Progress in Mathematics
Volume 202

European Congress of Mathematics

Barcelona, July 10–14, 2000
Volume II

Carles Casacuberta
Rosa Maria Miró-Roig
Joan Verdera
Sebastià Xambó-Descamps
Editors

Springer Basel AG

Editors:

Carles Casacuberta
Departament de Matemàtiques
Universitat Autònoma de Barcelona
08193 Bellaterra
Spain
casac@mat.uab.es

Joan Verdera
Departament de Matemàtiques
Universitat Autònoma de Barcelona
08193 Bellaterra
Spain
verdera@mat.uab.es

Rosa Maria Miró-Roig
Departament d'Algebra i Geometria
Facultat de Matemàtiques
Universitat de Barcelona
08007 Barcelona
Spain
miro@mat.ub.es

Sebastià Xambó-Descamps
Departament de Matemàtica Aplicada II
Universitat Politècnica de Catalunya
08028 Barcelona
Spain
sebastia.xambo@upc.es

2000 Mathematics Subject Classification 00B25

A CIP catalogue record for this book is available from the Library of Congress, Washington D.C., USA

Deutsche Bibliothek Cataloging-in-Publication Data

ECM <3, 2000, Barcelona>:
European Congress of Mathematics : Barcelona, July 10-14, 2000 /
Carles Casacuberta ... ed. – Basel ; Boston ; Berlin : Birkhäuser
ISBN 978-3-0348-9496-8 ISBN 978-3-0348-8266-8 (eBook)
DOI 10.1007/978-3-0348-8266-8

Vol. 2. - (2001)
(Progress in mathematics ; Vol. 202)
ISBN 978-3-0348-9496-8

Originally published by Birkhäuser Verlag, Basel- Boston - Berlin in 2001
Softcover reprint of the hardcover 1st edition 2001

Printed on acid-free paper produced of chlorine-free pulp. TCF ∞

ISBN 978-3-0348-9496-8

9 8 7 6 5 4 3 2 1

www.birkhauser-science.com

Table of Contents of Volume I

Articles by Parallel Speakers

Table of Contents of Volume II

Mini-Symposium on Curves over Finite Fields and Codes

Mini-Symposium on Free Boundary Problems

Mini-Symposium on Mathematical Finance: Theory and Practice

Mini-Symposium on Quantum Chaology

Mini-Symposium on Quantum Computing

Mini-Symposium on String Theory and M-Theory

Mini-Symposium on Symplectic and Contact Geometry and Hamiltonian Dynamics

Mini-Symposium on Wavelet Applications in Signal Processing

Classification Results on Valuations on Convex Sets

Semyon Alesker

Abstract. We discuss classification results on continuous valuations on convex sets obtained mostly during the last five years. Also we briefly describe some earlier results.

1. Introduction

In this short survey we will describe some of the classification results regarding continuous valuations on convex sets, which were obtained during approximately the last five years. For completeness we will state some earlier results, but we refer to the surveys [23] and [21] for a much more detailed discussion of previous works. Of course our survey and references do not pretend to be complete. Thus we do not discuss here the recent progress in description of *semicontinuous* valuations (see [18]).

First let us recall some basic definitions and examples. Let $\mathbb{R}^n$ denote the n-dimensional linear space over the reals. Let $\mathcal{K}^n$ denote the family of all convex compact subsets of $\mathbb{R}^n$. Equipped with the Hausdorff metric, $\mathcal{K}^n$ becomes a locally compact metric space.

Definition 1.1. *A scalar valued function*

$$\varphi\colon \mathcal{K}^n \longrightarrow \mathbb{C}$$

is called a valuation if for all convex compact sets $K_1, K_2 \in \mathcal{K}^n$ such that their union $K_1 \cup K_2$ is also convex one has

$$\varphi(K_1 \cup K_2) = \varphi(K_1) + \varphi(K_2) - \varphi(K_1 \cap K_2)\,.$$

A valuation φ is called continuous if it is continuous with respect to the Hausdorff metric on $\mathcal{K}^n$.

Let us make a few historical remarks on valuations, referring for more details and references to the surveys [23] and [21]. The class of translation invariant valuations is the most classical one. They played an important initial role in Dehn's solution in 1900 of Hilbert's third problem on non-equidecomposability of convex polytopes of equal volume in $\mathbb{R}^3$. A first attempt at classifying (in three-dimensional space) the rigid-motion invariant valuations (satisfying a suitable additional assumption) was made by Blaschke in the 1930s [see [8]). However his result was not satisfactory since in the course of the proof he had to introduce an additional invariance assumption. Probably the most famous result on valuations is Hadwiger's characterization of rigid motion invariant valuations, continuous with respect to the Hausdorff metric, as linear combinations of quermassintegrals (see [9]–[11], and below). It has numerous applications to integral geometry (see e.g., Chapter 9 of [17], or [11], or [24]).

The basic examples of valuations we would like to mention are as follows.

1) Any measure on $\mathbb{R}^n$ is a valuation.

2) The Euler characteristic χ is a valuation:

$$\chi(K) = 1 \text{ for every convex compact set } K \in \mathcal{K}^n .$$

For the next example let us recall the definition of mixed volumes. Let $K_1, \dots, K_n$ be convex compact subsets in $\mathbb{R}^n$. As it was observed by H. Minkowski, the volume of the Minkowski sum $\lambda_1 K_1 + \dots + \lambda_n K_n$ is a homogeneous polynomial of degree n in non-negative coefficients $\lambda_1, \dots, \lambda_n$ (here the sum $\lambda_1 K_1 + \dots + \lambda_n K_n$ denotes the set $\{\lambda_1 x_1 + \dots + \lambda_n x_n \mid x_i \in K_i\}$). The coefficient of $\lambda_1 \cdot \dots \cdot \lambda_n$ of this polynomial divided by $n!$ is called the *mixed volume* of $K_1, \dots, K_n$ and is denoted by $V(K_1, \dots, K_n)$. So the next example of valuations is as follows.

3) Fix a non-negative integer number $j \le n$. Fix convex compact sets $A_1, \dots, A_j$. Then the mixed volume

$$\varphi(K) = V(\underbrace{K, \dots, K}_{n-j \text{ times}}, A_1, \dots, A_j)$$

is a translation invariant continuous valuation (see [24]).

Note that if in this example one takes $j = 0$ one obtains the usual volume. If one takes $j = n$ one obtains the Euler characteristic multiplied by a constant.

Now let us state the classical result of H. Hadwiger. Let $O(n)$ denote the orthogonal group, and $SO(n)$ denote the special orthogonal group.

Theorem 1.2. *Every continuous translation invariant $SO(n)$-invariant valuation ϕ can be represented uniquely in the form*

$$\phi(K) = \sum_{i=0}^{n} c_i V_i(K) ,$$

where $V_i(K)$ denotes the mixed volume of the set K taken i times with the unit Euclidean ball taken $n - i$ times.

Moreover every expression of this form is a continuous translation invariant $O(n)$-invariant valuation.

Note now that the linear space of all continuous valuations has the natural topology given by a sequence of seminorms:

$$\|\phi\|_N = \sup_{K \subset N \cdot B} |\phi(K)|,$$

where $N \cdot B$ denotes the Euclidean ball of radius N. This topology defines the Frechet space structure on the space of continuous valuations.

2. Rotation Invariant Valuations

In this section we will present a description of $O(n)$- (resp. $SO(n)$-) invariant valuations generalizing Hadwiger's Theorem 1.2. We will need the following definition.

Definition 2.1. *The valuation ϕ is called polynomial of degree at most ℓ if $\phi(K+x)$ is a polynomial in $x \in \mathbb{R}^n$ of degree at most ℓ for every convex compact set K.*

Note that valuations polynomial of degree 0 are just translation invariant valuations, and valuations polynomial of degree 1 are called translation covariant (see [23]).

Definition 2.1 was introduced by Khovanski and Pukhlikov in [14]. They studied combinatorial-algebraic properties of polynomial valuations in [14] and [15]. The following result is the first step towards description of rotation invariant valuations.

Theorem 2.2. [1] *Every continuous $SO(n)$- (resp. $O(n)$-) invariant valuation can be approximated uniformly on compact subsets of $\mathcal{K}^n$ by continuous polynomial $SO(n)$- (resp. $O(n)$-) invariant valuations.*

It turns out that it is possible to describe explicitly *polynomial* continuous rotation invariant valuations. This description is not very interesting for the line $\mathbb{R}^1$; we will not discuss it here. So let us assume that $n \geq 2$. It is convenient to state the main result in terms of generalized curvature measures $\Theta_j(K, \cdot)$ of a convex set K. For the definition of it we refer to [24]. Below S^{n-1} denotes the unit $(n-1)$-dimensional sphere in $\mathbb{R}^n$.

Theorem 2.3. [1]

(i) *Let ϕ be a continuous polynomial valuation, which is $SO(n)$-invariant if $n \geq 3$ and $O(n)$-invariant if $n = 2$. Then there exist polynomials $p_0, \dots, p_{n-1}$ in two variables such that*

$$\phi(K) = \sum_{j=0}^{n-1} \int_{\mathbb{R}^n \times S^{n-1}} p_j(|s|^2, \langle s, n\rangle) d\Theta_j(K; s, n)$$

for every $K \in \mathcal{K}^n$, where $\Theta_j(K; \cdot)$ is the j-th generalized curvature measure of K, $|s|$ is the Euclidean norm of $s \in \mathbb{R}^n$, and $n \in S^{n-1}$. Moreover, any expression of this form is a continuous polynomial $O(n)$-invariant valuation.

(ii) *Let ϕ be a continuous polynomial $SO(2)$-invariant valuation on $\mathcal{K}^2$. Then there exist polynomials q_0, q_1 in two variables such that*

$$\phi(K)=\sum_{j=0}^{1}\int_{\mathbb{R}^2\times S^1} q_j(\langle s,n\rangle,\langle s,n'\rangle)d\Theta_j(K;s,n)\,,$$

where n' denotes the rotation of the vector n by $\frac{\pi}{2}$ counterclockwise. Moreover, any expression of this form is a continuous $SO(2)$-invariant polynomial valuation.

Note that Theorems 2.2 and 2.3 imply in particular that every continuous $SO(n)$-invariant valuation is $O(n)$-invariant provided $n\geq 3$; this is not true if $n\leq 2$.

The proof of these theorems uses a combination of methods of valuation theory and the representation theory of the orthogonal group. Note also that further extension of this method in [2] was applied to the classification of *isometry covariant* tensor valued valuations first introduced and studied by P. McMullen [22]. We refer for the discussion of this result to [3] or to [2] for the proof.

3. Translation Invariant Valuations

We are going to discuss in this section translation invariant continuous valuations without any additional symmetries. First of all let us state some well-known results about this class of valuations.

It is easy to see that every translation invariant continuous valuation ϕ on the line has the form $\phi([a,b])=\alpha(b-a)+\beta$, where α and β are constants.

Theorem 3.1. [19] *Every continuous translation invariant valuation φ on $\mathbb{R}^n$ can be presented uniquely as a sum*

$$\varphi=\sum_{i=0}^{n}\varphi_i\,,$$

where φ_i are homogeneous continuous translation invariant valuations of degree i, $0\leq i\leq n$, i.e., for every $K\in\mathcal{K}^n$ and a scalar $\lambda\geq 0$

$$\varphi_i(\lambda K)=\lambda^i\varphi_i(K)\,.$$

Theorem 3.2. *Let $\varphi_j\colon\mathcal{K}^n\to\mathbb{C}$ be a continuous translation invariant valuation, homogeneous of degree j. Then*

(a) *(trivial) φ_0 is proportional to the Euler characteristic;*
(b) [10] *φ_n is proportional to the standard volume, i.e., $\varphi_n(K)=a\cdot\mathrm{vol}_n K$;*
(c) [20] *For convenience let us fix a Euclidean metric on $\mathbb{R}^n$. Then*

$$\varphi_{n-1}(K)=\int_{S^{n-1}}f(\Omega)dS_{n-1}(K,\Omega)\,,$$

where $f\colon S^{n-1}\to\mathbb{R}$ is a continuous function, and $S_{n-1}(K,\cdot)$ is the surface area measure of K.

(d) [7] *A continuous translation invariant valuation* ϕ *homogeneous of degree* 1 *can be approximated uniformly on compact subsets of* $\mathcal{K}^n$ *by valuations of the form*

$$\phi(K) = V(K, L, \dots, L) - V(K, M, \dots, M),$$

where L, M *are fixed convex compact sets.*

Note that Theorems 3.1 and 3.2(a)-(c) provide an explicit description of translation invariant continuous valuations on the plane $\mathbb{R}^2$ (which was obtained earlier by Hadwiger [12]). Moreover these facts and Theorem 3.2(d) imply some description of translation invariant continuous valuations in 3-space $\mathbb{R}^3$, though not as explicit as on the plane.

It was conjectured by P. McMullen that *the mixed volumes span a dense subspace in the space of translation invariant continuous valuations.*

Clearly Theorems 3.1 and 3.2 imply this conjecture in $\mathbb{R}^n$, $n \leq 3$. We have shown that this conjecture has positive answer (see [5] for the general case; a particular case of even valuations in 4-dimensional space was proved in [4]). In fact a much stronger statement is true. In order to formulate it let us make two simple observations.

First of all observe that every valuation ϕ can be decomposed uniquely into even and odd parts

$$\phi = \phi^{\text{even}} + \phi^{\text{odd}},$$

where $\phi^{\text{even}}(-K) = \phi^{\text{even}}(K)$, $\varphi^{\text{odd}}(-K) = -\varphi^{\text{odd}}(K)$, for every $K \in \mathcal{K}^n$.

On the Frechet space of translation invariant continuous valuations we have the natural continuous representation π of the linear group $GL(n, \mathbb{R})$. Namely, for every $g \in GL(n, \mathbb{R})$, $K \in \mathcal{K}^n$,

$$(\pi(g)\varphi)(K) = \varphi(g^{-1}K).$$

Our next main result is

Theorem 3.3. *The natural representation of* $GL(n, \mathbb{R})$ *on the space of even (resp. odd) translation invariant continuous valuations of a given degree of homogeneity is irreducible.*

Note that this theorem immediately implies McMullen's conjecture. Indeed by Theorem 3.1 it is sufficient to prove it for valuations, homogeneous of a given degree, which are either even or odd. The linear subspace spanned by the mixed volumes is $GL(n, \mathbb{R})$-invariant, hence by Theorem 3.3 it must be either zero or dense everywhere. But obviously it is nonzero.

Very briefly the idea of the proof is as follows. Consider the space of translation invariant valuations of given degree of homogeneity and parity. One constructs two different embeddings of this $GL(n, \mathbb{R})$-module to two other $GL(n, \mathbb{R})$-modules, which are more standard from the representation theoretical point of view (they are induced from finite dimensional representation of certain parabolic subgroups of $GL(n, \mathbb{R})$). The injectivity of one of these embeddings is a quite non-trivial fact based on the Klain–Schneider characterization of simple translation invariant continuous valuations ([16, 25]). The next step is to show that these two $GL(n,\mathbb{R})$-modules can have at most one common irreducible subspace. This is a purely representation theoretical problem, and the proof is based on the Beilinson–Bernstein theory of D-modules ([6]). We refer to [5] for the details.

4. Unitarily Invariant Valuations

We will state here some results towards description of translation invariant continuous valuations on $\mathbb{C}^n$, which are invariant with respect to the unitary group $U(n)$.

Theorem 4.1. [5] *The space of translation invariant $U(n)$-invariant continuous valuations on $\mathbb{C}^n$ is finite dimensional. For every k, $0 \leq k \leq 2n$, the dimension of the space of such valuations, which are in addition homogeneous of degree k, is equal to*

$$1 + \left[\frac{\min(k, 2n-k)}{2}\right].$$

Note that if $n = 1$ then $U(1) = SO(2)$, and the explicit description of translation invariant $U(1)$-invariant continuous valuations just reduces to the Hadwiger Theorem 1.2 on the plane. For $n > 1$ Theorem 4.1 follows from the properties of the space of all even translation invariant continuous valuations as $GL(n,\mathbb{R})$-module discussed in the previous section and representation theoretical computations due to Howe and Lee [13].

We can give an explicit description of translation invariant $U(2)$-invariant continuous valuations on $\mathbb{C}^2$. Let us denote by $V_i(K)$ the mixed volume of K taken i times with the unit Euclidean ball in the remaining places.

Theorem 4.2. [4] *Every unitarily invariant translation invariant continuous valuation on $\mathbb{C}^2$ is a linear combination of V_i, $0 \leq i \leq 4$ and of the valuation*

$$\phi(K) := \int_{\mathbb{C}P^1} \mathrm{vol}_2(\mathrm{Pr}_\xi\, K) d\xi\,,$$

where $\mathbb{C}P^1$ denotes the complex projective space, and Pr_ξ is the orthogonal projection onto the complex line ξ. This valuation ϕ is linearly independent from the V_i's. In particular the dimension of the space of these valuations is equal to 6.

We believe that the explicit characterization of unitarily invariant valuations would be useful for obtaining new integral-geometric formulas in the complex affine space $\mathbb{C}^n$ and the complex projective space $\mathbb{C}P^n$.

References

[1] Alesker, S.; Continuous rotation invariant valuations on convex sets. Ann. of Math. (2) 149 (1999), no. 3, 977–1005.

[2] Alesker, S.; Description of continuous isometry covariant valuations on convex sets. Geom. Dedicata 74 (1999), no. 3, 241–248.

[3] Alesker, S.; Continuous valuations on convex sets. Geom. Funct. Anal. 8 (1998), no. 2, 402–409.

[4] Alesker, S.; On P. McMullen's conjecture on translation invariant valuations on convex sets. Adv. Math. 155 (2000), no. 2, 239–263.

[5] Alesker, S.; Description of translation invariant valuations with the solution of P. McMullen's conjecture. GAFA 11 (2001), no. 2, 244–272.

[6] Beĭlinson, A. and Bernstein, J.; Localisation de g-modules. (French) C. R. Acad. Sci. Paris Sér. I Math. 292 (1981), no. 1, 15–18.

[7] Goodey, P. and Weil, W.; Distributions and valuations. Proc. London Math. Soc. (3) 49 (1984), no. 3, 504–516.

[8] Blaschke, W.; Vorlesungen über Integralgeometrie. (German) 3te Aufl. Deutscher Verlag der Wissenschaften, Berlin, 1955. (First edition: 1937.)

[9] Hadwiger, H.; Beweis eines Funktionalsatzes für konvexe Körper. (German) Abh. Math. Sem. Univ. Hamburg 17 (1951), 69–76.

[10] Hadwiger, H.; Additive Funktionale k-dimensionaler Eikörper. I. (German) Arch. Math. 3 (1952), 470–478.

[11] Hadwiger, H.; Vorlesungen über Inhalt, Oberfläche und Isoperimetrie. (German) Springer-Verlag, Berlin-Göttingen-Heidelberg 1957.

[12] Hadwiger, H.; Translationsinvariant, additive und Stetige Eibereichfunktionale. Publ. Math. Debrecen 2 (1951), 81–94.

[13] Howe, Roger and Lee, Soo Teck; Degenerate principal series representations of $\mathrm{GL}_n(\mathbb{C})$ and $\mathrm{GL}_n(\mathbb{R})$. J. Funct. Anal. 166 (1999), no. 2, 244–309.

[14] Khovanskii, A. G. and Pukhlikov, A. V.; Finitely additive measures on virtual polyhedra. (Russian) Algebra i Analiz 4 (1992), no. 2, 161–85; translation in St. Petersburg, Math. J. 4 (1993), no. 2, 337–356.

[15] Khovanskii, A. G. and Pukhlikov, A. V.; A Riemann-Roch theorem for integrals and sums of quasipolynomials over virtual polytopes. (Russian) Algebra i Analiz 4 (1992) no. 4, 188–216; translation in St. Petersburg Math. J. 4 (1993), no. 4, 789–812.

[16] Klain, D. A.; A short proof of Hadwiger's characterization theorem. Mathematika 42 (1995), no. 2, 329–339.

[17] Klain, D. A. and Rota, G.-C.; Introduction to geometric probability. Lezioni Lincee. [Lincei Lectures] Cambridge University Press, Cambridge, 1997.

[18] Ludwig, M. and Reitzner, M.; A characterization of affine surface area. Adv. Math. 147 (1999), no. 1, 138–172.

[19] McMullen, P.; Valuations and Euler-type relations on certain classes of convex polytopes. Proc. London Math. Soc. (3) 35 (1977), no. 1, 113–135.

[20] McMullen, P.; Continuous translation-invariant valuations on the space of compact convex sets. Arch. Math. (Basel) 34 (1980), no. 4, 377–384.

[21] McMullen, P.; Valuations and dissections. Handbook of convex geometry, Vol. A, B, 933–988, North-Holland, Amsterdam, 1993, P. M. Gruber and J. M. Wills, eds.

[22] McMullen, P.; Isometry covariant valuations on convex bodies. II International Conference in "Stochastic Geometry, Convex Bodies and Empirical Measures" (Agrigento, 1996). Rend. Circ. Mat. Palermo (2) Suppl. No. 50, (1997), 259–271.

[23] McMullen, P. and Schneider, R.; Valuations on convex bodies. Convexity and its applications, 170–247, Birkhäuser, Basel-Boston, Mass., 1983.

[24] Schneider, R.; Convex bodies: the Brunn–Minkowski theory. Encyclopedia of Mathematics and its Applications, 44. Cambridge University Press, Cambridge, 1993.

[25] Schneider, R.; Simple valuations on convex bodies. Mathematika 43 (1996), no. 1, 32–39.

Department of Mathematics
Tel Aviv University
Ramat Aviv
69978 Tel Aviv, Israel
E-mail address: semyon@math.tau.ac.il

Towards a Microscopic Theory of Phase Coexistence

Raphaël Cerf

Abstract. One of the fundamental goals of statistical mechanics is to understand the macroscopic effects induced by random forces acting at the microscopic level. We illustrate this in the context of the Ising model in the phase coexistence regime: the most likely shapes of macroscopic droplets of one pure phase floating in the other pure phase are close to the Wulff crystal of the model. Furthermore, the law of configurations at equilibrium is governed by a minimal action principle. These results come from joint works with Agoston Pisztora. We list several related open problems.

1. Introduction

The aim of this text is to provide a gentle introduction to the subject and to review some results of the works [10, 11] in the context of the Ising model. For a summary of the development of this field, in particular for general references and historical remarks, we refer the reader to the introduction of [10]. Relevant and related results appeared in [2, 6, 7, 8, 9, 16, 17, 22, 23, 24, 25, 26, 29, 30, 31, 32, 33, 34].

Let us consider a volume of water in absence of gravity at ordinary temperature. We start to pour a very small quantity of oil in the water. First, nothing noticeable happens on the macroscopic scale, i.e., the oil is perfectly dissolved throughout the water and the oil molecules are homogeneously spread within the water: by observing the liquid at the macroscopic level, we cannot even tell that it is a mixture of two distinct types of particles, which nevertheless have the tendency to repell each other. Let us keep pouring oil into the water. We know that the solubility of the oil in the water is not infinite; at some density threshold (which increases with the temperature), we obtain a solution of water saturated with oil. This solution is still a pure phase, completely homogeneous on the macroscopic level, and it realizes a perfect tradeoff between entropy and energy; we call it the water phase. Let us pour more oil. The excess of oil is no longer dissolved and it precipitates: macroscopic droplets of oil emerge. These droplets are not regions where there are only oil molecules, rather in these regions we observe the symmetric pure phase consisting of oil saturated with water, which we call the oil phase. The droplets are delimited by an abrupt change of the local density of water and oil molecules. We wish to understand the law governing the evolution and the shapes of these droplets.

Classical phenomenological theory asserts the existence of a macroscopic surface free energy $\mathcal{I}$ and the evolution of droplets so as to minimize $\mathcal{I}$. For instance, at equilibrium, in case $\mathcal{I}$ is isotropic, one observes a unique spherical droplet of the oil phase floating in the sea of the water phase. Our aim is to confirm the predictions of phenomenological theory starting from a truly microscopic model. We wish to understand how random forces acting at the atomic level, more precisely the probabilistic repulsive effect between the two types of particles, can induce such deterministic macroscopic effects. Perhaps one of the most famous results in probability theory is the law of large numbers: If $(X_n)_{n\in\mathbb{N}}$ is a sequence of independent identically distributed random variables with mean m, then, with probability one,

$$\lim_{n\to\infty}\frac{1}{n}(X_1+\cdots+X_n)=m\,.$$

What we have in mind is a generalization of the law of large numbers, but in a fundamentally new context, which we could state informally as

$$\lim_{\text{number of particles}\to\infty}\begin{pmatrix}\text{total effect of the random}\\ \text{microscopic repulsive forces}\end{pmatrix}=\text{single droplet}\,.$$

The limiting deterministic object is the shape of the droplet at equilibrium and the problem is intrinsically geometric; we deal with spatially dependent random variables, leaving entirely the independent framework. Hence the geometry enters the problem in a decisive way, in random interactions and in the formulation of the result itself.

Let us try to set up a simple model for our experiment with water and oil. A convenient choice is a lattice model: each site of the lattice is occupied either by a water particle or by an oil particle, which we indicate by $+$ or $-$. The interaction between different particles is repulsive and occurs when the substances are in immediate contact. Hence a repulsive nearest neighbour interaction is a sensible choice. Since we focus only on the repulsive interaction between different molecules, we can assume that the two substances are symmetric and that their self-interactions are of equal magnitude, or equivalently, equal to zero. Thus the total energy of a configuration should be simply the sum of all nearest neighbour pairs with different signs. We end up exactly with the Hamiltonian of the famous Ising model (to be defined precisely in the next section). In our experiment the density of oil is fixed, therefore we have a constraint on the possible configurations: the proportion of pluses and minuses has to be fixed. This situation amounts to considering the Ising model with plus boundary conditions (guaranteeing the water dominance) conditioned on the event that the magnetization is equal to a fixed value smaller than the spontaneous magnetization at the given temperature.

The first mathematical proof of phase separation was achieved ten years ago by Dobrushin, Kotecký and Shlosman [17] in the context of the two dimensional Ising model. It started an intense research activity [2, 16, 22, 23, 24, 25, 29, 30, 32,

33, 34]. However all these works were confined to dimension 2. There was a bundle of important results for models in dimensions higher than 3 which indicated the occurrence of a phase separation phenomenon [1, 3, 4, 5, 14, 26, 31].

The first precise study of phase separation for a truly microscopic model in dimension 3 was achieved for the Bernoulli percolation model [9]. It was then quickly extended to the Ising model in two parallel, partially overlapping works: [6] which relies on [4, 9, 31] and is limited to sufficiently low temperatures on one hand and [10] which relies on [9, 31] on the other hand. Finally the Potts and FK percolation models were handled in [11]. The solution to this problem in dimensions higher than 3 requires a completely new strategy compared to the two-dimensional proofs. First a change of topology in the space of the shapes is needed in order to formulate adequately the result itself. Second, the two-dimensional proofs [2, 17, 23, 24, 25, 29] were all relying on a polygonalization procedure (called the skeleton) to approximate the shapes of the interfaces, together with a combinatorial bound on the number of admissible skeletons. Nobody has succeeded yet in designing an analogous procedure in dimension 3. The leading idea of the work [9] was to embed the analysis from the very beginning in a continuous world in order to replace the combinatorial argument, whose purpose is to control the explosion of the possible shapes for the droplet, by a compactness argument. In other terms, one should restate the problem in the correct variational framework.

Although the solution of the classical isotropic problem has been known since ancient times, it was formulated in a very satisfactory setting only fifty years ago: here, by satisfactory, we mean a setting where the space of candidate solutions is endowed with a convenient topology having the required closure and semi-continuity properties so that the direct methods of the calculus of variations can be applied. To that aim, it is necessary to extend the notion of perimeter or surface area to nonregular sets. This was achieved by Caccioppoli and developed afterwards by De Giorgi [12, 13]. Hence, at the geometric level, we use the current standard tools of geometric measure theory in codimension 1 [18, 19, 20, 35].

Let us point out that the idea to use these tools is also present in a series of works in the context of the Ising model with Kac potentials (a model which is intermediate between the nearest-neighbour Ising model and a mean-field model). Alberti, Bellettini, Cassandro, Presutti [1, 3] developed the general philosophy of embedding the problem in a continuum setting in order to use the BV framework (the equivalent functional formulation of sets of finite perimeter). Enhanced results were obtained in [4, 5], but still in some limit where the range of interactions goes to infinity.

On the probabilistic level, the proofs of [6, 9, 10, 11] rely on the coarse-graining estimates of Pisztora, which describe accurately the typical configurations on a scale intermediate between the microscopic and the macroscopic scale. These coarse-graining estimates are done with the help of the FK representation of the Ising model and a general renormalization procedure [14, 31]. The connection between the probabilistic world and the realm of the calculus of variations is made through the language of large deviations.

2. The Ising Model

For reasons of technical simplicity, it is easier to build our model on a lattice. We will work with the lattice $\mathbb{Z}^d$; each site of the lattice is occupied by one of the two types of particles, which we denote by $-$ and $+$. Let $\Lambda \subset \mathbb{Z}^d$ be a cubic box. A configuration in Λ is a map $\sigma\colon \Lambda \to \{-,+\}$ and for x in Λ, we denote by $\sigma(x)$ the type of the particle present at x. The energy or Hamiltonian $H_\Lambda(\sigma)$ of the configuration σ in Λ is, up to a constant, the number of interfaces between the minuses and the pluses, that is,

$$H_\Lambda(\sigma) = -\frac{1}{2} \sum_{\{x,y\}\subset\Lambda, |x-y|=1} \sigma(x)\sigma(y)\,.$$

We need further a mechanism to ensure the dominance of one type of particles. This is achieved through boundary conditions. We consider only two types of boundary conditions, by putting either a layer of pluses or of minuses around the box Λ. The energy or Hamiltonian $H^*_\Lambda(\sigma)$ with boundary conditions $*$ (where $*$ stands for $-$ or $+$) is given by

$$H^*_\Lambda(\sigma) = -\frac{1}{2} \sum_{\{x,y\}\in\Lambda, |x-y|=1} \sigma(x)\sigma(y) - \frac{1}{2} \sum_{x\in\partial^{in}\Lambda} \sigma(x)(*)$$

where $\partial^{in}\Lambda$ denotes the sites in Λ which are at distance less than 1 from the complement of Λ. We next add some randomness in the model. Let $T > 0$ be the temperature. We build a probability law on the space $\{-,+\}^\Lambda$ of the configurations. This space is huge but finite, hence to define the law we need to specify the individual probability of each possible configuration. The natural way to do this is to use the Boltzmann factor. So, the Gibbs measure $\mu^*_{\Lambda,T}$ in Λ at temperature T with boundary conditions $*$ is given by

$$\forall \sigma \in \{-,+\}^\Lambda \qquad \mu^*_{\Lambda,T}(\sigma) = \frac{1}{Z^*_{\Lambda,T}} \exp -\frac{H^*_\Lambda(\sigma)}{T}$$

where the normalizing factor $Z^*_{\Lambda,T}$, called the partition function, is given by

$$Z^*_{\Lambda,T} = \sum_{\sigma\in\{-,+\}^\Lambda} \exp -\frac{H^*_\Lambda(\sigma)}{T}\,.$$

Let us give a closer look at this formula. The elements Λ, T, $* \in \{-,+\}$ being fixed, the most likely configurations are those having a small energy, i.e., those for which the contacts between the minuses and the pluses are reduced. Thus we have built a complex probability law with spatial correlations between the sites. Let us play a bit with the elements controlling the Gibbs measures $\mu^\pm_{\Lambda,T}$. Imagine first that we fix the box Λ and the boundary conditions to $+$. If we send T to 0, then the measure $\mu^+_{\Lambda,T}$ concentrates on the configuration realizing the global minimum of the Hamiltonian H^+_Λ, in this case the configuration where all the sites are pluses. On the contrary, if we send T to ∞, the value of the Hamiltonian becomes irrelevant and $\mu^+_{\Lambda,T}$ converges towards the Bernoulli product law where all the sites are independent. The case of $\mu^-_{\Lambda,T}$ being symmetric, we see that

$$\begin{array}{ccccc} \text{dirac mass at "all pluses"} & \longleftarrow & \mu^+_{\Lambda,T} & \longrightarrow & \text{i.i.d. Bernoulli} \\ & T \downarrow 0 & & T \uparrow \infty & \\ \text{dirac mass at "all minuses"} & \longleftarrow & \mu^-_{\Lambda,T} & \longrightarrow & \text{i.i.d. Bernoulli} \end{array} .$$

Something remarkable has already happened: as $T \uparrow \infty$, the boundary conditions are forgotten, while as $T \downarrow 0$, they completely determine the limit. However, we wish to work at a fixed positive temperature T. In order to observe a sharp mathematical phenomenon, we must take another kind of limit, namely the thermodynamic limit where the number of particles goes to infinity. This is achieved by letting the box Λ grow and invade the whole lattice $\mathbb{Z}^d$. As Λ increases to $\mathbb{Z}^d$, the measure $\mu^+_{\Lambda,T}$ decreases and converges weakly towards the infinite volume Gibbs measure μ^+_T, which is a probability measure on the space of infinite volume configurations $\{-,+\}^{\mathbb{Z}^d}$. Similarly, $\mu^-_{\Lambda,T}$ increases weakly towards μ^-_T:

$$\lim_{\Lambda \uparrow \mathbb{Z}^d} \mu^-_{\Lambda,T} = \mu^-_T, \qquad \lim_{\Lambda \uparrow \mathbb{Z}^d} \mu^+_{\Lambda,T} = \mu^+_T .$$

Here is a heuristic explanation for this monotone convergence. Let us consider a huge box Λ and the site at the center of the box Λ. Without boundary conditions, the law of $\sigma(0)$ under $\mu_{\Lambda,T}$ is symmetric, hence it is the one of a fair coin, i.e.,

$$\mu_{\Lambda,T}(\sigma(0) = +) = 1/2 = \mu_{\Lambda,T}(\sigma(0) = -) .$$

If we put $+$ boundary conditions, these boundary conditions start to influence positively the sites at distance 1 from the boundary of the box, which themselves influence the sites at distance 2 from the boundary, hence this effect propagates and reaches the origin, so that the law of $\sigma(0)$ under $\mu^+_{\Lambda,T}$ is slightly biased towards $+$:

$$\mu^+_{\Lambda,T}(\sigma(0) = +) > 1/2 > \mu^+_{\Lambda,T}(\sigma(0) = -) .$$

The larger the box Λ is, the tinier is the resulting effect at the origin, hence the influence of the boundary conditions decreases as the box increases. The fundamental and basic question is whether something of this influence still remains after we have sent the boundary conditions to infinity. Equivalently, do the limiting measures μ^-_T and μ^+_T coincide?

We say that there is a phase transition at temperature T if $\mu^-_T \neq \mu^+_T$. The occurrence of a phase transition can be detected with the help of the spontaneous magnetization $m^*(T) = \mu^+_T(\sigma(0))$. This terminology stems from the fact that the Ising model was originally introduced as a model of ferromagnetism (under some adequate conditions, a magnet submitted to the influence of a magnetic field will remember the sign of the field even after it has disappeared). The fundamental result concerning the phase transition in the Ising model is the following. In any dimension $d \geq 2$, there exists a critical temperature $T_c(d)$ (depending on the dimension) such that the model exhibits a phase transition for $T < T_c(d)$ (i.e., $m^*(T) > 0$ and $\mu^-_T \neq \mu^+_T$ for $T < T_c(d)$) whereas the infinite volume Gibbs measure is unique for $T > T_c(d)$ (i.e., $m^*(T) = 0$ and $\mu^-_T = \mu^+_T$ for $T > T_c(d)$).

3. The Wulff Crystal

We can now mimic mathematically the initial experiment with the help of the Ising model. Let us consider a box $\Lambda(n)$ of diameter n full of pluses. We take n very large, of the order of the Avogadro number 6.02×10^{23}. We start deleting pluses and replacing them by minuses, first a small quantity of minuses. It is also possible to build a stochastic dynamics in the box which is conservative (i.e., the total numbers of minuses and pluses remain unchanged or equivalently the empirical magnetization $n^{-d} \sum_{x \in \Lambda(n)} \sigma(x)$ remains constant) and whose final equilibrium is the Gibbs measure $\mu^+_{\Lambda(n),T}$ conditioned to have this fixed magnetization. The most simple such dynamics is the so-called Kawasaki dynamics: at random exponential times, a pair of neighbouring particles might be exchanged according to a simple local probabilistic rule. As long as the empirical magnetization is larger than $m^*(T)$, the configuration in $\Lambda(n)$ at equilibrium is expected to be spatially homogeneous. If we keep pouring minuses in the box and removing pluses, we soon reach the value $m^*(T)$, and at this point we obtain the (saturated) pure phase μ^+_T, i.e., the configuration in $\Lambda(n)$ looks like a sample of the infinite volume Gibbs measure μ^+_T. We finally add some more minuses and we cross the threshold $m^*(T)$. We wish to understand the answer of the system and the most likely configurations inside the box when there is an excess of minuses. It turns out that this simple model indeed confirms the prediction of phenomenological theory. At equilibrium, with probability tending to 1 as n goes to ∞, one sees inside the box a region where the configuration looks like the minus phase μ^-_T, surrounded by a region filled with the plus phase μ^+_T. When rescaled by a factor n, the shape of this region converges in the limit $n \to \infty$ towards a deterministic shape, called the Wulff crystal of the Ising model. This crystal is convex, it depends on the temperature and on the initial lattice $\mathbb{Z}^d$; it bears the name of Wulff, who studied it one century ago [36]. In order to detect conveniently the Wulff region, we average locally the magnetization over an intermediate scale (this idea was originally introduced in [1]) and we send the box $\Lambda(n)$ of diameter n onto the unit cube. Let $f(n)$ be a fixed function from $\mathbb{N}$ to $\mathbb{N}$ such that both $n/f(n)^{d-1}$ and $f(n)/\ln n$ tend to ∞ as n goes to ∞. The locally averaged magnetization σ_n is the map from the open unit cube $\Omega =]-1/2, 1/2[^d$ to $[-1, 1]$ defined by

$$\forall x \in \Omega \qquad \sigma_n(x) = \frac{1}{f(n)^d} \sum_{y \in \Lambda(n), |y-nx|_\infty < f(n)/2} \sigma(y)$$

where $|\cdot|_\infty$ is the usual supremum norm. We define next the random set Ω_n^- where σ_n is negative and its mass center a_n, i.e.,

$$\Omega_n^- = \{x \in \Omega : \sigma_n(x) < 0\}, \qquad a_n = \frac{1}{\mathcal{L}^d(\Omega_n^-)} \int_{\Omega_n^-} x \, dx$$

where $\mathcal{L}^d$ is the d-dimensional Lebesgue measure. Let $d \geq 3$. There exists a subset $\mathcal{V}(d)$ of $\mathbb{R}^+$ which is at most countable and whose closure does not contain 0 such that for any $T \notin \mathcal{V}(d)$, there exist at most two extremal translation invariant Gibbs states, namely μ_T^- and μ_T^+. Let $\widehat{T}_c(d) = -1/\ln(1-\widehat{p}_c(d,2))$ where $\widehat{p}_c(d,2)$ is the limit of the slab critical points of the associated FK percolation model (see [31] for the precise definition). It is known that $0 < \widehat{T}_c(d) \le T_c(d)$. For $T < \widehat{T}_c(d)$, $T \notin \mathcal{V}(d)$, we can extract from the model a surface tension τ, which is a map from the $d-1$-dimensional unit sphere S^{d-1} into $\mathbb{R}^+ \setminus \{0\}$. The surface tension τ inherits automatically the basic symmetry properties of the model and it satisfies an important inequality called the weak triangle inequality [27, 28]. We define then the Wulff crystal $\mathcal{W}_\tau$ associated to τ as

$$\mathcal{W}_\tau = \left\{ x \in \mathbb{R}^d;\ x\cdot w \le \tau(w) \text{ for all } w \text{ in } S^{d-1} \right\}.$$

The Wulff crystal $\mathcal{W}_\tau$ is bounded, closed, convex and contains 0 in its interior. It is the solution to the anisotropic isoperimetric problem associated to the surface tension τ [18, 19, 35, 36].

Theorem 3.1. *Let $d \ge 3$, $T < \widehat{T}_c(d)$, $T \notin \mathcal{V}(d)$. For $\delta > 0$,*

$$\limsup_{n\to\infty} n^{1-d} \ln \mu^+_{\Lambda(n),T}\Big[\int_\Omega \Big| |\sigma_n(x)| - m^* \Big| \, dx > \delta \Big] = -\infty .$$

Let m be large enough so that the rescaled Wulff crystal

$$\mathcal{W}(m) = \left(\frac{m^* - m}{2m^* \mathcal{L}^d(\mathcal{W}_\tau)} \right)^{1/d} \mathcal{W}_\tau$$

fits completely into Ω; it is the case if $1 - 2(\operatorname{diam} \mathcal{W}_\tau)^{-d} \mathcal{L}^d(\mathcal{W}_\tau) < m/m^* < 1$. There exist constants $b = b(d,T,m,\delta)$, $c = c(d,T,m,\delta) > 0$ such that

$$\mu^+_{\Lambda(n),T}\Big[\mathcal{L}^d\Big(\Omega_n^- \Delta (a_n + \mathcal{W}(m)) \Big) < \delta \,\Big|\, \frac{1}{n^d} \sum_{x\in\Lambda(n)} \sigma(x) \le m \Big] \ge 1 - b\exp(-cn^{d-1})$$

where $\mathcal{L}^d$ is the Lebesgue measure in $\mathbb{R}^d$ and Δ is the symmetric difference.

Challenging open conjectures are the equality $\widehat{T}_c(d) = T_c(d)$ and the inclusion $\mathcal{V}(d) \subset \{T_c(d)\}$ (see [21, 31]). The way to prove Theorem 3.1 is rather long, but it is very likely that substantial simplifications will appear in the future. For instance, an intermediate step is to compute the asymptotics of the conditioning event

$$\lim_{n\to\infty} n^{1-d} \ln \mu^+_{\Lambda(n),T}\Big[\frac{1}{n^d} \sum_{x\in\Lambda(n)} \sigma(x) \le m \Big] = -d\Big(\frac{m^*-m}{2m^*} \Big)^{(d-1)/d} (\mathcal{L}^d(\mathcal{W}_\tau))^{1/d} .$$

In two dimensions the Wulff droplet can be identified with a random region surrounded by a minus spin cluster. Its external boundary is therefore a large contour separating plus and minus spins which follows closely the boundary of the Wulff crystal in the sense of the Hausdorff metric [17, 25]. In dimensions $d \ge 3$, it is widely believed that for low temperatures, the Wulff droplet can still be defined by a microscopic contour. However for temperatures close to T_c, a fundamentally new situation is expected. The dominant minus spin cluster of the Wulff droplet should percolate all the way to the boundary of the box. More precisely, there should exist two big spin clusters, one of pluses and one of minuses, and they should be both omnipresent in the entire box; the densities of these clusters should undergo an abrupt change at the boundary of the Wulff droplet. In which case the phase boundaries cannot be described directly with contours.

There are a lot of challenging open questions related to the surface tension and the Wulff crystal. A major problem is to understand the fluctuations of the interfaces. Dobrushin [15] proved that at low temperatures, interfaces perpendicular to the axis directions are localized. It is conjectured that in dimensions $d = 3$ there exists a temperature $T_R < T_c$ where a so-called "roughening" transition occurs: the nature of the fluctuations undergoes a fundamental change for $T > T_R$. As far as we know, there is still no proof or disproof for the existence of this transition in dimensions $d = 3$. It is also expected that the facets of the Wulff crystal should disappear at T_R (see [28]). This circle of questions is also linked with the delicate problem of the sharp large deviations, i.e., to obtain a more precise expansion of the asymptotics of the probability of a deviation for the empirical magnetization.

4. Minimal Surfaces

In the subsequent work [11], the phenomenon of phase separation was handled in a more general framework, namely, general large deviation principles and weak law of large numbers for the Ising, Potts and FK percolation models were proven in a rather general domain with general boundary conditions. We next restate the corresponding result for the case of the Ising model.

Consider a bounded open region Ω in $\mathbb{R}^d$ whose boundary Γ has locally a Lipschitz parametrization. Note that this hypothesis is satisfied when Ω is a bounded open set with a C^1 boundary or when Ω is a polyhedral domain. Let Γ^- and Γ^+ be two disjoint and relatively open subsets of Γ such that the relative boundary of $\Gamma \setminus \Gamma^- \setminus \Gamma^+$ in Γ has zero $\mathcal{H}^{d-1}$ measure (where $\mathcal{H}^{d-1}$ is the $d-1$ dimensional Hausdorff measure). We will study the Ising model in the region Ω. To obtain a discretized version of the region Ω, we define for $n \in \mathbb{N}$,

$$\mathbb{Z}^d_n = \mathbb{Z}^d/n \qquad \text{(the rescaled lattice)}$$

$$\Omega_n = \big\{x \in \mathbb{Z}^d_n;\ \exists y \in \Omega \quad |x-y|_\infty < 1/n\big\} \qquad \text{(the discrete counterpart of } \Omega)$$

$$\Gamma_n = \big\{x \in \Omega_n;\ \exists y \in \mathbb{Z}^d_n \setminus \Omega_n \quad |x-y| = 1/n\big\} \ \text{(the inner vertex boundary of } \Omega_n)$$

$$\Gamma^-_n = \big\{x \in \Gamma_n;\ \exists y \in \Gamma^- \quad |x-y|_\infty < 1/n\big\}$$

$$\Gamma^+_n = \big\{x \in \Gamma_n;\ \exists y \in \Gamma^+ \quad |x-y|_\infty < 1/n \quad \text{and} \quad \forall y \in \Gamma^- \quad |x-y|_\infty \geq 1/n\big\}$$

where $|\cdot|_\infty$ is the usual supremum norm. We use the sets Γ^-_n, Γ^+_n to specify the boundary conditions; namely we put pluses on Γ^+_n and minuses on Γ^-_n. We denote the corresponding Ising measure in Ω_n by μ_n. Let $f(n)$ be a fixed function from

$\mathbb{N}$ to $\mathbb{N}$ such that both $n/f(n)^{d-1}$ and $f(n)/\ln n$ tend to ∞ as n goes to ∞. The locally averaged magnetization σ_n is the map from Ω to $[-1,1]$ defined by

$$\forall x \in \Omega \qquad \sigma_n(x) = \frac{1}{f(n)^d} \sum_{y \in \Omega_n, |y-x|_\infty < f(n)/(2n)} \sigma(y)\,.$$

We partition Ω into the random sets Ω_n^-, Ω_n^0 and Ω_n^+ according to whether the value of the local magnetization is smaller, equal to or larger than 0. We consider the space of phase partitions $P(\Omega)$ consisting of 3-tuples (A^0, A^-, A^+) of Borel subsets of Ω forming a partition of Ω. We endow $P(\Omega)$ with the following metric:

$$\operatorname{dist}\Big((A^0,A^-,A^+),(B^0,B^-,B^+)\Big) = \mathcal{L}^d(A^0 \Delta B^0) + \mathcal{L}^d(A^- \Delta B^-) + \mathcal{L}^d(A^+ \Delta B^+)$$

where $\mathcal{L}^d$ is the Lebesgue measure in $\mathbb{R}^d$ and Δ is the symmetric difference. We recall that the perimeter $\mathcal{P}(E)$ of a Borel set E is

$$\mathcal{P}(E) = \sup\Big\{ \int_E \operatorname{div} f(x)\, dx : f \in C_0^\infty(\mathbb{R}^d, B(1)) \Big\}$$

where $C_0^\infty(\mathbb{R}^d, B(1))$ is the set of the compactly supported C^∞ vector functions from $\mathbb{R}^d$ to the unit ball $B(1)$ and div is the usual divergence operator. If E is a set of finite perimeter (i.e., $\mathcal{P}(E) < \infty$), one can define its reduced boundary $\partial^* E$; it is a subset of the topological boundary (it is exactly the topological boundary for a C^1 set) which can be used to extend several formulas of the classical differential calculus on surfaces [12]. The surface energy $\mathcal{I}$ on $P(\Omega)$ is given as follows. For any (A^0, A^-, A^+) with $A^0 = \emptyset$ and A^-, A^+ having finite perimeter we set

$$\begin{aligned}\mathcal{I}(A^0,A^-,A^+) = &\frac{1}{2}\int_{\partial^* A^- \cap \Omega} \tau(\nu_{A_-}(x))\, d\mathcal{H}^{d-1}(x) + \int_{\partial^* A^- \cap \Gamma^+} \tau(\nu_{A_-}(x))\, d\mathcal{H}^{d-1}(x)\\ &+\frac{1}{2}\int_{\partial^* A^+ \cap \Omega} \tau(\nu_{A_+}(x))\, d\mathcal{H}^{d-1}(x) + \int_{\partial^* A^+ \cap \Gamma^-} \tau(\nu_{A_+}(x))\, d\mathcal{H}^{d-1}(x)\,.\end{aligned}$$

If $A^0 \neq \emptyset$ or if $\mathcal{P}(A^-) + \mathcal{P}(A^+) = \infty$, we set $\mathcal{I}(A^0, A^-, A^+) = \infty$.

Theorem 4.1. *Let $d \geq 3$, $T < \widehat{T}_c(d)$, $T \notin \mathcal{V}(d)$. For $\delta > 0$,*

$$\limsup_{n\to\infty} n^{1-d} \ln \mu_n \Big[\int_\Omega \Big| |\sigma_n(x)| - m^* \Big|\, dx > \delta \Big] = -\infty\,.$$

The sequence $(\vec{\Omega}_n)_{n\in\mathbb{N}} = ((\Omega_n^0, \Omega_n^-, \Omega_n^+))_{n\in\mathbb{N}}$ of the empirical phase partitions of Ω satisfies a large deviation principle in $(P(\Omega), \operatorname{dist})$ with respect to μ_n with speed n^{d-1} and rate function $\mathcal{I} - \min_{P(\Omega)} \mathcal{I}$, i.e., for any Borel subset $\mathbb{E}$ of $P(\Omega)$,

$$\begin{aligned}-\inf_{\mathring{\mathbb{E}}} \mathcal{I} + \min_{P(\Omega)} \mathcal{I} &\leq \liminf_{n\to\infty} n^{1-d} \ln \mu_n \big[\vec{\Omega}_n \in \mathbb{E}\big]\\ &\leq \limsup_{n\to\infty} n^{1-d} \ln \mu_n \big[\vec{\Omega}_n \in \mathbb{E}\big] \leq -\inf_{\overline{\mathbb{E}}} \mathcal{I} + \min_{P(\Omega)} \mathcal{I}\,.\end{aligned}$$

The study of phase boundaries leads naturally to the theory of minimal surfaces, and for instance to the Plateau problem corresponding to anisotropic surface measures. Let Ω be a bounded open set in $\mathbb{R}^3$ with smooth boundary and let γ be a Jordan curve drawn on $\partial\Omega$ which separates $\partial\Omega$ into two disjoint relatively open sets Γ^+ and Γ^-. Typical configurations in the Ising model on a fine grid in Ω with boundary conditions plus on Γ^+ and minus on Γ^- will exhibit two phases separated by an interface close to a minimal surface which is a global solution to the following Plateau type problem:

$$\text{minimize} \int_S \tau(\nu_S(x))\, d\mathcal{H}^{d-1}(x) : \ S \text{ is a surface in } \Omega \text{ spanned by } \gamma$$

where $\nu_S(x)$ is the normal vector to S at x. It is conjectured that, as the temperature approaches T_c from below, the surface tension τ becomes more and more isotropic, thus the solution of the above minimization problem should converge to the solution of the classical Plateau problem.

References

[1] G. Alberti, G. Bellettini, M. Cassandro and E. Presutti: Surface tension in Ising systems with Kac potentials. J. Statist. Phys. 82, 743–796 (1996).

[2] K. S. Alexander, J. T. Chayes and L. Chayes: The Wulff construction and asymptotics of the finite cluster distribution for two-dimensional Bernoulli percolation. Comm. Math. Phys. 131, 1–50 (1990).

[3] G. Bellettini, M. Cassandro and E. Presutti: Constrained minima of nonlocal free energy functionals. J. Statist. Phys. 84, 1337–1349 (1996).

[4] O. Benois, T. Bodineau, P. Buttà and E. Presutti: On the validity of van der Waals theory of surface tension. Markov Process. Rel. Fields 3, 175–198 (1997).

[5] O. Benois, T. Bodineau and E. Presutti: Large deviations in the van der Waals limit. Stochastic Process. Appl. 75, 89–104 (1998).

[6] T. Bodineau: The Wulff construction in three and more dimensions. Comm. Math. Phys. 207 no. 1, 197–229 (1999).

[7] T. Bodineau, D. Ioffe and Y. Velenik: Rigorous probabilistic analysis of equilibrium crystal shapes. Probabilistic techniques in equilibrium and nonequilibrium statistical physics. J. Math. Phys. 41 no. 3, 1033–1098 (2000).

[8] R. Cerf: Large deviations of the finite cluster shape for two-dimensional percolation in the Hausdorff and L^1 metric, J. Theoret. Probab. 13 no. 2, 491–518 (2000).

[9] R. Cerf: Large deviations for three dimensional supercritical percolation. Astérisque 267 (2000).

[10] R. Cerf and A. Pisztora: On the Wulff crystal in the Ising model. Ann. Probab. 28 no. 3, 945–1015 (2000).

[11] R. Cerf and A. Pisztora: Phase coexistence in Ising, Potts and percolation models. Ann. Inst. Henri Poincaré, Probab. Stat., 2001, to appear.

[12] E. De Giorgi: Nuovi teoremi relativi alle misure $(r-1)$-dimensionali in uno spazio ad r dimensioni. Ricerche Mat. 4, 95–113 (1955).

[13] E. De Giorgi: Sulla proprieta isoperimetrica dell'ipersfera, nella classe degli insiemi aventi frontiera orientata di misura finita. Atti Accad. Naz. Lincei, Cl. Sci. Fis. Mat. Nat., VIII Ser. 5, 33–44 (1958).

[14] J.-D. Deuschel and A. Pisztora: Surface order large deviations for high-density percolation. Probab. Theory Related Fields 104, 467–482 (1996).

[15] R. K. Dobrushin: Gibbs state describing coexistence of phases for a three-dimensional Ising model. Theor. Probability Appl. 17 no. 4, 582–600 (1973).

[16] R. K. Dobrushin and O. Hryniv: Fluctuations of the phase boundary in the 2D Ising ferromagnet. Comm. Math. Phys. 189, 395–445 (1997).

[17] R. L. Dobrushin, R. Kotecký and S. B. Shlosman: Wulff construction: a global shape from local interaction. AMS translations series, Providence (Rhode Island) (1992).

[18] I. Fonseca: The Wulff theorem revisited. Proc. R. Soc. Lond. Ser. A 432 No. 1884, 125–145 (1991).

[19] I. Fonseca and S. Müller: A uniqueness proof for the Wulff theorem. Proc. R. Soc. Edinb. Sect. A 119 No. 1/2, 125–136 (1991).

[20] E. Giusti: Minimal surfaces and functions of bounded variation. Birkhäuser (1984).

[21] G. R. Grimmett: The stochastic random-cluster process and the uniqueness of random-cluster measures. Ann. Probab. 23, 1461–1510 (1995).

[22] O. Hryniv: On local behaviour of the phase separation line in the 2D Ising model. Probab. Theory Related Fields 110, 91–107 (1998).

[23] D. Ioffe: Large deviations for the 2D Ising model: a lower bound without cluster expansions. J. Statist. Phys. 74, 411–432 (1993).

[24] D. Ioffe: Exact large deviation bounds up to T_c for the Ising model in two dimensions. Probab. Theory Related Fields 102, 313–330 (1995).

[25] D. Ioffe and R. Schonmann: Dobrushin-Kotecký-Shlosman Theorem up to the critical temperature. Comm. Math. Phys. 199, 117–167 (1998).

[26] H. Kesten and Y. Zhang: The probability of a large finite cluster in supercritical Bernoulli percolation. Ann. Probab. 18, 537–555 (1990).

[27] A. Messager, S. Miracle-Solé and J. Ruiz: Convexity properties of the surface tension and equilibrium crystals. J. Statist. Phys. 67 nos. 3/4, 449–469 (1992).

[28] S. Miracle-Solé: Surface tension, step free energy, and facets in the equilibrium crystal. J. Statist. Phys. 79 nos. 1/2, 183–214 (1995).

[29] C. E. Pfister: Large deviations and phase separation in the two-dimensional Ising model. Helv. Phys. Acta 64, 953–1054 (1991).

[30] C. E. Pfister and Y. Velenik: Large deviations and continuum limit in the 2D Ising model. Probab. Theory Related Fields 109, 435–506 (1997).

[31] A. Pisztora: Surface order large deviations for Ising, Potts and percolation models. Probab. Theory Related Fields 104, 427–466 (1996).

[32] R. Schonmann and S. B. Shlosman: Constrained variational problem with applications to the Ising model. J. Statist. Phys. 83, 867–905 (1996).

[33] R. Schonmann and S. B. Shlosman: Complete analyticity for the 2D Ising model completed. Comm. Math. Phys. 170, 453–482 (1996).

[34] R. Schonmann and S. B. Shlosman: Wulff droplets and the metastable relaxation of kinetic Ising models. Comm. Math. Phys. 194, 389–462 (1998).

[35] J. E. Taylor: Crystalline variational problems. Bull. Am. Math. Soc. 84 no. 4, 568–588 (1978).

[36] G. Wulff: Zur Frage der Geschwindigkeit des Wachstums und der Auflösung der Kristallflächen. Z. Kristallogr. 34, 449–530 (1901).

Université Paris Sud
Mathématique
Bâtiment 425
91405 Orsay Cedex, France
E-mail address: Raphael.Cerf@math.u-psud.fr

Constructing Compact 8-Manifolds with Holonomy Spin(7) from Calabi–Yau Orbifolds

Dominic Joyce

Abstract. Compact Riemannian 7- and 8-manifolds with holonomy G_2 and Spin(7) were first constructed by the author in 1994–5, by resolving orbifolds T^7/Γ and T^8/Γ. This paper describes a new construction of compact 8-manifolds with holonomy Spin(7). We start with a Calabi–Yau 4-orbifold Y with isolated singularities of a special kind. We divide by an antiholomorphic involution σ of Y to get a real 8-orbifold $Z = Y/\langle\sigma\rangle$. Then we resolve the singularities of Z to get a compact 8-manifold M, which has metrics with holonomy Spin(7). Manifolds constructed in this way typically have large fourth Betti number $b^4(M)$.

Let M be an n-dimensional manifold and g a Riemannian metric on M. The *holonomy group* $\mathrm{Hol}(g)$ of g is the group of isometries of T_xM generated by parallel transport around closed loops based at x in M, using the Levi-Civita connection ∇ of g. We consider $\mathrm{Hol}(g)$ to be a subgroup of $O(n)$, defined up to conjugation by elements of $O(n)$. Then $\mathrm{Hol}(g)$ is independent of the base point x in M.

The classification of possible holonomy groups was due to Berger [1], although Alekseevskii later eliminated Spin(9) from Berger's list.

Theorem. (Berger) *Let M be a simply-connected, n-dimensional manifold, and g an irreducible, nonsymmetric Riemannian metric on M. Then either*

(i) $\mathrm{Hol}(g) = SO(n)$,
(ii) $n = 2m$ *and* $\mathrm{Hol}(g) = SU(m)$ *or* $U(m)$,
(iii) $n = 4m$ *and* $\mathrm{Hol}(g) = \mathrm{Sp}(m)$ *or* $\mathrm{Sp}(m)\,\mathrm{Sp}(1)$,
(iv) $n = 7$ *and* $\mathrm{Hol}(g) = G_2$, *or*
(v) $n = 8$ *and* $\mathrm{Hol}(g) = \mathrm{Spin}(7)$.

The exceptional cases G_2 and Spin(7) in this classification are called the *exceptional holonomy groups*. For some time after Berger's classification, they remained a mystery. In 1987, Bryant [2] used the theory of exterior differential systems to show that locally there exist many metrics with holonomy G_2 and Spin(7), and gave some explicit, incomplete examples. Then in 1989, Bryant and Salamon [3] found explicit, *complete* metrics with holonomy G_2 and Spin(7) on noncompact manifolds.

In 1994–5, the author found examples of metrics with holonomy G_2 and Spin(7) on *compact* manifolds [6, 7], by resolving the singularities of orbifolds T^7/Γ

and T^8/Γ. These are described at length in Chapters 11–14 of the author's book [9], which also makes the construction substantially more powerful by incorporating Calabi Conjecture methods, and gives many more examples.

The subject of this paper is a *second* construction of compact 8-manifolds with holonomy Spin(7), which was published by the author in [8] and [9, Chapter 15]. As in the first construction, we begin with a compact Spin(7)-orbifold, and resolve its singularities to get a Spin(7)-manifold.

However, in this case the initial orbifold is not T^8/Γ but a quotient $Y/\langle\sigma\rangle$ of a Calabi–Yau 4-orbifold Y by an antiholomorphic involution σ. It turns out that if Y and σ are carefully chosen so that the singularities of $Y/\langle\sigma\rangle$ are of a very special kind, then we can resolve $Y/\langle\sigma\rangle$ with holonomy Spin(7).

1. The Holonomy Group Spin(7)

Let $\mathbb{R}^8$ have coordinates $(x_1,\dots,x_8)$. Write $\mathrm{d}\mathbf{x}_{ijkl}$ for the 4-form $\mathrm{d}x_i\wedge\mathrm{d}x_j\wedge\mathrm{d}x_k\wedge\mathrm{d}x_l$ on $\mathbb{R}^8$. Define a 4-form Ω_0 on $\mathbb{R}^8$ by

$$\begin{aligned}\Omega_0 = {} & \mathrm{d}\mathbf{x}_{1234}+\mathrm{d}\mathbf{x}_{1256}+\mathrm{d}\mathbf{x}_{1278}+\mathrm{d}\mathbf{x}_{1357}-\mathrm{d}\mathbf{x}_{1368}\\ & -\mathrm{d}\mathbf{x}_{1458}-\mathrm{d}\mathbf{x}_{1467}-\mathrm{d}\mathbf{x}_{2358}-\mathrm{d}\mathbf{x}_{2367}-\mathrm{d}\mathbf{x}_{2457}\\ & +\mathrm{d}\mathbf{x}_{2468}+\mathrm{d}\mathbf{x}_{3456}+\mathrm{d}\mathbf{x}_{3478}+\mathrm{d}\mathbf{x}_{5678}\,.\end{aligned}\tag{1}$$

The subgroup of $\mathrm{GL}(8,\mathbb{R})$ preserving Ω_0 is the holonomy group Spin(7). This group also preserves the orientation on $\mathbb{R}^8$ and the Euclidean metric $g_0=\mathrm{d}x_1^2+\cdots+\mathrm{d}x_8^2$. It is a compact, semisimple, 21-dimensional Lie group, a subgroup of SO(8).

A Spin(7)-structure on an 8-manifold M gives rise to a 4-form Ω and a metric g on M, such that each tangent space of M admits an isomorphism with $\mathbb{R}^8$ identifying Ω and g with Ω_0 and g_0 respectively. By an abuse of notation we will refer to the pair (Ω,g) as a Spin(7)-structure.

Proposition 1.1. *Let M be a compact 8-manifold and (Ω,g) a* Spin(7)*-structure on M. Then the following are equivalent:*

(i) $\mathrm{Hol}(g)\subseteq\mathrm{Spin}(7)$, *and Ω is the induced 4-form,*
(ii) $\nabla\Omega=0$ *on M, where ∇ is the Levi-Civita connection of g, and*
(iii) $\mathrm{d}\Omega=0$ *on M.*

We call $\nabla\Omega$ the *torsion* of the Spin(7)-structure (Ω,g), and (Ω,g) *torsion-free* if $\nabla\Omega=0$. A triple (M,Ω,g) is called a Spin(7)-*manifold* (or Spin(7)-*orbifold*) if M is an 8-manifold (or 8-orbifold) and (Ω,g) a torsion-free Spin(7)-structure on M. If g has holonomy $\mathrm{Hol}(g)\subseteq\mathrm{Spin}(7)$, then g is Ricci-flat.

Here is a result on *compact* 8-manifolds with holonomy Spin(7).

Theorem 1.2. *Let (M,Ω,g) be a compact* Spin(7)*-manifold. Then* $\mathrm{Hol}(g)=\mathrm{Spin}(7)$ *if and only if M is simply-connected, and $b^3(M)+b^4_+(M)=b^2(M)+b^4_-(M)+25$. In this case the moduli space of metrics with holonomy* Spin(7) *on M, up to diffeomorphisms isotopic to the identity, is a smooth manifold of dimension $1+b^4_-(M)$.*

Next we sketch the construction in [7] of compact 8-manifolds with holonomy Spin(7). One begins with a flat Spin(7)-structure (Ω_0, g_0) on the 8-torus T^8, and a finite group Γ of isometries of T^8 preserving (Ω_0, g_0). Then T^8/Γ is an *orbifold*, a singular manifold with only quotient singularities.

For certain Γ one can resolve the singularities of T^8/Γ in a natural way, using complex geometry. This gives a nonsingular, compact 8-manifold M, and a projection $\pi\colon M \to T^8/\Gamma$. We write down a family of Spin(7)-structures (Ω_t, g_t) on M for $t \in (0, \epsilon)$, such that (Ω_t, g_t) has small torsion for small t, and converges to the singular Spin(7)-structure $\pi^*(\Omega_0, g_0)$ as $t \to 0$.

Finally, we prove using analysis that for small t, the Spin(7)-structure (Ω_t, g_t) can be deformed to a nearby Spin(7)-structure $(\tilde{\Omega}_t, \tilde{g}_t)$ on M, with zero torsion. If M satisfies the topological conditions of Theorem 1.2, then $\tilde{g}_t$ has holonomy Spin(7).

Here is the analytic result used in the last stage of this construction. It is proved in [7, Theorem A and Theorem B], and also in [9, Chapter 13].

Theorem 1.3. *Let λ, μ, ν be positive constants. Then there exist positive constants κ, K such that whenever $0 < t \le \kappa$, the following is true.*

Let M be a compact 8-manifold, and (Ω, g) a Spin(7)*-structure on M. Suppose that ϕ is a smooth 4-form on M with $\mathrm{d}\Omega + \mathrm{d}\phi = 0$, and*

(i) $\|\phi\|_{L^2} \le \lambda t^{9/2}$ *and* $\|\mathrm{d}\phi\|_{L^{10}} \le \lambda t$,
(ii) *the injectivity radius $\delta(g)$ satisfies $\delta(g) \ge \mu t$, and*
(iii) *the Riemann curvature $R(g)$ satisfies $\|R(g)\|_{C^0} \le \nu t^{-2}$.*

Then there exists a smooth, torsion-free Spin(7)*-structure $(\tilde{\Omega}, \tilde{g})$ on M with $\|\tilde{\Omega} - \Omega\|_{C^0} \le K t^{1/2}$.*

Here is how to interpret this theorem. As $\nabla\Omega = 0$ if and only if $\mathrm{d}\Omega = 0$, and $\mathrm{d}\phi + \mathrm{d}\Omega = 0$, the torsion $\nabla\Omega$ is determined by $\mathrm{d}\phi$. Thus we can think of ϕ as a *first integral of the torsion* of (Ω, g). So $\|\phi\|_{L^2}$ and $\|\mathrm{d}\phi\|_{L^{10}}$ are both measures of the torsion of (Ω, g). As t is small, part (i) of the theorem says that (Ω, g) has *small torsion* in a certain sense.

Parts (ii) and (iii) say that the injectivity radius of g should not be too small, and its curvature not too large. When a metric becomes singular, in general its injectivity radius becomes zero and its curvature infinite. So we can interpret (ii) and (iii) to mean that g is not too close to being singular.

Thus, the theorem as a whole says that if the torsion of (Ω, g) is small enough, and g is not too singular, then we can deform (Ω, g) to a nearby, torsion-free Spin(7)-structure $(\tilde{\Omega}, \tilde{g})$ on M.

We prove Theorem 1.3 using analysis: we write the condition that $(\tilde{\Omega}, \tilde{g})$ be torsion-free as a nonlinear elliptic p.d.e., which can be approximated by a linear elliptic p.d.e. when $\tilde{\Omega} - \Omega$ is small. Then we use tools such as Sobolev spaces, the Sobolev Embedding Theorem and elliptic regularity to show that this nonlinear elliptic p.d.e. has a smooth solution.

2. Calabi–Yau Orbifolds

We now give a brief introduction to Calabi–Yau geometry, and the relation between Calabi–Yau 4-folds and Spin(7)-manifolds. A suitable reference is [9, Chapter 6]. For convenience we work in the orbifold category.

A *Calabi–Yau orbifold* is a compact Kähler orbifold (Y, J, g) of dimension m, with $\mathrm{Hol}(g) \subseteq \mathrm{SU}(m)$. Calabi–Yau orbifolds are automatically Ricci-flat. Using Yau's proof of the Calabi conjecture [10], one can show that suitable complex orbifolds admit metrics making them Calabi–Yau.

Theorem 2.1. *Let (Y, J) be a compact complex orbifold admitting Kähler metrics, with trivial canonical bundle. Then there is a unique Ricci-flat Kähler metric in each Kähler class, making Y into a Calabi–Yau orbifold.*

By elementary properties of holonomy groups, every Calabi–Yau orbifold (Y, J, g) admits a Kähler form ω and a complex volume form θ, both constant under the Levi-Civita connection ∇, such that near every point $p \in Y$ we can choose complex coordinates $(z_1, \ldots, z_m)$ in which

$$\begin{aligned} g &= |\mathrm{d}z_1|^2 + \cdots + |\mathrm{d}z_m|^2, \\ \omega &= \frac{i}{2}(\mathrm{d}z_1 \wedge \mathrm{d}\bar{z}_1 + \cdots + \mathrm{d}z_m \wedge \mathrm{d}\bar{z}_m), \text{ and} \\ \theta &= \mathrm{d}z_1 \wedge \cdots \wedge \mathrm{d}z_m \quad \text{at } p. \end{aligned} \tag{2}$$

A Calabi–Yau 4-orbifold (Y, J, g) has $\mathrm{Hol}(g) \subseteq \mathrm{SU}(4)$. But $\mathrm{SU}(4) \subset \mathrm{Spin}(7) \subset \mathrm{SO}(8)$. Therefore $\mathrm{Hol}(g) \subset \mathrm{Spin}(7)$, and it follows from general principles of holonomy that g must be part of a torsion-free Spin(7)-structure (Ω, g) on Y. We shall show that we may take Ω to be $\frac{1}{2}\omega \wedge \omega + \mathrm{Re}(\theta)$.

Let $p \in Y$, and choose complex coordinates $(z_1, \ldots, z_4)$ near p satisfying (2). Define real coordinates $(x_1, \ldots, x_8)$ near p by $(z_1, \ldots, z_4) = (x_1 + ix_2, x_3 + ix_4, x_5 + ix_6, x_7 + ix_8)$. Then from (2) we see that at p we have

$$\begin{aligned} g &= \mathrm{d}x_1^2 + \cdots + \mathrm{d}x_8^2, \\ \omega &= \mathbf{dx}_{12} + \mathbf{dx}_{34} + \mathbf{dx}_{56} + \mathbf{dx}_{78} \quad \text{and} \end{aligned}$$

$$\mathrm{Re}(\theta) = \mathbf{dx}_{1357} - \mathbf{dx}_{1368} - \mathbf{dx}_{1458} - \mathbf{dx}_{1467} - \mathbf{dx}_{2358} - \mathbf{dx}_{2367} - \mathbf{dx}_{2457} + \mathbf{dx}_{2468},$$

where $\mathbf{dx}_{ij\ldots l} = \mathrm{d}x_i \wedge \mathrm{d}x_j \wedge \cdots \wedge \mathrm{d}x_l$.

It follows from this equation that $\Omega = \frac{1}{2}\omega \wedge \omega + \mathrm{Re}(\theta)$ coincides with the 4-form Ω_0 defined in (1). As this holds for all $p \in Y$, we see that (Ω, g) is a Spin(7)-*structure* on Y. But $\nabla\omega = \nabla\theta = 0$, so $\nabla\Omega = 0$, and (Ω, g) is torsion-free. This shows that Calabi–Yau 4-orbifolds are also Spin(7)-orbifolds.

3. ALE Spin(7)-Manifolds

Asymptotically Locally Euclidean manifolds, or *ALE manifolds* for short, are a class of noncompact Riemannian manifolds with one end modelled asymptotically on a quotient singularity $\mathbb{R}^n/G$.

Definition 3.1. *Let (X,Ω,g) be a* Spin(7)*-manifold, and G a finite subgroup of* Spin(7) *which acts freely on $\mathbb{R}^8\setminus\{0\}$. We call (X,Ω,g) an* ALE Spin(7)-manifold *asymptotic to $\mathbb{R}^8/G$ if there exists a surjective, continuous map $\pi\colon X\to\mathbb{R}^8/G$ such that $\pi\colon X\setminus\pi^{-1}(0)\to(\mathbb{R}^8/G)\setminus\{0\}$ is a diffeomorphism, and $\pi^{-1}(0)$ is compact, and*

$$\begin{aligned}\nabla^l(\pi_*(g)-g_0)&=O(r^{-8-l})\\ \text{and}\quad \nabla^l(\pi_*(\Omega)-\Omega_0)&=O(r^{-8-l})\end{aligned}\quad \text{on } \{x\in\mathbb{R}^8/G: r(x)>1\},\ \text{for all } l\geq 0.$$

Here $g_0=\mathrm{d}x_1^2+\cdots+\mathrm{d}x_8^2$ and ∇ its Levi-Civita connection, Ω_0 is the Spin(7) *form on $\mathbb{R}^8$ given in* (1)*, and r is the radius function on $\mathbb{R}^8/G$.*

The point of this is that if (X,Ω_X,g_X) is an ALE Spin(7)-manifold asymptotic to $\mathbb{R}^8/G$, and (Y,Ω_Y,g_Y) is a Spin(7)-orbifold with just one singular point p modelled on $\mathbb{R}^8/G$, then we can join X and Y together to get a nonsingular 8-manifold M. Moreover, it may be possible to glue the Spin(7)-structures (Ω_X,g_X) and (Ω_Y,g_Y) together to get a torsion-free Spin(7)-structure (Ω,g) on M.

We now study an example. Define $\alpha,\beta\colon\mathbb{R}^8\to\mathbb{R}^8$ by

$$\begin{aligned}\alpha&\colon(x_1,\dots,x_8)\mapsto(-x_2,x_1,-x_4,x_3,-x_6,x_5,-x_8,x_7),\\ \beta&\colon(x_1,\dots,x_8)\mapsto(x_3,-x_4,-x_1,x_2,x_7,-x_8,-x_5,x_6).\end{aligned}$$

Then α,β preserve the standard Spin(7)-structure (Ω_0,g_0) on $\mathbb{R}^8$ given in section 1, so they lie in Spin(7). Also α, β satisfy $\alpha^4=\beta^4=1$, $\alpha^2=\beta^2$ and $\alpha\beta=\beta\alpha^3$. Let $G=\langle\alpha,\beta\rangle$. Then G is a finite nonabelian subgroup of Spin(7) of order 8, which acts freely on $\mathbb{R}^8\setminus\{0\}$.

In the next two examples we shall construct two topologically distinct ALE Spin(7)-manifolds (X_1,Ω_1,g_1) and (X_2,Ω_2,g_2) asymptotic to $\mathbb{R}^8/G$.

Example 3.2. *Define complex coordinates $(z_1,\dots,z_4)$ on $\mathbb{R}^8$ by*

$$(z_1,z_2,z_3,z_4)=(x_1+ix_2,x_3+ix_4,x_5+ix_6,x_7+ix_8).$$

Then $g_0=|\mathrm{d}z_1|^2+\cdots+|\mathrm{d}z_4|^2$, and $\Omega_0=\frac{1}{2}\omega_0\wedge\omega_0+\mathrm{Re}(\theta_0)$, where ω_0 and θ_0 are the usual Kähler form and complex volume form on $\mathbb{C}^4$. In these coordinates, α and β are given by

$$\begin{aligned}\alpha&\colon(z_1,\dots,z_4)\mapsto(iz_1,iz_2,iz_3,iz_4),\\ \beta&\colon(z_1,\dots,z_4)\mapsto(\bar z_2,-\bar z_1,\bar z_4,-\bar z_3).\end{aligned}\tag{3}$$

Now $\mathbb{C}^4/\langle\alpha\rangle$ is a complex singularity, as $\alpha\in$ SU(4)*. Let (Y_1,π_1) be the blow-up of $\mathbb{C}^4/\langle\alpha\rangle$ at* 0*. Then Y_1 is the unique crepant resolution of $\mathbb{C}^4/\langle\alpha\rangle$. The action of β on $\mathbb{C}^4/\langle\alpha\rangle$ lifts to a* free *antiholomorphic map $\beta\colon Y_1\to Y_1$ with $\beta^2=1$. Define $X_1=Y_1/\langle\beta\rangle$. Then X_1 is a nonsingular* 8*-manifold, and the projection $\pi_1\colon Y_1\to\mathbb{C}^4/\langle\alpha\rangle$ pushes down to $\pi_1\colon X_1\to\mathbb{R}^8/G$.*

There exist ALE Kähler metrics g_1 on Y_1 with holonomy SU(4)*, which were written down explicitly by Calabi* [4, p. 285]*. Each such g_1 is invariant under the action of β on Y_1. Let ω_1 be the Kähler form of g_1, and $\theta_1=\pi_1^*(\theta_0)$ the holomorphic* *volume form on* Y_1. *Define* $\Omega_1 = \frac{1}{2}\omega_1 \wedge \omega_1 + \mathrm{Re}(\theta_1)$. *Then* (Ω_1, g_1) *is a torsion-free* Spin(7)*-structure on* Y_1.

As $\beta^*(\omega_1) = -\omega_1$ *and* $\beta^*(\theta_1) = \bar{\theta}_1$, *we see that* β *preserves* (Ω_1, g_1). *Thus* (Ω_1, g_1) *pushes down to a torsion-free* Spin(7)*-structure* (Ω_1, g_1) *on* X_1. *Then* (X_1, Ω_1, g_1) *is an* ALE Spin(7)-manifold *asymptotic to* $\mathbb{R}^8/G$. *The Betti numbers of* X_1 *are* $b^1 = b^2 = b^3 = 0$ *and* $b^4 = 1$, *and* $\pi_1(X_1) = \mathbb{Z}_2$.

Example 3.3. *Define new complex coordinates* $(w_1, \dots, w_4)$ *on* $\mathbb{R}^8$ *by*

$$(w_1, w_2, w_3, w_4) = (-x_1 + ix_3, x_2 + ix_4, -x_5 + ix_7, x_6 + ix_8).$$

Again we find that $g_0 = |\mathrm{d}w_1|^2 + \dots + |\mathrm{d}w_4|^2$ *and* $\Omega_0 = \frac{1}{2}\omega_0 \wedge \omega_0 + \mathrm{Re}(\theta_0)$. *In these coordinates,* α *and* β *are given by*

$$\begin{aligned} \alpha &: (w_1, \dots, w_4) \mapsto (\bar{w}_2, -\bar{w}_1, \bar{w}_4, -\bar{w}_3), \\ \beta &: (w_1, \dots, w_4) \mapsto (iw_1, iw_2, iw_3, iw_4). \end{aligned} \tag{4}$$

Observe that (3) *and* (4) *are the same, except that the rôles of* α, β *are reversed. Therefore we can use the ideas above again.*

Let Y_2 *be the crepant resolution of* $\mathbb{C}^4/\langle\beta\rangle$. *The action of* α *on* $\mathbb{C}^4/\langle\beta\rangle$ *lifts to a free antiholomorphic involution of* Y_2. *Let* $X_2 = Y_2/\langle\alpha\rangle$. *Then* X_2 *is nonsingular, and carries a torsion-free* Spin(7)*-structure* (Ω_2, g_2), *making* (X_2, Ω_2, g_2) *into an ALE* Spin(7)*-manifold asymptotic to* $\mathbb{R}^8/G$.

In effect, the two sets of complex coordinates z_j and w_j define two *different* complex structures on $\mathbb{R}^8$, and two *different* embeddings of SU(4) in Spin(7). In the first example $\alpha \in \mathrm{SU}(4)$ but $\beta \notin \mathrm{SU}(4)$, and we use complex geometry with respect to the first complex structure to resolve $\mathbb{C}^4/\langle\alpha\rangle$, in a β-equivariant way. In the second example $\beta \in \mathrm{SU}(4)$ but $\alpha \notin \mathrm{SU}(4)$, and we use the second complex structure to resolve $\mathbb{C}^4/\langle\beta\rangle$, in an α-equivariant way.

Now (X_1, Ω_1, g_1), (X_2, Ω_2, g_2) are clearly isomorphic as Spin(7)-manifolds, but they should be regarded as *topologically distinct* ALE manifolds, because the isomorphism between them acts nontrivially on $\mathbb{R}^8/G$. Thus, we have found two topologically distinct ALE Spin(7)-manifolds (X_1, Ω_1, g_1), (X_2, Ω_2, g_2) asymptotic to the same singularity $\mathbb{R}^8/G$.

Let i be 1 or 2, and let $t > 0$. Then $(t^4\Omega_i, t^2 g_i)$ is also a torsion-free Spin(7)-structure on X_i, which is homothetic to (Ω_i, g_i). That is, we obtain $(t^4\Omega_i, t^2 g_i)$ by shrinking distances on X_i by a factor t. Thus, $(X_i, t^4\Omega_i, t^2 g_i)$ is a Spin(7)-*manifold* for each $t > 0$.

Furthermore, it is an *ALE* Spin(7)-manifold. For if $\pi_i \colon X_i \to \mathbb{R}^8/G$ satisfies the conditions of Definition 3.1 for (Ω_i, g_i), then $t\pi_i \colon X_i \to \mathbb{R}^8/G$ satisfies the conditions of Definition 3.1 for $(t^4\Omega_i, t^2 g_i)$. This means that X_i carries not just one, but a whole 1-parameter family of ALE Spin(7)-structures $(t^4\Omega_i, t^2 g_i)$.

4. The Construction

Starting with a Calabi–Yau 4-orbifold Y with isolated singularities of a certain kind, and an antiholomorphic involution σ on Y, we will now explain how to construct a family of compact 8-manifolds $M_{\mathbf{i}}$ by resolving $Z = Y/\langle\sigma\rangle$, and show that there exist torsion-free Spin(7)-structures $(\tilde{\Omega}, \tilde{g})$ on $M_{\mathbf{i}}$, which have holonomy Spin(7) if $M_{\mathbf{i}}$ is simply-connected. The following conditions set out the ingredients for our construction.

Condition 4.1. *Let (Y, J) be a compact complex 4-orbifold with trivial canonical bundle, admitting Kähler metrics. Let σ be an antiholomorphic involution on Y. Define $\alpha\colon \mathbb{C}^4 \to \mathbb{C}^4$ by*

$$\alpha\colon (z_1, z_2, z_3, z_4) \longmapsto (iz_1, iz_2, iz_3, iz_4). \tag{5}$$

Then $\langle\alpha\rangle \cong \mathbb{Z}_4$, and $\mathbb{C}^4/\langle\alpha\rangle$ has an isolated singular point at 0. Suppose that the singular set of Y is k isolated points $p_1, \dots, p_k$ for some $k \geq 1$, each modelled on $\mathbb{C}^4/\langle\alpha\rangle$, and that the fixed set of σ in Y is exactly $\{p_1, \dots, p_k\}$. Suppose also that $Y \setminus \{p_1, \dots, p_k\}$ is simply-connected, and $h^{2,0}(Y) = 0$.

Assume these conditions hold. Choose a σ-invariant Kähler class on Y. By Theorem 2.1, there is a unique g_Y in this Kähler class making (Y, J, g_Y) into a Calabi–Yau 4-orbifold. Clearly g_Y is σ-invariant. Let ω_Y be the Kähler form of g_Y. Then $\sigma^*(\omega_Y) = -\omega_Y$, as σ is antiholomorphic.

As in Section 2, there exists a holomorphic volume form θ_Y on Y such that g_Y, ω_Y and θ_Y may be written in the form (2) at each point of Y. Multiplying θ_Y by a phase $e^{i\phi}$ if necessary, we can also ensure that $\sigma^*(\theta_Y) = \bar{\theta}_Y$.

Define $\Omega_Y = \frac{1}{2}\omega_Y \wedge \omega_Y + \mathrm{Re}(\theta_Y)$. Then (Ω_Y, g_Y) is a torsion-free Spin(7)-structure on Y, as in Section 2. Furthermore, the equations $\sigma^*(\omega_Y) = -\omega_Y$ and $\sigma^*(\theta_Y) = \bar{\theta}_Y$ imply that $\sigma^*(\Omega_Y) = \Omega_Y$. Thus (Ω_Y, g_Y) is σ-invariant.

Define $Z = Y/\langle\sigma\rangle$. Then Z is a compact, real 8-orbifold. As (Ω_Y, g_Y) is σ-invariant, it pushes down to give a torsion-free Spin(7)-structure (Ω_Z, g_Z) on Z. Thus, (Z, Ω_Z, g_Z) is a Spin(7)-*orbifold*.

Condition 4.1 implies that the singularities of Z are k points $p_1, \dots, p_k$. Furthermore, it can be shown that each singularity p_j is modelled on the singularity $\mathbb{R}^8/G$ considered in Section 3, in a way which identifies the Spin(7)-structures on $T_{p_j}Z$ and $\mathbb{R}^8/G$.

Now Examples 3.2 and 3.3 defined two distinct ALE Spin(7)-manifolds (X_i, Ω_i, g_i) asymptotic to $\mathbb{R}^8/G$. For $j = 1, \dots, k$, choose $i_j = 1$ or 2, and let $\mathbf{i} = (i_1, \dots, i_k)$. Define a compact, nonsingular 8-manifold $M_{\mathbf{i}}$ by resolving each singular point p_j in Z using the ALE 8-manifold X_{i_j}. We explain how to do this in [8, Section 5.3] and [9, Section 15.2.2]; the basic idea is to identify a small ball about p_j in Z with a small ball about 0 in $\mathbb{R}^8/G$, and then glue together open sets in Z and X_{i_j} along an annulus around 0 in $\mathbb{R}^8/G$.

As there are 2^k possible choices for the i_j, this gives 2^k 8-manifolds $M_{\mathbf{i}}$ resolving Z. Calculation shows that one of these 8-manifolds (with $i_j = 1$ for all

j) has $\pi_1(M_{\mathbf{i}}) = \mathbb{Z}_2$, and the rest are all simply-connected. Also, all the $M_{\mathbf{i}}$ have the same Betti numbers b^k, the sum of $b^k(Z)$ and the $b^k(X_{i_j})$ for $0 < k < 8$, which satisfy $b^3(M_{\mathbf{i}}) + b^4_+(M_{\mathbf{i}}) = b^2(M_{\mathbf{i}}) + b^4_-(M_{\mathbf{i}}) + 25$.

Next, we define a 1-parameter family of Spin(7)-structures (Ω^t, g^t) on $M_{\mathbf{i}}$ depending on $t \in (0, \epsilon]$, for some small $\epsilon > 0$, and a corresponding family of 4-forms ϕ^t on $M_{\mathbf{i}}$ such that $d\Omega^t + d\phi^t = 0$. The general idea is that ϕ^t is a measure of the *torsion* of (Ω^t, g^t), and the definitions are set up very carefully to ensure that ϕ^t is small when t is small. That is, when t is small, (Ω^t, g^t) has *small torsion.*

The details of the definition of (Ω^t, g^t) and ϕ^t are rather complicated, but here is a rough sketch. Recall from Section 3 that X_{i_j} carries a 1-parameter family of Spin(7)-structures $(t^4\Omega_{i_j}, t^2 g_{i_j})$ depending on $t > 0$. We make (Ω^t, g^t) by gluing together the torsion-free Spin(7)-structures (Ω_Z, g_Z) on Z and $(t^4\Omega_{i_j}, t^2 g_{i_j})$ on X_{i_j}, using a partition of unity on the $k+1$ patches used to define $M_{\mathbf{i}}$.

Because Ω^t must be a Spin(7) form, which does not hold for generic 4-forms but implies a pointwise restriction, we are unable to make Ω^t closed. Instead, we construct a 4-form ϕ^t such that $\Omega^t + \phi^t$ is closed, and try and arrange that ϕ^t is as small as possible.

We then prove the following estimates on (Ω^t, g^t) and ϕ^t.

Theorem 4.2. *In the situation above, there exist constants $\lambda, \mu, \nu > 0$ such that for all $t \in (0, \epsilon]$ we have*

(i) $\|\phi^t\|_{L^2} \le \lambda t^{24/5}$ *and* $\|d\phi^t\|_{L^{10}} \le \lambda t^{36/25}$*;*
(ii) *the injectivity radius $\delta(g^t)$ satisfies $\delta(g^t) \ge \mu t$; and*
(iii) *the Riemann curvature $R(g^t)$ satisfies $\|R(g^t)\|_{C^0} \le \nu t^{-2}$.*

Here all norms are calculated using the metric g^t on $M_{\mathbf{i}}$.

This should be compared with Theorem 1.3. Parts (ii) and (iii) of both results are the same, and the powers of t in part (i) of Theorem 1.3 are smaller than those in part (i) above, which means that part (i) of Theorem 4.2 is stronger for small t. So combining the two theorems shows that for small t there exists a torsion-free Spin(7)-structure $(\tilde{\Omega}, \tilde{g})$ on $M_{\mathbf{i}}$ close to (Ω^t, g^t).

Thus, $(M_{\mathbf{i}}, \tilde{\Omega}, \tilde{g})$ is a compact Spin(7)-manifold. We can then use Theorem 1.2 to determine the holonomy group of $\tilde{g}$. If $M_{\mathbf{i}}$ is simply-connected, which holds for $2^k - 1$ choices of $\mathbf{i}$, then $\mathrm{Hol}(\tilde{g}) = \mathrm{Spin}(7)$. If $\pi_1(M_{\mathbf{i}}) = \mathbb{Z}_2$, which holds for the remaining one choice of $\mathbf{i}$, it turns out that $\mathrm{Hol}(\tilde{g}) = \mathbb{Z}_2 \ltimes \mathrm{SU}(4)$. This proves our main result.

Theorem 4.3. *Suppose Condition 4.1 holds. Then there are $2^k - 1$ simply-connected, compact 8-manifolds $M_{\mathbf{i}}$ resolving $Z = Y/\langle\sigma\rangle$, which carry metrics with holonomy* Spin(7).

We do not claim here that the 8-manifolds $M_{\mathbf{i}}$ are distinct as smooth 8-manifolds, but only that they are topologically distinct as resolutions of Z.

5. An Example

Next we give an example of this construction. The main idea we shall use is borrowed from Candelas, Lynker and Schrimmrigk [5], who constructed a large number of Calabi–Yau 3-folds as crepant resolutions of hypersurfaces in weighted projective spaces $\mathbb{CP}^4_{a_0,\dots,a_4}$. Here *weighted projective spaces* are an important class of complex orbifolds.

Definition 5.1. *Let $m \geq 1$ be an integer, and $a_0, a_1, \dots, a_m$ positive integers with highest common factor* 1. *Let $\mathbb{C}^{m+1}$ have complex coordinates $(z_0, \dots, z_m)$, and define an action of the complex Lie group $\mathbb{C}^*$ on $\mathbb{C}^{m+1}$ by*

$$(z_0, \dots, z_m) \overset{u}{\longmapsto} (u^{a_0} z_0, \dots, u^{a_m} z_m), \qquad \textit{for } u \in \mathbb{C}^*.$$

The weighted projective space $\mathbb{CP}^m_{a_0,\dots,a_m}$ *is* $\big(\mathbb{C}^{m+1} \setminus \{0\}\big)/\mathbb{C}^*$.

Let Y be the hypersurface of degree 12 in $\mathbb{CP}^5_{1,1,1,1,4,4}$ given by

$$Y = \big\{[z_0, \dots, z_5] \in \mathbb{CP}^5_{1,1,1,1,4,4} : z_0^{12} + z_1^{12} + z_2^{12} + z_3^{12} + z_4^3 + z_5^3 = 0\big\}.$$

Calculation shows that Y has trivial canonical bundle and three singular points $p_1 = [0,0,0,0,1,-1]$, $p_2 = [0,0,0,0,1,e^{\pi i/3}]$ and $p_3 = [0,0,0,0,1,e^{-\pi i/3}]$, all modelled on $\mathbb{C}^4/\langle\alpha\rangle$, where α is given in (5).

Now define a map $\sigma\colon Y \to Y$ by

$$\sigma\colon [z_0, \dots, z_5] \longmapsto [\bar z_1, -\bar z_0, \bar z_3, -\bar z_2, \bar z_5, \bar z_4].$$

Note that $\sigma^2 = 1$, though this is not immediately obvious, because of the geometry of $\mathbb{CP}^5_{1,1,1,1,4,4}$. It can be shown that Condition 4.1 holds for Y and σ. So we can apply the construction of Section 4, and resolve $Z = Y/\langle\sigma\rangle$ to get seven compact 8-manifolds M_{i} admitting metrics with holonomy Spin(7). A long computation proves that the M_{i} have Betti numbers

$$b^0 = 1,\ b^1 = b^2 = b^3 = 0,\ b^4 = 2446,\ b^4_+ = 1639 \text{ and } b^4_- = 807,$$

and so by Theorem 1.2 the metrics with holonomy Spin(7) on each M_{i} form a smooth family of dimension 808.

6. Conclusions

Finding examples of complex orbifolds (Y, J) satisfying Condition 4.1 is not that easy, and requires a degree of ingenuity. The requirements that Y have only $\mathbb{Z}_4$ singularities of the given type, and that σ fixes these points and no others, are very restrictive. Nonetheless, by considering hypersurfaces and complete intersections in weighted projective spaces, and using various tricks, the author was able in [8] and [9, Chapter 15] to construct compact 8-manifolds with holonomy Spin(7) realizing 14 distinct sets of Betti numbers, which are given in Table 1. Probably there are many other examples which can be produced by the same methods.

Comparing these Betti numbers with those of the compact 8-manifolds constructed in [7] and [9, Chapter 14] by resolving torus orbifolds T^8/Γ, we see that

TABLE 1. Betti numbers (b^2, b^3, b^4) of compact Spin(7)-manifolds

$(4, 33, 200)$	$(3, 33, 202)$	$(2, 33, 204)$	$(1, 33, 206)$	$(0, 33, 208)$
$(1, 0, 908)$	$(0, 0, 910)$	$(1, 0, 1292)$	$(0, 0, 1294)$	$(1, 0, 2444)$
$(0, 0, 2446)$	$(0, 6, 3730)$	$(0, 0, 4750)$	$(0, 0, 11\,662)$	

in these examples the middle Betti number b^4 is much bigger, as much as 11 662 in one case.

Given that the two constructions of compact 8-manifolds with holonomy Spin(7) that we know appear to produce sets of 8-manifolds with rather different 'geography', it is tempting to speculate that the set of all compact 8-manifolds with holonomy Spin(7) may be rather large, and that those constructed so far are a small sample with atypical behaviour.

References

[1] M. Berger, *Sur les groupes d'holonomie homogène des variétés à connexion affines et des variétés Riemanniennes*, Bull. Soc. Math. France 83 (1955) 279–330.

[2] R. L. Bryant, *Metrics with exceptional holonomy*, Ann. of Math. 126 (1987) 525–576.

[3] R. L. Bryant and S. M. Salamon, *On the construction of some complete metrics with exceptional holonomy*, Duke Math. J. 58 (1989) 829–850.

[4] E. Calabi, *Métriques kähleriennes et fibrés holomorphes*, Ann. scient. éc. norm. sup. 12 (1979) 269–294.

[5] P. Candelas, M. Lynker and R. Schimmrigk, *Calabi–Yau manifolds in weighted* $\mathbb{P}_4^*$, Nucl. Phys. B341 (1990) 383–402.

[6] D. D. Joyce, *Compact Riemannian* 7*-manifolds with holonomy* G_2*. I* and *II*, J. Diff. Geom. 43 (1996) 291–328 and 329–375.

[7] D. D. Joyce, *Compact* 8*-manifolds with holonomy Spin*(7), Inventiones mathematicae 123 (1996) 507–552.

[8] D. D. Joyce, *A new construction of compact* 8*-manifolds with holonomy* Spin(7), J. Diff. Geom. 53 (1999), 89–130. math.DG/9910002.

[9] D. D. Joyce, *Compact Riemannian manifolds with special holonomy*, OUP Mathematical Monographs, Oxford, 2000.

[10] S.-T. Yau, *On the Ricci curvature of a compact Kähler manifold and the complex Monge–Ampère equations. I*, Comm. pure appl. math. 31 (1978) 339–411.

University of Oxford
Mathematical Institute
24-29 St Giles
Oxford OX1 3LB, United Kingdom
E-mail address: joyce@maths.ox.ac.uk

Banach KK-Theory and the Baum–Connes Conjecture

Vincent Lafforgue

Abstract. The report below describes the applications of Banach KK-theory to a conjecture of P. Baum and A. Connes about the K-theory of group C^*-algebras.

1. Introduction

There are many excellent articles and surveys on Kasparov's KK-theory and the Baum–Connes conjecture (see [2, 16, 22, 44, 48]) and on the Banach KK-theory ([44, 48]). Here we shall not define Banach KK-theory (see [33, 44, 48]) but only give, in a self-contained way, an idea of why it can be applied to the Baum–Connes conjecture.

Let G be a second countable locally compact group. The Baum–Connes conjecture predicts the bijectivity of a morphism of abelian groups from a topological K-theory group (which is computable in principle) to the K-theory of the reduced C^*-algebra of G. This C^*-algebra is non-commutative but in some way it is an algebra of continuous functions on the tempered unitary dual of G, and its K-theory groups are topological invariants of this dual, but they still exist when the topology of the dual is very bad. They are a receptacle for indices.

The injectivity in the Baum–Connes conjecture implies the Novikov conjecture about higher signatures ([37, 38, 23, 25, 2]). The surjectivity implies the Kadison–Kaplanski conjecture on idempotents and is tied to many questions of harmonic analysis (e.g., the exhaustion of discrete series of simple Lie groups).

The injectivity of the Baum–Connes map is known in many cases (e.g. for closed subgroups of Lie or p-adic groups). As far as surjectivity is concerned, Higson and Kasparov [15] pushed Kasparov KK-theoretic method probably as far as it can go. Banach KK-theory gives a few new cases: general Lie or p-adic reductive groups and some discrete groups with property (T) but the question of surjectivity remains largely open. In the last sections we will try to put the two methods in the same framework and to explain attempts for new methods.

2. Generalized Fredholm Operators

For us a Banach algebra is a (non-necessarily unital) $\mathbb{C}$-algebra A that is complete for a norm $\|.\|$ satisfying $\|ab\| \leq \|a\|\|b\|$ for any $a, b \in A$. If A and B are Banach algebras a morphism $u\colon A \to B$ is an algebra morphism such that $\|u(a)\| \leq \|a\|$ for any $a \in A$.

Each Banach algebra A has two K-theory groups $K_0(A)$ and $K_1(A)$ and any morphism $u\colon A \to B$ induces two morphisms of abelian groups $u_*\colon K_0(A) \to K_0(B)$ and $u_*\colon K_1(A) \to K_1(B)$. If X is a locally compact space and $C_0(X)$ the algebra of continuous functions vanishing at infinity, $K_*(C_0(X))$ are the Atiyah–Hirzebruch K-theory groups. We shall focus on K_0 because $K_1(A)$ is isomorphic to $K_0(C_0(\mathbb{R}, A))$ and for technical reasons we shall restrict ourselves to unital Banach algebras in this section.

Let A (and sometimes B) be unital Banach algebras.

A right A-module E is finitely generated projective (f.g.p.) if and only if it is a direct summand in A^n for some integer n. The set of isomorphism classes of right f.g.p. A-module is a semigroup because the direct sum of two right f.g.p. A-modules is a right f.g.p. A-module. Then $K_0(A)$ is the universal group associated to this semigroup (that is the group of formal differences of elements of the semigroup). If $u\colon A \to B$ is a morphism of Banach algebras and $u(1) = 1$, and E is a right f.g.p. A-module then $E \otimes_A B$ is a right f.g.p. B-module and this defines $u_*\colon K_0(A) \to K_0(B)$.

There is another definition of $K_0(A)$ for which the functoriality is even more obvious: $K_0(A)$ is the quotient of the free abelian group generated by all idempotents p in $M_k(A)$ for some integer k, by the relations $\left[\begin{pmatrix} p & 0 \\ 0 & q \end{pmatrix}\right] = [p] + [q]$ for any idempotents $p \in M_k(A)$ and $q \in M_l(A)$ and $[p] = [q]$ if p, q are idempotents of $M_k(A)$ and are connected by a path of idempotents in $M_k(A)$ and $[0] = 0$ where 0 is an idempotent in $M_k(A)$. The link with the former definition is that any idempotent $p \in M_k(A)$ acts on the left on A^k as a projector P and $\operatorname{Im} P$ is a right f.g.p. A-module (it is a direct summand in the right A-module A^k).

For us a continuous $\mathbb{C}$-linear map between two Banach spaces is said to be $\mathbb{C}$-compact if it is a uniform limit of finite rank operators. If E_0 and E_1 are Banach spaces and $u \in \mathcal{L}_{\mathbb{C}}(E_0, E_1)$ and $v \in \mathcal{L}_{\mathbb{C}}(E_1, E_0)$ are continuous $\mathbb{C}$-linear maps such that $\mathrm{Id}_{E_1} - uv$ and $\mathrm{Id}_{E_0} - vu$ are $\mathbb{C}$-compact, then $\operatorname{Ker} u$ has finite dimension and $\operatorname{Im} u$ is closed and has finite codimension in E_1, and the index of the Fredholm operator u is the formal difference of the finite dimensional vector spaces $\operatorname{Ker} u$ and $\operatorname{Coker} u = E_1/\operatorname{Im} u$ in $K_0(\mathbb{C}) = \mathbb{Z}$; in other words it is the rational integer $\dim \operatorname{Ker} u - \dim \operatorname{Coker} u$.

When E_0 and E_1 are Hilbert modules over a C^*-algebra A and u, v are A-linear (and have adjoints), and $\mathrm{Id}_{E_1} - uv$ and $\mathrm{Id}_{E_0} - vu$ are A-compact morphisms of Hilbert modules, the index of T lies in $K_0(A)$ ([38, 24]). It is possible to extend this construction: when A is a Banach algebra, and E_0 and E_1 are pairs of dual modules over A, we can also define an index in $K_0(A)$ ([33, 44]). Now we explain this construction, but we restrict ourselves to unital Banach algebras and in this way we avoid pairs.

Let A be a unital Banach algebra.

A right Banach A-module is a Banach space (with a given norm $\|.\|_E$) equipped with a right action of A such that $1 \in A$ acts by identity and $\|xa\|_E \leq \|x\|_E\|a\|_A$ for any $x \in E$ and $a \in A$. Let E and F be right Banach A-modules. A morphism $u\colon E \to F$ of A-modules is a continuous $\mathbb{C}$-linear map such that $u(xa) = u(x)a$ for any $x \in E$ and $a \in A$. The space $\mathcal{L}_A(E,F)$ of such morphisms is a Banach space with norm $\|u\| = \sup_{x\in E,\|x\|_E=1} \|u(x)\|_F$. A morphism $u \in \mathcal{L}_A(E,F)$ is said to be rank one if $u = w \circ v$ with $v \in \mathcal{L}_A(E,A)$ and $w \in \mathcal{L}_A(A,F)$. Let us notice that $\mathcal{L}_A(A,F)$ is equal to F. The space $\mathcal{K}_A(E,F)$ of A-compact morphisms is the closed vector span of rank one morphisms in $\mathcal{L}_A(E,F)$. If $E = F$, $\mathcal{L}_A(E) = \mathcal{L}_A(E,E)$ is a Banach algebra and $K_A(E) = K_A(E,E)$ is a closed ideal in it.

A generalized Fredholm operator over A is the data of a $\mathbb{Z}/2$ graded right Banach A-module E and an odd morphism $T \in \mathcal{L}_A(E)$ such that $T^2 - \mathrm{Id}_E \in \mathcal{K}_A(E)$. In other words $E = E_0 \oplus E_1$, and $T = \begin{pmatrix} 0 & v \\ u & 0 \end{pmatrix}$ and $u \in \mathcal{L}_A(E_0,E_1)$ and $v \in \mathcal{L}_A(E_1,E_0)$ satisfy $vu - \mathrm{Id}_{E_0} \in \mathcal{K}_A(E_0)$ and $uv - \mathrm{Id}_{E_1} \in \mathcal{K}_A(E_1)$.

If (E,T) is a generalized Fredholm operator on A and $u\colon A \to B$ a unital morphism, then $(E \otimes_A B, T \otimes 1)$ is a generalized Fredholm operator over B (here $E \otimes_A B$ is the completion of $E \otimes_A^{\mathrm{alg}} B$ for the maximal Banach norm such that $\|x \otimes b\| \leq \|x\|_E\|b\|_B$ for $x \in E$ and $b \in B$).

Let $A[0,1]$ be the Banach algebra of continuous functions from $[0,1]$ to A with the norm $\|f\| = \sup_{t\in[0,1]} \|f(t)\|_A$ and $u_0, u_1\colon A[0,1] \to A$ the evaluations at 0 and 1. Two generalized Fredholm operators on A are said to be homotopic if they are the images by u_0 and u_1 of a generalized Fredholm operator over $A[0,1]$.

Theorem 2.1. *There is a natural bijection between $K_0(A)$ and the set of homotopy classes of Fredholm operators over A, for any unital Banach algebra A.*

Let (E_0, E_1, u, v) be a generalized Fredholm operator over A. Its index, i.e., the corresponding element in $K_0(A)$, is constructed as follows. It is possible to find $n \in \mathbb{N}$ and $w \in \mathcal{K}_A(E_0 \oplus A^n, E_1)$ such that $(u,0) + w \in \mathcal{L}_A(E_0 \oplus A^n, E_1)$ is surjective. Its kernel is then f.g.p. and the index is the formal difference of $\mathrm{Ker}((u,0)+w)$ and A^n.

3. Statement of the Baum–Connes Conjecture

Let G be a second countable, locally compact group. We fix a left-invariant Haar measure dg on G. Denote by $C_c(G)$ the convolution algebra of complex-valued continuous compactly supported functions on G. The convolution of $f, f' \in C_c(G)$ is given by $f * f'(g) = \int_G f(h)f'(h^{-1}g)dh$ for any $g \in G$.

When G is discrete and dg is the counting measure, $C_c(G)$ is also denoted by $\mathbb{C}G$ and if e_g denotes the delta function at $g \in G$ (equal to 1 at g and 0 elsewhere), $(e_g)_{g\in G}$ is a basis of $\mathbb{C}G$ and the convolution product is given by $e_g e_{g'} = e_{gg'}$.

The completion of $C_c(G)$ for the norm $\|f\|_{L^1} = \int_G |f(g)|dg$ is a Banach algebra and is denoted by $L^1(G)$.

For any $f \in C_c(G)$ let $\lambda(f)$ be the operator $f' \mapsto f * f'$ on $L^2(G)$. The completion of $C_c(G)$ by the operator norm $\|f\|_{\mathrm{red}} = \|\lambda(f)\|_{\mathcal{L}_{\mathbb{C}}(L^2(G))}$ is called the reduced C^*-algebra of G and denoted by $C^*_{\mathrm{red}}(G)$. If G is discrete $(e_g)_{g\in G}$ is an orthonormal basis of $l^2(G)$ and $\lambda(e_g)\colon e_{g'} \mapsto e_{gg'}$.

For any continuous unitary representation π of G on a Hilbert space H and for any $f \in C_c(G)$, we write $\pi(f) = \int_G f(g)\pi(g)dg \in \mathcal{L}_{\mathbb{C}}(H)$. The completion of $C_c(G)$ for the norm $\|f\|_{\max} = \sup_{(H,\pi)} \|\pi(f)\|_{\mathcal{L}_{\mathbb{C}}(H)}$, where the supremum is taken over all continuous unitary representations of G, is called the maximal (or full) C^*-algebra of G and denoted by $C^*(G)$. For any continuous unitary representation π of G on H, $\pi\colon C_c(G) \to \mathcal{L}_{\mathbb{C}}(H)$ extends by continuity to $\pi\colon C^*(G) \to \mathcal{L}_{\mathbb{C}}(H)$.

For any $f \in C_c(G)$, $\|f\|_{L^1} \geq \|f\|_{\max}$ and $\|f\|_{\max} \geq \|f\|_{\mathrm{red}}$ because the left regular representation of G on $L^2(G)$ is continuous and unitary. This gives morphisms $L^1(G) \xrightarrow{i} C^*(G) \xrightarrow{r} C^*_{\mathrm{red}}(G)$.

The image of $L^1(G)$ is a dense subalgebra of $C^*(G)$ and $C^*_{\mathrm{red}}(G)$, the morphism r is surjective and is an isomorphism if and only if G is amenable.

Assume now that M is a smooth closed manifold, and $\tilde{M}$ a covering of M with group G (if $\tilde{M}$ is simply connected, $G = \pi_1(M)$). Let E_0 and E_1 be two smooth (hermitian) finite-dimensional vector bundles over M and u an order 0 elliptic pseudo-differential operator from $L^2(M, E_0)$ to $L^2(M, E_1)$. Since u is elliptic there is an order 0 pseudo-differential operator $v\colon L^2(M, E_1) \to L^2(M, E_0)$ such that $\mathrm{Id}_{L^2(M,E_0)} - vu$ and $\mathrm{Id}_{L^2(M,E_1)} - uv$ have order ≤ -1 and therefore are compact. Let $\mathcal{E}$ be the quotient of $\tilde{M} \times C^*_{\mathrm{red}}(G)$ by the diagonal action of G: $\mathcal{E}$ is a flat bundle over M, whose fibers are isomorphic to $C^*_{\mathrm{red}}(G)$. Then $L^2(M, E_0 \otimes \mathcal{E})$ and $L^2(M, E_1 \otimes \mathcal{E})$ are Hilbert modules over $C^*_{\mathrm{red}}(G)$ and it is possible to lift u and v to $\tilde{u}$ and $\tilde{v}$ so that $(L^2(M, E_0 \otimes \mathcal{E}), L^2(M, E_1 \otimes \mathcal{E}), \tilde{u}, \tilde{v})$ is a generalized Fredholm operator over $C^*_{\mathrm{red}}(G)$, whose index lies in $K_0(C^*_{\mathrm{red}}(G))$ and the index does not depend on the choice of the liftings.

In the last paragraph (M, u) defines an element of $K_{*,c}(BG)$, the K-homology with compact support of the classifying space BG, and for any discrete group G we have a morphism of abelian groups $K_{*,c}(BG) \to K_*(C^*_{\mathrm{red}}(G))$. The map $K_{*,c}(BG) \to K_*(C^*_{\mathrm{red}}(G))$ is the Baum–Connes assembly map when G is discrete and torsion free. When G is not discrete or has torsion, the index construction can be performed starting from a proper action of G (instead of the free action of G on $\tilde{M}$ in the last paragraph), and therefore we have to introduce the space $\underline{E}G$ that classifies the proper actions of G. Using Kasparov KK-theory the G-equivariant K-homology $K^G_*(\underline{E}G)$ (with suitable support requirements) may be defined, and there is an assembly map

$$\mu_{\mathrm{red}}\colon K_*^G(\underline{E}G) \to K_*(C^*_{\mathrm{red}}(G)).$$

In the same way we have assembly maps

$$\mu_{\max}\colon K_*^G(\underline{E}G) \to K_*(C^*(G)) \quad \text{and} \quad \mu_{L^1}\colon K_*^G(\underline{E}G) \to K_*(L^1(G))$$

and we have $\mu_{\max} = i_* \circ \mu_{L^1}$ and $\mu_{\mathrm{red}} = r_* \circ \mu_{\max}$.

Baum–Connes Conjecture. *If G is a second countable, locally compact group, then the assembly map $\mu_{\mathrm{red}}\colon K_*^G(\underline{E}G) \to K_*(C^*_{\mathrm{red}}(G))$ is an isomorphism.*

If G has Kazdhan's property (T) there is a projection $p \in C^*(G)$ such that for any continuous unitary representation π of G on a Hilbert space H, $\pi(p)$ is the orthogonal projection onto the subspace of G-invariant vectors in H ([14] is an excellent introduction to property (T)). If moreover G is not compact, the image of p in $C^*_{\mathrm{red}}(G)$ is zero (because $L^2(G)$ has no non-zero invariant vector under the left regular representation), therefore $r_*\colon K_0(C^*(G)) \to K_0(C^*_{\mathrm{red}}(G))$ is not injective and in particular $\mu_{\max}$ and μ_{red} cannot be both isomorphisms. Furthermore for most non-compact groups with property (T) it is possible to show that the class of p in $K_0(C^*(G))$ does not lie in the image of $\mu_{\max}$.

Starting from facts of harmonic analysis, Bost conjectured: If G is a second countable, locally compact group (and has reasonable geometric properties) then the assembly map $\mu_{L^1}\colon K_*^G(\underline{E}G) \to K_*(L^1(G))$ is an isomorphism.

In many cases $K_*^G(\underline{E}G)$ can be computed. For instance if G is a discrete torsion free subgroup of a reductive Lie group H and K is a maximal compact subgroup of H, then a possible $\underline{E}G$ is H/K and $K_*^G(\underline{E}G)$ is the K-homology with compact support of $G\backslash H/K$. This group may be computed thanks to Mayer–Vietoris sequences. If G is a connected Lie group, K a maximal compact subgroup and if G/K is even dimensional and admits a G-equivariant spin structure, $K_*^G(\underline{E}G)$ is equal to the representation ring of K (and the Baum–Connes map involves the index of a twisted Dirac operator on G/K).

Let G be a second countable, locally compact group and A a G-C^*-algebra (this is a C^*-algebra on which G acts continuously by C^*-algebra automorphisms). The space $C_c(G, A)$ of A-valued continuous compactly supported functions on G is endowed with the following convolution product:

$$f * f'(g) = \int_G f(h)\pi(h)(f'(h^{-1}g))dh$$

and the completion $L^1(G, A)$ of $C_c(G, A)$ for the norm $\|f\| = \int_G \|f(g)\|_A dg$ is a Banach algebra. We have also two C^*-algebras $C^*(G, A)$ and $C^*_{\mathrm{red}}(G, A)$ such that $L^1(G, A)$ is a dense subalgebra of these algebras and $C^*_{\mathrm{red}}(G, A)$ is a quotient of $C^*(G, A)$.

Then, using Kasparov's KK-theory a generalized assembly map

$$\mu_{\mathrm{red},A}\colon K_*^G(\underline{E}G, A) \to K_*(C^*_{\mathrm{red}}(G, A))$$

and similar maps $\mu_{\max,A}$ and $\mu_{L^1,A}$ can be defined. When $A = \mathbb{C}$ (and G acts trivially), $\mu_{\mathrm{red},A} = \mu_{\mathrm{red}}$, $\mu_{\max,A} = \mu_{\max}$ and $\mu_{L^1,A} = \mu_{L^1}$.

The Baum–Connes conjecture "with coefficients" claims that $\mu_{\mathrm{red},A}$ is an isomorphism and the Bost conjecture "with coefficients" claims that $\mu_{L^1,A}$ is an isomorphism. The surjectivity of the Baum–Connes conjecture with coefficients has been counter-exampled recently (Higson, Lafforgue, Ozawa, Skandalis, Yu) using a random group constructed by Gromov ([11]) but Bost conjecture with coefficients still stands. If the Baum–Connes conjecture with coefficients is true for a group, it is true also for all its closed subgroups; the Baum–Connes conjecture with coefficients is also stable under various kinds of extensions (Chabert [5], Chabert-Echterhoff [6], Oyono [39, 40], and Tu [45]).

4. Status of Injectivity and The Element γ

The injectivity of the Baum–Connes map μ_{red} (and therefore of μ_{L^1} and $\mu_{\max}$) is known in many cases: it is known, even with arbitrary coefficients, for the following classes of groups:

a) groups acting properly isometrically on a complete simply connected riemannian manifold with controlled non-positive sectional curvature, and in particular closed subgroups of reductive Lie groups ([26, 28]),

b) groups acting properly isometrically on an affine building and in particular closed subgroups of reductive p-adic groups ([29]),

c) groups acting properly isometrically on a discrete space with good properties at infinity (weakly geodesic, uniformly locally finite, and "bolic" [30]), and in particular hyperbolic groups (i.e., word-hyperbolic in the sense of Gromov),

d) groups acting amenably on a compact space ([17]).

In the cases a), b), c) above, the proof of injectivity provides an explicit idempotent endomorphism on $K_*(C^*_{\mathrm{red}}(G))$ whose image is the image of μ_{red} (and the same for $\mu_{\max}$ and μ_{L^1}).

To state this we need to understand a baby case of Kasparov's KK-groups. Let G be a second countable, locally compact group. A unitary Fredholm representation of G is a triple (H, π, T) where H is a $\mathbb{Z}/2$-graded Hilbert space, π a unitary representation of G and T an odd operator on H such $T^2 - \mathrm{Id}_H$ is compact and $\pi(g)T\pi(g^{-1}) - T$ is compact and depends norm continuously on $g \in G$. Then $KK_G(\mathbb{C}, \mathbb{C})$ is the quotient of the set of isomorphism classes of unitary Fredholm representations by the following equivalence relation: two unitary Fredholm representations are homotopic if they can be linked by a continuous field of unitary Fredholm representations on $[0,1]$, or, more precisely if there is a Hilbert module H' on $C([0,1])$, a unitary representation π' of G on H' and an odd morphism T', such that $T'^2 - \mathrm{Id}_{H'}$ is $C([0,1])$-compact and $\pi'(g)T'\pi'(g^{-1}) - T'$ is $C([0,1])$-compact and depends norm continuously on $g \in G$, and such that the two unitary Fredholm representations are the evaluations at 0 and 1 of (H', π', T'). Kasparov proved that

$KK_G(\mathbb{C},\mathbb{C})$ has a ring structure (using direct sum for the addition and a difficult construction with tensor products for the multiplication).

If π is a unitary representation of G on a Hilbert space H_0 and $H_1 = 0$, then $(H,\pi,0)$ is a unitary Fredholm representation if and only if H_0 has finite dimension. If moreover $H_0 = \mathbb{C}$ and π is the trivial representation of G, the class of $(H,\pi,0)$ is the unit of $KK_G(\mathbb{C},\mathbb{C})$ and is denoted by 1.

If G is compact, the unitary Fredholm representations $(H,\pi,0)$ with $H_1 = 0$ (and $\dim H_0 < +\infty$) generate $KK_G(\mathbb{C},\mathbb{C})$ and $KK_G(\mathbb{C},\mathbb{C})$ is equal to the representation ring of G.

The important fact is that there is a "descent morphism"

$$j_{\mathrm{red}}\colon KK_G(\mathbb{C},\mathbb{C}) \to \mathrm{End}(K_*(C^*_{\mathrm{red}}(G)))\,.$$

In fact it is a ring homomorphism and $j_{\mathrm{red}}(1) = \mathrm{Id}_{K_*(C^*_{\mathrm{red}}(G))}$. In Kasparov's theory it is defined as the composite of two maps $KK_G(\mathbb{C},\mathbb{C}) \to KK(C^*_{\mathrm{red}}(G), C^*_{\mathrm{red}}(G)) \to \mathrm{End}(K_*(C^*_{red}(G)))$.

Assume that G is discrete. We shall define j_{red} in this case. Since $C^*_{\mathrm{red}}(G)$ has a unit, $K_0(C^*_{\mathrm{red}}(G))$ is generated by classes of idempotents in $M_k(C^*_{\mathrm{red}}(G))$, $k \in \mathbb{N}^*$. For the sake of simplicity let p be an idempotent in $C^*_{\mathrm{red}}(G)$. Let (H,π,T) be a unitary Fredholm representation of G, and let us try to define $j_{\mathrm{red}}([(H,\pi,T)])([p]) \in K_0(C^*_{\mathrm{red}}(G))$, where $[(H,\pi,T)]$ is the class of (H,π,T) in $KK_G(\mathbb{C},\mathbb{C})$ and $[p]$ is the class of p in $K_0(C^*_{\mathrm{red}}(G))$.

If $H_1 = 0$ then $T = 0$, H_0 has finite dimension and if we identify H_0 with $\mathbb{C}^n$, $\sum_{g\in G} f(g)e_g \mapsto \sum_{g\in G} f(g)\pi(g)e_g$ is an algebra morphism $\mathbb{C}G \to M_n(\mathbb{C}G)$, this morphism extends by continuity to a morphism of Banach algebras (and even of C^*-algebras) $C^*_{\mathrm{red}}(G) \to M_n(C^*_{\mathrm{red}}(G))$, the image of p is an idempotent $P \in M_n(C^*_{\mathrm{red}}(G))$ and $j_{\mathrm{red}}([(H,\pi,T)])([p])$ is the class of P in $K_0(C^*_{\mathrm{red}}(G))$, i.e., the class of the f.g.p. module $\mathrm{Im}\,P \subset C^*_{\mathrm{red}}(G)^n$.

In the general case we define the Hilbert module $E = H \otimes C^*_{\mathrm{red}}(G)$ as the closed subspace of $\mathcal{L}_{\mathbb{C}}(l^2(G), H \otimes l^2(G))$ generated by the $h \otimes e_g\colon e_{g'} \mapsto h \otimes e_{gg'}$, $h \in H$, $g \in G$. The formula $(h \otimes e_g)e_{g'} = h \otimes e_{gg'}$ gives to E a right $C^*_{\mathrm{red}}(G)$-module structure and the formula $e_{g'}(h \otimes e_g) = \pi(g')(h) \otimes e_{g'g}$ gives to E a left $C^*_{\mathrm{red}}(G)$-module structure and the two structures commute. (The justification of the norm condition for the left module structure relies on the fact that if λ is the left regular representation of G on $l^2(G)$, the unitary representations $\pi \otimes \lambda$ and $1 \otimes \lambda$ on $H \otimes l^2(G)$ are conjugated by the unitary map $h \otimes e_g \mapsto \pi(g)(h) \otimes e_g$.) Moreover $T \otimes 1\colon h \otimes e_g \mapsto T(h) \otimes e_g$ is in $\mathcal{L}_{C^*_{\mathrm{red}}(G)}(E)$ (where E is considered as a right $C^*_{\mathrm{red}}(G)$-Hilbert module) and $(E, T \otimes 1)$ is a generalized Fredholm operator over $C^*_{\mathrm{red}}(G)$. The image P of p by the left module structure is a projector in $\mathcal{L}_{C^*_{\mathrm{red}}(G)}(E)$ and P commutes with $T \otimes 1$ up to a $C^*_{\mathrm{red}}(G)$-compact operator and therefore $(\mathrm{Im}\,P, P(T \otimes 1)P)$ is a generalized Fredholm operator over $C^*_{\mathrm{red}}(G)$ and $j_{\mathrm{red}}([(H,\pi,T)])([p])$ is its index in $K_0(C^*_{\mathrm{red}}(G))$.

We also have ring morphisms $j_{\max}\colon KK_G(\mathbb{C},\mathbb{C}) \to \mathrm{End}(K_*(C^*(G)))$ and $j_{L^1}\colon KK_G(\mathbb{C},\mathbb{C}) \to \mathrm{End}(K_*(L^1(G)))$.

Moreover if A is a G-C^*-algebra we have ring morphisms $j_{\mathrm{red},A}\colon KK_G(\mathbb{C},\mathbb{C})\to \mathrm{End}(K_*(C^*_{\mathrm{red}}(G,A)))$, $j_{\max,A}\colon KK_G(\mathbb{C},\mathbb{C}) \to \mathrm{End}(K_*(C^*(G,A)))$, and $j_{L^1,A}\colon KK_G(\mathbb{C},\mathbb{C}) \to \mathrm{End}(K_*(L^1(G,A)))$.

The following extremely important theorem also contains earlier works of Mishchenko and Solovjev.

Theorem 4.1. (Kasparov, Kasparov–Skandalis [28, 29, 30]**)** *If G belongs to one of the classes a), b), c) above, the geometric conditions in a), b) or c) allow us to construct an idempotent element $\gamma \in KK_G(\mathbb{C},\mathbb{C})$ such that μ_{red} is injective and its image is equal to the image of the idempotent $j_{\mathrm{red}}(\gamma) \in \mathrm{End}(K_*(C^*_{\mathrm{red}}(G)))$ and more generally for any G-C^*-algebra A, $\mu_{\mathrm{red},A}$ is injective and its image is equal to the image of the idempotent $j_{\mathrm{red},A}(\gamma) \in \mathrm{End}(K_*(C^*_{\mathrm{red}}(G,A)))$.*

Moreover the same results hold with $\mu_{\max}$, μ_{L^1} and $\mu_{\max,A}$, $\mu_{L^1,A}$.

5. Homotopies between γ and 1

We assume that G belongs to one of the classes a), b), c). Then the injectivity of $\mu_{\mathrm{red},A}$ is known for any G-C^*-algebra A and the surjectivity is equivalent with the equality $j_{\mathrm{red},A}(\gamma) = \mathrm{Id} \in \mathrm{End}(K_*(C^*_{\mathrm{red}}(G,A)))$. Unfortunately, when A is an arbitrary G-C^*-algebra this fact is known only in a few cases.

Theorem 5.1. *We have $\gamma = 1$ in $KK_G(\mathbb{C},\mathbb{C})$ if*

1. *G is a free group (Cuntz,* [8]*) or a closed subgroup of $SO(n,1)$ (Kasparov,* [27]*) or of $SU(n,1)$ (Julg-Kasparov,* [20]*) or of $SL_2(\mathbb{F})$ with $\mathbb{F}$ a local non-archimedian field (Julg-Valette,* [19]*),*
2. *G acts isometrically and properly on a Hilbert space (Higson–Kasparov,* [15]*, see also* [22, 46]*).*

In fact the second case contains the first one.

If $\gamma = 1 \in KK_G(\mathbb{C},\mathbb{C})$, then $\mu_{\mathrm{red},A}$ is a bijection for any G-C^*-algebra A but this is also true for $\mu_{\max,A}$. But, if G has property (T) and is not compact, $r_*\colon K_0(C^*(G)) \to K_0(C^*_{\mathrm{red}}(G))$ is not injective and therefore $\gamma \neq 1 \in KK_G(\mathbb{C},\mathbb{C})$. In fact it is impossible to deform 1 to γ among unitary Fredholm representations because the trivial representation is isolated in the unitary dual if G has property (T) and γ can be represented by (H,π,T) such that H has no invariant vector (and even H is tempered).

It is then natural to broaden the class of representations that are allowed in $KK_G(\mathbb{C},\mathbb{C})$ in order to break the isolation of the trivial one. In [21] Julg defined $KK_G^{\mathrm{ub}}(\mathbb{C},\mathbb{C})$ in the same way as $KK_G(\mathbb{C},\mathbb{C})$ but replacing unitary representations by uniformly bounded representations (a continuous representation π of G on a Hilbert space H is uniformly bounded if $\sup_{g\in G} \|\pi(g)\|_{\mathcal{L}_\mathbb{C}(H)} < +\infty$). Julg showed that the Baum–Connes conjecture is true for $Sp(n,1)$, with any coefficients, if γ is equal to 1 in $KK_G^{\mathrm{ub}}(\mathbb{C},\mathbb{C})$ but this is still unknown.

For any non compact group G the trivial representation is not isolated among isometric representations in Banach spaces (think of the left regular representation on $L^p(G)$, p going to infinity). A Banach Fredholm representation of G is a triple (E, π, T) with E a $\mathbb{Z}/2$-graded Banach space endowed with an isometric representation of G, and $T \in \mathcal{L}_{\mathbb{C}}(E)$ an odd operator such that $T^2 - \mathrm{Id}_E$ belongs to $\mathcal{K}_{\mathbb{C}}(E)$ and $\pi(g)T\pi(g^{-1}) - T$ belongs to $\mathcal{K}_{\mathbb{C}}(E)$ and depends norm continuously on $g \in G$. Then $KK_G^{\mathrm{ban}}(\mathbb{C}, \mathbb{C})$ is defined as the quotient of the set of isomorphism classes of Banach Fredholm representations by a homotopy relation involving Banach modules over $C([0,1])$.

Since any unitary representation of G on a Hilbert space H is an isometric representation on the Banach space H, there is a natural morphism of abelian groups

$$KK_G(\mathbb{C}, \mathbb{C}) \to KK_G^{\mathrm{ban}}(\mathbb{C}, \mathbb{C}) .$$

Now we introduce a subclass c') of c), which is essentially equal to c): in particular c') contains all hyperbolic groups.

Theorem 5.2. [33, 44] *For any group G in the classes a), b), or c'), we have $\gamma = 1$ in $KK_G^{\mathrm{ban}}(\mathbb{C}, \mathbb{C})$.*

In fact the statement is slightly incorrect, we should allow representations with a slow growth, but this adds no real difficulty. For general hyperbolic groups there is an additional subtlety. The proof of this theorem is quite technical. Let me just indicate some ingredients involved. If G is in class a), then G acts isometrically properly on a complete simply connected riemannian manifold X with controlled non-positive sectional curvature, and X is contractible (through geodesics) and the de Rham cohomology of X (without support) is $\mathbb{C}$ in degree 0 and 0 in other degrees. It is possible to put norms on the spaces of forms (on which G acts) and to build a parametrix for the de Rham operator (in the spirit of the Poincaré lemma) in order to obtain a resolution of the trivial representation, and in our language a Banach Fredholm representation that is equal to 1 in $KK_G^{\mathrm{ban}}(\mathbb{C}, \mathbb{C})$. The norms we use are essentially Sobolev L^∞ norms. Then it is possible to deform these norms to Hilbert norms (through L^p norms) and to reach γ.

If G belongs to class b) the de Rham complex is replaced by the simplicial homology complex (with l^1 norms) on the building. If G belongs to class c') a Rips complex plays the same role as the building in b).

It is not possible to apply this theorem directly to the Baum–Connes conjecture because there is no obvious descent map $KK_G^{\mathrm{ban}}(\mathbb{C}, \mathbb{C}) \to \mathrm{End}(K_*(C^*_{\mathrm{red}}(G)))$, and in the next section we shall see the difficulties encountered and how one bypasses them in a few cases.

We now show how one may apply this theorem to Bost's conjecture.

We assume that G is discrete (but we keep the notation $L^1(G)$ instead of $l^1(G)$). Let (E, π, T) be a Banach Fredholm representation of G. We denote by $L^1(G, E)$ the completion of $E \otimes \mathbb{C}G$ for the norm $\| \sum_{g \in G} x(g) \otimes e_g \| =$

$\sum_{g\in G}\|x(g)\|_E$. Then $L^1(G,E)$ is a right Banach $L^1(G)$-module and a left Banach $L^1(G)$-module by the same formulas as in the C^*_{red}-case. If p is an idempotent element in $L^1(G)$ the image P of p by the left module structure is the projection on a closed submodule $\operatorname{Im} P$ of $L^1(G,E)$. Then $(\operatorname{Im} P, P(T\otimes 1)P)$ is a generalized Fredholm operator over $L^1(G)$ and it has an index in $K_0(L^1(G))$. We may repeat the construction for any idempotent $p \in M_k(L^1(G))$ and if G is discrete this defines the descent map $j^{\text{ban}}_{L^1}\colon KK^{\text{ban}}_G(\mathbb{C},\mathbb{C}) \to \operatorname{End}(K_0(L^1(G)))$ with $j^{\text{ban}}_{L^1}(1) = \operatorname{Id}_{K_0(L^1(G))}$.

This map can be defined for any G and even with coefficients and for K_1 instead of K_0: for any G-C^*-algebra A, there is $j^{\text{ban}}_{L^1,A}\colon KK^{\text{ban}}_G(\mathbb{C},\mathbb{C}) \to \operatorname{End}(K_*(L^1(G,A)))$ and $j^{\text{ban}}_{L^1,A}(1) = \operatorname{Id}_{K_*(L^1(G,A))}$.

As a consequence we get that the Bost conjecture holds in many cases.

Theorem 5.3. *For any group G in the classes a), b) or c'), and for any G-C^*-algebra A, $\mu_{L^1,A}\colon K^G_*(\underline{E}G,A) \to K_*(L^1(G,A))$ is an isomorphism.*

6. Unconditional Completions

In the last section the main point in the construction of the descent map j_{L^1} was the two $L^1(G)$-modules structures on $L^1(G,E)$, when E is a Banach space with an isometric representation of G.

Now assume that $\mathcal{A}(G)$ is a Banach algebra containing $C_c(G)$ as a dense subalgebra, (for instance $\mathcal{A}(G)$ could be $L^1(G)$, or $C^*_{\text{red}}(G)$ but we have many other choices). We write $\mathcal{A}(G)$ instead of $\mathcal{A}$ for notational convenience. We ask for a necessary and sufficient condition so that there is a "natural" descent map $j_{\mathcal{A}}\colon KK^{\text{ban}}_G(\mathbb{C},\mathbb{C}) \to \operatorname{End}(K_*(\mathcal{A}(G)))$.

In order to simplify the argument below, we will assume G to be discrete. However no conceptual difficulty occurs in the non-discrete case.

Let E be a Banach space with an isometric representation of G. Then $E\otimes \mathbb{C}G$ has a right $\mathbb{C}G$-module structure given by $(x\otimes e_g)e_{g'} = x\otimes e_{gg'}$ and a left $\mathbb{C}G$-module structure given by $e_{g'}(x\otimes e_g) = \pi(g')(x)\otimes e_{g'g}$. We look for a completion $\mathcal{A}(G,E)$ of $E\otimes\mathbb{C}G$ by a Banach norm such that $\mathcal{A}(G,E)$ is a right and a left Banach $\mathcal{A}(G)$-module, i.e., $\|ff'\|_{\mathcal{A}(G,E)} \leq \|f\|_{\mathcal{A}(G,E)}\|f'\|_{\mathcal{A}(G)}$ and $\|f'f\|_{\mathcal{A}(G,E)} \leq \|f'\|_{\mathcal{A}(G)}\|f\|_{\mathcal{A}(G,E)}$ for any $f\in E\otimes\mathbb{C}G$ and $f'\in\mathbb{C}G$. Another condition is: for any $x\in E$ and $\xi\in\mathcal{L}_{\mathbb{C}}(E,\mathbb{C})$, we denote by $R_x\colon \mathbb{C}G\to E\otimes\mathbb{C}G$ the map $e_g\mapsto x\otimes e_g$ and by $S_\xi\colon E\otimes\mathbb{C}G\to\mathbb{C}G$ the map $y\otimes e_g\mapsto \xi(y)e_g$, and we impose $\|R_x(f)\|_{\mathcal{A}(G,E)} \leq \|x\|_E\|f\|_{\mathcal{A}(G)}$ for any $f\in\mathbb{C}G$ and $\|S_\xi(f)\|_{\mathcal{A}(G)} \leq \|\xi\|_{\mathcal{L}_{\mathbb{C}}(E,\mathbb{C})}\|f\|_{\mathcal{A}(G,E)}$ for any $f\in E\otimes\mathbb{C}G$ (this is to have enough $\mathcal{A}(G)$-compact operators acting on $\mathcal{A}(G,E)$). Now fix $x\in E$ and $\xi\in\mathcal{L}_{\mathbb{C}}(E,\mathbb{C})$ and denote by 1 the unit in G. For any $f=\sum_{g\in G}f(g)e_g\in\mathbb{C}G$, $S_\xi(f(R_x(e_1)))$ is $\sum_{g\in G}\xi(\pi(g)(x))f(g)e_g$ in $\mathbb{C}G$ (here f acts by the left $\mathbb{C}G$-module structure on $E\otimes\mathbb{C}G$). For any function c on G the Schur multiplication $\mathbb{C}G\to\mathbb{C}G$ is the pointwise product $\sum_{g\in G}f(g)e_g \mapsto \sum_{g\in G}c(g)f(g)e_g$. The necessary condition we obtained is therefore that for any $x \in E$ and $\xi \in \mathcal{L}_{\mathbb{C}}(E, \mathbb{C})$ the Schur multiplication by the matrix coefficient $g \mapsto \xi(\pi(g)(x))$ is bounded from $\mathcal{A}(G)$ to itself and its norm is less than $\|x\|_E \|\xi\|_{\mathcal{L}_{\mathbb{C}}(E,\mathbb{C})}$. But for any l^∞-function c on G we can find an isometric representation π of G on a Banach space E and $x \in E$ and $\xi \in \mathcal{L}_{\mathbb{C}}(E, \mathbb{C})$ such that $\|x\|_E \|\xi\|_{\mathcal{L}_{\mathbb{C}}(E,\mathbb{C})} \leq \|c\|_{l^\infty}$ and $c(g) = \xi(\pi(g)x)$ for any $g \in G$ (take $E = l^1(G)$, $x = \delta_1$, $\xi = c$). Therefore a necessary condition is that $\mathcal{A}(G)$ is an unconditional completion in the following sense.

Definition 6.1. *A Banach algebra $\mathcal{A}(G)$ (with a given norm $\|.\|_{\mathcal{A}(G)}$) containing $C_c(G)$ as a dense subalgebra is called an unconditional completion if for any $f_1, f_2 \in C_c(G)$ satisfying $|f_1(g)| \leq |f_2(g)|$ for any $g \in G$, we have $\|f_1\|_{\mathcal{A}(G)} \leq \|f_2\|_{\mathcal{A}(G)}$.*

Remark that $L^1(G)$ is always an unconditional completion of $C_c(G)$ but that $C^*_{\text{red}}(G)$ and $C^*(G)$ are not in general.

In fact this condition is also sufficient to construct the descent map. We still assume that G is discrete. If $\mathcal{A}(G)$ is an unconditional completion of $\mathbb{C}G$, and (E, π, T) is a Banach Fredholm representation of G, we define $\mathcal{A}(G, E)$ as the completion of $E \otimes \mathbb{C}G$ for the norm $\| \sum_{g \in G} x(g) \otimes e_g\| = \| \sum_{g \in G} \|x(g)\|_E \; e_g\|_{\mathcal{A}(G)}$ and $\mathcal{A}(G, E)$ is a right and a left Banach module over $\mathcal{A}(G)$ and if p is an idempotent in $\mathcal{A}(G)$ and P its image through the left action, $(\operatorname{Im} P, P(T \otimes 1)P)$ is a generalized Fredholm operator over $\mathcal{A}(G)$.

In this way, for any unconditional completion $\mathcal{A}(G)$ of $\mathbb{C}G$, we have a descent map $j^{\text{ban}}_{\mathcal{A}} \colon KK^{\text{ban}}_G(\mathbb{C}, \mathbb{C}) \to \operatorname{End}(K_*(\mathcal{A}(G))$. Moreover for any G-C^*-algebra B, the completion $\mathcal{A}(G, B)$ of $C_c(G, B)$ for the norm $\| \sum_{g \in G} f(g) e_g\| = \| \sum_{g \in G} \|f(g)\|_B \; e_g\|_{\mathcal{A}(G)}$ is a Banach algebra and we have a descent map $j^{\text{ban}}_{\mathcal{A},B} \colon KK^{\text{ban}}_G(\mathbb{C}, \mathbb{C}) \to \operatorname{End}(K_*(\mathcal{A}(G, B))$. If $\mathcal{A}(G)$ is an unconditional completion of $\mathbb{C}G$ and B a G-C^*-algebra, it is also possible to define an assembly map $\mu_{\mathcal{A},B} \colon K^G_*(\underline{E}G, B) \to K_*(\mathcal{A}(G, B))$. Moreover if $\mathcal{A}(G)$ is an involutive subalgebra of $C^*_{\text{red}}(G)$, $\mathcal{A}(G, B)$ is a subalgebra of $C^*_{\text{red}}(G, B)$ and if $i_{\mathcal{A},B} \colon \mathcal{A}(G, B) \to C^*_{\text{red}}(G, B)$ denotes the inclusion, $\mu_{\text{red},B} = i_{\mathcal{A},B,*} \circ \mu_{\mathcal{A},B}$. The same holds if G is not discrete.

Theorem 6.2. ([33]) *For any group G in the classes a), b) or c'), and for any unconditional completion $\mathcal{A}(G)$ of $C_c(G)$, and for any G-C^*-algebra B, $\mu_{\mathcal{A},B} \colon K^G_*(\underline{E}G, B) \to K_*(\mathcal{A}(G, B))$ is an isomorphism.*

7. RD Property

Let B be a Banach algebra, A a Banach algebra that is a subalgebra of B and $i \colon A \to B$ the inclusion. We say that A is stable under holomorphic functional calculus in B if any element of A has the same spectrum in A and in B. If A is dense and stable under holomorphic functional calculus in B, then $i_* \colon K_*(A) \to K_*(B)$ is an isomorphism (see the appendix of [4]).

Corollary 7.1. [33, 44] *For any group G in the classes a), b) or c'), if $C_c(G)$ admits an unconditional completion $\mathcal{A}(G)$ which is an involutive subalgebra of $C^*_{\text{red}}(G)$ and is stable under holomorphic functional calculus in $C^*_{\text{red}}(G)$, then $\mu_{\text{red}}: K^G_*(\underline{E}G) \to K_*(C^*_{\text{red}}(G))$ is an isomorphism.*

These conditions are fulfilled for

a) *hyperbolic groups,*
b) *cocompact lattices in $SL_3(\mathbb{F})$ with $\mathbb{F}$ a non-archimedian local field,*
c) *cocompact lattices in $SL_3(\mathbb{R})$, and $SL_3(\mathbb{C})$,*
d) *reductive groups over $\mathbb{R}$ or over a non-archimedian local field.*

In case d), $\mathcal{A}(G)$ is a variant of the Schwartz algebra on the group ([33]). In this case the Baum–Connes conjecture was already known for linear connected reductive groups (Wassermann [49]) and for the p-adic GL_n (Baum, Higson, Plymen [3]). In cases a), b), c) this result is based on a property first introduced by Haagerup for the free group and called (RD) (for rapid decay) by Jolissaint. In cases a), b), c) G has property (RD): this is due to Haagerup for free groups ([12]), Jolissant for "geometric hyperbolic groups", de la Harpe for general hyperbolic groups ([13]), Ramagge, Robertson and Steger in case b) ([42]) and the author in case c) ([34]). A discrete group G has property (RD) if there is a length function $\ell\colon G \to \mathbb{R}_+$ (satisfying $\ell(g^{-1}) = \ell(g)$ and $\ell(gh) \leq \ell(g) + \ell(h)$ for any $g, h \in G$) such that for $s \in \mathbb{R}_+$ big enough, the completion $H^s(G)$ of $\mathbb{C}G$ for the norm $\|\sum f(g)e_g\|_{H^s(G)} = \|\sum(1+\ell(g))^s f(g)e_g\|_{l^2(G)}$ is contained in $C^*_{\text{red}}(G)$. Then, for s big enough, $H^s(G)$ is a Banach algebra and an involutive subalgebra of $C^*_{\text{red}}(G)$ and is dense and stable under holomorphic functional calculus ([18, 34]); it is obvious that $H^s(G)$ is an unconditional completion of $\mathbb{C}G$.

8. Trying to Push the Method Further

KK-theory was generalized to groupoids in [28, 35, 36] and this generalized KK-theory was applied by Tu in [47] to the injectivity of the Baum–Connes map for hyperbolic foliations. It is possible to generalize also Banach KK-theory and unconditional completions. In this way we obtain the Baum–Connes conjecture for any hyperbolic group, with coefficients in any commutative C^*-algebra, and also for foliations with compact basis, admitting a negatively curved longitudinal riemannian metrics, and such that the holonomy groupoid is Hausdorff and has simply connected fibers (not yet published).

In [10, page 74], Valette conjectured that any cocompact lattice in a reductive group over $\mathbb{R}$ or over a non-archimedian local field has (RD) and this would imply the Baum–Connes conjecture (without coefficients) for these lattices. On the other hand $SL_3(\mathbb{Z})$ (a non-cocompact lattice in $SL_3(\mathbb{R})$) does not have property (RD) ([18]).

Hilbert and Banach approaches to prove $j_{\mathrm{red}}(\gamma) = \mathrm{Id}_{K_*(C^*_{\mathrm{red}}(G))}$ both look for a dense subalgebra $\mathcal{A}(G)$ of $C^*_{\mathrm{red}}(G)$ that is stable under holomorphic functional calculus and a homotopy between γ and 1 through (perhaps a special kind of) Banach Fredholm representations which all give a K-theory map $K_*(\mathcal{A}(G)) \to K_*(C^*_{\mathrm{red}}(G))$ by the descent construction (in the Hilbert approach $\mathcal{A}(G) = C^*_{\mathrm{red}}(G)$). Thanks to the discussion in Section 6 a necessary condition for this is that for any Banach Fredholm representation (E, π, T) in the homotopy between γ and 1, for any $x \in E$ and $\xi \in \mathcal{L}_{\mathbb{C}}(E, \mathbb{C})$, the Schur multiplication by the matrix coefficient $g \mapsto \xi(\pi(g)(x))$ is bounded from $\mathcal{A}(G)$ to $C^*_{\mathrm{red}}(G)$ and has norm $\leq \|x\|_E \|\xi\|_{\mathcal{L}_{\mathbb{C}}(E,\mathbb{C})}$. When we restrict the homotopy between γ and 1 at points $1/n, n \in \mathbb{N}^*$ (if 1 is at 0 and γ is at 1), we get a sequence π_n of isometric representations of G on Banach spaces E_n, which converges to the trivial one and therefore we can find $x_n \in E_n$ and $\xi_n \in \mathcal{L}_{\mathbb{C}}(E_n, \mathbb{C})$ of norm 1 such that the matrix coefficient $g \mapsto \xi_n(\pi_n(g)(x_n))$ converges to 1 uniformly on compact subsets of G. It is reasonable to assume that each of these coefficients vanish at infinity on G. Therefore we arrive at the following "necessary" condition for a proof along these lines of the Baum–Connes conjecture without coefficients for G:

Necessary condition

There is a subalgebra $\mathcal{A}(G)$ of $C^*_{\mathrm{red}}(G)$ which is a Banach algebra, contains $C_c(G)$, and is stable under holomorphic functional calculus in $C^*_{\mathrm{red}}(G)$, and a sequence of continuous functions $(c_n)_{n \in \mathbb{N}^*}$ from G to $\mathbb{C}$, all bounded by 1 and vanishing at infinity, such that c_n converges to 1 uniformly on compact subspaces of G and such that for any $n \in \mathbb{N}^*$, the Schur multiplication $M_{c_n} : \mathcal{A}(G) \to C^*_{\mathrm{red}}(G)$, $[g \mapsto f(g)] \mapsto [g \mapsto c_n(g) f(g)]$ is bounded of norm ≤ 1.

Of course this necessary condition is fulfilled (with $\mathcal{A}(G) = C^*_{\mathrm{red}}(G)$) when the trivial representation of G can be approximated by unitary representations whose coefficients vanish at infinity, i.e., when G acts isometrically properly on a Hilbert space (this is exactly the domain of the Hilbert method). It is also obviously fulfilled when $\mathcal{A}(G)$ is an unconditional completion and is a dense subalgebra of $C^*_{\mathrm{red}}(G)$ and is stable under holomorphic functional calculus in it, and this is essentially the domain of the Banach method. But this necessary condition is open for $SL_3(\mathbb{Z})$ where both Hilbert and Banach methods fail. It is open for all cocompact lattices in simple Lie or p-adic groups for which (RD) is not known, though it is much weaker than (RD).

In this necessary condition it is enough to ask that any element in $C_c(G)$ has the same spectral radius in $\mathcal{A}(G)$ and in $C^*_{\mathrm{red}}(G)$ (and the same with $M_k(C_c(G))$, $k \in \mathbb{N}^*$). In fact any reason for which the map $K_*(\mathcal{A}(G)) \to K_*(C^*_{\mathrm{red}}(G))$ is surjective would content us, and Oka's principle ([4]) should perhaps play an important role.

If A is a G-C^*-algebra a necessary condition for a proof along these lines of the surjectivity of $\mu_{\mathrm{red},A}$ is the same as above but with $\mathcal{A}(G)$ replaced by a subalgebra of $C^*_{\mathrm{red}}(G, A)$. This condition is fulfilled for any A if G is hyperbolic and this gives a hope for a proof of the Baum–Connes conjecture with any coefficients for hyperbolic groups. Unfortunately this necessary condition "with any coefficients" is not satisfied when G is a (cocompact) lattice in $SL_3(\mathbb{Q}_p)$.

It is also possible to state this necessary condition for groupoids.

References

[1] P. Baum and A. Connes, *K-theory for Lie groups and foliations*, Preprint, (1982).

[2] P. Baum, A. Connes and N. Higson, *Classifying space for proper actions and K-theory of group C^*-algebras*, C^*-algebras: 1943–1993 (San Antonio, TX, 1993), Contemp. Math., **167**, Amer. Math. Soc. (1994), 240–291.

[3] P. Baum, N. Higson and R. Plymen, *A proof of the Baum–Connes conjecture for p-adic $GL(n)$*, C. R. Acad. Sci. Paris Sér. I, **325**, (1997), 171–176.

[4] J.-B. Bost, *Principe d'Oka, K-théorie et systèmes dynamiques non commutatifs*, Invent. Math., **101**, (1990), 261–333.

[5] J. Chabert, *Baum–Connes conjecture for some semi-direct products*, J. Reine Angew. Math., **521**, (2000), 161–184.

[6] J. Chabert and S. Echterhoff, *Permanence properties of the Baume–Connes conjecture*, Documenta Math. **6**, (2001), 127–183.

[7] A. Connes, *Non-commutative geometry*, Academic Press, (1994).

[8] J. Cuntz, *K-theoretic amenability for discrete groups*, J. Reine Angew. Math., **344**, (1983), 180–195.

[9] J. Cuntz, *Bivariante K-theorie für lokalconvexe Algebren und der Chern-Connes Character*, Doc. Math., **2**, (1997), 139–182.

[10] *Novikov conjectures, Index theorems and Rigidity*, Edited by S. Ferry, A. Ranicki and J. Rosenberg, Volume 1, London Mathematical Society, LNS 226, (1993).

[11] M. Gromov, *Spaces and questions*, Preprint (1999).

[12] U. Haagerup, *An example of a nonnuclear C^*-algebra which has the metric approximation property*, Inv. Math., **50**, (1979), 279–293.

[13] P. de la Harpe, *Groupes hyperboliques, algèbres d'opérateurs et un théorème de Jolissaint*, C. R. Acad. Sci. Paris Sér. I, **307**, (1988), 771–774.

[14] P. de la Harpe and A. Valette, *La propriété (T) de Kazdhan pour les groupes localement compacts*, Astérisque, (1989).

[15] N. Higson and G. Kasparov, *Operator K-theory for groups which act properly and isometrically on Hilbert space*, Electron. Res. Announc. Amer. Math. Soc., **3**, (1997), 131–142.

[16] N. Higson, *The Baum–Connes conjecture*, Proc. of the Int. Cong. of Math., Vol. II (Berlin, 1998), Doc. Math., (1998), 637–646.

[17] N. Higson, *Bivariant K-theory and the Novikov conjecture*, Preprint, (1999).

[18] P. Jolissaint, *Rapidly decreasing functions in reduced C^*-algebra of groups*, Trans. Amer. Math. Soc., **317**, (1990), 167–196.

[19] P. Julg and A. Valette, *K-theoretic amenability for $SL_2(Q_p)$, and the action on the associated tree*, J. Funct. Anal., **58**, (1984), 194–215.

[20] P. Julg and G. Kasparov, *Operator K-theory for the group $SU(n,1)$*, J. Reine Angew. Math., **463**, (1995), 99–152.

[21] P. Julg, *Remarks on the Baum–Connes conjecture and Kazhdan's property T*, Operator algebras and their applications, Waterloo (1994/1995), Fields Inst. Commun., Amer. Math. Soc., **13**, (1997), 145–153.

[22] P. Julg, *Travaux de N. Higson et G. Kasparov sur la conjecture de Baum–Connes*, Séminaire Bourbaki. Vol. 1997/98, Astérisque, **252**, (1998), No. 841, 4, 151–183.

[23] G. G. Kasparov, *Topological invariants of elliptic operators. I. K-homology*, Izv. Akad. Nauk SSSR Ser. Mat., **39** no. 4, (1975), 796–838.

[24] G. G. Kasparov, *The operator K-functor and extensions of C^*-algebras*, Math. USSR Izv., **16**, (1980), 513–572.

[25] G. G. Kasparov, *Operator K-theory and its applications: elliptic operators, group representations, higher signatures, C^*-extensions*, Proc. of the Int. Cong. of Math., (Warsaw, 1983), 987–1000.

[26] G. G. Kasparov, *K-theory, group C^*-algebras, and higher signatures (conspectus)*, Novikov conjectures, index theorems and rigidity, Vol. 1 (Oberwolfach, 1993), 101–146, London Math. Soc. LNS 226, (1981).

[27] G. G. Kasparov, *Lorentz groups: K-theory of unitary representations and crossed products*, Dokl. Akad. Nauk SSSR, **275**, (1984), 541–545.

[28] G. G. Kasparov, *Equivariant KK-theory and the Novikov conjecture*, Invent. Math., **91**, (1988), 147–201.

[29] G. G. Kasparov and G. Skandalis, *Groups acting on buildings, operator K-theory, and Novikov's conjecture*, K-Theory, **4**, (1991), 303–337.

[30] G. Kasparov and G. Skandalis, *Groupes boliques et conjecture de Novikov*, Comptes Rendus Acad. Sc., **319**, (1994), 815–820.

[31] V. Lafforgue, *Une démonstration de la conjecture de Baum–Connes pour les groupes réductifs sur un corps p-adique et pour certains groupes discrets possédant la propriété (T)*, C. R. Acad. Sci. Paris Sér. I, **327**, (1998), 439–444.

[32] V. Lafforgue, *Compléments à la démonstration de la conjecture de Baum–Connes pour certains groupes possédant la propriété (T)*, C. R. Acad. Sci. Paris Sér. I, **328**, (1999), 203–208.

[33] V. Lafforgue, *KK-théorie bivariante pour les algèbres de Banach et conjecture de Baum–Connes*, Thèse de Doctorat de l'université Paris 11 (mars 1999).

[34] V. Lafforgue, *A proof of property (RD) for cocompact lattices in $SL_3(\mathbb{R})$ and $SL_3(\mathbb{C})$*, Journal of Lie Theory **10**, (2000), 255–267.

[35] P.-Y. Le Gall, *Théorie de Kasparov équivariante et groupoïdes*, C. R. Acad. Sci. Paris Sér. I, **324**, (1997), 695–698.

[36] P.-Y. Le Gall, *Théorie de Kasparov équivariante et groupoïdes. I*, K-Theory, **16**, (1999), 361–390.

[37] A. S. Mishchenko, *Homotopy invariants of multiply connected manifolds. I. Rational invariants*, Math. USSR Izv., **4**, (1970), 509–519, translated from Izv. Akad. Nauk SSSR Ser. Mat., **34**, (1970), 501–514.

[38] A. S. Mishchenko, *Infinite-dimensional representations of discrete groups, and higher signatures*, Math. USSR Izv., **8**, no. 1, (1974), 85–111.

[39] H. Oyono-Oyono, *La conjecture de Baum–Connes pour les groupes agissant sur les arbres*, C. R. Acad. Sci. Paris Sér. I Math., **326**, (1998), 799–804.

[40] H. Oyono-Oyono, *Baum–Connes conjecture and group actions on trees*, Preprint.

[41] M. V. Pimsner, *KK-groups of crossed products by groups acting on trees*, Invent. Math., **86**, (1986), 603–634.

[42] J. Ramagge, G. Robertson and T. Steger, *A Haagerup inequality for $\widetilde{A}_1 \times \widetilde{A}_1$ and $\widetilde{A}_2$ buildings*, Geom. Funct. Anal., **8**, (1998), 702–731.

[43] G. Skandalis, *Kasporov's bivariant K-theory and applications*, Expositiones Math., **9**, (1991), 193–250.

[44] G. Skandalis, *Progrès récents sur la conjecture de Baum–Connes, contribution de Vincent Lafforgue*, Séminaire Bourbaki, No. 869 (novembre 1999).

[45] J.-L. Tu, *The Baum–Connes conjecture and discrete group actions on trees*, K-Theory, **17**, (1999), 303–318.

[46] J.-L. Tu, *La conjecture de Baum–Connes pour les feuilletages moyennables*, K-Theory, **17**, (1999), 215–264.

[47] J.-L. Tu, *La conjecture de Novikov pour les feuilletages hyperboliques*, K-Theory, **16**, (1999), 129–184.

[48] A. Valette, *Introduction to the Baum–Connes conjecture*, to appear in the Series "Lectures in Mathematics – ETH Zürich", Birkhäuser.

[49] A. Wassermann, *Une démonstration de la conjecture de Connes–Kasparov pour les groupes de Lie linéaires connexes réductifs*, C. R. Acad. Sci. Paris Sér. I, **304**, (1987), 559–562.

Institut de Mathématiques de Jussieu
175 rue du Chevaleret
75013 Paris, France
E-mail address: `vlafforg@math.jussieu.fr`

An Introduction to Non-Commutative Mori Theory

Michael McQuillan

Abstract. The classification of algebraic surfaces of Enriques and Kodaira, and its conjectured higher dimensional generalisation by S. Mori et al, leaves as many questions unanswered as it solves; for example: which surfaces are hyperbolic? The highly conceptual nature of Mori's programme permits its extension to the birational classification of ordinary differential equations, and this in its full generality is the proper definition of non-commutative Mori theory. A remarkable observation of [2] is that not only can much of Mori's rational curve technology be brought to bear on this problem, but that many of the basic theorems of classification depend not on the dimension of the space on which the ODE is defined, but rather the dimension of its solutions. This suggests that, despite the difficulty of Mori's programme for higher dimensional varieties, its analogue for foliations by curves in any dimension will succeed. Whence a conjectural classification for ODEs on surfaces, and an exhaustive, yet wholly algebraic understanding of their hyperbolicity by way of the method of dynamic Diophantine approximation. A refined version of the programme leads similarly to the analogue of the Mordell conjecture for algebraic surfaces over function fields, which in [13] it had already succeeded in establishing under the hypothesis of positive index. Innovation in Diophantine approximation would lead to its applicability in arithmetic per se. Other applications are anticipated.

One of the most teleological of mathematical theorems is the profound relation between the topology, algebra, geometry and arithmetic of curves. In the first place an algebraic curve C may be considered over $\mathbb{C}$ as a Riemann surface, and as such is distinguished by a unique topological invariant, namely its genus g. The genus equally admits an algebraic, or more correctly algebro-geometric, definition as the dimension of the space of global sections of the bundle of holomorphic differentials, K_C, and three distinct types of behaviour emerge. In the first place should g be 0, the Riemann–Roch theorem establishes a conformal isomorphism with $\mathbb{P}^1$, and indeed the tangent bundle is even seen to be ample. Should the genus be 1, then we obtain an isomorphism of the curve with the variety of divisors of degree 0 on it, whence a group structure on the curve itself, i.e., C is an elliptic curve, and the tangent bundle is trivial. Finally should the genus be at least 2, then the bundle K_C is ample, and the linear system $|3K_C|$ yields an embedding of the curve into projective space. On the differential geometric side, this trichotomy manifests itself by way of the uniformisation theorem: $\mathbb{P}^1$ is simply connected and so indeed is not only its own universal cover but is positively curved; an elliptic curve is uniformised by $\mathbb{C}$ and admits a metric of zero curvature; while finally a curve of genus at least 2 is uniformised by the disc and has a metric of strictly negative sectional curvature. The arithmetic version of this trichotomy may be stated as soon as the curve is defined over a number field, then over sufficiently large number fields a curve is rational if and only if its number of rational points of big height at most H, say, grows like $O(H^2)$, while for elliptic curves the appropriate growth estimate is $O(\log H)$; finally by a celebrated theorem of G. Faltings [7], The Mordell Conjecture, a curve is of genus at least 2 if and only if the growth is $O(1)$.

It has been suggested that suitable generalisations of the above might hold in higher dimensions, where the genus is replaced by plurigenera, or more correctly by their rate of growth, i.e., by the Kodaira dimension. On the other hand this is not topologically invariant in dimension 3, and there is some evidence that this of itself might force the failure of a similar effort to relate, say, the algebraic and differential geometry of varieties. Nevertheless, the Kodaira dimension is, by a result of Friedman and Qin [9], a topological invariant in dimension 2, and thus suggests a quatrochotomy according to its possible values, as proposed by Lang and Vojta, [11] and [19].

Specifically, given an algebraic surface S, we once more consider the bundle, K_S, of top weight holomorphic forms, and study the maximum dimension $\kappa(S)$ of the image of S in projective space over all possible linear systems $|nK_S|$, for positive integers n, conventionally putting $\kappa(S)$ to be $-\infty$ if the image is always empty. The Kodaira dimension $\kappa(S)$ then exhibits four distinct types of behaviour. It is $-\infty$ if and only if S admits a pencil of rational curves. Otherwise it is zero for surfaces birational to a rather small list, viz: abelian surfaces, bi-elliptic surfaces, $K-3$ surfaces, or Enriques surfaces, all of which admit a metric of zero Ricci curvature, by Yau's famous theorem [21]; and it is 1 precisely for those elliptic surfaces which haven't already been noted. Finally, one puts everything else, i.e., those of Kodaira dimension 2, under the appellation "general type". From the point of view of differential geometry the situation for Kodaira dimension at most 1 is quite satisfactory with a clear understanding of the curvature, and parabolicity of such surfaces. However for surfaces of general type even Yau's celebrated theorem tells us very little; indeed no matter how we understand it, Enriques-Kodaira classification is very far from telling us to what extent is a surface of general type hyperbolic, and as for how many rational points they might have, we can say nothing beyond the small number of cases with high irregularity considered by G. Faltings, [8].

The original motivation of non-commutative Mori theory, hereby abbreviated NCMT, was to remedy this; it is however, as we shall see, in no way limited to its motivating example. It certainly cannot pretend to resolve the arithmetic nature of such a conjectured quatrochotomy per se, but there is every chance that it will succeed over function fields, and will contain all necessary geometric information for the arithmetic properly defined over number fields – this distinction results from a serious arithmetical gap in our understanding of Diophantine approximation.

Anyway, to illustrate the programme, let us concentrate on how we might wish to understand the hyperbolicity of higher dimensional varieties. A moment's thought shows that to ask for uniformisation by hyperbolic space is unreasonably strong; there are certainly many simply connected varieties which deserve the appellation "hyperbolic". The fundamental insight here is due to Kobayashi who defined a semi-distance on complex spaces to be a distance whenever length dominates area in all possible directions. Specifically given two points in a complex space one joins them by a chain of holomorphic discs, measures the length of this chain in the Poincaré metric, and takes the infimum over all such chains. Whence to relate the hyperbolicity of surfaces to their algebraic geometry, what is at stake is to show that on a surface of general type there is a Zariski open set on which the Kobayashi semi-distance dominates an actual distance, of which a weaker statement is the Green–Griffiths conjecture that the only parabolic curves on surfaces of general type are rational or elliptic. Now from the sense of the definition it is intuitively clear, and better still a theorem, that it is more than sufficient to classify all ordinary differential equations on an algebraic surface by the curvature of their solutions. It is this latter question that is the proper definition of NCMT.

One might argue that it would be better called the birational geometry of ODEs; however, just as Conne's non-commutative geometry [5], shows that even the strangest of spaces have much structure in common with the ordinary geometry of manifolds, so NCMT exhibits a remarkable parallel between differential and polynomial equations. The immediate question therefore is how even to define a differential equation algebraically. Fortunately this is not too hard. A variety X of dimension n comes equipped with a tower of jet spaces, each a $\mathbb{P}^{n-1}$ bundle over the previous one by adding in derivatives of higher order as introduced by Arrondo, Sols, and Speiser [1], and Demailly [6]. Whence an ODE of order m on a variety is a subvariety of the mth jet space. The ODE itself defines a possibly non-integrable foliation on this higher dimensional variety, whose solutions are in 1-1 correspondence to those of the original ODE by way of m-fold differentiation. Whence we have a canonical linearisation procedure which, provided we are prepared to replace our original space by a bigger one, allows us to restrict our attention to foliations.

In our motivating example the situation is particularly clean, since an mth order ODE on a surface corresponds to a foliation by curves on a variety of dimension $m+1$. Our objects of study are thus foliated varieties $(X, \mathcal{F})$, which really just amounts to equipping X with a saturated subsheaf $\mathcal{T}_{\mathcal{F}}$ of its tangent sheaf. Nevertheless this description permits the introduction of the main protagonist of NCMT, $K_{\mathcal{F}}$, as the dual of the top exterior power of $\mathcal{T}_{\mathcal{F}}$, which we may think of as the bundle of top weight holomorphic forms along the leaves. Not surprisingly then, NCMT wishes to classify foliations via $K_{\mathcal{F}}$. Now one observes that the case of the trivial foliation $X \to \operatorname{Spec} \mathbb{C}$ is nothing less than everyday Mori theory and a rather special case of NCMT, as is equivariant Mori theory, and as such the programme is ridiculously difficult. Even worse, supposing that such an absurdly difficult programme could be carried out, there is bound to be a large general type case about which one can say nothing. The first objection is wholly reasonable, but our real interest lies with foliations by curves, and there is already substantial evidence that regardless of the dimension of the ambient space this problem inherits many 1 dimensional features which render it remarkably tractable. Better still, foliations by curves of general type are exactly what we require to establish the hyperbolicity of their solutions, thanks to the method of dynamic Diophantine approximation applied on an appropriate foliation equivariant Zariski–Riemann surface. The main caveat here is singularities, since the possibility of their resolution is one of the fundamentals of ordinary Mori theory. Consequently NCMT may properly be divided into two distinct parts, not only that which may be considered as building on Mori theory proper, but also the establishment for foliated varieties of an analogue of Hironaka's resolution theorem.

Let us therefore concentrate on what resolution of foliation singularities might mean. Again our insight comes from the Mori programme where singularities are studied by way of the canonical sheaf. Naturally therefore we may extend the definitions of Mori, Kawamata et al [10], to the case of foliated varieties, and define foliated canonical, foliated terminal singularities, etc. One should note that being based on $K_{\mathcal{F}}$, these definitions have a priori nothing to do with the usual notions of the same on the underlying space X. They are, however, the wholly natural definitions for the bi-rational geometry of foliations, canonical being the largest class with respect to which the Kodaira dimension of $K_{\mathcal{F}}$ is a birational invariant. Needless to say another important notion is foliated Gorenstein, which while trivial when the underlying space is smooth, provides a fine illustration of the 1 dimensional nature of foliations by curves where one notes that a singularity is terminal and Gorenstein in the foliated sense iff it is smooth, i.e., admits a relatively smooth fibration as a local first integral. The key point however is that terminal resolutions do not exist in general, the most simple example being the equation $xdy + ydx$. This example is, nevertheless, canonical, and indeed rather typical of canonical singularities for foliations on algebraic surfaces, where the canonical singularities are classically referred to as reduced singularities, and Poincaré–Dulac singularities, and where a canonical resolution is known to exist.

The study of non-commutative Mori theory proper, naturally starts with a study of the closed cone of curves $\mathrm{NE}_1(X)$ on a foliated variety $(X, \mathcal{F})$. A complete study would seem impossibly difficult, but one of Mori's many insights was to realise that there is a clear distinction between the sub-cone on which the canonical class is positive, and the rest. Indeed away from the positive part, he proved the celebrated theorem that it is locally polyhedral, [17]. Naturally one wishes to pursue a similar theorem for $K_{\mathcal{F}}$, or specifically to find some parabolic structure for leaves through curves C on which $K_{\mathcal{F}}$ is negative. Substantial progress was already made in this direction by Y. Miyaoka [16], but his method required that the curve C should move in a family that covers the ambient space. The role of this assumption was brought out nicely by Shepherd-Barron in his reworking of

Miyoaka's theorem [10], where it is used to exhibit the foliation in positive characteristic as an inseparable scheme quotient. Extending this to arbitrary curves C therefore had two serious problems. Firstly to demonstrate the existence of an inseparable scheme quotient, and then to run Mori's bend and break argument even for curves passing through the singularities. This approach is technically feasible, but by no means easy [15], whereas with F. A. Bogomolov an alternative approach was discovered in which the ambient space X is replaced by a much smaller foliation invariant subvariety, cf. [2]. In consequence much stronger results are possible, for example:

Theorem 1. *Let $(X, \mathcal{F})$ be a normal (integrable) foliated variety such that $T_{\mathcal{F}}$ is a bundle and C a curve not contained in the singular locus of $\mathcal{F}$ on which $T_{\mathcal{F}}$ restricted to it is ample; then for all $x \in C$ there is a foliation invariant rationally connected subvariety V_x containing x of dimension the rank of $\mathcal{F}$.*

For foliations by curves the supposition that $T_{\mathcal{F}}$ is a bundle is nothing other than the definition of being Gorenstein, so it remains to examine curves in the singular locus. It is here that the hypothesis of canonical singularities is required. The situation is considered in full generality in [2], but for foliations by curves $K_{\mathcal{F}}$, or some power of it, always has a non-zero section over such a curve. Consequently, thanks to Mori type bounds on the rational curves implicit in the above theorem, we arrive at:

Theorem 2. *Let $(X, \mathcal{F})$ be a normal variety foliated by curves with (foliated) canonical Gorenstein singularities, and $\mathrm{NE}(X)_{K_{\mathcal{F}} \geq 0}$ the closed cone of curves on which $K_{\mathcal{F}}$ is non-negative; then there are $\mathcal{F}$ invariant rational curves L_i with $-2 \leq K_{\mathcal{F}}.L_i < 0$ such that,*

$$\mathrm{NE}_1(X) = \mathrm{NE}(X)_{K_{\mathcal{F}} \geq 0} + \mathbb{R}_+ L_i \,.$$

Better still the rays $\mathbb{R}_+ L_i$ are locally finite in the half space $\mathrm{NE}(X)_{K_{\mathcal{F}} < 0}$, any extremal ray is of this form, and should $K_{\mathcal{F}}.L_i = -2$ for some i, then $(X, \mathcal{F})$ is a foliation by rational curves.

At this point it is natural to discuss the contraction of extremal rays. However, unlike ordinary Mori theory, this is easily seen to be the wrong question. There are extremal rays on foliated 3-folds which are immobile, and as such admit no divisorial contraction, but are joined to another curve such that the combination admits a smooth contraction. At the other extreme, if the foliation is smooth then it is automatically its own minimal model. Whence in general one expects that the minimal model will enjoy a highly combinatorial description in terms of the integrable rational subvarieties through the singularities.

The situation for surfaces is of course relatively straightforward, and even better we have the existence of canonical resolutions. Whence one arrives at minimal models, [14] and [3]. The immediate question is then classification, and abundance. For foliations of numerical Kodaira dimension 0, abundance is easily established, and as one might expect such foliations admit wholly explicit descriptions in terms of group actions. For numerical Kodaira dimension 1, the situation is radically more complicated. Should the Kodaira dimension be non-negative, abundance holds, and we have a nice classification in terms of Ricatti foliations, elliptic fibrations, turbulent foliations, and products of curves. Abundance however fails in general, and it remains conjectural that the obvious examples of Shimura surfaces and Hilbert-modular surfaces with their natural foliations are exhaustive. Notwithstanding that we know that the foliations of general type do not have dense parabolic leaves, it would seem that NCMT must fail at this point to classify those foliations with parabolic leaves. Nevertheless, there is more than one way to skin a cat, and one may prove, cf. [12] or [4], that a dense parabolic leaf accumulates, in a suitable sense, at an interior point of the cone of curves, so in the presence of such a leaf, abundance holds. A flavour of the consequences of this analysis is,

Theorem 3. *Let X be an algebraic surface with $h^0(X, \mathrm{Sym}^k \Omega_X) \geq 2$ for some $k \in \mathbb{N}$; then X is hyperbolic iff it contains no rational curves or abelian varieties.*

This forcing of abundance in the case of studying rational points over function fields is rather more difficult, and even then one must still study integrable algebraic sub-curves of foliations by curves on 3-folds. The remarkable thing, once one manages to reduce to this later case, is that the abundance arguments in dimension 2 go through quasi verbatim, and one particularly attractive corollary is,

Theorem 4. *Let X be a non-isotrivial surface of general type and positive index over a field of functions in characteristic zero; then the rational points of X are not Zariski dense.*

It is tempting to hope that these considerations can be extended over number fields. There is, however, a serious technical problem. Specifically in Faltings and Vojta's work on Mordell type conjectures, [8] and [20], they take advantage of various linear systems on products of the original variety, or modifications of the same, which in some sense contract diagonals. In general NCMT predicts that contraction, modulo some well understood cases, i.e., classification, ought to remain possible but only analytically. One approach may therefore consist of making algebraic approximations of the analytic situation, a direction in which Siu is enjoying some success [18]. As ever a guide is provided by the dynamic counterpart, which in many cases enjoys an infinitesimal interpretation as a comparison of conformal structures. The goal of [14] is a definitive understanding of this relation, and whence some foundation for extending this circle of ideas to arithmetic proper.

References

[1] E. Arrondo, I. Sols and B. Speiser. 'Global moduli for contacts' *Ark. Math.* 35 (1997).

[2] F. A. Bogomolov and M. McQuillan. 'Rational curves on foliated varieties' *IHES pre-print* to appear Oct. 2000.

[3] M. Brunella. 'Minimal models of foliated algebraic surfaces' *Bull. SMF* 127 (1999) 289–305.

[4] M. Brunella. 'Courbes entières et feuilletages holomorphes' *Enseign. Math.* 45 (1999) 195–216.

[5] A. Connes. 'Non-commutative geometry' Academic Press 1994.

[6] J.-P. Demailly. 'Algebraic criteria for Kobayashi hyperbolic varieties and jet differentials' *Proc. Symp. Math.* 62 (1997).

[7] G. Faltings. 'Endlichkeitssatze fur abelsche Varietaten uber Zahlkorpen' *Invent. Math.* 73 (1983) 349–366.

[8] G. Faltings. 'Diophantine approximation on abelian varieties' *Ann. Math.* 133 (1991) 549–576.

[9] R. Friedman and Z. Qin. The smooth invariance of the Kodaira dimension of a complex surface *Math. Res. Lett.* 1 (1994) 369–376.

[10] J. Kollár et al. 'Flips and abundance for algebraic Threefolds' *Astérique* 211 (1992).

[11] S. Lang. 'Number Theory III' Springer-Verlag 1991.

[12] M. McQuillan. 'Diophantine approximation and foliations' *Inst. Hautes Études Sci. Publ. Math.* 87 (1999) 121–174.

[13] M. McQuillan. 'Non-commutative Mori theory' *IHES pre-print* March 2000, 100 pg.

[14] M. McQuillan. 'Comparisons of conformal structures via Macpherson's graph construction' *in preparation.*

[15] M. McQuillan. 'Deformation theory on inseparable scheme quotients' *in preparation.*

[16] Y. Miyaoka. 'Deformation of morphisms along a foliation' *Proc. Symp. Math.* 46 (1987).

[17] S. Mori. 'Threefolds whose canonical bundles are not numerically effective' *Ann. Math.* 116 (1982) 133–176.

[18] Y.-Y. Siu. Conversation at lunch, HKU 2000.

[19] P. Vojta. 'Diophantine approximation and value distribution theory' Lect. Notes. in Math. 1239 (1997).

[20] P. Vojta, P. 'Siegel's Theorem in the compact case' *Ann. Math.* 133 (1991) 509–548.

[21] S.-T. Yau. 'On the Ricci curvature of a complex Kahler manifold and the complex Monge-Ampère equation I' *Comm. Pure and App. Math.* 31 (1978) 339–411.

University of Oxford
Mathematical Institute
24-29 St Giles
Oxford OX1 3LB, United Kingdom
E-mail address: `mcquilla@maths.ox.ac.uk`

References

Geometric Methods in Complex Analysis

Stefan Nemirovski

Abstract. The talk surveys the applications of geometric topology to complex analysis in several complex variables.

1. Introduction

Complex analysis in several complex variables has always been one of the most omnivorous branches of mathematics. Sheaf theory and Banach algebras, algebraic topology and L^2-estimates for PDE's, all these and many other methods have been used to resolve various problems about holomorphic functions.

The purpose of this talk is to discuss complex analytic applications of geometric topology. Section 2 presents a rather personal and incomplete overview of complex analysis in $\mathbb{C}^n$. Section 3 explains how the topology of Stein manifolds is applied to analytic continuation of holomorphic functions and polynomial approximation. Section 4 concerns the connections between the differential topology of four-manifolds and analysis on complex surfaces.

My understanding of the subject has been largely formed at the complex analysis seminar conducted by Anatoliĭ Georgievich Vitushkin at the mekh-mat of the Moscow State University. I am deeply grateful to the participants of this seminar for many fruitful discussions.

2. Complex Analysis in $\mathbb{C}^n$

2.1. Analytic continuation

Let us consider a holomorphic function $f \in \mathcal{O}(U)$ in a domain $U \subset \mathbb{C}^n$. In many situations it is desirable to know what is the maximal analytic extension of f. More precisely, we want to describe the "Riemann surface" of the function f. An important point is that this extension may be multi-valued or, in other words, the Riemann surface of f may be a domain *over* $\mathbb{C}^n$.

In this setting, for individual functions or specific classes of functions, the problem can be extremely difficult. (Think of L-functions of algebraic varieties over $\mathbb{Q}$.) Therefore, one first looks for extension properties possessed by *all* holomorphic functions in a given domain. Technically, this is done by the notion of the envelope of holomorphy.

Definition 2.1. *Let $U \subset \mathbb{C}^n$ be a domain (an open connected subset). The envelope of holomorphy $\widetilde{U} \supset U$ is the maximal domain* over $\mathbb{C}^n$ *such that every $f \in \mathcal{O}(U)$ extends holomorphically to $\widetilde{U}$.*

It is well known that this notion becomes meaningful when $n \geq 2$. There are many examples of domains in $\mathbb{C}^2$ that are strictly contained in their envelopes of holomorphy. Perhaps the most classical example is given by (a neighbourhood of) the *Hartogs figure*

$$H = \{|z_1| = 1,\ |z_2| \leq 1\} \cup \{|z_1| \leq 1,\ z_2 = 0\} \subset \mathbb{C}^2 .$$

By the Hartogs Lemma, every function holomorphic in a neighbourhood of H extends to the unit bidisc $\{|z_1| \leq 1,\ |z_2| \leq 1\}$.

2.2. Polynomial and rational approximation

Let $f \in \mathcal{O}(K)$ be a holomorphic function in a neighbourhood of a compact set $K \Subset \mathbb{C}^n$. The problem is to find conditions (on f and K) under which f can be uniformly approximated by complex polynomials or rational functions. (This may be called "Runge-type" approximation problem. A much more subtle "Mergelyan-type" problem considers functions that are holomorphic in the interior of K only.)

Note that often every sequence of, say, polynomials converging uniformly on K converges on a larger set that does not depend on the sequence. This phenomenon can be observed already in $\mathbb{C}^1$. For instance, every sequence of polynomials converging on the unit circle converges on the entire unit disc by the maximum principle.

Definition 2.2. *Let $K \Subset \mathbb{C}^n$ be a compact subset. The convex hull of K with respect to a subalgebra of functions $\mathcal{P} \subset \mathcal{O}(K)$ is the set*

$$\widehat{K}_{\mathcal{P}} = \{z \in \mathbb{C}^n \mid |p(z)| \leq \max_{\xi \in K} |p(\xi)| \ \textit{for all}\ p \in \mathcal{P}\} .$$

Morally speaking, the larger is the $\mathcal{P}$-convex hull $\widehat{K}_{\mathcal{P}}$, the more obstructions there are to approximation by functions from $\mathcal{P}$.

2.3. The case of $\mathbb{C}^1$

In this case it is relatively easy to determine the envelopes and hulls defined above. Namely,

- the polynomially convex hull of $K \Subset \mathbb{C}^1$ is obtained by "filling in the holes" in K so that $\mathbb{C}^1 \setminus K$ becomes connected (Runge Theorem);
- every compact set $K \Subset \mathbb{C}^1$ is its own rationally convex hull;
- every domain $U \subset \mathbb{C}^1$ is its own envelope of holomorphy.

The last two statements follow from the fact that the singularities of rational functions of one complex variable are isolated points. Note also the first appearence of a topological condition in the Runge theorem.

2.4. Attaching analytic discs

In favourable circumstances, it is possible to approach higher dimensional problems by using essentially one-dimensional objects. The idea is to consider holomorphic maps $f\colon \Delta \to \mathbb{C}^n$ from the unit disc $\Delta \subset \mathbb{C}^1$ such that $f(\partial\Delta) \subset K$ where K is some fixed compact subset of $\mathbb{C}^n$.

The following properties of analytic discs make them an attractive tool for complex analysis:

- $f(\Delta)$ is contained in the polynomially convex hull of K (maximum principle);
- if $f(\partial\Delta)$ is homologous to zero in K, then $f(\Delta)$ is contained in the rationally convex hull of K (argument principle);
- if f can be included in a continuous family $f_t\colon \Delta \to \mathbb{C}^n$ starting from a disc $f_0(\Delta) \subset K$, then $f(\Delta)$ lifts to the envelope of holomorphy of K (continuity principle).

A (misleadingly) perfect example is the Hartogs figure $H \subset \mathbb{C}^2$. The obvious family of analytic discs

$$D_t = \{|z_1| \leq 1,\ z_2 = t\}, \quad |t| \leq 1$$

satisfies all the above requirements and, in fact, completely exhausts the polynomial and rational hulls and the envelope of holomorphy of H.

Unfortunately, in general it is rather hard to find analytic discs with boundary in a given set. Furthermore, the behaviour of families of analytic discs tends to be very complicated. Nonetheless, various techniques based on families of holomorphic discs (and more general Riemann surfaces with boundary) have found important applications in complex analysis, symplectic and contact geometry.

In this talk, however, we shall pursue a different approach, based on global properties of envelopes and hulls. The results are less explicit but the methods can be applied in a more general setting.

3. Stein Manifolds

3.1. Definition

A complex analytic manifold is Stein if it can be properly holomorphically embedded into $\mathbb{C}^N$ for some N. (This definition is good to work with but *not* to prove that one manifold or another is Stein.)

The following classical results show the crucial importance of Stein manifolds for several complex variables:

- envelopes of holomorphy are Stein manifolds spread over $\mathbb{C}^n$;
- polynomial and rational hulls have Stein neighbourhoods with the Runge property, i.e., holomorphic functions therein can be approximated by polynomials and rational functions, respectively.

Thus, every statement about Stein manifolds translates into a result on analytic continuation and approximation in $\mathbb{C}^n$.

3.2. Morse theory on Stein manifolds

Every Stein manifold X of complex dimension n admits a proper Morse function without critical points of index $> n$. To see this, consider the distance function to a sufficiently generic point in $\mathbb{C}^N$ for a proper holomorphic embedding $X \subset \mathbb{C}^N$. Standard Morse theory yields the following topological result:

Theorem 3.1. (Andreotti–Frankel, 1959) *A Stein manifold of complex dimension n is diffeomorphic to the interior of a (possibly infinite) handlebody without handles of index $> n$.*

In particular, Stein manifolds have no homology above the middle dimension, $H_p(X;\mathbb{Z}) = 0$ for all $p > n$. This has a strong analytic consequence:

Corollary 3.2. *The envelope $\widetilde{U} \supset U$ "fills in" all the cycles of dimension greater than n in U.*

For instance, if $V \subset \mathbb{C}^n$, $n \geq 2$, is a bounded domain with connected boundary ∂V and U is a neighbourhood of the boundary, then the envelope of U must contain V. In other words, holomorphic functions defined near ∂V extend to the entire domain V, which is the classical theorem of Hartogs. Needless to say, there are easier ways to prove this particular result. However, it is much harder to make these more direct methods work in the other situations covered by the above corollary.

3.3. Runge property and topology

More can be said about the topology of polynomially convex sets. A polynomially convex subset $K \Subset \mathbb{C}^n$ admits a base of Stein neighbourhoods of the form

$$U_\varphi = \{z \in \mathbb{C}^n \mid \varphi(z) < 0\}$$

where $\varphi\colon \mathbb{C}^n \to \mathbb{R}$ is a proper Morse function without critical points of index $> n$ (see [1]). It follows by Morse theory that

$$H_p(U_\varphi;\mathbb{Z}) = 0 \quad \text{for all } p \geq n\,.$$

Therefore, polynomial hulls "fill in" the cycles of dimension n as well. In particular, an n-dimensional closed submanifold in $\mathbb{C}^n$ is homologous to zero in its polynomially convex hull.

Somewhat surprisingly it is not clear what might be the analogue of this result for rationally convex sets. Applying Morse theory in the same way as for polynomially convex sets leads to problems about the topology of hypersurface complements in $\mathbb{C}^n$. Homology computations suggest that purely topological arguments could be insufficient to settle this question (see [18]).

3.4. Morse theory on Stein manifolds, continued

One may ask whether there exist other restrictions on the topology of Stein manifolds that are not captured by Morse theory. It turns out that the answer is different for $n \geq 3$ and $n = 2$. More precisely, the difference appears if we are interested in *differential* topology.

Let M be an open smooth manifold of real dimension $2n$ equipped with a Morse function without critical points of index $> n$ and an almost-complex structure. (These assumptions are clearly necessary for M to admit a Stein complex structure.)

Theorem 3.3. (Eliashberg, 1990) *If $n \geq 3$, then M is diffeomorphic to a Stein manifold.*

Hence, the Andreotti–Frankel theorem is sharp in dimensions ≥ 3. In particular, we cannot expect further general results for envelopes of holomorphy in $\mathbb{C}^n$, $n \geq 3$.

Theorem 3.4. (Gompf, 1996) *If $n = 2$, then M is homeomorphic to a Stein manifold.*

The assertion is weaker because the differential topology of Stein complex surfaces has some extra rigidity. In many important cases, this statement can be made precise by using gauge theoretic invariants. Consequently, interesting phenomena arise in complex analysis on $\mathbb{C}^2$ and other complex surfaces.

Remark 3.5. *The results of Eliashberg and Gompf are actually stronger. Namely, there exists a Stein complex structure realizing the homotopy type of the given almost-complex structure on M. In particular, the diffeomorphism (respectively, homeomorphism) to the Stein manifold is orientation preserving.*

4. Analytic Continuation and Embedded Real Surfaces in Complex Surfaces

4.1. Envelopes of holomorphy of real surfaces in $\mathbb{C}^2$

Before proceeding to the general statements, let us consider two typical analytic corollaries.

By Morse theory, Stein domains in $\mathbb{C}^2$ can have non-trivial homology up to real dimension 2. So, let $\Sigma \subset \mathbb{C}^2$ be a smoothly embedded closed real surface of genus g (a sphere with g handles) and let us try to use (a deformation of) Σ as the two-skeleton of a Stein domain. This turns out to be possible only if $g \geq 1$.

Theorem 4.1. (Forstnerič, 1992) *If $g \geq 1$, then there exists a C^0-close embedding with a base of Stein neighbourhoods.*

For instance, the Clifford torus $T = \{|z_1| = |z_2| = 1\} \subset \mathbb{C}^2$ has a fundamental system of Stein neighbourhoods

$$U_\varepsilon = \{1 - \varepsilon < |z_1|, |z_2| < 1 + \varepsilon\}.$$

Moreover, T is rationally convex. This is perhaps the simplest example to the Duval–Sibony theorem [5] on the rational convexity of Lagrangian submanifolds.

On the other hand, embedded two-spheres never have "small" Stein neighbourhoods:

Theorem 4.2. (Nemirovski, 1998) *A smoothly embedded 2-sphere $\Sigma \subset \mathbb{C}^2$ is homologous to zero in every Stein domain over $\mathbb{C}^2$ containing it.*

In particular, Σ is homologous to zero in the envelope of holomorphy of any neighbourhood $U \supset \Sigma$. In other words, envelopes of holomorphy of domains in $\mathbb{C}^2$ "fill in" two-dimensional cycles represented by smooth embeddings of the two-sphere.

For instance, the envelope of holomorphy of the standard two-sphere

$$S = \{|z_1|^2 + |z_2|^2 = 1, \operatorname{Im} z_2 = 0\} \subset \mathbb{C}^2$$

is the Levi-flat three-ball

$$B = \{|z_1|^2 + |z_2|^2 \leq 1, \operatorname{Im} z_2 = 0\}.$$

This follows from the continuity principle applied to the family of analytic discs

$$D_t = \{|z_1|^2 + t^2 \leq 1, z_2 = t\}, \quad t \in [-1, 1].$$

The last example was generalized by several authors (see [3, 13]). It was shown that the envelope of holomorphy of a generic two-sphere is given by a family of analytic discs provided that the sphere is contained in the boundary of a strictly pseudoconvex domain. Without the latter assumption, however, this geometric result is false [7].

4.2. Adjunction inequalities for real surfaces in Stein surfaces

The theorems stated in the previous subsection are a particular instance of general results on real surfaces in Stein complex surfaces [17].

Theorem 4.3. (Adjunction inequality) *Let $\Sigma \subset X$ be a smoothly embedded closed oriented real surface in a Stein complex surface. Then*

$$[\Sigma] \cdot [\Sigma] + |c_1(X) \cdot [\Sigma]| \leq 2g(\Sigma) - 2 \qquad (*)$$

if only Σ is not a null-homologous 2-sphere.

Here " $\cdot$ " denotes the intersection pairing, $[\Sigma] \in H^2_c(X;\mathbb{Z})$ is the cohomology class Poincaré dual to Σ, and $c_1(X) \in H^2(X;\mathbb{Z})$ is the first Chern class of X.

Inequality $(*)$ is called the "adjunction inequality" because it resembles the classical adjunction formula for complex curves. Inequalities of this type were proved for embedded real surfaces in compact complex surfaces by using the Seiberg–Witten invariants (see [12, 15], and [19]). The result for Stein surfaces can be deduced from this and the algebraic approximation theorem of Stout [21].

In the compact case, adjunction inequalities yield a proof of the so-called "generalized Thom conjecture." The latter asserts that a non-singular complex curve in a compact Kähler surface attains the minimal genus among all embedded real surfaces realizing the same homology class.

Stein manifolds do not contain compact complex curves, and hence this interpretation of the adjunction inequality does not make much sense for them. However, there is a different geometric way to see that inequality $(*)$ is sharp in the Stein case.

Theorem 4.4. (Existence of Stein neighbourhoods) *If a smoothly embedded real surface $\Sigma \subset Y$ in an arbitrary complex surface Y satisfies inequality $(*)$, then there exists a C^0-close embedded surface with a base of Stein neighbourhoods.*

This result is essentially a consequence of Gromov's h-principle for totally real embeddings. More precisely, a generic embedded real surface $S \subset Y$ admits a Stein neighbourhood base if and only if it has only hyperbolic complex tangencies. (A complex tangency is a point $p \in S$ such that the tangent plane T_pS is a complex line in T_pY.) The h-principle tells us that the surface Σ is isotopic to such a surface if there are no topological obstructions. However, by Lai's formulas for the number of complex tangencies of different types, the obstructions take on the form of inequality $(*)$.

4.3. Legendrian surgery and Bennequin's inequality

To emphasize the role of the adjunction inequality as an obstruction to an h-principle, we return briefly to the results of Eliashberg and Gompf on the existence of Stein manifolds with prescribed topology.

Let us restrict ourselves to the simplest situation. Suppose that we wish to build up a Stein complex surface diffeomorphic to the 4-ball with a two-handle attached to it along an unknotted circle in the 3-sphere. Note that to glue a two-handle, we must first choose a framing on this circle. In our case, the framing is determined by the integer n equal to the self-intersection index of the two-sphere formed by the core of the handle and a disc bounded by our circle in the ball.

The crucial observation of Eliashberg is that the resulting four-manifold can be endowed with a Stein complex structure if the following assumptions are satisfied:

(i) the gluing circle is Legendrian with respect to the standard contact structure on the 3-sphere,

(ii) the framing is the canonical Legendrian framing minus 1.

However, the framings realizable by Legendrian embeddings of the unknot are *negative* by the famous Bennequin inequality [2]. Therefore, the two-sphere obtained as the "core" of our Stein surface can have self-intersection at most -2, which is precisely the bound given by the adjunction inequality.

To overcome this difficulty Gompf uses infinite constructions of four-manifold topology (Casson handles). This leads to Stein manifolds that are only homeomorphic to the initial handlebody. On the other hand, there is no "Bennequin inequality" in higher dimensions, and the method works within the smooth category.

Let us also note that, in the opposite direction, the adjunction inequality for real surfaces in Stein surfaces was used by Lisca and Matić [14] to obtain generalizations of the Bennequin inequality.

4.4. More applications

Adjunction inequality $(*)$ has several other applications to analytic continuation. Sometimes it is necessary to leave $\mathbb{C}^2$ and consider holomorphic functions over other complex surfaces. We shall state two related results whose proofs use envelopes of holomorphy over $\mathbb{C} \times \mathbb{C}P^1$ and $\mathbb{C}P^2$.

The first theorem generalizes the Hartogs Lemma in $\mathbb{C}^2$ in the following way. The disc $\{|z_1| \leq 1, z_2 = 0\}$ in the Hartogs figure is replaced by a graph of a continuous function bounded by 1.

Theorem 4.5. (Generalized Hartogs Lemma) *Let $f: \overline{\Delta} \to \overline{\Delta}$ be a continuous self-map of the unit disc and let $\Gamma_f \subset \overline{\Delta} \times \overline{\Delta} \subset \mathbb{C}^2$ be its graph. Then every holomorphic function in a neighbourhood of the generalized Hartogs figure*

$$H_f = \{|z_1| = 1, |z_2| \leq 1\} \cup \Gamma_f \subset \mathbb{C}^2$$

extends to the unit bidisc.

This theorem was proved by Chirka [4] in 1997. A proof based on inequality $(*)$ is explained in [17]. However, the original argument was inspired by the approach developed by Ivashkovich–Shevchishin [11] using Gromov's theory of pseudoholomorphic curves.

The classical Hartogs Lemma holds true in $\mathbb{C}^n$ for all $n \geq 2$. On the other hand, Rosay showed in [20] that the generalization is false when $n \geq 3$. The counterexample is obtained by an "embedded version" of Eliashberg's Legendrian surgery. This shows once again the difference between complex dimension 2 and higher dimensions.

The next theorem (proved in [16]) was probably the first application of adjunction inequalities to complex analysis. This result was conjectured by Vitushkin in the 1980s as a by-product of his studies on the geometric aspects of the Jacobian conjecture in $\mathbb{C}^2$ (cf., for instance, [22]).

Theorem 4.6. (Vitushkin's Conjecture) *Let $S \subset \mathbb{C}P^2$ be an embedded 2-sphere that is not homologous to zero. Then every holomorphic function in a neighbourhood of S is constant.*

The proof uses R. Fujita's theorem [9] that the envelope of holomorphy of a domain in the complex projective space $\mathbb{C}P^n$ is either Stein or coincides with the entire $\mathbb{C}P^n$. If a domain in $\mathbb{C}P^2$ contains a homologically non-trivial two-sphere (which must have positive self-intersection), the adjunction inequality excludes the first possibility. Hence, every holomorphic function extends from this domain onto the entire $\mathbb{C}P^2$ and so is constant by the maximum principle.

The idea to regard Vitushkin's conjecture as an extension theorem is due to Sergeĭ Ivashkovich. A similar argument shows that every meromorphic function in a neighbourhood of S is rational, that is, extends meromorphically onto $\mathbb{C}P^2$.

On the other hand, it follows from Theorem 4.4 that there exists an embedded real surface $\Sigma \subset \mathbb{C}P^2$ of genus 3 homologous to the projective line and having a base of Stein neighbourhoods.

4.5. Selected problems: Fake Stein $\mathbb{R}^4$s

In [10], Gompf proves that there exist uncountably many exotic Stein $\mathbb{R}^4$s (i.e., Stein complex surfaces homeomorphic but not diffeomorphic to $\mathbb{R}^4$). However, he observes that the construction works only for some classes of exotic $\mathbb{R}^4$s. In particular, all Stein examples are actually diffeomorphic to domains in the standard $\mathbb{R}^4$.

It will be interesting to know which exotic $\mathbb{R}^4$s are Stein. For instance, it does not seem plausible that the "universal" exotic $\mathbb{R}^4$ of Freedman and Taylor admits a Stein complex structure.

Somewhat more precisely one may ask the following:

Question. *Are all exotic Stein $\mathbb{R}^4$s diffeomorphic to domains in $\mathbb{R}^4$?*

Question. *Are any (or all) of them biholomorphic to domains in $\mathbb{C}^2$?*

The last problem is related to the old question about the existence of Stein manifolds (of arbitrary dimension) that are homeomorphic to the $2n$-dimensional ball but cannot be realized as domains in $\mathbb{C}^n$.

In any case, a Stein domain in $\mathbb{C}^2$ that is homeomorphic to $\mathbb{R}^4$ but not diffeomorphic to it would be an interesting object for "hard" analysis.

References

[1] A. Andreotti and R. Narasimhan, *A topological property of Runge pairs*, Ann. of Math. (2) **76** (1962), 499–509.

[2] D. Bennequin, *Entrelacements et equation de Pfaff*, Asterisque **107/108** (1983), 87–161.

[3] E. Bedford and W. Klingenberg, *On the envelope of holomorphy of a 2-sphere in $\mathbb{C}^2$*, J. Am. Math. Soc. **4** (1991), 623–646.

[4] E. M. Chirka, *The generalized Hartogs lemma and the non-linear $\overline{\partial}$-equation*, in: Complex analysis in contemporary mathematics (dedicated to the memory of B. Shabat), Fasis, Moscow, 1998, pp. 19–30.

[5] J. Duval and N. Sibony, *Polynomial convexity, rational convexity, and currents*, Duke Math. J. **79** (1995), 487–513.

[6] Ya. Eliashberg, *Topological characterization of Stein manifolds of dimension > 2*, Int. J. Math. **1** (1990), 29–46.

[7] J. E. Fornaess and D. Ma, *A 2-sphere in $\mathbb{C}^2$ that cannot be filled in with analytic discs*, Int. Math. Res. Notices **1** (1995), 17–22.

[8] F. Forstnerič, *Complex tangents of real surfaces in complex surfaces*, Duke Math. J. **67** (1992), 353–376.

[9] R. Fujita, *Domaines sans point critique intérieur sur l'espace projectif complexe*, J. Math. Soc. Japan **15** (1963), 443–473.

[10] R. E. Gompf, *Handlebody construction of Stein surfaces*, Ann. of Math. (2) **148** (1998), 619–693.

[11] S. Ivashkovich and V. Shevchishin, *Structure of the moduli space in the neighbourhood of a cusp curve and meromorphic hulls*, Invent. Math. **136** (1999), 571–602.

[12] P. B. Kronheimer and T. Mrowka, *The genus of embedded surfaces in the projective plane*, Math. Res. Lett. **1** (1994), 797–808.

[13] N. G. Kruzhilin, *Two-dimensional spheres in the boundaries of strictly pseudoconvex domains in* $\mathbb{C}^2$, Izv. Acad. Nauk SSSR Ser. Mat. **55** (1991), 1194–1237; English transl. in *Math. USSR Izv.* **39** (1992), 1151–1187.

[14] P. Lisca and G. Matić, *Stein 4-manifolds with boundary and contact structures*, Topology Appl. **88** (1998), 55–66.

[15] J. W. Morgan, Z. Szabó and C. H. Taubes, *A product formula for Seiberg-Witten invariants and the generalized Thom conjecture*, J. Differential Geom. **44** (1996), 706–788.

[16] S. Nemirovski, *Holomorphic functions and embedded real surfaces*, Mat. Zametki **63** (1998), 599–606; English transl. in *Math. Notes* **63** (1998), 527–532.

[17] S. Nemirovski, *Complex analysis and differential topology on complex surfaces*, Uspekhi Mat. Nauk **54**:4 (1999), 47–74; English transl. in *Russ. Math. Surveys* **54**:4 (1999), 729–752.

[18] S. Nemirovski, in preparation.

[19] P. Ozsváth and Z. Szabó, *The symplectic Thom conjecture*, Ann. of Math. (2) **151** (2000), 93–124.

[20] J. P. Rosay, *A counterexample related to Hartogs' phenomenon (a question by E. Chirka)*, Michigan Math. J. **45** (1998), 529–535.

[21] E. L. Stout, *Algebraic domains in Stein manifolds*, in: Banach Algebras and Several Complex Variables, Proc. Conf. New Haven/Conn. 1983, *Contemp. Math.* **32** (1984), 259–266.

[22] A. G. Vitushkin, *Description of the homology of a ramified covering of* $\mathbb{C}^2$, Mat. Zametki **64** (1998), 839–846; English transl. in *Math. Notes* **64** (1998), no. 6.

Steklov Mathematical Institute
Russian Academy of Sciences
Gubkina str. 8
117966 Moscow GSP-1, Russia
E-mail address: `stefan@mccme.ru`

Vanishing Cycles and Mutation

Paul Seidel

Abstract. Using Floer cohomology, we establish a connection between Picard–Lefschetz theory and the notion of mutation of exceptional collections in homological algebra.

1. Introduction

This talk is about symplectic aspects of Picard–Lefschetz theory and the role of Floer cohomology in that context. I have relied on two main sources for inspiration, which are the ideas of Donaldson on vanishing cycles [3] respectively those of Kontsevich, partly in collaboration with Barannikov, on mirror symmetry for Fano varieties [8]. On a more technical level, the basis is provided by Fukaya's work on Floer cohomology [5]. I have tried to be as concrete as possible: not only are all objects mentioned rigorously defined, but they can be explicitly computed, and the assertions made about them checked, in many examples. This hands-on approach has its drawbacks, one of which will be mentioned after the summary of contents.

The basic geometric notion is that of an exact Morse fibration, a symplectic analogue of a holomorphic Morse function. One way of analyzing such fibrations is through the vanishing cycles in a fibre. When the base is a disc, one conveniently makes a choice of a finite family of vanishing cycles, forming a so-called distinguished basis. Any two such bases are connected by a sequence of Hurwitz moves. These are familiar concepts except that here vanishing cycles are considered as Lagrangian submanifolds, rather than only as homology classes.

Apart from their role as geometric objects worthy of study on their own, exact Morse fibrations are also relevant to Floer theory since, when equipped with suitable Lagrangian boundary conditions, they provide homomorphisms between Floer cohomology groups. As an application of these new maps we construct a long exact sequence analogous to that of Floer in gauge theory.

After that we return to exact Morse fibrations over a disc. The goal is to associate to each such fibration a triangulated category. An additional assumption is necessary: for simplicity, let us say that the total space of the fibration has zero first Chern class and zero first Betti number. The construction proceeds in several steps: first one chooses a distinguished basis of vanishing cycles. From that one obtains a Fukaya-type A_∞-category, unique up to quasi-isomorphism; our invariant is the derived category of this. It resembles derived categories of coherent sheaves on some Fano varieties, in that it is generated by an exceptional collection. The main point is to show that Hurwitz moves of the distinguished basis correspond to mutations of the exceptional collection, which leave the derived category unchanged. The explicit nature of mutations also allows them to be used for concrete computations of Floer cohomology.

We have to admit that our triangulated categories are only well-defined in a weak sense: different choices made during the construction lead to equivalent categories, but it has not been proved that the equivalences are canonical up to isomorphism (to have a completely satisfactory theory, it would further be necessary to establish coherence relations between functor isomorphisms). Rather than trying to do these improvements, it seems better to look for an alternative definition bypassing the choice of distinguished basis. The approach envisaged by Kontsevich is of this kind, but more work would be needed to put it on a rigorous footing.

2. Picard–Lefschetz Theory

(2A) Let (M, ω, θ) be an exact symplectic manifold of dimension $2n$. This means that M is compact with boundary, $\omega \in \Omega^2(M)$ is a symplectic form, and $\theta \in \Omega^1(M)$ satisfies $d\theta = \omega$. We will consider only symplectic automorphisms ϕ of M which are equal to the identity near ∂M. Such a ϕ is called exact if $[\phi^*\theta - \theta] \in H^1(M, \partial M; \mathbb{R})$ is zero. The exact symplectic automorphisms form a subgroup $\mathrm{Symp}^e(M) \subset \mathrm{Symp}(M)$. Note that any isotopy within this subgroup is Hamiltonian.

Let S be a smooth manifold with boundary. An exact symplectic fibration over S consists of data (E, π, Ω, Θ) as follows. $\pi\colon E \to S$ is a proper differentiable fibre bundle whose fibres are $2n$-dimensional manifolds with boundary. This means that E itself is a manifold with codimension two corners, with $\partial E = \partial_v E \cup \partial_h E$ consisting of two faces: $\partial_v E = \pi^{-1}(\partial S)$, while $\pi \mid \partial_h E\colon \partial_h E \to S$ is again a differentiable fibre bundle. $\Omega \in \Omega^2(E)$ is closed, its vertical part $\Omega \mid \ker(D\pi)$ nondegenerate at every point, and $\Theta \in \Omega^1(E)$ satisfies $d\Theta = \Omega$. We impose a final condition of triviality near $\partial_h E$. This means that there should be a neighbourhood $W \subset E$ of $\partial_h E$ and a diffeomorphism, for some $z \in S$, $\Xi\colon S \times (W \cap E_z) \to W$ lying over S, such that $\Xi^*\Omega = pr_2^*(\Omega \mid E_z)$ and $\Xi^*\Theta = pr_2^*(\Theta \mid E_z)$; here pr_2 is projection from $S \times (W \cap E_z)$ to the second factor. Clearly each fibre $(E_z, \omega_z = \Omega \mid E_z, \theta_z = \Theta \mid E_z)$ is an exact symplectic manifold. The form Ω defines a canonical connection on $\pi\colon E \to S$, with structure group $\mathrm{Symp}^e(E_z)$. In fact there is a bijective correspondence between fibre bundles with structure group $\mathrm{Symp}^e(E_z)$ in the usual sense, and cobordism classes of exact symplectic fibrations. We denote the parallel transport maps of the canonical connection by $\rho_c\colon E_{c(a)} \to E_{c(b)}$, for $c\colon [a;b] \to S$.

(2B) From now on assume that S is two-dimensional and oriented. An **exact Morse fibration** (this is shorthand for "exact symplectic fibration with Morse-type critical points") over S consists of data $(E, \pi, \Omega, \Theta, J_0, j_0)$. The properties of E, π, Ω, Θ are as before, except that π is allowed to have finitely many critical points. Each fibre may contain at most one of these points, and there should be none at all on ∂E. J_0 is an integrable complex structure defined in a neighbourhood of the set $E^{\text{crit}} \subset E$ of critical points, and Ω must be a Kähler form for it. Similarly j_0 is a positively oriented complex structure on a neighbourhood of the set $S^{\text{crit}} \subset S$ of critical values. They should be such that π is (J_0, j_0)-holomorphic, with nondegenerate second derivative at each critical point. We will usually denote exact Morse fibrations by (E, π) only.

One thing that needs explaining is why these are supposed to be analogues of holomorphic functions. For this one considers pairs (j, J) consisting of a positively oriented complex structure j on S and an almost complex structure J on E, such that $j = j_0$ near S^{crit}, $J = J_0$ near E^{crit}, π is (J, j)-holomorphic, and $\Omega(\cdot, J\cdot) \mid \ker(D\pi)$ is symmetric and positive definite everywhere. In this situation we say that J is compatible relative to j. The space of such pairs (j, J) is always contractible, and in particular nonempty. Moreover, for a fixed pair, by adding a positive two-form from S one can modify Ω such that it becomes symplectic and tames J.

Restricting any exact Morse fibration to $S \setminus S^{\text{crit}}$ yields an exact symplectic fibration. Before bringing the singular fibres into the picture, we need some more definitions. Let (M, ω, θ) be an exact symplectic manifold. A Lagrangian submanifold $L \subset M$, always assumed to be disjoint from ∂M, is called exact if $[\theta \mid L] \in H^1(L; \mathbb{R})$ is zero. A **framed Lagrangian sphere** is a Lagrangian submanifold L together with an equivalence class $[f]$ of diffeomorphisms $f \colon S^n \to L$. Here f_1, f_2 are equivalent if $f_2^{-1} f_1$ is isotopic to some element of $O(n+1) \subset \text{Diff}(S^n)$. One can associate to any $(L, [f])$ a Dehn twist $\tau_{(L,[f])} \in \text{Symp}(M)$ which is unique up to Hamiltonian isotopy. If L is exact, so is the Dehn twist along it. In future, we will often omit the framing $[f]$ from the notation.

To return to our discussion, let (E, π) be an exact Morse fibration. Take a path $c \colon [0; 1] \to S$ with $c^{-1}(S^{\text{crit}}) = \{1\}$ and $c'(1) \neq 0$. Let x be the unique critical point in $E_{c(1)}$. Then the stable manifold

$$B = \{y \in E_{c(s)},\ 0 \leq s < 1, \text{ with } \lim_{t \to 1} \rho_{c|[s,t]}(y) = x\} \cup \{x\}$$

is a smoothly embedded $(n+1)$-dimensional ball on which Ω vanishes identically, and therefore $V = \partial B = B \cap E_{c(0)}$ is an exact Lagrangian submanifold of $E_{c(0)}$ diffeomorphic to S^n. Moreover, V has a canonical structure of a framed Lagrangian sphere, constructed by first carrying V by parallel transport to $B \cap E_{c(s)}$, for some s close to 1, then projecting orthogonally in local Kähler coordinates to TB_x, and finally projecting radially to the unit sphere in that tangent space. The composition of these maps is a diffeomorphism $f^{-1} \colon V \to S^n$ whose inverse is the framing. $(V, [f])$ is called the **vanishing cycle** associated to c. A symplectic version of the Picard–Lefschetz theorem says that for l, c as in Figure 1, the monodromy $\rho_l \in \text{Symp}^e(E_{c(0)})$ is isotopic to $\tau_{(V,[f])}$.

(2C) Suppose now that $S = D$ is the closed unit disc in $\mathbb{C}$. Let (E, π) be an exact Morse fibration over it with m critical values. Let M be the fibre at some base point $z_0 \in \partial D$, say $z_0 = -i$. An admissible choice of paths is a family $(c_1, \dots, c_m)$

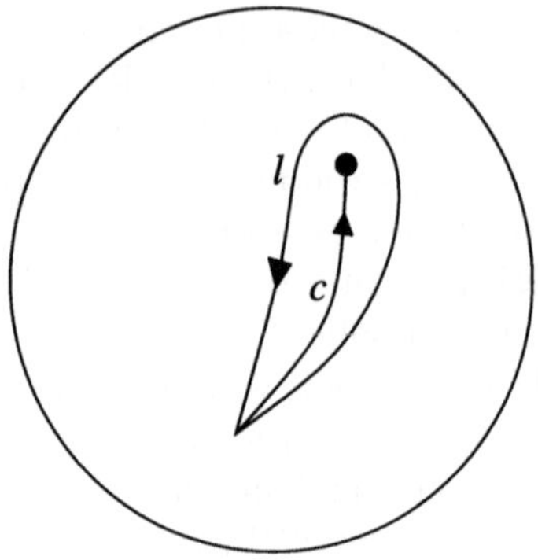

FIGURE 1.

looking as in Figure 2, ordered by their tangent directions at z_0. The collection of exact framed Lagrangian spheres in M arising from them is called a **distinguished basis** of vanishing cycles. Modifying the paths affects the distinguished basis in a way which can be determined using the Picard–Lefschetz theorem. The outcome is encoded in the following abstract notion:

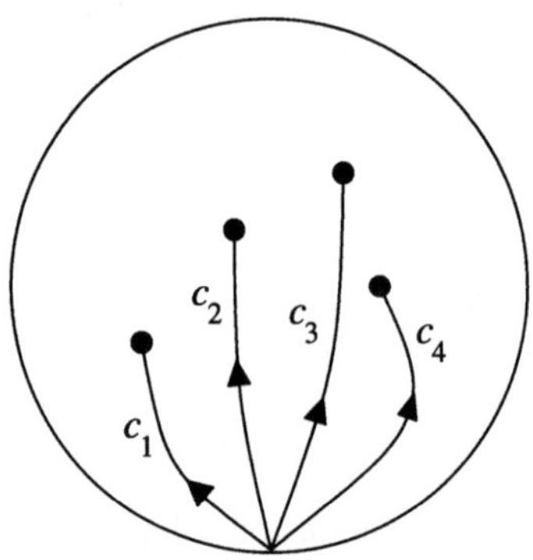

FIGURE 2.

Definition 2.1. *A Lagrangian configuration in an exact symplectic manifold M is an ordered family $\Gamma = (L_1, \dots, L_m)$ of exact framed Lagrangian spheres. Two configurations are Hurwitz equivalent if they can be connected by a sequence of the following moves and their inverses:*

- $\Gamma = (L_1, \dots, L_m) \rightsquigarrow (L_1, \dots, L_{i-1}, \phi(L_i), L_{i+1}, \dots, L_m)$ *for some* $1 \leq i \leq m$ *and some* $\phi \in \mathrm{Symp}^e(M)$ *isotopic to the identity;*
- $\Gamma \rightsquigarrow c\Gamma = (\tau_{L_1}(L_2), \tau_{L_1}(L_3), \dots, \tau_{L_1}(L_m), L_1)$;
- $\Gamma \rightsquigarrow r\Gamma = (L_1, \dots, L_{m-2}, \tau_{L_{m-1}}(L_m), L_{m-1})$.

Thus, the Hurwitz equivalence class of a distinguished basis is an invariant of the exact Morse fibration (E, π). Conversely, from M and that Hurwitz equivalence class one can reconstruct the fibration up to a suitable notion of deformation equivalence.

References. The symplectic nature of Dehn twists in all dimensions was first noticed by Arnol'd [1]. Global properties of these maps are discussed in [14, 15, 16]. I am not aware of any systematic exposition of symplectic Picard–Lefschetz theory in the literature.

3. Floer Cohomology

(3A) From now on, any exact symplectic manifold (M, ω, θ) is assumed to have contact type boundary, with $\theta \mid \partial M$ being the contact one-form (the same condition will be imposed on $\Theta \mid \partial E_z$, for E_z any fibre of a Morse fibration). Then for any pair (L_1, L_2) of exact Lagrangian submanifolds in M there is a well-defined Floer cohomology group $HF(L_1, L_2)$, which is a finite-dimensional vector space over the field $\mathbb{Z}/2$. We remind the reader that this is invariant under isotopies of L_1 or L_2, satisfies $HF(\phi L_1, \phi L_2) \cong HF(L_1, L_2)$ for any $\phi \in \mathrm{Symp}^e(M)$, and that there is a natural Poincaré duality $HF(L_1, L_2) \cong HF(L_2, L_1)^\vee$.

As a warm-up exercise, suppose that we have a compact oriented surface S with boundary, and an exact Morse fibration (E^{2n+2}, π) over it. A **Lagrangian boundary condition** for E is a closed submanifold $Q^{n+1} \subset \partial_v E \setminus \partial_h E$ such that $\pi \mid Q \colon Q \to \partial S$ is a smooth fibration, satisfying $\Omega \mid Q = 0$ and $[\Theta \mid Q] = 0 \in H^1(Q; \mathbb{R})$. Then the intersection $Q_z = Q \cap E_z$, for any $z \in \partial S$, is an exact Lagrangian submanifold in E_z, and parallel transport along ∂S takes these Lagrangian submanifolds into each other. Choose a complex structure j on S and an almost complex structure J on E which is compatible relative to j, as defined in the previous section. There is a Gromov type invariant $\Phi(E, \pi, Q) \in \mathbb{Z}/2$ which counts, in the familiar sense, the number of (j, J)-holomorphic sections $u \colon S \to E$ with $u(\partial S) \subset Q$. The exactness assumptions imply that there can be no bubbles (J-holomorphic spheres in a fibre E_z, or J-holomorphic discs in E_z with boundary in Q_z), so that the definition of the invariant is technically quite simple.

Example 3.1. *Let L be an exact framed Lagrangian sphere in an exact symplectic manifold M. Starting from a standard local model, one can construct an exact Morse fibration (E, π) over D with $E_{z_0} = M$ for $z_0 = -i \in \partial D$, having exactly one critical point, such that the monodromy around ∂D is τ_L. Because $\tau_L(L) = L$, there is a unique Lagrangian boundary condition $Q \subset E$ with $Q_{z_0} = L$. $\Phi(E, \pi, Q)$ vanishes because the expected dimension of the space of (j, J)-holomorphic sections is always odd, hence never zero.*

Now let S be as before but with a finite set of marked points $\Sigma \subset \partial S$. Suppose moreover that around each $\zeta \in \Sigma$ we have preferred local coordinates, given by an oriented embedding $\psi_\zeta \colon D^+ \to S$ of the half-disc $D^+ = D \cap \{\mathrm{im}(z) \geq 0\}$

with $\psi_\zeta(0) = \zeta$. Let (E, π) be an exact Morse fibration over $S^* = S \setminus \Sigma$ which is trivial near the marked points. This means that we have a fixed exact symplectic manifold M and preferred embeddings $\Psi_\zeta \colon (D^+ \setminus \{0\}) \times M \to E$ lying over ψ_ζ, satisfying some obvious conditions concerning Ω, Θ that we do not care to write down. If $Q \subset E$ is a Lagrangian boundary condition, there is for each $\zeta \in \Sigma$ a unique pair $L_{\zeta,\pm}$ of exact Lagrangian submanifolds of M such that $\Psi_\zeta^{-1}(Q) = [-1;0) \times L_{\zeta,-} \cup (0;1] \times L_{\zeta,+}$. After choosing a complex structure j on S^* such that the ψ_ζ become holomorphic, and an almost complex structure J on E which is compatible relative to j and satisfies some additional conditions with regard to Ψ_ζ, one can count pseudo-holomorphic sections with suitable behaviour near the marked points. The outcome is a relative invariant

$$\Phi_{rel}(E, \pi, Q) \in \bigotimes_{\zeta \in \Sigma} HF(L_{\zeta,+}, L_{\zeta,-}) \,. \tag{1}$$

These invariants satisfy the standard gluing law for a topological quantum field theory, which one can formulate in two parts as follows. First, if S is not connected then the relative invariant decomposes into the tensor product of relative invariants associated to its connected components. Second, suppose that there are two marked points $\zeta, \zeta' \in \Sigma$ with

$$L_{\zeta,\pm} = L_{\zeta',\mp} \,.$$

One can define a new surface $\overline{S}$ by removing small half-discs around ζ, ζ' and gluing together the resulting half-circles, as in Figure 3. There is a natural set of marked points $\overline{\Sigma} \subset \partial\overline{S}$ which is inherited from $\Sigma \setminus \{\zeta, \zeta'\}$. A similar process applied to (E, π) constructs a new exact Morse fibration over $\overline{S} \setminus \overline{\Sigma}$ with Lagrangian boundary conditions. The gluing rule says that on the level of the invariants Φ_{rel} this translates into contracting $HF(L_{\zeta,+}, L_{\zeta,-}) \otimes HF(L_{\zeta',+}, L_{\zeta',-})$ by Poincaré duality.

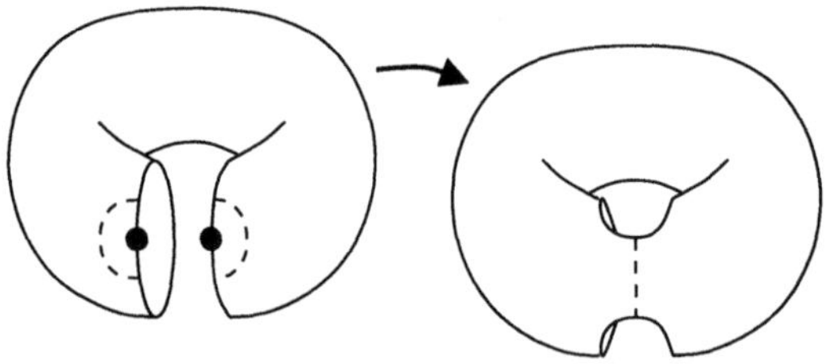

FIGURE 3.

Remark 3.2. *The reader is hereby warned of two possible misunderstandings: the gluing process does not take place along the boundaries ∂E_z of the fibres, and neither do we glue together two boundary circles of S.*

(3B) We next give three examples. Throughout, Poincaré duality will be used freely to write the invariants Φ_{rel} in various equivalent ways, e.g., as multilinear maps between Floer cohomology groups.

- Let M be an exact symplectic manifold and $L_1, L_2, L_3 \subset M$ exact Lagrangian submanifolds. Take $S = D$ and $\Sigma = \{\textit{three points}\} \subset \partial S$. Let I_1, I_2, I_3 be the three components of ∂S^*, ordered in positive sense. The trivial fibration $E = S^* \times M$ with Lagrangian boundary conditions $Q = \bigcup_{\nu=1}^{3} I_\nu \times L_\nu$ yields a relative invariant which can be written as a map $HF(L_2, L_3) \otimes HF(L_1, L_2) \to HF(L_1, L_3)$. This is just a reformulation of the usual "pair-of-pants" product. Reversing the orientation of S yields another relative invariant, which is the Poincaré dual coproduct $HF(L_1, L_3) \to HF(L_2, L_3) \otimes HF(L_1, L_2)$.
- Let M be an exact symplectic manifold, $L_1, L_2 \subset M$ exact Lagrangian submanifolds, and $\phi \in \mathrm{Symp}^e(M)$ an automorphism which is isotopic to the identity. Imagine $S = D$ as being constructed out of a smaller disc D_0 and two other pieces $D_1, D_2 = [0;1]^2$. The marked points are $\Sigma = \{\zeta_1, \zeta_2\}$, where ζ_1 is obtained by identifying $(0,1) \in D_1$ with $(0,0) \in D_2$, and ζ_2 is similarly $(1,1) \in D_1$ or $(1,0) \in D_2$. One can construct an exact symplectic fibration E_0 over D_0, which is topologically $D_0 \times M$ but has nontrivial forms Ω and Θ, such that the monodromy around ∂D_0 is ϕ. Assemble E_0 and the two trivial fibrations $E_1 = (D_1 \setminus \{(0,1),(1,1)\}) \times M$, $E_2 = (D_2 \setminus \{(0,0),(0,1)\}) \times M$ following the instructions in Figure 4. This gives an exact symplectic fibration over S^*; we equip it with the Lagrangian boundary condition which is the union of $(0;1) \times \{1\} \times L_1 \subset E_1$ and $(0;1) \times \{0\} \times L_2 \subset E_2$. The resulting relative invariant is a map $HF(L_1, L_2) \to HF(L_1, \phi(L_2))$, in fact just the familiar "continuation" isomorphism.

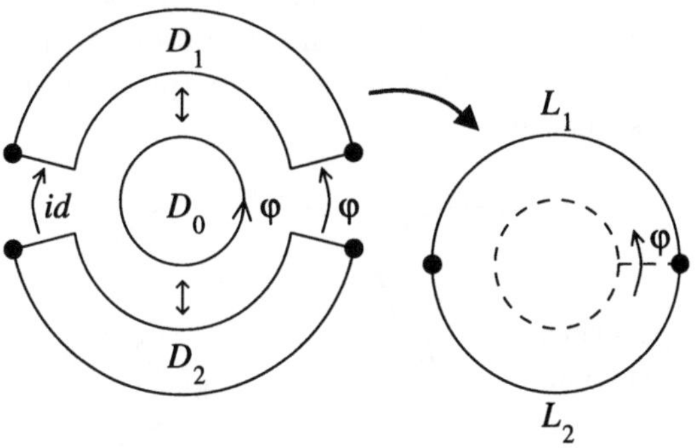

FIGURE 4.

- Let M, L_1, L_2 be as before, and $L \subset M$ an exact framed Lagrangian sphere. The central piece E_0 in the previous construction can be replaced by the Morse fibration from Example 3.1. This leads to a map $HF(L_1, L_2) \to HF(L_1, \tau_L(L_2))$, which is not an isomorphism in general. In fact there seems to be no way at all of getting from our formalism a map in the inverse direction; the reason is the presence of critical points, which prevents one from reversing the orientation of the base.

Theorem 3.3. *Let (M, ω, θ) be an exact symplectic manifold, $L_1, L_2 \subset M$ exact Lagrangian submanifolds, and $L \subset M$ an exact framed Lagrangian sphere. Suppose that $2c_1(M,L) \in H^2(M,L)$ is zero. Then there is a long exact sequence, with a being the pair of pants product and b the map defined in the last example above,*

$$\begin{array}{ccc} HF(L,L_2) \otimes HF(L_1,L) & \xrightarrow{\;a\;} & HF(L_1,L_2) \\ & \nwarrow & \big\downarrow b \\ & & HF(L_1, \tau_L(L_2))\,. \end{array} \tag{2}$$

To understand how this works, one needs to look at the Floer cochain complexes CF. To simplify, we suppress the dependence of these complexes on various additional choices, and write a, b for the maps between them inducing the Floer cohomology maps mentioned above. The central object in the proof is a map of complexes

$$\mathrm{Cone}(a) \xrightarrow{(h,b)} CF(L_1, \tau_L(L_2))\,. \tag{3}$$

Here $h\colon CF(L,L_2) \otimes CF(L_1,L) \to CF(L_1,\tau_L(L_2))$ is a chain homotopy $b \circ a \simeq 0$ defined as follows. Consider the exact Morse fibration with $S = D$, $\Sigma = \{\textit{three points}\}$, which is represented schematically, together with its Lagrangian boundary condition, in Figure 5. Moving the two leftmost marked points simultaneously along ∂D, in a way that preserves the symmetry of the picture with respect to the x-axis, yields a one-parameter family of fibrations, and h arises from the corresponding parametrized spaces of pseudo-holomorphic sections. Once (3) has been defined, an argument using the natural filtration of the Floer complexes by the action functional shows that it is a quasi-isomorphism; the long exact sequence is an immediate consequence.

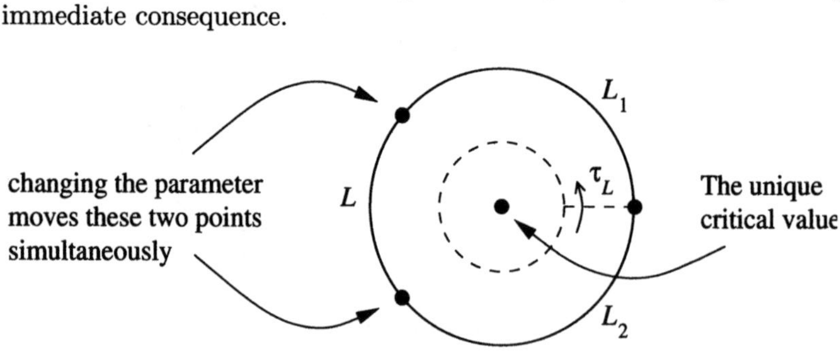

FIGURE 5.

As one can see from this sketch of the argument, the definition of the third arrow in (2) uses the inverse of the quasi-isomorphism (3). Poincaré duality yields a symmetry of the exact sequence, which suggests a more direct description of that arrow. Namely, it should be the composition

$$c\colon HF(L_1,\tau_L(L_2)) \to HF(L,\tau_L(L_2)) \otimes HF(L_1,L) \cong HF(L,L_2) \otimes HF(L_1,L)$$

of the coproduct and the isomorphism $HF(L,\tau_L(L_2)) \cong HF(\tau_L^{-1}(L),L_2) = HF(L,L_2)$. One can show that this is indeed the same map as that obtained by inverting (3); the proof uses invariants of the same kind as those defining h, for two-parameter families of exact Morse fibrations.

Remark 3.4. *The assumption $2c_1(M,L) = 0$ is a technical one. It implies that the space of (j,J)-holomorphic sections in Example 3.1 has expected dimension $2n-1$. From this it follows (except for $n=1$, which requires a separate treatment) that generically there is a positive-dimensional space of these sections u such that $u(z_0)$ is a specific point in L. That enters into the description of the limiting behaviour of sections of the family in Figure 5 when both moveable points of Σ go towards the fixed one, and through it into the proof that h is a homotopy $ba \simeq 0$. Still, it may be possible to substitute some other argument at this point, and thereby remove the assumption from Theorem 3.3.*

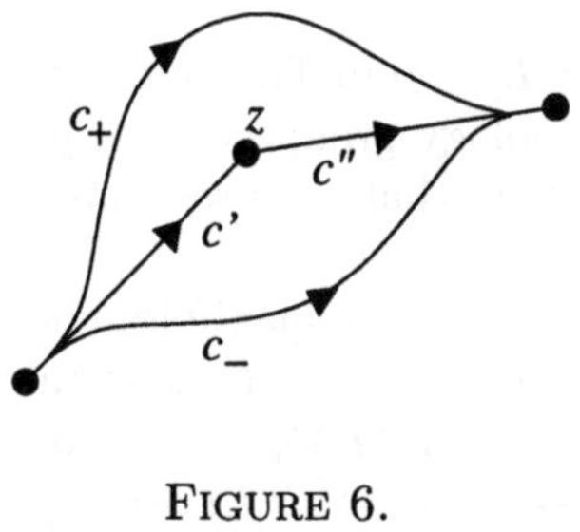

FIGURE 6.

There is an important special case in which Theorem 3.3 has a graphical interpretation, namely when all Lagrangian submanifolds involved are vanishing cycles. Take an exact Morse fibration (E,π) with arbitrary base S. To any path $c\colon [0;1] \to S$ with $c^{-1}(S^{\mathrm{crit}}) = \{0;1\}$ and $c'(0), c'(1) \neq 0$ one can associate the Floer cohomology group $HF(V_{c,0},V_{c,1})$ where $V_{c,0},V_{c,1} \subset E_{c(t)}$, for some $0 < t < 1$, are the vanishing cycles associated to $c \mid [0;t]$ resp. $c \mid [t;1]$. This group is independent of t, so we may write temporarily $HF(c)$ for it. Consider four paths as in Figure 6. As three of them run together for small t, we may assume that their vanishing cycles all lie in the same fibre, called M. Then $V_{c_-,0} = V_{c_+,0} = V_{c',0}$ and by the Picard–Lefschetz theorem, $V_{c_+,1} = \tau_{V_{c',1}}(V_{c_-,1})$. Taking the vanishing cycles of c'' and moving them to M by parallel transport shows that $HF(V_{c',1},V_{c_-,1}) = HF(V_{c',1},V_{c_+,1}) = HF(V_{c'',0},V_{c'',1})$. Applying Theorem 3.3 in this context yields a long exact sequence

$$\begin{array}{ccc} HF(c'') \otimes HF(c') & \longrightarrow & HF(c_-) \\ & \nwarrow & \downarrow \\ & & HF(c_+)\,. \end{array}$$

This can be thought of as a skein rule, with $HF(c'') \otimes HF(c')$ measuring the change to $HF(c_-)$ as the path moves over $z \in S^{\text{crit}}$ and becomes c_+. For $S = D$, allowing oneself for a moment to believe naively that "two terms in a long exact sequence determine the third one", one sees that all groups $HF(c)$ could be determined from knowing just finitely many of them, through a process of successively breaking up paths into shorter pieces.

References. Our TQFT formalism for Floer cohomology is a variation of that in [14], which in turn generalizes earlier work of Piunikhin–Salamon–Schwarz [11]. The pair-of-pants product and continuation map are defined in [10, Chapter 10] or [13] or [17], respectively [12]. The skein rule interpretation of the exact sequence is due to Donaldson.

4. Grading

(4A) The assumption $2c_1(M, L) = 0$ in Theorem 3.3 is very close to the condition needed to equip Floer cohomology groups with $\mathbb{Z}$-gradings, so one might just as well take advantage of it. Doing that requires some preliminaries, which we now go through.

Let M be an arbitrary symplectic manifold, and $\mathfrak{J}_M$ the space of all compatible almost complex structures. There is a canonical unitary line bundle $\Delta_M \to \mathfrak{J}_M \times M$ whose fibre at (J, x) is $\Lambda^n(TM_x, J)^{\otimes 2}$. To any Lagrangian submanifold $L \subset M$ one can associate a section $\det^2(TL)$ over $\mathfrak{J}_M \times L$ of the associated circle bundle $S(\Delta_M)$; and for any symplectic automorphism ϕ there is a canonical isomorphism $\det^2(D\phi)\colon \Delta_M \to \Delta_M$, covering the map $\phi_* \times \phi$ from $\mathfrak{J}_M \times M$ to itself. A Maslov map is a trivialization $\delta_M : \Delta_M \to \mathbb{C}$. Suppose that we have chosen such a δ_M. A grading of $L \subset M$ is a lift

$$\begin{array}{ccccc} & & \overset{\widetilde{L}}{\nearrow\!\!\!\longrightarrow} & & \mathbb{R} \\ & & & & \Big\downarrow {\scriptstyle \exp(2\pi i \cdot)} \\ \mathfrak{J}_M \times L & \xrightarrow{\det^2(TL)} & S(\Delta_M) & \xrightarrow{\delta_M} & S^1\,. \end{array}$$

Any two gradings differ by a locally constant function $L \to \mathbb{Z}$; we write $\widetilde{L}[\sigma]$ for $\widetilde{L} - \sigma$, $\sigma \in \mathbb{Z}$. Similarly, a grading of ϕ is a map $\tilde{\phi}\colon \mathfrak{J}_M \times M \to \mathbb{R}$ such that $\exp(2\pi i\, \tilde{\phi}(J, x)) = \delta_M(\det^2(D\phi)(w))/\delta_M(w)$ for any $w \in S(\Delta_M)$ in the fibre over (J, x). When dealing with connected manifolds M with boundary and maps ϕ

trivial near ∂M, there is often a preferred grading, characterized by being zero near $\mathfrak{J}_M \times \partial M$. Gradings of L and ϕ induce a grading of $\phi(L)$:

$$\tilde{\phi}(\widetilde{L}) \stackrel{\text{def}}{=} (\tilde{\phi} + \widetilde{L}) \circ (\phi_*^{-1} \times \phi^{-1}).$$

If $L \subset M$ is a framed Lagrangian sphere admitting gradings, there is a distinguished grading $\tilde{\tau}_L$ of τ_L, which is zero outside $\mathfrak{J}_M \times \{\text{neighbourhood of } L\}$ and satisfies

$$\tilde{\tau}_L(\widetilde{L}) = \widetilde{L}[1-n]. \tag{4}$$

Graded Lagrangian submanifolds and graded symplectic automorphisms are pairs consisting of an L resp. ϕ together with a choice of grading. For brevity we denote such pairs by $\widetilde{L}, \tilde{\phi}$ only. The Floer cohomology of a pair of graded Lagrangian submanifolds, whenever defined (e.g., when M is exact with contact type boundary, and the Lagrangian submanifolds are themselves exact), has a canonical $\mathbb{Z}$-grading with the properties

$$\begin{aligned} &HF^*(\widetilde{L}_1[-\sigma], \widetilde{L}_2) = HF^*(\widetilde{L}_1, \widetilde{L}_2[\sigma]) = HF^{*+\sigma}(\widetilde{L}_1, \widetilde{L}_2), \\ &HF^*(\tilde{\phi}(\widetilde{L}_1), \tilde{\phi}(\widetilde{L}_2)) \cong HF^*(\widetilde{L}_1, \widetilde{L}_2), \\ &HF^*(\widetilde{L}_2, \widetilde{L}_1) \cong HF^{n-*}(\widetilde{L}_1, \widetilde{L}_2)^\vee. \end{aligned} \tag{5}$$

(4B) Now consider an exact Morse fibration (E, π) over S, and the space $\mathfrak{J}_{E/S}$ of pairs (j, J) such that J is compatible relative to j. Let $\Delta_{E/S} \to \mathfrak{J}_{E/S} \times E$ be the line bundle with fibres $\Lambda^{n+1}(TE_x, J)^{\otimes 2} \otimes (TS_{\pi(x)}, j)^{\otimes -2}$. A **relative Maslov map** is a trivialization $\delta_{E/S}\colon \Delta_{E/S} \to \mathbb{C}$. Once such a map has been chosen, there is a canonical induced Maslov map δ_M on any regular fibre $M = E_z$. The vanishing cycles $V \subset M$ associated to paths $c\colon [0;1] \to S$, $c(0) = z$ and $c(1) \in S^{\text{crit}}$, admit gradings; and the monodromies ρ_l along closed loops l in $S \setminus S^{\text{crit}}$, $l(0) = z$, even have canonical gradings, which we denote by $\tilde{\rho}_l$. If $[l] \in H_1(S)$ vanishes, $\tilde{\rho}_l$ is zero near $\mathfrak{J}_M \times \partial M$. This implies a graded version of the Picard–Lefschetz theorem, saying that $\tilde{\rho}_l = \tilde{\tau}_V$ for c, l as in Figure 1.

(4C) Any Lagrangian boundary condition Q for (E, π) comes with a canonical section of $S(\Delta_{E/S}) \mid \mathfrak{J}_{E/S} \times Q$. We call $\delta_{E/S}$ adapted to Q if there is a diagram

$$\begin{array}{ccc} \mathfrak{J}_{E/S} \times Q & \xrightarrow{\text{canonical section}} & S(\Delta_{E/S}) \\ \Big\downarrow{\scriptstyle \mathrm{id}\times\pi} & & \Big\downarrow{\scriptstyle \delta_{E/S}} \\ \mathfrak{J}_{E/S} \times \partial S & \overset{\lambda}{\dashrightarrow} & S^1. \end{array}$$

For compact S, the existence of an adapted relative Maslov map means that all components of the space of (j, J)-holomorphic sections taking ∂S to Q have the same expected dimension $n\chi(S) + \deg(\lambda)$; we have already seen an instance of this in Remark 3.4. There is a generalization of this notion to the relative case, which ensures that the Floer groups in (1) have canonical gradings and that the invariant Φ_{rel} is of a specific degree, not necessarily zero. Instead of explaining this in general we will just look at an example, that of the long exact sequence.

Suppose then that we have an exact symplectic manifold M with a Maslov map δ_M, graded Lagrangian submanifolds $\widetilde{L}_1, \widetilde{L}_2$, and a framed exact Lagrangian sphere L which admits gradings. The map b: $HF^*(\widetilde{L}_1, \widetilde{L}_2) \rightarrow HF^*(\widetilde{L}_1, \tilde{\tau}_L(\widetilde{L}_2))$ has degree zero, and the same holds for the pair-of-pants product a: $HF^*(\widetilde{L}, \widetilde{L}_2) \otimes HF^*(\widetilde{L}_1, \widetilde{L}) \rightarrow HF^*(\widetilde{L}_1, \widetilde{L}_2)$. The coproduct $HF^*(\widetilde{L}_1, \tilde{\tau}_L(\widetilde{L}_2)) \rightarrow HF^*(\widetilde{L}, \tilde{\tau}_L(\widetilde{L}_2)) \otimes HF^*(\widetilde{L}_1, \widetilde{L})$ has degree n because of Poincaré duality, but then by (4) and (5) the isomorphism

$$HF^*(\widetilde{L}, \tilde{\tau}_L(\widetilde{L}_2)) = HF^*(\tilde{\tau}_L^{-1}(\widetilde{L}), \widetilde{L}_2) = HF^{*+1-n}(\widetilde{L}, \widetilde{L}_2)$$

has degree $1 - n$. Hence the third map c raises degrees by one, as in any other cohomological long exact sequence.

References. Floer's discussion of grading in [4] is essentially complete as it stands. Still, graded Lagrangian submanifolds, introduced by Kontsevich [9] somewhat later, allow a better formulation of the results.

5. Mutation

(5A) As in our discussion of Floer theory, we take the ground field to be $\mathbb{Z}/2$; all categories will be linear over it. Let $\mathcal{C}$ be a triangulated category such that the spaces $\mathrm{Hom}^*_{\mathcal{C}}(X, Y) = \bigoplus_r \mathrm{Hom}_{\mathcal{C}}(X, Y[r])$ are finite-dimensional for any X, Y. An exceptional collection in $\mathcal{C}$ is a finite family of objects $(X^1, \dots, X^m)$ satisfying

$$\mathrm{Hom}^*_{\mathcal{C}}(X^i, X^k) = \begin{cases} \mathbb{Z}/2 \cdot \mathrm{id}_{X^i} & i = k\,, \\ 0 & i > k\,. \end{cases}$$

Such a collection is called full if the X^i generate $\mathcal{C}$ in the triangulated sense. For $X, Y \in \mathrm{Ob}\,\mathcal{C}$ define $T_X(Y)$, up to isomorphism, as the object fitting into an exact triangle

$$T_X(Y)[-1] \rightarrow \mathrm{Hom}^*_{\mathcal{C}}(X, Y) \otimes X \xrightarrow{ev} Y \rightarrow T_X(Y)\,;$$

here the tensor product is just a finite sum of shifted copies of X, and ev the canonical evaluation map. If $(X^1, \dots, X^m)$ is a full exceptional collection then so are $(Y^1, \dots, Y^m)$, $(Z^1, \dots, Z^m)$ where

$$Y^i = \begin{cases} T_{X^1}(X^{i+1}) & i < m\,, \\ X^1 & i = m\,, \end{cases} \qquad (6)$$

respectively

$$Z^i = \begin{cases} X^i & i < m-1\,, \\ T_{X^{m-1}}(X^m) & i = m-1\,, \\ X^{m-1} & i = m\,. \end{cases} \qquad (7)$$

(5B) There is a slightly different version of the same story for A_∞-categories. Call an A_∞-category $\mathcal{A}$ **directed** if it has finitely many objects numbered $1,\dots,m$, say $\mathrm{Ob}\,\mathcal{A} = \{X^1,\dots,X^m\}$, such that

$$hom_{\mathcal{A}}(X^i,X^k) = \begin{cases} \text{finite-dimensional} & i<k\,, \\ \mathbb{Z}/2\cdot \mathrm{id}_{X^i} & i=k\,, \\ 0 & i>k\,. \end{cases}$$

Note that $\mu^d_{\mathcal{A}} = 0$ is necessarily zero for $d > \max\{m-1,2\}$. We recall the definition of the (bounded) derived category $D^b(\mathcal{A})$. The first step is to embed $\mathcal{A}$ into a larger A_∞-category $\mathcal{A}^\oplus$ which has finite sums and shifts. Thus, an object of $\mathcal{A}^\oplus$ is a formal sum $\bigoplus_{e\in E} X_e[\sigma_e]$ with E a finite set, $X_e \in \mathrm{Ob}\,\mathcal{A}$, and $\sigma_e \in \mathbb{Z}$. Next, a twisted complex in $\mathcal{A}$ is a pair (C,δ_C) consisting of $C \in \mathrm{Ob}\,\mathcal{A}^\oplus$ and $\delta_C \in hom^1_{\mathcal{A}^\oplus}(C,C)$, such that the "generalized Maurer–Cartan equation"

$$\sum_{d\geq 1} \mu^d_{\mathcal{A}^\oplus}(\delta_C,\dots,\delta_C) = 0 \tag{8}$$

holds. Twisted complexes form an A_∞-category $\mathrm{Tw}\,\mathcal{A}$ which again has direct sums and shifts. It also contains a cone, denoted by $\mathrm{Cone}(a)$, for any morphism a such that $\mu^1_{\mathrm{Tw}\,\mathcal{A}}(a) = 0$, and the cohomological category $D^b(\mathcal{A}) = H^0(\mathrm{Tw}\,\mathcal{A})$ inherits a triangulated structure from this. The objects of $\mathcal{A}$, seen as twisted complexes with zero differential, form a full exceptional collection in $D^b(\mathcal{A})$.

Remark 5.1. *The definition of the derived category of a general A_∞-category $\mathcal{A}$ uses pairs (C,δ_C) such that δ_C is strictly decreasing with respect to some finite filtration of C, which has the effect of making the sum* (8) *finite. The fact that this is not necessary in our case is just one of several technical simplifications which directedness brings with it.*

Any A_∞-functor $F\colon \mathcal{A} \to \mathcal{B}$ between directed A_∞-categories induces another one $\mathrm{Tw}\,F\colon \mathrm{Tw}\,\mathcal{A} \to \mathrm{Tw}\,\mathcal{B}$ taking cones to cones, and therefore an exact functor $D^b(F) = H^0(\mathrm{Tw}\,F)\colon D^b(\mathcal{A}) \to D^b(\mathcal{B})$. Call F a quasi-isomorphism if $H(F)\colon H(\mathcal{A}) \to H(\mathcal{B})$ is an isomorphism; since the objects of $\mathcal{A}$ generate $D^b(\mathcal{A})$, it follows that in this case $D^b(F)$ is an equivalence.

More interestingly, suppose that $\mathcal{A}$ is some directed A_∞-category, and $(Y^1,\dots,Y^m)$ an exceptional collection in $D^b(\mathcal{A})$. One can then define a new directed A_∞-category $\mathcal{B}$, called the directed A_∞-subcategory of $\mathrm{Tw}\,\mathcal{A}$ generated by the Y^i, as follows. Objects of $\mathcal{B}$ are the Y^i, and $hom_{\mathcal{B}}(Y^i,Y^k) = hom_{\mathrm{Tw}\,\mathcal{A}}(Y^i,Y^k)$ for $i<k$, with the same composition maps between these groups as in $\mathrm{Tw}\,\mathcal{A}$. All other morphism groups and compositions are as dictated by the directedness condition. The embedding $\mathcal{B} \hookrightarrow \mathrm{Tw}\,\mathcal{A}$ can be extended to an A_∞-functor $\iota\colon \mathrm{Tw}\,\mathcal{B} \to \mathrm{Tw}\,\mathcal{A}$, which induces an exact functor $H^0(\iota)\colon D^b(\mathcal{B}) \to D^b(\mathcal{A})$. For the same reason as before, $H^0(\iota)$ is full and faithful; it is even an equivalence if $(Y^1,\dots,Y^m)$ is a full exceptional collection.

(5c) Given $X \in \operatorname{Ob}\operatorname{Tw}\mathcal{A}$ and a finite-dimensional complex V of vector spaces, one can form the tensor product $V \otimes X \in \operatorname{Ob}\operatorname{Tw}\mathcal{A}$, which is a direct sum of shifted copies of X with a differential combining those on V and X. Taking $V = (hom_{\operatorname{Tw}\mathcal{A}}(X,Y), \mu^1_{\mathcal{A}})$ for some $Y \in \operatorname{Ob}\operatorname{Tw}\mathcal{A}$, one has a canonical evaluation morphism $ev \in hom^0_{\operatorname{Tw}\mathcal{A}}(hom_{\operatorname{Tw}\mathcal{A}}(X,Y) \otimes X, Y)$ with $\mu^1_{\operatorname{Tw}\mathcal{A}}(ev) = 0$. Let $T_X(Y) \in \operatorname{Ob}\operatorname{Tw}\mathcal{A}$ be the cone of ev. This is isomorphic in $D^b(\mathcal{A})$ to the object of the same name introduced above, but it is now unique in a strict sense, not just up to isomorphism. Taking the exceptional collection formed by the objects of $\mathcal{A}$ and applying (6) or (7) yields another full exceptional collection, hence a directed A_∞-subcategory $\mathcal{B}$ of $\operatorname{Tw}\mathcal{A}$ with $D^b(\mathcal{B}) \cong D^b(\mathcal{A})$. This process can be repeated indefinitely and leads to the following notion:

Definition 5.2. *Two directed A_∞-categories with m objects are mutations of each other if they can be related by a sequence of the following moves and their inverses:*

- *$\mathcal{A} \rightsquigarrow \mathcal{B}$ if there is a quasi-isomorphism between them.*
- *It is allowed to shift each object by some degree, which means changing the grading of each group $hom_{\mathcal{A}}(X^i, X^k)$ by $(\sigma_i - \sigma_k)$ for some $\sigma_1, \dots, \sigma_m \in \mathbb{Z}$, while keeping the same composition maps.*
- *$\mathcal{A} \rightsquigarrow c\mathcal{A}$ where, if the objects of $c\mathcal{A}$ are denoted by $\{Y^1, \dots, Y^m\}$, the nontrivial morphism spaces are*

$$hom_{c\mathcal{A}}(Y^i, Y^k) = \begin{cases} hom_{\mathcal{A}}(X^{i+1}, X^{k+1}) & i < k < m\,, \\ hom_{\mathcal{A}}(X^1, X^{i+1})^\vee[-1] & i < m,\ k = m\,. \end{cases}$$

Here $^\vee$ denotes the dual of a graded vector space. The compositions

$$\mu^d_{c\mathcal{A}} \colon \textstyle\prod_{\nu=1}^d hom_{c\mathcal{A}}(Y^{i_\nu}, Y^{i_{\nu+1}}) \to hom_{c\mathcal{A}}(Y^{i_1}, Y^{i_{d+1}}),\ i_1 < \dots < i_{d+1},$$

are equal to those in $\mathcal{A}$ except when $i_{d+1} = m$, in which case one has $\langle \mu^d_{c\mathcal{A}}(a^d, \dots, a^1), b\rangle = \langle a^d, \mu^d_{\mathcal{A}}(a^{d-1}, \dots, a^1, b)\rangle$ with $\langle \dots \rangle$ the dual pairing.
- *$\mathcal{A} \rightsquigarrow r\mathcal{A}$ with $\operatorname{Ob} r\mathcal{A} = \{Z^1, \dots, Z^m\}$ and the following nontrivial morphism spaces $hom_{r\mathcal{A}}(Z^i, Z^k)$:*

$$\begin{cases} hom_{\mathcal{A}}(X^i, X^k) & i < k \leq m-2\,, \\ hom_{\mathcal{A}}(X^i, X^{m-1}) & i \leq m-2,\ k = m\,, \\ hom_{\mathcal{A}}(X^{m-1}, X^m)^\vee[-1] & i = m-1,\ k = m\,, \\ (hom_{\mathcal{A}}(X^{m-1}, X^m) \otimes hom_{\mathcal{A}}(X^i, X^{m-1}))[1] & \\ \quad \oplus\ hom_{\mathcal{A}}(X^i, X^m) & i \leq m-2,\ k = m-1\,. \end{cases} \tag{9}$$

$\mu^1_{r\mathcal{A}}$ is given by $\mu^1_{\mathcal{A}}$ in the first two cases, its dual $(\mu^1_{\mathcal{A}})^\vee$ in the third, and in the final case by

$$\begin{pmatrix} \mu^1_{\mathcal{A}} \otimes \mathrm{id} + \mathrm{id} \otimes \mu^1_{\mathcal{A}} & 0 \\ \mu^2_{\mathcal{A}} & \mu^1_{\mathcal{A}} \end{pmatrix}.$$

Most of the higher order maps

$$\mu^d_{r\mathcal{A}} \colon \textstyle\prod_{\nu=1}^d hom_{r\mathcal{A}}(Z^{i_\nu}, Z^{i_{\nu+1}}) \to hom_{r\mathcal{A}}(Z^{i_1}, Z^{i_{d+1}}) \tag{10}$$

for $i_1 < \cdots < i_{d+1}$ are taken from those of $\mathcal{A}$ in a straightforward way, but there are two exceptions. One is when $i_{d+1} = m-1$, in which case

$$\mu^d_{r\mathcal{A}} = \begin{pmatrix} \mathrm{id} \otimes \mu^d_{\mathcal{A}} & 0 \\ \mu^{d+1}_{\mathcal{A}} & \mu^d_{\mathcal{A}} \end{pmatrix}$$

with respect to the obvious splittings on both sides of (10)*. The second exceptional case is when $i_d = m-1$, $i_{d+1} = m$: then $\mu^2_{r\mathcal{A}}$ is*

$$\begin{array}{c} hom_{\mathcal{A}}(X^{m-1}, X^m)^\vee \otimes hom_{\mathcal{A}}(X^{m-1}, X^m) \otimes hom_{\mathcal{A}}(X^{i_1}, X^{m-1}) \\ \oplus\ (hom_{\mathcal{A}}(X^{m-1}, X^m)^\vee \otimes hom_{\mathcal{A}}(X^{i_1}, X^m))[-1] \\ \Big\downarrow \\ hom_{\mathcal{A}}(X^{i_1}, X^{m-1}), \end{array}$$

$$\mu^2_{r\mathcal{A}}(a^3 \otimes a^2 \otimes a^1, b^2 \otimes b^1) = \langle a^3, a^2 \rangle a^1,$$

while the compositions of order $d \geq 3$ vanish.

Actually, while $r\mathcal{A}$ is precisely the directed A_∞-subcategory of $\mathrm{Tw}\,\mathcal{A}$ generated by the collection (7), $c\mathcal{A}$ is only canonically quasi-isomorphic to that generated by (6). But it is still true that two directed A_∞-categories which are mutations of each other have equivalent derived categories.

References. This section makes no claim to originality. Exceptional collections in triangulated categories are discussed in the work of Bondal, Gorodentsev, Kapranov, and others. Most relevant for us is [2] which emphasizes the role of dg categories. The short step from there to A_∞-categories was made by Kontsevich [8], who is also responsible for introducing $\mathrm{Tw}\,\mathcal{A}$ and $D^b(\mathcal{A})$ [9].

6. Fukaya Categories

(6A) Throughout this section M is a fixed exact symplectic manifold, with a Maslov map δ_M. A graded Lagrangian configuration in M is a family $\tilde{\Gamma} = (\tilde{L}_1, \dots, \tilde{L}_m)$ of graded, exact, framed Lagrangian spheres. Hurwitz equivalence for graded configurations is defined as in the ungraded case, with the following adaptations: for the isotopy invariance one wants to take a grading $\tilde{\phi}$ which is zero near $\partial_M \times \partial M$, so that $\tilde{\phi}$ is isotopic to the identity in the group of graded symplectic automorphisms; the moves $c\tilde{\Gamma}, r\tilde{\Gamma}$ use the canonical gradings $\tilde{\tau}_L$ of Dehn twists; and there is an additional shift move,

- $\tilde{\Gamma} \rightsquigarrow (\tilde{L}_1[\sigma_1], \dots, \tilde{L}_m[\sigma_m])$ for any $\sigma_1, \dots, \sigma_m \in \mathbb{Z}$.

This is something of an anticlimax, since it cancels out the extra information contained in the grading; but in fact, the whole notion of graded configuration has been introduced only for notational convenience.

We will associate to $\tilde{\Gamma}$ a **directed Fukaya A_∞-category** $Lag^{\rightarrow}(\tilde{\Gamma})$, unique up to quasi-isomorphism. Suppose first that the configuration is in general position, meaning that any two L_i are transverse and there are no triple intersections.

Objects of $\mathcal{A} = Lag^{\to}(\widetilde{\Gamma})$ are the graded Lagrangian submanifolds $\widetilde{L}_i$, in the given order, and

$$hom_{\mathcal{A}}(\widetilde{L}_i, \widetilde{L}_k) = \begin{cases} CF^*(\widetilde{L}_i, \widetilde{L}_k) = (\mathbb{Z}/2)^{L_i \cap L_k} & i < k\,, \\ \mathbb{Z}/2 \cdot \mathrm{id}_{\widetilde{L}_i} & i = k\,, \\ 0 & i > k\,. \end{cases}$$

Roughly speaking, $\mu^1_{\mathcal{A}}$ is the Floer boundary map, $\mu^2_{\mathcal{A}}$ the pair-of-pants product, and $\mu^3_{\mathcal{A}}, \mu^4_{\mathcal{A}}, \dots$ Fukaya's generalizations of that product. Each $\mu^d_{\mathcal{A}}$ depends on the choice of a family $\mathbf{J}^{d+1}$ of almost complex structures on M, and these choices have to obey certain consistency conditions, which we will now outline (this is joint work of Lazzarini and the author).

In a first step one takes, for each $1 \leq i_1 < i_2 \leq m$, a generic family of almost complex structures $\mathbf{J}^2(i_1, i_2)\colon [0;1] \to \mathfrak{J}_M$. As is well known, this causes all solutions of Floer's equations

$$u\colon \mathbb{R} \times [0;1] \to M, \qquad \begin{cases} \mathbf{J}^2(i_1, i_2, t) \circ du(s,t) = du(s,t) \circ j\,, \\ u(\mathbb{R} \times \{1\}) \subset L_{i_1},\ u(\mathbb{R} \times \{0\}) \subset L_{i_2}\,, \\ \int u^*\omega < \infty\,, \end{cases} \tag{11}$$

where j is the standard complex structure on $\mathbb{R} \times [0;1]$, to be regular. From the one-dimensional solution spaces (zero-dimensional after dividing by translation) one builds $\mu^1_{\mathcal{A}}\colon CF^*(\widetilde{L}_{i_1}, \widetilde{L}_{i_2}) \to CF^{*+1}(\widetilde{L}_{i_1}, \widetilde{L}_{i_2})$.

Next take $S = D$, three cyclically ordered marked points $\zeta^3_\nu \in \partial S$, $1 \leq \nu \leq 3$, local coordinates $\psi^3_\nu\colon D^+ \to S$ around them, and set $S^* = S \setminus \{\zeta^3_1, \zeta^3_2, \zeta^3_3\}$, all as in the definition of the invariants Φ_{rel}. Choose for each $1 \leq i_1 < i_2 < i_3 \leq m$ a generic family $\mathbf{J}^3(i_1, i_2, i_3)\colon S^* \to \mathfrak{J}_M$, such that for $s \ll 0$ and $t \in [0;1]$,

$$\mathbf{J}^3(i_1, i_2, i_3, \psi^3_\nu(e^{\pi(s+it)})) = \begin{cases} \mathbf{J}^2(i_\nu, i_{\nu+1}, t) & \nu = 1, 2\,, \\ \mathbf{J}^2(i_1, i_3, 1-t) & \nu = 3\,. \end{cases} \tag{12}$$

Denote by I^3_ν, $1 \leq \nu \leq 3$, the connected components of ∂S^*, ordered cyclically such that I^3_1 lies between ζ^3_3 and ζ^3_1. The equation defining $\mu^2_{\mathcal{A}}$ is

$$u\colon S^* \to M, \qquad \begin{cases} \mathbf{J}^3(i_1, i_2, i_3, z) \circ du(z) = du(z) \circ j\,, \\ u(I^3_\nu) \subset L_{i_\nu} \quad \text{for } \nu = 1,2,3\,, \\ \int u^*\omega < \infty\,. \end{cases}$$

Condition (12) causes this to agree with a suitable equation (11) in the tubular coordinates $\psi^3_\nu(e^{\pi(s+it)})$ on each end of S^*.

The additional ingredient in the definition of the products of order $d > 2$ are "moduli parameters" as the complex structure of the domain changes. Let $\mathcal{C}^{d+1} \subset (\partial D)^{d+1}$ be the configuration space of $d+1$ distinct, numbered and cyclically ordered points on ∂D. The moduli space $\mathcal{R}^{d+1}$ and the universal disc bundle $\mathcal{S}^{d+1}$ over it are defined as

$$\mathcal{S}^{d+1} = \mathcal{C}^{d+1} \times_{\mathrm{Aut}(D)} D \longrightarrow \mathcal{R}^{d+1} = \mathcal{C}^{d+1}/\mathrm{Aut}(D)\,,$$

where $\mathrm{Aut}(D) \cong PSL(2,\mathbb{R})$ is the holomorphic automorphism group. Each fibre $\mathcal{S}^{d+1}_r$ carries a canonical complex structure, and there are canonical sections $\zeta^{d+1}_\nu : \mathcal{R}^{d+1} \to \partial\mathcal{S}^{d+1}$, $1 \leq \nu \leq d+1$, such that the points $\zeta^{d+1}_\nu(r) \in \partial\mathcal{S}^{d+1}_r$ are distinct and cyclically ordered for each r. Write

$$\mathcal{S}^{d+1,*} = \mathcal{S}^{d+1} \setminus (\textstyle\bigcup_\nu \mathrm{im}\, \zeta^{d+1}_\nu), \qquad \mathcal{S}^{d+1,*}_r = \mathcal{S}^{d+1,*} \cap \mathcal{S}^{d+1}_r .$$

For each $1 \leq i_1 < \cdots < i_{d+1} \leq m$ one has to choose a family

$$\mathbf{J}^{d+1}(i_1,\dots,i_{d+1}) : \mathcal{S}^{d+1,*} \to \mathcal{J}_M ,$$

subject to two kinds of conditions.

- Compatibility with $\mathbf{J}^2$. This requires a preliminary choice of maps
$$\psi^{d+1}_\nu : \mathcal{R}^{d+1} \times D^+ \to \mathcal{S}^{d+1}, \quad 1 \leq \nu \leq d+1,$$
such that $\psi^{d+1}_\nu(r,\cdot)$ provides local coordinates around $\zeta^{d+1}_\nu(r)$ for each $r \in \mathcal{R}^{d+1}$. Then the conditions are similar to (12), requiring
$$\mathbf{J}^{d+1}(i_1,\dots,i_{d+1}, \psi^{d+1}_\nu(r,z))$$
to be determined by the previously chosen $\mathbf{J}^2$ for small $|z|$.
- Compatibility with $\mathbf{J}^{e+1}$ for $2 \leq e < d$. For this it is necessary to consider the Deligne–Mumford compactification of $\mathcal{R}^{d+1}$. Each stratum at infinity is a product of lower order spaces $\mathcal{R}^{e+1}$, and for a point $r \in \mathcal{R}^{d+1}$ sufficiently close to one such stratum, the fibre $\mathcal{S}^{d+1,*}_r$ is built by gluing together fibres of $\mathcal{S}^{e+1,*}$ for the various occurring e. The precise condition on $\mathbf{J}^{d+1}(i_1,\dots,i_{d+1}) \mid \mathcal{S}^{d+1,*}_r$ is too complicated to be written down here, but informally it says that this should be built up from the $\mathbf{J}^{e+1}$ in a corresponding way.

Let $I^{d+1}_{r,\nu}$, $1 \leq \nu \leq d+1$, be the connected components of $\partial\mathcal{S}^{d+1,*}_r$, ordered cyclically so that $I^{d+1}_{r,1}$ lies between $\zeta^{d+1}_{d+1}(r)$ and $\zeta^{d+1}_1(r)$. The consistency conditions leave enough freedom to make solutions of the equation

$$r \in \mathcal{R}^{d+1},\ u\colon \mathcal{S}^{d+1,*}_r \to M, \quad \begin{cases} \mathbf{J}^{d+1}(i_1,\dots,i_{d+1},z) \circ du(z) = du(z) \circ j , \\ u(I^{d+1}_{r,\nu}) \subset L_{i_\nu} \quad \text{for } \nu = 1,\dots,d+1 , \\ \int u^*\omega < \infty , \end{cases}$$

regular for generic $\mathbf{J}^{d+1}(i_1,\dots,i_{d+1})$. Counting such solutions defines

$$\mu^d_{\mathcal{A}} : CF^*(\widetilde{L}_{i_d}, \widetilde{L}_{i_{d+1}}) \otimes \cdots \otimes CF^*(\widetilde{L}_{i_1}, \widetilde{L}_{i_2}) \longrightarrow CF^{*+2-d}(\widetilde{L}_{i_1}, \widetilde{L}_{i_{d+1}}) .$$

The dependence of $Lag^{\rightarrow}(\widetilde{\Gamma})$ on the choice of almost complex structure can be analyzed using a one-parameter family argument. Since only finitely many moduli spaces are involved, the A_∞-structure is subject to a finite number of changes in the family. At each of these exceptional times one can produce a quasi-isomorphism relating the old A_∞-structure with the new one.

Remark 6.1. *As a technical point, note that directedness allows us to bypass some problems which plague Fukaya's original setup, having to do with the chain complexes underlying $HF(L,L)$ and unit elements in them.*

The next step is isotopy invariance which, as always in Floer theory, is also used to extend the definition to configurations which are not in general position.

Proposition 6.2. *Let $\widetilde{\Gamma} = (\widetilde{L}_1, \dots, \widetilde{L}_m)$ be a graded Lagrangian configuration in general position. Take $l \in \{1, \dots, m\}$, a symplectic automorphism ϕ isotopic to the identity, and a grading $\tilde{\phi}$ which is zero near $\mathfrak{J}_M \times \partial M$, such that $\widetilde{\Xi} = (\widetilde{L}_1, \dots, \widetilde{L}_{l-1}, \tilde{\phi}(\widetilde{L}_l), \dots, \tilde{\phi}(\widetilde{L}_m))$ is again in general position. Then there is a quasi-isomorphism $F\colon Lag^{\rightarrow}(\widetilde{\Gamma}) \to Lag^{\rightarrow}(\widetilde{\Xi})$.*

We will spend a moment discussing the structure of the proof, since it is a good example of arguments involving directed Fukaya categories. Recall that an A_∞-functor $F\colon \mathcal{A} \to \mathcal{B}$ consists of a map $F\colon \mathrm{Ob}\,\mathcal{A} \to \mathrm{Ob}\,\mathcal{B}$, chain maps $F^1\colon hom_{\mathcal{A}}(X,Y) \to hom_{\mathcal{B}}(FX, FY)$ for $X, Y \in \mathrm{Ob}\,\mathcal{A}$, and multilinear "higher order terms" F^d, $d \geq 2$. In the present case, the map on objects is the obvious one, and the nontrivial chain maps

$$F^1\colon CF^*(\widetilde{L}_i, \widetilde{L}_k) \longrightarrow CF^*(\widetilde{L}_i, \tilde{\phi}(\widetilde{L}_k)), \quad i < l \text{ and } k \geq l,$$

are those underlying the continuation homomorphisms, so that F is automatically a quasi-isomorphism. The main effort goes into defining higher order terms which satisfy the equations for an A_∞-functor.

(6B) We will now describe the relation between Hurwitz moves of $\widetilde{\Gamma}$ and mutations of $D^b Lag^{\rightarrow}(\widetilde{\Gamma})$. Proposition 6.2 says that isotopies of $\widetilde{\Gamma}$ result in a quasi-isomorphism. Shifting the gradings $\widetilde{L}_i$ obviously corresponds to the first mutation in Definition 5.2.

Lemma 6.3. *$Lag^{\rightarrow}(c\widetilde{\Gamma})$ is quasi-isomorphic to $cLag^{\rightarrow}(\widetilde{\Gamma})$.*

The proof relies on the $\mathbb{Z}/(d+1)$-action on $\mathcal{R}^{d+1}$ given by a cyclic shuffle of the marked points (this symmetry had not been used in the definition of directed Fukaya categories).

Theorem 6.4. *$rLag^{\rightarrow}(\widetilde{\Gamma})$ is quasi-isomorphic to $Lag^{\rightarrow}(r\widetilde{\Gamma})$.*

To see why this is plausible, set $\mathcal{A} = Lag^{\rightarrow}(\widetilde{\Gamma})$, and let $(Z^1, \dots, Z^m)$ be the objects of $r\mathcal{A}$. The cohomology $H(hom_{r\mathcal{A}}(Z^i, Z^k), \mu^1_{r\mathcal{A}})$, $i < k$, is

$$\begin{cases} HF^*(\widetilde{L}_i, \widetilde{L}_k) & i < k \leq m-2, \\ HF^*(\widetilde{L}_i, \widetilde{L}_{m-1}) & i \leq m-2,\ k = m, \\ HF^*(\tilde{\tau}_{L_{m-1}}(\widetilde{L}_m), \widetilde{L}_{m-1}) & i = m-1,\ k = m, \\ H(\mathrm{Cone}(\mu^2_{\mathcal{A}}\colon CF^*(\widetilde{L}_{m-1}, \widetilde{L}_m) \otimes CF^*(\widetilde{L}_i, \widetilde{L}_{m-1}) & \\ \qquad \to CF^*(\widetilde{L}_i, \widetilde{L}_m))) & i \leq m-2,\ k = m-1. \end{cases}$$

This is just (9) except that in writing down the third case we have used Poincaré duality and (4). As we saw when discussing Theorem 3.3, the cone in the last line is isomorphic to $HF^*(\widetilde{L}_i, \tilde{\tau}_{L_{m-1}}(\widetilde{L}_m))$. Therefore all cohomology groups are in fact isomorphic to the corresponding ones in $Lag^{\to}(r\widetilde{\Gamma})$. The remainder of the proof, as in Proposition 6.2, consists in extending this to a full-fledged A_∞-functor.

By the general theory of mutations, what we have shown implies that if two graded Lagrangian configurations in M are Hurwitz equivalent, their directed Fukaya categories have equivalent derived categories. Combining this with Picard–Lefschetz theory yields the following consequence:

Corollary 6.5. *Let (E,π) be an exact Morse fibration over D, with a relative Maslov map $\delta_{E/D}$. Make an admissible choice of paths, and let Γ be the corresponding distinguished basis of vanishing cycles in a fibre. Choose any gradings $\widetilde{\Gamma}$ and form $D^b Lag^{\to}(\widetilde{\Gamma})$. This is independent of all choices up to equivalence, and hence is an invariant of (E,π) and $\delta_{E/D}$.*

(6C) There is a computational aspect which Corollary 6.5 fails to convey, and which we will explain by giving an example. Let M be an exact symplectic four-manifold with $2c_1(M) = 0$, and (L_1, L_2) two Lagrangian spheres in M. $L_2' = \tau_{L_1}^2(L_2)$ and L_2 are always isotopic as smooth submanifolds. There are two cases, $L_1 = L_2$ and $L_1 \cap L_2 = \emptyset$, in which L_2' is also Lagrangian isotopic to L_2 for obvious reasons, but in general this is false:

Proposition 6.6. *Suppose that L_1, L_2 intersect transversally, with $|L_1 \cap L_2| \geq 3$, and that the local intersection numbers at all points are the same. Then L_2' is not Lagrangian isotopic to L_2.*

The proof goes as follows. Choose a Maslov map δ_M and gradings $\widetilde{L}_1, \widetilde{L}_2$. The directed Fukaya category $\mathcal{A} = Lag^{\to}(\widetilde{\Gamma})$ of the configuration $\widetilde{\Gamma} = (\widetilde{L}_1, \widetilde{L}_1, \widetilde{L}_2, \widetilde{L}_2)$ is determined up to quasi-isomorphism by the graded vector space $R = HF^*(\widetilde{L}_1, \widetilde{L}_2)$ together with the degree two maps $q_1, q_2 \in \mathrm{End}(R)$ given by the pair-of-pants product with the unique nontrivial element in $HF^2(\widetilde{L}_1, \widetilde{L}_1)$ resp. $HF^2(\widetilde{L}_2, \widetilde{L}_2)$. These satisfy $q_1 \circ q_2 = q_2 \circ q_1$ and $q_1^2 = q_2^2 = 0$. Lemma 6.3 and Theorem 6.4 give an explicit sequence of mutations which transforms $\mathcal{A}$ into the directed Fukaya category $\mathcal{B}$ associated to the Hurwitz equivalent configuration

$$ccrc^{-1}rc^{-1}\widetilde{\Gamma} = (\tilde{\tau}_{L_1}^2(\widetilde{L}_2), \widetilde{L}_1, \widetilde{L}_1, \widetilde{L}_2).$$

In particular this determines $HF^*(\tilde{\tau}_{L_1}^2(\widetilde{L}_2), \widetilde{L}_2)$, since that is a morphism space in $H(\mathcal{B})$. We will not write down the actual computation; the outcome is the total cohomology of the complex

$$\mathbb{Z}/2 \oplus \mathbb{Z}/2[-2] \xrightarrow{(\mathrm{id}, q_2)} \mathrm{End}(R) \xrightarrow{\psi} \mathrm{End}(R)[2],$$

where the second arrow is $\psi(x) = q_1 \circ x - x \circ q_1$. Since $\psi^2(x) = 2q_1 \circ x \circ q_1 = 0$, linear algebra tells us that the dimension of $\mathrm{coker}(\psi)$ is $\geq (\dim R)^2/2$. With the assumption $\dim R \geq 3$ this implies that $HF(L_2', L_2)$ is bigger than $HF(L_2, L_2)$, which completes the argument.

References. The presence of A_∞-structures in Floer theory was discovered by Fukaya [5, 6, 7]; at the time of writing, there are still no published proofs of the basic analytic results. Our approach to transversality is joint work with Lazzarini. The formal resemblance between Hurwitz moves and mutations was pointed out by Kontsevich. With the benefit of hindsight one can see that Theorem 6.4 is, in conjectural form, implicit in his discussion of that phenomenon [8]. Proposition 6.6 takes up a topic discussed in [15] and [16]. However, the result itself is new, and can apparently not be proved by the elementary methods used in those earlier papers.

Acknowledgments

Thanks go to Ivan Smith for reading the manuscript and suggesting several improvements.

References

[1] V. I. Arnol'd, *Some remarks on symplectic monodromy of Milnor fibrations*, The Floer Memorial Volume (H. Hofer, C. Taubes, A. Weinstein, and E. Zehnder, eds.), Progress in Mathematics, vol. 133, Birkhäuser, 1995, pp. 99–104.

[2] A. I. Bondal and M. M. Kapranov, *Enhanced triangulated categories*, Math. USSR Sbornik **70** (1991), 93–107.

[3] S. K. Donaldson, *Polynomials, vanishing cycles, and Floer homology*, Preprint.

[4] A. Floer, *A relative Morse index for the symplectic action*, Comm. Pure Appl. Math. **41** (1988), 393–407.

[5] K. Fukaya, *Morse homotopy, A_∞-categories, and Floer homologies*, Proceedings of GARC workshop on Geometry and Topology (H. J. Kim, ed.), Seoul National University, 1993.

[6] ———, *Floer homology for three-manifolds with boundary I*, Preprint, 1997.

[7] K. Fukaya and Y.-G. Oh, *Zero-loop open strings in the cotangent bundle and Morse homotopy*, Asian J. Math. **1** (1998), 96–180.

[8] M. Kontsevich, *Lectures at ENS Paris, Spring 1998*, set of notes taken by J. Bellaiche, J.-F. Dat, I. Marin, G. Racinet and H. Randriambololona.

[9] ———, *Homological algebra of mirror symmetry*, Proceedings of the International Congress of Mathematicians (Zürich, 1994), Birkhäuser, 1995, pp. 120–139.

[10] D. McDuff and D. Salamon, *J-holomorphic curves and quantum cohomology*, University Lecture Notes Series, vol. 6, Amer. Math. Soc., 1994.

[11] S. Piunikhin, D. Salamon and M. Schwarz, *Symplectic Floer–Donaldson theory and quantum cohomology*, Contact and symplectic geometry (C. B. Thomas, ed.), Cambridge Univ. Press, 1996, pp. 171–200.

[12] D. Salamon and E. Zehnder, *Morse theory for periodic solutions of Hamiltonian systems and the Maslov index*, Comm. Pure Appl. Math. **45** (1992), 1303–1360.

[13] M. Schwarz, *Cohomology operations from S^1-cobordisms in Floer homology*, Ph.D. thesis, ETH Zürich, 1995.

[14] P. Seidel, *Floer homology and the symplectic isotopy problem*, Ph.D. thesis, Oxford University, 1997.

[15] ———, *Lagrangian two-spheres can be symplectically knotted*, J. Differential Geom. **52** (1999), 145–171.

[16] ———, *Graded Lagrangian submanifolds*, Bull. Soc. Math. France **128** (2000), 103–146.

[17] V. De Silva, *Products in the symplectic Floer homology of Lagrangian intersections*, Ph.D. thesis, Oxford University, 1998.

Centre de Mathématiques
Ecole Polytechnique
91128 Palaiseau, France
E-mail address: `seidel@math.polytechnique.fr`

Critical Exponents, Conformal Invariance and Planar Brownian Motion

Wendelin Werner

Abstract. In this review paper, we first discuss some open problems related to two-dimensional self-avoiding paths and critical percolation. We then review some closely related results (joint work with Greg Lawler and Oded Schramm) on critical exponents for two-dimensional simple random walks, Brownian motions and other conformally invariant random objects.

1. Introduction

The conjecture that the scaling limits of many two-dimensional systems in statistical physics exhibit conformally invariant behaviour at criticality has led to striking predictions by theoretical physicists concerning, for instance, the values of exponents that describe the behaviour of certain quantities near (or at) the critical temperature. Some of these predictions can be reformulated in elementary terms (see for instance the conjectures for the number of self-avoiding walks of length n on a planar lattice).

From a mathematical perspective, even if the statement of the conjectures are clear, the understanding of these predictions and of the non-rigorous techniques (renormalisation group, conformal field theory, quantum gravity, the link with highest-weight representation of some infinite-dimensional Lie algebras, see e.g., [20, 11]) used by physicists has been limited. Our aim in the present review paper is to present some results derived in joint work with Greg Lawler and Oded Schramm [33, 34, 29, 30, 31, 32] that proves some of these conjectures, and improves substantially our understanding of others. The systems that we will focus on (self-avoiding walks, critical percolation, simple random walks) correspond in the language of conformal field theory to zero central charge.

We structure this paper as follows: In order to put our results into perspective, we start by very briefly describing two models (self-avoiding paths and critical percolation) and some of the conjectures that theoretical physicists have produced and that are, at present, open mathematical problems. Then, we state theorems derived in joint work with Greg Lawler and Oded Schramm [29, 30, 32] concerning critical exponents for simple random walks and planar Brownian motions (these had been also predicted by theoretical physics). We then show how all these problems are mathematically related, and, in particular, why the geometry of critical percolation in its scaling limit should be closely related to the geometry of a planar Brownian path via a new increasing set-valued process introduced by Schramm in [38].

2. Review of Some Predictions of Theoretical Physics

2.1. Predictions for self-avoiding walks

We first very briefly describe some predictions of theoretical physics concerning self-avoiding paths in a planar lattice. For a more detailed mathematical account on this subject, see for instance [35].

Consider the square lattice $\mathbb{Z}^2$ and define the set Ω_n of nearest-neighbour paths of length n started at the origin that are self-avoiding. In other words, Ω_n is the set of injective functions $\{0,\dots,n\} \to \mathbb{Z}^2$, such that $w(0) = (0,0) = 0$ and $|w(1) - w(0)| = \cdots = |w(n) - w(n-1)| = 1$.

The first problem is to understand the asymptotic behaviour of the number $a_n := \#\Omega_n$ of such self-avoiding paths when $n \to \infty$. A first trivial observation is that for all $n, m \geq 1$, $a_{n+m} \leq a_n a_m$ because the first n steps and the last m steps of an $n+m$ long self-avoiding path are self-avoiding paths of length n and m respectively. Furthermore, $a_n \geq 2^n$ because if the path goes only upward or to the right, then it is self-avoiding. This leads immediately to the existence of a constant $\mu \in [2,3)$ (called the connectivity constant of the lattice $\mathbb{Z}^2$) such that

$$\mu := \inf_{n\geq 1} (a_n)^{1/n} = \lim_{n\to\infty} (a_n)^{1/n} .$$

Note that if one counts the number a'_n of self-avoiding paths of length n on a triangular lattice, the same argument shows that $(a'_n)^{1/n}$ converges when $n \to \infty$ to some limit $\mu' \geq 3$. The connectivity constant is lattice-dependent.

One can also look at other regular planar lattices, such as the honeycombe lattice. We will say that a property is 'lattice-independent' if it it holds for all these three 'regular' lattices (square lattice, triangular lattice, honeycombe lattice). One possible way to describe a larger class of 'regular' lattices for which our 'lattice-independent' properties should hold, could (but we do not want to discuss this issue in detail here) be that they are transitive (i.e., for each pair of points, there exists a euclidean isometry that maps the lattice onto itself and one of the two points onto the other) and that rescaled simple random walk on this lattice converges to planar Brownian motion. For instance, the lattice $\mathbb{Z} \times 2\mathbb{Z}$ is not allowed.

When L denotes such a planar lattice, we denote by $a_{n,L}$ the number of self-avoiding paths of length n in the lattice starting from a fixed point, and by $\mu_L := \inf_{n\geq 1}(a_{n,L})^{1/n}$ its connectivity constant.

A first striking prediction from theoretical physics is the following:

Prediction 2.1. **(Nienhuis** [37]**)** *For any regular planar lattice L, when $n \to \infty$,*

$$a_n(L) = (\mu_L)^n n^{11/32+o(1)} .$$

The first important feature is the rational exponent 11/32. The second one is that this result does not depend on the lattice i.e., the first-order term is lattice-dependent while the second is 'universal'.

The following statement is almost equivalent to the previous prediction. Suppose that we define under the same probability P_n two independent self-avoiding paths w and w' of length n on the lattice L (the law of w and w' is the uniform probability on Ω_n).

Prediction 2.2. **(Intersection exponents version)** *When $n \to \infty$,*

$$P_n[w\{1,2,\dots,n\} \cap w'\{0,1,\dots,n\} = \emptyset] = n^{-11/32+o(1)} .$$

Indeed, $w\{1,2,\dots,n\}$ and $w'\{0,1,\dots,n\}$ are disjoint if and only if the concatenation of the two paths w and w' is a self-avoiding path of length $2n$ so that the non-intersection probability is exactly $a_{2n}/(a_n)^2$.

A second question concerns the typical behaviour of a long self-avoiding path, chosen uniformly in Ω_n when n is large. Let $d(w)$ denote the diameter of w. Theoretical physics predicts that the typical diameter is of order $n^{3/4}$. This had already been predicted using a different ('very non-rigorous') argument by Flory [18] in the late 1940s. A formal way to describe this prediction is the following:

Prediction 2.3. **(Nienhuis** [37]**)** *For all $\epsilon > 0$ and all regular lattices, when $n \to \infty$,*

$$P_n\left[d(w) \in [n^{3/4-\epsilon}, n^{3/4+\epsilon}]\right] \to 1 .$$

One of the underlying beliefs that led to these conjectures is that the measure on long self-avoiding paths, suitably rescaled, converges when the length goes to infinity, towards a measure on continuous curves that posesses some invariance properties under conformal transformations. The counterparts of the previous predictions in terms of this limiting measure then go as follows: Take two independent paths defined under the limiting measure, started at distance ϵ from each other. Then, the probability that the two paths are disjoint decays like $\epsilon^{11/24}$ when $\epsilon \to 0$. For the second prediction: the Hausdorff dimension of a path defined under the limiting measure is almost surely 4/3.

2.2. Predictions for critical planar percolation

We now review some results predicted by theoretical physics concerning critical planar percolation. A more detailed acount on these conjectures for mathematicians can be found for instance in [25]. See [19] for a general introduction to percolation.

Let $p \in (0,1)$ be fixed. For each edge between neighbouring points of the lattice, erase the edge with probability $1-p$ and keep it (and call the edge open) with probability p independently for all edges. In other words, for each edge we toss a biased coin to decide whether it is erased or not. This procedure defines a random subgraph of the square grid. It is not difficult to see that the large-scale geometry of this subgraph depends a lot on the value of p. In particular, there exists a critical value p_c (called the critical probability), such that if $p < p_c$, there exists almost surely no unbounded connected component in the random subgraph, while if $p > p_c$, there exists almost surely a unique unbounded connected component of open edges. It is not very difficult to see that the value p_c of the critical probability is lattice-dependent. Kesten has shown that for $L = \mathbb{Z}^2$, $p_c = 1/2$. We are going to be interested in the geometry of large connected components when $p = p_c$.

In regular planar lattices at $p = p_c$, it is known that almost surely no infinite connected component exists. However, a simple duality argument shows that in the square grid, at $p = p_c = 1/2$, for any $n \geq 1$, with probability $1/2$, there exists a path of open edges joining (in that rectangle) the bottom and top boundaries of a fixed $n \times (n+1)$ rectangle. This loosely speaking shows that in a big box, with large probability, there exist connected components of diameter comparable to the size of the box.

Theoretical physics predicts that large-scale properties of the geometry of critical percolation (i.e., percolation on a planar lattice at its critical probability) are lattice-independent (even though the value of the critical probability is lattice-dependent), and, in the scaling limit, invariant under conformal transformations; see e.g., [1, 25]. Furthermore, physicists have produced explicit formulas that describe some of its features.

A first prediction is the following: Consider critical percolation restricted to an $n \times n$ square (in the square lattice, say), and choose the connected component C with largest diameter (among all connected components). The previous observation shows that the diameter of C is of the order of magnitude of n. Define the rescaled discrete outer perimeter of C, $\partial_n = \partial C/n$, where ∂C is the boundary of the unbounded connected component of the complement of C in the plane.

Prediction 2.4. (Cluster boundaries [16, 12, 4]) *The law of ∂_n converges when $n \to \infty$ towards a law μ on continuous paths ∂. Moreover, μ-almost surely: the Hausdorff dimension of ∂ is $7/4$, the path ∂ is not self-avoiding, and the outer boundary of ∂ has Hausdorff dimension* $4/3$.

A weaker version of the first part of this prediction is that for all $\epsilon > 0$, when $n \to \infty$, $P[\#\partial C \in (n^{7/4-\epsilon}, n^{7/4+\epsilon})] \to 1$.

A second prediction concerns the crossing probabilities of a quadrilateral. Suppose that $L > 0$ and $l > 0$, and perform critical percolation in the rectangle $[0, a_n] \times [0, b_n]$ where a_n and b_n are the respective integer parts of Ln and ln. Let $x(L, l)$ denote the cross-ratio between the four corners of the $L \times l$ rectangle (more precisely, it is the value x such that there exists a conformal mapping from the rectangle onto the upper half-plane, such that the left and right-hand side of the rectangle are mapped onto the intervals $(-\infty, 0]$ and $[1 - x, 1]$).

Prediction 2.5. (Cardy's formula [10]) *When $n \to \infty$, the probability that there exists a path of open edges in the rectangle $[0, a_n] \times [0, b_n]$ joining the left- and right-hand sides of the boundary of the rectangle converges to*

$$F(x) = \frac{3\Gamma(2/3)}{\Gamma(1/3)^2} x^{1/3} {}_2F_1(1/3, 2/3, 4/3; x)$$

where ${}_2F_1$ is the usual hypergeometric function.

These results are believed to be lattice-independent. This second statement has been predicted by Cardy [9, 10], using and generalising conformal field theory considerations and ideas introduced in [6, 7]. Note that this prediction is of a different nature than the previous ones. It gives an exact formula for an event in the scaling limit rather than just an exponent. Assuming conformal invariance, this prediction can be reformulated in a half-plane as follows:

Prediction 2.6. (Cardy's formula in a half-plane) *For all $a, b > 0$, the probability that there exists a crossing (a path of open edges) joining $(-\infty, -an]$ to $[0, bn]$ in the upper half-plane converges when $n \to \infty$ towards $F(b/(a+b))$.*

Carleson was the first to note that Cardy's formula takes on a very simple form in an equilateral triangle. Suppose A, B, C are the vertices of an equilaterial triangle, say $A = 0$, $B = e^{2i\pi/3}$, $C = e^{i\pi/3}$.

Prediction 2.7. (Cardy's formula in an equilateral triangle) *In the scaling limit (performing critical percolation on the grid $\epsilon\mathbb{Z}^2$, say, and letting $\epsilon \to 0$), the law of the left-most point on $[B, C]$ that is connected (in the triangle) to $[A, C]$ is the uniform distribution on $[B, C]$.*

3. Brownian Exponents

We now come to the core of the present paper and state some of the results derived in the series of papers [29]–[32] concerning critical exponents for planar Brownian motions and simple random walks. These are mathematical results (as opposed to the predictions reviewed in the previous section) that had been predicted some 15 or 20 years ago. The proofs of these theorems (that we shall briefly outline in the coming sections) do use conformal invariance, complex analysis and univalent functions.

3.1. Intersection exponents

Suppose that $(S_n, n \geq 0)$ and $(S'_n, n \geq 0)$ are two independent simple random walks on the lattice $\mathbb{Z}^2$ that are both started from the origin. We are interested in the asymptotic behaviour (when $n \to \infty$) of the probability that the traces of S and S' are disjoint.

Theorem 3.1. ([30]) *When $n \to \infty$,*

$$P[S\{1, 2, \dots, n\} \cap S'\{0, 1, \dots, n\} = \emptyset] = n^{-5/8+o(1)}.$$

This result had been predicted by Duplantier–Kwon [15], see also [13, 14] for another non-rigorous derivation based on quantum gravity ideas and predictions by Khnizhnik, Polyakov and Zamolodchikov.

Note that, as opposed to self-avoiding walks and percolation cluster boundaries, the scaling limit of planar simple random walk is well understood mathematically: It is planar Brownian motion (this is lattice-independent) and it is invariant under conformal transformations (modulo time-change). This is what makes it possible to prove Theorem 3.1 as opposed to the analogous prediction for self-avoiding walks. The scaling limit analog of Theorem 3.1 is the following:

Theorem 3.2. [30] *Let B and B' denote two independent planar Brownian motions started at distance 1 from each other. Then, when $t \to \infty$,*

$$P[B[0,t] \cap B'[0,t] = \emptyset] = t^{-5/8+o(1)}.$$

In [29]–[32], analogous results concerning non-intersection exponents between more than two Brownian motions (or simple random walks), in the plane or in the half-plane, are derived.

In fact, Theorem 3.1 is a consequence of Theorem 3.2 via an invariance principle argument (see [28] and the references therein for the connection between the two results).

3.2. Mandelbrot's conjecture

Let $(B_t, t \geq 0)$ denote a planar Brownian motion. Define the hull of $B[0,1]$ as the complement of the unbounded connected component of $\mathbb{C} \setminus B[0,1]$ and define the outer frontier of $B[0,1]$ as the boundary of the hull of $B[0,1]$.

Theorem 3.3. ([29, 30, 32]) *Almost surely, the Hausdorff dimension of the outer boundary of $B[0,1]$ is $4/3$.*

This result had been conjectured by Mandelbrot [36] based on simulations and the analogy with the conjectures for self-avoiding walks. See also [14] for a physics approach based on quantum gravity. This theorem is in fact a consequence (using results derived by Lawler in [26]) of the determination of the following critical exponent (called disconnection exponent):

Theorem 3.4. ([29, 30, 32]) *If B and B' are two independent planar Brownian motions started from 0, then, when $t \to \infty$,*

$$P[B[0,t] \cup B'[0,t] \text{ does not disconnect } 1 \text{ from } \infty] = t^{-1/3+o(1)}.$$

Similarly, a consequence of Theorem 3.2 is that the Hausdorff dimension of the set of cut points of the path $B[0,1]$ is almost surely $3/4$. Analogously, we get that the set of pioneer points (i.e., points B_t that are on the outer boundary of $B[0,t]$) in a planar Brownian path has Hausdorff dimension $7/4$, which – together with Theorem 3.3 – is reminiscent of Prediction 2.4. Also, the determination of more general exponents gives the multifractal spectrum of the outer frontier of a Brownian path. See [27] and the references therein (papers by Lawler) for the link between critical exponents and Hausdorff dimensions. See also [5].

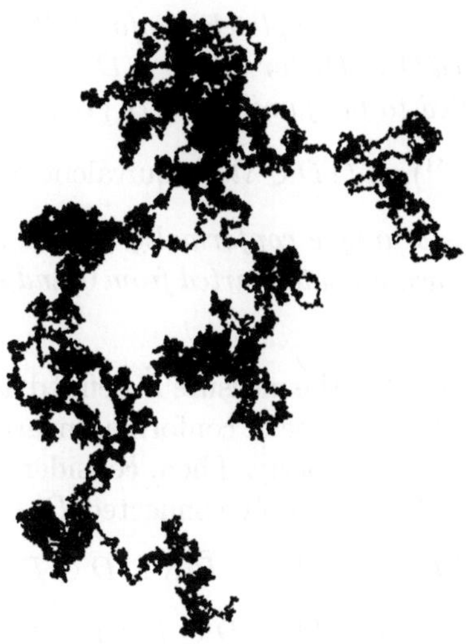

FIGURE 1. Simulation of a Brownian motion

4. Universality

4.1. In the plane

A first important step in the proof of Theorems 3.2 and 3.3 is the observation that some conformally invariant random probability measures with a special 'restriction' property are identical. Our presentation of universality here differs slightly from that of the preprints [34, 29, 30, 31]. A more extended version of the present approach and of its consequences is in preparation.

More precisely, let D denote the open unit disc, and suppose that P^0 is a rotationally invariant probability measure defined on the set of all simply connected compact subsets K of the closed unit disc that contain the origin and such that $K \cap \partial D$ is just one single point $e(K)$ (we endow this set with a well-chosen σ-field). Note that the law of $e(K)$ is the uniform probability λ on ∂D.

For all $x \in D$, the probability measure P^x is defined as the image probability measure of P^0 under a Möbius transformation Φ from D onto D such that $\Phi(0) = x$ (rotational invariance of P^0 shows that P^x is independent of the actual choice of Φ). Similarly, for any simply connected open set $\Omega \subset \mathbb{C}$ (that is not identical to the whole plane) and $x \in \Omega$, we define the probability $P^{x,\Omega}$ as the image measure of P^0 under a conformal transformation Φ from D onto Ω with $\Phi(0) = x$. When Φ does not extend continuously to the boundary of D, one can view $\Phi(K)$ as the union of $\Phi(K \setminus \{e(K)\})$ and the prime end $\Phi(e(K))$.

Definition 4.1. *We say that P^0 is completely conformally invariant (in short: CCI) if for any simply connected $\Omega \subset D$, for any $x \in D$, the measures $P^{x,D}$ and $P^{x,\Omega}$ are identical when restricted to the family of sets $\{K : K \cap D \subset \Omega\}$.*

Note that (under $P^{x,\Omega}$), $K \cap D \subset \Omega$ is equivalent to $\Phi(e(K)) \subset \partial D$.

Theorem 4.2. *There exists a unique conformally invariant probability measure. It is given by the hull of a Brownian path started from 0 and stopped at its first hitting of the unit circle.*

Idea of the proof. First, note that the measure $\tilde{P}$ defined using the hull of a stopped Brownian path is indeed CCI, because of conformal invariance of planar Brownian motion and the strong Markov property. Then, consider another CCI measure P. It is sufficient to show that for all simply connected $D' \subset D$ that contains 0,

$$P[K \cap D \subset D'] = \tilde{P}[K \cap D \subset D'] \tag{1}$$

since the family of all such events $\{K \cap D \subset D'\}$ is a generating π-system of the σ-field on which we define the measure. Let Φ denote a conformal map from D' onto D with $\Phi(0) = 0$. Then, because of the CCI property, both sides of (1) are equal to $\lambda(\Phi(\partial D \cap \partial D'))$ and hence equal.

4.2. In the half-plane

Analogous (slightly more complicated) arguments can be developed for subsets K of a domain that join two parts of the boundary of this domain (as opposed to joining a point in the interior to the boundary, as in the previous subsection). For convenience, we consider subsets of the equilateral triangle $A = 0$, $B = e^{2i\pi/3}$, $C = e^{i\pi/3}$.

We study probability measures P on the set of all simply connected compact subsets K of $\overline{\mathcal{T}}$ such that

$$K \cap \partial \mathcal{T} = [A, A_1] \cup [A, A_2] \cup \{e(K)\}$$

where $A_1 = A_1(K) \in [A, B]$, $A_2 = A_2(K) \in [A, C]$ and $e(K) \in (B, C)$, and such that $\mathcal{T} \setminus K$ consists of exactly two connected components (having respectively $[A_1, B] \cup [B, e(K)]$ and $[e(K), C] \cup [C, A_2]$ on their boundary).

We say that P satisfies Cardy's formula if the law of $e(K)$ is the uniform probability measure on $[B, C]$.

We say that P satisfies Property I, if for all non-empty $(B', C') \subset [B, C]$, if Φ denotes the conformal map from $\mathcal{T}$ onto $\mathcal{T}$ with $\Phi(A) = A$, $\Phi(B') = B$ and $\Phi(C') = C$, the image measure of P (restricted to $\{e(K) \in (B', C')\}$) under the mapping Φ is exactly the measure P restricted to $\{A_1 \in [A, \Phi(B)]\} \cap \{A_2 \in [A, \Phi(C)]\}$.

Suppose that K' is a simply connected compact subset of $\overline{\mathcal{T}}$ such that $K' \cap [B, C] \neq \emptyset$, $K' \cap ([A, B] \cup [A, C]) = \emptyset$, and $\mathcal{T}' := \mathcal{T} \setminus K'$ is simply connected. Let Φ denote the conformal mapping from $\mathcal{T}'$ onto $\mathcal{T}$ with $\Phi(A) = A$, $\Phi(B) = B$ and $\Phi(C) = C$. We say that P satisfies Property II, if for all such K', the image of P

(restricted to $\{K : K \cap K' = \emptyset\}$) under Φ is identical to the measure P restricted to the set $\{K : K \cap \Phi(\partial \mathcal{T}' \setminus \partial \mathcal{T}) = \emptyset\}$.

We then say that P is invariant under restriction if it satisfies Property I and Property II. There are other equivalent definitions and formulations of this restriction property.

Theorem 4.3. *There exists a unique probability measure P that satisfies Cardy's formula and is invariant under restriction. It is given by the hull in $\mathcal{T}$ (see the precise definition below) of Brownian motion, started from A, reflected with oblique angle $\pi/3$ (pointing 'away' from A) on $[A, B]$ and $[A, C]$, and stopped when reaching $[B, C]$.*

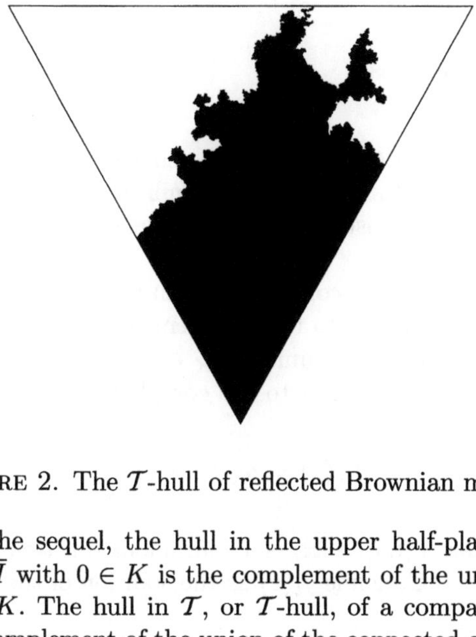

FIGURE 2. The $\mathcal{T}$-hull of reflected Brownian motion

Here and in the sequel, the hull in the upper half-plane (or H-hull) of a compact set $K \subset \overline{H}$ with $0 \in K$ is the complement of the unbounded connected component of $\overline{H} \setminus K$. The hull in $\mathcal{T}$, or $\mathcal{T}$-hull, of a compact subset of $\overline{\mathcal{T}}$ that contains A, is the complement of the union of the connected components of $\overline{\mathcal{T}} \setminus K$ that have B or C on their boundaries.

Idea of the proof. Uniqueness follows easily from a similar argument as in Theorem 4.2, by identifying $P(K \cap \mathcal{T} \subset \mathcal{T}')$ for a certain class of simply connected subsets $\mathcal{T}'$ of $\mathcal{T}$. For existence, one needs to verify that such a reflected Brownian motion satisfies Cardy's formula, and the restriction property follows from conformal invariance and the strong Markov property for such a reflected Brownian motion.

Let us briefly indicate how this reflected Brownian motion is defined (see for instance [40] for details). We first define it in the upper half-plane. Define for any $x \in \mathbb{R}$, the vector $u(x) = \exp(i\pi/3)$ if $x \geq 0$ and $u(x) = \exp(2i\pi/3)$ if $x < 0$. Suppose that $B(t)$ is an ordinary planar Brownian path started from 0. Then, there exists a unique pair (Z_t, ℓ_t) of continuous processes such that Z_t takes its values in $\overline{H}$, ℓ_t is a non-decreasing real-valued function with $\ell_0 = 0$ that increases only when $Z_t \in \mathbb{R}$, and

$$Z_t = B_t + \int_0^t u(Z_s) d\ell_s .$$

The process $(Z_t, t \geq 0)$ is called the reflected Brownian motion in H with reflection vector field $u(\cdot)$. At each time t, we define V_t as the H-hull of $Z[0, t]$. In order to define the $\mathcal{T}$-hull of reflected Brownian motion in the triangle that is refered to in the theorem, consider for instance a conformal mapping Φ from H onto the triangle, such that $\Phi(0) = A$, $\Phi(\infty) = B$, $\Phi(M) = C$ for some real $M > 0$. Then the $\mathcal{T}$-hull is $\Phi(V_T)$ where T is the first time at which Z hits $[M, \infty]$.

5. Schramm's Processes

5.1. In the plane

It is possible to contruct a CCI measure P using Loewner's differential equation (that encodes a certain class of growing families of compact sets) driven by a Brownian motion. See, for instance, [17] for a general introduction to Loewner's equation.

For any simply connected compact K such that the complement of K in the plane is conformally equivalent to the complement of a disc, Riemann's mapping theorem shows that there exists a unique $\alpha_K \in \mathbb{R}$ and a unique conformal map $\hat{f}_K$ that maps the complement of K onto the complement of the unit disk in such a way that

$$\hat{f}_K(z) = z e^{-\alpha_K} + O(1)$$

when $z \to \infty$.

Suppose now that $(\zeta(t), t \in \mathbb{R})$ is a continuous function taking values on the unit circle. For any $z \in \mathbb{C}$, define $f_t(z)$ as the solution of the ordinary differential equation

$$\partial_t f_t(z) = -f_t(z) \frac{f_t(z) + \zeta(t)}{f_t(z) - \zeta(t)}$$

such that

$$\lim_{t \to -\infty} e^{-t} f_t(z) = z .$$

For any fixed $z \neq 0$, the mapping $t \mapsto f_t(z)$ is well defined up to a possibly infinite 'explosion time' T_z, at which $f_t(z)$ hits the singularity $\zeta(t)$ (and we put $T_0 = -\infty$). Simple considerations show that for any time $t \in \mathbb{R}$, $\alpha_{K_t} = t$, and $f_t = \hat{f}_{K_t}$, where

$$K_t = \{z \in \mathbb{C} : T_z \leq t\}$$

i.e., f_t is the conformal mapping from the complement of K_t onto the complement of the unit disc such that $f_t(z) = z e^{-t} + o(z)$ when $z \to \infty$.

We now take $\zeta(t) = \exp(i\sqrt{\kappa}W_t)$, $t \in \mathbb{R}$, where $\kappa > 0$ is a fixed constant and $(W_t, t \in \mathbb{R})$ denotes a one-dimensional Brownian motion such that the law of $\sqrt{\kappa}W(0)$ is the uniform distribution on $[0, 2\pi]$. We call $(K_t, t \in \mathbb{R})$ Schramm's radial process with parameter κ (in [38, 30], it is refered to as SLE_κ for Stochastic Loewner Evolution process). Then, let T denote the first time at which K_t intersects the unit circle. The set K_T is a (random) simply connected compact set that intersects the unit circle at just one point.

Theorem 5.1. [29, 30] *If $\kappa = 6$, the law of K_T is a CCI probability measure.*

In fact, a more general result (complete conformal invariance of SLE_6 as a process) also holds, see [29, 30]. This theorem may seem quite surprising. Note that it fails for all other values of κ. The idea of the proof is to prove 'invariance' of the law under infinitesimal deformations of D.

A direct consequence of Theorems 5.1 and 4.2 is that the law of K_T is identical to the law of the hull of a stopped Brownian path. This is a rather surprising result since the two processes (hulls of planar Brownian motion and SLE_6) are a priori very different. For instance, the joint distribution at the first hitting times of the circles of radius 1 and 2 are not the same for SLE_6 and for the hull of a planar Brownian path.

5.2. In the half-plane

Similarly, one can construct natural processes of compact subsets of the closed upper-half plane $\overline{H} = \{z : \Im(z) \geq 0\}$ that are 'growing from the boundary'.

For any simply connected compact subset K of $\overline{H}$ such that $0 \in K$ and $H \setminus K$ is simply connected, there exists a unique number $\beta_K \geq 0$ and a unique conformal map $\hat{g}_K$ from $H \setminus K$ onto H such that $\hat{g}_K(\infty) = \infty$ and

$$\hat{g}_K(z) = z + \frac{\beta_K}{z} + o\left(\frac{1}{z}\right)$$

when $z \to \infty$. β_K which is increasing in K, is a half-space analog of capacity.

Suppose now that $(a(t), t \geq 0)$ is a continuous real-valued function, and define for all $z \in \overline{H}$ the solution $g_t(z)$ to the ordinary differential equation

$$\partial_t g_t(z) = \frac{2}{g_t(z) - a(t)}$$

with $g_0(z) = z$. This equation is well defined up to the (possibly infinite) time T_z at which $g_t(z)$ hits $a(t)$. Then, define

$$K_t := \{z \in \overline{H} : T_z \leq t\}.$$

$(K_t, t \geq 0)$ is an increasing family of compact sets, and it is not difficult to see that for each time $t \geq 0$, $\beta_{K_t} = 2t$ and $g_t = \hat{g}_{K_t}$ i.e., g_t is the unique conformal mapping from $H \setminus K_t$ onto H such that $g_t(\infty) = \infty$, and $g_t(z) = z + 2t/z + o(1/z)$ when $z \to \infty$.

When $a(t) = \sqrt{\kappa}W_t$, where W is one-dimensional Brownian motion with $W(0) = 0$, we get a random increasing family $(K_t, t \geq 0)$ which we call Schramm's chordal process with parameter κ (it is refered to as chordal SLE_κ in [38, 29, 30, 31]).

Note that if we put $Z_t(z) = W_t - \kappa^{-1/2} g_t(z)$, then

$$Z_t(z) = W_t + \int_0^t \frac{2}{\kappa Z_s(z)} ds$$

so that Z can be interpreted as a (translation of a) complex Bessel flow of dimension $1 + (4/\kappa)$. If $\kappa < 4$, it is easy to see, by comparison with a Bessel process, that almost surely $T_z < \infty$ for all $z \in \overline{H}$, in other words, $\cup_{t \geq 0} K_t = \overline{H}$.

The scaling properties of Brownian motion easily show that it is possible to define (modulo increasing time-reparametrization) the law of an increasing process of hulls in any simply connected set. More precisely, if Φ is a conformal mapping from H onto some simply connected domain Ω, then we say that $(\Phi(K_t), t \geq 0)$ is a chordal Schramm process (with parameter κ) started from $\Phi(0)$ aiming at $\Phi(\infty)$ in Ω. This process is well defined modulo increasing time-reparametrization (i.e., if there exists a (random) continuous increasing ψ such that for all t, $K_t = K'_{\psi(t)}$, then we say that the two processes K and K' are equal).

For the remainder of this paper, we will assume that $\kappa = 6$ as this case exhibits many very interesting properties. It is possible to compute explicitly certain probabilities. For instance, if $a, b > 0$,

$$P(T_{-a} < T_b) = F(b/(a+b))$$

where F is the same as in Cardy's formula In fact, much more is true:

Theorem 5.2. ([29]) *Consider SLE_6 in the equilateral triangle, started from A and aiming at some point $M \in [B,C]$. Let T denote the first time t at which $K_t \cap [B,C] \neq \emptyset$. Then, the law of $K_{T-} = \overline{\cup_{t<T} K_t}$ is independent of the choice of $M \in [B,C]$, it satisfies Cardy's formula and it is invariant under restriction.*

Again, this property is only valid when $\kappa = 6$ and there exists a more general version in terms of processes [29].

5.3. Link with reflected Brownian motion

From Theorem 5.2, Theorem 4.3 and the strong Markov property, it follows that the chordal process $(K_t, t \geq 0)$ can be reinterpreted in terms of reflected Brownian motions. In particular, the law of K_{T-} is that of the $\mathcal{T}$-hull of a stopped Brownian motion with oblique reflection. More generally, finite-dimensional marginals of the chordal SLE_6 process can for instance be constructed as follows (other more general statements with other stopping times hold as well).

Suppose that $J_1, \dots, J_p$ is a decreasing family of closed subsets of $\overline{H}$ such that $H \setminus J_1, \dots, H \setminus J_p$ are simply connected. When $(K_t, t \geq 0)$ is chordal SLE_6 in $\overline{H}$ (started at 0 and aiming at infinity), we put, for all $j \leq p$,

$$T_j = \inf\{t > 0 : K_t \cap J_j \neq \emptyset\}.$$

Let V denote a simply connected compact subset of the upper half-plane such that $H' = H \setminus V$ is simply connected and let $x \in \partial H'$. Define a reflected Brownian motion $(B_s, s \geq 0)$ in H' with oblique reflection angle (angle $\pi/3$ on the part of the boundary between x and $+\infty$ and reflection $2\pi/3$ between $-\infty$ and x) as the conformal image in H' of reflected Brownian motion in H. If J is a compact set, define the stopping time $S = S(J)$ at which $V \cup B[0, S]$ intersects J for the first time, and the hull $\mathcal{V} = \mathcal{V}(V, x, J)$ of $V \cup B[0, S(J)]$ in H.

Now, define recursively (using each time independent Brownian motions), $V_0 = \{0\}$, $x_0 = 0$,

$$V_{j+1} = \mathcal{V}(V_j, x_j, J_j) \text{ and } x_{j+1} = B(S(J_j)) .$$

Theorem 5.3. *The laws of* $(V_1, \dots, V_p)$ *and of* $(K_{T_1}, \dots, K_{T_p})$ *are identical.*

In particular, this implies that the law of V_p is identical to that of K_{T_p} and therefore is independent of p and $J_1, \dots, J_{p-1}$. This can be viewed as a way to reformulate the restriction property for SLE_6 in terms of reflected Brownian motions only. So far, there is no direct proof of this fact that does not use the link with Schramm's process.

5.4. Relation between radial and chordal SLE_6

Radial and chordal Schramm processes with parameter 6 are very closely related. The CCI property and the restriction property in their 'process' versions can be very non-rigorously described as follows: The evolution of K_t (for radial and chordal SLE_6) depends on K_t only in a local way i.e., suppose that at time t_0, K_t is increasing near some point $x \in \partial K_{t_0}$. Then the evolution of K_t immediately after t_0 depends only on how K_{t_0} looks in the neighbourhood of x (for a rigorous version of this statement, see [29, 30]). The link between radial and chordal SLE_6 (and this is only true when $\kappa = 6$) is that this local evolution is the same for radial and chordal SLE_6, see [30]. This also leads to a description of the finite-dimensional marginal laws of radial SLE_6 in terms of hulls of reflected Brownian motion.

6. Computation of the Exponents

We now give a very brief outline of the proof of Theorem 3.1. As a consequence of the results stated in the last two sections, hulls of Brownian paths at certain stopping times can be also constructed in an a priori completely different way using SLE_6. The latter turns out to be much better suited to compute probabilities involving only the shape of its complement, as SLE_6 is a process that is 'continuously growing to the outside', whereas Brownian motion does incursions inside its own hull, so that for instance the point at which the Brownian hull 'grows' makes a lot of jumps.

It is very easy to show that Theorem 3.1 is a consequence of the following fact that we shall now derive: Let B and B' denote two independent planar Brownian motions started from 0 and $\epsilon > 0$, and killed when they hit the unit circle. Then, when $\epsilon \to 0$,

$$P[B \cap B' = \emptyset] = \epsilon^{5/4+o(1)}. \tag{2}$$

To derive (2), note first that if $h_{K(B)}$ denotes the harmonic measure of ∂D at ϵ in $D \setminus K(B)$ (here $K(B)$ is the hull of B), then

$$P[B \cap B' = \emptyset \mid B] = h_{K(B)}.$$

Hence,

$$P[B \cap B' = \emptyset] = E[h_{K(B)}] = E[h_{K_T}]$$

where K_T is radial SLE_6 stopped as in Theorem 5.1. The construction of K_T via Loewner's equation makes it possible to study explicitly the asymptotic behaviour of $E[h_{K_T}]$ when $\epsilon \to 0$ by computing the highest eigenvalue computation of a differential operator, see [30]. From this, (2) follows.

Similar (though more involved) arguments lead to Theorem 3.3 and to the determination of many other such critical exponents that are defined in terms of planar Brownian paths [29]–[32].

7. The Conjecture for Percolation

Suppose that one performs critical bond percolation ($p = p_c = 1/2$) in the discrete half-plane $\mathbb{Z} \times \mathbb{N}$. Decide that all edges of the type $[x, x+1]$ when x is on the real axis are erased when $x \geq 0$ and open when $x < 0$. Then, there exists a unique infinite cluster C_- formed by the negative half-axis and the union of all clusters that are attached to it. We now explore the outer boundary of this cluster, starting from the point $(0,0)$. In other words, we follow the left-most possible path that is not allowed to cross open edges as shown in the picture below (for convenience, we draw the line that stays at 'distance' 1/4 from C_-). The open edges are the thin plain lines, and the exploration process is the thick plain line.

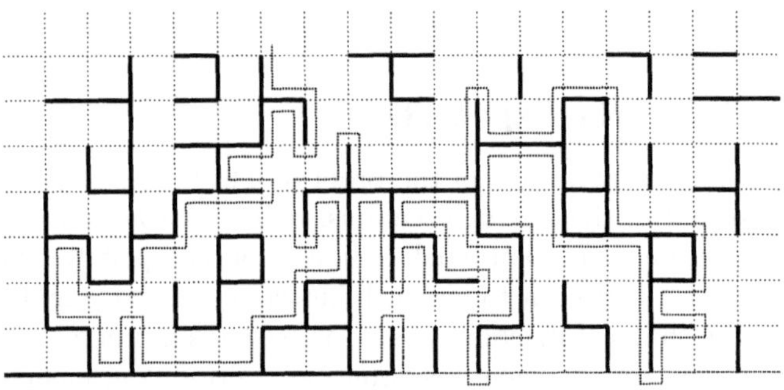

FIGURE 3. The discrete exploration process

Note that this exploration process is almost symmetric because of the self-duality property of the planar lattice: It is (almost) the same as exploring the outer boundary of the cluster of 'open' edges in the dual lattice (i.e., duals of closed edges in the original lattice) attached to the 'positive half-line' $(1-i)/2+\mathbb{N}$ (other choices of the lattice make this exploration process perfectly symmetric).

The conformal invariance conjecture leads naturally to the conjecture [38, 39] that in the scaling limit, this exploration process can be described using Loewner's differential equation in the upper half-plane, and that the driving process $a(t)$ is a continuous symmetric Markov process with stationary increments i.e., $a(t) = \sqrt{\kappa}W_t$ for some parameter κ. It is then easy to identify $\kappa = 6$ as the only possible candidate (looking for instance at the probability of crossing a square, see [39]), and the restriction property gives additional support to this conjecture. In particular, note that Cardy's formula for Schramm's process with parameter 6 would indeed correspond to the crossing probability.

So far, there is no mathematical proof of the fact that this exploration process converges to Schramm's process. However, Theorem 5.3 and the fact that the exploration process can be viewed as a discrete random walk reflected on its past hull gives at least a heuristic hand-waving justification.

One can also recover non-rigorously the exponents predicted for self-avoiding walks (see e.g., [34]) using the exponents derived (rigorously) for Brownian motions and Schramm processes.

The only discrete models that have been mathematically shown to exhibit some conformal invariance properties in the scaling limit — apart from simple random walks — are those studied by Kenyon in [21]–[24], namely loop-erased walks and uniform spanning trees. These are conjectured to correspond to Schramm's processes with parameters 2 and 8, see [38]. For partial results concerning percolation scaling limit and its conformal invariance, see [2, 3, 8].

For some other critical exponents related to Hausdorff dimensions of conformally invariant exceptional subsets of the planar Brownian curve (such as pivoting cut points for instance) that we do not know (yet?) the value of, see [5].

Acknowledgements

Without my coauthors Greg Lawler and Oded Schramm, this paper would of course not exist. I also use this opportunity to thank my Orsay colleagues Richard Kenyon and Yves Le Jan for many very stimulating and inspiring discussions, as well as Vincent Beffara for the nice picture (Figure 2).

References

[1] M. Aizenman, *The geometry of critical percolation and conformal invariance*, Statphys19 (Xiamen, 1995), 104–120 (1996).

[2] M. Aizenman and A. Burchard, *Hölder regularity and dimension bounds for random curves*, Duke Math. J. **99**, 419–453 (1999).

[3] M. Aizenman, A. Burchard, C. Newman, and D. Wilson, *Scaling limits for minimal and random spanning trees in two dimensions*, Random Str. Algo. **15**, 319–367 (1999).

[4] M. Aizenman, B. Duplantier, and A. Aharony, *Path crossing exponents and the external perimeter in 2D percolation*, Phys. Rev. Let. **83**, 1359–1362 (1999).

[5] V. Beffara, *On some conformally invariant subsets of the planar Brownian curve*, preprint (2000).

[6] A. A. Belavin, A. M. Polyakov, and A. B. Zamolodchikov, *Infinite conformal symmetry of critical fluctuations in two dimensions*, J. Stat. Phys. **34**, 763–774 (1984).

[7] A. A. Belavin, A. M. Polyakov, and A. B. Zamolodchikov, *Infinite conformal symmetry in two-dimensional quantum field theory*, Nucl. Phys. B **241**, 333–380 (1984).

[8] I. Benjamini and O. Schramm, *Conformal invariance of Voronoi percolation*, Comm. Math. Phys. **197**, 75–107 (1998).

[9] J. L. Cardy, *Conformal invariance and surface critical behavior*, Nucl. Phys. B **240**, 514–532 (1984).

[10] J. L. Cardy, *Critical percolation in finite geometries*, J. Phys. A **25**, L201–L206 (1992).

[11] J. L. Cardy, *Scaling and renormalization in Statistical Physics*, Cambridge University Press, 1996.

[12] J. L. Cardy, *The number of incipient spanning clusters in two-dimensional percolation*, J. Phys. A **31**, L105 (1998).

[13] B. Duplantier, *Random walks and quantum gravity in two dimensions*, Phys. Rev. Let. **82**, 5489–5492 (1998).

[14] B. Duplantier, *Two-dimensional copolymers and exact conformal multifractality*, Phys. Rev. Let. **82**, 880–883 (1999).

[15] B. Duplantier and K.-H. Kwon, *Conformal invariance and intersection of random walks*, Phys. Rev. Let. 2514–2517 (1988).

[16] B. Duplantier and H. Saleur, *Exact determination of the percolation hull exponent in two dimensions*, Phys. Rev. Lett. **58**, 2325 (1987).

[17] P. Duren, *Univalent functions*, Springer, 1983.

[18] P. J. Flory, *The configuration of a real polymer chain*, J. Chem. Phys. **17**, 303–310 (1949).

[19] G. Grimmett, *Percolation*, Springer-Verlag, 1989.

[20] C. Itzykon and J.-M. Drouffe, *Statistical Field Theory*, Vol. 2, Cambridge University Press, 1989.

[21] R. Kenyon, *Conformal invariance of domino tiling*, Ann. Probab. **28**, 759–795 (2000).

[22] R. Kenyon, *The asymptotic determinant of the discrete Laplacian*, Acta Math. **185**, 239–286 (2000).

[23] R. Kenyon, *Long-range properties of spanning trees in* $\mathbb{Z}^2$, J. Math. Phys. **41**, 1338–1363 (2000).

[24] R. Kenyon, *Dominos and the Gaussian free field*, preprint (2000).

[25] R. Langlands, Y. Pouillot, and Y. Saint-Aubin, *Conformal invariance in two-dimensional percolation*, Bull. A.M.S. **30**, 1–61, (1994).

[26] G. F. Lawler, *The dimension of the frontier of planar Brownian motion*, Electron. Comm. Probab. **1** 29–47 (1996).

[27] G. F. Lawler, *Geometric and fractal properties of Brownian motion and random walks paths in two and three dimensions*, in *Random Walks, Budapest 1998*, Bolyai Society Mathematical Studies **9**, 219–258 (1999).

[28] G. F. Lawler and E. E. Puckette, *The intersection exponent for simple random walk*, Comb. Probab. Comput., **9**, 441–464 (2000).

[29] G. F. Lawler, O. Schramm, and W. Werner, *Values of Brownian intersection exponents I: Half-plane exponents*, Acta. Math., to appear.

[30] G. F. Lawler, O. Schramm, and W. Werner, *Values of Brownian intersection exponents II: Plane exponents*, Acta Math., to appear.

[31] G. F. Lawler, O. Schramm, and W. Werner, *Values of Brownian intersection exponents III: Two-sided exponents*, Acta Math., to appear.

[32] G. F. Lawler, O. Schramm, and W. Werner, *Analyticity of planar Brownian intersection exponents*, Acta Math., to appear.

[33] G. F. Lawler and W. Werner, *Intersection exponents for planar Brownian motion*, Ann. Probab. **27**, 1601–1642 (1999).

[34] G. F. Lawler and W. Werner, *Universality for conformally invariant intersection exponents*, J. European Math. Soc. **2**, 291–328 (2000).

[35] N. Madras and G. Slade, *The self-avoiding walk*, Birkhäuser, Boston, 1993.

[36] B. B. Mandelbrot, *The fractal geometry of nature*, Freeman, 1982.

[37] B. Nienhuis, *Critical behavior of two-dimensional spin models and charge asymmetry in the Coulomb gas*, J. Stat. Phys. **34**, 731–761 (1983).

[38] O. Schramm, *Scaling limits of loop-erased random walks and uniform spanning trees*, Israel J. Math. **118**, 221–288 (2000).

[39] O. Schramm, *Conformal invariant scaling limits*, in preparation.

[40] S. R. S. Varadhan and R. J. Williams, *Brownian motion in a wedge with oblique reflection*, Comm. Pure Appl. Math. **38** 405–443 (1985).

Laboratoire de Mathématiques
Université Paris-Sud
Bât. 425
91405 Orsay cedex, France
E-mail address: `wendelin.werner@math.u-psud.fr`

Computer Algebra Algorithms for Linear Ordinary Differential and Difference equations

Manuel Bronstein

Abstract. Galois theory has now produced algorithms for solving linear ordinary differential and difference equations in closed form. In addition, recent algorithmic advances have made those algorithms effective and implementable in computer algebra systems. After introducing the relevant parts of the theory, we describe the latest algorithms for solving such equations.

1. Introduction

Linear ordinary differential equations are equations (resp. systems) of the form

$$\sum_{i=0}^{n} a_i(x)\frac{d^i y(x)}{dx^i} = 0 \quad \left(\text{resp. } \frac{dY(x)}{dx} = A(x)Y(x)\right),$$

while linear ordinary difference equations are equations (resp. systems) of the form

$$\sum_{i=0}^{n} a_i(x)y(x+i) = 0 \quad (\text{resp. } Y(x+1) = A(x)Y(x)),$$

where in both cases the unknown(s) and the coefficients are functions of the continuous or discrete variable x. Similarities between those equations have been noticed and used for a long time, to the point that algebraic algorithms based on the underlying linear operators allow large common parts of differential or difference equations solvers to be described, and indeed programmed, in the same algebraic setting (see e.g., [1, 8]). With the recent discovery of a difference Galois theory [30] with effective algorithms [10], both problems of deciding whether differential or difference equations have closed-form solutions are now solved. Furthermore, a recent reformulation of differential Galois theory [22] allows both cases to be presented using the same algebraic framework, and its interpretation using invariants [12, 13, 29] has led for the first time to effective implementations in computer algebra systems. After describing the common algebraic setting (below) and outlining the Galois theory required (Section 2), we describe the computation of invariants of differential equations (Section 3) and of Liouvillian solutions of difference equations (Section 4). All fields in this paper are commutative, rings are not necessarily commutative, and all rings and fields have characteristic 0.

We briefly outline the common algebraic setting for differential and difference equations. This formalism will be used in the rest of this paper when describing constructions common to both cases. A *σ-differential ring (resp. field)* is a triple (R, σ, δ) where R is a ring (resp. field), σ is an automorphism of R and δ is an additive map of R satisfying the modified Leibniz rule $\delta(ab) = (\sigma a)(\delta b) + (\delta a)b$ for all $a, b \in R$. The set

$$\mathrm{Const}_{\sigma,\delta}(R) = \{a \in R \text{ such that } \sigma a = a \text{ and } \delta a = 0\}$$

is a ring (resp. field), which is called the *constant subring (resp. subfield)* of R. Given a left R-module M, a map $\theta\colon M \to M$ is called *pseudo-linear* if it is additive and if $\theta(am) = (\sigma a)\theta m + (\delta a)m$ for any $a \in R$ and $m \in M$. The set of all the pseudo-linear maps of M is denoted $\mathrm{End}_{R,\sigma,\delta}(M)$. A *$\sigma$-differential extension of* (R, σ, δ) is a σ-differential ring (S, σ', δ') such that $R \subseteq S$, $\sigma' = \sigma$ on R and $\delta' = \delta$ on R. When R is commutative, $\mathrm{End}_{R,\sigma,\delta}(R) = \{\gamma\sigma + \delta \text{ for } \gamma \in R\}$ [5], which means that any $\theta \in \mathrm{End}_{R,\sigma,\delta}(R)$ has a unique extension to any commutative σ-differential extension (S, σ', δ') of R, namely $\theta' = \gamma\sigma' + \delta'$ where $\gamma \in R$ is such that $\theta = \gamma\sigma + \delta$. Given a σ-differential ring (R, σ, δ), the *univariate skew-polynomial ring* over R, denoted $(R[X]; \sigma, \delta)$, is the ring of univariate polynomials with the usual addition, but with the multiplication given by $Xa - \sigma(a)X = \delta a$ for any $a \in R$. This ring was introduced by Ore [19] who studied in particular its factorization properties. A key property that we use in this paper is that when R is a field, it has both left and right Euclidean divisions, which implies the existence of left and right greatest common divisors and least common multiples (see [8] for additional properties and algorithms). When σ is the identity on R, (R, δ) is a differential ring and $(R[X]; 1_R, \delta)$ is the ring of linear ordinary differential operators with coefficients in R. When $\delta r = 0$ for every $r \in R$, (R, σ) is a difference ring and $(R[X]; \sigma, 0_R)$ is the ring of linear ordinary difference operators with coefficients in R.

2. Galois Groups and Liouvillian Solutions

We summarize in this section the basic definitions and results that allow us to describe and justify the algorithms of the next sections. See [10, 22, 30] for proofs and a complete treatment of this subject.

Definition 2.1. ([22, 30]) *Let (k, σ, δ) be a σ-differential field, $\theta \in \mathrm{End}_{k,\sigma,\delta}(k)$ and $A \in GL_n(k)$. A* Picard–Vessiot ring *over k for the equation $\theta Y = AY$ is a commutative σ-differential extension ring R of k satisfying:*

(i) *The only ideals of R closed under σ and δ are (0) and (1).*
(ii) *$\exists U \in GL_n(R)$ such that $\theta U = AU$.*
(iii) *R is generated as a ring by k, the coefficients of U and $1/\det(U)$.*

For any σ-differential extension ring S of k, the σ-differential Galois group *of S over k is the group of automorphisms of S over k that commute with σ and δ.*

While the above definition is stated in terms of systems of equations, it applies to scalar equations as well: we associate to the scalar equation $Ly = 0$ where $L = \theta^n + \sum_{i=0}^{n-1} a_i \theta^i$, the companion system $\theta Y = A_L Y$ where

$$A_L = \begin{pmatrix} 0 & 1 & & \\ & \ddots & \ddots & \\ & & 0 & 1 \\ -a_0 & -a_1 & \dots & -a_{n-1} \end{pmatrix}.$$

The map $y \to (y, \theta y, \dots, \theta^{n-1} y)^T$ is an isomorphism between the solution spaces of $Ly = 0$ and $\theta Y = A_L Y$, so we define the Picard–Vessiot ring and Galois group for L to be the ones of the associated companion system.

If $(k, 1_k, \delta)$ is a differential field with an algebraically closed constant field C, then for any $A \in GL_n(k)$, there exist a Picard–Vessiot ring R, for $\delta Y = AY$, and it is unique up to differential isomorphism [22]. Furthermore, R is an integral domain, whose field of fractions K is called the *Picard–Vessiot field* for the equation $\delta Y = AY$, and $\text{Const}_\delta(K) = \text{Const}_\delta(R) = C$ [22]. In that case, the σ-differential Galois group of R over k is called the *differential Galois group* of $\delta Y = AY$ over k, and it coincides with the group of field-automorphisms of K over k that commute with the derivation.

If $(k, \sigma, 0_k)$ is a difference field with an algebraically closed constant field C, then for any $A \in GL_n(k)$, there exist a Picard–Vessiot ring R, for $\sigma Y = AY$, it is unique up to difference isomorphism and $\text{Const}_\sigma(R) = C$ [30]. In that case, the σ-differential Galois group of R over k is called the *difference Galois group* of $\sigma Y = AY$ over k.

In both of the above cases, the σ-differential Galois group G has a natural structure of a linear algebraic group over C, *i.e.*, it is an algebraic subgroup of $GL_n(C)$. Furthermore, the Picard–Vessiot extension R is normal over k, *i.e.* for any $t \in R \setminus k$, there is an element g of the group such that $g(t) \neq t$ [22, 30]. Since G is an algebraic group, it is the disjoint union of finitely many connected components in the Zariski topology. The component containing the identity will be denoted G^0.

A major success of differential and difference Galois theory has been the discovery of effective group-theoretic criteria for the existence of closed-form solutions to linear ordinary differential and difference equations. Before presenting those criteria, we need to formalize the notions of closed-form solutions that they use, starting with the differential case: let $(k, 1_k, \delta)$ be a differential field and K a differential extension of k. We say that K is *Liouvillian over* k if there are $t_1, \dots, t_m \in K^*$ such that $K = k(t_1, \dots, t_m)$ and for each i, either t_i is algebraic over $K(t_1, \dots, t_{i-1})$, or $\delta t_i \in K(t_1, \dots, t_{i-1})$ or $\delta(t_i)/t_i \in K(t_1, \dots, t_{i-1})$. We say that a differential equation with coefficients in k has a *Liouvillian solution* if it has a nonzero solution in a Liouvillian extension of k. The main criterion for the differential case is then:

Theorem 2.2. ([17]) *Let $L \in k[D; 1_k, \delta]$ be a linear ordinary differential operator with coefficients in a differential field k whose constant subfield is algebraically closed, and G be the differential Galois group of L over k.*

(i) *All the solutions of $Ly = 0$ are Liouvillian over k if and only if G^0 is a solvable group.*

(ii) *$Ly = 0$ has a Liouvillian solution over k if and only if it has a solution y such that y'/y is algebraic over k.*

Even though Theorem 2.2 relates the existence of Liouvillian solutions to the existence of a solution of a very special form, it does not yield an algorithm because it gives no bound on the degree of y'/y as an algebraic function. Singer produced such a bound for n arbitrary as well as an algorithm for determining the coefficients of the minimal polynomial of y'/y over k.

Theorem 2.3. ([25]) *There exists a function $F\colon \mathbb{N} \to \mathbb{N}$ such that $Ly = 0$ has a Liouvillian solution over k if and only if it has a solution y such that y'/y is algebraic over k and $[k(y'/y) : k] \leq F(n)$.*

The function $F(n)$ is defined by $F(0) = 1$ and $F(n) = \max(f(n), n!F(n-1))$ where $f(n)$ is such that every finite subgroup of $GL_n(\mathbb{C})$ has a normal Abelian subgroup of index at most $f(n)$. Jordan's Theorem [15] implies the existence of such a function and there are explicit formulas. This general upper bound can be improved in many cases, in particular for specific values of n [28, 31, 32].

A criterion also exists for a restricted class of difference equations [10], namely when $k = C(x)$ for some algebraically closed constant field C, and σ is the automorphism of k over C that maps x to $x+1$. In that case, Picard–Vessiot rings over k can be viewed as subrings of the commutative ring $\mathcal{S}$ of germs of sequences defined as $\mathcal{S} = C^{\mathbb{N}}/J$ where J is the ideal of sequences having finitely many nonzero terms. The map $\sigma(a_0, a_1, a_2, \dots) = (a_1, a_2, \dots)$ is a well-defined automorphism of $\mathcal{S}$ and k can be embedded in $\mathcal{S}$ by the difference embedding that maps $f \in k$ to the sequence $a_n = 0$ if n is a pole of f, $f(n)$ otherwise. A sequence $a \in \mathcal{S}$ is called *hypergeometric* if $\sigma a = fa$ for some $f \in k$. For a sequence $a = (a_n)_{n \geq 0}$ in $\mathcal{S}$, and $m > 0$, define the mth *spread of* a to be the sequence $a^{\vec{m}}$ given by

$$\left(a^{\vec{m}}\right)_n = \begin{cases} a_{n/m} & \text{if } n \equiv 0 \pmod{m} \\ 0 & \text{if } n \not\equiv 0 \pmod{m} \end{cases} . \tag{1}$$

The *interlacing* of m sequences $a^{(1)}, \dots, a^{(m)}$ is the sequence

$$\biguplus_{j=1}^{m} a^{(j)} = \sum_{j=1}^{m} \sigma^{1-j} \left(a^{j\vec{m}}\right) = (a_0^{(1)}, \dots, a_0^{(m)}, a_1^{(1)}, \dots).$$

Finally, the ring $\mathcal{L}$ of the *Liouvillian sequences* is the smallest difference subring of $\mathcal{S}$ containing k such that

1. $a \in k$, $b \in \mathcal{S}$, $\sigma b = ab \implies b \in \mathcal{L}$.
2. $a \in \mathcal{L}$, $b \in \mathcal{S}$, $\sigma b = b + a \implies b \in \mathcal{L}$.
3. $a \in \mathcal{L} \implies \forall i, m$ such that $0 \leq i < m$, $\sigma^{-i}(a^{\vec{m}}) \in \mathcal{L}$.

We say that a difference equation with coefficients in k has a *Liouvillian solution* if it has a nonzero solution in $\mathcal{L}$. The main criterion for the difference case is then:

Theorem 2.4. ([10]) *Let $L \in k[E;\sigma,0_k]$ be a linear ordinary differential operator with coefficients in the difference field $k = C(x)$, where C is algebraically closed and σ is the automorphism of k over C mapping x to $x+1$. Let G be the difference Galois group of L over k.*

(i) *All the solutions of $Ly = 0$ are Liouvillian if and only if G is a solvable group.*
(ii) *$Ly = 0$ has a Liouvillian solution if and only if it has a solution y such that y is the interlacing of m hypergeometric sequences where $1 \le m \le n$.*

3. Invariants and Differential Equations

Let W be a vector space over a field F and $G \subseteq GL(W)$ be a group of automorphisms of W. We say that $w \in W$ is an *invariant of* G if $g(w) = w$ for every $g \in G$. The set of all the invariants of G in W is denoted W^G and is a subspace of W. We say that w is a *semi-invariant of* G if for every $g \in G$, there exists $f_g \in F$ such that $g(w) = f_g w$. Invariants and semi-invariants of differential Galois groups play an important role in algorithms for solving differential equations. For the rest of this section, let k be a differential field with algebraically closed constant field C, L be a linear ordinary differential operator of order $n > 0$ with coefficients in k, K be its Picard–Vessiot extension, $V \subset K$ be the solution space of $Ly = 0$, G be its Galois group and G^0 be its connected component of the identity. The (faithful) action of G on V is extended to the symmetric algebra $S(V)$ via

$$g\left(\sum_{e=(e_1,\dots,e_{m_e})} c_e v_{e_1} \otimes \cdots \otimes v_{e_{m_e}}\right) = \sum_e c_e g(v_{e_1}) \otimes \cdots \otimes g(v_{e_{m_e}}) \qquad (2)$$

for any $c_e \in C$ and $v_i \in S^1(V) = V$. Note that $S(V) = \bigoplus_{m\ge 0} S^m(V)$ is a graded C-algebra, and that the action of G given by (2) preserves the grading, so in particular $S(V)^G = \bigoplus_{m\ge 0} S^m(V)^G$ and we can restrict our attention to *homogeneous* invariants and semi-invariants in $S^m(V)$. Define the *degree* of a nonzero $w \in S(V)$ to be the smallest $d \ge 0$ such that $w \in \bigoplus_{m=0}^{d} S^m(V)$. We say that $w \in S(V)$ *factors into linear forms* if $w = w_1 \otimes \cdots \otimes w_m$ for some $w_1, \dots, w_m \in V$. Liouvillian solutions of $Ly = 0$ and semi-invariants of G are related by the following result, in which $F(n)$ is as in Theorem 2.3.

Theorem 3.1. ([29]) *$Ly = 0$ has a Liouvillian solution over k if and only if G has a semi-invariant $I \in S(V)$ of degree at most $F(n)$ and which factors into linear forms.*

It turns out that looking for semi-invariants can be reduced to looking for invariants of the same degree of transformed operators [12]. In addition, "semi-invariant" can be replaced by "invariant" in the above theorem for a large class of equations[1] (see the discussion at the end of this section for details), so we consider only the problem of searching for invariants in $S^d(V)^G$ where d ranges over a given subset of $\{1,\ldots,F(n)\}$. The basic idea to is to map them via G-morphisms to G-modules where their images can be computed. Singer and Ulmer [29] used the C-algebra evaluation homomorphism $\psi\colon S(V) \to K$ given by

$$\psi\left(\sum_{e=(e_1,\ldots,e_{m_e})} c_e v_{e_1} \otimes \cdots \otimes v_{e_{m_e}}\right) = \sum_e c_e v_{e_1} \ldots v_{e_{m_e}}$$

for any $c_e \in C$ and $v_i \in S^1(V) = V \subset K$. It is easily checked that ψ commutes with the actions of G on $S(V)$ and K, hence that $\psi(S(V)^G) \subseteq K^G$. Since Picard–Vessiot extensions are normal, $K^G = k$, so $I \in S(V)^G \implies \psi(I) \in k$. Define the dth *symmetric power* of L, denoted $L^{ⓢd}$, to be a differential operator of minimal order whose solution space is $\psi(S^d(V))$. Symmetric powers can be computed from L using only linear algebra over k [7]. Then $I \in S^d(V)^G \implies \psi(I) \in k$ and $L^{ⓢd}(\psi(I)) = 0$. If in addition $L^{ⓢd}$ has order $\dim_C(S^d(V)) = \binom{n+d-1}{n-1}$, then ψ_d (the restriction of ψ to $S^d(V)$) must be injective, so in that case

$$I \in S^d(V)^G \setminus \{0\} \iff \psi(I) \in k^* \text{ and } L^{ⓢd}(\psi(I)) = 0\,.$$

Such solutions in k^* can be computed for a large class of fields [4, 6, 26], and a basis $(f_1,\ldots,f_r)$ for them implies that $I_1,\ldots,I_r$ is a basis for $S^d(V)^G$ where $I_j = \psi_d^{-1}(f_j)$. Inverting ψ_d is possible whenever k is a finitely generated differential extension of $\mathbb{Q}$: Seidenberg's Embedding Theorem [23, 24] states that any such field is isomorphic to a differential field of meromorphic functions on an open region of $\mathbb{C}$. This means that given any finite subset S of k, there exists infinitely many $x_0 \in \mathbb{C}$ such that the elements of S can be seen as analytic functions at x_0, and hence evaluated at $x = x_0$. We now say that $x_0 \in \mathbb{C}$ is an *ordinary point* of $L = \sum_{i=0}^n a_i D^i$ if each a_i is analytic in an open neighborhood of x_0 (including x_0) and if $a_n(x_0) \neq 0$. Since meromorphic functions have isolated singularities, it follows from the Embedding Theorem that L has infinitely many ordinary points. To compute $I = \psi_d^{-1}(f)$ for $f \in k$, pick an ordinary point $c \in \mathbb{C}$ of L and compute a basis $(y_1,\ldots,y_n)$ of the formal Taylor series solutions of $Ly = 0$ around c. Writing I as a homogeneous polynomial of degree d in $Y_1,\ldots,Y_n$ with undetermined constant coefficients and equating the first $\binom{n+d-1}{n-1}$ coefficients of $I(y_1,\ldots,y_n)$ with those of the Taylor series of f yields a nonsingular linear algebraic system for the coefficients of I, whose solution yields I. If $L^{ⓢd}$ has order strictly smaller than $\binom{n+d-1}{n-1}$, then ψ_d is not injective and one needs to perform a generic change of variable to ensure that ψ_d will be injective for the transformed operator [29]. Alternatively, one can use the above series method to compute $\ker \psi_d$, since its dimension is known.

The above method is particularly well adapted for second-order operators, because $L^{ⓢd}$ is straightforward to compute via a simple iteration in that case [7].

[1] At the cost of increasing the degree bound.

Furthermore, its order is always $d+1$ so ψ_d is always injective. Finally, any homogeneous polynomial in $C[Y_1, Y_2]$ factors into linear forms, implying that any homogeneous invariant factors into linear forms. The bound $F(2)$ can be improved and an optimal result for $n = 2$, which holds for arbitrary coefficients fields, even when the constant field is not algebraically closed, is that $Ly = 0$ has a Liouvillian solution if and only if G has either a semi-invariant of degree 1 or a homogeneous invariant of degree $d \in \{2, 4, 6, 8, 12\}$ [33].

Example 3.2. *Consider the equation* $x^2y''(x) - (36x^6e^{4x^3} + 9x^6 + 2)y(x) = 0$, *whose operator* $L = x^2D^2 - 36x^6t^4 - 9x^6 - 2$ *has coefficients in* $k = \mathbb{Q}(x,t)$ *with* $\delta x = 1$ *and* $\delta t = 3x^2t$. *Its symmetric square is*

$$L^{\circledS 2} = x^3D^3 - 4x(36x^6t^4 + 9x^6 + 2)D - 8(1 + 3x^3)(36x^6t^4 + 3x^3 - 1),$$

whose solution space in k *is generated by* $f = x^{-2}t^{-2}$. *A basis of local solutions of the equation around* $x = 1$ *is*

$$y_1 = 1 + \left(18e^4 + \frac{11}{2}\right)(x-1)^2 + \cdots,$$

$$y_2 = (x-1) + \left(6e^4 + \frac{11}{6}\right)(x-1)^3 + \cdots,$$

and equating the first three terms of $c_{11}y_1^2 + c_{12}y_1y_2 + c_{22}y_2^2$ *with the first three terms of* $f = e^{-2} - 8e^{-2}(x-1) + 27e^{-2}(x-1)^2 + \cdots$ *yields* $c_{11} = e^{-2}, c_{12} = -8e^{-2}$ *and* $c_{22} = 16e^{-2} - 36e^2$. *It follows that*

$$I = e^{-2}(Y_1^2 - 8Y_1Y_2 + (16 - 36e^4)Y_2^2) = e^{-2}(Y_1 + (6e^2 - 4)Y_2)(Y_1 - (6e^2 + 4)Y_2)$$

is an invariant of the Galois group of L *that factors into linear forms.*

For equations of higher order, the above method suffers from the cost of computing $L^{\circledS d}$ as well as from the occasionally required generic transformation. To avoid those problems, van Hoeij and Weil used a different C-algebra homomorphism [13]. For any commutative ring R, write $N_d = \binom{n+d-1}{n-1}$ and define $\sigma_d \colon R^n \to R^{N_d}$ by $\sigma_d((r_1, \dots, r_n)^T) = (r_1^d, r_1^{d-1}r_2, r_1^{d-1}r_3, \dots, r_n^d)^T$ where the N_d homogeneous monomials of degree d in $r_1, \dots, r_n$ are ordered lexicographically with $r_1 < r_2 < \cdots < r_d$. Let $R = F[Y_1, \dots, Y_n]$ be a polynomial ring over a field F and $Y = (Y_1, \dots, Y_n)^T \in R^n$. For any $M \in GL_n(F)$, the entries of MY are linear forms in the Y_i's, which implies that the entries of $\sigma_d(MY)$ are homogeneous of degree d. Since $\sigma_d(Y)$ is a basis for the homogeneous polynomials of degree d, this defines a matrix $\mathrm{Sym}^d(M)$ given by $\sigma_d(MY) = \mathrm{Sym}^d(M)\sigma_d(Y)$. The map $\mathrm{Sym}^d \colon GL_n(F) \to GL_{N_d}(F)$ is then a group-homomorphism. When F is the constant field C of k, Sym^d describes the extension (2) of the action of the

Galois group G from V to $S^d(V)$ as it makes the following diagram commute:

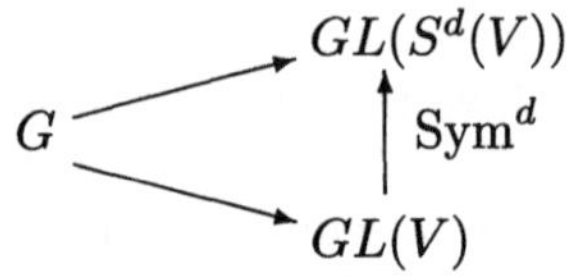

Let now $T \in GL_n(\mathbb{Q})$ be the upper triangular matrix given by $T_{ij} = 1$ for $i \leq j$ and $\Delta_d \in GL_{N_d}(\mathbb{Q})$ be the diagonal matrix whose diagonal is the first row of $\mathrm{Sym}^d(T)$. One can check that the diagonal element of Δ_d corresponding to the monomial $Y_1^{e_1} \dots Y_n^{e_n}$ is $d!/\prod_{i=1}^n e_i!$. Let $\mathcal{Y} = (y_1, \dots, y_n)$ be a basis for V and $W \in GL_n(K)$ be the corresponding Wronskian matrix. Using the notation v^e for the e-fold tensor $v \otimes \dots \otimes v$ of $S(V)$, let $\mathcal{Y}_d$ be the basis $(y_1^d, y_1^{d-1} \otimes y_2, y_1^{d-1} \otimes y_3, \dots, y_n^d)$ of $S^d(V)$ where the symmetric tensors are ordered lexicographically with $y_1 < y_2 < \dots < y_d$. The *injective* morphism used by van Hoeij and Weil is then $\lambda_d \colon S^d(V) \to K^{N_d}$ given by $\lambda_d(w) = \mathrm{Sym}^d(W)\Delta_d^{-1}\overline{w}$ where $\overline{w} \in C^{N_d}$ is the vector of coefficients of w with respect to $\mathcal{Y}_d$. If G acts pointwise on the coordinates in K^{N_d}, then they prove in [13] that λ_d commutes with the actions of G on $S^d(V)$ and K^{N_d}, hence that $\lambda_d\left(S^d(V)^G\right) \subseteq {K^{N_d}}^G$. As earlier, the normality of the Picard–Vessiot extension implies that ${K^{N_d}}^G = k^{N_d}$, hence that $I \in S^d(V)^G \implies \lambda_d(I) \in k^{N_d}$. They then define the *dth symmetric power system* of L to be the differential system $Z' = S^d(L)Z$ where $S^d(L)$ is the $N_d \times N_d$ matrix with entries in k given by $\sigma_d(\mathcal{Y}^T)' = S^d(L)\sigma_d(\mathcal{Y}^T)$. The entries of $S^d(L)$ can be easily computed from the coefficients of L [13] and they prove that the solution space of $Z' = S^d(L)Z$ is generated by the columns of $\mathrm{Sym}^d(W)$, which implies that the solution space is exactly $\lambda_d(S^d(V))$. Therefore,

$$I \in S^d(V)^G \setminus \{0\} \iff \lambda_d(I) \in k^{N_d} \setminus \{0\} \text{ and } \lambda_d(I)' = S^d(L)\lambda_d(I)\,. \qquad (3)$$

Their algorithm, which is applicable when k is the rational function field $C(x)$ with derivation d/dx, proceeds as follows: compute first from the Newton polygons of L at all its poles denominators of the entries of a solution in $C(x)^{N_d}$ of $Z' = S^d(L)Z$ as well as bounds on the degrees of the corresponding numerators. Compute then a local fundamental solution matrix $\widehat{W}$ of $Ly = 0$ around some point $x_0 \in \mathbb{P}^1(\overline{C})$. Let $\overline{I}$ be a vector of N_d indeterminates, then $F = \mathrm{Sym}^d(\widehat{W})\Delta_d^{-1}\overline{I}$ is a local formal solution of $Z' = S^d(L)Z$ around x_0. Multiply each row by its corresponding denominator and reduce the remaining series modulo x^{n_i+1} where n_i is the corresponding degree bound. This yields a vector Z of rational functions in x where the indeterminates of $\overline{I}$ appear linearly. Equating Z' with $S^d(L)Z$ yields a linear algebraic system of the form $M\overline{I} = 0$ whose kernel is isomorphic to $S^d(V)^G$ ([13] contains several heuristics to reduce the number of unknowns, down to the dimension of $S^d(V)^G$ in many cases).

Their algorithm is quite effective for operators of order 3 or more when the coefficient field is of the form $(C(x), d/dx)$. Furthermore, once they find an invariant (or semi-invariant) I that factors linearly, the entries of $\lambda_d(I)$ can be used to produce a polynomial $P \in C[x,u]$ such that $P(x,u) = 0 \implies Le^{\int u dx} = 0$ [12].

In the case of more general coefficient fields, even though $\widehat{W}$ can still be computed, it is not known how to convert Taylor series to elements of k, so the above method is not applicable. However, we can invert λ_d without knowing a fundamental solution matrix, so we can still use the symmetric power system to compute invariants:

Theorem 3.3. *Suppose that k is a finitely generated differential extension of $\mathbb{Q}$ and let $F \in k^{N_d}$ be a nonzero solution of $Z' = S^d(L)Z$. Then, for any $M \in GL_n(\mathbb{C})$ and any ordinary point x_0 of L, $\Delta_d \operatorname{Sym}^d(M)F(x_0)$ is the coefficient vector of an invariant of G with respect to a basis that depends on M. In particular $\Delta_d F(x_0)$ is an invariant of G with respect to the basis $(y_1, \dots, y_n)$ satisfying $y_j^{(i-1)}(x_0) = \delta_{ij}$.*

Proof. By the classical existence and uniqueness theorems [14], there exists for each $1 \le j \le n$ a unique solution y_j of $Ly = 0$, analytic at x_0 and satisfying $y_j^{(i-1)}(x_0) = M_{ij}^{-1}$ for $1 \le i \le n$. Let W be the Wronskian matrix of $y_1, \dots, y_n$. Since $W(x_0) = M^{-1}$ is invertible, W is invertible so it is a fundamental matrix for L. Since the solution space of $Z' = S^d(L)Z$ is $\lambda_d(S^d(V))$, it follows from (3) that $F = \lambda_d(I)$ for some invariant $I \in S^d(V)^G$. Writing $\overline{I}$ for the coefficients of I, we have $F = \operatorname{Sym}^d(W)\Delta_d^{-1}\overline{I}$. Since the entries of W are analytic at x_0, the entries of $\operatorname{Sym}^d(W)$ are also analytic at x_0, so evaluating at x_0 yields

$$F(x_0) = \operatorname{Sym}^d(W)(x_0)\Delta_d^{-1}\overline{I} = \operatorname{Sym}^d(W(x_0))\Delta_d^{-1}\overline{I} = \operatorname{Sym}^d(M)^{-1}\Delta_d^{-1}\overline{I}$$

which implies that $\overline{I} = \Delta_d \operatorname{Sym}^d(M)F(x_0)$. The last sentence of the theorem follows from taking M to be the identity matrix in the proof. □

Using cyclic vectors [9], we can compute the solutions in k^{N_d} of symmetric power systems for a large class of fields [4, 6, 26]. Theorem 3.3 then implies that $\Delta_d F(x_0)$ is an invariant of G for any solution, and this yields an alternative to the algorithm of Singer & Ulmer that does not require ψ_d to be injective. Prototypes implementations of those algorithms exist both in the MAPLE system and the BERNINA[2] server.

Example 3.4. *Consider the same equation as in Example 3.2. Its second symmetric power system is*

$$Z' = \begin{pmatrix} 0 & 2 & 0 \\ 36x^4e^{4x^3} + 9x^4 + 2x^{-2} & 0 & 1 \\ 0 & 72x^4e^{4x^3} + 18x^4 + 4x^{-2} & 0 \end{pmatrix} Z$$

[2] `http://www.inria.fr/cafe/Manuel.Bronstein/sumit/bernina.html.`

whose solution space in k^3 *is generated by*

$$F = (x^{-2}t^{-2}, -(1+3x^3)x^{-3}t^{-2}, (1+6x^3+9x^6-36x^6t^4)x^{-4}t^{-2})^T. \quad (4)$$

We have $\Delta_3 F(1) = (e^{-2}, -8e^{-2}, 16e^{-2} - 36e^2)^T$, *so we get the same invariant as in Example 3.2.*

Once an invariant that factors linearly has been found, a Liouvillian solution of the equation can be computed: using $\sigma_d((Z_1,\dots,Z_n)^T)$ as a basis of the ring $k[Z_1,\dots,Z_n]_d$ of homogeneous polynomials of degree d with coefficients in k, we can identify k^{N_d} with $k[Z_1,\dots,Z_n]_d$. Under this identification, Theorem 2.1 of [12] implies that if $F \in k^{N_d}$ is a nonzero solution of $Z' = S^d(L)Z$, then $Q = (\Delta_d F)(U,-1,0,\dots,0) \in k[U]$ has degree d and is such that $Q(u) = 0$ implies that $L(e^{\int u}) = 0$. Such a polynomial can also be computed from a semi-invariant that factors into linear forms [12].

Example 3.5. *Consider the same equation as in Examples 3.2 and 3.4. From the solution (4) of its second symmetric power system, we get*

$$Q = (\Delta_3 F)(U,-1) = x^{-4}t^{-2}\left(x^2U^2 + (2x+6x^4)U + (1+6x^3+9x^6-36x^6t^4)\right)$$

whose roots are $u = -(1+3x^3 \pm 6x^3t^2)/x$. *It follows that a basis of the solution space of* $Ly = 0$ *is* $x^{-1}e^{-x^3 \pm e^{2x^3}}$.

We conclude this section with a discussion of when it is sufficient to look for invariants rather than semi-invariants. The results in the rest of this section are all from M. Singer (personal communications). Lemma 3.6 was independently discovered by D. Boucher and appears as a non-integrability criterion in [3].

Lemma 3.6. *If* G *is reductive, then every connected solvable normal subgroup of* G *is diagonalizable.*

Proof. Every group is diagonal for $n = 1$, so suppose that $n > 1$ and that the lemma holds for $1 \le m < n$, and let H be a connected solvable normal subgroup of G. H is triangularizable by the Lie–Kolchin Theorem [16, 18], so it has a semi-invariant $v \in V$. Let $g \in G$ and $h \in H$ be arbitrary. Since H is normal in G, $hg = gh'$ for some $h' \in H$. Then, $h(g(v)) = g(h'(v)) = g(c_{h'}v) = c_{h'}g(v)$ where $c_{h'} \in C$. This implies that $g(v)$ is a semi-invariant of H for every $g \in G$. Let $(g_1(v),\dots,g_m(v))$ be a basis for V', the C-span of Gv. If $m = n$, then H is diagonal with respect to that basis. Otherwise, since G is reductive, $V = V' \oplus W$ where $GW \subseteq W$. Let $(w_1,\dots,w_{n-m})$ be a basis of W. Relative to the basis $(g_1(v),\dots,g_m(v),w_1,\dots,w_{n-m})$ of V, every $g \in G$ has the block-diagonal form

$$g = \begin{pmatrix} A_g & \\ & B_g \end{pmatrix}$$

where $A_g \in GL_m(C)$ and $B_g \in GL_{n-m}(C)$. The map $g \to B_g$ is a rational morphism from G into $GL_{n-m}(C)$. It is continuous in the Zariski topology [16,

Lemma 4.3], so the image of H is connected and solvable. By our induction hypothesis, there is a basis $(b_1, \ldots, b_{n-m})$ of W which diagonalizes the image of H. H is then diagonal with respect to the basis $(g_1(v), \ldots, g_m(v), b_1, \ldots, b_{n-m})$. □

Lemma 3.7. *If G has a diagonalizable unimodular normal subgroup of finite index, then G has an invariant $I \in S(V)$ that factors into linear forms.*

Proof. Let H be a diagonalizable unimodular normal subgroup of G of finite index, $(y_1, \ldots, y_n)$ be a basis of V that diagonalizes H and $z = y_1 \otimes \cdots \otimes y_n$. Then, for each $h \in H$, $h(z) = h(y_1) \otimes \cdots \otimes h(y_n) = \det(h)z = z$ since H is unimodular. Let $g_1H, \ldots, g_mH$ be the distinct cosets of H in G. Any $g \in G$ can be written as $g = g_jh$ for some $h \in H$, which implies that $g(z) = g_j(h(z)) = g_j(z)$. Therefore, $Gz \subseteq \{g_1(z), \ldots, g_m(z)\}$ is finite, so $Gz = \{f_1(z), \ldots, f_s(z)\}$ where $s \leq m$ and each f_j is in G. Then $I = f_1(z) \otimes \cdots \otimes f_s(z)$ factors into linear forms and since each $g \in G$ acts as a permutation on Gz, $g(I) = I$ so I is an invariant of G. □

Since G^0 is always a connected normal subgroup of G of finite index, combining Theorem 2.2 with Lemmas 3.6 and 3.7 we get:

Theorem 3.8. *If G is reductive, G^0 is unimodular and all the solutions of $Ly = 0$ are Liouvillian over k, then G has an invariant that factors into linear forms.*

Remark that $L = \text{llcm}(D - x^2, D^2 + x^2D - 1)$ has a reductive unimodular Galois group, as well as the Liouvillian solution $e^{\int x^2 dx}$, but no invariant, so the above theorem does not hold when some solutions are Liouvillian and some not. However, having one Liouvillian solution is equivalent to all solutions being Liouvillian in the case of irreducible equations, so a consequence of Theorem 3.8 is that an irreducible equation with a unimodular G^0 has a Liouvillian solution if and only if G has an invariant that factors into linear forms. If we can find rational and exponential solutions over k, then we can factor operators over k into irreducibles [8, 11, 27], which ensures that we only solve operators with irreducible Galois groups. In addition, the transformation $z = ye^{\int a_{n-1}/na_n}$ where $L = a_nD^n + a_{n-1}D^{n-1} + \cdots + a_0$ transforms $Ly = 0$ into $\bar{L}z = 0$ where $\bar{L}$ has order n and a unimodular Galois group, which remains irreducible if L was irreducible. In addition to factoring, it is thus sufficient to look for homogeneous invariants of prescribed degrees in order to check for Liouvillian solutions (although it may be preferable in some cases to look for semi-invariants of lower degrees).

4. Difference Equations

We describe in this section the algorithm of [10] for computing all the Liouvillian solutions of ordinary linear difference equations with polynomial coefficients. Based on Theorem 2.4, this algorithm looks for solutions of $Ly = 0$ that are interlacings of m hypergeometric sequences for $1 \leq m \leq n$.

Definition 4.1. *Let F be a field, $F[t;\sigma,\delta]$ be a skew-polynomial ring over F and $p \in F[t;\sigma,\delta]\setminus\{0\}$. For any $m > 0$, the* least m-sparse left multiple of p *is a monic $\overleftarrow{p^m} \in F[t;\sigma,\delta]$ of minimal degree such that $\overleftarrow{p^m} = qp$ for some $q \in F[t;\sigma,\delta]$ and such that only powers of t^m appear in $\overleftarrow{p^m}$.*

Such multiples can be computed by linear algebra over F: let θ be the image of t^m in $V = F[t;\sigma,\delta]/F[t;\sigma,\delta]p$, which is a vector space of dimension $n = \deg(p)$ over F, and N be the largest integer such that $\theta^0, \theta^1, \dots, \theta^{N-1}$ are linearly independent over F (note that $1 \le N \le n$). Then $\theta^N = \sum_{i=0}^{N-1} c_i\theta^i$ for some $c_0, \dots, c_{N-1} \in F$ and $\overleftarrow{p^m} = t^{Nm} - \sum_{i=0}^{N-1} c_i t^{im}$ is the desired multiple.

Let now $L = \sum_{i=0}^{n} a_i E^i$ be a linear ordinary difference operator with coefficients in (k,σ) where $k = C(x)$ and $\sigma x = x+1$. Let $\overleftarrow{L^m} = E^{Nm} + \sum_{i=0}^{N-1} b_i E^{im}$ be its least m-sparse left multiple and for $1 \le j \le m$ let

$$\overleftarrow{L_j{}^m} = E^N + \sum_{i=0}^{N-1} \left(\tau_m \sigma^{j-1} b_i\right) E^i$$

where $\tau_m \colon C(x) \to C(x)$ is the automorphism over C mapping x to mx.

Theorem 4.2. [10] *For any $a^{(1)}, \dots, a^{(m)} \in \mathcal{S}$,*

$$L(\biguplus_{j=1}^{m} a^{(j)}) = 0 \implies \overleftarrow{L_j{}^m}\left(a^{(j)}\right) = 0 \text{ for } 1 \le j \le m.$$

Proof. Let $a = \biguplus_{j=1}^{m} a^{(j)}$. Since $La = 0$ and $\overleftarrow{L^m}$ is a left-multiple of L, $\overleftarrow{L^m} a = 0$, so $a_{s+Nm} + \sum_{i=0}^{N-1} b_i(s) a_{s+im} = 0$ for all $s \ge 0$. For $1 \le j \le m$ and any $t \ge 0$, applying that equality to $s = tm + j - 1$ and remarking that $a_{s+im} = a_{(t+i)m+j-1} = a_{t+i}^{(j)}$, we obtain $a_{t+N}^{(j)} + \sum_{i=0}^{N-1} b_i(tm+j-1) a_{t+i}^{(j)} = 0$ for all $t \ge 0$. □

We therefore need to look for all the hypergeometric solutions of $\overleftarrow{L_j{}^m}$ for $1 \le j \le m$. We can use the algorithm `Hyper` [20, 21], which, given a linear ordinary difference operator R with coefficients in $C(x)$ returns $\gamma_1, \dots, \gamma_t \in C(x)^*$ and finitely many polynomials $p_{rs} \in C[x] \setminus \{0\}$ such that the hypergeometric solutions of $Ry = 0$ are exactly all the sequences satisfying

$$\frac{\sigma y}{y} = \gamma_r \frac{\sum_s c_{rs}\sigma p_{rs}}{\sum_s c_{rs} p_{rs}} \tag{5}$$

for some r, where the c_{rs}s are arbitrary constants. Let now $1 \le j \le m$ be given and let $\mathcal{H}_j$ be $\{0\}$ if $\overleftarrow{L_j{}^m}$ has no nonzero hypergeometric solution, the finite set of all the $\gamma_r \sigma(p_{rs})/p_{rs}$ for all the γ_r and p_{rs} returned by `Hyper` on $\overleftarrow{L_j{}^m}$ otherwise. For each $h \in \mathcal{H}_j$, let z_h be 0 if $h = 0$, a nonzero hypergeometric solution of $\sigma y = hy$ otherwise. Then, (1) implies that $\sigma^m(\overrightarrow{z_h{}^m}) = \overrightarrow{h^m}\,\overrightarrow{z_h{}^m}$, hence that $(E^m - \tau_m^{-1}h)\overrightarrow{z_h{}^m} = 0$, which implies in turn that $(E^m - \sigma^{1-j}\tau_m^{-1}h)\sigma^{1-j}(\overrightarrow{z_h{}^m}) = 0$. Let now $a^{(j)}$ be any hypergeometric solution of $L_j^{\overleftarrow{m}} y = 0$. It follows from (5) that $a^{(j)} = \sum_{h \in \mathcal{H}_j} c_h z_h$ for some constants $c_h \in C$, hence, by linearity of all the operations involved, that $\uplus_{j=1}^m a^{(j)} = \sum_{j=1}^m \sum_{h \in \mathcal{H}_j} c_h \sigma^{1-j}(z_h{}^{\vec{m}})$. Therefore, $L^{\vec{m}}(\uplus_{j=1}^m a^{(j)}) = 0$ where

$$L^{\vec{m}} = \operatorname*{llcm}_{\substack{1 \le j \le m \\ h \in \mathcal{H}_j}} \left(E^m - \sigma^{1-j} \tau_m^{-1} h\right).$$

Since $L^{\vec{m}}$ annihilates all the interlacings of hypergeometric solutions of $L_1^{\overleftarrow{m}}, \ldots, L_m^{\overleftarrow{m}}$, Theorems 2.4 and 4.2 imply that $Ly = 0$ has a Liouvillian solution if and only if $\operatorname{gcrd}(L, L^{\vec{m}})$ is nontrivial for some $1 \le m \le n$. The Hendriks–Singer algorithm proceeds as follows: if $\operatorname{gcrd}(L, L^{\vec{m}})$ has degree 0 for all $1 \le m \le n$, then $Ly = 0$ has no Liouvillian solution. Otherwise, let $m \ge 1$ be the smallest integer such that $R = \operatorname{gcrd}(L, L^{\vec{m}})$ is nontrivial. Then, $L^{\vec{m}} = QR$ for some operator Q of degree $q \ge 0$. Since $\{\uplus_{j=1}^m z_{h_j}, (h_1, \ldots, h_m) \in \mathcal{H}_1 \times \cdots \times \mathcal{H}_m\}$ generates the solution space of $L^{\vec{m}}$, a basis $(y^{(1)}, \ldots, y^{(t)})$ of that space can be extracted from that set. If $q = 0$, then $(y^{(1)}, \ldots, y^{(t)})$ is a basis of the solution space of R. Otherwise, let $g^{(i)} = Ry^{(i)}$ for $1 \le i \le t$, $N \ge 0$ be an integer such $L^{\vec{m}}, Q$ and R have no singularities at $x = N + s$ for any integer $s \ge 0$, and M be the $(q+1) \times t$ matrix given by $M_{ij} = g_{N+i-1}^{(j)}$ for $1 \le i \le q+1$ and $1 \le j \le t$. Since $Qg^{(i)} = 0$ for each i and $\deg(Q) = q$, the map $(c_1, \ldots, c_t)^T \to \sum_{i=1}^t c_i y^{(i)}$ is an isomorphism between the kernel of M and the solution space of R, so we obtain a basis of the solution space of R composed of interlacings of hypergeometric sequences, and those are Liouvillian solutions of $Ly = 0$. Writing $L = \tilde{L}R$ and applying the algorithm recursively to $\tilde{L}$ eventually yields a basis of all the Liouvillian solutions of $Ly = 0$.

A prototype implementation of the above algorithm is in the SHASTA[3] server. While the above algorithm is complete, remark that it really needs to compute the hypergeometric solutions of the $L_j^{\overleftarrow{m}}$'s over the algebraic closure of the constant field C. A rational alternative based on eigenring computations is developed [2], which computes only in the constant field C of the equation to solve, even if it is not algebraically closed.

Acknowledgements

I would like to thank R. Bomboy, M. Singer, F. Ulmer and J.-A. Weil for their many useful comments and remarks during the preparation of this paper, as well as Prof. W. Decker for inviting me to present this survey at 3ecm.

[3] `http://www.inria.fr/cafe/Manuel.Bronstein/sumit/shasta.html`.

References

[1] S. A. Abramov, M. Bronstein, and M. Petkovšek. On polynomial solutions of linear operator equations. In *Proceedings of ISSAC'95*, pages 290–296. ACM Press, 1995.

[2] R. Bomboy. *Solutions Liouvilliennes des équations aux différences finies linéaires.* Thèse de mathématiques, Université de Nice, 2001.

[3] D. Boucher. *Sur les équations différentielles paramétrées, une application aux systèmes hamiltoniens.* Thèse de mathématiques, Université de Limoges, 2000.

[4] M. Bronstein. On solutions of linear ordinary differential equations in their coefficient field. *J. Symbolic Computation*, 13(4):413–440, April 1992.

[5] M. Bronstein. On solutions of linear ordinary difference equations in their coefficient field. *Journal of Symbolic Computation*, 29(6):841–877, June 2000.

[6] M. Bronstein and A. Fredet. Solving linear ordinary differential equations over $C(x, e^{\int f(x)dx})$. In S. Dooley, editor, *Proceedings of ISSAC'99*, pages 173–179. ACM Press, 1999.

[7] M. Bronstein, T. Mulders, and J.-A. Weil. On symmetric powers of differential operators. In *Proceedings of ISSAC'97*, pages 156–163. ACM Press, 1997.

[8] M. Bronstein and M. Petkovšek. An introduction to pseudo-linear algebra. *Theoretical Computer Science*, 157:3–33, 1996.

[9] F. T. Cope. Formal solutions of irregular linear differential equations II, *American Journal of Mathematics*, 58:130–140, 1936.

[10] P. A. Hendriks and M. F. Singer. Solving difference equations in finite terms. *J. Symbolic Computation*, 27(3):239–260, March 1999.

[11] M. van Hoeij. Factorization of differential operators with rational functions coefficients. *J. Symbolic Computation*, 24(5):537–562, November 1997.

[12] M. Van Hoeij, J.-F. Ragot, F. Ulmer, and J.-A. Weil. Liouvillian solutions of linear differential equations of order three and higher. *J. Symbolic Computation*, 28(4 and 5):589–610, October/November 1999.

[13] M. van Hoeij and J.-A. Weil. An algorithm for computing invariants of differential galois groups. *Journal of Pure and Applied Algebra*, 117 & 118:353–379, 1997.

[14] E. L. Ince. *Ordinary Differential Equations.* Dover Publications Inc., 1956.

[15] C. Jordan. Mémoire sur les équations différentielles linéaires à intégrale algébrique. *Journal für Mathematik*, 84:89–215, 1878.

[16] Irving Kaplansky. *An introduction to differential algebra.* Hermann, Paris, 1957.

[17] E. R. Kolchin. Algebraic matrix groups and the Picard–Vessiot theory of homogeneous linear ordinary differential equations. *Annals of Mathematics*, 49:1–42, 1948.

[18] E. R. Kolchin. *Differential algebra and algebraic groups.* Academic Press, New York and London, 1973.

[19] O. Ore. Theory of non-commutative polynomials. *Annals of Mathematics*, 34:480–508, 1933.

[20] M. Petkovšek. Hypergeometric solutions of linear recurrences with polynomial coefficients. *J. Symbolic Computation*, 14(2 and 3):243–264, August&September 1992.

[21] M. Petkovšek, H. S. Wilf, and D. Zeilberger. $A = B$. A. K. Peters, Wellesley, 1996.

[22] M. van der Put. Galois theory of differential equations, algebraic groups and Lie algebras. *J. Symbolic Computation*, 28(4 and 5):441–472, October/November 1999.

[23] A. Seidenberg. Abstract differential algebra and the analytic case. *Proceedings of the American Mathematical Society*, 9:159–164, 1958.

[24] A. Seidenberg. Abstract differential algebra and the analytic case II. *Proceedings of the American Mathematical Society*, 23:689–691, 1969.

[25] M. F. Singer. Liouvillian solutions of nth order homogeneous linear differential equations. *American Journal of Mathematics*, 103:661–682, 1981.

[26] M. F. Singer. Liouvillian solution of linear differential equations with Liouvillian coefficients. *J. Symbolic Computation*, 11(3):251–274, March 1991.

[27] M. F. Singer. Testing reducibility of linear differential operators: a group theoretic perspective. *Applicable Algebra in Engineering, Communication and Computing*, 7:77–104, 1996.

[28] M. F. Singer and F. Ulmer. Liouvillian and algebraic solutions of second and third order linear differential equations. *J. Symbolic Computation*, 16(1):37–74, July 1993.

[29] M. F. Singer and F. Ulmer. Linear differential equations and products of linear forms. *Journal of Pure and Applied Algebra*, 117 & 118:549–563, 1997.

[30] M. F. Singer and M. van der Put. *Galois Theory of Difference Equations*. LNM 1666. Springer, 1997.

[31] F. Ulmer. On Liouvillian solutions of linear differential equations. *Applicable Algebra in Engineering, Communication and Computing*, 2:171–193, 1992.

[32] F. Ulmer. Irreducible linear differential equations of prime order. *J. Symbolic Computation*, 18(4):385–401, October 1994.

[33] F. Ulmer and J.-A. Weil. Note on Kovacic's algorithm. *J. Symbolic Computation*, 22(2):179–200, August 1996.

INRIA
2004 route des Lucioles - B.P. 93
F-06902 Sophia Antipolis Cedex, France
E-mail address: `Manuel.Bronstein@sophia.inria.fr`

Some Introductory Remarks on Computer Algebra

Wolfram Decker

Abstract. Computer algebra is a relatively young but rapidly growing field. In this introductory note to the mini-symposium on computer algebra organized as part of the third European Congress of Mathematics, I will not even attempt to address all major streams of research and the many applications of computer algebra. I will concentrate on a few aspects, mostly from a mathematical point of view, and I will discuss a few typical applications in mathematics. I will present a couple of examples which underline the fact that computer algebra systems provide easy access to powerful computing tools. And, I will quote from and refer to a couple of survey papers, textbooks and web-pages which I recommend for further reading.

1. Some Historical Remarks

Most of mathematics is concerned at some level with setting up and solving equations, for example to model applications in science and engineering. In many cases this involves tedious computations which are difficult to get right or too extensive to be carried through by hand. Numerical analysis and, more recently, computer algebra originated from this problem. That mechanized computing is not restricted to numerical computation was already evident to Charles Babbage and Lady Ada Augusta in the 19th century (see [63] for the following quotations and for more on the story of these pioneers of computer science). Besides working on his Difference Engines (see [94]) Babbage spent many years of his life with designing a more universal mechanical machine (Analytical Engine) to be programmed with punch-cards invented by J.-M. Jacquard for the control of his automatic looms. Babbage's earliest ideas on how such a machine could also perform mechanized algebraic manipulations are documented in his notebook of July 1836 [5]:

> "This day I had for the first time a general but very indistinct conception of the possibility of making an engine work out algebraic developments – I mean without any reference to the value of the letters. My notion is that as the cards (Ja[c]quards) of the calc. engine direct a series of operations and then recommence with the first, so it might perhaps be possible to cause the same cards to punch others equivalent to any given

number of repetitions. But these hole[s] might perhaps be small pieces of formulae previously made by the first cards ... "

In Lady Ada's extensive notes accompanying her translation from French into English of L. F. Menabrea's 1842 paper on the Analytical Engine [70] we find, for example, the following passage:

"There are many ways in which it may be desired in special cases to distribute and keep separate the numerical values of different parts of an algebraic formula; and the power of effecting such distributions to any extent is essential to the algebraic character of the Analytical Engine. Many persons who are not conversant with mathematical studies imagine that because the business of the engine is to give its results in numerical notation, the nature of its process must consequently be arithmetical rather than algebraic and analytical. This is an error. The engine can arrange and combine its numerical quantities exactly as if they were letters or any other general symbols; and, in fact, it might bring out its results in algebraic notation were provisions made accordingly. It might develop three sets of results simultaneously, viz. symbolic results ... ; numerical results (its chief and primary object); and algebraic results in literal notation."

The Analytical Engine, however, was never built, and symbolic computations were not carried through in an automated way until electronic computers were available. The first documented computer algebra programs, written for analytic differentiation, were described in two 1953 master's theses [53, 73]. According to the historical account by K. Geddes, S. R. Czapor, and G. Labahn in their introductory textbook [42] the beginning 1960's saw increasing activities in the field, in particular, after the LISP language,

"a major advancement on the road to languages for symbolic computation"

had been developed. It was the period between 1961 and 1971 in which

"the field progressed from birth through adolescence to at least some level of maturity."

I would like to mark this progress by some celebrated algorithmic breakthroughs, by the first releases of some well-known LISP-based general purpose computer algebra systems (see [42, 99] for some important systems not mentioned here), and by the beginning organization of researchers in interest groups, most notably in the Special Interest Group on Symbolic and Algebraic Manipulation (SIGSAM) of the Association for Computing Machinery (ACM).

1965	B. Buchberger's algorithm for computing Gröbner bases [16, 17]
1966	SYMSAM'66. First SYMposium on Symbolic and Algebraic Manipulation by ACM [38]
1967, 1970	E. R. Berlekamp's algorithm for factoring univariate polynomials over finite fields [7, 8]
1967	First issue of the SIGSAM Bulletin
1968–1970	R. H. Risch's algorithm for the indefinite integration of elementary functions [81, 82, 83, 84]
1969	H. Zassenhaus' algorithm for factoring univariate polynomials over the integers [100]
1968, 1970	First releases of REDUCE respectively REDUCE2 (A. C. Hearn)
1970	First release of MACSYMA (J. Moses, W. Martin)
1971	First release of SCRATCHPAD (J. Griesmer, R. Jenks)
1971	SYMSAM'71 [76]

A good impression of the speed with which computer algebra has been evolving since its beginnings can be obtained by checking

- the home-pages of its main interest groups and its leading researchers,
- recent textbooks such as [40, 90] (with extensive bibliographies),
- the Journal of Symbolic Computation and other journals such as Mathematics of Computation and Applicable Algebra in Engineering, Communication and Computing, and
- the proceedings of the leading conferences of computer algebra.

SYMSAM'66 was followed by several North American and European conferences which have been amalgamated into the annual International Symposium on Symbolic and Algebraic Computation (ISSAC) since 1988. ISSAC2000, chaired by T. Recio, is the 25th edition in this series. There are various other series of conferences either being organized by continental, national or local interest groups, or gathering together users of one of the computer algebra systems, or being devoted to special features such as COCOA (COmputational COmmutative Algebra), DISCO (Design and Implementation of Symbolic Computation Systems), MEGA (Effective Methods in Algebraic Geometry), or PASCO (PArallel Symbolic COmputation). See [99] for a much larger list.

2. Some Features of Computer Algebra

Calculations in computer algebra are carried through exactly, that is, no approximation is applied at any step. Infinite precision arithmetic allows us to actually compute in the *ring* of integers and in the *field* of rationals, in finite prime fields, algebraic number fields, and in arbitrary Galois fields. In fact, there is a much larger variety of algebraic structures in which algebraic algorithms allow us to manipulate algebraic objects or the structures itself. And, exact computer algebra methods allow us to create algorithms that decide, for example, the solvability of systems of polynomial equations or the solvability of indefinite summation and integration problems in certain specified classes of functions. Let us look at a few examples and introduce some computer algebra systems as well (the home-pages of these and other systems can be easily found on the net).

Example 2.1. *GAP (Groups, Algorithms and Programming), until recently developed under the direction of J. Neubüser, is a system "for computational discrete algebra with particular emphasis on computational group theory." In the following GAP4 session we define the Heisenberg group H_3 of level 3 in its Schrödinger representation, check that H_3 is a finite group of order 27, and compute its character table. Here Schrödinger representation means that $H_3 \subset \mathrm{GL}(3, \mathbf{Q}(\xi))$, $\xi = e^{\frac{2\pi i}{3}}$ a primitive 3rd root of unity, is given as the matrix group generated by*

$$\sigma = \begin{pmatrix} 0 & 0 & 1 \\ 1 & 0 & 0 \\ 0 & 1 & 0 \end{pmatrix} \quad \textit{and} \quad \tau = \begin{pmatrix} 1 & 0 & 0 \\ 0 & \xi & 0 \\ 0 & 0 & \xi^2 \end{pmatrix}.$$

```
gap> m1:=[[0,0,1],[1,0,0],[0,1,0]];;
gap> m2:=[[1,0,0],[0,E(3),0],[0,0,E(3)^2]];;
gap> G:=Group(m1,m2);;
gap> Size(G);
27
gap> Display(CharacterTable(G));
CT1

      3  3  2  2  2  2  2  2  2  2  3  3

        1a 3a 3b 3c 3d 3e 3f 3g 3h 3i 3j

X.1      1  1  1  1  1  1  1  1  1  1  1
X.2      1  A  A  A /A /A /A  1  1  1  1
X.3      1 /A /A /A  A  A  A  1  1  1  1
X.4      1  1 /A  A  1 /A  A /A  A  1  1
X.5      1  A  1 /A /A  A  1 /A  A  1  1
X.6      1 /A  A  1  A  1 /A /A  A  1  1
X.7      1  1  A /A  1  A /A  A /A  1  1
X.8      1  A /A  1 /A  1  A  A /A  1  1
X.9      1 /A  1  A  A /A  1  A /A  1  1
X.10     3  .  .  .  .  .  .  .  .  B /B
X.11     3  .  .  .  .  .  .  .  . /B  B

A = E(3)
  = (-1+ER(-3))/2 = b3
B = 3*E(3)^2
  = (-3-3*ER(-3))/2 = -3-3b3
```

Via the online GAP user manual we easily find out how to read the output.

Example 2.2. *MAGMA, developed under the direction of J. Cannon, is a computer algebra system "for Algebra, Number Theory and Geometry". In the following MAGMA session we define again H_3 in its Schrödinger representation. Then H_3, as a subgroup of* $\mathrm{GL}(3, \mathbf{Q}(\xi))$, *acts on the polynomial ring* $\mathbf{Q}(\xi)[x_1, x_2, x_3]$ *by linear substitution. All polynomials which are invariant under this action form the invariant ring* $\mathbf{Q}(\xi)[x_1, x_2, x_3]^{H_3}$. *We compute a corresponding fundamental system of invariants, that is, a minimal, finite set of homogeneous invariants generating* $\mathbf{Q}(\xi)[x_1, x_2, x_3]^{H_3}$ *as a ring.*

```
> K<g>:=NumberField(x^2+x+1);
> m1:=[0,0,1,1,0,0,0,1,0];
> m2:=[1,0,0,0,g,0,0,0,g^2];
> G:=MatrixGroup<3,K|m1,m2>;
> R:=InvariantRing(G);
> FundamentalInvariants(R);

[

    x1^3 + x2^3 + x3^3,

    x1*x2*x3,

    x1^6 + x2^6 + x3^6,

    x1^6*x3^3 + x1^3*x2^6 + x2^3*x3^6

]
```

Example 2.3. *NTL, by V. Shoup, "is a high-performance, portable C++ library providing data structures and algorithms for manipulating signed, arbitrary length integers, and for vectors, matrices, and polynomials over the integers and over finite fields." The following is taken from the online Tour of NTL.*

"NTL provides extensive support for very fast polynomial arithmetic. In fact, this was the main motivation for creating NTL in the first place, because existing computer algebra systems and software libraries had very slow polynomial arithmetic. The class ZZX represents univariate polynomials with integer coefficients. The following program reads a polynomial, factors it, and prints the factorization.

```
#include <NTL/ZZXFactoring.h>
main() {
   ZZX f;

   cin >> f;
```

```
   vec_pair_ZZX_long factors;
   ZZ c;

   factor(c, factors, f);

   cout << c << "\n";
   cout << factors << "\n";
}
```

When this program is compiled and run on input

```
[2 10 14 6]
```

which represents the polynomial $2 + 10 * X + 14 * x^2 + 6 * X^3$, the output is

```
2
[[[1 3] 1] [[1 1] 2]]
```

The first line of output is the content of the polynomial, which is 2 in this case as each coefficient of the input polynomial is divisible by 2. The second line is a vector of pairs, the first member of each pair is an irreducible factor of the input, and the second is the exponent to which is appears in the factorization. Thus, all of the above simply means that

```
2 + 10*X + 14*x^2 +6*X^3 =2 * (1 + 3*X) * (1 + X)^2
```

Admittedly, I/O in NTL is not exactly user friendly, but then NTL has no pretensions about being an interactive computer algebra system: it is a library for programmers."

Example 2.4. *MAPLE, founded by G. Gonnet and K. Geddes and firstly released in 1983, is a general purpose system with a kernel written in C, and with most higher level functions or packages written in the MAPLE user language. In the following MAPLE session we factor a multivariate polynomial over the integers, namely*

$$p = x^2y^4z - xy^9z^2 + xyz^3 + 2x - y^6z^4 - 2y^5z \in \mathbf{Z}[x, y, z].$$

```
> p:=x^2*y^4*z - x*y^9*z^2 + x*y*z^3 + 2*x - y^6*z^4 - 2*y^5*z:
> factor(p);
                          4        3             5
                        (y  z x + y z  + 2) (x - y  z)
```

Example 2.5. *SCHUBERT, written by S. Katz and S. A. Strømme, is a "MAPLE package for Intersection Theory." Intersection rings of algebraic varieties are represented by generators and relations, that is, as residue class rings of polynomial rings. The following two SCHUBERT sessions, taken from the SCHUBERT manual, compute the number of lines respectively conics on a general quintic hypersurface in complex projective 4-space (see also* [58]*).*

```
# Lines on a quintic threefold.  This is the top Chern class of the
# 5th symmetric power of the universal quotient bundle on the
# Grassmannian of lines.
>
> grass(2,5,c):          # Lines in P^4.
> B:=symm(5,Qc):         # Qc is the rank 2 quotient bundle, B its
>                        # 5th symmetric power.
> c6:=chern(rank(B),B):     # the 6th Chern class of this
                            # rank 6 bundle.
> integral(c6);
                                    2875

#-----------------------------------------------------------------

# Conics on a quintic threefold. This is the top Chern class of the
# quotient of the 5th symmetric power of the universal quotient on
# the Grassmannian of 2 planes in P^5 by the subbundle of quintics
# containing the tautological conic over the moduli space of
# conics.
>
> grass(3,5,c):           # 2-planes in P^4.
> B:=Symm(2,Qc):          # The bundle of conics in the 2-plane.
> Proj(X,dual(B),z):      # X is the projective bundle of all conics.
> A:=Symm(5,Qc)-Symm(3,Qc)&*o(-z):       # The rank 11 bundle of
>                                        # quintics restricted to the
                                         # universal conic.
> c11:=chern(rank(A),A):  # its top Chern class.
> lowerstar(X,c11):       # push down to G(3,5).
> integral(Gc,");         # and integrate there.
                                   609250
```

Example 2.6. *SINGULAR, developed under the direction of G.-M. Greuel, G. Pfister, and H. Schönemann, "is a Computer Algebra System for polynomial computations with special emphasis on the needs of commutative algebra, algebraic geometry, and singularity theory." In the following SINGULAR session we check that the system of polynomial equations*

$$-x^2+xy-xz+y+1 = x^2+xy-y^2-2 = -x^3-x^2y+xyz+xz^2-xy-y$$
$$= -x^3+xy^2+x^2z+xyz+xz^2-z^2 = 0$$

has no solution over the complex numbers. Indeed, by Hilbert's Nullstellensatz, it suffices to compute that the reduced Gröbner basis of the ideal $I \subset \mathbf{Q}[x,y,z]$ defined by the polynomials is 1.

```
> ring r=0,(x,y,z),dp;
> ideal i1=-x2+xy-xz+y+1, x2+xy-y2-2;
```

```
> ideal i2=-x3-x2y+xyz+xz2-xy-y, -x3+xy2+x2z+xyz+xz2-z2;
> ideal i=i1+i2;
> std(i);
_[1]=1
```

For the very simple examples considered so far "almost no" computing time is needed. Just a little bit more involved is the following example.

Example 2.7. *LIDIA, developed under the direction of J. Buchmann, is "a C++ Library for Computational Number Theory". In the following LIDIA session we factor the 8th Fermat number* $F_8 = 2^{2^8} + 1$.

```
lc> factor(2^(2^8)+1);
$0 =  [(1238926361552897,1)
(93461639715357977769163558199606896584051237541638188580280321,1)]
>
```

We will discuss further examples later-on. Now, let us again quote Geddes, Czapor, and Labahn [42]:

> "There are three recognizable, yet interdependent, forces in the development of this field. We may classify them under the headings *systems*, *algorithms*, and *applications*."

People are concerned with developing (algebraic) algorithms and analyzing the complexity of algorithms, they provide the necessary surroundings for implementing algorithms and for allowing users to work with them, they implement algorithms and test their practical performance, and they apply algorithms. A more detailed impression is obtained by looking at the home-page of ISSAC2000 (which can be easily found on the net).

> "The conference topics include, but are not limited to:
> *Algorithmic mathematics*: Algebraic, symbolic, and symbolic-numeric algorithms including: simplification, polynomial and rational function manipulations, algebraic equations, summation and recurrence equations, integration and differential equations, linear algebra, number theory, group computations, and geometric computing.
> *Computer science*: Theoretical and practical problems in symbolic mathematical manipulation including: computer algebra systems, data structures, computational complexity, problem solving environments, programming languages and libraries for symbolic-numeric-geometric computation, user interfaces, visualization, software architectures, parallel or distributed computing, mapping algorithms to architectures, analysis and benchmarking, automatic differentiation and code generation, automatic theorem proving, mathematical data exchange protocols.

> *Applications*: Problem treatments incorporating algebraic, symbolic, symbolic-numeric and geometric computation in an essential or novel way, including engineering, economics and finance, architecture, physical and biological sciences, computer science, logic, mathematics, statistics, and use in education."

Let me add a couple of observations:.

- The different branches of computer algebra are cross-linked in numerous ways.
- There are increasing efforts to bring the symbolic and numerical communities together.
- There is an increasing use of computer algebra systems in education.

I would like to illustrate these observations by giving some examples. J. Cannon and D. Holt are the guest editors of the Special issue on computational algebra and number theory: Proceedings of the first MAGMA conference of the Journal of Symbolic Computation. Let me quote from their foreword [20]:

> " ... even if one wanted to, it would no longer be possible to treat a particular area of computational algebra, such as computational group theory, as an isolated and self-sufficient branch of mathematics. As the various different branches of computational algebra mature they are seen to rely on a common set of fundamental tools. To take a simple example, if we wish to compute chief factors of a group G, we may perhaps quickly find elementary abelian sections M/N of G (where M and N are normal in G). But we then need to refine the section, and to do this, we need to consider M/N as a module for G over a finite field, and to find a composite series of the module. Currently, the best method known of achieving this involves factorizing polynomials over finite fields, which is drawing us much closer to the realms of traditional symbolic computation. Another example concerns techniques for the efficient computation of Hermite and Smith normal forms for integral matrices (and the LLL algorithm) which are key tools in both computational group theory and algebraic number theory."

H. M. Möller [72] gives several examples of how to use Gröbner bases in numerical analysis. In his introduction we also find the following general remarks:

> " In the nineties, there is an increasing interest in combining symbolic and numerical methods. This can be seen at diverse instances. There are now international symposia supported by organizations from both sides, and the number of contributes displaying the symbolic-numerical interplay is increasing. Other examples are the facilities of using floating point arithmetics and simple numerical procedures in Computer Algebra Systems on the one hand and the (eventually partly) integration of Computer Algebra Systems into numerical packages on the other hand. The most prominent example is here the migration of the Computer Algebra System Axiom to NAG, the Numerical Algorithm Group.

Many interesting results have been obtained by combining symbolic and numerical methods, like in polynomial continuation the avoiding of solution paths diverging to infinity by means of concepts from toric ideals or like the numerical solving of systems of polynomial equations using resultants, see for instance Canny and Manocha (1993) [21]."

For more information in this direction I refer to the articles by G. Gonnet [43] and G.-M. Greuel [48] in these proceedings, to the Special Issues of the Journal of Symbolic Computation on Validated numerical methods and computer algebra [59] respectively Symbolic numeric algebra for polynomials [98], and to the homepage of CASC2000, the Third International Workshop on Computer Algebra in Scientific Computing.

The Special Issue [62] of the Journal of Symbolic Computation presents a couple of articles which deal with the use of computer algebra in courses on group theory, abstract algebra, computational non-associative algebra, differential equations and interpolation. The paper by B. Amrhein, O. Gloor and R. E. Maeder [2] in this special issue is concerned with Visualizations for mathematics courses based on a computer algebra system. Let me quote from the introduction:

> "Computer-algebra systems (CAS) provide the necessary algorithms needed to compute mathematics visualizations. Furthermore, in the last years, the graphics capabilities of computer-algebra systems have improved considerably and now satisfy the needs of visualization in education.
>
> CAS also give teachers and students another and more direct approach to using the computer. Applying a CAS, much less effort is needed to treat a simple practical problem than with the classical approach, learning a full programming language first. Therefore, the focus moves from computer handling to the application. This enables the possibility of applying the computer in education not as a teaching object but as a tool to solve problems in other disciplines. In addition, CAS became of increasing importance and are widely used in the industry. Hence, an introduction to CAS will soon belong to a modern curriculum."

There are in fact numerous textbooks and tutorials on using computer algebra in high-school and university courses.

Example 2.8. *The book by D. Cox, J. Little, and D. O'Shea* [29] *gives an introduction into both Gröbner bases and algebraic geometry, starting from scratch. This approach allows us to work out first examples in algebraic geometry without developing too much of the abstract theory behind algebraic geometry, that is, it allows us to introduce students into algebraic geometry at a very early stage of their studies. On the other hand, the geometric point of view nicely motivates the need for a variety of algorithms manipulating ideals in polynomial rings.*

3. Some Remarks on Systems and Algorithms

One reason for the great success of computer algebra is, as indicated several times above, that modern computer algebra systems (CAS) provide easy access to powerful computing tools. Nowadays there is a large variety of systems suiting different needs. In addition, a modern CAS comes with a programming language which allows us to extend the system (for some systems even the source code is available). Some of these extensions are publicly available, and, typically, a CAS which has been around for a little while is supplemented by a variety of user-written packages or libraries such as the MAPLE package SCHUBERT introduced in Example 2.5.

There are general purpose and special purpose CAS. Typically, one has to pay for a general purpose system whereas many of the special purpose systems are for free. From a general purpose system we expect that it provides tools for *symbolic computations*, for *numeric computations*, and for *visualization*, and we expect that there is a large variety of such tools allowing attacks on many different problems. Information on such systems can be found on the net and in numerous introductory textbooks. The practical guide [99] contains articles which compare the capabilities of systems such as REDUCE, MACSYMA, MAPLE, DERIVE, MATHEMATICA, MUPAD, AXIOM (the successor of SCRATCHPAD) and TI-92.

For many of the more special and advanced applications, general purpose systems are not powerful enough. Therefore a researcher with a desperate need for computational power in the context of certain problems actually may decide to create his own system. Some of todays' special purpose systems started out in this way. A pioneering and prominent example is SCHOONSCHIP by M. J. G. Veltman which helped to win a Nobel prize in physics in 1999 (awarded to Veltman and G. t'Hooft "for having placed particle physics theory on a firmer mathematical foundation", see http://nobelprizes.com/nobel/physics/physics.html). From a special purpose system we expect highly tuned implementations of the algorithms needed for the area in which the system is specializing. As pointed out in the last section this includes special algorithms such as those for computing fundamental systems of invariants in Example 2, and it includes algorithms which are common ground to every system such as algorithms for polynomial arithmetic. Again, information on such systems can be found on the net and in recent textbooks.

Most people who want to apply computer algebra are actually not interested in computer algebra itself. They will find "their" system(s) already on the market and they are going to use such a system as a kind of a black box machine without being forced to understand details of the algorithms or of their implementations. In some cases, however, information on which variant of an algorithm is implemented is necessary for understanding the meaning of the output (see the discussion of Risch's algorithm below). Unfortunately, detailed information is often difficult or even impossible to find. Setting a flag might allow one to obtain at least partial information.

Example 3.1. *Let us compute*

$$\sum_{0\leq k\leq m}(-1)^k\binom{n}{k}=(-1)^m\binom{n-1}{m}$$

with MAPLE V, Release 5.1.

```
> infolevel[sum]:=3:
> sum((-1)^k*binomial(n,k), k=0..m);
sum/indefnew:    indefinite summation
sum/extgosper:    applying Gosper algorithm to
                  a(   k   ):=   (-1)^k*binomial(n,k)
sum/gospernew:    a(   k   )/a(   k   -1):=    (-n+k-1)/k
sum/gospernew:    Gosper's algorithm applicable
sum/gospernew:    p:=   1
sum/gospernew:    q:=   -n+k-1
sum/gospernew:    r:=   k
sum/gospernew:    degreebound:=   0
sum/gospernew:    solving equations to find f
sum/gospernew:    Gosper's algorithm successful
sum/gospernew:    f:=   -1/n
sum/indefnew:     indefinite summation finished

                                 (m + 1)
                  (m + 1) (-1)            binomial(n, m + 1)
                - --------------------------------------
                                     n
```

Here we understand that there is an algorithm due to Gosper which does the job. Indeed, R. W. Gosper's algorithm [46] solves the indefinite summation problem for hypergeometric functions. This problem and Gosper's solution to it can be formulated rigorously by using difference fields (see, for example, [40]).

Differential fields provide the algebraic setting for indefinite symbolic integration. Symbolic integration has a long history starting with I. Newton and G. W. Leibniz, including work by J. Bernoulli in the 18th century, by J. Liouville, M. W. Ostrogradsky and E. Hermite in the 19th century, by J. F. Ritt in the middle of the 20th century, and stunning modern work subsumed under the name Risch's algorithm (see M. F. Singer's survey article [91] and M. Bronstein's book [14] for details and references). As is Gosper's algorithm for the summation of hypergeometric functions, Risch's algorithm is a decision algorithm for integrating elementary functions. Given such a function f, the algorithm computes an elementary function which is an anti-derivative of f or decides that no such anti-derivative exists. An algorithm for the case of transcendental elementary functions was given by R. H. Risch in [83]. Subsequent improvements are due to many people (see again [91]). An algorithm for the general case, which is much more involved than the algorithm for transcendental functions, was outlined by Risch in [81, 82, 84]. New ideas and improvements in the case of algebraic elementary functions are due to J. H. Davenport [30] and B. Trager [97]. M. Bronstein [13] generalized particularly Trager's ideas to give an algorithm for the general case. He implemented a great part of this algorithm in AXIOM, but to the best of my knowledge, no complete implementation exists (I refer to [13] and [24] for a discussion of some of the problems in this direction). As a consequence, if a CAS can't integrate a given elementary function, this might mean that no elementary anti-derivative exists, or that the variant of Risch's algorithm which is implemented can't handle the particular case (but this might not be evident for the user).

Example 3.2. *Bronstein's ISSAC'98 tutorial* [15] *outlines Risch's algorithm for various classes of elementary functions and gives corresponding examples such as*

$$\int \frac{(x^2+2x+1)\sqrt{x+\log x}\ +(3x+1)\log x\ +3x^2+x}{(x^2+x\log x)\sqrt{x+\log x}\ +x^2\log x\ +x^3}\,dx$$

$$= 2\sqrt{x+\log x}\ +2\log(x+\sqrt{x+\log x}\,)$$

in the algebraic logarithmic case. From among the systems which I tested with this example only AXIOM 2.2 gave an honest answer:

```
>> Error detected within library code:
integrate: implementation incomplete (constant residues)
```

The interested reader is invited to test his favorite CAS more systematically.

A central area of computer algebra, whose history also started in classical papers, and which saw dramatic modern developments with numerous research papers, is polynomial factorization.

There is a classical algorithm for factoring univariate polynomials over the integers due to F. T. von Schubert [89] who generalized ideas of I. Newton. This algorithm was rediscovered by L. Kronecker [60] and implemented in early computer algebra programs. Modern, highly practical algorithms rely on ideas of E. R. Berlekamp [7, 8] and D. G. Cantor and H. Zassenhaus [22] for factoring univariate polynomials over finite fields, and of Zassenhaus [100], who used Hensel-lifting for rediscovering the factors of a (squarefree and primitive) polynomial $f \in \mathbf{Z}[x]$ from those of f reduced modulo a suitable prime p. During the last three decades enormous progress has been made, with new ideas and improvements of the basic algorithms. Now one can handle problems of a size which was inconceivable a couple of years ago.

Example 3.3. *P. Roelse* [86] *reports on an implementation of Niederreiter's algorithm for factoring univariate polynomials over* $\mathbf{F}_2$*. With this implementation he is able to factor a pseudo-randomly chosen polynomial of degree* 300 000.

Factoring univariate integer polynomials as described above works well "on the average example in praxis". It has, however, exponential running time in the worst case (see, for example, [40]). A polynomial-time algorithm was given by A. K. Lenstra, H. W. Lenstra, Jr. and L. Lovász [66] who replaced the factor combination step at the very end of Zassenhaus' algorithm by a step involving their celebrated LLL-algorithm for lattice basis reduction. The LLL-algorithm together with its more recent variants has many applications, for example, to integer programming (see H. W. Lenstra's paper in these proceedings [68]).

Kronecker [60] also showed how to reduce multivariate factorization to the univariate case. Again, there are more practical, modern algorithms (for different representations of the multivariate polynomials). A polynomial-time reduction from dense multivariate to univariate integer polynomial factorization is due to E. Kaltofen [54]. Kaltofen is also one of the protagonists of the further development. I refer to his three survey papers [55, 56, 57] for the history of univariate and multivariate polynomial factorization over various coefficient domains until 1991. Further remarks can be found in [40, 90] (according to the notes on univariate factorization in [40] some of the basic ideas of the modern algorithms can actually be traced back to classical work, most notably to work of Gauss).

4. Some Remarks on Applications and Algorithms

Some of the fields in which computer algebra has applications are mentioned above (see also the home-page of ACA2000, the 6th conference on Applications of Computer Algebra by IMACS (International Association for Mathematics and Computers in Simulation)). Some examples for industrial applications are given in the paper by L. Gonzalez-Vega and T. Recio in these proceedings [44]. I will not comment further on applications outside of mathematics. Instead I will illustrate the impact of computer algebra on mathematical research by giving some examples, mostly from group theory, number theory, and algebraic geometry. Needless to say that there are applications in other areas of mathematics as well. Exact computer algebra methods may help

- to correct and supplement old mathematical tables or to create new tables (databases),
- to revive classical problems,
- to solve enumerative problems,
- to verify theorems whose proof has been reduced to straightforward but tedious calculations,
- to construct interesting examples such as counterexamples to conjectures,
- to find mathematical evidence by experiments.

Example 4.1. *The GAP Data Library contains "a number of databases with GAP language interfaces allowing them to be searched and studied efficiently", see http://www-history.mcs.st-and.ac.uk/~gap/. Among others one can find, "in various databases, all groups of order up to* 1,000, *excluding* 512 *and* 768″ *(see the papers by H. U. Besche and B. Eick* [9, 10], *who also announced a solution for the order 768 case). The case of order 512 is dealt with in* [35]. *Whereas all groups of* *order* 512 *with p-class at least three can be explicitly generated, the computations in the case of p-class* 2 *are extremely large. Therefore the authors decided to only enumerate these groups. In fact, they showed that there are* 8 785 772 *such groups. Altogether there exist* 10 494 213 *groups of order* 512.

Example 4.2. *In a tour de force of enumerative geometry G. Ellingsrud and S.-A. Strømme* [37] *computed that the number of twisted cubic curves on a general quintic hypersurface in complex projective* 4*-space is* 317 206 375 *(this was one of the motivations for creating the package SCHUBERT). The number* 317 206 375 *had been predicted by theoretical physicists* [19] *with a method originating from string theory (this method allows, in fact, prediction of the number of rational curves of any degree on a general quintic).*

Example 4.3. *The Number Theory Website, maintained by Keith Matthews at http://www.maths.uq.edu.au/~krm/web.html, provides plenty of information on number theory including links to the home-pages of CAS and to tables and databases (numbers, elliptic curves, number fields, modular forms). In particular, there is a link to the Primes Pages by C. K. Caldwell with tables such as Mersenne primes, or largest primes, and with recent records such as*

> *"On 1 June 1999, the team of Nayan Hajratwala, George Woltman, Scott Kurowski et. al. discovered a new record prime:* $2^{6972593} - 1$*. This is the 38th known Mersenne prime ... "*

(see http://www.utm.edu/research/primes/index.html). On the page on Fermat numbers (see http://vamri.xray.ufl.edu/proths/fermat.html) compiled by W. Keller one can find, for example, the following:

> *"On May 9, 2000, Ray Ballinger discovered a new factor of a Fermat number using Yves Gallot's program Proth.exe:* $591909 \cdot 22063+1$ *divides* F_{2059}*. This is the eighth factor found with Gallot's program."*

H. W. Lenstra, Jr. and A. K. Lenstra [64] give a theoretical and a practical reason for the renewed interest in primality testing and factoring integers in recent years: the introduction of complexity theory, which

> "enabled workers in the field to phrase the fruits of their intellectual labors in terms of theorems that apply to more than a finite number of cases",

and

> "the discovery, by Rivest, Shamir and Adleman [85], that the difficulty of factorization can be applied for cryptographical purposes ... we note that for the construction of the cryptographic scheme that they proposed it is important that primality testing is easy, and that for the unbreakability of the scheme it is essential that factorization is hard. Thus, as far as factorization is concerned, this is a *negative* application ... "

We refer to [64] for a discussion of modern algorithms for primality testing and factoring including, for factoring, the quadratic sieve by C. Pomerance [77, 78] and the elliptic curve method by H. W. Lenstra, Jr. [67]. Various articles on the number field sieve can be found in [65]. An introduction to computational number theory in general is given in H. Cohen's books [25, 26]. These books also contain a short discussion of some of the CAS suitable for number theory including MAGMA, PARI-GP, developed under the direction of H. Cohen, KANT/KASH, developed under the direction of M. Pohst, LIDIA, and SIMATH, developed under the direction of H. G. Zimmer.

Classical invariant theory (see [71] and [47]) is a 19th century field involving tedious calculations (see, for example, the many pages of extensive explicit tables in A. Cayley's collected works [23]). Invariants such as the discriminant $b^2 - ac$ of a quadratic binary form $ax^2 + 2bxy + cy^2$ come into play when one asks for properties of sets of solutions of polynomial equations which are invariant under certain classes of transformations.

Example 4.4. *J. J. Sylvester* [95, 96] *enumerated the fundamental invariants (more generally covariants) of binary forms of some low degrees. Together with Franklin* [39] *he arrived at his figures by manipulating the generating functions of the rings of covariants under consideration following ideas of Cayley and Sylvester. In fact, this approach yields for each given degree only a lower bound on the number of fundamental covariants. In order to conclude that this lower bound is actually the correct number a*

> *" ... fundamental postulate still awaiting demonstration is necessary .... The validity of the fundamental postulate ... is verified by its conducting to results which have been proved to be accurate for single binary quantics up to sixth order inclusive ... "*

Unfortunately, the fundamental postulate proved to be wrong in the next case, that is, in degree 7 [49]. *That there are indeed four more fundamental invariants in this degree than those enumerated by Sylvester was computed by Dixmier and Lazard* [34] *in 1986 with the help of computer algebra. The correct number of fundamental covariants, however, seems to be still unknown. Whereas nowadays every reader can easily recheck Sylvester's manipulations with his favorite computer algebra system, the contemporaries of Sylvester had to travel to Baltimore:*

> *"The manuscript sheets containing the original calculations ... are deposited in the iron safe of the John Hopkins University Baltimore, where they can be seen and examined ... "*

Classical invariant theory culminated in two papers of Hilbert [50, 51] which contain several "lemmas" which deeply influenced the development of modern abstract algebra and algebraic geometry. The first paper, for example, contains the basis theorem, the syzygy theorem and the theorem on the structure of generating functions (Hilbert functions). In our context these results play a crucial role in the theory of Gröbner bases. An interesting side remark is that the concepts of monomial orders, normal-form reductions and Gröbner bases were already introduced in 1899 by P. Gordan [45] in order to give another proof of Hilbert's basis theorem. Gordan's paper does not contain, however, Buchberger's celebrated algorithm for computing Gröbner bases [16, 17]. This algorithm contributes as one of its numerous applications (see [18]) to nowadays' revival of computational invariant theory (see [93, 32]) which in turn has many new applications ranging from topology and geometry, to physics, continuum mechanics, and computer vision (see also [75, 41, 92]). Let us quote J. P. S. Kung and G.-C. Rota [61]:

> "Like the Arabian phoenix rising out of its ashes, the theory of invariants, pronounced dead at the turn of the century, is once again at the forefront of mathematics. During its long eclipse, the language of modern algebra was developed, a sharp tool now at last being applied to the very purpose for which it was invented."

Let me quickly mention some further examples. A classical example of finishing a proof with the help of computers is the solution of the four-color problem by K. Appel and W. Haken [3]. The celebrated conjectures of B. J. Birch and H. P. F. Swinnerton-Dyer [11, 12] are based on extensive computer calculations. The same holds for the heuristics on class groups by H. Cohen and H. W. Lenstra, Jr. [27, 28]. An example from algebraic geometry is the systematic treatment by D. Eisenbud and S. Popescu [36] of some counterexamples to the minimal resolution conjecture [69]. The authors are grateful to D. Bayer and M. Stillman respectively D. Grayson and M. Stillman for the systems MACAULAY respectively MACAULAY2

> "which have been extremely useful for us; without them we would probably never have been bold enough to guess the existence of the structure that we explain here."

The first counterexamples to the minimal resolution conjecture had been found by F.-O. Schreyer (unpublished, compare [52]) with the help of MACAULAY. Schreyer's approach also yielded some interesting examples of projective varieties of low codimension [87, 88]. The construction method of W. Decker, L. Ein and F.-O. Schreyer combines syzygy theory, the theorem of Beilinson [6] and various computational techniques to construct smooth surfaces in projective 4-space and to find out where these surfaces stand in the Enriques-Kodaira classification (see [33, 31, 4, 80, 79, 1]). These authors also used MACAULAY. For some applications in algebraic geometry and singularity theory obtained with the help of the more recent system SINGULAR we refer to G.-M. Greuel's article in these proceedings [48].

Finally, let me come back to number theory and mention the indirect disproof of Merten's conjecture by A. M. Odlyzko and H. J. J. Riele [74] for which the LLL-algorithm

> "was the main new ingredient that allowed us to obtain much stronger results than those of previous authors."

References

[1] H. Abo, W. Decker, N. Sasakura, *An elliptic conic bundle in* $\mathbf{P}^4$ *arising from a stable rank-3 vector bundle,* Math. Z. **229** (1998), 725–741.

[2] B. Amrhein, O. Gloor, R. E. Maeder, *Visualizations for mathematics based on a computer algebra system,* J. Symbolic Computation **23** (1997), 447–452.

[3] K. Appel, W. Haken, *Every planar map is four colorable,* AMS, Providence, 1989.

[4] A. Aure, W. Decker, K. Hulek, S. Popescu, K. Ranestad, *Syzygies of abelian and bielliptic surfaces in* $\mathbf{P}^4$, International J. Math. **8** (1997), 849–919.

[5] C. Babbage, *Scribbling books,* volume 2 (1836), Science Museum Library, London.

[6] A. Beilinson, *Coherent sheaves on* $\mathbf{P}^N$ *and problems of linear algebra,* Funkt. Anal. Appl. **12** (1978), 214–216.

[7] E. R. Berlekamp, *Factoring polynomials over finite fields,* Bell System Tech. J. **46** (1967), 1853–1859.

[8] E. R. Berlekamp, *Factoring polynomials over large finite fields,* Math. Comp. **24** (1970), 713–735.

[9] H. U. Besche, B. Eick, *Construction of finite groups,* J. Symbolic Computation **27** (1999), 387–404.

[10] H. U. Besche, B. Eick, *The groups of order at most 1000 except 512 and 768,* J. Symbolic Computation **27** (1999), 405–413.

[11] B. J. Birch, H. P. F. Swinnerton-Dyer, *Notes on elliptic curves. I,* J. Reine Angew. Math. **212** (1963), 7–25.

[12] B. J. Birch, H. P. F. Swinnerton-Dyer, *Notes on elliptic curves. II,* J. Reine Angew. Math. **218** (1965), 79–108.

[13] M. Bronstein, *Integration of elementary functions,* J. Symbolic Computation **9** (1990), 117–173.

[14] M. Bronstein, *Symbolic integration I – transcendental functions,* Springer-Verlag, Berlin, 1997.

[15] M. Bronstein, *Symbolic integration tutorial, ISSAC'98,* downloadable from http://www-sop.inria.fr/cafe/Manuel.Bronstein/bronstein-eng.html

[16] B. Buchberger, *Ein Algorithmus zum Auffinden der Basiselemente des Restklassenrings nach einem nulldimensionalen Polynomideal,* PdH thesis, Lepold-Franzens-Universität, Innsbruck, 1965.

[17] B. Buchberger, *Ein algorithmisches Kriterium für die Lösbarkeit eines algebraischen Gleichungssystems,* Aequationes mathematicae **4** (1970), 374–383, English translation by M. Ambramson and R. Lumbert in [18], 535–545.

[18] B. Buchberger, F. Winkler (eds.), *Gröbner bases and applications, Linz 1999,* Cambridge University Press, Cambridge, 1999.

[19] P. Candelas, X. C. de la Ossa, P. S. Green, L. Parkes, *A pair of Calabi-Yau manifolds as an exactly soluble superconformal theory,* Nuclear Phys. B. **359** (1991), 21–74.

[20] J. Cannon, D. Holt, Foreword of the guest editors to the Special issue on computational algebra and number theory: Proceedings of the first MAGMA conference, J. Symbolic Computation **24** (1997), 233–234.

[21] J. F. Canny, D. Manocha, Multipolynomial resultant algorithms, J. Symbolic Computation **15** (1997), 99–122.

[22] D. G. Cantor, H. Zassenhaus, *A new algorithm for factoring polynomials over finite fields,* Math. Comp. **36** (1981), 587–592.

[23] A. Cayley, *The collected mathematical papers,* vols 1–12, Cambridge University Press, Cambridge, 1889.

[24] A. M. Cohen, J. H. Davenport, J. P. Heck, *An overview of computer algebra,* in A.M. Cohen (ed.), Computer algebra in industry, Wiley, Chichester, 1993.

[25] H. Cohen, *A course in computational algebraic number theory (3rd corrected printing),* Springer-Verlag, New York, 1996.

[26] H. Cohen, *Advanced topics in computational number theory,* Springer-Verlag, New York, 2000.

[27] H. Cohen, H. W. Lenstra, Jr., *Heuristics on class groups,* in D.V. Chudnovsky et al. (eds.), *Number theory, New York 1982,* Springer-Verlag, Berlin, 1984.

[28] H. Cohen, H. W. Lenstra, Jr., *Heuristics on class groups of number fields,* in H. Jager (ed.), *Number theory, Noordwijkerhout, 1983,* Springer-Verlag, Berlin, 1984.

[29] D. Cox, J. Little, D. O'Shea, *Ideals, varieties, and algorithms, second edition,* Springer-Verlag, New York, 1997.

[30] J. H. Davenport, *On the integration of algebraic functions,* Springer-Verlag, New York, 1981.

[31] W. Decker, L. Ein, F.-O. Schreyer, *Construction of surfaces in* $\mathbf{P}^4$, J. Algebraic Geometry **2** (1993), 185–237.

[32] W. Decker, T. de Jong, *Gröbner bases and invariant theory,* in [18], 61–89.

[33] W. Decker, F.-O. Schreyer, *Non-general type surfaces in* $\mathbf{P}^4$*: some remarks on bounds and constructions,* J. Symbolic Computation **29** (2000), 545–582.

[34] J. Dixmier, D. Lazard, *Minimum number of fundamental invariants for the binary form of degree 7,* J. Symbolic Computation **6** (1988), 113–115.

[35] B. Eick, E. A. O'Brien, *The groups of order 512,* in B.H. Matzat et al. (eds.), *Algorithmic algebra and number theory,* 379–380, Springer-Verlag, Berlin, 1999.

[36] D. Eisenbud, S. Popescu, *Gale duality and free resolutions of ideals of points,* Invent. Math. **136** (1999), 419–449.

[37] G. Ellingsrud, S.-A. Strømme, *The number of twisted cubic curves on the general quintic threefold,* Math. Scand. **76** (1995), 5–34.

[38] R. W. Floyd (ed.), *Proc. of ACM Symposium on Symbolic and Algebraic Manipulation (SYMSAM'66), Washington D.C.,* Comm. ACM **9** (1966), 574–643.

[39] F. Franklin, *On the calculation of the generating functions and tables of groundforms for binary quantics,* Amer. J. of Math. **3** (1883), 128–153.

[40] J. von zur Gathen, J. Gerhard, *Modern computer algebra,* Cambridge University Press, Cambridge, 1999.

[41] K. Gatermann, *Computer algebra methods for equivariant dynamical systems,* Springer-Verlag, Berlin, 2000.

[42] K. Geddes, S. R. Czapor, G. Labahn, *Algorithms for computer algebra,* Kluwer Academic Publishers, Boston, 1992.

[43] G. Gonnet, *A study of iteration formulas for root finding, where mathematics, computer algebra and software engineering meet,* in these proceedings.

[44] L. Gonzalez-Vega, T. Recio, *Industrial applications of computer algebra: climbing up a mountain, going down a hill,* in these proceedings.

[45] P. Gordan, *Neuer Beweis des Hilbertschen Satzes über homogene Funktionen,* Nachrichten König. Ges. der Wiss. zu Gött., 1899, 240–242, English translation by M. Abramson in SIGSAM Bulletin **32**, Number 2 (1998), 47–48.

[46] R. W. Gosper, Jr., *Decision procedure for indefinite hypergeometric summation,* Proc. Nat. Acad. Sci. USA **75** (1978), 40–42.

[47] J. H. Grace, A. Young, *The algebra of invariants,* Cambridge University Press, Cambridge, 1903.

[48] G.-M. Greuel, *Applications of computer algebra to algebraic geometry, singularity theory and symbolic-numerical solving,* in these proceedings.

[49] J. Hammond, *On the solution of the differential equation of sources,* Amer. J. of Math. **5** (1883), 218–227.

[50] D. Hilbert, *Über die Theorie der algebraischen Formen,* Math. Ann. **36** (1890), 473–534.

[51] D. Hilbert, *Über die vollen Invariantensysteme,* Math. Ann. **42** (1893), 313–373.

[52] A. Hirschowitz, C. Simpson, *La résolution minimale de l'idéal d'un arrangement général d'un grand nombre de points dans* $\mathbf{P}^n$, Invent. Math. **126** (1996), 467–503.

[53] H. G. Kahrimanian, *Analytical differentiation by a digital computer,* Master's thesis, Temple University, Philadelphia, 1953.

[54] E. Kaltofen, *Polynomial-time reductions from multivariate to bi- and univariate integral polynomial factorization,* SIAM J. Comp. **14** (1985), 469–489.

[55] E. Kaltofen, *Polynomial factorization,* in B. Buchberger et al. (eds.), *Computer algebra,* 95–113, Springer-Verlag, Wien, 1982.

[56] E. Kaltofen, *Polynomial factorization 1982-1986,* in I. Simon (ed.), *Computers in mathematics,* 285–309, Marcel Dekker, New York, 1990.

[57] E. Kaltofen, *Polynomial factorization 1987–1991,* in D.V. Chudnovsky, R.D. Jenks (eds.), *Proceedings of LATIN'92, Sao Paulo,* 294–313, Springer-Verlag, New York, 1992.

[58] S. Katz, *On the finiteness of rational curves on quintic threefolds,* Composito Math. **60** (1986), 151–162.

[59] W. Krandick, S. Rump (eds.), *Special issue on Validated numerical methods and computer algebra,* J. Symbolic Computation **24** (1997), 649–803.

[60] L. Kronecker, *Grundzüge einer arithmetischen Theorie algebraischer Grössen,* J. Reine Angew. Math. **92** (1882), 1–122.

[61] J. P. S. Kung, G.-C. Rota, *The Invariant Theory of Binary Forms,* Bull. Am. Math. Soc. **10** (1984), 27–85.

[62] L. A. Lambe (ed.), *Special issue,* J. Symbolic Computation **23** (1997), 445–623.

[63] P. L. Larcombe, *On Lovelace, Babbage and the origins of computer algebra,* in [99].

[64] A. K. Lenstra, H. W. Lenstra, Jr., *Algorithms in number theory,* in J. van Leeuwen (ed.), *Algorithms and complexity,*, Volume A, 673–716, Elsevier, Amsterdam, 1990.

[65] A. K. Lenstra, H. W. Lenstra, Jr. (eds), *The development of the number field sieve,* Springer-Verlag, Berlin, 1993.

[66] A. K. Lenstra, H. W. Lenstra, Jr., L. Lovász, *Factoring polynomials with rational coefficients,* Math. Ann. **261** (1982), 515–534.

[67] H. W. Lenstra, Jr., *Factoring integers with elliptic curves,* Ann. of Math. **126** (1987), 649–673.

[68] H. W. Lenstra, Jr., *Flags and lattice basis reduction,* in these proceedings.

[69] A. Lorenzini, *The minimal resolution conjecture,* J. Algebra **156** (1993), 5–35.

[70] L. F. Menabrea, *Sketch of the analytical engine invented by Charles Babbage. With notes upon the memoir by the translator, Ada Augusta, Countess of Lovelace,* in P. Morrison and E. Morrison (eds.), *Charles Babbage and his calculating engines,* part II, Dover Publications, New York, 1961.

[71] W. F. Meyer, *Invariantentheorie,* in *Encyklopädie der mathematischen Wissenschaften, Erster Band,* Teubner, Leipzig, 1898–1904.

[72] H. M. Möller, *Gröbner bases and numerical analysis,* in [18], 159–178.

[73] J. F. Nolan, *Analytical differentiation on a digital computer,* Master's thesis, MIT, Cambridge, 1953.

[74] A. M. Odlyzko, H. J. J. te Riele, *Disproof of the Mertens conjecture,* J. Reine Angew. Math. **357** (1985), 138–160.

[75] P. J. Olver, Classical invariant theory, Cambridge University Press, Cambridge, 1999.

[76] S. R. Petrick (ed.), *Proc. of the Second Symposium on Symbolic and Algebraic Manipulation (SYMSAM'71),* Los Angeles, ACM Press, New York, 1971.

[77] C. Pomerance, *Analysis and comparism of some factoring algorithms,* in H.W. Lenstra, Jr., R. Tijdeman (eds.), *Computational methods in number theory,* 89–139, Mathematical Centre Tracts **154/155**, Math. Centrum, Amsterdam, 1982.

[78] C. Pomerance, *The quadratic sieve factoring algorithm,* in T. Beth et al. (eds.), *Advances in cryptology,* Springer-Verlag, Berlin, 1985.

[79] S. Popescu, On smooth surfaces of degree ≥ 11 in $\mathbf{P}^4$, PhD thesis, Universität des Saarlandes, Saarbrücken, 1993.

[80] S. Popescu, K. Ranestad, Surfaces of degree 10 in the projective fourspace via linear systems and linkage, J. Algebraic Geometry **5** (1996), 13–76.

[81] R. Risch, *On the integration of elementary functions which are built up using algebraic operations,* Report SP-2801/002/00, System Development Corp., Santa Monica, 1968.

[82] R. Risch, *Further results on elementary functions,* Report RC-2402, IBM Corp., Yorktown Heights, 1969.

[83] R. Risch, *The problem of integration in finite terms,* Trans. A. M. S. **139** (1969), 167–189.

[84] R. Risch, *The solution of the problem of integration in finite terms,* Bull. A. M. S. **76** (1970), 605–608.

[85] R. L. Rivest, A. Shamir, L. Adleman, *A method for obtaining digital signatures and public-key cryptosystems,* Comm. ACM **21** (1978), 120–126.

[86] P. Roelse, *Factoring high-degree polynomials over* $\mathbf{F}_2$ *with Niederreiter's algorithm on the IBM SP2,* Math. Comp. **68** (1999), 869–880.

[87] F.-O. Schreyer, *Small fields in constructive algebraic geometry,* 221–228, in *Moduli of vector bundles, Sanda 1994,* 221–228, Dekker, New York, 1996.

[88] F.-O. Schreyer, F. Tognoli, *Constructions and investigations with small finite fields,* to appear in D. Eisenbud et al. (eds.), *Mathematical computations with Macaulay2,* Springer-Verlag, New York.

[89] F. T. von Schubert, *De inventione divisorum,* Nova Acta Academiae Scientiarum Imperalis Petropolitanae **11** (1793), 172–182.

[90] I. E. Shparlinski, *Finite fields: theory and computations,* Kluwer Academic Publishers, Dordrecht, 1999.

[91] M. F. Singer, *Formal solutions of differential equations,* J. Symbolic Computation **10** (1990), 59–94.

[92] L. Smith, *Polynomial Invariants of Finite Groups,* A.K. Peters, Wellesley, 1995.

[93] B. Sturmfels, *Algorithms in invariant theory,* Springer-Verlag, Wien, 1993.

[94] D. D. Swade, *Der mechanische Computer des Charles Babbage,* Spektrum der Wissenschaft, April 1993.

[95] J. J. Sylvester, *Tables of the generating functions and groundforms for the binary quantics of the first ten orders,* Amer. J. Math. **2** (1879), 223–251.

[96] J. J. Sylvester, *Tables of the generating functions and groundforms for the binary duodecimic, with some general remarks, and tables of the irreducible syzygies of certain quantics,* Amer. J. Math. **4** (1881), 41–61.

[97] B. Trager, *Integration of algebraic functions,* PhD thesis, MIT, Boston, 1984.

[98] S. M. Watt, H. Stetter (eds.), *Special issue on Symbolic numeric algebra for polynomials,* J. Symbolic Computation **26** (1998), 649–652.

[99] M. Wester (ed.), *Computer algebra systems. A practical guide,* Wiley, Chichester, 1999.

[100] H. Zassenhaus, *On Hensel factorization,* J. Number Theory **1** (1969), 291–311.

Department of Mathematics
Universität des Saarlandes
D-66041 Saarbrücken, Germany
E-mail address: `decker@math.uni-sb.de`

A Study of Iteration Formulas for Root Finding, Where Mathematics, Computer Algebra and Software Engineering Meet

Gaston H. Gonnet

Abstract. For various reasons, including speed code simplicity and symbolic approximation, it is still very interesting to analyze simple iteration formulas for root finding. The classical analysis of iteration formulas concentrates on their convergence near a root. We find experimentally, that this information is almost useless. The (apparently) random walk followed by iteration formulas before reaching convergence is the dominating factor in their performance. We study a set of 29 iteration formulas from a theoretical and a practical point of view. We define a new property of the formulas, their far-convergence, in an effort to explain their behaviours. Extensive experimentation finding polynomial roots, shows that there are extreme differences in performance of seemingly similar iterators. This is a surprising result. We use this experimental approach to select the most effective performer, which is Laguerre's method. The best companion (second method) to handle the failures of Laguerre's is a new method which is an adaptation of Halley's method to multipoint computation. The little-known Ostrowski's method comes out with one of the best performances. We also find that an unknown simple variant of Newton's method behaves much better than Newton's method itself, which behaves very poorly. This shows that sometimes it pays to modify a method to improve its far-convergence. Various performance curiosities cannot be explained in terms of neither order of convergence and are probably caused by the paths that the methods force on the iteration values. The study of these random paths is an open problem, probably beyond our present tools.

1. Introduction

We consider the problem of finding a numerical approximation to a root. For simplicity and uniformity reasons we will approximate roots of polynomials. We use $p(x)$ to denote the univariate polynomial of degree d. A root α, $p(\alpha) = 0$, may be real or complex. To avoid trivial cases, we will assume that $d \geq 3$ and $p(0) \neq 0$. A single point iteration formula, can be written as

$$x_{n+1} = x_n - \phi(x_n)$$

where $\phi(\alpha) = 0$. The error of the iteration at step n is

$$\varepsilon_n = x_n - \alpha\,.$$

We have noticed that all single-point iteration formulas that we studied can be expressed in terms of two values:

$$\delta = p(x)/p'(x) \qquad \text{and} \qquad \eta = \frac{p(x)p''(x)}{p'(x)^2}\,.$$

For example, Newton's method has $\phi = \delta$ and Halley's method has $\phi = \delta/(1-\eta/2)$. Both $\delta \to 0$ and $\eta \to 0$ when we are converging to a simple root.

1.1. Near- and far-convergence

The standard definition of order of convergence [4, 7, 6] is based on the behaviour of the iteration in the neighbourhood of a root. That is

$$\varepsilon_{n+1} = x_{n+1} - \alpha = O((x_n - \alpha)^\gamma) = O(\varepsilon_n^\gamma)$$

where α is the searched root, x_n is the nth approximation to the root and ε_n is the error on the nth iteration. A method which verifies the above for maximal γ is said to have order of convergence γ. We will call this order of convergence, *near-convergence*, as it refers to the behaviour at the limit when $x_n \to \alpha$, or when x_n is near α.

We define *far-convergence*, as the convergence order, the same as above, but in the limit when $|x_n - \alpha| \to \infty$. For example, the Newton iteration for the approximate computation of $\sqrt{N}$

$$x_{n+1} = \frac{N/x_n + x_n}{2}$$

has near-convergence of order 2 (quadratic convergence)

$$x_{n+1} - \sqrt{N} = \frac{(x_n - \sqrt{N})^2}{2\sqrt{N}} + O((x_n - \sqrt{N})^3)$$

and far-convergence of order 1 (linear convergence with constant factor 1/2) since

$$\varepsilon_{n+1} = \varepsilon_n/2 + O(1)$$

when $|x_n - \sqrt{N}| = |\varepsilon_n| \to \infty$. The Newton iteration for square roots can be considered very reliable as it converges to a root from any starting point due to its properties of far- and near-convergence (although not at the same speeds).

In this paper we are studying iteration formulas to be used for approximating complex roots of polynomials. In this context, the following can be asserted.

- Any number of derivatives of the function can be computed, each one costing about the same to compute.
- We are interested in working with multiple precision in the context of a computer algebra system, and hence the cost of evaluating $p(x)$ and its derivatives dominates the cost of each iteration.

- When convergence is achieved, an order $\gamma = 1.6$ or $\gamma = 2$ or $\gamma = 3$ makes little difference in the total number of iterations required. However, when the method fails to converge and the values wander around the complex plane, tens or hundreds of iterations can be wasted. From this we conclude that far-convergence is more important than near-convergence.
- Our view is pragmatic. We are searching for an iteration which delivers the best overall efficiency over a sufficiently large sample of polynomials.

2. Iteration Generators

It is possible to generate an infinite number of iteration formulas with a given order of near-convergence. These iterations can be single-point or multi-point. Methods for generating iteration formulas are quite straightforward, we will describe one for single-point direct methods.

Let $A(a_1, a_2, \ldots, a_k, x)$ be an arbitrary function of x with parameters a_1, $a_2, \ldots, a_k$. E.g.

$$A(a_1, a_2, a_3, x) = a_1 + \frac{a_2}{x - a_3}.$$

For most functions A with k parameters, it is possible to generate an iteration formula with order of convergence k. This is done through the following steps. Let x_n be an approximation of a zero of $p(x)$. At this point we compute k equations

$$\begin{aligned} A(a_1, a_2, \ldots, a_k, x_n) &= p(x_n) \\ A'_x(a_1, a_2, \ldots, a_k, x_n) &= p'_x(x_n) \\ &\vdots \\ A_x^{(k-1)}(a_1, a_2, \ldots, a_k, x_n) &= p_x^{(k-1)}(x_n). \end{aligned}$$

These equations are viewed as equations in the parameters $a_1, a_2, \ldots, a_k$. We solve these equations and obtain the parameters in terms of x_n, i.e. $a_1(x_n)$, $a_2(x_n)$, etc. Plugging these parameters in A we obtain a function $A(a_1(x_n), a_2(x_n), \ldots, a_k(x_n), x)$ which is a k-order approximation to $p(x)$ at x_n. Let x_{n+1} be the zero of $A(\ldots, x)$, i.e. $A(a_1(x_n), a_2(x_n), \ldots, a_k(x_n), x_{n+1}) = 0$, then x_{n+1} becomes our next approximation to the root. The iteration formula is defined by the computation producing x_{n+1} from x_n. Normally we can find a closed form expression of x_{n+1} in terms of $p(x)$ and x_n. To achieve this, the function A is chosen so that the solution of the system of equations can be expressed in closed form and that the zero of A is easy to compute, i.e. $A(\ldots, x) = 0$ is easy to solve. For the above example,

$$\begin{aligned} A(a_1, a_2, a_3, x_n) &= a_1 + \frac{a_2}{x - a_3} = p(x_n), \\ A'_x(a_1, a_2, a_3, x_n) &= -\frac{a_2}{(x_n - a_3)^2} = p'_x(x_n) \end{aligned}$$

and

$$A''_x(a_1, a_2, a_3, x_n) = \frac{2a_2}{(x_n - a_3)^3} = p''_x(x_n)$$

from which we derive

$$a_2 = -\frac{4p'(x_n)^3}{p''(x_n)^2}, \qquad a_3 = \frac{2p'(x_n)}{p''(x_n)} + x_n$$

and

$$a_1 = p(x_n) - \frac{a_2}{x_n - a_3}$$

$$A(a_1, a_2, a_3, x_{n+1}) = a_1 + \frac{a_2}{x_{n+1} - a_3} = 0 \implies x_{n+1} = a_3 - \frac{a_2}{a_1}.$$

Simplifying and rearranging we obtain

$$x_{n+1} = x_n - \frac{p(x_n)}{p'(x_n)} \frac{1}{(1 - \frac{p(x_n)p''(x_n)}{2p'(x_n)^2})} = x_n - \frac{\delta}{1 - \eta/2}$$

which is Halley's method [1]. We can say that approximating $p(x)$ by $a_1 + \frac{a_2}{x - a_3}$ generates Halley's iteration.

Table 1 shows the iteration formulas tested and the approximation formulas which generate them. Iterations 2 to 6 are all special cases of $\delta(1 + \frac{\eta}{2a})^a$. This iteration gives third order near-convergence for any value of a. In all cases, it corresponds to an inverse interpolation approximation

$$x \approx (a_1 + a_2 p(x))(1 + a_3 p(x))^a.$$

Halley's, Ostrowski's [5] and inverse quadratic interpolation are special cases of this formula. Methods 10 to 17 are derived from Laguerre's method by assuming that the approximation is of a fixed degree (2 to 9). For degree 2, this coincides with Euler's method or direct quadratic interpolation. Many variations of Laguerre's method were tested because all of them have a very good behaviour, in particular the optimal number of successes is achieved by iterator 15. Methods 18 to 25 were artificially created. 18 and 19 are truncated Taylor series in η of method 10. Methods 20 to 25 are perturbations of other methods in their $O(\eta^2)$ term (in $O(\delta^2)$ for 25) so that the order of far-convergence becomes $O(1)$ instead of $O(\eta)$. Methods 26 to 29 are multipoint methods based on similar approximations. The Hansen-Patrick methods [2], generated by

$$\frac{\delta(a+1)}{a + \sqrt{1 - (a+1)\eta}}$$

for fixed a, are not included since they can be generated from Laguerre's formula, by setting $a = 1/(d-1)$.

Table 2 shows the near- and far-convergence for each of the methods. For both last columns, the expression represents the asymptotic value of ε_{n+1} in terms of ε_n. Only the most significant term is shown. For the near convergence, the exponent of ε_n is the classical order of convergence. $p' = p'_x(\alpha)$, $p'' = p''_x(\alpha)$ and

	Iteration formula	Generating Approximation	Notes
1	δ	$p(x) \approx a_2\,(x-a_1)$	Newton
2	$\frac{\delta}{1-1/2\eta}$	$p(x) \approx \frac{x-a_1}{a_2+a_3x}$	Halley
3	$\frac{\delta}{\sqrt{1-\eta}}$	$x \approx \frac{a_1+a_2p(x)}{\sqrt{1+a_3p(x)}}$	Ostrowski
4	$\frac{\delta}{\sqrt[3]{1-3/2\eta}}$	$x \approx \frac{a_1+a_2p(x)}{\sqrt[3]{1+a_3p(x)}}$	
5	$\delta\sqrt{1+\eta}$	$x \approx$ $(a_1+a_2p(x))\sqrt{1+a_3p(x)}$	
6	$\delta\,(1+1/2\eta)$	$x \approx a_1+a_2p(x)+a_3p(x)^2$	Inverse quadratic interpolation
7	$\frac{\delta(1-\delta+1/2\eta)}{1-\delta+\delta^2-1/2\eta\delta}$	$x \approx$ $1-\frac{1}{(a_1+a_2p(x)+a_3p(x)^2)}$	
8	$\frac{\delta\left(2+\sqrt{1+\eta}\right)}{3-\eta}$	$p(x)^2 \approx \frac{(x-a_1)^2}{a_2+a_3x}$	
9	$\frac{\delta d}{1+\sqrt{(d-1)^2-d(d-1)\eta}}$	$p(x) \approx$ $(x-a_1)\,(a_2x+a_3)^{d-1}$	Laguerre
10	$\frac{2\delta}{1+\sqrt{1-2\eta}}$	$p(x) \approx (x-a_1)\,(a_2x+a_3)$	Euler, direct quadratic interpolation
11-17	$\frac{\delta k}{1+\sqrt{(k-1)^2-k(k-1)\eta}}$	$p(x) \approx$ $(x-a_1)\,(a_2x+a_3)^k$	Laguerre with fix degree $k=3..9$
18	$\delta\,(1+\eta/2+\eta^2/2)$	artificial	matching 10 to order 2 in η
19	$\frac{\delta}{1-\eta/2-\eta^2/4}$	artificial	inverse matching inverse of 10 to order 2 in η
20	$\frac{\delta}{1-\eta/2-\eta^2/8}$	artificial	inverse matching inverse of 3 to order 2 in η
21	$\delta/\left(1-\eta/2-\frac{d\eta^2}{2(d-1)}\right)$	artificial	using degree to force 2 to have far convergence
22	$\frac{\delta}{\sqrt{1-\eta-\frac{\eta^2}{d-1}}}$	artificial	using degree to force 3 to have far convergence
23	$\delta\sqrt{1+\eta+\frac{d(d^2+d-1)\eta^2}{d-1}}$	artificial	using degree to force 5 to have far convergence
24	$\delta\left(1+\frac{\eta}{2}+\frac{(2d-1)d\eta^2}{2(d-1)}\right)$	artificial	using degree to force 6 to have far convergence
25	$\delta/\left(1-\frac{(d-1)\delta}{x}\right)$	artificial	Perturbing Newton with degree for good far convergence
26	$\frac{p_n\Delta x_{01}}{\Delta p_{01}}$	$p(x) \approx a_2(x-a_1)$	Secant
27	$\frac{p_n}{\Delta p_{12}}\left(\frac{p_{n-1}\Delta x_{02}}{\Delta p_{02}}-\frac{p_{n-2}\Delta x_{01}}{\Delta p_{01}}\right)$	$x \approx a_1+a_2p(x)+a_3p(x)^2$	Multipoint inverse quadratic interpolation

TABLE 1. Iteration formulas.

28	$\frac{2b_0}{b_1(1+\sqrt{1+4b_2b_0/b_1^2})}$ $b_0 = p_n \Delta x_{02} \Delta x_{12} \Delta x_{01}$ $b_1 = \Delta p_{01} \Delta x_{02}^2 - \Delta p_{02} \Delta x_{01}^2$ $b_2 = p_{n-1} \Delta x_{02} - p_n \Delta x_{12} - p_{n-2} \Delta x_{01}$	$p(x) \approx (x-a_1)(a_2 x + a_3)$	Muller
29	$\frac{p_n \Delta p_{02} \Delta x_{02} \Delta x_{01}}{p_n \Delta p_{12} x - p_{n-1} \Delta p_{02} x_{n-1} + p_{n-2} \Delta p_{01} x_{n-2}}$	$p(x) \approx \frac{x-a_1}{a_2+a_3 x}$	Multipoint Halley

TABLE 1. Iteration formulas (Cont.).

	near-convergence		far-convergence	
	order	ε_{n+1}	order	ε_{n+1}
1	2	$\frac{p''}{2p'}\varepsilon_n^2$	1	$\frac{d-1}{d}\varepsilon_n$
2-6	3	$\frac{3(3a+1)p''^2-4ap'p'''}{24ap'^2}\varepsilon_n^3$	1	$\left(1-\frac{\left(1+\frac{d-1}{2da}\right)^a}{d}\right)\varepsilon_n$
7	3	$\frac{6p'^2-6p'p''+3p''^2-p'p'''}{6p'^2}\varepsilon_n^3$	1	ε_n
8	3	$\frac{9p''^2-4p'p'''}{24p'^2}\varepsilon_n^3$	1	$\frac{2d-1-\sqrt{2-1/d}}{2d+1}\varepsilon_n$
9	3	$\frac{3(d-2)p''^2-4(d-1)p'p'''}{24(d-1)p'^2}\varepsilon_n^3$	0	$-\alpha+\frac{-c_{d-1}}{dc_d}$ $+\frac{\sqrt{(d-1)\left((d-1)c_{d-1}^2-2dc_dc_{d-2}\right)}}{dc_d}$
10-17	3	$\frac{3(k-2)p''^2-4(k-1)p'p'''}{24(k-1)p'^2}\varepsilon_n^3$	1	$\left(1-\frac{k}{d+\sqrt{-d(d-k)(k-1)}}\right)\varepsilon_n$
18	3	$-\frac{p'''}{6p'}\varepsilon_n^3$	1	$\frac{(d-1)\left(2d^2-2d+1\right)}{2d^3}\varepsilon_n$
19	3	$-\frac{p'''}{6p'}\varepsilon_n^3$	1	$\frac{d^2-1}{d^2+4d-1}\varepsilon_n$
20	3	$\frac{3p''^2-4p'p'''}{24p'^2}\varepsilon_n^3$	1	$\frac{3d^2-2d-1}{3d^2+6d-1}\varepsilon_n$
21	3	$\frac{-3(d+1)p''^2-2(d-1)p'''p'}{12p'^2(d-1)}\varepsilon_n^3$	0	$-\alpha-\frac{c_{d-1}}{c_d}$
22	3	$\frac{3(d-5)p''^2-4(d-1)p'''p'}{24p'^2(d-1)}\varepsilon_n^3$	0	$-\alpha-\frac{c_{d-1}}{c_d}$
23	3	$\frac{-(12d^3+12d^2-27d+15)p''^2-4(d-1)p'''p'}{24p'^2(d-1)}\varepsilon_n^3$	0	$-\alpha-\frac{c_{d-1}}{c_d}$
24	3	$\frac{-(d-1)p'p'''-(6d^2-6d+3)p''^2}{6p'^2(d-1)}\varepsilon_n^3$	0	$-\alpha-\frac{c_{d-1}}{c_d}$
25	2	$\frac{\alpha p''-2p'(d-1)}{2p'\alpha}\varepsilon_n^2$	0	$-\alpha-\frac{c_{d-1}}{c_d}$
26	1.618 ...	$\frac{p''}{2p'}\varepsilon_n\varepsilon_{n-1}$		
27	1.839 ...	$\frac{3p''^2-p'p'''}{6p'^2}\varepsilon_n\varepsilon_{n-1}\varepsilon_{n-2}$		
28	1.839 ...	$-\frac{p'''}{6p'}\varepsilon_n\varepsilon_{n-1}\varepsilon_{n-2}$		
29	1.839 ...	$-\frac{3p''^2-2p'p'''}{12p'^2}\varepsilon_n\varepsilon_{n-1}\varepsilon_{n-2}$		

TABLE 2. near- and far-convergence of iteration formulas.

$p''' = p_x'''(\alpha)$ denote the derivatives of $p(x)$ at the root. $\Delta x_{ij} = x_{n-i} - x_{n-j}$, $\Delta p_{ij} = p(x_{n-i}) - p(x_{n-j})$, p_n stands for $p(x_n)$ and c_i is the coefficient of x^i in $p(x)$. For both columns, a smaller coefficient for ε_n is better. For the first column, a higher exponent of ε_n is better, as $\varepsilon_n \to 0$, while for the last column a lower exponent is better, as $\varepsilon_n \to \infty$. Far-convergence cannot be easily defined for multistep methods and hence was not computed.

3. The Search

Although the method described here could be used for any type of problems, or for any order of convergence, we concentrate on methods with cubic convergence ($\gamma = 3$) used on polynomials. As we said earlier, we will take a very pragmatic approach and try as many formulas as possible against a sufficiently large number of problems and select the best performing one(s).

Since the target application of this method is in Maple, we will test the algorithms using the same system so that the complexity measures (mainly time) are compatible.

3.1. The problem set

The problem set should be large and open-ended, so that we avoid selecting methods which happen to be lucky with a particular sample set. This is relevant if we are going to test many iteration formulas. We tested each method with 100000 polynomials so that critiques with respect to sample size may be silenced. The best performing formulas (3,9-17,22 and 25) were tested against 200000 polynomials. The sample is generated at random, so that it is open-ended, i.e. we could test these methods for 10 times more problems if desirable. In all cases the polynomials have integer coefficients. We generate the polynomials in four different groups. In all cases, a random variable distributed as $U(a, b)$ denotes a uniform random distribution between the values a and b inclusive. All polynomials are generated with a random degree distributed as $U(3, 20)$.

(a) 40% of the polynomials with random $U(-10^{10}, 10^{10})$ coefficients. The number of non-zero coefficients is also random $U(3, deg(p) + 1)$. This is an example of such a polynomial.

$$422434762x^7 - 8643406254x^6 + 6868525159x^5 + 9284921461x^3 + 3505658871$$

(b) 30% of the polynomials with coefficients which are random integers $U(-10^k, 10^k)$ and k is $U(0, 40)$ for each coefficient. The number of non-zero coefficients is $U(2, deg(p) + 1)$. E.g.

$$\begin{aligned}
&- 708032792x^{19} + 77361635429252588371 0x^{17} \\
&+ 9069436236132014192781 7x^{15} + 287243520412466889x^{13} + 990598934x^{12} \\
&+ 6420995340618263067643600578251x^{10} - 54255029639x^9 \\
&- 539443581303941066250263794671941160274
\end{aligned}$$

(c) 10% of the polynomials as in (b) above, but with random gaussian integers (complex integers) as coefficients. This is done by multiplying each coefficient by i^k where k is a random integer $U(0, 1)$. E.g.

$$\begin{aligned}
&972019526003356644271014109899146206803 8ix^8 \\
&\qquad - 6779x^3 + 33787241729985 + 606i
\end{aligned}$$

(d) 20% of the polynomials which are a minor perturbation of a product of powers of simpler polynomials. Although the degree of the polynomial is $U(3,20)$, each factor of these polynomials has random $U(1,3)$ degree and is powered to a random exponent so that it does not exceed the expected degree of the final polynomial. The perturbation is obtained by adding $\pm x^k$ where k is $U(0, deg(p))$. E.g.

$$(9887x+446)^6\,(-273x+12)^4\,\left(-3230x^3-20\right)^2(3178x+6)-x^{13}.$$

The polynomials in group (a) are truly random polynomials but are fairly easy to resolve by all methods and present a lesser challenge. Thus we have weighted them only 40% in the sample. The groups (b) and (c), represent cases where the large differences in the magnitude of the coefficients could lead to extreme placements of the roots. Roots could lie very close together, or have very small imaginary or real parts, or have very different magnitudes. In our experience these polynomials reveal the weaknesses of the iteration formulas, and consequently we have weighted these two groups as much as group (a). Notice that (c) should not behave too differently from (b), it was included just to insure that polynomials with complex coefficients do not cause additional problems. The last group, (d), contains polynomials which will have roots very close together, i.e. very small perturbations of non square-free polynomials. This is not so uncommon in practice, and again, most iteration formulas have difficulties in this situation.

4. Test for each Polynomial

For each polynomial, and for each formula, the iteration is started at the same point.

$$x_0=(1+i)\sqrt{2}\max_k \sqrt[k]{\left|\frac{c_{d-k}}{c_d}\right|}$$

where d is the degree of the polynomial and c_i is the coefficient of x^i in $p(x)$. This is a point at 45 degrees on a complex circle which encloses all roots of the polynomial.

All computations were done with increasing precision, the initial precision is identical to the number of digits in the largest coefficient and it is increased by one decimal digit after each iteration. This reflects the accepted practice in variable precision computer algebra systems. One of the reasons for which iterations fail is the ill-conditioning of the polynomial which may require additional precision to resolve.

Convergence to a root is decided from the computed values and estimates of the errors of the last step. For multipoint methods with error formulas $\varepsilon_{n+1} \approx$

$\alpha\varepsilon_n\varepsilon_{n-1}$ or $\varepsilon_{n+1} \approx \alpha\varepsilon_n\varepsilon_{n-1}\varepsilon_{n-2}$, in the neighbourhood of a root and with convergence, the estimate for ε_{n+1} is

$$\varepsilon_{n+1} \approx \frac{\varepsilon_n^2}{\varepsilon_{n-2}} \quad \text{or} \quad \varepsilon_{n+1} \approx \frac{\varepsilon_n^2}{\varepsilon_{n-3}} .$$

The convergence criteria, once x_{n+1} has been computed, is to test whether the relative error of x_{n+1} is small enough, or

$$\left| \frac{\varepsilon_{n+1}}{x_{n+1}} \right| < M_\varepsilon$$

which gives the complete condition

$$|\varepsilon_n| < |\varepsilon_{n-1}| < |\varepsilon_{n-2}| < |\varepsilon_{n-3}| \text{ and } \left| \frac{\varepsilon_n^2}{\varepsilon_{n-2}x_{n+1}} \right| < M_\varepsilon$$

where M_ε is the machine epsilon, or the relative precision at which we want to compute the roots. We do not know the exact ε_{n-i}, but we can approximate them with $\varepsilon_{n-i} \approx x_{n-i} - x_{n+1}$. In the neighbourhood of a root and with superlinear convergence, this approximation is sufficient.

For single-point methods, with error term of the form $\varepsilon_{n+1} \approx \alpha\varepsilon_n^\gamma$ we compute $\varepsilon_{n+1} \approx \varepsilon_n^{\gamma+1}/\varepsilon_{n-1}^\gamma$ and obtain the first condition. For these iterations we already computed the derivative of the polynomial at each step, so we can use these derivatives to estimate the error at x_{n+1}

$$p'(x_n)(x_n - \alpha) \approx p(x_n) - p(\alpha) = p(x_n)$$

$$\varepsilon_n = p(x_n)/p'(x_n) = \delta_n .$$

We can use both estimates ($\varepsilon_{n-i} = \delta_{n-i}$ and $\varepsilon_{n-i} = x_{n-i} - x_{n+1}$) to derive a safer criteria for convergence

$$|\varepsilon_n| < |\varepsilon_{n-1}| \text{ and } \left| \frac{\varepsilon_n^{\gamma+1}}{\varepsilon_{n-1}^\gamma x_{n+1}} \right| < M_\varepsilon \text{ and } \left| \frac{\delta_n^{\gamma+1}}{\delta_{n-1}^\gamma x_{n+1}} \right| < M_\varepsilon .$$

No acceleration of the iterations is attempted (as is common practice when the method shows linear convergence), as this is one of the features which we want to measure. We want to evaluate the iterators on their exclusive own merits.

We compute only one root for each polynomial, as it is cheaper and safer to compute a new polynomial than to deflate one.

The best iteration schemes use an average of about 10 iterations per root. We let each method run at most 50 iterations. If the algorithm runs for 50 iterations and does not converge, we consider it has failed. Failures to compute the iteration formula (like divisions by zero, very large floating exponents and repeated values) are also counted as failures. In the case of failures, the number of iterations and time consumed is accounted up to the point of failure. Each failure due to lack of convergence is very costly in time, and hence total time becomes an excellent compound measure for the iteration formulas.

We measure, for each polynomial and each method, the number of iterations, the time spent and the failures. Notice that the relation between number of iterations and total time spent is not necessarily linear due to the increasing precision of the computation. Time and number of failures are the most important measures.

	iterations					time					failures (%)				
	a	b	c	d	all	a	b	c	d	all	a	b	c	d	all
1	18.7	33.5	32.3	40.4	28.8	1.00	3.19	3.13	4.14	2.50	5.10	48.21	45.35	57.54	32.55
2	10.1	26.7	25.6	31.9	21.0	.66	3.90	3.82	5.24	2.86	1.87	34.03	32.33	28.65	19.92
3	*5.7*	19.1	18.4	22.9	14.4	*.36*	3.04	2.97	3.94	2.14	*.00*	19.48	18.99	5.95	8.93
4	15.9	19.3	19.2	35.5	21.2	2.56	5.24	5.23	9.96	5.11	1.36	8.68	8.38	51.05	14.20
5	15.2	30.6	29.4	36.4	25.5	1.27	4.81	4.70	6.19	3.66	4.58	42.02	40.00	43.82	27.20
6	15.0	30.9	29.3	35.9	25.4	1.13	4.40	4.25	5.64	3.32	5.49	42.76	40.12	41.71	27.38
7	13.9	38.8	38.7	42.7	29.6	1.10	6.16	6.24	6.77	4.27	7.08	68.60	68.90	71.41	44.59
8	11.5	28.0	27.0	33.8	22.5	.89	4.66	4.54	6.11	3.43	1.86	36.76	35.08	34.76	22.23
9	6.2	13.6	13.7	*15.0*	*10.9*	.49	2.20	2.30	*2.59*	1.60	4.30	13.60	13.00	5.18	8.14
10	11.4	15.0	15.7	29.0	16.4	.94	2.19	2.34	5.25	2.32	.33	7.82	9.51	29.62	9.35
11	10.3	14.1	14.5	27.4	15.3	.84	2.05	2.17	5.06	2.18	.28	7.49	8.70	25.91	8.41
12	9.6	13.4	13.2	25.9	14.3	.78	1.95	1.92	4.83	2.05	.25	7.20	6.66	22.44	7.41
13	9.0	*12.8*	12.7	24.8	13.7	.72	1.86	1.84	4.62	1.95	.21	*7.01*	*6.54*	20.26	6.89
14	8.4	13.0	12.7	23.8	13.3	.67	1.88	1.83	4.43	1.90	.18	7.78	7.06	17.68	6.65
15	7.9	13.0	12.6	22.7	12.9	.62	1.86	1.81	4.23	1.83	.16	7.84	7.25	15.25	*6.19*
16	7.5	13.4	13.0	22.1	12.7	.58	1.95	1.90	4.09	1.82	.14	9.20	8.29	13.82	6.41
17	7.1	13.4	13.3	21.4	12.5	.53	1.96	1.95	3.94	1.78	.08	9.59	9.53	12.07	6.28
18	14.3	29.0	27.8	32.9	23.8	1.12	4.27	4.16	5.40	3.22	6.58	38.20	36.10	31.88	24.08
19	9.7	24.0	22.9	26.1	18.6	.67	3.58	3.50	4.38	2.57	6.03	28.25	27.27	14.54	16.52
20	9.4	24.9	24.0	29.3	19.5	.63	3.77	3.70	4.98	2.75	3.63	30.88	29.59	22.05	18.08
21	30.1	31.8	31.7	27.6	30.3	3.06	4.53	4.63	4.94	4.03	60.97	51.71	51.24	38.19	52.66
22	9.1	22.6	22.3	17.3	16.1	.71	3.51	3.58	2.68	2.23	4.75	13.12	13.22	*4.99*	8.16
23	31.9	35.8	36.0	45.5	36.2	2.72	5.62	6.12	7.49	4.88	89.81	88.89	88.09	93.44	90.09
24	30.4	34.7	34.8	42.4	34.5	2.19	4.95	5.36	6.55	4.21	85.91	85.23	84.96	87.04	85.84
25	17.1	21.2	21.6	25.7	20.5	1.08	2.24	2.30	3.29	1.99	10.42	19.94	21.17	24.58	17.18
26	19.2	35.4	33.8	42.7	30.2	.82	2.59	2.50	3.14	1.98	14.90	58.20	54.81	67.08	42.32
27	19.6	13.3	*12.5*	27.3	18.5	1.21	*1.07*	*1.00*	2.82	*1.47*	47.46	63.56	63.01	56.61	55.68
28	13.1	25.9	24.8	39.2	23.3	1.35	5.02	4.81	7.07	3.94	2.56	38.14	35.68	59.68	27.97
29	15.9	17.2	16.6	31.5	19.5	1.04	1.96	1.88	3.56	1.90	17.23	54.98	55.26	45.39	37.99

TABLE 3. Average values for each method (row) and type of polynomial.

5. Conclusions

The whole computations in this paper used about 11,984 hours of CPU time on a small subnetwork of HP and Sun workstations. The results are shown in Table 3 from which we can reach several conclusions.

The first conclusion is that there is a substantial difference in performances. Not only in the time consumed, but also in the number of failures. The effectiveness of the best method compared to the effectiveness of the worst is in a ratio of approximately 1:30. We define effectiveness as the fraction of solved polynomials divided by time. Methods with apparently very similar characteristics perform dissimilarly. Order of near-convergence does not appear to be a criteria on which to base any consistent judgement. In particular the order of the worst performer is equal to the best performer and is better than the one of the third best performer!

Not surprisingly, the overall winner is Laguerre's iteration. It is a winner in effectiveness and in three of the 15 categories (columns of the table). Entries in italics indicate the best value for the given column. It only loses by 2% to the best in number of failures. Other methods of the Laguerre family are also very effective, in particular for high fixed degree. One of these methods, 15, is the one producing the fewest failures.

The relatively unknown Ostrowski's method, which comes next in efficiency, is remarkable in one way. It does not fail with *any* of the polynomials of group a. We continued to run Ostrowski's method against polynomials of type a, and we ran it with over one million polynomials without failure. There is something quite remarkable about its convergence which escapes our analysis, and which places it in a unique category with respect to all other methods.

Method 25 ranks next in effectiveness. This is a big surprise. Method 25 is a modification of Newton's method, which improves the far-convergence order and outperforms Newton's in 13 of the 15 categories. The number of failures is cut in half. This shows that modifying a method to improve its far-convergence, while keeping its near-convergence, could improve the method significantly.

The artificially generated method 22 (a modification of Ostrowski's method to improve far-convergence) comes next in efficiency, still outperforming all other more famous methods. It produces a very small number of failures, and it appears to be the best suited for problems in the group d.

Again, for reasons unknown to us, methods 23 and 24, which were supposed to be improvements on methods 5 and 6, behave disastrously. In particular much worse than 5 and 6. Notice that the same type of change was done for 22 (from 3), in which case they remained about the same and for 25 (from 1) which produced a significant improvement.

All these curiosities and oddities indicate that the overall performance of these methods is not tied to the convergence part alone, but rather to the apparently random walks over the complex plane. These walks are the most expensive part of the computation and depend on subtleties of the iterations, at present beyond our understanding. Hence our selection method here is empirical.

The effectiveness of these walks is dramatically different for different methods. Our methods of analysis cannot explain this behaviour which is probably one of the most difficult open problems in the area.

For samples of 100,000 or 200,000 polynomials, the confidence intervals of the values in the tables are very small. For example, the totals for Laguerre's method have 95% confidence intervals $10.94 \pm .12$, $1.604 \pm .027$ and $8.14 \pm .24$ respectively. The difference in efficiency with the second most effective method (17) are well outside these confidence intervals.

5.1. The best companion for Laguerre

Once that we have observed that the best performer is Laguerre's, and that this method still fails on one out of 12 polynomials, the immediate next question is which method will handle the largest portion of problems where Laguerre fails. If there is a pattern in the problems being solved, then there is a pattern (complementary) on the failed ones, and hence a particular method may be well suited to handle them.

	iterations					time					failures (%)				
	a	b	c	d	all	a	b	c	d	all	a	b	c	d	all
1	22.7	30.8	32.3	48.6	31.7	.87	2.58	2.85	6.35	2.75	19.11	41.58	45.69	85.06	43.17
2	12.9	23.4	24.7	42.5	23.9	.61	3.03	3.40	9.11	3.37	10.78	33.03	36.70	57.99	32.18
3	*5.4*	18.1	20.7	*32.2*	17.7	*.25*	2.79	3.36	7.86	3.00	*.00*	26.90	32.94	*23.57*	*21.78*
4	17.7	27.0	27.7	49.7	28.1	2.38	7.25	7.58	16.87	7.53	*.00*	26.41	28.62	96.38	30.30
5	18.1	26.9	28.2	46.1	27.8	1.09	3.79	4.16	9.99	4.09	14.22	37.09	40.00	73.16	37.43
6	20.0	28.2	30.1	45.7	29.1	1.12	3.56	4.02	9.24	3.86	19.61	41.24	45.78	71.18	41.32
7	11.8	29.5	31.2	47.8	28.4	.69	4.29	4.80	10.29	4.39	.07	44.72	49.54	86.11	41.47
8	14.0	24.5	25.8	44.1	25.0	.78	3.62	4.02	10.23	3.95	10.42	34.84	37.61	64.18	33.96
9	39.8	45.5	37.6	46.1	43.0	4.43	8.52	7.41	10.24	7.70	75.22	93.84	76.97	83.20	85.73
10	12.3	22.6	24.9	46.1	23.8	.74	3.43	3.99	10.32	3.85	*.00*	25.64	31.93	70.71	27.13
11	11.5	22.2	24.1	46.0	23.4	.69	3.36	3.85	10.61	3.82	*.00*	25.76	30.83	73.51	27.37
12	10.9	21.7	22.6	44.5	22.5	.64	3.29	3.52	10.46	3.70	*.00*	*25.45*	27.43	70.95	26.32
13	10.5	21.4	22.2	44.8	22.3	.61	3.24	3.46	10.45	3.66	*.00*	25.92	27.52	74.45	27.02
14	10.1	21.2	21.8	43.8	21.9	.58	3.20	3.39	10.28	3.60	*.00*	25.73	*27.16*	73.86	26.79
15	9.6	20.8	21.6	41.6	21.3	.55	3.15	3.37	10.10	3.54	*.00*	25.73	27.98	68.61	26.24
16	9.0	20.4	21.2	39.6	20.6	.51	3.10	3.30	9.87	3.47	.14	25.92	27.71	63.71	25.68
17	8.6	20.0	21.1	38.4	20.1	.49	3.04	3.32	9.72	3.42	.22	25.76	28.17	61.14	25.36
18	16.9	26.0	27.6	43.2	26.6	.95	3.43	3.85	9.27	3.74	16.95	38.97	42.48	62.19	37.92
19	14.6	24.5	26.2	36.2	24.2	.86	3.39	3.82	8.44	3.58	19.54	39.74	43.85	36.76	35.76
20	11.9	22.6	24.1	40.0	22.9	.59	3.10	3.50	9.09	3.41	11.35	34.04	37.98	49.01	31.85
21	6.8	33.8	34.4	46.8	29.9	.36	5.21	5.65	7.89	4.61	100.0	84.09	81.38	89.26	87.67
22	45.3	45.6	45.7	44.2	45.4	5.00	7.47	7.90	6.55	6.90	76.29	67.07	67.16	72.58	69.75
23	10.3	24.3	26.8	48.2	24.9	.37	3.13	4.56	8.00	3.42	100.0	96.95	97.16	99.77	98.00
24	12.0	24.9	27.1	48.3	25.6	.41	2.93	4.09	7.59	3.20	99.93	95.81	96.06	99.42	97.19
25	47.8	36.5	35.6	42.0	39.4	3.73	4.33	4.29	7.07	4.55	83.69	63.16	63.58	64.06	67.68
26	19.7	29.0	31.0	49.7	30.1	.72	1.96	2.18	*4.62*	2.08	36.42	51.06	55.14	95.22	54.39
27	17.2	15.8	13.6	43.7	19.4	.90	1.26	1.08	5.90	1.76	37.00	47.98	50.92	82.26	50.61
28	7.8	20.3	22.7	49.6	21.9	.60	3.71	4.28	10.99	4.09	.86	25.82	31.93	95.57	30.63
29	14.9	*13.4*	*11.7*	42.3	*17.2*	.82	*1.16*	*1.01*	6.13	*1.71*	12.72	34.44	37.89	70.83	35.15

TABLE 4. Average values for each method (row) and type of polynomial over the failures of method 9.

Table 4 contains the results of running problems which were not resolved by Laguerre. For Laguerre itself, the runs were done starting at a different point ($x_0 e^{2\pi i\phi}$, where ϕ is the golden ratio). This gives an indication of how dependent the failures were from an initial unlucky value. For this run we are also interested in the effectiveness, that is number of problems solved divided by total time.

The best companion for Laguerre is method 29, a multipoint version of Halley's iteration, which fails on 35% of the residual polynomials. Notice that the statistics for method 29 as a second method are *better*, in all three categories, than as a first method. This means that it is very complementary to Laguerre's method, the polynomials which were harder for Laguerre, are easier for 29.

Using Laguerre a second time, with a different starting value, does not appear to be a good idea at all. It fails 86% of the time, i.e. a different starting value is successful about $1/7^{\text{th}}$ of the time. This somehow *proves* that the failure set has a pattern.

Ostrowski's method (3) is the most successful companion, but it takes more time than method 29, and hence its effectiveness is lower.

5.2. Best companion of Laguerre and 29

Following the same idea, we ran all the methods against the residuals of Laguerre and method 29. The best method for handling these residuals is method 25, the modified Newton method. We omit the table showing these results. Method 25 solves 56% of the remaining polynomials, in a very competitive time. It is by far the most effective method for these residual polynomials.

We ran this process six times, to find the most effective sequence of methods to handle these polynomials. When a method is used more than once with the same polynomial, we start the iteration at a different value. The starting value for the jth run of a method is set to $x_0 e^{2\pi i(j-1)\phi}$. Table 5 summarizes the results of these simulations. Each method was selected to have highest effectiveness, that is number of problems solved divided by total time. It is obvious from the first three methods already, that each one of them *specializes* for some type of polynomials and that their combined use is much more effective than their repeated use.

prev methods	meth	iter	time	failures (%)	solved/ time	cumulative fail (%)	sample size
	9	10.901±.086	1.597±.019	7.99±.17	57.61	7.99±.17	92450
9	29	17.16±.40	1.710±.060	35.2±1.2	37.92	2.824±.093	82000
9,29	25	24.70±.55	3.398±.098	44.3±1.1	16.39	1.298±.034	246000
9,29,25	3	21.87±.42	5.22±.16	15.32±.89	16.22	.196±.011	492000
9,29,25,3	22	35.85±.43	11.24±.19	20.7±1.4	7.05	.0403±.0027	1640000
9,29,25,3,22	9	36.0±1.1	14.22±.62	52.9±3.1	3.31	.0210±.0012	2460000

TABLE 5. Performance of each method used on the residuals previous methods.

5.3. Conclusions

From this experimental/theoretical study we can derive the following conclusions.

The effectiveness of different iteration formulas cannot be explained by orders or convergence or any other form of analysis presently available.

Different iterations have very different performances, up to a factor of 30 in effectiveness. This cannot be ignored. It appears that each method has its own *personality*.

Newton's iteration formula is one of the poorest performers. An interesting observation on the most popular iteration formula. Laguerre's, Ostrowski's and multipoint-Halley's are the best iterations and should be given special consideration.

Pairs of companion methods are much more effective than one method applied repeatedly.

References

[1] W. Gander. On Halley's iteration method. *Americam Mathematical Monthly.* 92(2):131–134, Feb. 1985.

[2] E. Hansen and M. Patrick. *A family of root finding methods.* Numer. Math., 27:257–269, 1977.

[3] P. Henrici. *Elements of Numerical Analysis.* John Wiley, New York, 1964.

[4] P. Henrici. *Essentials of Numerical Analysis.* John Wiley, New York, 1982.

[5] A. M. Ostrowski. *Solution of Equations and Systems of Equations.* Academic Press, New York, 1973.

[6] A. Ralston. *A First Course in Numerical Analysis.* McGraw-Hill, 1965.

[7] J. F. Traub. *Iterative Methods for Solution of Equations.* Prentice-Hall, 1964.

Institut für Wissenschaftliches Rechnen
Eidgenössische TH Zürich-Zentrum
8092 Zürich, Switzerland
E-mail address: `gonnet@inf.ethz.ch`

Industrial Applications of Computer Algebra: Climbing Up a Mountain, Going Down a Hill

Laureano Gonzalez-Vega and Tomas Recio

Abstract. In this paper we present some personal experiences with Computer Algebra applications to industrial problems. In many cases the involved Computer Algebra problems seem as challenging as climbing up a difficult peak. Then one finds out that the trail leads up to a quite rugged hill ... This point of view will be illustrated with "real" examples coming from robot kinematics and path planning, parametric CAD and shape design in automotive industry.

1. Considerations about the Nature of Computer Algebra

Computer Algebra bounces back and forth, from Computer Science to Mathematics. As a scientific discipline Computer Algebra arises, in the 50's, as a response to the difficulties posed by the first attempts to implement computer programs for analytical integration or differentiation (see the historical chapter in [10]). These programs were, in many cases, application oriented. The addressed difficulties were of various sorts, ranging from very mathematical – for instance, purely algebraic – to very computational, such as the need for a specific memory management policy.

No matter how far from the implementation step could happen to be our research in Computer Algebra, we should not forget this: the success of so many symbolic computation programs that have been (and that are being) used to solve so many different problems, is ultimately responsible for the existence, today, of Computer Algebra as a scientific discipline. There is not a "purely mathematical" Computer Algebra subfield. There are different levels of applicability, i.e. of proximity to an externally given goal ... Thus, its achievement should measure the success of the application. This applies, in particular, to Computer Algebra industrial applications.

Whether we agree totally, or just in part, or whether we disagree openly with the above statements, we will probably agree that the interest of Computer Algebra industrial applications should be primarily evaluated from the industrial partners' side of the picture. But they tend to be, for good or for bad, rather silent. They rarely spend time and energy writing papers, and they usually do not attend Computer Algebra conferences. A negative consequence of this state of things is that we do not have an objective way to provide a solid overview of industrial applications, and that we do not readily learn about successful/failed cases. A positive consequence is that we can claim whatever we aim to claim, without much precaution.

It was not always like this. In earlier times, a handful of successful Computer Algebra stories made their way to the communication media. We can recall articles published in the Scientific News [18], New York Times [23], Nature [22], Scientific American [17], etc. One could argue that it was, perhaps, just because of the scientific novelty of symbolic computation – performing in a few minutes or seconds some computations that required, previously, a titanic effort for humans – and that some of these were news on non-industrial aspects of Computer Algebra. Anyway, it has been quite some time since Computer Algebra methods and tools were last echoed in the scientific news for the general public (with few exceptions such as [7]). And we all realize that today, in many scientific fields, success stories with relevant consequences for industry make their way rather easily into the newspapers. We could ask, then, what is the matter today with Computer Algebra industrial applications?

The goal of this paper is to present some personal considerations around this question. The next section will describe some of the external circumstances that face academic/industrial cooperation. In Section 3, we will summarily describe the more intrinsic cooperation problems posed by the (in days gone by) promising field of Computer Algebra applications in Robotics: certainly, we were trying to climb up a high mountain ... Section 4 is devoted to presenting the more modest aims of an ongoing cooperation project which is effectively changing the practice of a concrete enterprise: we can say that we are now exploring the challenging top of a rugged hill.

2. Working with Industrial Partners

Working with industrial partners is quite uncomfortable for academics[1]. They obstinately care about solving a problem, but only in most cases, or at least in some cases, or even in "this" particular case, instead of caring about solving it in general. Some times the solution they search for is conceptually rather simple, but tiresome to execute in practice. Other times they do not care about the problem they have just posed, if they see that you do not progress fast enough to solve it; and then they merely switch their mathematical model to a different and simpler one, asking you to forget about the interesting question you have just started to think about ... It is difficult to get a paper properly done under these changing conditions, it seems impossible to bind in a Ph.D. thesis based on such "rush hour" solutions ... As a result, only very few and very obstinate scholars persist in working in this near-zero atmosphere (i.e. extremely poor in academic oxygen: publications and dissertations).

[1]The abstract of this paper explicitly states that we are telling here some **personal** experiences: the reader must have this in mind, when reading what follows.

Even if we assume this environement and if we try our best to achieve "costumer satisfaction", we can run into other kind of problems. The following example comes from the *FRISCO* ("A FRamework for Integrated Symbolic/Numeric COmputation", ESPRIT LTR Program, 1996–1999) project. This project arose as an industrial application of another ESPRIT project (the *POSSO* project: "Polynomial System Solving"). The FRISCO consortium was leaded by a software company, NAG Ltd (Oxford, UK) and included the universities of Cantabria (Spain), Pisa (Italy), Rennes I and the research institute INRIA at Sophia-Antipolis (both in France).

After the POSSO experience, shared by some of the FRISCO teams, the FRISCO partners carefully started addressing the problem of industrial cooperation through an opinion poll to R&D directors in different industries. The conclusions of this survey, published under the name "The Needs of Industry on Polynomial System Solving" (see [11]), were discussed at a Workshop (Barcelona, fall 96). The questionnaire attempted to identify symbolic polynomial system solving problems appearing in each of the addressed industries, and it included:

1. Information about the corporation itself.
2. A specific item on: how did the corresponding polynomial system appear in the industrial context?
3. What kind of system was it (number of variables, coefficient type, etc.)?
4. What kind of solutions were looked for?
5. When had it to be solved (real time, off-line)?

More than 60 enterprises were contacted, in several European countries, representing very different fields of economic activity: EDF, Alcatel, FIAT, REIS, Daimler Benz, VW, PEGOP, TECNATOM, CANDEMAT, SIMULOG, CASA, APIA XXI, LABEIN, ... From this large group, only about a dozen replied (mostly Spanish, also EDF, CCETT, PEGOP, REIS ROBOTICS, etc.). It must be said that the coordinator of this activity was one of the coauthors of this paper, and this could explain the greater success with Spanish firms. Finally, just a half dozen of these industries made the effort to attend the Workshop (invited by FRISCO). A first conclusion we can extract from these data is that it is also very hard for engineers working in industries to distract some time and energy from daily occupations to participate in long-term cooperation projects with Academia.

On the positive side, the information collected by the enquiry was quite interesting. The polynomial system was, in most cases, generated by means of standard Symbolic/Numeric software. Usually the system had an equal number of equations and of unknowns (from 16 to several thousands). The system was quite sparse, but this fact was due, perhaps, to the way it was generated. Its degree was quite low, 2 or 3, in general. The usual coefficients were real numbers, a few cases had coefficients in $\mathbb{Z}_3$, or they were exact rational numbers. If the system contained parameters, they were first specialized to numerical values, then the system was solved. The sought solutions were always real numbers, with a 10^{-7} accuracy requirement, both for solutions found on or off-line. If problems arose in the solving process (for Newton-like methods) the mathematical model was then simplified. Usually, parametric solutions had not been regarded (but they would have been welcome, in principle).

We can also extract some general conclusions. One, is that it is hard to communicate with industries, because of the lack of common language and because of the divergent short-term interests from both sides. Moreover, technically speaking, we can say that cooperation with industry requires,

1. to develop Symbolic/Numerical integrated packages,
2. to adapt symbolic solvers to the kind of systems appearing in industry (which are not of general type), and
3. to disseminate the existence of symbolic solvers that deal with parameters.

Moreover we observed that many industrially relevant situations are not so (mathematically speaking) challenging. Sometimes, to explore the capabilities of Computer Algebra for a particular problem involves changing the mathematical model, not simply solving it in the given model. Finally, we can remark that interesting contexts for cooperation seem to arise mainly with those industries which are in the academic vicinity.

3. Robot Kinematics and Motion Planning

There are of course other, more intrinsic, reasons that explain the difficulties of Computer Algebra for industrial cooperation.

Few subjects have attracted more interest as an application domain for Computer Algebra than robot kinematics and motion planning problems. We can say that Robotics entered into Computer Algebra through the seminal paper [21]. The motion planning problems were, subsequently, reduced to a kind of quantifier elimination problem, in as many variables as the robot's degrees of freedom. Luckily, Computer Algebra experts had successfully developed and implemented a Quantifier Elimination algorithm by that time (see [4] and [2]). In the next years, a number of prestigious scientists contributed to this problem. For instance J. Canny obtained the Young Scientist Presidential Award in 1987 for his dissertation ([3]) on this topic. R. Alami and J.-P. Laumond went even further (see [1]), providing a general solution for the task planning problem (i.e. algorithmically describing how a robot should deal with a given task, including non self-movable objects and fixed obstacles).

Yet, we can apply to this happy picture some statements taken from the Foreword to the chapter "Motion Planning", in the book "Autonomous Robot Vehicles" (see [5])[2]:

> "*Although the theoretical approach has the* **drawback that the algorithms produced tend to be far too complicated to implement**, *the heuristic approach suffers from the problem that there is no guarantee that the*

[2]The emphasis is ours.

methods will work in all cases. However, the hope is that the theoretical results can reveal the fundamental structure of the problem, thus providing a framework to assist in making decisions as to what compromises should be made when developing a heuristic approach."

And, more clearly, in the chapter on "Prospects" from his book [14], Latombe states:

"*Although* **these results have still made few inroads in industrial applications,** *this should change soon. Several graphic simulators already include collision checking capabilities and it will not take long before they also offer path planning tools."*

The obvious reason is that the Quantifier Elimination algorithms of general purpose offered by Computer Algebra are still too far from being able to deal with real size motion problems, for the incredible amounts of time and space they currently require. One can hope that "this should change soon", of course ..., although complexity results [6] apparently say that it could take some time to become reality.

A similar analysis applies to the long standing problem of solving robot kinematic equations (those giving the position and orientation of each robot arm for a desired position and orientation of its hand) [16]. As McCarthy [15] mentions, the interesting case is that in which the equations include parameters, each corresponding to the possible geometric data for the different parts of a robot:

"*At the core of both the analysis of existing machines and the design of new ones is the solution of sets of nonlinear algebraic equations with parameters obtained form either the dimensions of the machine or the designed set of output positions ....*
... the designer is less often interested in a particular solution than in ranges of solutions and their relation to the prescribed features of the design."

We can track this problem for the 6R manipulator (with six degrees of freedom) from 1969 till the final computational solution in 1989 [20]. Essentially, the problem deals with solving symbolically a collection of non-linear equations and Computer Algebra provides many different algorithms to cope with this situation. Yet, after 20 years of research in this issue, one of the persons who had contributed more to its solution, Professor Roth, from U. Stanford, confessed (in a personal communication) that he did not recall any immediate industrial reaction to his discovery. The reason, again, is that the obtained characteristic polynomials are by far too large to be handled with current computers.

Summarizing, we can say that the field that has attracted most interest as a challenging application for Computer Algebra in the past decades, seems to have made little industrially oriented progress in such a high-tech field as Robotics. For a more down to earth and promising approach see [13].

4. CAD and Shape Design in the Automotive Industry

To be less negative, we would like to present some ongoing work on an issue of modest theoretical challenge that is producing a concrete impact in the output of the industrial partner. The activity of the project "*Integrating new algebraic-numerical techniques in Computer Aided Geometric Design (CAGD): Developing Problem Solving Environments and Implementation into an industrial CAD/CAM framework*" funded by the Spanish Ministery of Education (through European Union support for regional development) is directed to the resolution, in a more efficient way, of the mathematical problems arising when manipulating curves and surfaces into the CSIS software. This software is developed and maintained by the company CANDEMAT S.A. in order to be used for design purposes, production control and quality verification of the final product (shape design and cast construction in the automotive industry).

In order to check the possibility of using Computer Algebra techniques to improve the CSIS software, firstly three concrete problems were isolated:

- the computation of the implicit equation of a surface in $\mathbb{R}^3$ presented by a rational parametrization,
- the problem of conversion between the VDA and IGES formats (see 4.2), and
- the tracing, algebraically guided, of algebraic curves implicitly defined.

The reasons justifying the choice of the first problem rely on the fact that, when available, the implicit equation of a parametric surface is very useful to deal with point-surface position problems, surface intersections, sectioning, trimmed surfaces and sculptured solids, etc. For the second problem, to remark that the manipulation of trimmed surfaces (surfaces with holes) involves the exact topology computation of several real algebraic plane curves defined in the parameter space.

4.1. Generic implicitation applied to sectioning and offsetting

Two main difficulties were encountered when trying to use inside CSIS the usual elimination technics to deal with implicitation problems: first, it is usually a very costly algebraic operation and, second, the coefficients of the parametrization are usually floating-point numbers.

The second difficulty was overcome by taking into account that, in general, a concrete object to model (and then to construct) is made up of several hundreds (or thousands) of small patches, all of them sharing the same algebraic structure. For such an object a database has been constructed containing the implicit equation of every class of patches appearing in its definition. This database must also contain the inversion formulae (giving the parameters in terms of the cartesian coordinates) and must be pruned to avoid specialization problems. Moreover the database for a specific object is kept in a bigger and more general database for further use.

In the particular case of the objects provided by CANDEMAT S.A., the implicitation process was made by using Sylvester resultants, Grobner Basis computations and ad-hoc techniques for specific cases. Since the implicitation procedure was a preprocessing step, no time is spent when the CAD/CAM user is working.

Moreover the solution to the floating-point coefficients problem is rather simple under this approach, requiring just the specialization in the precomputed implicit equation.

Example 4.1. *In the database, the implicit equation of the parametric patch*

$$x = x_{00}\frac{t_2 - t}{t_2 - t_1} + x_{01}\frac{t - t_1}{t_2 - t_1}$$

$$y = y_{00}\frac{s_2 - s}{s_2 - s_1} + y_{11}\frac{s - s_1}{s_2 - s_1}$$

$$z = \left(z_{00}\frac{s_2 - s}{s_2 - s_1} + z_{10}\frac{s - s_1}{s_2 - s_1}\right)\frac{t_2 - t}{t_2 - t_1} + \left(z_{10}\frac{s_2 - s}{s_2 - s_1} + z_{11}\frac{s - s_1}{s_2 - s_1}\right)\frac{t - t_1}{t_2 - t_1}$$

appears as:

$$\begin{aligned}&(z_{00} - 2z_{10} + z_{11})xy + (y_{00}z_{10} + y_{11}z_{10} - y_{00}z_{11} - y_{11}z_{00})x \\ &+ (x_{00}z_{10} + x_{01}z_{10} - x_{00}z_{11} - x_{01}z_{00})y + (x_{00}y_{11} - x_{00}y_{00} + x_{01}y_{00} - x_{01}y_{11})z \\ &+ x_{01}y_{11}z_{00} - x_{00}y_{11}z_{10} + x_{00}y_{00}z_{11} - x_{01}y_{00}z_{10}\,.\end{aligned}$$

The sectioning by the plane $X = k$ of a B-spline proceeds, with this new strategy, by performing the following steps:

- First, we consider all the implicit equations $H_i(X, Y, Z)$ of the different patches integrating the considered B-spline surface.
- Second, the starting points in the parametric domain are determined and these are used to determine the interesting part of every curve

  $$H_i(k, Y, Z) = 0$$

  through a convenient discretisation.
- The obtained points are interpolated by a spline curve which represents the desired section.

We should mention, finally, that for specific objects, the time for sectioning, by using this strategy, has already been divided by a factor of 3 when compared with the previous approach followed in CSIS.

As a future work we regard the resolution of some problems arising in the implicitation problem:

- Some algebraic structures arising in the database construction are very complicated and the implicit equation has not been generated. Namely:

  $$x = \frac{f_1(s,t)}{f_0(s,t)}, \quad y = \frac{f_2(s,t)}{f_0(s,t)}, \quad z = \frac{f_3(s,t)}{f_0(s,t)}$$

  with ($i \in \{0, 1, 2, 3\}$):

  $$\begin{aligned}f_i = {}& s^3(\alpha_3^{(i)}t^2 + \beta_3^{(i)}t + \gamma_3^{(i)}) + s^2(\alpha_2^{(i)}t^2 + \beta_2^{(i)}t + \gamma_2^{(i)}) \\ &+ s(\alpha_1^{(i)}t^2 + \beta_1^{(i)}t + \gamma_1^{(i)}) + \alpha_0^{(i)}t^2 + \beta_0^{(i)}t + \gamma_0^{(i)}\,.\end{aligned}$$

- We would like to deal in advance with the specialization problems: up to this moment these are detected "a posteriori" when substituting the parametrization into the candidate to be the implicit equation.

If the sectioning by using the implicitation procedure has been a first successful application, this will be much more important when dealing with the offsetting by taking into account that the offset of a B-spline is no longer a B-spline or, in general, a parametric surface (but an algebraic surface) and thus the use of implicit equations is a direct method not currently used in any CAD/CAM software.

4.2. Format conversion: from rational to polynomial parameterizations

VGA and IGES are two of the formats most commonly used in CAD/CAM systems to represent and deal with the data required to describe and communicate the essential engineering characteristics of physical objects such as manufactured products. The VDA (Verband der Automobilundustrie) format was created in 1982 by the VDA Committee founded mainly by several German automobile and automotive supply companies. The IGES (Initial Graphics Exchange Specification) was created by the IGES/PDES Organisation (USA).

The main difference between these two formats lies in the fact that VDA accepts only surfaces defined by polynomial parameterizations while IGES accepts both rational and polynomial. This is a very important problem in many companies in the automobile industry, since they get the information in the IGES format but the specific CAD/CAM software they use allows the use of only the VDA format. The only way of solving this problem up to now has been by means of a method based upon a uniform subdivision of the parameter domain plus a polynomial interpolation procedure whose efficiency depends strongly on the degree of the polynomials to appear in the required polynomial parameterization (see [19]). The Computer Algebra behind this problem is still not completely well understood and new algorithms for solving this problem have been developed and initially checked by using new Hermite-Birkhoff interpolation schemes for multivariate polynomials (see [9]). These techniques also allow the reduction of the degree of the considered surface which produce a simplification of the implicitation procedure described in 4.1, since the degree of the parametric equations is smaller (see [8]).

4.3. Algebraically guided tracing of implicitly defined algebraic curves and surfaces

Many important problems in Computer Aided Geometric Design are reduced to the computation of the graph of a planar algebraic curve implicitly presented. For example if we want to section the surface

$$x = \frac{X(s,t)}{W(s,t)}, \quad y = \frac{Y(s,t)}{W(s,t)}, \quad z = \frac{Z(s,t)}{W(s,t)}; \qquad s,t \in [0,1]$$

with respect to the plane $X = x_0$, then we have two possibilities: either we draw "into the square unit" $[0,1] \times [0,1]$ the planar algebraic curve defined by

$$x_0 W(s,t) - X(s,t) = 0$$

and then this picture is lifted to the considered surface or, if the implicit equation $H(x,y,z)$ of the considered surface is available, the lifting procedure can be avoided by merely computing the graph of the planar curve $H(x_0,y,z)=0$ as it was shown in Section 4.1.

The problem of computing the graph (even topologically) of a planar algebraic curve defined implicitly has received special attention from Computer Algebra since it has been responsible for many advances regarding sub-resultants, real root counting, infinitesimal computations, etc. These algorithms have been successfully implemented in several Computer Algebra Systems (AXIOM and Maple) and now we are involved in the process of integrating the algorithms into the CSIS software (but some new practical problems have arisen since CSIS has just moved to Visual Basic under Windows 98 ...).

The main reason for including these algorithms in CSIS, apart from its use in the sectioning procedure, is found in the manipulation of trimmed surfaces: parametric surfaces with holes defined as parametric curves in the parametric domain of the surface. For example, the sectioning of this kind of surfaces is not currently available in the CSIS software.

5. Conclusions

It is clear at this moment that the practical algebraic resolution of nonlinear systems with thousands of equations and unknowns is far from the capabilities of Computer Algebra but probably, in many cases, Computer Algebra could help to reformulate the mathematical problem in such a way that the system to solve gets much smaller (and simpler to solve).

For example the following areas have already been identified as cornerstones when trying to apply algebraic techniques to an industrial practice:

- To make explicit the capabilities of Computer Algebra when dealing with the resolution of polynomial systems involving parameters: in many cases final users did not imagine that such a possibility existed and is already available for some non-trivial problems.
- To determine some specific needs in the CAD area with respect to polynomial system solving.
- To create/investigate links between Computer Algebra software and nonlinear optimization packages, or between Computer Algebra techniques and the theory of linear systems and control.
- To develop Computer Algebra facilities to deal with the nonlinear systems of equations which are produced by discretization schemes or finite elements.
- To establish a good collection of test suites, not only with a statement of the problem and results, but also containing calling sequences allowing the user to experiment with every system on a WWW server providing access to Computer Algebra software. A test suites collection is already available from

  `http://www-sop.inria.fr/saga/POL/`

and a very interesting collection of highly specialized symbolic software is available at the following web addresses:

1. The PoSSo Library:

   `http://janet.dm.unipi.it/posso_demo.html`

2. FGb+RS Servers:
   `http://posso.lip6.fr/~jcf/` and `https://calfor.lip6.fr/`
3. Gb+RealSolving through MuPAD:

   `http://www.loria.fr/~rouillie/software.html`

4. The ALP Library:

   `http://www-sop.inria.fr/safir/WHOSWHO/Bernard.Mourrain/ALP/`.

A very easy introduction to the use of almost all the libraries before mentioned can be found in [12].

Acknowledgements

The first author acknowledges partial support by the grant DGESIC PB 98–0713–C02–02 (Ministerio de Educación y Cultura).

The second author acknowledges partial support by the grant DGESIC PB 98–0756–C02–02 (Ministerio de Educación y Cultura).

References

[1] R. Alami and J. P. Laumond, *A geometrical approach to planning manipulation tasks in robotics,* First Canadian Conference on Computational Geometry (Montreal, 1989).

[2] B. Buchberger, G. Collins, M. Encarnación, H. Hong, J. Johnson, W. Krandick, R. Loos and A. Neubacher, *A SACLIB Primer,* Tech. Rep. 92–34, RISC-Linz, Johannes Kepler University (Linz, Austria), **1992**.

[3] J. F. Canny, *The complexity of robot motion planning,* ACM Doctoral Dissertation Series, MIT Press, Cambridge Mass., **1988**.

[4] G. E. Collins, *Quantifier elimination for real closed fields by Cylindrical Algebraic Decomposition,* Lecture Notes in Computer Science (Second GI Conference on Automata Theory and Formal Languages), **33** (1975), 134–183.

[5] I. J. Cox and G. T. Wilfong, *Motion Planning,* Autonomous Robot Vehicles (I. J. Cox and G. T. Wilfong eds), Springer-Verlag, **1990**.

[6] J. H. Davenport and J. Heintz, *Real Quantifier Elimination is Doubly Exponential,* Journal of Symbolic Computation, **5** (1988), 29–36.

[7] D. Duval, *Calcul symbolique: automatisation en cours,* La Recherche, **291** (1996), 64–71.

[8] J. Espinola, L. Gonzalez-Vega and I. Necula, *Generic implicitation of low degree rational surfaces,* Preprint available at `http://www.frisco.matesco.unican.es` (1999).

[9] J. Espinola, L. Gonzalez-Vega and I. Necula, *Algebraic approximation in CAGD,* Preprint available at `http://www.frisco.matesco.unican.es` (2000).

[10] K. Geddes, S. R. Czapor and G. Labhan, *Algorithms for computer algebra,* Kluwer Academic Publishers, Boston, **1992**.

[11] L. Gonzalez-Vega, *The Needs of Industry for Polynomial System Solving,* The SAC Newsletter, **3** (1998), 21–46.

[12] L. Gonzalez-Vega, *FRISCO Software Overview,* Preprint available at `http://www.frisco.matesco.unican.es` (1999).

[13] P. Kovacs, *Rechnergestützte Symbolische Roboterkinematik,* Vieweg Verlag, **1993**.

[14] J. C. Latombe, *Robot Motion Planning,* The Kluwer International Series Series in Engineering and Computer Sceince, Kluwer Academic Publishers, **1991**.

[15] J. M. McCarthy, *Kinematics of Robot Manipulators,* International Journal of Robotics Research, **5(2)** (1986).

[16] R. P. Paul, *Robot manipulators: Mathematics, Programming and Control,* The MIT Press Series in Artificial Intelligence, **1981**.

[17] R. Pavelle, M. Rothstein and J. Fitch, *Algebra por ordenador,* Scientific American (Spanish edition), **1981**.

[18] L. Steen, *Computer calculus,* Science News, **119** (1981), 250–251.

[19] N. Patrikalakis, *Approximate Conversion of Rational B-spline Patches,* Computer Aided Geometric Design, **6** (1989), 189–204.

[20] M. Raghavan and B. Roth, *Kinematic analysis of the 6R manipulator of general geometry.* Proceedings of the International Symposium on Robotics Research (Tokyo), 314–320 (1989).

[21] J. T. Schwartz and M. Sharir, *On the "piano movers" problem. II. General techniques for computing topological properties of real algebraic manifolds,* Advances in Applied Mathematics, **4(3)** (1983), 298–351.

[22] *Algebra Made Mechanical,* Nature, **290** (1981), 198–200.

[23] *Liberating the prose of math from its grammar,* New York Times, july 19 (1988), 21.

Departamento de Matemáticas, Estadística y Computación
Facultad de Ciencias
Universidad de Cantabria
Santander, Spain
E-mail address: `gvega@matesco.unican.es`
E-mail address: `recio@matesco.unican.es`

Applications of Computer Algebra to Algebraic Geometry, Singularity Theory and Symbolic-Numerical Solving

Gert-Martin Greuel

1. Introduction

Although computer algebra is a research field in its own, its main driving force comes from various fields of applications. These applications range from mathematics and computer science over physics and engineering up to applications to technical and industrial problems.

In this article I should like to show, by means of examples, the impact of computer algebra on two branches of pure mathematics: algebraic geometry and singularity theory. Today, algorithms, programmes and systems in computer algebra have reached a stage where it is possible to compute highly sophisticated mathematical objects such as moduli spaces and objects related to mixed Hodge structures (Sections 2 and 3). Moreover, computer algebra has been and is still successfully used in testing or disproving conjectures, or in computing interesting examples (Section 4). Finally, I shall report on recent experiments where different methods of computer algebra have been applied to symbolic-numerical solving of polynomial equations (Section 5), an important application of computer algebra to real life problems. Each of the sections contains a few unsolved problems, respectively projects, for further research.

The examples were chosen either from diploma theses of some of my students (T. Bayer, M. Schulze, M. Wenk), respectively from joint research projects together with C. Lossen and E. Shustin. All algorithms are implemented in the computer algebra system SINGULAR [30]. They are mainly based on Gröbner basis methods which were foundationally developed by Buchberger [6, 7] for polynomial rings. Subsequently, they have been extended to local and "mixed" rings in [28] for use in singularity theory.

Gröbner basis computations are, nowadays, implemented in all major general purpose computer algebra systems such as the big-M-systems (`Magma, Maple, Mathematica, MuPad`) but also in special systems designed for use in commutative algebra and algebraic geometry (`CoCoA, Macaulay,` SINGULAR). However, having only the possibility to compute Gröbner bases (w.r.t. a few monomial orderings) is, for applications to mathematical research problems, not much more than having the elementary numerical operations on a calculator for applications to engineering problems. Hence, I should like to emphasize the necessity to further develop packages and libraries to make the systems still more useful for the "working mathematician". Today, new and advanced algorithms can be built on already existing powerful procedures for computing, for example, free resolutions, Ext and Tor groups, sheaf cohomology, primary decomposition, ring normalization, versal deformations, and many more (cf. [30]). The development of new algorithms provides, in addition, a better understanding and often even produces new theoretical insight, as has been the case, just to mention one example, for primary decomposition (cf. [33, 20]). This has also been the case for some topics treated in the present article.

For more applications of computer algebra to algebraic geometry and singularity theory see [32].

We assume the reader is familiar with the main notions of Gröbner bases (cf. [12, 9]).

2. Monodromy and Gauß-Manin Connection

The monodromy of a morphism $f\colon X \to S$ between complex spaces or algebraic schemes/$\mathbb{C}$, which we suppose to be a differentiable fibre bundle outside the discriminant $\Delta \subset S$, describes the action of the fundamental group of $S \setminus \Delta$ on the cohomology $H^*(X_t, \mathbb{C})$ of the general fibre. The Gauß-Manin connection may be considered as an algebraic description of the monodromy action by means of differential forms. Finally, the mixed Hodge structure is an analytic structure on $H^*(X_t, \mathbb{C})$ generalizing the Hodge decomposition of compact, smooth algebraic varieties. These concepts have many applications and were widely studied in the global situation for proper maps as well as in the local situation for isolated singularities, for a survey see [35]. Here we shall consider only the local case.

Let $f \in \langle x \rangle \subset \mathbb{C}\{x_0, \ldots, x_n\}$ be a convergent power series (in practice a polynomial) with isolated singularity at 0 and $\mu = \dim_{\mathbb{C}} \mathbb{C}\{x\}/\langle f_{x_0}, \ldots, f_{x_n}\rangle$ the Milnor number of f. Then f defines in an ε-ball B_ε around 0 a holomorphic function $f\colon B_\varepsilon \to \mathbb{C}$, and, by a theorem of Milnor, there exists a small δ-disc S_δ in $\mathbb{C}$ around 0 such that $f\colon B_\varepsilon \setminus X_0 \to S_\delta \setminus \{0\}$ is a C^∞-fibre bundle so that the general fibre $X_t = f^{-1}(t)$, $t \neq 0$, is homotopy equivalent to a bouquet of μ n-dimensional spheres.

The simple, counterclockwise path γ in S_δ around 0 induces a C^∞-diffeomorphism of X_t $(t \neq 0)$ and an automorphism T of the singular cohomology group $H^n(X_t, \mathbb{C})$ which is a μ-dimensional $\mathbb{C}$-vector space. The automorphism T is called the local **Picard-Lefschetz monodromy** of f. We address the problem of computing the eigenvalues and the Jordan normal form of T.

The first important theorem is the monodromy theorem, due to Deligne in the global and to Brieskorn in the local situation which says that the eigenvalues of T are roots of unity, that is, we have $T = e^{2\pi i M}$, where M is a complex matrix with eigenvalues in $\mathbb{Q}$.

Hence, we are left with the problem of computing the eigenvalues and the Jordan normal form of M. Since X_t is a complex Stein manifold, its complex cohomology can be computed, via the holomorphic de Rham theorem, by using holomorphic differential forms, which is the starting point of Brieskorn's algorithm for computing the monodromy. To cut a long story short, we just mention that the Brieskorn lattices (cf. [5])

$$H' = \Omega^n / \left(df \wedge \Omega^{n-1} + d\Omega^{n-1}\right), \quad H'' = \Omega^{n+1}/df \wedge d\Omega^{n-1}$$

are free $\mathbb{C}\{t\}$-modules of rank μ. Here $(\Omega^\bullet, d)$ denotes the complex of holomorphic differential forms in $(\mathbb{C}^n, 0)$. We define the local **Gauß-Manin connection** of f as

$$\nabla : df \wedge H' = df \wedge \Omega^n / df \wedge d\Omega^{n-1} \longrightarrow H'', \quad [df \wedge \omega] \longmapsto [d\omega] .$$

Extending ∇ to an endomorphism of $H'' \otimes_{\mathbb{C}\{t\}} \mathbb{C}(t)$ and describing it with respect to a basis, we see immediately that the kernel of ∇, together with a basis of H'', is the same as the solutions of a rank μ system of ordinary differential equations

$$\frac{dy}{dt} = -Ay, \qquad A = (a_{ij}) = \sum_{i \geq -p} A_i t^i \in \mathrm{Mat}\big(\mu \times \mu, \mathbb{C}(t)\big) ,$$

in a neighbourhood of 0 in $\mathbb{C}$. The connection matrix A has a pole at $t = 0$ and is holomorphic for $t \neq 0$. If $\phi_t = (\phi_1, \dots, \phi_\mu)$ is a fundamental system of solutions at a point $t \neq 0$, then the analytic continuation of ϕ_t along the path γ transforms ϕ_t into another fundamental system ϕ_t' which satisfies $\phi_t' = T_\nabla \phi_t$ for some matrix $T_\nabla \in \mathrm{GL}(\mu, \mathbb{C})$.

It is a fundamental fact that the Picard-Lefschetz monodromy T coincides with the monodromy T_∇ of the Gauß-Manin connection.

Brieskorn [5] used this fact to describe the essential steps for an algorithm to compute the characteristic polynomial of T. Results of Gerard and Levelt [23] allowed the extension of this algorithm to compute the Jordan normal form of T. An implementation of Schulze in SINGULAR is able to compute interesting examples (including the uni- and bimodal singularities, [45]).

The algorithm uses the regularity theorem which says that there exists a basis of some lattice in $H'' \otimes \mathbb{C}(t)$ such that the connection matrix A has a pole of order 1.

Basically, if $A = A_{-1}t^{-1} + A_0 + A_1 t + \cdots$ has a simple pole, then $T = e^{2\pi i A_{-1}}$ is the monodromy (this holds if the eigenvalues of A_{-1} do not differ by integers which can be achieved algorithmically).

SINGULAR example for computing the monodromy (omitting the output):

```
> LIB "mondromy.lib";
> ring R = 0,(x,y),ds;
> poly f = x2y2+x6+y6;      //example of A'Campo (monodromy is not
> matrix M = monodromy(f); //diagonalisable)
```

```
> print(jordanform(M));     //prints Jordan normalform of monodromy
```

Ingredients for the implementation of Brieskorn's algorithm:

1. Computation of standard bases and normal forms for local orderings;
2. find k so that $f^k \in \langle f_{x_0}, \ldots, f_{x_n}\rangle$ and express f^k as linear combination of $f_{x_0}, \ldots, f_{x_n}$;
3. computation of the connection matrix on increasing lattices in $H'' \otimes \mathbb{C}(t)$ up to sufficiently high order (until saturation) by linear algebra over $\mathbb{Q}$;
4. computation of the transformation matrix to a simple pole by linear algebra over $\mathbb{Q}$.

The most expensive parts are certain normal form computations for a local ordering and the linear algebra part because here one has to deal iteratively with matrices with several thousand rows and columns.

In the remaining part of this section we describe a new algorithm, developed by M. Schulze, based on the theory of D-modules ($\mathrm{D} = \mathbb{C}\{t\}[\partial_t]$): the complex $\Omega^\bullet[D]$ with differential $\mathbf{d}$ defined by

$$\mathbf{d}(\omega D^k) := d\omega D^k - df \wedge \omega D^{k+1}$$

is a complex of D-modules with D-action

$$\partial_t \omega D^k = \omega D^{k+1}, \qquad t\omega D^k = f\omega D^k - k\omega D^{k-1}.$$

The D-module $\mathrm{H} := H^{n+1}(\Omega^\bullet[D], \mathbf{d}) = \Omega^{n+1}[D]/\mathbf{d}\Omega^n[D]$ is called the **Gauß-Manin system** of f. The operator ∂_t is invertible on H. For $k \geq 0$, let

$$F_k \Omega^{n+1}[D] := \bigoplus_{i=0}^{k} \Omega^{n+1} D^i$$

and $F_k\mathrm{H}$ be the image of $F_k\Omega^{n+1}[D]$ under the canonical map $\Omega^{n+1}[D] \to \mathrm{H}$. This defines a filtration F on H called the **Hodge filtration**. By the De Rham and Poincaré lemma, $df \wedge H' = \partial_t^{-1} F_0 \mathrm{H} \subset F_0\mathrm{H} = H''$. We denote by

$$\mathbb{C}\{\{\partial_t^{-1}\}\} := \left\{ \sum_{i \geq 0} a_i \partial_t^{-i} \in \mathbb{C}[[\partial_t^{-1}]] \;\middle|\; \sum_{i\geq 0} \frac{a_i}{i!} t^i \in \mathbb{C}\{t\} \right\},$$

the ring of **micro-differential operators** with constant coefficients and abbreviate $s := \partial_t^{-1}$. Then H'' is a free $\mathbb{C}\{\{s\}\}$-module of rank μ. By definition, $df \wedge \Omega^n \subset \Omega^{n+1}$ is isomorphic to the Jacobian ideal of f, and $\Omega_f = \Omega^{n+1}/df \wedge \Omega^n$ to the Milnor algebra. Using Gröbner basis methods, one can compute a monomial $\mathbb{C}$-basis $m = \{m_1, \ldots, m_\mu\}$ of Ω_f, inducing a section $v \in \mathrm{Hom}_{\mathbb{C}}(\Omega_f, H'')$ of the projection π and an isomorphism $\mathbb{C}\{\{s\}\}^\mu \cong H''$, by Nakayama's lemma. We define the matrix

$$B = \sum_{k\geq 0} B_k s^k \in \mathrm{Mat}\big(\mu \times \mu, \mathbb{C}\{\{s\}\}\big)$$

of multiplication by t with respect to m, i.e., $Bm := tm$. An easy computation shows that $B + s^2\partial_s$ is the basis representation of t with respect to m.

By definition of the differential $\mathbf{d}$, computing B up to order $k-1$ amounts to expressing k times an element of Ω^{n+1} in the basis m and $df \wedge \Omega^n$, which is the Jacobian ideal of f. This can be done using Gröbner basis methods. To do the k-th step in the computation of the saturation H''_∞ of H'', one has to compute B up to order k. To compute the residue of ∂_t on H''_∞, whose eigenvalues are the eigenvalues of monodromy, one has to compute B up to sufficiently high order and compute a $\mathbb{C}\{\{s\}\}$-basis of H''_∞ as well as the basis representation of the images of this basis under $\partial_t t$ with respect to this basis. This can also be done using Gröbner basis methods.

Compared to the Brieskorn algorithm, we have interchanged the roles of ∂_t^{-1} and t. The ∂_t^{-1}-structure of H'' is much more natural and there are many advantages of this new algorithm: There are no problems with estimations, no huge linear algebra problems, we need not lift a power of f in the Jacobian ideal, the basis of H'' is easier to compute, and so on. The main point is that we can continue the computation when we have to increase the order of B. In the Brieskorn algorithm, we have to start again almost from the beginning. Nevertheless, the three components of this new algorithm explained above also require difficult computations, especially the first one. The new algorithm can be extended to compute the Jordan normal form of the monodromy in a similar way as it was done in [45].

Problems

1. Generalize the algorithm of M. Schulze to isolated complete intersection singularities.
2. Find an algorithm to compute the V-filtration of the mixed Hodge structure of an isolated hypersurface singularity.
3. Compute the spectrum, resp. the spectral pairs, of an isolated hypersurface singularity.

The last problem was solved (and implemented in SINGULAR) by S. Endraß for nondegenerate singularities. M. Schulze has made progress in attacking 2. and 3.

3. Moduli Spaces and Invariants

When classifying objects in algebraic geometry, one usually fixes discrete invariants, such as the genus of a projective curve, and then one would like to have a distinct view on the set of objects with fixed invariants with respect to some equivalence relation. For small invariants it is sometimes possible to enumerate the equivalence classes and to provide normal forms. For bigger invariants this usually fails and a way to describe the objects is to construct a classifying space such that each point of this space corresponds to a unique equivalence class. In algebraic geometry this classifying space should again be an algebraic variety, together with certain functorial properties. These ideas lead to the notion of a fine, respectively coarse, **moduli space** ([40, 42]).

Classically, moduli spaces have been constructed for global algebraic objects such as projective varieties, or for vector bundles on a fixed projective variety. During the past years there has also been some progress in constructing moduli spaces for singularities (cf. [21]) and for Cohen-Macaulay modules on a fixed local ring of a curve singularity ([29], see also [31] for a survey). Indeed, the methods of proof are constructible and can be transferred to algorithms and finally to programmes.

In the following, we describe an algorithm to compute a moduli space for isolated hypersurface singularities, following [21]. The algorithm has been developed and implemented in SINGULAR by T. Bayer ([3]).

Let $\mathbf{w} = (w_1, \ldots, w_n) \in \mathbb{Z}^n$, $w_i > 0$, be a weight vector and $f \in \mathbb{C}\{x_1, \ldots, x_n\}$ a semiquasihomogeneous power series, i.e.,

$$f = f_0 + \sum_{\langle \mathbf{w},\alpha\rangle > d} c_\alpha x^\alpha, \quad f_0 = \sum_{\langle \mathbf{w},\alpha\rangle = d} c_\alpha x^\alpha$$

such that the quasi-homogeneous (or weighted homogeneous) principal part f_0 has an isolated singularity at the origin. We denote the class of all such power series (resp. singularities) by $\mathcal{C}_{f_0}$.

Two power series f, g are called **right equivalent**, $f \overset{r}{\sim} g$, if there exists a holomorphic coordinate change $\phi^\# : (\mathbb{C}^n, 0) \to (\mathbb{C}^n, 0)$ such that $f = g \circ \phi^\#$, or, equivalently, $f = \phi(g)$, where $\phi \in \mathrm{Aut}(\mathbb{C}\{x_1, \ldots, x_n\})$ is the algebra automorphism corresponding to $\phi^\#$.

In a series of papers, V. I. Arnold classified all isolated hypersurface singularities w.r.t. right equivalence up to modality 2, by giving normal forms [1].

Here we should like to present an algorithm to compute a moduli space for semiquasihomogeneous power series with fixed principal part w.r.t. right equivalence. In [21], also a moduli space for contact equivalence was constructed, but that construction is more involved and not treated here.

To start with we need an algebraic variety which parametrizes all semiquasihomogeneous power series (up to right equivalence) and then to identify equivalent objects. Indeed, equivalent objects belong to the same orbit of an algebraic group action and the aim is to compute explicitly the group, the action of the group and, finally, the quotient space.

Giving f_0, we compute the set of exponents $B \subset \mathbb{N}^n$ so that $\{x^\alpha \mid \alpha \in B\}$ is a monomial basis of the Milnor algebra $M_{f_0} = \mathbb{C}\{x_1, \ldots, x_n\}/\langle f_{0,x_1}, \ldots, f_{0,x_n}\rangle$. This requires a standard basis computation for a local ordering (cf. [27]). Then we select $B_- = \{\alpha \in B | \langle \mathbf{w}, \alpha\rangle > d\}$ and set $T_- = \mathbb{C}^k$ with $k = |B_-|$.

The polynomial

$$F_t(x) = f_0(x) + \sum_{\alpha \in B_-} t_\alpha x^\alpha$$

is the miniversal μ-constant unfolding of f_0. By a theorem of Arnold ([1]), for any $f \in \mathcal{C}_{f_0}$ there exists a $t \in T_-$ such that $f \overset{r}{\sim} F_t$. The next step in the algorithm is to compute, for a given $f \in \mathcal{C}_{f_0}$, a coordinate change ϕ and a $t \in T_-$ such that $\phi(f) = F_t$. The computation follows Arnold's proof, constructing ϕ degree by degree until the maximal weighted degree $\langle \mathbf{w}, \alpha\rangle$, $\alpha \in B_-$.

Usually there exist $t \neq t' \in T_-$ such that $F_t \overset{r}{\sim} F_{t'}$. However, we have the following fact (proved in [21] by using the Gauß-Manin connection): let $f, g \in \mathcal{C}_{f_0}$, $\varphi \in \operatorname{Aut}\mathbb{C}\{x\}$ and assume $\varphi(f) = g$. Then $\operatorname{ord}_{\mathbf{w}}(\varphi) \geq 0$, that is $\operatorname{ord}_{\mathbf{w}}\big(\varphi(x_i) - x_i\big) \geq w_i$ for $i = 1, \dots, n$.

In the theorem of Arnold, $\operatorname{ord}_{\mathbf{w}}(\phi) > 0$, which implies that $t \in T_-$ is unique. Moreover, $\operatorname{Aut}_{>0}\mathbb{C}\{x\} = \{\varphi \in \operatorname{Aut}\mathbb{C}\{x\} \mid \operatorname{ord}_{\mathbf{w}}(\varphi) > 0\}$ is a normal subgroup of $\operatorname{Aut}_{\geq 0}\mathbb{C}\{x\} = \{\varphi \in \operatorname{Aut}\mathbb{C}\{x\} \mid \operatorname{ord}_{\mathbf{w}}(\varphi) \geq 0\}$, and the quotient

$$G^{\mathbf{w}} = \operatorname{Aut}_{\geq 0}\mathbb{C}\{x\} / \operatorname{Aut}_{>0}\mathbb{C}\{x\}$$

acts algebraically on T_-. Let $G^{\mathbf{w}}_{f_0} \subset G^{\mathbf{w}}$ denote the subgroup which fixes f_0 and denote by $E_{f_0} \subset \operatorname{Aut}(T_-)$ the image of $G^{\mathbf{w}}_{f_0}$. Then E_{f_0} is a finite group acting algebraically on T_- and the geometric quotient T_-/E_{f_0} is the desired coarse moduli space for unfoldings in $\mathcal{C}_{f_0}$ modulo right equivalence (cf. [21]).

The following steps are needed for computing the moduli space:

0. Compute miniversal μ-constant unfolding,
1. compute $G^{\mathbf{w}}_{f_0}$,
2. compute the action of $G^{\mathbf{w}}_{f_0}$ on T_- using Arnold's theorem,
3. compute E_{f_0} and linearize to get E'_{f_0} acting linearly on some $\mathbb{C}^\ell$, $\ell \geq k$, and compute an equivariant embedding $i \colon T_- \hookrightarrow \mathbb{C}^\ell$,
5. determine generating invariant polynomials for E'_{f_0},
6. determine the relations between the invariants to get the equations for $i(T_-)/E'_{f_0} \cong T_-/E_{f_0}$, which is the desired moduli space.

SINGULAR example for computing the moduli space (we omit intermediate commands):

```
> LIB "qhmoduli.lib";
> ring R = 0, (x,y,z), ls;        // define a local ring
> poly f = x2y + x2z + y5 - z5; // principal part
```

Step 0. Compute a basis for the semi-universal unfolding.

```
> ideal B = UpperMonomials(f); B;
B[1]=y3z3, B[2]=x2y3, B[3]=x2y2
```

Hence, $F = f + t_1 y^3 z^3 + t_2 x^2 y^3 + t_3 x^2 y^2$ is the miniversal μ-constant unfolding. The dimension of the moduli space is 3.

Step 1, 2 and 3. Compute the equations of the stabilizer of f, compute the induced action on $T_- = \mathbb{C}^3$, linearize the action with equivariant embedding $T_- \hookrightarrow \mathbb{C}^4$

```
> list stab = StabEqn(f);  ....  // commands omitted
> actionid; //linearized action of E'_f on C^4 ⊃ T_- = C^3
actionid[1]=s(1)*t(1), actionid[2]=-s(3)*t(2)+s(3)*t(4)+s(5)*t(2)
actionid[3]=s(4)*t(3), actionid[4]=s(5)*t(4)
```

Step 4. Compute generators for the invariant ring of E'_f

```
> def T = InvariantRing(groupid,actionid); setring T;
> invars;              //there are 21 invariants of degree 3 to 10
invars[1]=t(1)*t(4)^2, invars[2]=t(2)*t(3)*t(4)-t(3)*t(4)^2, ...
```

Step 5. Compute equations of the moduli space

```
> def R4 = ImageVariety(V, invars);  //V is the ideal of T_ ⊂ C^4
> setring R4; imageid;        //simplified equation of moduli space
imageid[1]=Y(5)^2-Y(4)*Y(6), imageid[2]=Y(3)*Y(5)-Y(2)*Y(6), ...
imageid[55]=9*Y(1)^5+2816*Y(2)^2*Y(6)*Y(9)+296*Y(6)*Y(7)^2
  -152*Y(2)*Y(6)*Y(8)-960*Y(6)*Y(7)*Y(9)-9*Y(6)*Y(10)
```

Hence, the moduli space for $x^2y+x^2z+y^5-z^5$ is a 3-dimensional affine subvariety of $\mathbb{C}^{10}$ defined by 55 equations of degrees between 2 and 5.

This shows already that moduli spaces have a complicated structure, even for relatively small examples.

Problems

1. Extend the algorithms to construct moduli spaces for singularities with respect to contact equivalence. This will contain completely new parts since we need not only handle finite groups but unipotent groups.
2. Moduli spaces for torsion free modules on curve singularities have been constructed in [29] with constructive proofs. Again unipotent group actions come into play. It would be desirable to develop and implement algorithms and test conjectures related to the structure of these moduli spaces.

4. Curves with Prescribed Singularities

It is a classical and interesting problem, which is still in the centre of theoretical research, to study the variety $V = V_d(S_1,\dots,S_r)$ of (irreducible) curves $C \subset \mathbb{P}^2_{\mathbb{C}}$ of degree d having exactly r singularities of prescribed (topological or analytical) types $S_1,\dots,S_r$. Among the most important questions are:

- Is $V \neq \emptyset$ (existence problem)?
- Is V irreducible (irreducibility problem)?
- Is V smooth of expected dimension (T-smoothness problem)?

A complete answer is known only in the special case of nodal curves, that is, for $V_d(r) = V_d(S_1,\dots,S_r)$ with S_i ordinary nodes (A_1-singularities): $V_d(r) \neq \emptyset$ and T-smooth $\iff r \leq \frac{(d-1)(d-2)}{2}$ (Severi, 1921), $V_d(r)$ is irreducible (if $\neq \emptyset$) (Harris, 1985). Even for cuspidal curves there is no sufficient and necessary answer to any of the above questions and one can hardly expect such an answer.

Clearly, one can easily give an upper bound for the number of singular points that may occur on a plane irreducible curve C of degree d: by the genus formula C can have, at most, $(d-1)(d-2)/2$ singularities. Another upper bound for the (weighted) number of singularities arises from applying Bézout's Theorem to the intersection of two generic polars of C:

$$\sum_{z \in C} \mu(C, z) \leq (d-1)^2,$$

$\mu(C, z) = \dim_{\mathbb{C}} \mathbb{C}\{x, y\}/(f_x, f_y)$ the Milnor number of C at z.

On the other hand, in the case of arbitrary topological types S_i, we have the following existence theorem, which is asymptotically optimal (with respect to the occurring invariants and the exponent of d)

Theorem 4.1. ([24, 37]) *$V_d(S_1, \dots, S_r) \neq \emptyset$ if $\sum_{i=1}^{r} \mu(S_i) \leq \frac{1}{46}(d+2)^2$ and two additional conditions for the five "worst" singularities hold true.*

In case of only one singularity we have the slightly better sufficient condition for existence, $\mu(S_1) \leq \frac{1}{29}(d-5)^2$.

The theorem is just an existence statement, the proof gives no hint how to produce any equation. To produce explicit equations one needs some constructive method. Then the computer can be used in order to check the construction, or even, to improve the results. The following is a prominent example (actually, it belongs to a series of "world record" examples):

Example 4.2. ([26]) *The irreducible curve with affine equation*

$$y^2 - 2y\bigl(x^{28} + 2x^{21}y^{16} - 2x^{14}y^{32} + 4x^7y^{48} - 10y^{64}\bigr) + x^{56} + 4x^{49}y^{16} = 0$$

has degree 65 and an A_{2260}-singularity ($x^2 - y^{2261} = 0$) and a semiquasihomogeneous singularity $S_{9,16}$ with principal part $f_0 = x^9 + y^{16}$ as only singularities. In particular, it is an element of the variety $V_{65}(A_{2260}, S_{9,16})$ which has negative expected dimension (hence is not T-smooth).

In order to verify this, one may proceed, using Singular, as follows:

```
> ring s = 0,(x,y),ds;
> poly f = y2-2x28y-4x21y17+4x14y33-8x7y49+20y65+x56+4x49y16;
> matrix Hess = jacob(jacob(f));      //the Hessian matrix of f
> vdim(std(jacob(f))); //the Milnor number of f
2260
```

Since the rank of the Hessian at 0 is checked to be 1, f has an A_k singularity at 0; it is an A_{2260}-singularity since the Milnor number is 2260. In the following we show that the projective curve defined by f has no further singularities in the affine part. This follows from

$$\dim_{\mathbb{C}}(\mathbb{C}[x,y]_{\langle x,y\rangle}/\langle \mathrm{jacob}(f), f\rangle = \dim_{\mathbb{C}}(\mathbb{C}[x,y]/\langle \mathrm{jacob}(f), f\rangle,$$

confirmed by Singular:

```
> vdim(std(jacob(f)+f));
2260          // multiplicity of Sing(C) at 0 (local ordering)
> ring r = 0,(x,y),dp;
> poly f = fetch(s,f);
```

```
> vdim(std(jacob(f)+f));
2260            // multiplicity of Sing(C)     (global ordering)
```

Finally, we have to consider the singularities at infinity:

```
> ring sh = 0,(x,y,z),dp;
> poly f = fetch(s,f);
> poly F = homog(f,z); F;   // homogeneous polynomial defining C
4x49y16+20y65+x56z9-8x7y49z9+4x14y33z18-4x21y17z27-2x28yz36+y2z63
> ring r1 = 0,(y,z),dp;
> map phi = sh,1,y,z;
> poly g = phi(F);         // F in affine chart (x=1)
> vdim(std(jacob(g)+g));
120
> ring r2 = 0,(y,z),ds;  // local ring at (1:0:0)
> poly g = fetch(r1,g); g;
z9+4y16-2yz36-4y17z27+4y33z18-8y49z9+20y65+y2z63
> vdim(std(jacob(g)+g));
120
```

As before, we can conclude that there is precisely one singularity of C on the line at infinity, situated at $(1:0:0)$, being semiquasihomogeneous of type $S_{9,16}$. (Note that in our computation we have considered all points at infinity except $(0:1:0)$. The latter is obviously not a point of C).

In the following we should like to mention a few **problems and conjectures** which are currently in the centre of research in connection with singular curves in $\mathbb{P}^2_{\mathbb{C}}$.

Computing zero-dimensional ideals

Many of the questions concerning plane projective curves with prescribed singularities can be translated to properties of zero-dimensional (homogeneous) ideals $I \subset \mathbb{C}[x,y,z]$, e.g.,

- existence of curves with (ordinary) multiple points *in prescribed position*, or, more generally, existence of curves with prescribed position of infinitely near points (clusters),
- T-smoothness of the varieties $V_d(S_1,\ldots,S_r)$,
- existence of (global) deformations of projective curves.

For instance, consider the following problem: given points $p_1,\ldots,p_n \in \mathbb{P}^2_{\mathbb{C}}$ and positive integers $m_1,\ldots,m_n$. Determine the dimension of the variety of curves of any degree d passing through each of the points p_i with multiplicity (at least) m_i, $i=1,\ldots,n$. The equivalent formulation would be: determine the ideal

$$I = \mathfrak{m}_{p_1}^{m_1} \cap \cdots \cap \mathfrak{m}_{p_n}^{m_n} \subset \mathbb{C}[x,y,z]\,,$$

$\mathfrak{m}_{p_i}$ the maximal ideal at p_i, and compute the Hilbert function H_I of I.

Conjecture 4.3. (Harbourne-Hirschowitz) *Let $n > 9$, $p_1, \ldots, p_n \in \mathbb{P}^2_{\mathbb{C}}$ in general position, m a positive integer, and let $I = \mathfrak{m}_{p_1}^m \cap \cdots \cap \mathfrak{m}_{p_n}^m$. Then the Hilbert function satisfies*

$$H_I(d) = \max\left\{0, \frac{(d+1)(d+2)}{2} - n \cdot \frac{m(m+1)}{2}\right\}$$

for all $d > 0$. In other words, the variety of curves with n singular points of multiplicity (at least) m at the prescribed (generic) points has the expected dimension.

There are several special cases where this conjecture is known to hold true; in particular, C. Ciliberto and R. Miranda [14] have proven that it always holds for $m \leq 12$. Nevertheless the general conjecture is still far from being proven.

Conjecture 4.4. (Nagata) *Let $n>9$, $p_1, \ldots, p_n \in \mathbb{P}^2_{\mathbb{C}}$ in general position, $m_1, \ldots, m_n$ positive integers, and let $a(m_1, \ldots, m_n)$ denote the minimal degree of a curve passing through each of the points p_i with multiplicity (at least) m_i, $i = 1, \ldots, n$. Then*

$$a(m_1, \ldots, m_n) > \frac{m_1 + \cdots + m_n}{\sqrt{n}} .$$

N. Nagata [41] has proven the statement to be true for any $n > 9$ being a square. There are many people working to prove this conjecture for other integers n ([43]), or, at least, to improve the known lower bounds for $a(m_1, \ldots, m_n)$ (the best known general bound is probably given in [44]). But the general question is still widely open.

Computer algebra could be used to provide evidence for such conjectures (or, to produce counter examples) provided one can solve the following **problems:**

1. find algorithms to compute the 0-dimensional ideals (related to the above problems). In many cases this is easy but for others this is unknown (e.g., to compute the equisingularity ideal for a sufficiently general singularity);
2. find *fast* algorithms to compute the intersection of zero-dimensional ideals. The general method for computing intersections via syzygies or elimination is too slow, due to the high complexity of the algorithms involved. There is already some considerable progress made, by the so-called Buchberger-Möller algorithm and further generalizations (cf. [2]), but certainly this is not yet sufficient.

5. Symbolic-Numerical Polynomial Solving

Algebraic geometry is concerned with investigating the structure of the set of solutions of finitely many polynomial equations. Solving such a system is considered to be part of numerical analysis rather than of algebraic geometry. However, knowing something about the structure of the solution set can actually help in finding the solutions.

Given polynomials $f_1, \ldots, f_k \in K[x_1, \ldots, x_n]$, K being $\mathbb{R}$ or $\mathbb{C}$, numerical solving means determining the coordinates $(p_1, \ldots, p_n)$, up to a given precision of all (respectively some, respectively one) points of the variety

$$V = \{p = (p_1, \ldots, p_n) \in K^n \mid f_1(p) = \cdots = f_k(p) = 0\}.$$

Algebraically, we are interested in describing the structure of V, in computing its dimension, in the number of solutions (if finite), in the decomposition into irreducible varieties (e.g., by primary decomposition), in the radical of the ideal I generated by $f_1, \ldots, f_k$ or the normalization of the ring $K[x_1, \ldots, x_n]/I$. Since V, the set of solutions of $f_1 = \cdots = f_k = 0$ depends only on I, even only on the radical of I, all the above mentioned methods can be used in preparing the given polynomial system for better numerical solving.

Hence, algebraic geometry and computer algebra may be used as symbolic preprocessing for easy numerical postprocessing. In particular, the following algorithms and methods may be applied (all being implemented in SINGULAR).

- Find other generators of I, or of ideals with the same solution set, for example, triangular sets, based on Gröbner basis computations (see below), which allow better numerical solving: the numerical algorithms become more stable, we can solve overdetermined systems and find all solutions.
- Create other ideals, having the same solution set, for example, the *radical* for obtaining only simple zeros, *primary decomposition* for splitting the system into several smaller ones.
- Compute a parametrization of the solution set V, which is only possible for rational V, sometimes it is achieved by the *normalization* of $K[x_1, \ldots, x_n]/I$.
- Reduce higher dimensional solving to 0-dimensional solving by applying a *Noether normalization.*

Pure numerical solving has the advantage of being fast, flexible in accuracy by using iterative methods and being applicable not only to polynomial systems. Indeed, the big success of numerical methods during the past years seems to show that symbolic methods are of little use in solving systems coming from real life problems. However, due to rounding errors, numerical methods are principally uncertain, often unstable in an unpredictable way, sometimes do not find all solutions and have problems with overdetermined systems. Moreover, they can hardly treat underdetermined systems (sometimes curves, at most surfaces, as solution sets) and certainly get into trouble near singularities.

On the other hand, symbolic methods are principally exact and stable. However they have a high complexity, are, therefore, slow, and, in practice, are applicable only to small systems (this is the case, in particular, for radical computation, primary decomposition and normalization). Nevertheless, they are applicable to any polynomial system of any dimension and for zero-dimensional systems they can predict precisely the number of complex solutions (counted with multiplicities). Moreover, as is well known, symbolic preprocessing of a system of polynomials (even of ordinary and partial differential equations) may not only lead to better conditions for the system to be solved numerically but can help to find all solutions or even make numerical solving possible (see below).

There is continuous progress in applying symbolic methods to numerical solving, cf. the various articles in the ISSAC Proceedings, the survey article by Möller [39], the textbook by Cox, Little and O'Shea [13] or the recent paper by Verschelde [49]. Besides Gröbner basis many other methods have been used. Recently, resultant methods have been re-popularized, in particular in connection with numerical solving (cf. [11, 51]), partly due to the new sparse resultants by Gelfand, Kapranov and Zelevinsky [22]. I should also mention the work of Stetter (cf. [47, 48]), which is not discussed in this paper.

In the following I shall describe roughly how Gröbner bases and resultants can be applied to prepare for numerical solving of zero-dimensional systems. Moreover, I shall present experimental material for comparing the performance of the two methods, which seems to be the first practical comparison of resultants and Gröbner bases in connection with numerical solving under equal conditions. The motivation for doing this came from a collaboration with electrical engineers, aiming at symbolic analysing and sizing micro-electric analog circuits.

Let $f_1, \ldots, f_k \in K[x_1, \ldots, x_n]$, and assume that the system $f_1 = \cdots = f_k = 0$ has only finitely many complex solutions. The problem is to find all solutions up to a given precision. We present two methods, one by computing a lexicographical Gröbner basis and then splitting this into triangular sets, the second by computing the sparse u-resultants and the determinants of the partly evaluated resultant matrix. Both methods end up with the problem of solving univariate polynomial equations for which we use Laguerre's method.

Solving polynomial systems using Gröbner bases and triangular sets:

Input: Zero-dimensional system $f_1, \ldots, f_k \in K[x_1, \ldots, x_n]$, $k \geq n$.
Output: Complex roots of $f_1 = \cdots = f_k = 0$ in $\mathbb{C}^n$.

- Compute a reduced lexicographical Gröbner basis $G = \{g_1, \ldots, g_s\}$ of the ideal $I = \langle f_1, \ldots, f_k \rangle$ with $s \geq n$.
- Compute a triangular system: a *triangular basis* is a reduced lexicographical Gröbner basis $G = \{g_1, \ldots, g_n\}$ (as many polynomials as variables) with g_i of the form $g_i = x_i^{p_i} + g_i'(x_i, \ldots, x_n)$ with $\deg_{x_i} g_i' < p_i$. A *triangular system* for I consists of triangular bases $T_1, \ldots, T_s$ such that $V(I) = V(T_1) \cup \cdots \cup V(T_s)$. Triangular systems can be computed effectively, basically by two different methods, one due to Lazard [36, 18], the other due to Möller [39]. Choose any of these methods to compute a triangular system $T_1, \ldots, T_s$ for I.
- Use a numerical solver (e.g. Laguerre's method) to find all zeros of T_i, $i = 1, \ldots, s$. The union of these zero-sets is the desired solution set.

There are several variations on how to compute triangular sets. The $V(T_i)$ need not be disjoint (but can be made disjoint). The T_i need not define maximal ideals (but this can be achieved), we may use the factorising Gröbner, etc. Some of these have been implemented in SINGULAR by D. Hillebrand, a former student of M. Möller (cf. [34]).

SINGULAR example (the output has been changed to save space):

```
> ring s = 0,(x,y,z),lp;
> ideal i = x2+y+z-1,x+y2+z-1,x+y+z2-1;
> option(redSB); //option for computing a reduced Groebner basis
> ideal j = groebner(i); j;
j[1]=z6-4z4+4z3-z2, j[2]=2yz2+z4-z2, j[3]=y2-y-z2+z, j[4]=x+y+z2-1
> LIB "triang.lib";
> triangMH(j);     //triangular system with Moeller's method
>                  //(fast, but not necessarily disjoint)
[1]:                    [2]:
   _[1]=z2                 _[1]=z4-4z2+4z-1
   _[2]=y2-y+z             _[2]=2y+z2-1
   _[3]=x+y-1              _[3]=2x+z2-1
> triangMH(j,2); //triangular system (with Moeller's method,
>                //improved by Hillebrand) and factorisation
[1]:            [2]:            [3]:                [4]:
   _[1]=z          _[1]=z          _[1]=z2+2z-1        _[1]=z-1
   _[2]=y          _[2]=y-1        _[2]=y-z            _[2]=y
   _[3]=x-1        _[3]=x          _[3]=x-z            _[3]=x
```

We can now solve the system easily by recursively finding roots of univariate polynomials, SINGULAR commands are:

```
> LIB "solve.lib";
> triang_solve(triangMH(j,2),30);    //accuracy of 30 digits
```

or applying `triangLf_solve(i);` directly to i.

Resultant methods have recently become popular again, due to new sparse resultants invented by Gelfand, Kapranov and Zelevinsky [22]. Indeed, they beat by far the classical Macaulay resultants as is shown, for example, in [50]. The following algorithm to use sparse resultants for polynomial solving is due to Canny and Emiris [11], it has been implemented in SINGULAR by M. Wenk [50].

For computing resultants, we have to start with a zero-dimensional polynomial system with as many equations as variables, and have to compute the u-resultant (named so by Van der Waerden) where u_i are new variables which have to be specialized later. The construction of the sparse resultant matrix uses a mixed polyhedral subdivision of the Minkowski sum of the Newton polytopes. Specializations of the u-coordinates are used then to reduce the problem to the univariate case. The determinants of the specialized u-resultant matrices are univariate polynomials, the roots are determined by Laguerre's algorithm.

The main advantage of the sparse resultants against Macaulay resultants is that the sizes of the resultant matrices depend on the Newton polytopes and not just on the degrees of the input polynomials, hence is much smaller. Note that by the resultant method we can only determine roots in $(\mathbb{C} \setminus \{0\})^n$.

Here is a more detailed description of the algorithm (for details see [11, 50]):

Solving polynomial systems using resultants

(After Gelfand, Kapranov, Zelevinsky (1994) and Canny, Emiris (1997)):

Input: Zero-dimensional system $f_1, \dots, f_n \in \mathbb{C}[x_1, \dots, x_n]$.
Output: Complex roots of $f_1 = \cdots = f_n = 0$ in $(\mathbb{C} \setminus \{0\})^n$.

- Add $f_0 = u_0 + u_1 x_1 + \cdots + u_n x_n \in \mathbb{K}[u_0, \dots, u_n, x_1, \dots, x_n]$ and compute the Newton polytopes $Q_i \subset \mathbb{R}^n$ of f_i, $i = 0, \dots, n$.
- Compute, using linear programming, a polyhedral subdivision Δ of the Minkowski sum $Q = Q_0 + \cdots + Q_n$. Translate Δ by a small vector in $\mathbb{R}^n$ such that lattice points are interior points.
- Construct from Δ the square resultant matrix $M(u_0, \dots, u_n)$, which has as entries either a number or a variable u_i.
- Set $M^i(u_0) := M(u_0, 0, \dots, 0, -1, 0, \dots, 0)$, -1 at the i-th place. The set L_i of all roots of $\det\bigl(M^i(u_0)\bigr)$, $i = 1, \dots, n$, contains the i-th components of all complex solutions of the system $f_1 = \cdots = f_n = 0$ (in an unordered manner).
- For each solution of $f_1 = \cdots = f_n = 0$ identify the components which were computed in the previous step. This is done by substituting $u_1, \dots, u_i$ by random numbers and $u_{i+1}, \dots, u_n$ by 0 in $M(u_0, u_1, \dots, u_n)$, computing the determinant and solve this for u_0.

Most of the time is spent in the second last and, in particular, in the last step.

As an example, we show the computation of the complex zero-set of the ideals $I_1, \dots, I_5$ (with precision of 30 digits, see below) which represent more than 60 examples. On average our resultant solver could manage the same examples as Mathematica and MAPLE (but MAPLE found fewer roots). The problems for the resultant solver occurred either because of too big matrices or because of numerical problems for the subsequent Laguerre solver (no convergence). Sometimes not all solutions were found, sometimes the system returned too many solutions because multiple solutions were interpreted as different ones.

Our experiments do not confirm the claim made in [51] that resultant methods are best suited to polynomial systems solving, at least for bigger examples. On the contrary, Gröbner bases and triangular sets showed the best performance, identified most precisely the correct number of simple roots, could treat many more examples and had the least numerical difficulties. The most expensive part was usually the computation of a lexicographical Gröbner basis (computed through FGLM); the triangular set computation was less expensive and the numerical solving depended very much on the example.

Of course, the Gröbner bases in SINGULAR are highly tuned and the resultant computations can certainly be improved. The examples show that resultant methods are a good alternative for small examples. But for many variables even the sparse resultants become huge and we have to compute several determinants of these matrices. This is the main bottleneck for the resultant method. Nevertheless, still more research has to be done.

Interpretation of the table: vars: number of variables, mult: number of complex solutions with multiplicity, # roots: number of different roots without multiplicity, time: total time, degree of res.: number of (not necessarily simple roots) in $(\mathbb{C}\setminus\{0\})^n$, matrix size: number of rows of (square) resultant matrix.
Commands used:

- `triang_solve(I);` and `ures_solve(I);` (SINGULAR),
- `NSolve(eqns,vars,30);` (Mathematica),
- `evalf(solve(eqns,vars),30);` (MAPLE),
- `GROESOLVE(eqns,vars);`, respectively `solve(I,var);` with options `on complex; on rounded; precision 30;` (REDUCE).

No	vars	mult	#roots	SINGULAR 1-3-7				
				Triang. systems		Resultant method		
				roots found	time	degree of res.	time	matrix size
1	3	53	32	32	24 sec	31	~ 40 sec	~ 160
2	4	16	16	16	2 sec	16	~ 5 sec	~ 70
3	5	70	70	70	5 sec	70	~ 50 sec	~ 880
4	6	156	156	156	22 sec	156	> 5000 sec	~ 5460
5	10	27	27	27	66 sec	30	> 5000 sec	~ 10000

No	Mathematica 4.0		MAPLE V.5		REDUCE 3.7	
	roots found	time	roots found	time	roots found	time
1	37	100 sec	—	> 5000 sec	—	> 5000 sec
2	16	9 sec	1	1 sec	16	22 sec

$I_1 = \langle x^3z+6x^2-2xz^3+y^4+yz,\ 5x^2y^2+y^3z+3yz^3+z^3,\ -x^2z-2xyz^2+4y^4+2y^2z\rangle \subset \mathbb{C}[x,y,z]$,

$I_2 =$ (cf. [51]) $\langle 5x^2+6x+3y+6z+2w+3, 4x+4y^2+3z+2w+4, x-11z^2+3z+7w-9, x+3y+2z-3w^2+13\rangle \subset \mathbb{C}[x,y,z,w]$,

$I_3 =$ (cyclic 5) $\langle a+b+c+d+e,\ ab+ae+bc+cd+de,\ abc+abe+ade+bcd+cde,\ abcd+abce+abde+acde+bcde,\ abcde-1\rangle \subset \mathbb{C}[a,b,c,d,e]$,

$I_4 =$ (Arnborg 6) $\langle a+b+c+d+e+f,\ ab+af+bc+cd+de+ef,\ abc+abf+aef+bcd+cde+def,\ abcd+abcf+abef+adef+bcde+cdef,\ abcde+abcdf+abcef+abdef+acdef+bcdef,\ abcdef-1\rangle \subset \mathbb{C}[a,b,c,d,e,f]$,

$I_5 =$ (POSSO, Methan6_1) $\langle -10ai-320000a+64bh+10hj+11hk,\ 160000a-32bh-5bi-5bk-5000b,\ -ci+gi+210g+jk+1300000,\ -ei+700000,\ -2f+k^2,\ -gi-210g+hj,\ 320000a-64bh-10hj-11hk-16h+7000000,\ ei-hj-jk-410j,\ -10ai-10bi+10bk+20000b-10ci-10ei+14f-10gi+11hk,\ 10bi-10bk+10ci+10gi+1400g+10hj-11hk-10jk-10k^2-4200k\rangle \subset \mathbb{C}[a,b,c,d,e,f,g,h,i,j,k]$.

The computations, with precision of 30 digits, were performed on a Pentium Pro 200 with 128 MB.

MAPLE and REDUCE could not solve examples 3–5 within our time limit of 5000 seconds, Mathematica stopped with an error.

MAPLE offers also the opportunity to preprocess a set of ideal generators in view of solving, by using the command `gsolve;`. But within our time limit of 5000 seconds, this also lead only to a result for example 2. However, in this case, MAPLE is able to compute all roots, by successively applying `fsolve(eqn,var,complex);` and substituting.

Acknowledgement

I should like to thank T. Bayer, C. Lossen und M. Schulze for helping to prepare this article and M. Möller for useful comments concerning symbolic-numerical solving.

Development and implementation of algorithms on which this paper is based were supported by the DFG-Schwerpunkt “Effiziente Algorithmen für diskrete Probleme und ihre Anwendungen“ and by the “Stiftung Rheinland-Pfalz für Innovation”, which we kindly acknowledge.

References

[1] V. I. Arnold, S.M. Gusein–Zade and A. N. Varchenko, *Singularities of Differential Maps*, Volume I. Birkhäuser, 1985.

[2] J. Abbott, M. Kreuzer and L. Robbiano, *Computing zero-dimensional schemes*, Preprint in preparation (2000).

[3] T. Bayer, *Computation of moduli spaces for semiquasihomogeneous singularities and an implementation in* SINGULAR, Diplomarbeit, Kaiserslautern, 2000.

[4] A. Borel, *Linear Algebraic Groups*, 2nd Edition, Springer-Verlag, 1991.

[5] E. Brieskorn, *Die Monodromie der isolierten Singularitäten von Hyperflächen*, Manuscripta Math., **2**, 103–161 (1970).

[6] B. Buchberger, *Ein Algorithmus zum Auffinden der Basiselemente des Restklassenringes nach einem nulldimensionalen Polynomideal*, PhD Thesis, University of Innsbruck, Austria, 1965.

[7] B. Buchberger, *Ein algorithmisches Kriterium für die Lösbarkeit eines algebraischen Gleichungssystems*, Äqu. Math., **4**, 374–383 (1970).

[8] B. Buchberger, *Gröbner bases: an algorithmic method in polynomial ideal theory*, in: Recent trends in multidimensional system theory, N.B. Bose, ed., Reidel (1985).

[9] B. Buchberger and F. Winkler, *Gröbner Bases and Applications*, LNS **251**, 109–143, CUP (1998).

[10] T. Becker and V. Weispfenning, *Gröbner Bases, A Computational Approach to commutative Algebra*, Graduate Texts in Mathematics 141, Springer-Verlag, 1993.

[11] J.F. Canny and I.Z. Emiris, *A Subdivision-Based Algorithm for the Sparse Resultant*, Preprint, Berkeley, 1997.

[12] D. Cox, J. Little and D. O'Shea, *Ideals, Varieties and Algorithms*, Springer-Verlag, 1992.

[13] D. Cox, J. Little and D. O'Shea, *Using Algebraic Geometry*, Springer-Verlag, 1998.

[14] C. Ciliberto and R. Miranda, *Linear Systems of Plane Curves with Base Points of Equal Multiplicity*, Duke preprint, no. math.AG/9804018 (1998).

[15] H. Derksen, *Constructive Invariant Theory and the Linearization Problem*, PhD Thesis, University of Basel, 1997.

[16] W. Decker, G.-M. Greuel, T. de Jong and G. Pfister, *The normalization: a new algorithm, implementation and comparisons*, in: Proc. EUROCONFERENCE Computational Methods for Representations of Groups and Algebras (1.4.–5.4.1997), 177–185, Birkhäuser, 1998.

[17] W. Decker, G.-M. Greuel and G. Pfister, *Primary decomposition: algorithms and comparisons*, in: G.-M. Greuel, B. H. Matzat and G. Hiß (Eds.), *Algorithmic Algebra and Number Theory*, 187–220, Springer-Verlag, 1998.

[18] J. Della Dorca, C. Dicrescenzo and D. Duval, *About a new method for computing in algebraic number fields*, EUROCAL 1985, LN in CS 204, 289–290 (1985).

[19] D. Eisenbud, *Commutative Algebra with a view toward Algebraic Geometry.* Springer-Verlag, 1995.

[20] D. Eisenbud, C. Huneke and W. Vasconcelos, *Direct methods for primary decomposition*, Invent. Math., **110**, 207–235 (1992).

[21] G.-M. Greuel, C. Hertling and G. Pfister, *Moduli Spaces of Semiquasihomogeneous Singularities with fixed Principal Part.* J. of Alg. Geom. **6**, no. 1, 169–199 (1997).

[22] I. Gelfand, M. Kapranov and A. Zelevinski, *Discriminants, Resultants and Multidimensional Determinants*, Birkhäuser, 1994.

[23] R. Gerard and A.H.M. Levelt, *Invariants mesurant l'irrégularité en un point singulier des systèmes d'équations différentielles linéaires*, Ann. Inst. Fourier Grenoble, **23**, 157–195 (1973).

[24] G.-M. Greuel, C. Lossen and E. Shustin, *Plane curves of minimal degree with prescribed singularities*, Invent. Math. **133**, 539–580 (1998).

[25] G.-M. Greuel, B.H. Matzat and G. Hiß (Eds.), *Algorithmic Algebra and Number Theory*, 187–220, Springer-Verlag, 1998.

[26] S.M. Gusein–Zade and N.N. Nekhoroshev, *On A_k-singularity on a plane curve of fixed degree*, Duke preprint, no. math.AG/9906147 (1999).

[27] G.-M. Greuel and G. Pfister, *Advances and improvements in the theory of standard bases and syzygies*, Arch. Math. **66**, 163–176 (1996).

[28] G.-M. Greuel and G. Pfister, *Gröbner bases and algebraic geometry*, in: B. Buchberger and F. Winkler (Eds.): Gröbner Bases and Applications, LNS **251**, 109–143. CUP (1998).

[29] G.-M. Greuel and G. Pfister, *Moduli spaces for torsion free modules on curve singularities I*, J. Algebraic Geometry **2**, 81–135 (1993).

[30] G.-M. Greuel, G. Pfister and H. Schönemann. SINGULAR, *A System for Polynomial Computations*, version 1.2 User Manual, in: Reports On Computer Algebra, number 21. Centre for Computer Algebra, University of Kaiserslautern, June 1998. Available via `http://www.singular.uni-kl.de`.

[31] G.-M. Greuel, *Deformation und Klassifikation von Singularitäten und Moduln*, Jahresber. Deutsche Math.–Verein Jubiläumstagung 1990, 177–238 (1992).

[32] G.-M. Greuel, *Computer algebra and algebraic geometry – achievements and perspectives*, to appear in JSC, 2000.

[33] P. Gianni, B. Trager and G. Zacharias, *Gröbner Bases and Primary Decomposition of Polynomial Ideals*, J. Symbolic Computation **6**, 149–167 (1988).

[34] D. Hillebrand, *Triangulierung nulldimensionaler Ideale – Implementierung und Vergleich zweier Algorithmen*, Diplomarbeit, Dortmund, 1999.

[35] V. Kulikov, *Mixed Hodge Structures and Singularities*, Cambridge Tracts in Mathematics **132**, Cambridge University Press, 1998.

[36] D. Lazard, *Solving Zero-dimensional Algebraic Systems*, J. Symbolic Computation **13**, 117–131, (1992).

[37] C. Lossen, *The geometry of equisingular and equianalytic families of curves on a surface*, PhD Thesis, University of Kaiserslautern, 1998.

[38] J. Milnor, *Singular Points of Complex Hypersurfaces*, Ann. of Math. Studies **61**, Princeton (1968).

[39] H. M. Möller, *Gröbner Bases and Numerical Analysis*, in: B. Buchberger and F. Winkler (Eds.): *Gröbner Bases and Applications.* LNS 251, 159–178, CUP (1998).

[40] D. Mumford and J. Fogarty, *Geometric Invariant Theory*, Springer-Verlag, 1982.

[41] N. Nagata, *On the fourteenth problem of Hilbert*, Amer. J. Math. **81**, 766–772, (1959).

[42] P. E. Newstead, *Introduction to moduli problems and orbit spaces*, Lecture Notes, Tata Institute of Fundamental Research, Springer-Verlag, 1978.

[43] Z. Ran, *On the Nagata Problem*, Duke preprint, no. math.AG/9809101 (1998).

[44] J. Roé, *On the existence of plane curves with prescribed multiple points*, Duke preprint, no. math.AG/9807066 (1998).

[45] M. Schulze, *Computation of the Monodromy of an Isolated Hypersurface Singularity*, Diploma Thesis, Kaiserslautern, 1999.

[46] B. Sturmfels, *Algorithms in Invariant Theory*, Springer-Verlag, 1993.

[47] H. Stetter, *Matrix eigenproblems are at the heart of polynomial systems solving*, Sigsam Bulletin **30**, 22–25 (1996).

[48] H. Stetter, *Stabilization of polynomial systems solving with Groebner bases*, in Proceedings of ISSAC 97, 117–124 (1997).

[49] J. Verschelde, *Polynomial homotopies for dense, sparse and determinantal systems*, Duke preprint, no. math.NA/9907060 (1999).

[50] M. Wenk, *Resultantenmethoden zur Lösung algebraischer Gleichungssysteme implementiert in* SINGULAR, Diplomarbeit, Kaiserslautern, 1999.

[51] A. Wallack, I. Z. Emiris and D. Monacho, *MARS: A Maple/Matlab/C Resultant-Based Solver*, in Proceedings of ISSAC 98, 244–251 (1998).

Department of Mathematics
University of Kaiserslautern
Erwin-Schrödinger-Straße
D-67663 Kaiserslautern, Germany
E-mail address: `greuel@mathematik.uni-kl.de`

Explicit Towers of Drinfeld Modular Curves

Noam D. Elkies

Abstract. We give explicit equations for the simplest towers of Drinfeld modular curves over any finite field, and observe that they coincide with the asymptotically optimal towers of curves constructed by Garcia and Stichtenoth.

1. Introduction

Fix a finite field k_1 of size q_1. It has been known for almost twenty years (see [16, 18, 17]) that any curve C/k_1 of genus $g = g(C)$ has at most $(q_1^{1/2} - 1 + o(1))g$ points rational over k_1 as $g \to \infty$ (the *Drinfeld-Vlăduţ bound* [1]), and that if q_1 is a square then there are various families of classical, Shimura, or Drinfeld modular curves C/k_1 of genus $g \to \infty$ with $\#(C(k_1)) \geq (q_1^{1/2} - 1)(g-1)$. Thus such curves are "asymptotically optimal": they attain the lim sup of $\#(C(k_1))/g(C)$. Moreover, asymptotically optimal curves yield excellent linear error-correcting codes over k_1 [12, 17].

To actually construct and use these Goppa codes one needs explicit equations for asymptotically optimal C. Now the definitions of modular curves are in principle constructive, but it is usually not feasible to actually exhibit a given modular curve. For certain Drinfeld modular curves, [17, pp. 453ff.] gives explicit, albeit unpleasant, models as plane curves with two complicated singularities. Garcia and Stichtenoth gave nice formulas for asymptotically optimal families of curves in [6] for $q_1 = 4, 9$ (see also [7]), and in [4] for all square prime powers q_1. In each case they construct a sequence of curves $C_1, C_2, C_3, \ldots$ forming what we'll call a *recursive tower.* A "tower" $\{C_n\}$ is a sequence of curves in which each C_n is given as a low-degree cover of C_{n-1}. We shall call a tower "recursive" if C_2 is given as a curve in $C_1 \times C_1$, and C_n is the curve in C_1^n consisting of n-tuples $(P_1, \ldots, P_n)$ of points in C_1 such that $(P_j, P_{j+1}) \in C_2$ for each $j = 1, 2, \ldots, n-1$. That is, C_n is obtained by iterating $n-1$ times the correspondence from C_1 to itself given by C_2. In each of the cases discovered by Garcia and Stichtenoth, analysis of this correspondence yields the genus of each C_n and enough k_1-rational points on C_n to prove that $\#(C_n(k_1)) \geq (q_1^{1/2} - 1)g(C_n)$. The genus grows exponentially with n, yet only $O(n)$ equations of bounded degree are needed to exhibit C_n.

Now while equations for most modular curves cannot be reasonably exhibited, modular curves whose conductors are products of small primes form towers, and modular curves with a single repeated factor even form recursive towers. For instance, for any integers $N_0 \geq 1$ and $l > 1$, the classical modular curves $X_0(N_0 l^n)$ $(n = 1, 2, 3, \dots)$ form a recursive tower of curves related by maps of degree l. There are analogous towers of Shimura and Drinfeld modular curves. We observed in [3] that this can be used to exhibit asymptotically optimal towers, and carried out the computation of explicit equations for eight such towers, six of classical modular curves and two of Shimura curves. Moreover, we showed that the towers of classical modular curves $\mathrm{X}_0(2^n)$ and $\mathrm{X}_0(3^n)$ in characteristics 3 and 2 respectively are the same as the towers constructed in [6].

In the same paper we noted that similar methods can be used to exhibit towers of Drinfeld modular curves, and announced that one such tower recovers the curves constructed in [4] for arbitrary square q_1. In the present paper we specify this tower and perform the computations that determine its equations and thus identify it with the Garcia-Stichtenoth tower. We do the same for a closely related tower obtained by Garcia and Stichtenoth in [5].

In the next section we recall basic definitions of Drinfeld modules, isogenies, and supersingularity. The following section describes certain Drinfeld modular curves. In both sections we simplify the exposition by describing only the modules and curves that arise in the interpretation of the Garcia-Stichtenoth towers; for a thorough treatment of the general case we refer to [9]. In the final section we give explicit equations for several of the simplest towers of Drinfeld modular curves and observe that two of them coincide with asymptotically optimal towers obtained by Garcia and Stichtenoth.

2. Drinfeld Modules and Isogenies

Fix a finite field k of size q. (In our application, k_1 will be the quadratic extension of k, so $q_1 = q^2$; the use of "k" instead of "$\mathbf{F}_q$" is our only divergence from the notations of [9].) For any field $L \supseteq k$, we denote by $L\{\tau\}$ the non-commutative L-algebra generated by τ and satisfying the relation $\tau a = a^q \tau$ for all $a \in L$. Equivalently, $L\{\tau\}$ is the ring of endomorphisms of $\mathbf{G}_a$ defined over L and linear over k. Explicitly, the polynomial $\sum_{i=0}^{n} l_i \tau^i$ acts on $\mathbf{G}_a$ as the endomorphism taking any X to the q-linearized polynomial $\sum_{i=0}^{n} l_i X^{q^i}$.

We only consider Drinfeld modules of rank 2, and usually only ones associated to a function field of genus 0 with a place at infinity of degree 1. Let then $K = k(T)$ and $A = k[T]$. (See [9] for the general case, in which A is the ring of functions on a curve over k with poles at most at a fixed place ∞ of the curve.) In general, a Drinfeld module is defined as a k-algebra homomorphism $\phi\colon a \mapsto \phi_a$ from A to $L\{\tau\}$ satisfying certain technical conditions. In our case, $A = k[T]$, so specifying ϕ is equivalent to choosing ϕ_T. The rank of the resulting Drinfeld module is then simply the degree of ϕ_T as a polynomial in τ. Thus for us

$$\phi_T = l_0 + l_1\tau + l_2\tau^2 = l_0 + g\tau + \Delta\tau^2\,, \tag{1}$$

with nonzero *discriminant* $\Delta = \Delta(\phi)$. In general the map $\gamma\colon A \to L$ taking any $a \in A$ to the "constant term" (τ^0 coefficient) of ϕ_a is a ring homomorphism; in our case γ is determined by $\gamma(T) = l_0$. If ϕ, ψ are two Drinfeld modules, an *isogeny* from ϕ to ψ is a $u \in \bar{L}\{\tau\}$ such that

$$u \circ \phi_a = \psi_a \circ u \tag{2}$$

for all $a \in A$. For our A, (2) holds for all $a \in A$ if and only if it holds for $a = T$. The *kernel* of the isogeny is[1]

$$\ker(u) := \{x \in \bar{L} : \phi_u(x) = 0\}\,. \tag{3}$$

This is a k-vector subspace of $\bar{L}$, which is of finite dimension unless $u = 0$. By (2), $\phi_a(x) \in \ker(u)$ for all $x \in \ker(u)$, so $\ker(u)$ in fact has the structure of an A-module. Conversely, for every finite $G \subset \bar{L}$ which is an A-submodule of $\bar{L}$ for the ϕ-action of A on $\bar{L}$, one may define $u \in \bar{L}\{\tau\}$ of degree $\dim_k G$ by

$$u(X) = \prod_{x \in G} (X - x)\,, \tag{4}$$

and then u is an isogeny with kernel G from ϕ to some Drinfeld module ψ_a. In particular, if $u = \phi_{a_1}$ then (2) holds with $\psi = \phi$ for any $a_1 \in A$; thus ϕ_{a_1} is an isogeny from ϕ to itself, called *multiplication by* a_1. If $\gamma(a_1) \neq 0$, the kernel of this isogeny is isomorphic with $(A/a_1A)^2$ as an A-module [9, Proposition I.1.6]. Elements of $\ker(\phi_{a_1})$ are called a_1*-division points* or a_1*-torsion points* of the Drinfeld module ϕ. In particular, the T-torsion points are the roots in $x \in \bar{L}$ of

$$\phi_T(X) = \gamma(T)X + gX^q + \Delta X^{q^2}\,. \tag{5}$$

If γ is not injective then $\ker\gamma = Aa_0$ for some irreducible $a_0 \in A$. Then the a_0-torsion points of ϕ constitute a vector space of dimension 1 or 0 over the field A/a_0A. The Drinfeld module ϕ is then said to be *supersingular* if $\ker(\phi_{a_0}) = \{0\}$, *ordinary* otherwise. We shall use the case $\deg(a_0) = 1$, when $\phi_{a_0} = g\tau + \Delta\tau^2$ and thus ϕ is supersingular if and only if $g = 0$. Note that a_0 is of degree 1 if and only if $\gamma(T) \in k$, and that Drinfeld modules over $\bar{k}$ with $\gamma(T) \in k$ may arise as the "reduction mod $(T - \gamma(T))$" of Drinfeld modules over $\bar{k}(T)$ with $\gamma(T) = T$.

[1] When u is not separable, i.e. has τ^0 coefficient zero, it is for many purposes better to consider $\ker(u)$ not as a subgroup of $\bar{L}$ but as a group subscheme of $\mathbf{G}_a$. We shall not need this refinement here. Note that the condition $\gamma(a) \neq 0$ occurring later in this paragraph is equivalent to the separability of ϕ_a.

3. Drinfeld Modular Curves

An *isomorphism* between Drinfeld modules is an invertible isogeny, i.e. some $u \in \bar{L}^*$ satisfying (2). This isomorphism multiplies each coefficient l_i in (1) by u^{1-q^i}. We define the *J-invariant* of a Drinfeld module ϕ given by (1) as follows:[2]

$$J(\phi) = \frac{g^{q+1}}{\Delta}. \tag{6}$$

Two Drinfeld modules with the same γ are isomorphic (over $\bar{L}$) if and only if their J-invariants are equal. Thus, in analogy with the case of the classical modular curves parametrizing elliptic curves, we refer to the J-line as the *Drinfeld modular curve* $\mathrm{X}(1)$ for Drinfeld modules with a given γ. Likewise, for $N \in A$ such that $\gamma(N) \neq 0$, we have Drinfeld modular curves $\mathrm{X}_0(N)$, $\mathrm{X}_1(N)$, $\mathrm{X}(N)$ parametrizing Drinfeld modules with a given γ and a choice of torsion subgroup $G \cong A/NA$, or such a subgroup G together with a generator of G as an A-module, or an identification of the group of N-torsion points with $(A/NA)^2$. These are finite separable covers of $\mathrm{X}(1)$, all of which except $\mathrm{X}(N)$ ($N \notin k^*$) are geometrically irreducible; $\mathrm{X}(N)$ is a normal cover of $\mathrm{X}(1)$ with Galois group $\mathrm{GL}_2(A/NA)$, and $\mathrm{X}_1(N)$ is an abelian normal cover of $\mathrm{X}_0(N)$ with Galois group $(A/NA)^*$. Note that, unlike $\mathrm{X}(1)$, these curves $\mathrm{X}_0(N)$, $\mathrm{X}_1(N)$, $\mathrm{X}(N)$ generally depend on the choice of γ. If $\gamma(T) \in k$, we may regard the curves $\mathrm{X}(1)$, $\mathrm{X}_0(N)$, $\mathrm{X}_1(N)$, $\mathrm{X}(N)$ as the "reduction mod $(T - \gamma(T))$" of the corresponding modular curves for $\gamma(T) = T$. More generally, reducing the $\gamma(T) = T$ curves modulo any irreducible $a_0 \in A$ yields the curves parametrizing Drinfeld modules for which $\gamma(T)$ is a root of a_0.

If γ is not injective, we say that a point on a Drinfeld modular curve is *ordinary* or *supersingular* according as the Drinfeld modules it parametrizes are ordinary or supersingular. It is known that in this case the supersingular points constitute a nonempty finite set. For instance, we have seen in effect that if $\gamma(T) \in k$ then $\mathrm{X}(1)$ has the unique supersingular point $J = 0$. The supersingular points on $\mathrm{X}_0(N)$, $\mathrm{X}_1(N)$ and $\mathrm{X}(N)$ are then the preimages of $J = 0$ under the natural maps from those Drinfeld modular curves to $\mathrm{X}(1)$. It is known that each supersingular point on $\mathrm{X}_0(N)$ is defined over the quadratic extension k_1 of k, and that the same is true for certain twists of $\mathrm{X}_1(N)$ and of each component of $\mathrm{X}(N)$.

We shall relate the Garcia-Stichtenoth curves to certain Drinfeld modular curves with $N = T^n$ and $\gamma(T) = 1$. We shall find that in some cases the modules parametrized by these curves must satisfy the additional condition $\Delta = -1$, i.e.

$$\phi_T = 1 + g\tau - \tau^2. \tag{7}$$

We call such Drinfeld modules *normalized.* A Drinfeld module ϕ with $\gamma(T) = 1$ is isomorphic to a normalized one if and only $-\Delta(\phi)$ is a (q^2-1)-st power. This

[2]Usually a lower-case j is used for this. We use a capital J to forestall confusion with the integer variable j appearing in the next section. For elliptic curves one sometimes sees $J = j/12^3$; in the Drinfeld modular setting, no factor analogous to 12^3 is needed, so we might plausibly claim that the invariant of a Drinfeld module corresponds to J as well as j in the classical theory of modular invariants of elliptic curves ...

condition is invariant under isogeny. One thus expects that there would be an equivalent condition in terms of the torsion structure of ϕ. Such a condition cannot be given in completely elementary terms, because $\mathrm{GL}_2(A)$ does not have a large enough cyclic quotient. But[3] one can give an equivalent condition in terms of $\wedge^2\phi$. For a general Drinfeld module ϕ given by (1), "$\wedge^2\phi$" is the rank-1 module $T \mapsto l_0 - l_2\tau$, whose Tate modules are the discriminants of those of ϕ [14, Theorem 4.1]. Thus ϕ is normalized if and only if $\wedge^2\phi$ takes T to $1-\tau$.

In terms of the coordinate J on $X(1)$, the condition that ϕ be equivalent to a normalized module is that $-J(\phi)$ be a $(q+1)$-st power. Thus normalized modules, or normalized modules with suitable level-N structure, are parametrized by curves we shall call $\dot{\mathrm{X}}(1)$, $\dot{\mathrm{X}}_0(N)$, $\dot{\mathrm{X}}_1(N)$, $\dot{\mathrm{X}}(N)$, whose function fields are obtained from those of $\mathrm{X}(1)$, $\mathrm{X}_0(N)$, $\mathrm{X}_1(N), \mathrm{X}(N)$ by adjoining a $(q+1)$-st root of $(-J)$.

Now by analogy with the case of classical and Shimura curves, one might expect that the curves $\mathrm{X}_0(N)$ are asymptotically optimal over k_1, and that the same is true of the twists mentioned earlier of $\mathrm{X}_1(N)$ and of the components of $\mathrm{X}(N)$, with supersingular points already providing $(q-1+o(1))g$ rational points in each case. One might even hope that the same is true with X_0, X_1, X replaced by $\dot{\mathrm{X}}_0$, $\dot{\mathrm{X}}_1$, $\dot{\mathrm{X}}$. It would be enough to prove this for $\dot{\mathrm{X}}(N)$, since all the other curves listed are quotients of $\dot{\mathrm{X}}(N)$ by subgroups of $\mathrm{GL}_2(A/NA)$. This is stated explicitly in the literature only for components of $\mathrm{X}(N)$ in the case that N is an irreducible polynomial of odd degree (see for instance [17, pp. 449ff.]). We expect that the same is true for $\dot{\mathrm{X}}(N)$ and arbitrary N, and indeed even for modular curves for Drinfeld modules over rings other than $k[T]$. In each case the supersingular points are readily enumerated, and the main technical challenge is computing the genus of the curve, since the covering maps to $\mathrm{X}(1)$ are highly and wildly ramified above the "cusp" $J=\infty$. [This was also the most difficult part of Garcia and Stichtenoth's direct construction in [4].]

Even though explicit statements of asymptotic optimality have not been made, the genera of $\mathrm{X}_0(N)$, $\mathrm{X}_1(N)$, and the components of $\mathrm{X}(N)$ are known. They were computed in [13, Theorem 4.4] for $\mathrm{X}(N)$, and also in Gekeler's thesis [8, Satz 3.4.8], which also deals with the case of $\mathrm{X}_0(N)$ (Satz 3.4.18). The curves $\mathrm{X}_1(N)$ were treated in [11]. These results, combined with the enumeration of supersingular points, should yield the asymptotic optimality of all these curves over k_1. Alternatively, M. Zieve suggests, and Gekeler confirms by e-mail, that one may be able to entirely avoid the genus computation and the enumeration of supersingular points by adapting an earlier proof by Ihara [15, pp. 292–3] that a classical modular curve of genus g has at least $(p-1)(g-1)$ points rational over the field of p^2 elements. That argument uses the reduction mod p of certain Hecke correspondences on the curve, such as $\mathrm{X}_0(Np)$ considered as a correspondence on $\mathrm{X}_0(N)$ via its map to $\mathrm{X}_0(N) \times \mathrm{X}_0(N)$ (when $\gcd(p,N)=1$); similar correspondences are available in the Drinfeld modular setting. We are not aware of a published analysis along the same lines of the curves $\dot{\mathrm{X}}_0(N)$ etc.; but as Gekeler

[3] Thanks to Bjorn Poonen for pointing this out, and to David Goss for the reference to Hamahata.

points out it should be straightforward to obtain their genera from those of $\mathrm{X}_0(N)$ etc., of which they are cyclic covers of degree prime to q. At any rate the results of the next section, together with the genus calculations in [4, 5], show at least that the curves $\dot{\mathrm{X}}_0(T^n)$ are asymptotically optimal.

4. Some Drinfeld Modular Curves of Conductor T^n

If x_1 is a nonzero torsion point of ϕ then we can solve for g by setting (5) equal to zero, obtaining

$$0 = T(x_1) = x_1 + gx_1^q - x_1^{q^2}, \quad \text{i.e.} \quad g = x_1^{-q}(x_1^{q^2} - x_1). \tag{8}$$

Thus x_1 may be regarded as a coordinate for the rational curve $\dot{\mathrm{X}}_1(T)$ parametrizing normalized Drinfeld modules ϕ with a T-torsion point. For any nonzero $x \in \bar{L}$ we let t_x be the corresponding linearized polynomial

$$t_x(X) := X + \frac{x^{q^2} - x}{x^q} X^q - X^{q^2}. \tag{9}$$

The supersingular x_1 are those for which the X^q coefficient of t_{x_1} vanishes, i.e. the units of the quadratic extension k_1 of k. (The points $x_1 = 0, \infty$ are the cusps of $\dot{\mathrm{X}}_1(T)$.) We next construct the curves we shall call $\dot{\mathrm{X}}'_0(T^n)$, parametrizing ϕ with such a torsion point x_1 as well as a torsion group $G_n \cong A_0[T]/T^n$ containing x_1, and find the supersingular points on these curves. Note that $\dot{\mathrm{X}}'_0(T) = \dot{\mathrm{X}}_1(T)$, but for $n \geqslant 2$ the curve $\dot{\mathrm{X}}'_0(T^n)$ is only a quotient of $\dot{\mathrm{X}}_1(T^n)$, because, as in the case of elliptic curves, we demand not that each point of G_n be rational, only that G_n be permuted by Galois.

For $j \leqslant n$ let G_j be the group $T^{n-j}G_n$ of T^j-torsion points in G_n. Of course $G_1 = kx_1$. For any $x \neq 0$ let P_x be the linearized polynomial

$$P_x(X) = x^{q-1}X - X^q \tag{10}$$

vanishing on kx. Since t_x vanishes on kx we expect it to factor through P_x (see for instance [2, Proposition 3]), and indeed we find

$$t_x(X) = Q_x(P_x(X)) \tag{11}$$

where Q_x is the linearized polynomial

$$Q_x(X) = x^{1-q}X + X^q. \tag{12}$$

Let $t'_x(X)$ be the reverse composition $P_x \circ Q_x$, given by

$$t'_x(X) := P_x(Q_x(X)) = X + (x^{q-1} - x^{q-q^2})X^q - X^{q^2}. \tag{13}$$

Note that t'_x again gives a normalized Drinfeld module of rank 2, namely the module T-isogenous[4] to ϕ obtained as the quotient of ϕ by G_1. Now suppose y_2 is a generator of G_2 such that $t_{x_1}(y_2) = x_1$. There are q such y_2, all differing by multiples of x_1, so $x_2 := P_{x_1}(y_2)$ does not depend on the choice of t_2; conversely

[4] i.e. with an isogeny (here P_x or Q_x) whose kernel is isomorphic with A/TA as an A-module.

x_2 determines G_2. Since $t_{x_1} = Q_{x_1} \circ P_{x_1}$, the condition $t_{x_1}(y_2) = x_1$ is equivalent to $Q_{x_1}(x_2) = x_1$. Recalling the definition (12) and multiplying by x_1^q we find

$$x_2 = x_1^{-1} z_2 \quad \text{where} \quad z_2^q + z_2 = x_1^{q+1}. \tag{14}$$

Thus the curve $\dot{X}_0'(T^2)$ is just $z_2^q + z_2 = x_1^{q+1}$, which is known to be k_1-isomorphic with the Fermat curve of degree $q+1$ (a.k.a. the "Hermitian curve" over k_1). Note that G_2 is the k-vector space of zeros of the linearized polynomial $P_{x_2} \circ P_{x_1}$. Moreover,

$$0 = P_{x_1}(Q_{x_1}(x_2)) = t_{x_1}'(x_2). \tag{15}$$

That is, x_2 is a T-torsion point on our T-isogenous module. It follows that $t_{x_1}' = t_{x_2}$. By induction we can now determine the tower of curves $\dot{X}_0'(T^n)$ explicitly: the function field of $\dot{X}_0'(T^n)$ is generated by $x_1, x_2, \ldots, x_{n-1}$ with relations $Q_{x_{j-1}}(x_j) = x_{j-1}$, or equivalently

$$x_j = x_{j-1}^{-1} z_j \quad \text{where} \quad z_j^q + z_j = x_{j-1}^{q+1} \tag{16}$$

$(1 < j < n)$; the point $(x_1, x_2, \ldots, x_{n-1})$ parametrizes the Drinfeld module with $\phi_T = t_{x_1}$, with T^n-torsion subgroup generated by any of the q^{n-1} solutions of

$$(P_{x_{n-1}} \circ P_{x_{n-2}} \circ \cdots \circ P_{x_2} \circ P_{x_1})(y_n) = x_n. \tag{17}$$

This works because by applying

$$Q_{x_j} \circ P_{x_j} = T_{x_j} = P_{x_{j-1}} \circ Q_{x_{j-1}} \tag{18}$$

$(n-2)$ times we find

$$\begin{aligned} x_{n-1} = Q_{x_{n-1}}(x_n) &= (Q_{x_{n-1}} \circ P_{x_{n-1}} \circ P_{x_{n-2}} \circ \cdots \circ P_{x_1})(y_n) \\ &= (P_{x_{n-2}} \circ Q_{x_{n-2}} \circ P_{x_{n-2}} \circ \cdots \circ P_{x_1})(y_n) \\ = \cdots &= (P_{x_{n-2}} \circ \cdots \circ P_{x_2} \circ Q_{x_1} \circ P_{x_1})(y_n) \\ &= (P_{x_{n-2}} \circ \cdots \circ P_{x_2} \circ P_{x_1} \circ T_{x_1})(y_n) \\ &= (P_{x_{n-2}} \circ \cdots \circ P_{x_2} \circ P_{x_1})(T_{x_1}(y_n)), \end{aligned} \tag{19}$$

i.e. $T_{x_1}(y_n)$ satisfies the equation for y_{n-1}. As with G_2 we see that G_n is the k-vector space of zeros of the linearized polynomial

$$P_{x_n} \circ P_{x_{n-1}} \circ \cdots \circ P_{x_2} \circ P_{x_1}. \tag{20}$$

In [4] Garcia and Stichtenoth obtained many k_1-rational points on these curves as follows. For each of the $q^2 - 1$ points on $\dot{X}_0'(T)$ with $x_1 \in k_1^*$ we have $x_1^{q+1} \in k_1^*$, so the q solutions z_2 of (14) are just the q elements of k_1 whose trace to k is x_1^{q+1}. Clearly none of these z_2 vanish, so $x_2 = x_1 z_2$ is again in k_1^*. Inductively we see that x_1 lies under q^{n-1} rational points of $\dot{X}_0'(T^n)$ defined over k_1. Garcia and Stichtenoth use the resulting $(q^2-1)q^{n-1}$ rational points to confirm that the $\dot{X}_0'(T^n)$ form an asymptotically optimal tower over k_1. From the Drinfeld modular viewpoint we recognize $x_1 \in k_1^*$ as the condition that a point on $\dot{X}'_0(T^n)$ be supersingular. Thus the $(q^2-1)q^{n-1}$ points of $\dot{X}'_0(T^n)$ lying above k_1^* are precisely the supersingular points on $\dot{X}'_0(T^n)$.

The group k_1^* acts on $\dot{X}'_0(T^n)$ by

$$c(x_1,\dots,x_n)=(cx_1,\bar{c}x_2,cx_3,\bar{c}x_4,\dots,c^{q^{n-1}}x_n)\quad(c\in k_1^*,\bar{c}:=c^q)\,. \tag{21}$$

If $c\in k^*$ then the automorphism (21) preserves ϕ (see (8)) and each G_j, but changes the generator x_1 of G_1 to cx_1. Thus we recover $\dot{X}_0(T^n)$ as the quotient of $\dot{X}'_0(T^n)$ by the action of k^*. The tower of curves $\dot{X}_0(T^n)$ was obtained in this way (again without mention or use of Drinfeld modules) in [5]. For arbitrary $c\in k_1^*$, the automorphism (21) multiplies g by the $(q+1)$-st root of unity $c/\bar{c}$. This takes ϕ to a Drinfeld module ϕ_c with the same J-invariant but a different choice of $(-J)^{1/(q+1)}$. The isomorphism c from ϕ to ϕ_c respects our choice of subgroups G_j. Thus the quotient of $\dot{X}_0(T^n)$ by k_1^*/k^*, or equivalently of $\dot{X}'_0(T^n)$ by $k)1^*$, is the Drinfeld modular curve $X_0(T^n)$.

We next obtain an explicit description of $\{X_0(T^n)\}$ as a recursive tower. As with several of the examples in [3], it will be convenient to start the tower at $n=2$. [To start at $n=1$ we would have to use $x_j^{q^2-1}$ as the j-th coordinate, and then all the supersingular points would be on the normalization of the highly singular point $(x_1^{q^2-1},\dots,x_n^{q^2-1})=(1,\dots,1)$ on the resulting curve.] The curve $X_0(T)$ is the quotient of the x_1-line $\dot{X}'_0(T)$ by k_1^*, so has genus zero and coordinate $x_1^{q^2-1}$. To obtain $X_0(T^2)$, raise both sides of the equation (14) for $\dot{X}'_0(T^2)$ to the power $q-1$ to obtain

$$Z_2(1+Z_2)^{q-1}=x_1^{q^2-1},\quad\text{where } Z_2:=z_2^{q-1}=(x_1x_2)^{q-1}\,. \tag{22}$$

Now Z_2 is invariant under k_1^*, and (22) shows that the Z_2-line is a degree-q cover of the $x_1^{q^2-1}$-line $X_0(T)$; since the cover $X_0(T^2)/X_0(T)$ is also of degree q, we conclude that Z_2 generates the function field of $X_0(T^2)$.

Thus for $n\geq 2$ the function field of $X_0(T^n)$ is generated by

$$Z_j:=z_j^{q-1}\quad(2\leq j\leq n)\,. \tag{23}$$

For each $j=2,\dots,n-1$, we have $Z_{j+1}(1+Z_{j+1})^{q-1}=x_j^{q^2-1}$ as in (22), which in turn equals

$$(z_j/x_{j-1})^{q^2-1}=Z_j^{q+1}/x_{j-1}^{q^2-1}=Z_j^q/(1+Z_j)^{q-1}\,. \tag{24}$$

Thus

$$Z_{j+1}(1+Z_{j+1})^{q-1}=Z_j^q/(1+Z_j)^{q-1}\,. \tag{25}$$

This gives Z_{j+1} as an algebraic function of degree q in Z_j (and vice versa). Thus the relations (25) for $j=2,\dots,n-1$ determine the function field of $X_0(T^n)$. From our description of the supersingular points on $\dot{X}'_0(T^n)$ we see that the q^{n-1}

supersingular points on $X_0(T^n)$ are the points for which each Z_j is in

$$\begin{aligned} &\{Z \in k_1 : Z^{q+1} = 1, Z \neq -1\} \\ =&\{Z \in k_1 : Z(1+Z)^{q-1} = 1\} \\ =&\{Z \in k_1 : Z^q = (1+Z)^{q-1}\}. \end{aligned} \tag{26}$$

Acknowledgements

I thank Bjorn Poonen for introducing me to Drinfeld modules, and Henning Stichtenoth for information on his and Garcia's asymptotically optimal towers. Thanks also to E.-U. Gekeler, D. Goss, B. H. Gross, B. Poonen, and M. Zieve for much enlightening communication concerning Drinfeld modules and modular curves.

This work was made possible in part by funding from the Packard Foundation.

References

[1] S. G. Drinfeld and S. G. Vlăduţ: The number of points of an algebraic curve. *Functional Anal. Appl.* **17** (1983), #1, 53–54 (translated from the Russian paper in *Funktsional. Anal. i Prilozhen*).

[2] N. D. Elkies: Linearized algebra and finite groups of Lie type, I: Linear and symplectic groups. Pages 77–108 in *Applications of Curves over Finite Fields* (1997 AMS-IMS-SIAM Joint Summer Research Conference, July 1997, Washington, Seattle; M. Fried, ed.; Providence: AMS, 1999) = *Contemp. Math.* **245**.

[3] N. D. Elkies: Explicit modular towers. Pages 23–32 in *Proceedings of the Thirty-Fifth Annual Allerton Conference on Communication, Control and Computing* (1997, T. Basar, A. Vardy, eds.), Univ. of Illinois at Urbana-Champaign 1998.

[4] A. Garcia and H. Stichtenoth: A tower of Artin-Schreier extensions of function fields attaining the Drinfeld-Vlăduţ bound. *Invent. Math.* **121** (1995), #1, 211–233.

[5] A. Garcia and H. Stichtenoth: On the asymptotic behaviour of some towers of function fields over finite fields. *J. Number Theory* **61** (1996), #2, 248–273.

[6] A. Garcia and H. Stichtenoth: Asymptotically good towers of function fields over finite fields. *C. R. Acad. Sci. Paris I* **322** (1996), #11, 1067–1070.

[7] A. Garcia, H. Stichtenoth and M. Thomas: On towers and composita of towers of function fields over finite fields. *Finite Fields and their Appl.* **3** (1997), #3, 257–274.

[8] E.-U. Gekeler: *Drinfeld-Moduln und modulare Formen über rationalen Funktionenkörpern.* Bonner Math. Schriften 119, 1980.

[9] E.-U. Gekeler: *Drinfeld Modular Curves.* Berlin: Springer-Verlag, 1980 (Lecture Notes in Math. 1231).

[10] E.-U. Gekeler: Über Drinfeld'sche Modulkurven vom Hecke-Typ. *Compositio Math.* **57** (1986), #2, 219–236.

[11] E.-U. Gekeler and U. Nonnengardt: Fundamental domains of some arithmetic groups over function fields. *International J. Math.* **6** (1995), #5, 689–708.

[12] V. D. Goppa: Codes on algebraic curves. *Soviet Math. Dokl.* **24** (1981), #1, 170–172.

[13] D. Goss: π-adic Eisenstein series for function fields. *Compositio Math.* **41** (1980), #1, 3–38.

[14] Y. Hamahata: Tensor products of Drinfeld modules and v-adic representations. *Manusc. Math.* **79** (1993), #3–4, 307–327.

[15] Y. Ihara: Congruence relations and Shimura curves. Pages 291–311 of *Automorphic Forms, Representations, and L-functions* (A. Borel and W. Casselman, eds.; Providence: AMS, 1979; Part 2 of Vol. 33 of Proceedings of Symposia in Pure Mathematics).

[16] Y. Ihara: Some remarks on the number of rational points of algebraic curves over finite fields. *J. Fac. Sci. Tokyo* **28** (1981), #3, 721–724.

[17] M. A. Tsfasman and S. G. Vlăduţ: *Algebraic-Geometric Codes.* Dordrecht: Kluwer, 1991.

[18] M. A. Tsfasman, S. G. Vlăduţ and T. Zink: Modular curves, Shimura curves and Goppa codes better than the Varshamov-Gilbert bound. *Math. Nachr.* **109** (1982), 21–28.

Department of Mathematics
Harvard University
Cambridge, MA 02138 USA
E-mail address: elkies@math.harvard.edu

Curves over Finite Fields Attaining the Hasse-Weil Upper Bound

Arnaldo Garcia

Abstract. Curves over finite fields (whose cardinality is a square) attaining the Hasse-Weil upper bound for the number of rational points are called *maximal curves.* Here we deal with three problems on maximal curves:

1. Determination of the possible genera of maximal curves.
2. Determination of explicit equations for maximal curves.
3. Classification of maximal curves having a fixed genus.

1. Introduction

The theory of equations over finite fields (or the theory of congruences) is in the basis of classical number theory. Its foundations were laid, among others, by mathematicians like Fermat, Euler, Lagrange, Gauss, and Galois (see Dickson's book [6]). Historically, the object of the first investigations in this theory were the congruences of the special form

$$y^2 \equiv f(x) \quad \text{(modulo a prime number)}, \tag{1}$$

where $f(x)$ is a polynomial (or rational function) with integer coefficients. Such congruences were used to get results such as the representability of integers as sums of four squares, or the distribution of pairs of quadratic residues, or even the estimation of the sum of Legendre's quadratic residues symbols.

E. Artin constructed a quadratic extension of the field $\mathbb{F}_p(x)$, p a prime, by adjoining the roots of the congruence (1) and he introduced a zeta-function for this field, in analogy with Dedekind's zeta-function for quadratic extensions of the field of rational numbers. Assuming that Riemann's hypothesis was valid for his zeta-function, Artin conjectured an upper bound for the number of solutions of congruences such as the one in (1) above. Artin's conjecture was then proved by Hasse for polynomials $f(x)$ of degrees 3 and 4 over arbitrary finite fields, and widely generalized by A. Weil (see [29]) as follows. Let X be a projective geometrically irreducible nonsingular algebraic curve of genus g, defined over a finite field $\mathbb{F}_\ell$ with ℓ elements. Then,

$$|\#X(\mathbb{F}_\ell) - (\ell + 1)| \leq 2g\sqrt{\ell}, \tag{2}$$

where $X(\mathbb{F}_\ell)$ denotes the set of $\mathbb{F}_\ell$-rational points of the curve X. Inequality (2) is equivalent to the validity of Riemann's hypothesis for the zeta-function associated to the curve X. Bombieri [2] gave an elementary proof of Inequality (2) following ideas of Stepanov, Postnikov and Manin that were used to treat the special case of hyperelliptic curves; see Chapter 5 in [25].

The interest in curves over finite fields with many rational points was renewed after Goppa's construction of codes with good parameters from such curves, see [15]. The number of solutions of congruences in two variables has other applications such as estimates of exponential sums over finite fields (see [22]), finite geometries (see [16]), correlations of shift register sequences (see [21]).

Here we will be interested in *maximal curves over* $\mathbb{F}_\ell$ with $\ell = q^2$, that is, we will consider curves X attaining Hasse-Weil's upper bound:

$$\#X(\mathbb{F}_{q^2}) = q^2 + 1 + 2gq\,.$$

It is often the case that maximal curves are special (and interesting) from other points of view. For example, it is often the case that they have large automorphism groups, see [18] and [26]. Also, they are always nonclassical for the canonical linear series if $g \geq q-1$, see Proposition 1.7 in [7].

We will consider here three important problems on maximal curves over $\mathbb{F}_{q^2}$:

1. Determination of the *possible genera* of maximal curves over $\mathbb{F}_{q^2}$.
2. Determination of *explicit equations* for maximal curves over $\mathbb{F}_{q^2}$.
3. *Classification* of maximal curves over $\mathbb{F}_{q^2}$ of a given genus.

The methods used to deal with these three problems are: the action of the Frobenius morphism on the Jacobian of a maximal curve (see the fundamental equation (3) here), Weierstrass Point Theory (including Stöhr-Voloch theory of Frobenius orders of a morphism), Castelnuovo's genus bound for curves in projective spaces and Riemann-Hurwitz genus formula for separable coverings of algebraic curves. It is also crucial that a subcovering of a maximal curve is also a maximal curve.

2. The Genera of Maximal Curves

There are only finitely many possibilities for the genus g of a maximal curve X over $\mathbb{F}_{q^2}$, since Ihara [17] has shown that

$$g \leq q(q-1)/2\,.$$

The proof of the inequality above uses the knowledge of the zeta-function associated to a maximal curve and the following trivial observation:

$$\#X(\mathbb{F}_{q^4}) \;\geq\; \#X(\mathbb{F}_{q^2})\,.$$

Not every positive integer g with $g \leq q(q-1)/2$ can be the genus of a maximal curve over $\mathbb{F}_{q^2}$. It was shown in [8] that the genus g of a maximal curve over $\mathbb{F}_{q^2}$

satisfies (see also [27]):

$$\text{if} \quad g < q(q-1)/2, \quad \text{then} \quad g \le (q-1)^2/4 .$$

In other terms, the second largest genus g of a maximal curve over $\mathbb{F}_{q^2}$ is given by

$$g = [(q-1)^2/4], \quad \text{where the brackets mean the integer part}.$$

The proof of this fact (i.e., that there are no genera of maximal curves over $\mathbb{F}_{q^2}$ in the open interval $\left(\frac{(q-1)^2}{4}, \frac{q(q-1)}{2}\right)$) uses the Castelnuovo genus bound for curves in projective spaces and the following "*fundamental linear equivalence of divisors on a maximal curve over* $\mathbb{F}_{q^2}$":

$$qP + FrP \sim (q+1)P_0 , \tag{3}$$

where P is any point on the curve, FrP is the image of P by the $\mathbb{F}_{q^2}$-Frobenius morphism and P_0 is any $\mathbb{F}_{q^2}$-rational point on the curve.

It should be pointed out here that the exact value of the third largest genus of a maximal curve is still unknown (see [5]).

It is a result of Serre (see [19]) that if X is maximal over $\mathbb{F}_{q^2}$ and Y is covered by X over $\mathbb{F}_{q^2}$ (i.e., there is a surjective morphism $X \to Y$ defined over $\mathbb{F}_{q^2}$), then Y is also maximal over $\mathbb{F}_{q^2}$. This leads one to the consideration of quotient curves X/G of a maximal curve X over $\mathbb{F}_{q^2}$ under the action of subgroups G of the automorphism group $\mathrm{Aut}(X)$ of X. One then may hope to get several genera of maximal curves by applying the Riemann-Hurwitz formula to the covering $X \twoheadrightarrow X/G$ in order to determine the genus of the curve X/G. These ideas were systematically used in [11], where the curve X is taken as the Hermitian curve; i.e., the curve X given by the affine equation:

$$y^q + y = x^{q+1} \quad \text{over} \quad \mathbb{F}_{q^2} . \tag{4}$$

This is a maximal curve over $\mathbb{F}_{q^2}$ with the biggest genus possible (i.e., with genus given by $g = q(q-1)/2$). The advantage in taking the Hermitian curve X is its huge automorphism group (see [26] and [20])

$$|\mathrm{Aut}(X)| = (q^2-1)\cdot q^3 \cdot (q^3+1) .$$

Here it is worth mentioning that there is no example of a maximal curve for which it is known that it cannot be covered by the Hermitian curve.

By a systematic use of subgroups of $\mathrm{Aut}(X)$, it is determined in [11] lots of possible genera of maximal curves over $\mathbb{F}_{q^2}$. In particular it is shown that for a fixed integer $g \ge 1$, there are maximal curves over $\mathbb{F}_{q^2}$ of genus g for infinitely many values of q (see Remark 6.2 in [11]). Another interesting result of this paper (writing $q = p^n$ and assuming that the characteristic p is odd) is the existence of maximal curves over $\mathbb{F}_{q^2}$ with genus g given by:

$$g = \frac{1}{2}p^{n-v}\cdot(p^{n-w}-1) , \tag{5}$$

for each $0 \le v \le n$ and for each $0 \le w \le (n-1)$. The genera above are obtained by considering p-subgroups G of $\mathrm{Aut}(X)$, where X denotes the Hermitian curve (see also [13], [14]).

3. Explicit Equations for Maximal Curves

For applications to Coding Theory it is necessary that the curves are explicitly given by equations. In Example 6.3 of [11] it is pointed out that (see also Example 6.4 in [11])

$$z^n = t(t+1)^{q-1}, \qquad \text{with } n \text{ a divisor of } (q^2-1)\,, \tag{6}$$

is the equation of a maximal curve over $\mathbb{F}_{q^2}$ with genus given by $g = (n-\delta)/2$, where $\delta = gcd(n, q-1)$. It is with those maximal curves in equation (6) that it is shown in [11] that, for a fixed integer $g \ge 1$, there are maximal curves over $\mathbb{F}_{q^2}$ of genus g for infinitely many values of q (see also [12]).

Another interesting instance of an explicit equation for a maximal curve over $\mathbb{F}_{q^2}$ is given in Remark 4.4 of [10]. Denoting by $\varphi_n(T)$ the reduction modulo p of the (normalized) Chebyshev polynomial (i.e., the (monic) polynomial expressing $\cos n\theta$ in terms of $\cos\theta$), it is shown in [10] that (see also Remark 5.2 in [10])

$$v^{q+1} = \varphi_n(u) + 2, \qquad \text{with } n \text{ odd}\,, \tag{7}$$

is the equation of a maximal curve over $\mathbb{F}_{q^2}$. It is interesting to point out that properties of Chebyshev polynomials were deduced from the fact that equation (7) defines a maximal curve (see Section 6 of [10]).

We now mention a situation showing that sometimes it is hard to get such explicit equations. In Theorem 5.1 of [11] it is shown that there are maximal curves over $\mathbb{F}_{q^2}$ having genus g given by:

$$g = \frac{s-1}{2}, \qquad \text{for each divisor } s \text{ of } (q^2-q+1)\,. \tag{8}$$

This is shown by considering subgroups G of a certain cyclic subgroup of order (q^2-q+1) of the automorphism group of the Hermitian curve. The determination of explicit equations for the maximal curves over $\mathbb{F}_{q^2}$ with genera as in formula (8) above is not so easy (see [4]).

We end up this section with the following explicit equation for a maximal curve over $\mathbb{F}_{q^{2k}}$ with $k \ge 2$ (see [9]):

$$\sum_{j=0}^{k-1} y^{q^j} = w \cdot x^{q^k+1}, \qquad \text{with} \quad w^{q^k-1} = -1. \tag{9}$$

This curve has genus $g = q^k(q^{k-1}-1)/2$ and, in particular, its genus appears among those given in formula (5). It can be shown that this curve is Galois covered by the Hermitian curve with a Galois group G of order q. In the particular case when $q = p$, this curve also appears in Theorem 2.1 of [5].

4. Classification of Maximal Curves

Since it is not known whether all maximal curves are covered by the Hermitian curve, one sees that the classification problem for maximal curves is a wide open problem. General results on this problem have been obtained (so far) when the genus of the maximal curve is large compared with the cardinality of the finite field.

The first result on the classification of maximal curves (see [23]) asserts that the Hermitian curve over $\mathbb{F}_{q^2}$ given by equation (4) is the **unique maximal curve** over $\mathbb{F}_{q^2}$ with genus $g = q(q-1)/2$. The main ingredient of the proof of this uniqueness of the Hermitian curve is the fundamental linear equivalence in equation (3) above.

Consider now the curve over $\mathbb{F}_{q^2}$ given by

$$y^q + y = x^m, \quad m \text{ a divisor of } (q+1). \tag{10}$$

Since the curve in equation (10) is covered by the Hermitian curve, we have that it is a maximal curve. Its genus g is given by $g = (m-1)(q-1)/2$. For q odd and for $m = (q+1)/2$, we get a maximal curve over $\mathbb{F}_{q^2}$ with the second largest possible genus $g = (q-1)^2/4$. It was shown in [7] that this curve (equation (10) with q odd and $m = (q+1)/2$) is the **unique maximal curve** over $\mathbb{F}_{q^2}$ with genus $g = (q-1)^2/4$. In Theorem 2.3 of [7] is given a characterization of the maximal curves in equation (10) above, but this characterization requires an extrahypothesis on Weierstrass nongaps at a rational point over $\mathbb{F}_{q^2}$.

Write $q = p^t$ and consider the curve over $\mathbb{F}_{q^2}$ given by the affine equation:

$$\sum_{i=1}^{t} y^{q/p^i} + w \cdot x^{q+1} = 0, \qquad \text{with } w^{q-1} = -1. \tag{11}$$

Equation (11) defines a maximal curve and its genus g is given by $g = q(q-p)/2p$. In the case of characteristic $p = 2$, one gets the second largest genus possible and this curve (i.e., equation (11) with $p = 2$) is characterized in [1] with a similar extrahypothesis on Weierstrass nongaps at a rational point. Also, equation (11) appears in Theorem 2.1 of [5] where is classified the **Galois subcoverings of prime degrees** of the Hermitian curve. Besides the fundamental linear equivalence in equation (3), the other main ingredient for the classification problem of maximal curves over $\mathbb{F}_{q^2}$ of a given genus is the Stöhr-Voloch theory of Frobenius-orders (see [28]). This theory is similar to Weierstrass Point Theory in prime characteristics (see [24]) and provides also a proof of Weil's theorem (i.e., a proof of Inequality (2)). Roughly speaking, instead of counting rational points on the curve (i.e., the fixed points for the Frobenius action on the curve) Stöhr-Voloch's approach counts the number of points such that the image under the Frobenius action of a point lies on the osculating hyperplane to the curve at that point.

The first example of nonisomorphic maximal curves over $\mathbb{F}_{q^2}$ with the same genus was given in [3]. For $q \equiv 3(\text{modulo } 4)$, it is shown in [3] that the following two curves of genus $g = (q-1)(q-3)/8$ are not isomorphic:

$$y^q + y = x^{\frac{q+1}{4}} \quad \text{and} \quad x^{\frac{q+1}{2}} + y^{\frac{q+1}{2}} = 1\,. \tag{12}$$

Both curves in equation (12) are maximal because they are Galois covered by the Hermitian curve, the first one with a cyclic group $G \simeq \mathbb{Z}/4\mathbb{Z}$ and the second one with a group $G \simeq \mathbb{Z}/2\mathbb{Z} \times \mathbb{Z}/2\mathbb{Z}$.

We end up this paper by saying that the second curve in equation (12) above was characterized in [3] as the **unique maximal curve** over $\mathbb{F}_{q^2}$ with genus given by $g = (q-1)(q-3)/8$ that has a nonsingular plane model over $\mathbb{F}_{q^2}$.

References

[1] M. Abdón and F. Torres, *On maximal curves in characteristic two,* Manuscripta Math., **99** (1999), 39–53.

[2] E. Bombieri, *Hilbert's 8th problem: An analogue,* Proc. Symp. Pure Math., **28** (1976), 269–274.

[3] A. Cossidente, J. W. P. Hirschfeld, G. Korchmáros and F. Torres, *On plane maximal curves,* Compositio Math., **121** (2000), 163–181.

[4] A. Cossidente, G. Korchmáros and F. Torres, *On curves covered by the Hermitian curve,* J. Algebra, **216** (1999), 56–76.

[5] A. Cossidente, G. Korchmáros and F. Torres, *Curves of large genus covered by the Hermitian curve,* Comm. Algebra, **28** (2000), 4707–4728.

[6] L. E. Dickson, *History of the theory of numbers,* Vol. II, Chelsea Publ. Comp., New York, (1971).

[7] R. Fuhrmann, A. Garcia and F. Torres, *On maximal curves,* J. Number Theory, **67**(1) (1997), 29–51.

[8] R. Fuhrmann and F. Torres, *The genus of curves over finite fields with many rational points,* Manuscripta Math., **89** (1996), 103–106.

[9] A. Garcia and L. Quoos, *A construction of curves over finite fields,* to appear in Acta Arithmetica.

[10] A. Garcia and H. Stichtenoth, *On Chebyshev polynomials and maximal curves,* Acta Arithmetica, **90** (1999), 301–311.

[11] A. Garcia, H. Stichtenoth and C. P. Xing, *On subfields of the Hermitian function field,* Compositio Math., **120** (2000), 137–170.

[12] A. Garcia and F. Torres, *On maximal curves having classical Weierstrass gaps,* Contemporary Math., **245** (1999), 49–59.

[13] G. van der Geer and M. van der Vlugt, *Fibre products of Artin-Schreier curves and generalized Hamming weights of codes,* J. Comb. Theory A, **70** (1995), 337–348.

[14] G. van der Geer and M. van der Vlugt, *Generalized Hamming weights of codes and curves over finite fields with many points,* Israel Math. Conf. Proc., **9** (1996), 417–432.

[15] V. D. Goppa, *Geometry and Codes,* Mathematics and its applications, **24**, Kluwer Academic Publishers, Dordrecht-Boston-London (1988).

[16] J. W. P. Hirschfeld, *Projective geometries over finite fields,* second edition, Oxford University Press, Oxford (1998).

[17] Y. Ihara, *Some remarks on the number of rational points of algebraic curves over finite fields,* J. Fac. Sci. Tokio, **28** (1981), 721–724.

[18] A. I. Kontogeorgis, *The group of automorphisms of function fields of the curve $x^n + y^m + 1 = 0$,* J. Number Theory, **72** (1998), 110–136.

[19] G. Lachaud, *Sommes d'Eisenstein et nombre de points de certaines courbes algébriques sur les corps finis,* C.R. Acad. Sci. Paris, **305**, Série I (1987), 729–732.

[20] H. W. Leopoldt, *Über die Automorphismengruppe des Fermatkörpers,* J. Number Theory, **56** (1996), 256–282.

[21] R. Lidl and H. Niederreiter, *Finite fields,* Encyclopedia of mathematics and its applications, **20**, Addison-Wesley (1983).

[22] C. J. Moreno, *Algebraic curves over finite fields,* Cambridge University Press, **97** (1991).

[23] H. G. Rück and H. Stichtenoth, *A characterization of Hermitian function fields over finite fields,* J. Reine Angew. Math., **457** (1994), 185–188.

[24] F. K. Schmidt, *Zur arithmetischen Theorie der algebraischen Funktionen II. Allgemeine Theorie der Weierstrasspunkte,* Math. Z., **45** (1939), 75–96.

[25] S. A. Stepanov, *Arithmetic of algebraic curves,* Monographs in contemporary mathematics, Consultants Bureau, New York - London (1994).

[26] H. Stichtenoth, *Über die Automorphismengruppe eines algebraischen Funktionenkörpers von Primzahlcharakteristik,* Arch. Math., **24** (1973), 527–544 and 615–631.

[27] H. Stichtenoth and C. P. Xing, *The genus of maximal function fields over finite fields,* Manuscripta Math., **86** (1995), 217–224.

[28] K. O. Stöhr and J. F. Voloch, *Weierstrass points and curves over finite fields,* Proc. London Math. Soc. (3) **52** (1986), 1–19.

[29] A. Weil, *Courbes algébriques et variétés abeliennes,* Hermann, Paris, (1971).

IMPA,
Estrada Dona Castorina 110
22.460–320, Rio de Janeiro, RJ
Brasil
E-mail address: `garcia@impa.br`

[illegible] Number Theory [illegible]

[illegible]

[illegible]

[illegible]

Asymptotically Good Towers of Global Fields

Farshid Hajir and Christian Maire

Abstract. The study of the maximal p-extension of a global field k unramified everywhere and totally split at a finite set of places of k has at least two important applications: it gives information on the asymptotic behavior of discriminants versus degree in the number field case (as measured by the Martinet constant $\alpha(t)$), and on the relationship between genus and the number of places of degree one (for large genus) in the function field case (as measured by the Ihara constant $A(q)$). We survey recent work on class-field-theoretical constructions of towers of global fields which are optimal for the study of these phenomena, including best known examples in both settings; these contain, among others, an infinite unramified tower of totally complex number fields with small root discriminant improving Martinet's record. We show that allowing wild ramification to limited depth leads to asymptotically good towers. However, we demonstrate also that the investigation of the infinitude of these towers involves difficulties absent in the tame case.

1. Introduction and Definitions

An infinite tower of global fields $K_0 \subset K_1 \subset K_2 \cdots$ is "asymptotically good" if the relationship between certain of its layers' invariants is in some sense optimal. The precise condition is: (1) for number fields, that $\mathrm{rd}_{K_i} := |\operatorname{disc} K_i|^{1/[K_i:\mathbb{Q}]}$ remain bounded from above, and (2) for function fields with fixed finite constant field $\mathbb{F}_q$, that N_{K_i}/g_{K_i} remain bounded away from 0, where g_K, N_K are, respectively, the genus, and the number of places of degree 1 of K.

The relationships between these invariants (disc_K vs. $[K:\mathbb{Q}]$, and g_K vs. N_K) are governed by important general bounds (Stark-Odlyzko for number fields and Hasse-Weil for function fields). The interest of asymptotically good towers is that they measure the sharpness of the leading terms of these bounds.

In the number field case, currently the only method for constructing asymptotically good towers is that of p-class field towers, for a fixed prime p. The same method works well for function fields; however, at least when q is a square, there

F. Hajir was partially supported by a Mathematical Sciences Postdoctoral Fellowship from the NSF.

are other rich sources of asymptotically good towers as well: modular curve constructions [14], certain types of explicit equations [7] (which potentially are always modular! cf. [4]), and a new "rigid" construction of Frey, Kani, and Völklein [6].

In this paper, we survey the p-class field tower constructions. One fixes a prime p and a finite non-empty set S of places of a base field K and studies the maximal p-extension of K which is everywhere unramified and totally split at S. For these applications, it is also possible to allow a finite number of places to ramify, as long as we can obtain a bound for the minimal number of relations of the Galois groups which appear. In §2 we recall all of this; we give the best known examples for totally complex and totally real number fields (§3) and for function fields over fields with 2, 3 and 5 elements (§4).

In §5 we consider extensions where wild ramification can occur but only to a limited depth; we show first that these lead to asymptotically good towers. We state a theorem (Theorem 5.5) which shows that allowing wild ramification leads to complications in the study of the minimal number of relations of the resulting Galois groups.

We conclude with some open problems which come up naturally in the search for asymptotically good towers.

1.1. Martinet's constant

For a number field k of degree $n = r_1 + 2r_2$ over $\mathbb{Q}$, with signature (r_1, r_2), let $t = t_k := r_1/n$ be its "infinity type," i.e. the proportion of its embeddings into $\mathbb{C}$ which factor through $\mathbb{R}$. We will write disc_k, rd_k for its discriminant and root discriminant, respectively. Thanks to the work of Stark [32], Odlyzko [22, 23], Poitou [25] and Serre [30], we have a very good lower bound for rd_k, an asymptotic version of which reads: for a number field k of infinity type t and large enough degree,

$$\mathrm{rd}_k \geq A^t B^{1-t}, \tag{1}$$

with $A = 60$, $B = 22$; under GRH, one may take $A = 215$, $B = 44$.

For fixed $t \in \mathbb{Q} \cap [0,1]$, and integers n such that number fields of degree n and infinity type t exist, we let $\alpha_n(t)$ be the minimal root discriminant attained by number fields of degree n and infinity type t and define

$$\alpha(t) = \liminf_n \alpha_n(t).$$

For more details see [18, 10]. From (1), we see that

$$\alpha(t) \geq A^t B^{1-t}, \quad t \in \mathbb{Q} \cap [0,1].$$

A nested sequence of distinct number fields $K_0 \subset K_1 \subset \cdots$ is "asymptotically good" (Tsfasman-Vladut [34]) if rd_{K_j} is bounded from above. An asymptotically good tower with fixed infinity type t and root discriminant bounded above by R gives an upper bound $\alpha(t) \leq R$. In §4, we will present the best current upper bounds for $\alpha(0)$ and $\alpha(1)$, namely (see [11])

$$B \leq \alpha(0) < 82.2, \quad A \leq \alpha(1) < 954.3.$$

1.2. Function field case

We will fix a finite field $\mathbb{F}_q$ and ask how large the group of rational points $X(\mathbb{F}_q)$ of a smooth absolutely irreducible algebraic curve X of genus g over $\mathbb{F}_q$ can be as g tends to infinity. Actually, to maintain notational consistency with number fields, we will consider the corresponding function fields $K = \mathbb{F}_q(X)$ and count the places of degree 1. Let $N_q(g)$ be the maximum number of degree 1 places of a genus g function field with contant field $\mathbb{F}_q$. By the celebrated theorem of Hasse-Weil,

$$N_q(g) \leq q + 1 + 2g\sqrt{q}\,.$$

Various improvements of this bound have been obtained in the last two decades. To measure the asymptotically optimal bound, Ihara introduced

$$A(q) = \limsup_{g\to\infty} \frac{N_q(g)}{g}\,.$$

In the 80's, Ihara [14] and, Tsfasman, Vladut and Zink [35] independently showed by using modular curves that $A(q) \geq \sqrt{q} - 1$ if q is a square. Shortly thereafter, Drinfeld and Vladut [3] proved that for all q,

$$A(q) \leq \sqrt{q} - 1\,,$$

and so $A(q) = \sqrt{q} - 1$ for square q. Using class field towers, Serre showed that there is an absolute constant $c > 0$ such that $A(q) > c \log q$ for all q. Variations on Serre's proof [31] can be found in Neiderreiter-Xing [21], Temkine [33] and Li-Maharaj [17]. Recently, Elkies, Kresch, Poonen, Wetherell, and Zieve [5] have shown that $\liminf_g N_q(g)/g \geq (\sqrt{q} - 1)/3$ for square q and $\liminf_g N_q(g)/g \geq c' \log q$ for an absolute constant $c' > 0$ (all q). When q is not prime, it can be shown that the growth of $A(q)$ is faster than logarithmic (e.g. [21, 33, 17]). The major outstanding problem here is, then, to improve Serre's lower bound for $A(q)$ when q is prime.

We will be contented here with describing what is known for three small prime values of q, namely $q = 2$, 3 and 5. The best known bounds are: $A(2) \geq 81/317$ (see [19]), $A(3) \geq 12/25$ (see [1]) and $A(5) \geq 8/11$ (see [2, 33]): we will present these examples in §4.

2. Tamely Ramified Situation

We fix a prime p and two finite sets S and T of places of k such that:

1. In the function field case, S is non-empty and contains only degree 1 places.
2. In the number field case, S contains all infinite places S_∞ of k.
3. $S \cap T = \emptyset$.
4. For all places $\mathfrak{p} \in T$, p divides $\mathbb{N}\mathfrak{p} - 1$, where $\mathbb{N}\mathfrak{p}$ is the absolute norm. In the function field case, that means that p divides $q^{\deg(\mathfrak{p})} - 1$ for $\mathfrak{p} \in T$.

Now we define k_T^S to be the maximal p-extension (inside a fixed algebraic closure) of k unramified ouside T in which S splits completely. By our assumptions, the ramification in k_T^S/k is tame; put $G_T^S = \mathrm{Gal}(k_T^S/k)$. One has to introduce two quantities:

Definition 2.1. *Let G be a finitely generated pro-p-group. Then*

1. *$d(G)$ is the minimal number of generators of G: $d(G) = \dim_{\mathbb{F}_p} H^1(G, \mathbb{F}_p)$.*
2. *$r(G)$ is the minimal number of relations of G: $r(G) = \dim_{\mathbb{F}_p} H^2(G, \mathbb{F}_p)$.*

By the Burnside Basis Theorem, the generator rank of a group is the same as that of its maximal abelian quotient, thus $d(G_T^S)$ can be understood in terms of class field theory (it is the p-rank of the S-ray class group mod T). The deepest known fact about these groups was first established by Shafarevich (see [15], or [2]):

Theorem 2.2. *With the above assumptions,*

$$r(G_T^S) - d(G_T^S) \leq |S| - 1 + \theta_{k,T},$$

where $\theta_{k,T} = 1$ when k contains μ_p and $T = \emptyset$, 0 otherwise.

Remark 2.3. *One has the trivial inequalities: $d(G_\emptyset^S) \leq d(G_T^S) \leq d(G_\emptyset^S) + |T|$.*

The famous Theorem of Golod-Shafarevich, as improved by Vinberg and Gaschütz, says that for a non-trival finite p-group G, $r(G) > d(G)^2/4$. Thus,

Theorem 2.4. *If*

$$d(G_T^S) \geq 2 + 2\sqrt{|S| + \theta_{k,T}},$$

then G_T^S is infinite.

The last ingredient we need is a standard genus theory bound for the p-rank of the S-class group in a degree p Galois extension.

Theorem 2.5. *Suppose k/k' is a Galois extension of degree p. Let $S' = S \cap k'$ be the set of places of k' lying under the places in S. Suppose r places of k' ramify in k. Then*

$$d(G_\emptyset^S) \geq r - |S'| - \delta_{k'},$$

where $\delta_{k'} = 1$ when k' contains μ_p, and $\delta_{k'} = 0$ otherwise.

For this and more refined genus theory bounds, see e.g. [28, 27].

3. Number Fields

A first observation is that the layers of an infinite tamely ramified tower form an asymptotically good family, (i.e. they have bounded root discriminant) [10]:

Theorem 3.1. *Let k be a number field of degree n over $\mathbb{Q}$ of infinity type t such that G_T^S is infinite. Then*

$$\alpha(t) \leq \mathrm{rd}_k \prod_{\mathfrak{p} \in T} \left(N_{k/\mathbb{Q}}\mathfrak{p}\right)^{1/n}.$$

3.1. Martinet's example

The first idea was to construct a number field k with small root discriminant admitting an infinite *unramified* 2-extension ($S = S_\infty$, $T = \emptyset$). The layers of such a tower comprise a family with *constant* root discriminant. Since 1978, the best such example known has been that of Martinet [18]: he proved that the field $\mathbb{Q}(\cos(2\pi/11), \sqrt{2}, \sqrt{-23})$ with root discriminant $2^{3/2}11^{4/5}23^{1/2}$ has an infinite unramified 2-tower ($T = \emptyset$, $S = S_\infty$), and so $\alpha(0) < 92.4$. Martinet also provided a totally real infinite unramified tower: $\mathbb{Q}(\sqrt{2}, \sqrt{29}, \sqrt{3 \cdot 5 \cdot 7 \cdot 23 \cdot 29})$ has an infinite unramified 2-tower, giving $\alpha(1) < 1058.6$.

3.2. An infinite unramified tower which improves Martinet's record

In [10], we found that class field towers over non-Galois base fields seem to yield asymptotically good towers. We now apply that idea to give an unramified tower with root discriminant smaller than the previous example.

Let $k' = \mathbb{Q}(\xi)$ where ξ is a root of $f = x^5 - 2x^4 + 3x^3 - 3x^2 - x + 1$. The discriminant of f is -31391, a prime; thus, this is also the discriminant of k', and $\mathcal{O}_{k'} = \mathbb{Z}[\xi]$. Since $\operatorname{disc}_{k'}$ is negative, k' has signature $(3, 1)$. Since $\operatorname{disc}_{k'}$ is a quadratic discriminant, it follows (see Kondo [16]) that the Galois group of f is S_5; indeed, the Galois closure of k' is an unramified A_5-extension of $\mathbb{Q}(\sqrt{-31391})$. We will not need this fact, however.

The element $\eta = -36\xi^4 + 125\xi^3 - 221\xi^2 + 182\xi - 80 \in \mathcal{O}_{k'}$ is negative at all three real places of k'. Its minimal polynomial is $g(y) = y^5 + 223y^4 + 18336y^3 + 10907521y^2 + 930369979y + 18559139599$. The $\mathcal{O}_{k'}$-ideal it generates factors into nine prime ideals of $\mathcal{O}_{k'}$: $\eta = \pi_7\pi_7'\pi_{11}\pi_{11}'\pi_{13}\pi_{19}\pi_{19}'\pi_{23}\pi_{29}$ where π_r generates an ideal of norm r. We let $k = k'(\sqrt{\eta})$, a totally complex field of degree 10. A defining polynomial for k is $g(y^2)$. We note that η is congruent to a square modulo $4\mathcal{O}_k$; explicitly, $\eta = \beta^2 - 4\gamma$ with $\beta = \xi^4 + \xi + 1$ and $\gamma = 11\xi^4 - 31\xi^3 + 56\xi^2 - 45\xi + 20$. Thus, k/k' is ramified at the three real places and at the nine primes dividing η and nowhere else. Thus, the root discriminant of k is $rd_k = 31391^{1/5}(7^2 \cdot 11^2 \cdot 13 \cdot 19^2 \cdot 23 \cdot 29)^{1/10} = 84.375\ldots$ By Theorem 2.5 ($p = 2$), the 2-rank of the ideal class group of k is at least $9 + 3 - (3+1) - 1 = 7$. (This is confirmed by a Pari calculation which gives $\mathrm{Cl}_k = \mathbb{Z}/3 \oplus (\mathbb{Z}/2)^7$.) By Theorem 2.4, k admits an infinite everywhere unramified 2-tower since $2 + 2\sqrt{5+1} = 2 + \sqrt{24} < 7$. To our best knowledge, this tower gives the least root discriminant for an *unramified* tower which is known to be infinite.

3.3. The best known bounds for $\alpha(0)$ and $\alpha(1)$

Tamely ramified towers and asymmetric (non-Galois) constructions of the base (such as the one presented above) were two ideas introduced in [10] for improving Martinet's constant. We briefly present the best known current estimates for $\alpha(0)$ and $\alpha(1)$ (see [11] for details).

3.3.1. Totally complex situation The totally imaginary field $k = \mathbb{Q}(\theta)$ where θ is a root of

$$x^{12} + 339x^{10} - 19752x^8 - 2188735x^6 + 284236829x^4 + 4401349506x^2 + 15622982921$$

has discriminant $7 \cdot 13 \cdot 19^2 \cdot 23^4 \cdot 29 \cdot 31 \cdot 35509^2$; it admits an infinite 2-extension ramified at a prime $\mathfrak{p}_9$ with absolute norm 9 and unramified everywhere else. Thus $\alpha(0) \leq \mathrm{rd}_k \cdot 9^{1/12} < 82.2$.

3.3.2. Totally real situation The totally real field $k = \mathbb{Q}(\theta)$ where θ is a root of

$$x^{12} - 56966x^{10} + 959048181x^8 - 5946482981439x^6 + 14419821937918124x^4 - 12705425979835529941x^2 + 3527053069602078368989$$

has discriminant $7^{10} \cdot 13^7 \cdot 29^4 \cdot 41^4 \cdot 97 \cdot 113^2$; it admits an infinite 2-extension ramified at a prime $\mathfrak{p}_{13}$ with absolute norm 13 and unramified elsewhere. Thus $\alpha(1) \leq \mathrm{rd}_k \cdot 13^{1/12} < 954.3$.

4. Function Fields

As in the number field case, a tamely ramified p-extension of function fields is asymptotically good, meaning N_K/g_K is bounded from below for the layers K of the p-extension. (This was used in [2] to give improved lower bounds for $A(3)$ and $A(5)$.) To be precise, [2]:

Theorem 4.1. *Fix a prime p not dividing q. Let k be a genus g function field with constant field $\mathbb{F}_q$, S a non-empty set of degree 1 places of k, T a (possibly empty) set of places of k disjoint from S. If G_T^S is infinite, then*

$$A(q) \geq \frac{|S|}{g - 1 + \frac{1}{2}\sum_{\mathfrak{p} \in T} \deg \mathfrak{p}}.$$

4.1. Unramified towers

The best lower bound for $A(2)$ has been given by Niederreiter and Xing [19]:

Let $k = \mathbb{F}_2(x)$, $N_0 = x^4$ and $N_1 = (x^2 + x + 1)(x^6 + x^3 + 1) \in F_2[x]$. Let K_i be the subfield of the cyclotomic function field k_{N_i} associated to N_i for $i = 0, 1$ (for more details see [13]). Consider now the subfield F_i of K_i fixed by $\langle \overline{x + i} \rangle$. Put $F = F_0F_1$. Then $[F : k] = 84$. Now let S be the set of places of F lying over ∞ together with one place lying over x. Then $|S| = 81$. For this example, and $p = 2$, $G_\emptyset^S$ is infinite, and so

$$0.255 < \frac{81}{317} \leq A(2) \leq \sqrt{2} - 1 < 0.414.$$

4.2. Tamely ramified towers

The following example is from [1]: Let $k = \mathbb{F}_3(x, \sqrt{D})$, where $D = (x^{27} - x)(x^9 - x)(x+1)(x^3 - x)^{-2}(x^3 - x^2 + x + 1)^{-1} \in \mathbb{F}_3[x]$; the polynomial D has 11 prime factors over $\mathbb{F}_3$. If we take $p = 2$, S to be the set of k-places above x, $x - 1$ and $1/x$, T to consist of the unique k-place above $x + 1$, Golod-Shafarevich implies that k_T^S is infinite, hence by Theorem 4.1,

$$0.48 = 12/25 \leq A(3) \leq \sqrt{3} - 1 < 0.74\,.$$

The following example is from [2]. For the field $k = \mathbb{F}_5(x, \sqrt{D})$ where $D = (x-1)(x-2)(x-3)(x-4)(x^2+x+1)(x^2+3)(x^2+2)(x^2+x+2)(x^2+2x+3)$ with $p = 2$, S all places above x and $1/x$, and T the place above $x - 1$, one has k_T^S/k is infinite and then

$$0.72 < 8/11 \leq A(5) \leq \sqrt{5} - 1 < 1.24\,.$$

Note that by using unramified extensions, Temkine [33] has recovered the same bound for $A(5)$.

5. Wild Ramification

We fix a prime p. Now we suppose (for simplicity) that T consists solely of places dividing p in the number field case, and in the function field case we suppose that $p \mid q$. To each prime $\mathfrak{p} \in T$ we associate $i_\mathfrak{p} \in [0, \infty]$. We call $T(I) = \{(\mathfrak{p}, i_\mathfrak{p}), \mathfrak{p} \in T\}$, and define $k_{T(I)}^S$ as being the maximal p-extension of k unramified outside T, totally decomposed for all places in S, such that $D_\mathfrak{p}^{(i_\mathfrak{p})}$ is trivial, for all $\mathfrak{p}$ in T, where $D_\mathfrak{p}^{(i_\mathfrak{p})}$ is the ramification group with upper numbering (see for example [29] for more details). The condition $i_\mathfrak{p} = \infty$ means that there is no restriction for ramification at $\mathfrak{p}$. Note that we can assume without loss of generality that $i_\mathfrak{p} > 1$ as the following proposition demonstrates.

Proposition 5.1. *If $i_\mathfrak{p} \leq 1$, $\mathfrak{p}$ is unramified in $k_{T(I)}^S/k$.*

Proof. Fix $\mathfrak{p} \in T$. Let K be a field such that $k \subset K \subset k_{T(I)}^S$. By the restriction property of ramification groups, $D_\mathfrak{p}^{(i_\mathfrak{p})}(K/k)$ is trivial. Put $n = \psi_{K/k}(i_\mathfrak{p})$ where $\psi_{K/k}$ is the Herbrand function associated to $\mathfrak{p}$ in K/k: $D_\mathfrak{p}^{(i_\mathfrak{p})}(K/k) = \{1\} \Rightarrow D_{\mathfrak{p},(n)}(K/k) = \{1\}$ where $D_{\mathfrak{p},(j)}(K/k)$ is the ramification group with lower numbering. If $n \leq 1$, then since $D_{\mathfrak{p},(1)}(K/k) \subset D_{\mathfrak{p},(n)}(K/k)$, we find that $D_{\mathfrak{p},(1)}(K/k) = \{1\}$. But $D_{\mathfrak{p},(0)}(K/k)/D_{\mathfrak{p},(1)}(K/k)$ has order prime to p and $D_{\mathfrak{p},(0)}(K/k)$ is a p-group, hence $D_{\mathfrak{p},(0)}(K/k) = \{1\}$.

Now we want to show that $n \leq 1$. Suppose $n \geq 1$; $m = \lfloor n \rfloor \geq 1$. We have

$$i_\mathfrak{p} = \psi_{K/k}^{-1}(n) = \frac{g_1 + \cdots + g_m + (n-m)g_{m+1}}{g_0}$$

where $g_i = |D_{\mathfrak{p},(i)}(K/k)|$. Since $D_{\mathfrak{p},(0)}(K/k)/D_{\mathfrak{p},(1)}(K/k)$ is trivial, $g_0 = g_1$. So one obtains ([29, Chapter IV §3]):

$$1 + \frac{m-1}{g_0} + \frac{(n-m)g_{m+1}}{g_0} \leq i_p \leq 1\,,$$

giving $n = m = 1$. □

The main question of this section is the following:

Problem 5.2. *What is the relation rank of the group $G^S_{T(I)}$?*

Before looking more closely at this question we explain why it is interesting for the problem of finding asymptotically good towers.

Theorem 5.3. *Assume that for all $\mathfrak{p} \in T$, $i_\mathfrak{p} > 1$ is finite. Suppose that $k^S_{T(I)}/k$ is infinite. Then*

1) *In the number field case, if k has degree n and infinity type t, one has:*

$$\alpha(t) \leq \mathrm{rd}_k \cdot \left(\prod_{\mathfrak{p}\in T} N_{k/\mathbb{Q}}\mathfrak{p}^{i_\mathfrak{p}+1}\right)^{1/n}.$$

2) *In the function field case, suppose $p \mid q$, k is a genus g function field with constant field $\mathbb{F}_q$, S is a non-empty set of degree 1 places of k disjoint from T; then one has:*

$$A(q) \geq \frac{|S|}{g - 1 + \frac{1}{2}\sum_{\mathfrak{p}\in T}(i_\mathfrak{p}+1)\deg \mathfrak{p}}\,.$$

Proof. Let K be such that $k \subset K \subset k^S_{T(I)}$. By restriction, for all $\mathfrak{p} \in T$, $D^{(i_\mathfrak{p})}_\mathfrak{p}(K/k)$ is trivial. Let $\mathfrak{p} \in T$. By definition one has:

$$D^{(i_\mathfrak{p})}_\mathfrak{p}(K/k) = D_{\mathfrak{p},(\psi_{K/k}(i_\mathfrak{p}))}(K/k)\,,$$

where $\psi_{K/k}$ is the Herbrand function. Put $n = \psi_{K/k}(i_\mathfrak{p})$ and $m = \lfloor n \rfloor$. Then for all $\mathfrak{P} \mid \mathfrak{p}$, $\mathfrak{P}$ a prime of K, one knows the $\mathfrak{P}$-valuation $v_\mathfrak{P}(\mathcal{D}_{K/k})$ of the different of K/k [29]:

$$v_\mathfrak{P}(\mathcal{D}_{K/k}) = g_0 + g_1 + \cdots + g_m - (m+1)\,,$$

where $g_j = |D_{\mathfrak{p},(j)}(K/k)|$; $g_j = 1$ for all $j \geq m+1$. If we use the definition of ψ, and the fact that $\varphi = \psi^{-1}$ is the reciprocal function, one gets:

$$i_p = \varphi_{K/k}(n) = \frac{g_1 + \cdots + g_m + (n-m)}{g_0}\,,$$

and then

$$\begin{aligned} v_\mathfrak{P}(\mathcal{D}_{K/k}) &= g_0 + g_1 + \cdots + g_m - (m+1) \\ &= g_0(i_\mathfrak{p}+1) - (m+1) - (n-m) \\ &\leq e_\mathfrak{P}(K/k)(i_\mathfrak{p}+1) \end{aligned}$$

because $g_0 = e_\mathfrak{P}(K/k)$. The rest follows easily as in [2]. □

5.1. A sub-extension of $k^S_{T(I)}$

For a finite extension K/k we consider T_K and S_K the set of places of K above T and S. For all places $\mathfrak{P} \in T_K$ we let $i_{\mathfrak{P}} = i_{\mathfrak{p}}$ where $\mathfrak{P} \cap \mathcal{O}_k = \mathfrak{p} \in T$, and we write simply $K^S_{T(I)}$ instead of $K^{S_K}_{T_K(I_K)}$.

So now we can define $k_\infty \subseteq k^S_{T(I)}$ inductively as follows: start with $k_0 = k$; for each $i \geq 0$, let k_{i+1} be the maximal abelian extension of k_i contained in $(k_i)^S_{T(I)}$; for the whole tower, put $k_\infty = \cup k_i$. Then $k_\infty \subseteq k^S_{T(I)}$. Note that in the tamely ramified situation, $k_\infty = k^S_{T(I)}$. Put $G = \mathrm{Gal}(K_\infty/k)$. In [24] Perret proposed a bound for $r(G) - d(G)$ when G is finite. Niederreiter and Xing [20] showed that Perret's conjecture would imply the infinitude of a certain tower over $\mathbb{F}_2$ violating the Drinfeld-Vladut bound.

We note that if k_∞/k is infinite, it gives better estimates for $\alpha(t)$ and $A(q)$ than those of Theorem 5.3, namely:

- $\alpha(t) \leq \mathrm{rd}_k \cdot \left(\prod_{\mathfrak{p} \in T} (N_{k/\mathbb{Q}})\, \mathfrak{p}^{\lfloor i_\mathfrak{p} \rfloor^* + 1} \right)^{1/n}$, in the number field case and
- $A(q) \geq \dfrac{|S|}{g - 1 + \frac{1}{2} \sum_{\mathfrak{p} \in T} (\lfloor i_\mathfrak{p} \rfloor^* + 1) \deg \mathfrak{p}}$ in the function field case,

where $\lfloor i_\mathfrak{p} \rfloor^* = \lfloor i_\mathfrak{p} \rfloor$ if $i_\mathfrak{p}$ is not an integer, $\lfloor i_\mathfrak{p} \rfloor^* = i_\mathfrak{p} - 1$ otherwise.

This comes from the following observation: If $\mathfrak{p}$ is ramified in K/k, let

$$n(i_\mathfrak{p}) = \sup_j \{ D^{(j)}_\mathfrak{p}(K/k) \neq \{e\} \}.$$

Then $n(i_\mathfrak{p}) \leq i_\mathfrak{p}$. But if K/k is an abelian extension we know that $n(i_\mathfrak{p})$ is an integer and so $n(i_\mathfrak{p}) \leq \lfloor i_\mathfrak{p} \rfloor^*$: this is the Hasse-Arf Theorem (see [29] for example). To conclude we use the proof of Perret [24].

5.2. Iwasawa theory

Suppose that for all $\mathfrak{p} \in T$, $i_\mathfrak{p} = \infty$. Then the difference $r - d$ is well understood (see [15, 9]). In particular, if T contains all places of k above p with $i_\mathfrak{p} = \infty$, then $r - d = -(r_2 + 1)$ (for $p > 2$).

One has the following natural question:

Problem 5.4. *Can one give explicitly a function f depending on S and on $T(I)$ with* **values in** $\mathbb{R}$ *such that*

$$r(G^S_{T(I)}) - d(G^S_{T(I)}) \leq f(S, T(I))?$$

When all the indices $i_\mathfrak{p} = \infty$, Shafarevich has given a very satisfactory answer. However, at least when some of the indices in I are finite and others are not, the groups in question are not even necessarily finitely presentable! For example, we have the following theorem [12]:

Theorem 5.5. *Let $p = 2$. Let ℓ be a prime such that $\ell \equiv 7 \pmod{16}$ and put $k = \mathbb{Q}(\sqrt{-\ell})$. Let $\mathfrak{p}_1$ and $\mathfrak{p}_2$ be the two primes of k above 2. Take $1 < i_{\mathfrak{p}_1} < \infty$ and $i_{\mathfrak{p}_2} = \infty$. Then $G_{T(I)}^S$ is a finitely generated pro-2-group with $r(G_{T(I)}^S) = \infty$.*

Proof. We give only the two crucial points of the proof:

1) The condition on ℓ forces the decomposition group of $\mathfrak{p}_1$ in $k_{T(I)}^S/k$ to be exactly the absolute Galois group of the maximal p-extension $\mathcal{K}_{\mathfrak{p}_1}$ of $k_{\mathfrak{p}_1}$ (the completion of k at $\mathfrak{p}_1$): this is an application of a result of Wingberg [36].
2) Let $\mathcal{G}$ be the Galois group of the absolute p-extension of a local field k, and let $\mathcal{G}^i$ be the ramification group of $\mathcal{G}$ with upper numbering. Then for $i > 1$ the number of relations of $\mathcal{G}/\mathcal{G}^i$ is infinite: this is a result of Gordeev [8]. □

In view of Theorem 5.3, it would be very interesting to investigate the above problem when all the indices are finite.

6. Two Further Questions

To finish we want to mention two questions.

Problem 6.1. *Does every infinite T-ramified p-tower k_T^S/k contain an intermediate field K (of finite degree over k) such that K has an infinite* unramified *p-tower $K_\emptyset^S/K$?*

In a forthcoming work, we investigate this question, and show, in particular, that it has a positive answer when T has large cardinality with respect to the unit rank of k.

Problem 6.2. *Consider, for simplicity, the number field situation and $p = 2$. Suppose that for all primes $\mathfrak{p}$ of k not dividing 2, the maximal $\{\mathfrak{p}\}$-ramified 2-extension $k_{\{\mathfrak{p}\}}$ over k is infinite. Does this imply that the maximal unramified 2-extension of k is infinite?*

The second can give a very nice application for bounding $\alpha(t)$ and is, in essence, a refinement of the Golod-Shafarevich criterion. For instance, it has long been conjectured that the imaginary quadratic field k of discriminant $-5460 = -4 \cdot 3 \cdot 5 \cdot 7 \cdot 13$ whose class group has exponent 2 and rank 4 has an infinite unramified 2-class field tower. It is easy to see that this field satisfies the hypothesis of problem 6.2, a positive answer to which would then yield $\alpha(0) \leq \sqrt{5460} < 74$.

References

[1] W. Aitken and F. Hajir, *Some asymptotically good towers of function fields*, in preparation.

[2] B. Angles and C. Maire, *A note on tamely ramified towers of global functions fields*, Preprint, 1999.

[3] V. G. Drinfeld and S. G. Vladut, *Number of points of an algebraic curve*, Funct. Anal. **17** (1983), 53–54.

[4] N. Elkies, *Explicit Modular Towers*, Proceedings of the Thirty-Fifth Annual Allerton Conference on Communication, Control and Computing (1997, T. Basar and A. Vardy, eds), Univ. of Illinois at Urbana-Champaign 1998, pp. 23–32.

[5] N. Elkies, A. Kresch, B. Poonen, J. Wetherell and M. Zieve, *Curves of every genus with many points II: asymptotically good families*, in preparation.

[6] G. Frey, E. Kani and H. Völklein, *Curves with infinite K-rational geometric fundamental group* in Aspects of Galois Theory (Gainesville, FL, 1996), 85–118, London Math. Soc. Lecture Note Ser., 256, Cambridge Univ. Press, Cambridge, 1999, **46** (1994), 467–476.

[7] A. Garcia and H. Stichtenoch, *A tower of Artin-Schreier extensions of function fields attaining the Drinfeld-Vladut bound,* Invent. Math. **121** (1995), 211–222.

[8] N. L. Gordeev, *Infinitude of the number of relations in the Galois group of the maximal p-extension of a local field with restricted ramification,* Math. USSR **18** (1982), 513–524.

[9] K. Haberland, Galois cohomology of algebraic number fields, V.E.B. Deutscher Verlag der Wissenschaften, 1970.

[10] F. Hajir and C. Maire, *Tamely ramified towers and discriminants bounds for number fields,* Comp. Math., to appear.

[11] F. Hajir and C. Maire, *Tamely ramified towers and discriminants bounds for number fields II,* Preprint, 2000.

[12] F. Hajir and C. Maire, *Extensions of number fields with wild ramification of bounded depth,* Preprint, 2000.

[13] D. R. Hayes, *Explicit class field theory for rational function fields*, Trans. Amer. Math. Soc. **189** (1974), 77–91.

[14] Y. Ihara, *Some remarks on the number of rational points of algebraic curves over finite fields,* J. Fac. Sci. Tokyo **28** (1981), 721–724.

[15] H. Koch, Galoissche Theorie der p-Erweiterungen, VFB Deutscher Verlag der Wissenschaften, Berlin, 1970.

[16] T. Kondo, *Algebraic number fields with the discriminant equal to that of quadratic number field*, J. Math. Soc. Japan **47** (1995), 31–36.

[17] W. W. Li and H. Maharaj, *Coverings of curves with asymptotically many rational points*, Preprint, 1999.

[18] J. Martinet, *Tours de corps de classes et estimations de discriminants,* Invent. Math. **44** (1978), 65–73.

[19] H. Niederreiter and C. Xing, *Towers of global function fields with asymptotically many rational places and an improvement on the Gilbert-Varshamov bound,* Math. Nach **195** (1998), 171–186.

[20] H. Niederreiter and C. Xing, *A counterexample to Perret's conjecture on infinite class field towers for global function fields,* Finite Fields Appl. **5** (1999), 240–245.

[21] H. Niederreiter and C. Xing, *Curve sequences with asymptotically many rational points* in Applications of Curves Over Finite Fields (ed. M. Fried) Contemp. Math. vol. 245, American Math. Soc., 1999.

[22] A. M. Odlyzko, *Lower bounds for discriminants of number fields II*, Tokoku Math. J. **29** (1977), 209–216.

[23] A. M. Odlyzko, *Bounds for discriminants and related estimates for class numbers, regulators and zeros of zeta functions: a survey of recent results*, Sém. de Théorie des Nombres, Bordeaux **2** (1990), 119–141.

[24] M. Perret, *Tours ramifiées infinies de corps de classes,* J. Number Theory **38** (1991), 300–322.

[25] G. Poitou, *Minorations de discriminants (d'aprés A. M. Odlyzko)*, Séminaire Bourbaki, Vol. 1975/76, 28ème année, Exp. No. 479, pp. 136–153, Lecture Notes in Math. 567, Springer-Verlag, 1977.

[26] P. Roquette, On Class field towers, in Algebraic Number Theory, ed. J. Cassels, A. Fröhlich, Academic Press, 1980.

[27] R. Schoof, *Algebraic curves over* $\mathbb{F}_2$ *with many rational points,* J. Number Theory **41** (1992), 6–14.

[28] R. Schoof, *Infinite class field towers of quadratics fields*, J. Reine Angew. Math. **372** (1986), 209–220.

[29] J.-P. Serre, Corps locaux, Hermann, Paris, 1962.

[30] J.-P. Serre, *Minorations de discriminants*, note of October 1975, published on pp. 240–243 in vol. 3 of Jean-Pierre Serre, Collected Papers, Springer-Verlag, 1986.

[31] J.-P. Serre, *Rational Points on Curves Over Finite Fields*, Harvard Course Notes by F. Gouvea (unpublished), 1985.

[32] H. M. Stark, *Some effetive cases of the Brauer-Siegel theorem*, Invent. Math. **23** (1974), 135–152.

[33] A. Temkine, *Hilbert class field towers of function fields over finite fields and lower bounds for* $A(q)$, Preprint, 1999.

[34] M. A. Tsfasman and S. G. Vladut, Asymptotic properties of global fields and generalized Brauer-Siegel Theorem, Prétirage 98–35, Institut Mathématiques de Luminy, 1998.

[35] M. A. Tsfasman, S. G. Vladut and T. Zink, Modular curves, *Shimura curves and Goppa codes better than the Varshamov-Gilbert bound,* Math. Nach. **109** (1982), 21–28.

[36] K. Wingberg, *Galois groups of local and global type,* J. reine angew. Math. **517** (1999), 223–239.

Farshid Hajir
Department of Mathematics
California State University, San Marcos
San Marcos CA 92096, U.S.A.
E-mail address: `fhajir@csusm.edu`

Christian Maire
Department A2X
University of Bordeaux I
351, cours de la Libération
33400 Talence, France
E-mail address: `maire@math.u-bordeaux.fr`

Explicit Constructions of Towers of Function Fields with Many Rational Places

Henning Stichtenoth

Abstract. We discuss several examples of function field towers $F_0 \subseteq F_1 \subseteq F_2 \subseteq \ldots$ over a finite field $\mathbb{F}_l$, for which the limit (number of rational places of F_n)/(genus of F_n) is positive.

1. Introduction

We consider algebraic function fields (of one variable) $F/\mathbb{F}_l$ whose constant field is the finite field $\mathbb{F}_l$ of cardinality l. By $g(F)$ (resp. $N(F)$) we denote the genus (resp. the number of rational places, i.e. places of degree one) of $F/\mathbb{F}_l$. A tower of function fields over $\mathbb{F}_l$ is a sequence

$$\mathcal{F} = (F_0, F_1, F_2, \ldots)$$

of function fields $F_i/\mathbb{F}_l$ with the following properties:

i) $F_0 \subseteq F_1 \subseteq F_2 \subseteq \ldots$
ii) For each $n \geq 0$, the extension F_{n+1}/F_n is separable of degree $[F_{n+1} : F_n] > 1$.
iii) $g(F_j) \geq 2$, for some $j \geq 0$.

It is easily seen that $g(F_n) \to \infty$ for $n \to \infty$, and that the limit

$$\lambda(\mathcal{F}) := \lim_{n\to\infty} N(F_n)/g(F_n)$$

exists [5]. The Drinfeld-Vladut bound [1] gives an upper bound for $\lambda(\mathcal{F})$,

$$\lambda(\mathcal{F}) \leq \sqrt{l} - 1.$$

A tower $\mathcal{F}$ is said to be asymptotically good if $\lambda(\mathcal{F}) > 0$; it is asymptotically optimal if it attains the Drinfeld-Vladut bound $\lambda(\mathcal{F}) = \sqrt{l} - 1$. These notions are motivated by applications to coding theory: asymptotically good towers of function fields yield asymptotically good sequences of (algebraic geometric) codes over $\mathbb{F}_l$, see [9, 10].

In general, it is hard to find asymptotically good towers of function fields. A famous result of Ihara [7] and Tsfasman, Vladut and Zink states that certain (reductions of) modular towers are asymptotically optimal, for $l = q^2$ being a square. As a consequence, there exist sequences of codes having excellent error-correcting properties: they improve the so-called Gilbert-Varshamov bound [10].

Another approach to constructing asymptotically good towers is the method of class field towers which was introduced by Serre [8]. Class field towers are, however, not explicit, and modular towers require deep results from algebraic geometry.

It is therefore desirable to give a more elementary construction of asymptotically good towers of function fields. In the following I will present some joint results with A. García.

2. The Method

We begin with some simple observations on the asymptotic behaviour of a tower of function fields.

Proposition 2.1. *A tower $\mathcal{F} = (F_0, F_1, F_2, \dots)$ of function fields $F_i/\mathbb{F}_l$ is asymptotically good if and only if there exists constants $c_1, c_2 > 0$ such that*

$$g(F_n) \le c_1 \cdot [F_n : F_0], \quad \textit{and} \tag{A}$$

$$N(F_n) \ge c_2 \cdot [F_n : F_0] \tag{B}$$

hold for all $n \ge 0$.

Proof. Obviously conditions (A) and (B) imply that $\mathcal{F}$ is asymptotically good. Conversely we assume now that $\mathcal{F}$ is asymptotically good. For $n \ge j \ge 0$, the Hurwitz genus formula yields a relation between the genera $g(F_n)$ and $g(F_j)$

$$g(F_n) - 1 = [F_n : F_j] \cdot (g(F_j) - 1) + \frac{1}{2}\deg(\mathrm{Diff}(F_n/F_j))$$
$$\ge [F_n : F_j] \cdot (g(F_j) - 1),$$

where $\mathrm{Diff}(F_n/F_j)$ denotes the different divisor of F_n/F_j. On the other hand, we have the trivial estimate

$$N(F_n) \le [F_n : F_0] \cdot N(F_0).$$

Conditions (A) and (B) follow easily from these two inequalities. □

The simplest way to ensure condition (B) is as follows:

Lemma 2.2. *Suppose that $\mathcal{S}$ is a non-empty set of rational places of F_0 such that any $P \in \mathcal{S}$ splits completely in all extensions F_n/F_0. Then*

$$N(F_n) \ge \#(\mathcal{S}) \cdot [F_n : F_0].$$

Condition (A) is more delicate: we need to know the behaviour of the degree $\deg(\mathrm{Diff}(F_n/F_0))$ of the different divisor of F_n/F_0. The following notion is useful: A place P of F_0 is said to be unramified in $\mathcal{F}$ if P is unramified in all extensions F_n/F_0; otherwise P is called ramified in $\mathcal{F}$. The set

$$V(\mathcal{F}) = \{P \mid P \text{ is a place of } F_0 \text{ which is ramified in } \mathcal{F}\}$$

is called the ramification locus of $\mathcal{F}$. For a place P of F_0 and a place Q_n of F_n lying above P, let $d(Q_n \mid P)$ be the different exponent of Q_n over P, and let

$$d_n(P) := \sum_{Q_n|P} d(Q_n \mid P) \cdot \deg Q_n.$$

Then we have

$$\deg(\mathrm{Diff}(F_n/F_0)) = \sum_{P\in V(\mathcal{F})} d_n(P).$$

Lemma 2.3. *Assume that the ramification locus $V(\mathcal{F})$ is finite, and that there exists a constant $c_3 \geq 0$ such that*

$$d_n(P) \leq c_3 \cdot [F_n : F_0] \qquad (\mathrm{A}_1)$$

for all $P \in V(\mathcal{F})$ and all $n \geq 0$. Then condition (A) *holds.*

Proof. This follows immediately from the Hurwitz genus formula. □

From here on we will consider towers $\mathcal{F} = (F_0, F_1, F_2, \ldots)$ over $\mathbb{F}_l$ of the following specific form: There are two rational functions $f(Z), h(Z) \in \mathbb{F}_l(Z)$ of the same degree

$$\deg f(Z) = \deg h(Z) = m,$$

such that $F_n = \mathbb{F}_l(x_0, x_1, \ldots, x_n)$ with

$$h(x_{n+1}) = f(x_n) \quad \text{and} \quad [F_{n+1} : F_n] = m \qquad (*)$$

for all $n \geq 0$ (as usual, the degree of a rational function $f(Z) = f_0(Z)/f_1(Z)$ with relatively prime polynomials $f_0(Z), f_1(Z) \in \mathbb{F}_l[Z]$ is defined as $\deg f(Z) = \max\{\deg f_0(Z), \deg f_1(Z)\}$). Note that $F_0 = \mathbb{F}_l(x_0)$ is a rational function field, and $[F_n : F_0] = m^n$ for all $n \geq 0$.

We need a criterion to determine whether the ramification locus of this tower is finite. Since ramification does not change in constant field extensions, we may replace $\mathbb{F}_l$ by its algebraic closure $\bar{\mathbb{F}}$ and the fields F_n by $\bar{F}_n := F_n \cdot \bar{\mathbb{F}}$, and we consider the tower $\bar{\mathcal{F}} = (\bar{F}_0, \bar{F}_1, \bar{F}_2, \ldots)$ of function fields over $\bar{\mathbb{F}}$. For $\gamma \in \bar{\mathbb{F}} \cup \{\infty\}$ we denote by "$x_i = \gamma$" the unique place of the rational function field $\bar{\mathbb{F}}(x_i)$ which is a zero of $x_i - \gamma$ (resp. the pole of x_i if $\gamma = \infty$). Let

$$R_0 := \{\gamma \in \bar{\mathbb{F}} \cup \{\infty\} \mid x_0 = \gamma \text{ ramifies in } \bar{F}_1/\bar{F}_0\}.$$

This is a finite set since $\bar{F}_1/\bar{F}_0$ is separable.

Lemma 2.4. *Notation as above. Assume in addition that $R \subset \bar{\mathbb{F}} \cup \{\infty\}$ is a finite set which contains R_0 and has the following property:*

$$\gamma \in R \Rightarrow \text{all roots of the equation } f(Z) = h(\gamma) \text{ are in } R. \qquad (\mathrm{A}_2)$$

Then all places $x_0 = \alpha$ with $\alpha \notin R$ are unramified in $\bar{\mathcal{F}}$. In particular the ramification locus $V(\bar{\mathcal{F}})$ and, a forteriori, the ramification locus $V(\mathcal{F})$ are finite.

Proof. Suppose that $x_0 = \gamma \in \bar{\mathbb{F}} \cup \{\infty\}$ is ramified in $\bar{\mathcal{F}}$. Then there is some $n \geq 0$ and a place Q_n of $\bar{F}_n$ lying above $x_0 = \gamma$ which ramifies in the extension $\bar{F}_{n+1}/\bar{F}_n$. The restriction of Q_n to $\bar{\mathbb{F}}(x_n)$ is of the form $x_n = \alpha$ with $\alpha \in \bar{\mathbb{F}} \cup \{\infty\}$. Since Q_n ramifies in $\bar{F}_{n+1}/\bar{F}_n$, the place $x_n = \alpha$ ramifies in the extension $\bar{\mathbb{F}}(x_n, x_{n+1})/\bar{\mathbb{F}}(x_n)$, hence $\alpha \in R_0$. By condition (A_2) we conclude that $\gamma \in R$. □

We summarize: For a tower $\mathcal{F}$ as defined in $(*)$, the two conditions (A_1) and (A_2) imply condition (A).

3. Examples

In this section we describe some examples of towers of the form $(*)$ explicitly.

3.1. Tame towers

We say that a tower $\mathcal{F} = (F_0, F_1, F_2, \ldots)$ over $\mathbb{F}_l$ is tame if for all $n \geq 0$ and all places Q_n of F_n, the ramification index $e(Q_n)$ in F_n/F_0 is relatively prime to the characteristic of $\mathbb{F}_l$. This implies that

$$d_n(P) \leq [F_n : F_0] \cdot \deg P$$

for all places P of F_0, i.e. condition (A_1) holds in tame towers.

Example 3.1. (See [6]) *Let $l = p^e$ with $e \geq 2$ and $m = (l-1)/(p-1)$. Consider the tower $\mathcal{F} = (F_0, F_1, F_2, \ldots)$ with $F_n = \mathbb{F}_l(x_0, \ldots, x_n)$ and*

$$x_{i+1}^m = 1 - (1 + x_i)^m \quad \text{for} \quad 0 \leq i \leq n-1,$$

i.e. the functions $f(Z)$ resp. $h(Z)$ in $(*)$ are here

$$f(Z) = 1 - (1+Z)^m,\ h(Z) = Z^m.$$

This is a tame tower, and it is easily seen that the place $x_0 = \infty$ splits completely in $\mathcal{F}$. Condition (A_2) from Lemma 2.4 holds for $R := \mathbb{F}_l$, and we obtain

$$\lambda(\mathcal{F}) \geq \frac{2}{l-2} > 0.$$

For $l = 4$ the tower is optimal, i.e. $\lambda(\mathcal{F}) = \sqrt{l} - 1$.

Example 3.2. (See [6]) *Let $l = q^2 > 4$ be a square and $F_n = \mathbb{F}_l(x_0, \ldots, x_n)$ with*

$$x_{i+1}^{q-1} = 1 - (1 + x_i)^{q-1} \quad \text{for} \quad 0 \leq i \leq n-1.$$

Here again the pole of x_0 splits completely, and we can take $R := \mathbb{F}_q$ in this case. It follows that

$$\lambda(\mathcal{F}) \geq \frac{2}{q-2} > 0.$$

The tower is optimal for $l = 9$.

3.2. Wild towers

If some ramification index in the tower is divisible by the characteristic of $\mathbb{F}_l$, the tower $\mathcal{F}$ is said to be wild. In this case one does not have an obvious bound (A_1) for the different degrees.

Example 3.3. (See [5]) *Let $l = q^2$ be a square and $F_n = \mathbb{F}_l(x_0, \dots, x_n)$ with*

$$x_{i+1}^q + x_{i+1} = \frac{x_i^q}{x_i^{q-1} + 1} \quad \text{for} \quad 0 \le i \le n-1,$$

i.e. $f(Z) = Z^q/(Z^{q-1} + 1)$ and $h(Z) = Z^q + Z$. Condition (A_2) holds for $R = \{\gamma \in \mathbb{F}_l \mid \gamma^q + \gamma = 0\} \cup \{\infty\}$ (this is easily checked), and condition (A_1) holds with $c_3 = 2$ (this is non-trivial). All places $x_0 = \alpha$ with $\alpha \in \mathbb{F}_l \setminus R$ split completely in this tower, and we conclude that

$$\lambda(\mathcal{F}) = q - 1.$$

Hence the tower is optimal: it attains the Drinfeld-Vladut bound $\sqrt{l} - 1$.

Example 3.4. (See [4]) *This example is similar to Example 3.3. Again let $l = q^2$ be a square. Define $F_n = \mathbb{F}_l(x_0, \dots, x_n)$ by*

$$x_{i+1}^q x_i^{q-1} + x_{i+1} = x_i^q \quad \text{for} \quad 0 \le i \le n-1.$$

Although this is not precisely of the form (*), one can determine ramification and rational places in an analogous manner. One obtains that $\lambda(\mathcal{F}) = q-1$, i.e. this tower is also optimal. It can be shown that Example 3.3 is in fact a subtower of example 3.4 (see [5]).

For more information about the examples of this section see N. Elkies' paper [2] in this volume.

References

[1] V. G. Drinfeld and S. G. Vladut, *Number of points of an algebraic curve,* Functional Anal. Appl., **17** (1983), 53–54.

[2] N. Elkies, *Explicit towers of Drinfeld modular curves,* this volume.

[3] G. van der Geer, *Curves over finite fields and codes,* this volume.

[4] A. García and H. Stichtenoth, *A tower of Artin-Schreier extensions of function fields attaining the Drinfeld-Vladut bound,* Invent. Math., **121** (1995), 211–222.

[5] A. García and H. Stichtenoth, *On the asymptotic behaviour of some towers of function fields over finite fields,* J. Number Theory, **61** (1996), 248–273.

[6] A. García, H. Stichtenoth and M. Thomas, *On towers and composita of towers of function fields over finite fields,* Finite Fields Appl., **3** (1997), 257–274.

[7] Y. Ihara, *Some remarks on the number of rational points of algebraic curves over finite fields,* J. Fac. Sci. Univ. Tokyo Sect. IA Math., **28** (1981), 721–724.

[8] J. P. Serre, *Rational points on curves over finite fields,* Lecture Notes, Harvard University, 1985.

[9] H. Stichtenoth, *Algebraic Function Fields and Codes,* Universitext, Springer-Verlag, Berlin, 1993.

[10] M. A. Tsfasman, S. G. Vladut and T. Zink, *Modular Curves, Shimura curves and Goppa codes, better than the Varshamov-Gilbert bound,* Math. Nachr., **109** (1982), 21–28.

FB 6
Mathematik und Informatik
Universität GH Essen
45117 Essen, Germany
E-mail address: `stichtenoth@uni-essen.de`

Curves over Finite Fields and Codes

Gerard van der Geer

Abstract. This paper gives a review of recent developments in this field and discusses some questions.

1. Introduction

The history of counting points on curves over finite fields goes back at least to C. F. Gauss who counted the number of points on several types of curves defined over the prime field $\mathbf{Z}/p\mathbf{Z}$. For example, in §358 of his Disquisitiones of 1801 he counts the number of points on the Fermat curve

$$x^3 + y^3 + z^3 \equiv 0 \pmod{p} \tag{1}$$

with p a prime greater than 3 and gives a beautiful answer: if $p \not\equiv 1 \pmod{3}$ then the number $\#C(\mathbf{F}_p)$ of $\mathbf{F}_p$-rational points on the projective curve C defined by (1) is $p+1$, while if $p \equiv 1 \pmod{3}$ there is a unique way of writing $4p = a^2 + 27b^2$ with a and b integers and $a \equiv 1 \pmod{3}$ and then $\#C(\mathbf{F}_p) = p + 1 + a$. Note that one has $|a| < 2\sqrt{p}$. Also Jacobi worked on the number of solutions of such congruences because he wanted to obtain estimates for Gauss sums. After them the problem sunk into oblivion for a long time.

In his 1924 thesis E. Artin introduced a zeta function $\zeta_F(s)$ for hyperelliptic function fields $F = \mathbf{F}_q(x, y)$ over a finite field $\mathbf{F}_q$, with q odd, where y satisfies $y^2 = f(x)$, in analogy with the Dedekind zeta function $\zeta_K(s) = \sum_{\mathfrak{a}} N(\mathfrak{a})^{-s}$ for a number field K. Artin noticed that by substituting $t = q^{-s}$ this zeta function became a rational function $Z_F(t)$ of t and that it satisfied a functional equation relating $\zeta_F(s)$ and $\zeta_F(1-s)$. He also advanced a conjecture as analogue of the Riemann hypothesis saying that the zeros of $Z_F(t)$ satisfy $|t| = q^{-1/2}$. Artin formulated everything in terms of ideals and ideal classes, but shortly afterwards F. K. Schmidt introduced a more geometric point of view and wrote the zeta function for a smooth (absolutely) irreducible projective curve X defined over $\mathbf{F}_q$ in the form

$$Z_X(t) = \exp(\sum_{r=1}^{\infty} c(r) \frac{t^r}{r})$$

with $c(r) = \#X(\mathbf{F}_{q^r})$ and observed that the theorem of Riemann-Roch implied that for a curve of genus g the function$Z_X(t)$ is of the form

$$Z_X(t) = \frac{P_X(t)}{(1-t)(1-qt)}, \tag{2}$$

with $P(t)$ a polynomial of degree $2g$, and that it satisfies a functional equation

$$Z_X(1/qt) = q^{1-g} t^{2-2g} Z_X(t).$$

Around 1932 Hasse noticed that the conjecture by Artin implied

$$|\#X(\mathbf{F}_q) - (q+1)| \leq 2g\sqrt{q} \tag{3}$$

and proved it for $g = 1$ using correspondences. Deuring observed then that in order to extend this proof to higher genus one needed a theory of correspondences. This theory was developed by A. Weil and within 16 years after Artin put forward his conjecture it was proved by Weil. He showed that the polynomial $P_X(t)$ in (2) is a polynomial with integral coefficients of the form $P_X(t) = \prod_{i=1}^{2g}(1-\alpha_i t)$, where the α_i are algebraic integers with $|\alpha_i| = \sqrt{q}$ and this implies the famous Hasse-Weil bound (3). One also finds the formula

$$\#X(\mathbf{F}_{q^r}) = q^r + 1 - \sum_{i=1}^{2g} \alpha_i^r . \tag{4}$$

Note that there exist curves which attain this bound over $\mathbf{F}_{q^2}$; for example the so-called hermitian curves defined by

$$x^{q+1} + y^{q+1} + z^{q+1} = 0$$

have genus $g = q(q-1)/2$ and satisfy $\#X(\mathbf{F}_{q^2}) = q^3 + 1 = q^2 + 1 + 2gq$.

After Weil obtained his result interest waned again, but it was brought back by coding theory.

2. Error Correcting Codes

The theory of error-correcting codes was born out of frustration about the frequent stops made by early computers every time these observed an inaccuracy when a parity check failed. Early pioneers like Hamming, Golay and others invented schemes to add redundancy with which inaccuracies due to noise could be repaired and tried to do this efficiently. They found some beautiful mathematical structures while doing this. For a review of the history and the results we refer to [24].

Fix a finite field $\mathbf{F}_q$ of cardinality q and consider the $\mathbf{F}_q$-vector space $\mathbf{F}_q^n$. The set $\mathbf{F}_q$ is called the alphabet, the elements of $\mathbf{F}_q^n$ are called *words* and the integer n is called the word length. The Hamming distance on $\mathbf{F}_q^n$ is the distance function with $d(x,y) = \#\{i\colon 1 \leq i \leq n, x_i \neq y_i\}$. The weight $w(x)$ of a word x is its distance to the origin.

By a linear code we shall mean a linear subspace C of $\mathbf{F}_q^n$. We say that $C \subseteq \mathbf{F}_q^n$ is a (n,k,d)-code if its dimension is k and the minimum weight of a non-zero word in C is d. A rough gauge of the quality of a code is provided by two invariants: the transmission rate $R = k/n$ and the relative distance $\delta = d/n$. In essence coding theory is a game where one tries to find codes that optimize these invariants. We shall restrict ourselves to linear codes.

If $C \subseteq \mathbf{F}_q^n$ is a linear code, then

$$C^{\perp} = \{x \in \mathbf{F}_q^n : \langle x, y\rangle = 0, \quad \text{for all} \quad y \in C\}$$

with $\langle x, y\rangle = \sum_{i=1}^{n} x_i y_i$ is called the dual code.

An important problem of coding theory is the determination of the weight distribution of a code, that is, of the polynomial

$$\sum_{c \in C} X^{w(c)} \in \mathbf{Z}[X].$$

The weight distribution and that of the dual code can be expressed in each other by the MacWilliams identities.

The notion of weight of a word admits an extension to the weight of a subspace of C. If $D \subseteq C$ is a linear subspace of dimension r we define the weight of D as the number of coordinate places for which C contains a codeword with a non-vanishing coordinate at that place, or equivalently as

$$w(D) = \frac{1}{q^r - q^{r-1}} \sum_{c \in D} w(c).$$

The weight hierarchy of C is the set $\{d_r(C) : 1 \leq r \leq n\}$ of generalized Hamming weights $d_r(C)$ defined by

$$d_r(C) = \min\{w(D) : D \subseteq C, \dim(D) = r\}.$$

The numbers $d_r(C)$ are a measure for the reliability of a transmission channel where part of the data fall prey to hostile eavesdroppers, cf. [39].

An important class of codes is given by the generalized Reed-Muller codes. Consider the set of polynomials

$$P_s = \{f \in \mathbf{F}_q[X_1, \ldots, X_r] : \deg(f) \leq s\}.$$

Here the degree is the total degree. We can evaluate elements of P_s at the points of affine r-space over $\mathbf{F}_q$:

$$\beta : P_s \to \mathbf{F}_q^n, \quad f \mapsto (f(P)_{P \in \mathbf{F}_q^r}),$$

where $n = q^r$. The image $\beta(P_s)$ of this map is the generalized q-ary Reed-Muller code $R_q(s, r)$. These codes were studied intensively from the middle of the 20th century. The weight of a code word $\beta(f)$ is $n - \#H_f(\mathbf{F}_q)$, with $H_f(\mathbf{F}_q)$ the set of $\mathbf{F}_q$-rational points on the hypersurface H_f given by $f = 0$. Determining the weight distribution of this code is equivalent to determining the distribution of the number of points in the family of all hypersurfaces of degree $\leq s$ in $\mathbf{F}_q^r$. This makes the difficulty of the problem apparent and it will be no surprise that the weight distribution of $R_q(s, m)$ in general is not known for $s \geq 3$.

3. Asymptotically Good Sequences of Codes

There are classical bounds on the parameters of codes, like the simple Singleton bound $k+d \leq n+1$ for codes with dimension k and minimum distance d, which show that the parameters of codes underlie certain restrictions.

A central problem of coding theory from its early beginnings has been to find long codes which can correct a fixed percentage of errors per codeword and which have a strictly positive transmission rate. To a linear code one can associate the pair $(\delta = d/n, R = k/n) \in [0,1]^2$ of relative parameters. Let U_q^{lin} be the set of limit points of the set of all such pairs coming from linear codes. The region U_q^{lin} is called the domain of codes. It is bounded in the unit square by the sides of the unit squares on the δ- and R-axis and by the graph of a function $\alpha_q^{\mathrm{lin}}\colon [0,1] \to [0,1]$ defined by

$$\alpha_q^{\mathrm{lin}}(\delta) = \sup\{R\colon (\delta, R) \in U_q^{\mathrm{lin}}\}.$$

A sequence of codes (C_i) with parameters (n_i, k_i, d_i) such that the ratios d_i/n_i and k_i/n_i converge to δ and R with $\delta R > 0$ is called an *asymptotically good sequence.*

Define the entropy function $H_q(\delta)$ by

$$H_q(\delta) = \begin{cases} 0 & \delta = 0\,, \\ \delta \log_q(q-1) - \delta\log_q(\delta) - (1-\delta)\log_q(1-\delta) & 0 < \delta \leq (q-1)/q\,. \end{cases}$$

We denote by $b_q(n,d)$ the number of points in a ball of radius d in $\mathbf{F}_q^n$. Gilbert made the simple observation that, if $q^n > q^{k-1} b_q(n, d-1)$, there exists a (n,k,d)-code over $\mathbf{F}_q$. This together with the fact that $\lim_{n\to\infty}(\log_q b_q(n,[\delta n]))/n$ equals the entropy function for $0 \leq \delta \leq (q-1)/q$ implies the following bound.

Proposition 3.1. (Gilbert-Varshamov bound) *For* $0 \leq \delta \leq (q-1)/q$ *the function* α_q^{lin} *satisfies* $\alpha_q^{\mathrm{lin}}(\delta) \geq 1 - H_q(\delta)$.

Another elementary bound from coding theory, the Plotkin bound, implies

$$\alpha_q^{\mathrm{lin}}(\delta) \leq 1 - (\frac{q}{q-1})\delta \quad \text{for} \quad 0 \leq \delta \leq (q-1)/q\,,$$

and

$$\alpha_q^{\mathrm{lin}}(\delta) = 0 \quad \text{for} \quad (q-1)/q \leq \delta \leq 1\,.$$

Thus on the interval $[0,(q-1)/q]$ the function α_q^{lin} is bounded from below and above by the graphs of $1 - H_q(\delta)$ and $1 - (q/(q-1))\delta$. Manin showed moreover that α_q^{lin} is a continuous function of δ which is decreasing on $[0,(q-1)/q]$.

For a long time coding theorists were unable to construct explicit sequences of codes with limit points on or above the Gilbert-Varshamov bound and they were thus led to suspect that $\alpha_q^{\mathrm{lin}}(\delta) = 1 - H_q(\delta)$ for $0 \leq \delta \leq (q-1)/q$.

4. Goppa Codes

In 1973 V. D. Goppa succeeded in constructing sequences of codes attaining the Gilbert-Varshamov bound. The codes were obtained by taking as entries of their parity check matrix the values of rational functions. When he tried to improve upon this he had the idea of taking values of rational functions on algebraic curves and thus discovered around 1980 an unexpected relationship between algebraic curves and codes, see [14, 15].

If X is a smooth irreducible projective curve of genus g defined over a field $\mathbf{F}_q$ with a given set $P = \{P_1, \ldots, P_n\}$ of n distinct $\mathbf{F}_q$-rational points and if $L \subset \mathbf{F}_q(X)$ is a $\mathbf{F}_q$-linear subspace of the function field $\mathbf{F}_q(X)$ such that no $f \in L$ has a pole in the points P_i, we can define an evaluation map

$$\alpha\colon L \to \mathbf{F}_q^n, \quad f \mapsto (f(P_i)_{i=1}^n).$$

The image of L under α is then a code. In case

$$L = L(D) = \{f \in k(X)^*\colon (f) + D \geq 0\} \cup \{0\}$$

with D a $\mathbf{F}_q$-divisor we find the code $C(D, P)$. Invariants of these codes, like the dimension k and minimum distance d, can be expressed in terms of properties of the curve; for example, one has the elementary result:

Theorem 4.1. *If D is a divisor defined over $\mathbf{F}_q$ of degree $g \leq \deg D \leq n$ and with* $\mathrm{supp}(D) \cap P = \emptyset$, *then we have $k \geq \deg(D) + 1 - g$ with equality if* $\deg(D) \geq 2g - 1$. *Moreover, $d \geq n - \deg(D)$.*

If we fix the ratio $\deg(D)/n$ then the transmission rate $R = k/n$ increases with the ratio n/g. Therefore to obtain good codes one has to construct curves with as many points as possible.

If X_ℓ is a sequence of curves defined over $\mathbf{F}_q$ such that their genera g_ℓ tend to ∞ and such that $\lim_{\ell\to\infty} \#X_\ell(\mathbf{F}_q)/g_\ell = \gamma > 0$, then the part of the line $\delta + R = 1 - 1/\gamma$ in the positive quadrant is contained in the domain U_q^{lin}. This follows by taking divisors D_ℓ of degree $[\#X_\ell(\mathbf{F}_q)(1-\delta)]$ and evaluating the functions in $L(D_\ell)$ in the rational points $X_\ell(\mathbf{F}_q)$. Then Theorem 4.1 tells us that for the geometric Goppa codes $C_\ell = C((D_\ell, X_\ell(\mathbf{F}_q))$ we have

$$R_\ell + \delta_\ell \geq 1 + (1 - g_\ell)/\#X_\ell(\mathbf{F}_q)$$

which tends to $1 - 1/\gamma$. Hence this sequence of codes has a limit point on or above the line $R + \delta = 1 - 1/\gamma$.

The fact that for q a square there exists a sequence of curves X_ℓ defined over $\mathbf{F}_q$ of genus g_ℓ with the ratio $\#X_\ell(\mathbf{F}_q)/g_\ell$ tending to $\sqrt{q} - 1$ was observed by Ihara (see [16]) and independently by Tsfasman, Vladuts and Zink who applied it to coding theory, see [37]. They used modular curves $X_0(N)$ and the rational points on them provided by the supersingular points.

Theorem 4.2. (Tsfasman, Vladuts, Zink) *Let q be a square. There exists a sequence of geometric Goppa codes over $\mathbf{F}_q$ with limit point on or above the line $R + \delta = 1 - 1/(\sqrt{q} - 1)$.*

For $q \geq 49$ this line comes above the Gilbert-Varshamov bound and one thus gets asymptotically good sequences of codes above the Gilbert-Varshamov bound, and this came at that time as quite a surprise for coding theorists.

Goppa's discovery led to renewed interest in curves over finite fields and drew attention to new aspects of these curves.

5. Upper Bounds for the Number of Points on a Curve

The new interest in curves over finite fields soon paid its dividends. It was noted by Ihara (see [17]) that the Hasse-Weil bound (3) could be improved. He did this by comparing $\#X(\mathbf{F}_q)$ and $\#X(\mathbf{F}_{q^2})$ using (4) and the Cauchy-Schwartz inequality. His bound

$$\#X(\mathbf{F}_q) \leq q+1+[(\sqrt{(8q+1)g^2+4(q^2-q))g}-g)/2]$$

is better than the Hasse-Weil bound for $g > (q-\sqrt{q})/2$. Drinfeld and Vladuts generalized this idea by using all extensions $\mathbf{F}_{q^r}/\mathbf{F}_q$ instead of just the quadratic one (see [38]). Let us define

$$N_q(g) := \text{maximum value of } \#X(\mathbf{F}_q)\,, \tag{5}$$

where X runs through all curves of genus g defined over $\mathbf{F}_q$. Moreover, we define

$$A(q) := \limsup_{g\to\infty} N_q(g)/g\,.$$

The result of Drinfeld and Vladuts says that $A(q) \leq \sqrt{q}-1$.

Serre started the study of the actual value of $N_q(g)$, see [32]. He improved the Hasse-Weil bound slightly by applying some arithmetic to the α_i (cf. (4)):

$$N_q(g) \leq q+1+g[2\sqrt{q}]\,. \tag{6}$$

Serre also transplanted the 'formules explicites' from number theory. He takes a trigonometric polynomial

$$f = 1+2\sum_{n\geq 1} u_n \cos n\theta$$

with real coefficients $u_n \geq 0$ such that $f(\theta) \geq 0$ for all $\theta \in \mathbb{R}$ and sets $\psi = \sum_{n\geq 1} u_n t^n$. Then he gets the estimate

$$N_q(g) \leq a_f g + b_f$$

with

$$a_f = \frac{1}{\psi(1/\sqrt{q})} \quad \text{and} \quad b_f = 1+\frac{\psi(\sqrt{q})}{\psi(1/\sqrt{q})}\,.$$

Finding the optimal choices for such functions$f$ is a linear programming problem which was solved by Oesterlé, see [34], or [30] for an exposition. The resulting bounds are called Oesterlé bounds.

6. Asymptotics

Combining the result of Drinfeld and Vladuts $A(q) \leq \sqrt{q}-1$ with the result of Ihara shows that

$$A(q) = \sqrt{q}-1 \quad \text{for } q \text{ a square}.$$

In 1996 Garcia and Stichtenoth constructed an explicit tower of curves X_ℓ defined over $\mathbf{F}_{q^2}$. This tower starts with the rational line X_1 with coordinate x_1 and consists of successive Artin-Schreier extensions defined by

$$y_{\ell+1}^q + y_{\ell+1} = x_\ell^{q+1},$$

where x_ℓ is defined recursively by

$$x_\ell = y_\ell / x_{\ell-1}$$

and for which the ratio $\#X_\ell(\mathbf{F}_{q^2})/g(X_\ell)$ tends to $q-1$. Later Elkies showed that these towers are modular towers, cf. [2].

For q not a square the situation is yet unclear. Zink used in [40] degenerations of Shimura surfaces to show that if $q=p^3$ then there exists a sequence of curves over $\mathbf{F}_q$ with $\gamma = \lim X_\ell/g_\ell \geq 2(p^2-1)/(p+2)$. In certain cases one can construct explicit towers of curves reaching this bound γ, cf. [13].

In any case we know since the 1980s by work of Serre, who used class field towers, that $A(q) > c \log q$ for an absolute constant $c>0$. We refer to the paper by Hajir and Maire in this volume for a survey of asymptotically good towers.

These results make use of towers where the genera that occur may be rather sparse. If one insists on using all sufficiently large genera the problem becomes more difficult.

In [18] it is shown that $N_q(g)$ goes to ∞ with g:

Theorem 6.1. ([18]) *For fixed q we have $\lim_{g\to\infty} N_q(g) = \infty$.*

The following question suggests itself:

Question 6.2. *What is $\liminf_{g\to\infty} N_q(g)/g$?*

It follows from the work [3] that $\liminf_{g\to\infty} N_q(g)/g \geq (\sqrt{q}-1)/3$ if q is a square.

The following result gives restrictions for the number $N_q(g)$ for all g:

Theorem 6.3. ([18]) *For fixed q there are constants e_q and f_q depending on q with $0 < e_q < f_q$ such that for every $g > 0$ one has $e_q g < N_q(g) < f_q g$.*

7. Maximal Curves

We shall call a curve X defined over $\mathbf{F}_q$ *maximal* if $\#X(\mathbf{F}_q) = q+1+2g\sqrt{q}$, i.e. if the curve attains the Hasse-Weil bound. Then q is a square if the genus is not zero and since the Ihara bound improves the Hasse-Weil bound if $g > (q-\sqrt{q})/2$, this can only happen if $g \leq (q-\sqrt{q})/2$. If X is a maximal curve and Y a curve dominated by X then Y is also a maximal curve, since the Jacobian of Y is an isogeny factor of the Jacobian of X.

For a maximal curve X the action of Frobenius F on the Jacobian is by $-\sqrt{q}$. Choose a $\mathbf{F}_{q^2}$-rational point P_0 on X and map X to the Jacobian $\mathrm{Jac}(X)$ by $P \mapsto [P - P_0]$. If $F(P)$ is the Frobenius image of P on X then one has

$$-\sqrt{q}[P - P_0] = [F(P) - P_0]$$

in the Jacobian, so for any point P the divisor $\sqrt{q}P + F(P)$ is linearly equivalent to $(\sqrt{q}+1)P_0$ and this gives a canonically defined linear system on such curves. Fuhrmann and Torres and Garcia applied ideas of Stöhr and Voloch ([36]; see also Section 8 below) to this linear system to prove the following restrictions on the genera of maximal curves. Let $g_0 = g_0(q) = (q - \sqrt{q})/2$ and $g_1 = g_1(q) = (\sqrt{q}-1)^2/4$.

Theorem 7.1. ([4, 5]) *If X is a maximal curve of genus g over $\mathbf{F}_q$ then either $g = g_0$ or $g \leq g_1$. Moreover, if q is odd and $(\sqrt{q}-1)(\sqrt{q}-2)/4 < g \leq g_1$ then $g = g_1$.*

A natural question arises:

Question 7.2. *Determine the genera for which there are maximal curves. Characterize these curves.*

Maximal curves with $g = g_0$ or $g = g_1$ have been characterized: for genus $g = g_0$ the curve is the hermitian curve, for $g = g_1$ it is given by $y^q + y = x^{(\sqrt{q}+1)/2}$, which is dominated by a hermitian curve. For curves of genus $g_2 = (\sqrt{q}-1)(\sqrt{q}-3)/8$ there are two non-isomorphic types of maximal curves for $\sqrt{q} \equiv 3 \pmod 4$, the Fermat curve of degree $(\sqrt{q}+1)/2$ and the Artin-Schreier curve $y^{\sqrt{q}} + y = x^{(\sqrt{q}+1)/4}$. It is not known whether there are other types. We refer to the paper by Garcia in this volume for more details.

The maximal curves that we know all seem to come from the hermitian curve, although this has not been verified for all of them, which could motivate the following question, but the evidence is limited.

Question 7.3. (Stichtenoth) *Is every maximal curve the image under a dominant map of a hermitian curve?*

8. The Function $N_q(g)$

For each pair (q, g) there is the constant of nature $N_q(g)$ defined by (5). Besides the asymptotic behavior the actual value of $N_q(g)$ for relatively small q and g is interesting, just as a test for our knowledge on curves over finite fields, but also with an eye to practical applications in coding theory or cryptography.

The Hasse-Weil bound and its improvements (Ihara, Serre, Oesterlé) give us an upper bound for $N_q(g)$. To see how good this upper bound is one has to construct curves with as many points as possible. In practice this results in an interval $[a, b]$ to which $N_q(g)$ is constrained. Here b the best upper bound we know and a is the largest number of points we know to occur for a curve of genus g over $\mathbf{F}_q$.

Serre determined the actual value of $N_q(g)$ for a number of small values of g and q. For example he determined $N_q(2)$ for all q and $N_q(g)$ for certain pairs (q, g) with $q = 2$.

In some cases for small genera certain values of $N_q(g)$ can be eliminated by listing all the possibilities of the zeta function and showing that zeta functions in this list imply a decomposition of the Jacobian as a product of principally polarized abelian varieties, which contradicts the irreducibility of the theta divisor of the curve, cf. [34, 20, 21].

Furthermore, a result of Serre says that for a curve of genus ≥ 3 with $\#X(\mathbf{F}_q) < q + 1 + g[2\sqrt{q}]$ one has $\#X(\mathbf{F}_q) \leq q - 1 + g[2\sqrt{q}]$.

Sometimes one can rule out that $N_q(g)$ equals the Serre bound (6) by a specific argument, like Galois descent. This works e.g. for $(q = 27, g = 3)$ and $(q = 8, g = 4)$, cf. [34, 21].

We give a sample of results thus obtained.

Proposition 8.1. *One has the following explicit results:*

1) $N_2(7) = 10$;
2) $N_3(5) \leq 13$ *and* $N_3(7) = 16$;
3) $N_4(4) = 15$ *and* $N_9(4) = 30$;
4) $N_{27}(3) = 56$.

It seems not unlikely that with such methods many more results can be obtained, maybe even of a general nature.

A look at tables for $N_q(g)$ (see [11]) suggests the following question.

Question 8.2. *Is the function $N_q(g)$ a non-decreasing function of g for fixed q?*

Instead of studying the maximum value of $\#X(\mathbf{F}_q)$ for all curves of genus g one could restrict to curves of a specific type. For example, for hyperelliptic curves one has the obvious bound $\#X(\mathbf{F}_q) \leq 2(q + 1)$. One could try to generalize this to bounds for the maximum number of points on a curve of genus g with given gonality. The gonality vector $\gamma(X) = (\gamma_1, \gamma_2, \dots)$ of a curve X over an algebraically closed field k is given by

$$\gamma_r(X) = \min\{d\colon 1 \leq d \leq g - 1, \text{ there exists a } \mathfrak{g}^r_d \text{ on } X\},$$

where $\mathfrak{g}^r_d$ stands for a linear system of degree d and dimension r. A curve of genus g admits a map of degree $\leq [(g + 3)/2]$ to the projective line, so the geometric gonality $\gamma_1(X)$ is bounded by $\leq [(g + 3)/2]$. But this map need not be defined over our ground field. For example, a non-hyperelliptic curve of genus 4 over $\mathbf{F}_q$ has a map of degree 3 to $\mathbb{P}^1$ over $\mathbf{F}_{q^2}$, but not necessarily over $\mathbf{F}_q$, depending on whether the two rulings of the quadric containing the canonical curve are defined over $\mathbf{F}_q$ or not.

Question 8.3. *What is the maximum number of rational points on a curve of genus g and gonality γ defined over $\mathbf{F}_q$?*

In [36] Stöhr and Voloch present a sort of answer to this question, namely an upper bound for the maximum number of points on a curve over $\mathbf{F}_q$ which does not only depend on the genus g, but also on a given linear system defined over $\mathbf{F}_q$. It uses an infinitesimal approach which counts points on a curve, embedded in projective space with a linear system such that the Frobenius image of a point lies in the osculating hyperplane of the curve at that point. As a special case this provides a new proof of the Hasse-Weil bound.

Quite a lot of people have tried to construct curves with many points in order to test how good the upper bounds on $N_q(g)$ are. A variety of methods have been used for this, like methods from class field theory (Serre, Schoof, Lauter, Niederreiter, Xing and Auer), methods from Drinfeld modules (Niederreiter and Xing), fibre products of Artin-Schreier curves (Van der Geer and Van der Vlugt, Shabat), Kummer curves (Van der Geer and Van der Vlugt) and various other methods. We refer to [9] and [11] for a summary of the methods and results.

9. Another Relationship between Curves and Codes

A relationship between curves and codes quite different from the one discovered by Goppa, but not less important, arises from the observation that many classical codes are trace codes of the form $\mathrm{Tr}(C')$, where C' is a code over a field $\mathbf{F}_{q^s}$ obtained from evaluating functions f at points P_i of the affine line and $\mathrm{Tr} = \mathrm{Tr}_{q^s/q}$ is the trace from $\mathbf{F}_{q^s}$ to $\mathbf{F}_q$. To give an example, the classical dual Melas codes $M(q)^{\perp}$ over $\mathbf{F}_p$ are codes of length $q-1$ with words of the form

$$c_{a,b} = (\mathrm{Tr}_{\mathbf{F}_q/\mathbf{F}_p}(ax + b/x))_{x\in\mathbf{F}_q^*} \quad \text{with} \quad a, b \in \mathbf{F}_q\,.$$

Since $\mathrm{Tr}_{\mathbf{F}_q/\mathbf{F}_p}(u) = 0$ if and only if there exists an element $v \in \mathbf{F}_q$ with $v^p - v = u$ we see that the weight of a word $c_{a,b}$ equals

$$q - 1 - \frac{1}{p}(\#X_{a,b}(\mathbf{F}_q) - 2)\,,$$

where $X_{a,b}$ is the (smooth projective) curve given by $y^p - y = ax + b/x$.

More generally, we consider a Goppa code $C' = C(D, P)$ over $\mathbf{F}_{q^s}$ associated to some triple (X, D, P) as above. Since the words of $\mathrm{Tr}_{q^s/q}(C')$ are of the form

$$\mathrm{Tr}(f(P_i))_{i=1}^{n} \quad \text{for} \quad f \in L(D)$$

we see that there is a simple relation between the weight of a word $\mathrm{Tr}(\alpha(f)) = \mathrm{Tr}(f(P_i))_{i=1}^{n}$ and the number of rational points on the curve defined by $y^q - y = f$. In this way we obtain a one-to-one correspondence between words in a trace code of a Goppa code and curves in a k-dimensional family of curves. Linear subspaces correspond then to fibre products of curves. The weight distribution in the code is directly related to the distribution of the number of points in this family. For example, the weight distribution of the quadratic Reed-Muller code $R_q(2, r)$ is related to a family of supersingular curves given by equations of the form $y^p - y = xR(x)$ with R running through a family of linearized polynomials of the form $R = \sum_{i=0}^{h} a_i x^{p^i}$. These curves are thus related to quadric hypersurfaces and the fact that we are dealing with quadrics enables one to determine the weight distribution.

This relationship between codes and families of curves leads to difficult questions on the behavior of the zeta function in families of curves.

10. Distribution of Traces of Frobenius and Weights

Although most of the attention so far focused on determining or bounding the function $N_q(g)$ one may ask more generally:

Question 10.1. *For given pair (q, g) which values can the number of points on a smooth projective irreducible curve of genus g over $\mathbf{F}_q$ assume?*

In a given family of curves over $\mathbf{F}_q$ we can ask for the frequencies with which a given number of points is assumed. Here we count the contribution of each curve X defined over $\mathbf{F}_q$ with multiplicity $1/\#\,\mathrm{Aut}_{\mathbf{F}_q}(X)$. The most basic families are the universal families over a cover of the moduli space M_g of smooth irreducible complete curves of genus g.

Question 10.2. *Determine the frequencies of the number of points in such (universal) families.*

Note that this is a difficult question in general since the information given by these frequencies suffices to determine the number of points on $M_{g,n}(\mathbf{F}_q)$, where $M_{g,n}$ is the universal n-pointed curve for all n. In joint work with S. del Baño and C. Faber we have determined the frequencies for the genus 2 moduli spaces $M_2 \otimes \mathbf{F}_p$ for all primes $p \leq 181$.

11. Strata on the Moduli Spaces

To an abelian variety in characteristic $p > 0$ we can associate the characteristic polynomial of Frobenius (acting on $H^1_{et}(X, \mathbf{Q}_\ell)$ with $p \neq \ell$) and the Newton polygon of this characteristic polynomial. As the Newton polygon goes up under specialization this leads to a stratification on moduli spaces of (polarized) abelian varieties. For an abelian variety over a finite field the isogeny class of X is determined by the characteristic polynomial as Tate showed and by work of Tate and Honda it is known which characteristic polynomials can occur. We restrict ourselves to principally polarized abelian varieties. The two extreme cases in this stratification on the moduli space $A_g \otimes \mathbf{F}_p$ of principally polarized abelian varieties of dimension g are the ordinary case and the supersingular case; the corresponding strata have dimension $g(g+1)/2$ and $[g^2/4]$, respectively, see [22]. Intermediate cases are the strata given by the condition that the p-rank equals r. These strata have codimension $g - r$ in A_g.

By associating to a curve its Jacobian variety this induces also a stratification on the moduli space $M_g \otimes \mathbf{F}_p$ of curves in characteristic $p > 0$. But unlike the case of abelian varieties it is not known which Newton polygons can occur. We know that the generic curve is ordinary, but for example for general p it is not known whether for every genus g there exists a curve defined over a finite field of characteristic p which is supersingular. For $p = 2$ it was proven in [12] that for every g there exists a supersingular curve of genus g defined over $\mathbf{F}_2$. A similar construction in characteristic $p > 2$ gives supersingular curves for genera whose p-adic expansion uses 0 and $(p-1)/2$ only. This motivates the following questions, which can also be specialized to finite fields, but one may ask them as well in a more general setting.

Question 11.1. *For which genera does there exist a supersingular curve in characteristic p?*

Question 11.2. *Which Newton polygons occur for Jacobians?*

Since the characteristic polynomial of Frobenius is an isogeny invariant we can ask more generally the following important question that belongs to the folklore of the field:

Question 11.3. *Which isogeny classes of abelian varieties contain a jacobian variety?*

It is proved in [6] that the stratum of curves of p-rank r has codimension $g-r$ in the moduli space of curves $M_g \otimes \mathbf{F}_p$.

It might well be that the following final question admits unexpected answers over finite fields.

Question 11.4. *Can one give an effective procedure for deciding whether or not a polarized abelian variety is a jacobian?*

References

[1] R. Auer: Ray class fields of global function fields with many rational places. Report University of Oldenburg, 1998.

[2] N. Elkies: Explicit Modular Towers. *Proceedings of the Thirty-Fifth Annual Allerton Conference on Communication, Control and Computing* (1997, T. Basar and A. Vardy, eds), Univ. of Illinois at Urbana-Champaign 1998, pp. 23–32.

[3] N. Elkies, A. Kresch, B. Poonen, J. Wetherell and M. Zieve: Curves of every genus with many points II: asymptotically good families, in preparation.

[4] R. Fuhrmann and F. Torres: The genus of curves over finite fields with many rational points. *Manuscripta Math.* **89** (1996), pp. 103–106.

[5] R. Fuhrmann, A. Garcia and F. Torres: On maximal curves. *J. Number Theory* **67** (1997), pp. 29–51.

[6] C. Faber and G. van der Geer: Complete subvarieties of the moduli space of curves. Manuscript in preparation.

[7] F. Garcia and H. Stichtenoth: A tower of Artin-Schreier extensions of function fields attaining the Drinfeld-Vladut bound. *Invent. Math.* **121** (1995), pp. 211–22.

[8] C. F. Gauss: Disquisitiones Arithmeticae 1801.

[9] G. van der Geer and M. van der Vlugt: How to construct curves over finite fields with many points. In: *Arithmetic Geometry*, (Cortona 1994), F. Catanese Ed., Cambridge Univ. Press, Cambridge, 1997, pp. 169–189.

[10] G. van der Geer and M. van der Vlugt: On generalized Reed-Muller codes and curves with many points. *J. of Number Theory* **72** (1998), pp. 257–268.

[11] G. van der Geer and M. van der Vlugt: Tables for the function $N_q(g)$. Math. of Computation. Regularly updated tables at: http://www.science.uva.nl~/geer.

[12] G. van der Geer and M. van der Vlugt: On the existence of supersingular curves of given genus. *Journal für die Reine und angewandte Math.* **458** (1998), pp. 53–61.

[13] G. van der Geer and M. van der Vlugt: An asymptotically good tower of curves over the field with eight elements. Preprint. arXiv math.AG/0102158.

[14] V. D. Goppa: Codes associated with divisors. (Russian) *Problemy Peredavci Informacii* **13** (1977), pp. 33–39.

[15] V. D. Goppa: Codes on algebraic curves. (Russian) *Dokl. Akad. Nauk SSSR* **259** (1981), pp. 1289–1290.

[16] Y. Ihara: Congruence relations and Shimura curves. II. *J. Fac. Sci. Univ. Tokyo* **25** (1979), pp. 301–361.

[17] Y. Ihara: Some remarks on the number of points of algebraic curves over finite fields. *J. Fac. Sci. Tokyo* **28** (1982), pp. 721–724.

[18] A. Kresch, J. Wetherell and M. Zieve: Curves of every genus with many points, I: Abelian and toric families. Preprint 2000.

[19] K. Lauter: Ray class field constructions of curves over finite fields with many rational points. In: *Algorithmic Number Theory* (Talence 1996), H. Cohen Ed., Lecture Notes in Computer Science 1122, Springer-Verlag Berlin, 1996, pp. 187–195.

[20] K. Lauter: Non-existence of a curve over $\mathbf{F}_3$ of genus 5 with 14 rational points. *Proc. Amer. Math. Soc.* **128** (2000), pp. 369–374.

[21] K. Lauter: Improved upper bounds for the number of rational points on algebraic curves over finite fields. *Comptes Rend. Acad. Sci. Paris Sér. I Math.* **328** (1999), pp. 1181–1185.

[22] K.-Z. Li and F. Oort: Moduli of supersingular abelian varieties. Springer Lecture Notes in Mathematics, 1680. Springer-Verlag, Berlin, 1998.

[23] J. van Lint: Introduction to Coding Theory. Graduate Texts in Mathematics. Springer-Verlag, 1998.

[24] F. MacWilliams and N. Sloane: The theory of error-correcting codes. North-Holland Publishing Company 1977.

[25] Yu. I. Manin: What is the maximum number of points on a curve over $\mathbf{F}_2$? *J. Fac. Sci. Tokyo* **28** (1981), pp. 715–720.

[26] H. Niederreiter and C. P. Xing: Drinfeld modules of rank 1 and algebraic curves with many rational points II. *Acta Arithm.* **81** (1997), pp. 81–100.

[27] H. Niederreiter and C. P. Xing: Global function fields with many rational points over the ternary field. *Acta Arithm.* **83** (1998), pp. 65–86.

[28] H. Niederreiter and C. P. Xing: Algebraic curves with many rational points over finite fields of characteristic 2. In: *Number Theory in Progress* (Zakopane 1997), pp. 359–380, de Gruyter, Berlin, 1999.

[29] H. Niederreiter and C. P. Xing: A general method of constructing global function fields with many rational places. To appear in: *Algorithmic Number Theory* (Portland 1998), Lecture Notes in Comp. Science **1423** (1998), Springer-Verlag, Berlin.

[30] R. Schoof: Algebraic curves and coding theory. UTM 336, University of Trento, 1990.

[31] J.-P. Serre: Sur le nombre des points rationnels d'une courbe algébrique sur un corps fini. *Comptes Rendus Acad. Sci. Paris* **296** (1983), pp. 397–402.

[32] J.-P. Serre: Nombre de points des courbes algébriques sur $\mathbf{F}_q$. Sém. de Théorie des Nombres de Bordeaux, 1982/83, exp. no. 22. (= Oeuvres III, No. 129, pp. 664–668).

[33] J.-P. Serre: Quel est le nombre maximum de points rationnels que peut avoir une courbe algébrique de genre g sur un corps fini $\mathbf{F}_q$? Résumé des Cours de 1983–1984. (=Œuvres III, No. 132, pp. 701–705).

[34] J.-P. Serre: Rational points on curves over finite fields. Notes of lectures at Harvard University, 1985.

[35] V. Shabat: Curves with many points. Thesis, University of Amsterdam, 2001.

[36] K. O. Stöhr and J. F. Voloch: Weierstrass points and curves over finite fields. *Proc. London Math. Soc.* **52** (1986), pp. 1–19.

[37] M. A. Tsfasman, S. G. Vladuts and Th. Zink: On Goppa codes which are better than the Varshamov-Gilbert bound. *Math. Nachrichten* **109** (1982), pp. 21–28.

[38] S. G. Vladuts and V. G. Drinfeld: Number of points of an algebraic curve. *Funct. Anal.* **17** (1983), pp. 68–69.

[39] V. K. Wei: Generalized Hamming weights for linear codes. *IEEE Trans. Inform. Theory* **37** (1991), pp. 1412–1418.

[40] Th. Zink: Degeneration of Shimura surfaces and a problem in coding theory. Fundamentals of computation theory (Cottbus, 1985), 503–511, Lecture Notes in Comput. Sci., 199, Springer-Verlag, Berlin-New York, 1985.

Korteweg-de Vries Instituut
Universiteit van Amsterdam
Plantage Muidergracht 24
1018 TV Amsterdam, The Netherlands
E-mail address: `geer@science.uva.nl`

Authentication Codes and Algebraic Curves

Chaoping Xing

Abstract. We survey a recent application of algebraic curves over finite fields to the constructions of authentication codes.

1. Introduction

Authentication codes were invented by Gilbert, MacWilliams and Sloane [5]. The general theory of unconditional authentication has been developed by Simmons ([10, 11]) and has been extensively studied in recent years.

In the conventional model for unconditional authentication, there are three participants: a *transmitter*, a *receiver* and an *opponent*. The transmitter wants to communicate some information to the receiver using a public channel which is subject to active attack. That is, the opponent can either impersonate the transmitter and insert a message on the channel, or replace a transmitted message with another. To protect against these threats, the transmitter and the receiver share a secret key, the key is then used in an authentication code (A-code for short).

A *systematic* A-code (or A-code *without secrecy*) is a code where the *source state* (i.e. plain text) is concatenated with an *authenticator* (or a *tag*) to obtain a *message* which is sent through the channel. Such a code is a triple $(\mathcal{S}, \mathcal{E}, \mathcal{T})$ of finite sets together with a (authentication) mapping $f\colon \mathcal{S} \times \mathcal{E} \to \mathcal{T}$. Here $\mathcal{S}$ is the set of source states, $\mathcal{E}$ is the set of keys and $\mathcal{T}$ is the set of authenticators. When the transmitter wants to send the information $s \in \mathcal{S}$ using a key $e \in \mathcal{E}$, which is secretly shared with the receiver, he transmits the message $m = (s,t)$, where $s \in \mathcal{S}$ and $t = f(s,e) \in \mathcal{T}$. When the receiver receives a message $m = (s,t)$, she checks the authenticity by verifying whether $t = f(s,e)$ or not, using the secret key $e \in \mathcal{E}$. If the equality holds, the message m is called *valid*.

Suppose the opponent has the ability to insert messages into the channel and/or to modify existing messages. When the opponent inserts a new message $m' = (s', t')$ into the channel, this is called *impersonation attack*. When the opponent sees a message $m = (s,t)$ and changes it to a message $m' = (s', t')$ where $s \neq s'$, this is called *substitution attack*.

We assume that there is a probability distribution on the source states, which is known to all the participants. Given the probability distribution on the source

Research supported by the MOE-ARF research grant R-146-000-018-112.

states, the receiver and the transmitter will choose a probability distribution for $\mathcal{E}$. We will denote the probability of success for the opponent when trying impersonation attack and substitution attack, by P_I and P_S, respectively, and $P(\cdot)$ and $P(\cdot|\cdot)$ specify probability and conditional probability distribution on the message space $\mathcal{M} := \mathcal{S} \times \mathcal{E}$. Then we have

$$P_I = \max_{(s,t)\in\mathcal{M}} P(m = (s,t) \text{ valid}) \qquad \text{and}$$

$$P_S = \max_{(s,t),(s',t')\in\mathcal{M}, s\neq s'} P(m' = (s',t') \text{ valid} \mid m = (s,t) \text{ observed}).$$

If we further assume that the keys and the source states are uniformly distributed, then the deception probabilities can be expressed as

$$P_I = \max_{(s,t)\in\mathcal{M}} \frac{|\{e \in \mathcal{E}:\ t = f(s,e)\}|}{|\mathcal{E}|},$$

$$P_S = \max_{(s,t),(s',t')\in\mathcal{M}, s\neq s'} \frac{|\{e \in \mathcal{E}:\ t = f(s,e), t' = f(s',e)\}|}{|\{e \in \mathcal{E}:\ t = f(s,e)\}|}.$$

In the remainder of the paper, we will always assume that the keys and the source states are uniformly distributed.

We observe that each parameter of an A-code $(\mathcal{S}, \mathcal{E}, \mathcal{T})$ plays a role:

- the size of $\mathcal{S}$ indicates how large the plain text could be;
- the size of $\mathcal{E}$ is the number of keys, which represents the number of users;
- the size of $\mathcal{T}$ is the number of authenticators, which represents the transmission rate;
- P_I is the security measure against the impersonation attack;
- P_S is the security measure against the substitution attack.

It is clear that for fixed sizes of $\mathcal{S}, \mathcal{E}$ and $\mathcal{T}$, we want P_i and P_S to be as small as possible. In other words, if P_I, P_S and $|\mathcal{T}|$ are fixed, we are interested in A-codes $(\mathcal{S}, \mathcal{E}, \mathcal{T})$ with $|\mathcal{S}|$ and $|\mathcal{E}|$ as large as possible. In particular, the study of the asymptotic behaviour of $(\log|\mathcal{S}|)/|\mathcal{E}|$ for fixed $|\mathcal{T}|$, $P_I = 1/|\mathcal{T}|$ and P_S is one of the most important topics for A-codes. For a review of different bounds and constructions for A-codes, we refer to [7, 13, 8].

In this survey paper, we present an explicit construction of A-codes based on algebraic curves over finite fields.

2. Constructions

In this section, we describe a construction of authentication codes based on algebraic curves over finite fields.

Before starting our construction, we need to introduce some concepts and notations that are essential for the construction. For further results on algebraic curves over finite fields, we refer to [12, 16].

We fix some notation for this section.

ℓ: power of a prime;
$\mathbf{F}_\ell$: the finite field of ℓ elements;
$\mathcal{X}$: a projective, absolutely irreducible, complete algebraic curve defined over $\mathbf{F}_\ell$. We simply say that $\mathcal{X}$ is an algebraic curve;
$g = g(\mathcal{X})$: the genus of $\mathcal{X}$;
$\mathbf{F}_\ell(\mathcal{X})$: the function field of $\mathcal{X}$;
$\mathcal{X}(\mathbf{F}_\ell)$: the set of all $\mathbf{F}_\ell$-rational points on $\mathcal{X}$.

A *divisor* G of $\mathcal{X}/\mathbf{F}_q$ is a formal sum

$$G = \sum_{P \in S} \nu_P(G)P,$$

where S is a finite non-empty set of points of $\mathcal{X}$, and $\nu_P(G) \in \mathbf{Z}$ for all $P \in S$. A divisor G of $\mathcal{X}/\mathbf{F}_\ell$ is called *$\mathbf{F}_\ell$-rational* if

$$G^\sigma = G$$

for all automorphisms $\sigma \in \mathrm{Gal}(\overline{\mathbf{F}}_\ell/\mathbf{F}_\ell)$, where $\overline{\mathbf{F}}_\ell$ is a fixed algebraic closure of $\mathbf{F}_\ell$ and $\mathrm{Gal}(\overline{\mathbf{F}}_\ell/\mathbf{F}_\ell)$ is the Galois group of $\overline{\mathbf{F}}_\ell/\mathbf{F}_\ell$. In this paper we always mean a rational divisor whenever a divisor is mentioned.

We write ν_P for the normalized discrete valuation corresponding to the point P of $\mathcal{X}$.

For a divisor G we form the vector space

$$L(G) = \{x \in \mathbf{F}_\ell(\mathcal{X})\backslash\{0\} : \mathrm{div}(x) + G \geq 0\} \cup \{0\}.$$

Then $L(G)$ is a finite-dimensional vector space over $\mathbf{F}_\ell$, and we denote its dimension by $l(G)$. By the Riemann-Roch theorem (see [12, 16]), we have

$$l(G) \geq \deg(G) + 1 - g,$$

and equality holds if $\deg(G) \geq 2g - 1$.

Now we are ready to describe the construction.

Let $\mathcal{P}$ be a subset of $\mathcal{X}(\mathbf{F}_\ell)$, i.e., $\mathcal{P}$ is a set of $\mathbf{F}_\ell$-rational points of $\mathcal{X}$. Let D be a positive divisor with $\mathcal{P} \cap \mathrm{Supp}(D) = \emptyset$. Choose an $\mathbf{F}_\ell$-rational point R in $\mathcal{P}$ and put $G = D - R$. Then $\deg(G) = \deg(D) - 1$, $L(G) \subset L(D)$ and $\mathbf{F}_\ell \cap L(G) = \{0\}$. Moreover, we have

$$L(D) = \mathbf{F}_\ell \oplus L(G) = \{\alpha + f | f \in L(G), \alpha \in \mathbf{F}_\ell\}.$$

Put

$$\mathcal{S} = L(G), \quad \mathcal{E} = \mathcal{P} \times \mathbf{F}_\ell, \quad \mathcal{T} = \mathbf{F}_\ell,$$

and consider the map f

$$\mathcal{S} \times \mathcal{E} \to \mathcal{T}, \quad (s, (P, \alpha)) \mapsto s(P) + \alpha.$$

It can be proved that $(\mathcal{S}, \mathcal{E}, \mathcal{T})$ constructed above together with f forms an A-code with the deception probabilities

$$P_I = \frac{1}{\ell}, \quad P_S = \frac{\deg(D)}{|\mathcal{P}|}$$

provided $\deg(D) \geq 2g + 1$. More precisely, we have the following result.

Theorem 2.1. *Let $\mathcal{X}$ be an algebraic curve and $\mathcal{P}$ a set of $\mathbf{F}_\ell$-rational points on $\mathcal{X}$. Suppose that D is a positive divisor with $\deg(D) \geq 2g + 1$ and $\mathcal{P} \cap \mathrm{Supp}(D) = \emptyset$. Then there exists an A-code $(\mathcal{S}, \mathcal{E}, \mathcal{T})$ with*

$$|\mathcal{S}| = \ell^{l(D)-1} = \ell^{\deg(D)-g}, \quad |\mathcal{E}| = \ell|\mathcal{P}|, \quad |\mathcal{T}| = \ell$$

$$P_I = \frac{1}{\ell}, \quad P_S = \frac{\deg(D)}{|\mathcal{P}|}.$$

Theorem 2.1 gives a construction of A-codes based on general algebraic curves over finite fields. In the examples below, we apply Theorem 2.1 to some special curves to obtain an A-code with nice parameters.

Example 2.2. *Consider the projective line $\mathcal{X}/\mathbf{F}_\ell$. Then $g = g(\mathcal{X}) = 0$.*

(a) Let d be an integer between 1 and ℓ, and P an $\mathbf{F}_\ell$-rational point of $\mathcal{X}$. Put

$$D = dP, \quad \mathcal{P} = \mathcal{X}(\mathbf{F}_\ell) - \{P\}.$$

Then $\deg(D) = d \geq 2g + 1$, $|\mathcal{P}| = \ell$ and $\mathcal{P} \cap \mathrm{Supp}(D) = \emptyset$. By Theorem 2.1, we obtain an A-code $(\mathcal{S}, \mathcal{E}, \mathcal{T})$ with

$$|\mathcal{S}| = \ell^d, \quad |\mathcal{E}| = \ell^2, \quad |\mathcal{T}| = \ell$$

$$P_I = \frac{1}{\ell}, \quad P_S = \frac{d}{\ell}.$$

The A-code $(\mathcal{S}, \mathcal{E}, \mathcal{T})$ with the above parameters can also be found in [2]. It can be proved that the above A-code is optimal in the sense that

(b) Let d be an integer between 2 and ℓ. Put $\mathcal{P} = \mathcal{X}(\mathbf{F}_\ell)$. As there always exists an irreducible polynomial of degree d over $\mathbf{F}_\ell$, we can find a positive divisor D such that $\deg(D) = d$ and $\mathcal{P} \cap \mathrm{Supp}(D) = \emptyset$. Then $\deg(D) = d \geq 2g + 1$, $|\mathcal{P}| = \ell + 1$. By Theorem 4.1, we obtain an A-code $(\mathcal{S}, \mathcal{E}, \mathcal{T})$ with

$$|\mathcal{S}| = \ell^d, \quad |\mathcal{E}| = \ell(\ell + 1), \quad |\mathcal{T}| = \ell$$

$$P_I = \frac{1}{\ell}, \quad P_S = \frac{d}{\ell + 1}.$$

These A-codes had not been known before the construction in Theorem 2.1 was introduced (see [19]).

Example 2.3. *Let ℓ be a square and put $r = \sqrt{\ell}$. Consider a sequence of algebraic curves $\mathcal{X}_m/\mathbf{F}_\ell$ given in [4] as follows. Let $\mathcal{X}_1$ be the projective line with the function field $\mathbf{F}_\ell(\mathcal{X}) = \mathbf{F}_\ell(x_1)$. Let $\mathcal{X}_m$ be obtained by adjoining a new equation:*

$$x_m^r + x_m = \frac{x_{m-1}^r}{x_{m-1}^{r-1}+1},$$

for all $m \geq 2$. Then the number of $\mathbf{F}_\ell$-rational points of $\mathcal{X}_m$ is more than $(r^2 - r)r^{m-1}$, and the genus g_m of $\mathcal{X}_m$ is less than r^m for all $m \geq 1$. Choose an integer c between 2 and $\sqrt{\ell}-1$ (c is independent of m) and an $\mathbf{F}_\ell$-rational point P_m of $\mathcal{X}_m$ and put $D_m = c\ell^{m/2}P_m$. Let $\mathcal{P}_m$ be a subset of $\mathcal{X}_m(\mathbf{F}_\ell) - \{P_m\}$ with

$$|\mathcal{P}_m| = (r^2 - r)r^{m-1} = \ell^{m/2}(\sqrt{\ell}-1).$$

By Theorem 2.1, we obtain a sequence of A-code $(\mathcal{S}_m, \mathcal{E}_m, \mathcal{T}_m)$ with

$$|\mathcal{S}_m| = \ell^{(c-1)\ell^{m/2}}, \quad |\mathcal{E}_m| = \ell^{m/2}(\ell\sqrt{\ell_m} - \ell), \quad |\mathcal{T}_m| = \ell,$$

and with deception probabilities

$$P_I = \frac{1}{\ell}, \qquad P_S = \frac{c}{\sqrt{\ell}-1}.$$

The above example provides the first explicit construction of A-codes with $\lim_{|\mathcal{E}|\to 0}(\log|\mathcal{S}|)/|\mathcal{E}| > 0$ for fixed $|\mathcal{T}|$, $P_I = 1/|\mathcal{T}|$ and P_S.

References

[1] J. Bierbrauer, "Universal hashing and geometric codes", *Designs, Codes and Cryptography,* Vol.11, pp. 207–221, 1997.

[2] J. Bierbrauer, T. Johansson, G. Kabatianskii and B. Smeets, "On families of hash functions via geometric codes and concatenation", *Advances in Cryptology–CRYPTO'93, Lecture Notes in Computer Science,* **773**, pp. 331–342, 1994.

[3] J. L. Carter and M. N. Wegman, "Universal classes of hash functions", *J. Computer and System Sci.*, Vol.18, pp. 143–154, 1979.

[4] A. Garcia and H. Stichtenoth, "On the asymptotic behavior of some towers of function fields over finite fields", *J. of Number Theory,* Vol.61, pp. 248–273, 1996.

[5] E. N. Gilbert, F. J. MacWilliams and N. J. A. Sloane, "Codes which detect deception", *The Bell System Technical Journal,* Vol.33, No.3, pp. 405–424, 1974.

[6] T. Helleseth and T. Johansson, "Universal hash functions from exponential sums over finite fields and Galois Rings", *Advances in Cryptology–Crypto'96,* Lecture Notes in Computer Science, **1109**, pp. 31–44, 1996.

[7] T. Johansson, *Contributions to unconditionally secure authentication,* Ph.D. thesis, Lund, 1994.

[8] G. Kabatianskii, B. Smeets, and T. Johansson, "On the cardinality of systematic authentication codes via error correctings", *IEEE Trans. Inform. Theory,* Vol. 42, pp. 566–578, 1996.

[9] J.-P. Serre, *Rational points on curves over finite fields,* lecture notes, Harvard University, 1985.

[10] G. J. Simmons, "Authentication theory/coding theory", *Advances in Cryptology-Crypto '84*, Lecture Notes in Computer Science, **196**, pp. 411–431, 1984.

[11] G. J. Simmons, "A survey of information authentication", in *Contemporary Cryptology, The Science of Information Integrity*, G.J. Simmons, ed., IEEE Press, pp. 379–419, 1992.

[12] H. Stichtenoth, *Algebraic function fields and codes*, Berlin: Springer-Verlag, 1993.

[13] D. R. Stinson, " Combinatorial characterization of authentication codes", *Designs, Codes and Cryptography*, Vol. 2, pp. 175–187, 1992.

[14] D. R. Stinson, "Universal hashing and authentication codes", *Designs, Codes and Cryptography*, Vol. 4, pp. 377–346, 1994. (also *Advances in Cryptology–CRYPTO '91*, Lecture Notes in Computer Sci. **576**, pp. 74–85, 1992.)

[15] D. R. Stinson, "On the connection between universal hashing, combinatorial designs and error-correcting codes, *Congressus Numerantium*, Vol. 114, pp. 7–27, 1996.

[16] M. A. Tsfasman and S. G. Vladut, *Algebraic-geometric codes*, Dordrecht: Kluwer, 1991.

[17] W. C. Waterhouse, "Abelian varieties over finite fields", *Ann. Sci. Ecole Norm. Sup.*, Vol. 2, pp. 521–560, 1969.

[18] M. N. Wegman and J. L. Carter, "New hash functions and their use in authentication and set equality", *Journal of Computer and System Sciences*, Vol. 22, pp. 265—279, 1981.

[19] C. P. Xing and H. X. Wang, and K. Y. Lam, "Constructions of authentication codes from algebraic curves over finite fields," IEEE Trans. Inform. Theory, Vol. 48, pp. 886–892, 2000.

Department of Mathematics
National University of Singapore
2 Science Drive 2
117543, Singapore
E-mail address: matxcp@nus.edu.sg

Some Aspects of Mean Curvature Flow in Presence of Nonsmooth Anisotropies

Giovanni Bellettini

Abstract. We discuss some aspects of motion by mean curvature of hypersurfaces in presence of nonsmooth anisotropies. We include the crystalline case in three dimensions.

1. Introduction

In this short note we discuss some aspects of anisotropic motion by mean curvature. One of the most interesting examples is when the anisotropy is crystalline in three dimensions (i.e. the Wulff shape is a polytope). In this respect, pioneeristic papers have been written by J. Taylor, see for instance [20, 21, 22]. Our main idea is to work in relative geometry, looking at the anisotropy as a norm inducing a new geometry in the ambient space. A number of difficulties arise, due to the nonstrict convexity and nonsmoothness of the Wulff shape. The starting paper leading to this approach is [12]. The reference list is largely incomplete. We refer to [22, 15, 24, 16, 17] for a more detailed bibliography.

It is a pleasure to thank Matteo Novaga and Maurizio Paolini for enlightening discussions on the arguments considered in this paper.

2. Notation

Anisotropies on $\mathbb{R}^n$. We indicate by $\mathcal{M}(\mathbb{R}^n)$ the class of all anisotropies of $\mathbb{R}^n$ (Finsler metrics), i.e. $\phi \in \mathcal{M}(\mathbb{R}^n)$ if $\phi\colon \mathbb{R}^n \to [0, +\infty[$ is convex and satisfies the properties

$$\phi(\xi) \geq \lambda|\xi|, \qquad \phi(a\xi) = a\phi(\xi), \qquad \xi \in \mathbb{R}^n, \, a \geq 0,$$

for a suitable constant $\lambda \in]0, +\infty[$. The dual function $\phi^o\colon \mathbb{R}^n \to [0, +\infty[$ of ϕ is defined as $\phi^o(\xi^*) := \sup \{\xi^* \cdot \xi \ : \ \phi(\xi) \leq 1\}$ for $\xi^* \in \mathbb{R}^n$, and belongs to $\mathcal{M}(\mathbb{R}^n)$; ϕ^o plays the rôle of the surface tension, see (1). We set

$$\mathcal{B}_\phi := \{\xi \in \mathbb{R}^n : \phi(\xi) \leq 1\}, \qquad \mathcal{B}_\phi^o := \{\xi^* \in \mathbb{R}^n : \phi^o(\xi^*) \leq 1\}.$$

$\mathcal{B}_\phi$ and $\mathcal{B}_\phi^o$ are sometimes called the Wulff shape and the Frank diagram, respectively. We say that ϕ is smooth if $\mathcal{B}_\phi$ and $\mathcal{B}_\phi^o$ are of class C^2 and strictly convex.

We say that ϕ is crystalline if $\mathcal{B}_\phi$ (and therefore $\mathcal{B}_\phi^o$) is a polytope. In general, we say that ϕ is nonsmooth if $\mathcal{B}_\phi$ is not at the same time of class C^2 and strictly convex. By a facet of $\partial\mathcal{B}_\phi$ we mean a facet of $\partial\mathcal{B}_\phi$ of dimension $(n-1)$.

$\mathbb{R}^n$ endowed with the norm ϕ becomes a finite dimensional Banach space (sometimes called Minkowski space); our approach consists of properly introducing a notion of ϕ-surface measure, of ϕ-regular boundary (this is necessary only in the nonsmooth case), of ϕ-normal vector and of ϕ-mean curvature, in order to define (intrinsically) and to study the associated geometric evolution problems.

Duality mappings. Let $\phi \in \mathcal{M}(\mathbb{R}^n)$. By T and T^o we denote the possibly multivalued duality mappings defined by

$$T := \frac{1}{2}D^-(\phi^2), \qquad T^o := \frac{1}{2}D^-((\phi^o)^2),$$

where we drop the explicit dependence on the anisotropy in the notation, and D^- denotes the subdifferential. T, T^o are maximal monotone operators, and T (resp. T^o) takes $\partial\mathcal{B}_\phi$ (resp. $\partial\mathcal{B}_\phi^o$) onto $\partial\mathcal{B}_\phi^o$ (resp. onto $\partial\mathcal{B}_\phi$). The geometric properties of the maps $T_{|\partial\mathcal{B}_\phi}$, $T^o_{|\partial\mathcal{B}_\phi^o}$ are of basic importance for describing the geometry of ϕ-regular boundaries in the nonsmooth case: if $\xi \in \partial\mathcal{B}_\phi$, $T(\xi)$ is the intersection of the closed outward normal cone to $\partial\mathcal{B}_\phi$ with $\partial\mathcal{B}_\phi^o$.

ϕ-distance function. Given a nonempty set $E \subset \mathbb{R}^n$ and $x \in \mathbb{R}^n$, we set

$$\mathrm{dist}_\phi(x,E) := \inf_{y\in E} \phi(x-y), \qquad \mathrm{dist}_\phi(E,x) := \inf_{y\in E} \phi(y-x),$$

$$d_\phi^E(x) := \mathrm{dist}_\phi(x,E) - \mathrm{dist}_\phi(\mathbb{R}^n \setminus E, x).$$

The function d_ϕ^E is therefore the oriented ϕ-distance function negative inside E; since in general $\mathcal{B}_\phi$ is not symmetric with respect to the origin, $-d_\phi^E$ does not necessarily coincide with $d_\phi^{\mathbb{R}^n\setminus E}$. At each point x where d_ϕ^E is differentiable, there holds $\nabla d_\phi^E(x) \in \partial\mathcal{B}_\phi^o$ (eikonal equation at x).

The surface energy. The surface energy functional P_ϕ is defined as follows: if $E \subset \mathbb{R}^n$ is a finite perimeter set, then

$$P_\phi(E) := \int_{\partial E} \phi^o(\nu^E)\, d\mathcal{H}^{n-1}, \tag{1}$$

where $\mathcal{H}^{n-1}$ is the $(n-1)$-dimensional Hausdorff measure and ν^E is the outward unit normal to the (reduced) boundary of E.

P_ϕ coincides with the Minkowski content of ∂E (at least if ∂E is sufficiently smooth) defined by means of the distance dist_ϕ and does not necessarily coincide with the $(n-1)$-dimensional Hausdorff measure with respect to dist_ϕ [12]. The functional P_ϕ can be written as $\sup\{\int_E \mathrm{div}\sigma\, dx \colon \sigma \in C_0^1(\mathbb{R}^n;\mathcal{B}_\phi)\}$ [3], and coincides with the $\Gamma - L^2(\mathbb{R}^n)$ limit, as $\epsilon \to 0^+$, of the sequence of functionals

$$M_\epsilon(u) := \frac{1}{2c}\int_{\mathbb{R}^n} \left[\epsilon(\phi^o(\nabla u))^2 + \frac{1}{\epsilon}W(u)\right] dx, \qquad u \in W^{1,2}(\mathbb{R}^n), \tag{2}$$

where $W(s) := (1-s^2)^2$, $c := \int_{-1}^{1} \sqrt{W(s)}\, ds$, and $M_\epsilon := +\infty$ in $L^2(\mathbb{R}^n) \setminus W^{1,2}(\mathbb{R}^n)$ [12].

In the sequel, the function $g\colon \mathbb{R}^n \times [0,+\infty[\to \mathbb{R}$ denotes a fixed bounded function, which plays (depending on the context) either the rôle of prescribed mean curvature or of the forcing term (external field).

3. Definitions and Results in the Smooth Case

In this section $\phi \in \mathcal{M}(\mathbb{R}^n)$ is smooth.

ϕ-normal vectors, ϕ-mean curvature: smooth case. Let E be a bounded open set of class C^∞. We set $\nu_\phi^E := \nabla d_\phi^E$ on ∂E (which has unit ϕ^o-norm) and we define [12] the vector field $n_\phi^E\colon \partial E \to \partial\mathcal{B}_\phi$ as

$$n_\phi^E := T^o(\nu_\phi^E) \qquad \text{on } \partial E\,. \tag{3}$$

n_ϕ^E is the natural ϕ-normal vector field to ∂E, and $n_\phi^E \cdot \nu_\phi^E = 1$. It is sometimes called the Cahn-Hoffman vector field; it can be defined in a suitable neighbourhood of ∂E, keeping the fundamental constraint $\phi(n_\phi^E) = 1$, as $n_\phi^E := T^o(\nabla d_\phi^E)$. The ϕ-mean curvature κ_ϕ^E of ∂E is then defined [12, 11] as

$$\kappa_\phi^E := \operatorname{div}\, n_\phi^E \qquad \text{on } \partial E\,. \tag{4}$$

It turns out that κ_ϕ^E coincides with the tangential divergence of n_ϕ^E and that $\partial\mathcal{B}_\phi$ has ϕ-mean curvature equal to $n-1$.

First variation of P_ϕ: smooth case. Let $E \subset \mathbb{R}^n$ be a bounded open set of class C^∞. Let $\Psi \in C^\infty(\mathbb{R}^{n+1};\mathbb{R}^n)$ and define $\Psi_\lambda(x) := \Psi(x,\lambda)$ for any $(x,\lambda) \in \mathbb{R}^{n+1}$. Assume that $\Psi_0 = \mathrm{Id}$ and that $\Psi_\lambda - \mathrm{Id}$ has compact support, and let $X := \frac{\partial \Psi_\lambda}{\partial \lambda}_{|\lambda=0}$. Then one can prove [11] that

$$\frac{d}{d\lambda} P_\phi\left(\Psi_\lambda(E)\right)_{|\lambda=0} = \int_{\partial E} \kappa_\phi^E\, \nu_\phi^E \cdot X\, d\mathcal{P}_\phi\,, \tag{5}$$

where, here and in the following, $d\mathcal{P}_\phi := \phi^o(\nu^E)\, d\mathcal{H}^{n-1}$. Therefore, if $\mathcal{F}$ is the functional defined by

$$\mathcal{F}(E) := P_\phi(E) + \int_E g\, dx\,, \tag{6}$$

we have

$$\frac{d}{d\lambda} \mathcal{F}\left(\Psi_\lambda(E)\right)_{|\lambda=0} = \int_{\partial E} \left(\kappa_\phi^E - g\right) \nu_\phi^E \cdot X\, d\mathcal{P}_\phi\,. \tag{7}$$

Denoting by dP_ϕ the variation of P_ϕ as an element of the normed space $L^2_\phi(\partial E;\mathbb{R}^n)$, and denoting by $\langle\cdot,\cdot\rangle$ the duality, one can also prove that a scalar multiple of the vector field $\kappa_\phi^E n_\phi^E$ is a solution of the problem

$$\min\left\{\langle dP_\phi, X\rangle : X \in L^2_\phi(\partial E;\mathbb{R}^n),\ \|X\|^2_{L^2_\phi(\partial E;\mathbb{R}^n)} := \int_{\partial E} \phi(X)^2 d\mathcal{P}_\phi \le 1\right\}.$$

For the computation of the second variation of P_ϕ see [5].

Geometric evolution law: smooth case. Assume that g is smooth. Let $E(t)$ be a family of smooth bounded open sets, varying smoothly with $t \in [0,T]$. Set $d_\phi(x,t) := d_\phi^{E(t)}(x)$. We say that $t \in [0,T] \to E(t)$ is a ϕ-smooth flow on $[0,T]$ with forcing term g (and initial set $E(0)$) if

$$\frac{\partial d_\phi}{\partial t}(x,t) = \kappa_\phi^{E(t)}(x) + g(x,t), \qquad x \in \partial E(t), \quad t \in [0,T]. \tag{8}$$

Remark 3.1.

(i) *Under the evolution law (8), $\mathcal{B}_\phi$ shrinks self-similarly.*

(ii) *No mobility factor is present in (8).*

(iii) *If $n = 2$, setting $\phi^o(\xi) = |\xi|\phi^o(\frac{\xi}{|\xi|}) =: \rho\psi(\theta)$ (polar coordinates), it turns out that the velocity of the front along $n_\phi^{E(t)}$ is equal to $\kappa^{E(t)}(\psi + \psi'')$ (where $\kappa^{E(t)}$ is the euclidean curvature of $\partial E(t)$), while along the euclidean direction is equal to $\kappa^{E(t)}\psi(\psi + \psi'')$, see* [11].

(iv) *(8) admits local existence and uniqueness of a smooth solution, and the comparison principle holds. Different notions of weak solutions (giving sense to the evolution after the onset of singularities) can be defined, see for instance* [2, 14].

(v) *Smooth evolutions under (8) can be approximated, as $\epsilon \to 0^+$, with the solutions of the reaction-diffusion type equation*

$$u_t = \operatorname{div}(T^o(\nabla u)) - \frac{1}{2\epsilon^2}W'(u) - \frac{1}{2c\epsilon}g, \tag{9}$$

obtained as the gradient flow of the functionals M_ϵ in (2), after a time rescaling, see [11, 4] *for more details.*

We conclude this section with some comments. The definitions of P_ϕ, n_ϕ^E and κ_ϕ^E, as well as the convergence result of solutions of (9), can be extended for smooth inhomogeneous anisotropies $\phi(x,\xi)$. The results do not cover the case when $\phi(x,\xi)$ vanishes (or becomes infinite) at some point x, or when $\phi(x,\xi)$ depends in a discontinuous way on x. Finally, the extension of our approach to the evolution of manifolds with arbitrary codimension is still an open problem.

4. Extension to the Nonsmooth Case

In this section $\phi \in \mathcal{M}(\mathbb{R}^n)$ is nonsmooth. Unless otherwise stated, we assume also $g = 0$ for simplicity. Our aim is to extend the approach of Section 3. The first step is to define the vector field n_ϕ^E. It is clear that the definition in (3) must be modified, since $T^o(\nu_\phi^E)$ is now a (convex) set, whose dimension may vary from point to point on ∂E. Moreover (as it clearly happens in the crystalline case) we must be able to let evolve nonsmooth (for instance polyhedral) sets, which do not admit a normal vector field everywhere defined. These (and other) considerations lead to redefine the concept of smooth boundary [6, 7].

Definition 4.1. *Let $\phi \in \mathcal{M}(\mathbb{R}^n)$ be nonsmooth. Let $E \subset \mathbb{R}^n$ be a bounded open set. We say that E is Lipschitz ϕ-regular, and we write $E \in \mathcal{R}_\phi(\mathbb{R}^n)$, if ∂E is Lipschitz continuous and there exists a vector field $\eta \in \mathrm{Lip}_{\nu,\phi}(\partial E;\mathbb{R}^n)$, where*

$$\mathrm{Lip}_{\nu,\phi}(\partial E;\mathbb{R}^n) := \{v \in \mathrm{Lip}(\partial E;\mathbb{R}^n) : v(x) \in T^o(\nu_\phi^E(x)) \text{ for } \mathcal{H}^{n-1} - \text{a.e. } x \in \partial E\}.$$

Remark 4.2.

(i) *Lipschitz ϕ-regular sets are the analog, in the euclidean case, of the sets of class $C^{1,1}$.*

(ii) *It has been shown in [8] that, if $\phi \in \mathcal{M}(\mathbb{R}^3)$ is crystalline, there is a Lipschitz ϕ-regular set E, which is polyhedral, convex and very close to the Wulff shape $\mathcal{B}_\phi$, whose evolution, under crystalline mean curvature flow, does not remain a polyhedral set (i.e. a facet of ∂E bends instantly). See also [19] for numerical simulations. This is why, even for a crystalline ϕ, we impose ∂E to be Lipschitz: we cannot restrict Definition 4.1 to polyhedral sets ∂E. It must be said that, if $n = 2$, polygonal curves remain polygonal during the crystalline flow.*

(iii) *The lipschitzianity of η is a regularity requirement making more difficult the proof of a short time existence result of a ϕ-mean curvature flow in the class $\mathcal{R}_\phi(\mathbb{R}^n)$ (which is an open problem). In this respect, one could relax the regularity of η, for instance by requiring η to be only bounded with divergence in $L^2(\partial E)$ or, alternatively, with divergence in $L^\infty(\partial E)$, see [8]. A fourth definition can be given by imposing that η admits an extension in a suitable neighbourhood of ∂E which is bounded with bounded divergence [7, 9]. The three last definitions are expected to coincide for a rather large class of ϕ and E, and hopely to coincide with Definition 4.1 for some choice of ϕ and E.*

(iv) *$\mathcal{B}_\phi$ belongs to $\mathcal{R}_\phi(\mathbb{R}^n)$: take $\eta(x) := x/\phi(x)$.*

(v) *Let $n = 2$, $\phi(\xi) := \max\{|\xi_1|, |\xi_2|\}$ and $B := \{\xi \in \mathbb{R}^2 : |\xi| < 1\}$ be the euclidean ball. Then there exists no $\eta \in \mathrm{Lip}_{\nu,\phi}(\partial B;\mathbb{R}^2)$ (and no η in one of the other three classes introduced in (iii)). Therefore B is not Lipschitz ϕ-regular. The regularity of B, in our relative approach, is analogous to the regularity of the square in the euclidean geometry.*

(vi) *The structure and classification of Lipschitz ϕ-regular sets in $n = 2$ dimensions is essentially known: for instance, in the crystalline case, roughly speaking the boundary of E is a closed Lipschitz curve which is a sequence (with a precise order) of segments which are parallel to some edge of $\partial\mathcal{B}_\phi$ and of segments or arcs which correspond to vertices of $\partial\mathcal{B}_\phi$ [23, 18]. In $n = 2$ dimensions, there is a natural choice of a special vector field, which is the one which makes the edges and the arcs of ∂E of constant ϕ-curvature.*

(vii) *The structure and classification of Lipschitz ϕ-regular sets in $n = 3$ dimensions, even for special ϕ's, is an open problem. Some of their basic properties are studied in [9]. If ϕ is crystalline and E is a polyhedron with the property that at any of its vertices v, the intersection of $T^o(\nu_\phi^E(x))$ ($x \in \mathrm{int}(Q)$) over all facets Q containing v is non-empty, then $E \in \mathcal{R}_\phi(\mathbb{R}^3)$.*

(viii) *It is easily seen, for instance in $n = 2$ dimensions and for ϕ crystalline, that ϕ-convex sets E (in the sense that given $x, y \in E$ there exists at least a ϕ-geodesic connecting x and y and lying inside E) are not necessarily convex in the usual sense.*

First variation of P_ϕ: nonsmooth case. Let $E \in \mathcal{R}_\phi(\mathbb{R}^n)$. In [9] the first variation of the functional P_ϕ at E is rigorously computed. Roughly speaking, it turns out that the expression of (minus) the norm of the gradient of P_ϕ at E is given by

$$- \inf_{N \in Y} \left(\int_{\partial E} (\mathrm{div}_{\phi,\tau} N)^2 \, dP_\phi \right)^{1/2} . \tag{10}$$

The operator $\mathrm{div}_{\phi,\tau}$ denotes the tangential divergence with respect to ϕ; here it suffices to say that it is the natural definition of tangential divergence in relative geometry, and coincides, on flat regions of ∂E, with the usual tangential divergence. The space Y is the class of all vector fields in $L^2(\partial E; \mathbb{R}^n)$ satisfying the constraint to belong to $T^o(\nu_\phi^E)$ and having ϕ-tangential divergence in $L^2(\partial E)$. It can be proved that the minimum problem (10) has a solution, which we denote by $N_{\min}^E$, which is *unique in the divergence.* It can then be proved that the direction of minimal slope for the functional P_ϕ at E is given by $N_{\min}^E$: this is one of the motivations for studying problem (10) in connection with the geometric evolution problem. It is clear that $\mathrm{div}_{\phi,\tau} N_{\min}^E$ is expected to identify the initial velocity of the front.

Definition 4.3. *Let $E \in \mathcal{R}_\phi(\mathbb{R}^n)$. We define [9] the ϕ-mean curvature κ_ϕ^E of E as $\kappa_\phi^E := \mathrm{div}_{\phi,\tau} N_{\min}^E \in L^2(\partial E)$.*

It turns out that the ϕ-mean curvature of $\mathcal{B}_\phi$ is constantly equal to $n - 1$.

Let $E \in \mathcal{R}_\phi(\mathbb{R}^2)$. If C is an edge of ∂E and W_C is the corresponding edge in the Wulff shape, then κ_ϕ^E is constant on C and is equal to $\delta_C \frac{|W_C|}{|C|}$, where $\delta_C \in \{0, \pm 1\}$ is a convexity factor. $N_{\min}^E$ is, on C, the linear combination of the vector η at the vertices of C (all $\eta \in \mathrm{Lip}_{\nu,\phi}(\partial E; \mathbb{R}^2)$ coincide on the vertices of ∂E).

One of the main results is the following global regularity result on the minimizers of (10), see [9, 24].

Theorem 4.4. *Let $E \in \mathcal{R}_\phi(\mathbb{R}^n)$. Then $\kappa_\phi^E \in L^\infty(\partial E)$. Moreover, κ_ϕ^E has bounded variation on all facets of ∂E corresponding to facets of $\partial \mathcal{B}_\phi$.*

Open problem. Is there a solution $\overline{N}$ of problem (10) with $\overline{N} \in \mathrm{Lip}_{\nu,\phi}(\partial E; \mathbb{R}^n)$?

Theorem 4.4 makes it possible to speak of the jump set of κ_ϕ^E on the facets of ∂E corresponding to facets of $\partial \mathcal{B}_\phi$. If $F \subset \partial E$ is such a facet, it is of particular interest to find necessary and sufficient conditions on E and F ensuring that the jump set of κ_ϕ^E on F is empty: that is, to prove that $\mathrm{div}_{\phi,\tau} N_{\min}^E$ is continuous on F. For small times in the evolution problem, F is expected to translate parallel to itself, possibly changing its shape, if κ_ϕ^E is constant on F (in this case we say that F is ϕ-calibrable), or to bend if κ_ϕ^E is continuous but not constant on F. In the first example of [8] is shown a crystalline mean curvature flow of a Lipschitz ϕ-regular set having a facet F which instantly subdivides (κ_ϕ^E is piecewise constant on F). In the second example of [8] the bending phenomenon of a facet of a Lipschitz ϕ-regular polyhedral set is described. These two evolutions are ϕ-regular evolutions, i.e. they are not examples of singularities of the flow.

In general, it is interesting to find the structure and the properties of the jump set of κ_ϕ^E on F, see [9] for some results in this direction. A characterization of ϕ-calibrability for convex facets F of convex sets E will appear in [10].

Geometric evolution problem: nonsmooth case. We now define the notion of ϕ-regular flow: to avoid technical difficulties, we skip some details. Let $E(t)$, for $t \in [0,T]$, be a family of bounded open sets. We say that $t \in [0,T] \to E(t)$ is a ϕ-regular flow on $[0,T]$ if $E(t)$ is Lipschitz ϕ-regular and there exists a family $t \in [0,T] \to N(\cdot,t)$ of vector fields $N(\cdot,t)\colon \partial E(t) \to \mathbb{R}^n$, such that $N(x,t) \in T^o(\nu_\phi^{E(t)}(x))$ for $\mathcal{H}^{n-1}$-almost every $x \in \partial E(t)$, $\mathrm{div}_{\phi,\tau} N(\cdot,t) \in L^2(\partial E(t))$ and such that, setting $d_\phi(x,t) := d_\phi^{E(t)}(x)$, there holds

$$\frac{\partial d_\phi}{\partial t}(x,t) = \mathrm{div}_{\phi,\tau} N(x,t)\,, \qquad \mathcal{H}^{n-1} - \text{a.e. } x \in \partial E(t)\,, \quad \text{a.e. } t \in [0,T]\,. \tag{11}$$

By analogy with semigroup theory (see [13, Theorem 3.1]), an open problem is to prove that

$$\frac{\partial d_\phi}{\partial t}(x,t) = \mathrm{div}_{\phi,\tau} N_{\min}^{E(t)}(x)\,, \qquad \mathcal{H}^{n-1} - \text{a.e. } x \in \partial E(t)\,, \quad t \in [0,T] \tag{12}$$

(possibly $\frac{\partial d_\phi}{\partial t}$ being the right derivative), where $N_{\min}^{E(t)}$ solves (10) with $E(t)$ in place of E.

Remark 4.5.

(i) *In [6] it is proved that, in two dimensions, the solutions of the reaction-diffusion type inclusion*

$$u_t - \mathrm{div}(T^o(\nabla u)) + \frac{1}{2\epsilon^2} W'(u) \ni 0 \tag{13}$$

approximate, as $\epsilon \to 0^+$, the ϕ-curvature flow with a quasi-optimal error estimate of order $\epsilon^2 |\log \epsilon|^2$. A similar result holds in any space dimensions, with a sub-optimal error estimate of order $\epsilon |\log \epsilon|^2$ [7].

(ii) *In [7] is proved a comparison principle between ϕ-regular flows (see [18] for a comparison result in two dimensions), where the notion of ϕ-regularity used is the fourth one in (iii) of Remark 4.2: this result implies that, if a ϕ-regular* *flow exists, then it is* unique *(in that class). This result is also sufficient to ensure that the first example constructed in* [8] *is the unique crystalline evolution starting from its initial datum.*

(iv) *Another notion of flow can be given by imposing that each $E(t)$ is Lipschitz ϕ-regular, and $\partial E(t)$ admits a vector field $\eta(\cdot, t) \in \mathrm{Lip}_{\nu,\phi}(\partial E(t); \mathbb{R}^n)$ such that (11) holds with η in place of N (Lipschitz ϕ-regular flow). Finding conditions ensuring that a ϕ-regular flow is also a Lipschitz ϕ-regular flow is an open problem, strictly related to the problem addressed after Theorem 4.4. We do not know whether the first evolutionary example constructed in* [8] *(subdivision of a facet at the initial time) is a Lipschitz ϕ-regular flow.*

(v) *The extension of the above results in the presence of a forcing term g depending on x is a largely open problem.*

References

[1] F. J. Almgren and J. E. Taylor, *Flat flow is motion by crystalline curvature for curves with crystalline energies,* J. Diff. Geom. **42** (1995), 1–22.

[2] F. J. Almgren, J. E. Taylor and L. Wang, *Curvature-driven flows: a variational approach,* SIAM J. Control Optim., **31** (1993), 387–437.

[3] M. Amar and G. Bellettini, *A notion of total variation depending on a metric with discontinuous coefficients,* Ann. Inst. H. Poincaré Anal. Non Linéaire, **11** (1993), 91–133.

[4] G. Bellettini, P. Colli Franzone and M. Paolini, *Convergence of front propagation for anisotropic bistable reaction-diffusion equations,* Asymp. Anal., **15** (1997), 325–358.

[5] G. Bellettini and I. Fragalà, *Elliptic approximations to prescribed mean curvature surfaces in Finsler geometry,* Asymp. Anal., **22** (2000), 87–111.

[6] G. Bellettini, R. Goglione and M. Novaga, *Approximation to driven motion by crystalline curvature in two dimensions,* Adv. Math. Sci. and Appl., **10** (2000), 467–493.

[7] G. Bellettini and M. Novaga, *Approximation and comparison for non-smooth anisotropic motion by mean curvature in R^N,* Math. Mod. Meth. Appl. Sc., **10** (2000), 1–10.

[8] G. Bellettini, M. Novaga and M. Paolini *Facet-breaking for three-dimensional crystals evolving by mean curvature,* Interfaces and Free Boundaries **1** (1999), 39–55.

[9] G. Bellettini, M. Novaga and M. Paolini, *On a crystalline variational problem,* Arch. Rational Mech. Anal., to appear.

[10] G. Bellettini, M. Novaga and M. Paolini, *Characterization of facets breaking for nonsmooth mean curvature flow in 3D in the convex case,* Interfaces and Free Boundaries, to appear.

[11] G. Bellettini and M. Paolini, *Anisotropic motion by mean curvature in the context of Finsler geometry,* Hokkaido Mathematical J., **89** (1996), 537–566.

[12] G. Bellettini, M. Paolini and S. Venturini, *Some results on surface measures in the calculus of variations,* Ann. Mat. Pura Appl., **CLXX** (1996), 329–359.

[13] H. Brezis, Operateurs Maximaux Monotones et Semi-groupes de Contractions Dans les Espaces de Hilbert, North-Holland Mathematics Studies, 1973.

[14] Y.-G. Chen, Y. Giga and S. Goto, *Uniqueness and existence of viscosity solutions of generalized mean curvature flow equations,* J. Diff. Geom., **3** (1991), 749–786.

[15] M. E. Gage, *Evolving plane curves by curvature in relative geometries. II.,* Duke Math. J., **75** (1994), 79–98.

[16] M.-H. Giga and Y. Giga, *Evolving graphs by singular weighted curvature,* Arch. Rational Mech. Anal., **141** (1998), 117–198.

[17] M.-H. Giga and Y. Giga, *Crystalline and level set flow-convergence of a crystalline algorithm for a general anisotropic curvature flow in the plane,* Free boundary problems: theory and applications, I (Chiba 1999), 64–79, GAKUTO Internat. Ser. Math. Sc. Appl., 13, Gakkutosho, Tokyo, 2000.

[18] Y. Giga and M. E. Gurtin, *A comparison theorem for crystalline evolutions in the plane,* Quarterly of Applied Mathematics, **IV** (1996), 727–737.

[19] M. Paolini and F. Pasquarelli, *Numerical simulations of crystalline curvature flow in 3D by interface diffusion,* in: Free Boundary Problems: theory and applications II, GAKUTO Intern. Ser. Math. Sci. Appl. **14** (N. Kenmochi ed.), Gakkötosho (2000), 376–389, to appear.

[20] J. E. Taylor, *Crystalline variational problems,* Bull. Amer. Math. Soc. (N.S.), **84** (1978), 568–588.

[21] J. E. Taylor, *Complete catalog of minimizing embedded crystalline cones,* Proc. Symposia Math., **44** (1986), 379–403.

[22] J. E. Taylor, *II-Mean curvature and weighted mean curvature,* Acta Metall. Mater., **40** (1992), 1475–1485.

[23] J. E. Taylor, *Motion of curves by crystalline curvature, including triple junctions and boundary points,* Proc. of Symposia in Pure Math., **54** (1993) 417–438.

[24] J. Yunger, *Facet stepping and motion by crystalline curvature,* Ph.D. Thesis, Rutgers University (1998).

Dipartimento di Matematica
Università di Roma "Tor Vergata"
via della Ricerca Scientifica
I–00133 Roma, Italy
E-mail address: `belletti@mat.uniroma2.it`

A Phase-Field Model for Diffusion-Induced Grain Boundary Motion

Klaus Deckelnick

Abstract. We consider a phase-field model for diffusion-induced grain boundary motion. This model couples a parabolic variational inequality to a degenerate diffusion equation. We summarize recent results on existence and uniqueness, sharp interface limits and numerical discretization.

1. Introduction

The aim of this note is to review recent work on the mathematical study of diffusion-induced grain boundary motion. This kind of motion is observed when a thin polycrystalline film of metal is exposed to vapor consisting of another metal. Atoms from the vapor diffuse into the film along the grain boundaries that separate the crystals inducing the migration of the boundary. As it advances, the solute atoms are left behind the boundary, which changes the concentration of the growing crystal (see [6] and the references in [2]). The following thermodynamically consistent phase-field system was suggested in [2] (see also [5]) in order to describe this phenomenon:

$$\rho\epsilon\phi_t - \epsilon\Delta\phi - \frac{1}{\epsilon}\phi + \beta(\phi) + p_\phi(\phi,u) \ni 0 \tag{1}$$

$$\epsilon u_t - \nabla\cdot\big(D(\phi)\nabla w\big) = 0\,. \tag{2}$$

The system (1), (2) is in nondimensionalized form, with the constants ρ and ϵ satisfying

$$0<\rho\le 1, 0<\epsilon\ll 1\,.$$

Furthermore, ϕ is an order parameter, which has the value $+1$ in one crystal and -1 in the other. Within the grain boundary we have $|\phi|<1$. The constraint $|\phi|\le 1$ is realized in (1) by the use of the subdifferential β of

$$I_{[-1,1]}(s) := \begin{cases} 0 & \text{if } s\in[-1,1]\,, \\ +\infty & \text{otherwise}\,. \end{cases}$$

Thus,

$$\beta(s) = \partial I_{[-1,1]}(s) = \begin{cases} (-\infty,0] & \text{if } s=-1\,, \\ 0 & \text{if } |s|<1\,, \\ [0,\infty) & \text{if } s=1\,. \end{cases}$$

The function $\Psi(s) := I_{[-1,1]}(s) + \frac{1}{2}(1-s^2)$ is the well-known double obstacle potential (see [1]).

The variable u denotes the concentration of the solute atoms, while

$$w = u + \frac{\hat{\epsilon}}{\pi} p_u(\phi,u) \qquad (0 < \hat{\epsilon} \ll 1)\,.$$

The coupling term p is given by

$$p(\phi,u) = \frac{\pi}{8}(1+\phi)^2 u^2$$

and models elastic interaction. Finally, the diffusivity is much larger in the grain boundary than in the crystals, so that a reasonable ansatz for D is

$$D(\phi) = \frac{2}{\pi}(1-\phi^2)\,.$$

Note that $D(\phi)$ vanishes within the grains. Thus, (1), (2) couples a parabolic variational inequality to a degenerate diffusion equation leading to challenging mathematical problems.

Let us briefly outline the plan of this note: in § 2 we shall review an existence and uniqueness result for the above system in a two-dimensional slab. § 3 is concerned with results on the sharp-interface limit $\epsilon \to 0$. Finally, in § 4 we briefly introduce a numerical method for the problem studied in § 2 and present some computations.

2. Existence and Uniqueness for the Phase-Field System

We consider the following two-dimensional geometry, which models a metal plate that is surrounded by vapor: for $H>0$ let $\Omega = \mathbb{R}\times(-H,H)$, where we think of the vapor as being above $x_2 = H$ and below $x_2 = -H$. Instead of (1), (2) we consider the following slightly modified system (which still retains all the mathematical difficulties):

$$\epsilon\phi_t - \epsilon\Delta\phi - \frac{1}{\epsilon}\phi + \beta(\phi) + \frac{\pi}{4}u^2 \ni 0 \tag{3}$$

$$\epsilon u_t - \nabla\cdot\big(D(\phi)\nabla u\big) = 0 \tag{4}$$

together with the boundary and initial conditions

$$\frac{\partial\phi}{\partial n} = 0 \qquad \text{on } \partial\Omega\times(0,T), \tag{5}$$

$$D(\phi)\frac{\partial u}{\partial n} + \alpha D(\phi)^2(u-1) = 0 \qquad \text{on } \partial\Omega\times(0,T), \tag{6}$$

$$\phi(.,0)=\phi_0, u(.,0) = u_0 \qquad \text{in } \Omega\,. \tag{7}$$

The fact that $D(\phi)^2$ (rather than $D(\phi)$) appears in (6) has a technical reason, namely to ensure an a priori estimate needed for the existence proof. Nevertheless, (6) approximates for large α the condition $u = 1$ on $\{(x_1, x_2) \mid D(\phi) \neq 0, x_2 = \pm H\}$ which is used in [2].

We assume that the initial functions (ϕ_0, u_0) satisfy

$$\phi_0 \in W^{2,2}_{\text{loc}}(\bar{\Omega}), \qquad -1 \leq \phi_0 \leq 1 \quad \text{in} \quad \Omega, \quad \frac{\partial \phi_0}{\partial n} = 0 \quad \text{on} \quad \partial\Omega,$$
$$u_0 \in H^1_{\text{loc}}(\bar{\Omega}), \qquad 0 \leq u_0 \leq 1 \quad \text{in} \quad \Omega$$

and that there exists $R_0 > 0$ such that

$$\begin{aligned} \phi_0(x) &= 1 \quad \text{for} \quad x_1 \geq R_0, \quad \phi_0(x) = -1 \quad \text{for} \quad x_1 \leq -R_0, \\ u_0(x) &= 0 \quad \text{for} \quad |x_1| \geq R_0. \end{aligned}$$

Furthermore, we define

$$K := \{v \in H^1_{\text{loc}}(\bar{\Omega}) \mid |v| \leq 1 \quad \text{a.e. in } \Omega, \quad \exists R = R(v) \quad v(x) = \pm 1, \quad \pm x_1 \geq R\}$$

as well as the space-time function spaces

$$X_1 := \{\phi \in L^\infty(\Omega_T) | \nabla\phi \in L^2(0,T; H^1(\Omega) \cap L^\infty(\Omega)), \ \phi_t \in L^2(0,T; H^1(\Omega))\},$$
$$X_2 := L^2(0,T; H^1(\Omega)) \cap W^{1,2}(0,T; L^2(\Omega)).$$

Definition 2.1. *The pair $(\phi, u) \in X_1 \times X_2$ is called a solution of (3)–(7) provided that*

$$\phi(0) = \phi_0, \quad u(0) = u_0 \textit{ in } \Omega,$$
$$\phi(t) \in K \textit{ for } t \in (0,T) \quad \textit{and} \quad \frac{\partial \phi}{\partial n} = 0 \textit{ on } \partial\Omega \times (0,T),$$
$$\epsilon \int_\Omega \phi_t(\zeta - \phi) + \epsilon \int_\Omega \nabla\phi \cdot \nabla(\zeta - \phi) - \frac{1}{\epsilon} \int_\Omega \phi(\zeta - \phi) + \frac{\pi}{4} \int_\Omega u^2(\zeta - \phi) \geq 0,$$
$$\epsilon \int_\Omega u_t \eta + \int_\Omega D(\phi) \nabla u \cdot \nabla \eta + \alpha \int_{\partial\Omega} D(\phi)^2 (u - 1)\eta = 0,$$

for all $\zeta \in K$, all $\eta \in H^1(\Omega)$ and for almost all $t \in (0,T)$.

Theorem 2.2. *Under the above assumptions there exists a solution of (3)–(7) in the sense of Definition 2.1.*

Proof. We briefly outline the main ideas and difficulties refering the reader to [3] for a detailed proof.

1. One introduces a regularized strictly parabolic system, in which the subdifferential β and the diffusivity D are replaced by functions β_δ and D_δ in such a way that $\beta_\delta \to \beta, D_\delta \to D$ as $\delta \to 0$ in a suitable sense. In order to avoid difficulties related to the unboundedness of Ω, the system is initially considered on $(-L, L) \times (-H, H) \times (0, T)$, where L is chosen sufficiently large depending on ϵ, R_0 and T.
2. The procedure in step 1 yields a family (ϕ_δ, u_δ) of approximate solutions, for which $0 \leq u_\delta \leq 1$ and higher norms of ϕ_δ can be estimated independently of

δ (by the maximum principle and parabolic regularity theory respectively). The main difficulty consists in obtaining uniform bounds on derivatives of u_δ. Note that the usual energy estimate for (2) formally reads

$$\sup_{0\le t\le T} \|u(t)\|^2_{L^2} + \int_0^T \int_\Omega D(\phi)|\nabla u|^2 + \alpha \int_0^T \int_{\partial\Omega} D(\phi)^2|u|^2 \le C\,,$$

which only provides information on ∇u in the interior of the boundary layer. To obtain a bound on ∇u_δ it is therefore natural – although rather complicated – to work with the evolution equations for $u_{\delta,x_k}, k=1,2$.

3. Using the a priori estimates of step 2 and standard compactness results one obtains a subsequence $(\phi_{\delta_k}, u_{\delta_k}), (\delta_k \to 0)$, which converges to a pair (ϕ, u) satisfying the variational identities of Definition 2.1 for $(-L,L)\times(-H,H)$ instead of Ω. A comparison argument shows that

$$\phi(x,t) = \pm 1, \quad u(x,t) = 0 \quad \text{for } \pm x_1 > \frac{L}{2},\ 0<t<T\,,$$

provided that L is chosen large enough. By extending (ϕ, u) appropriately for $|x_1| \ge L$ one obtains a solution in the sense of Definition 2.1. □

Unfortunately we are not able to show that the above solution is unique within the class $X_1 \times X_2$. However, if we assume slightly more regularity we can prove a uniqueness result, which relies on a duality argument applied to the diffusion equation. Details can again be found in [3].

Theorem 2.3. *Let $(\phi_1,u_1),(\phi_2,u_2) \in X_1 \times X_2$ be two solutions in the sense of Definition 2.1 which in addition satisfy*

$$\Delta\phi_i \in L^1(0,T;L^\infty(\Omega)), \quad \nabla u_i \in L^2(0,T;L^p(\Omega)), \quad i=1,2$$

for some $p>2$. Then $(\phi_1,u_1)\equiv(\phi_2,u_2)$.

3. Sharp-Interface Limits

In [5] formal asymptotics are used to derive the following moving free boundary problem for the sharp interface $\Gamma(t)$ as $\epsilon \to 0$:

$$\begin{aligned} \rho v &= \kappa + u^2 && \text{on } \Gamma(t) \\ u_{ss} &= vu && \text{on } \Gamma(t)\,. \end{aligned} \tag{8}$$

Here, v is the normal velocity of $\Gamma(t)$, κ is its curvature and s denotes arclength. The system (8) couples forced curve shortening flow to an elliptic equation on $\Gamma(t)$. If one considers the limit problem for the geometry treated in § 2, boundary conditions have to be added to (8), namely that the curve $\Gamma(t)$ meets the lines $\mathbb{R}\times\{\pm H\}$ orthogonally and that $u=1$ at these contact points. At $t=0$, $\Gamma(0)$ is prescribed, while the values $u_0 = u(.,0)$ have to satisfy the compatibility condition

$$u_{0,ss} - \frac{1}{\rho}\big(\kappa(.,0) + u_0^2\big)u_0 = 0 \quad \text{on} \quad \Gamma(0)\,.$$

Apart from deriving the sharp interface limit, [5] also investigates the existence of traveling wave solutions to (8) for the infinite slab. Two types of waves are considered: the first type of solution connects the two faces $x_2 = H$ and $x_2 = -H$, while the second type does not reach the other side of the plate and trails behind.

Including terms of order $O(\epsilon)$ in the derivation of the sharp interface limit, one obtains the following more accurate system for $\Gamma(t)$ (see again [5]):

$$\begin{aligned} \rho v &= \kappa(1+\epsilon C) + u^2\big(1+\epsilon(B+Au^2)\big) && \text{on} \quad \Gamma(t) \\ \epsilon^* \frac{du}{dt} &= \big(u_s(1+\epsilon\alpha u^2)\big)_s - vu - \epsilon^* uv\kappa && \text{on} \quad \Gamma(t) \end{aligned} \tag{9}$$

for certain constants A, B, C, α, g_1 and $\epsilon^* = \frac{\epsilon\pi}{1+\epsilon g_1}$. The symbol $\frac{du}{dt}$ denotes differentiation along flow lines that are perpendicular to $\Gamma(t)$ (see [7] for a precise definition). Due to the presence of $\epsilon^* uv\kappa$ in the second equation of (9), the system is fully nonlinear. In [7], [8] local existence and uniqueness of classical solutions to (9) is proved for a geometry, in which the grain boundary is modeled by a closed curve in the plane (with vapor in the third dimension).

4. Numerical Discretization

In this final section we briefly present a numerical method, which was used in [4] to calculate solutions of (3)–(7). We consider the problem on the rectangle $\Omega_L = (-L, L) \times (-H, H)$ (cf. step 1 of the Proof of Theorem 2.2) and introduce a triangulation $\mathcal{T}_h$ of Ω_L as well as the corresponding space of linear finite elements

$$X_h := \{v_h \in C^0(\bar{\Omega}_L) \mid v_h \in P_1(T) \text{ for all } T \in \mathcal{T}_h\}.$$

Furthermore, let

$$K_h := \{v_h \in X_h \mid |v_h(x)| \leq 1 \text{ for all } x \in \Omega_L\}$$

and $\Delta t > 0$ the timestep. Denoting by $\phi_h^0 = I_h(\phi_0)$ and $u_h^0 = I_h(u_0)$ the Lagrange interpolants of ϕ_0 and u_0 the numerical algorithm reads as follows: for $0 \leq n \leq [\frac{T}{\Delta t}]$ find $(\phi_h^n, u_h^n) \in K_h \times X_h$ such that

$$\begin{aligned} \frac{\rho\epsilon}{\Delta t} \int_{\Omega_L} (\phi_h^{n+1} - \phi_h^n)(\zeta_h - \phi_h^{n+1}) + \epsilon \int_{\Omega_L} \nabla\phi_h^n \cdot \nabla(\zeta_h - \phi_h^{n+1}) \\ - \frac{1}{\epsilon} \int_{\Omega_L} \phi_h^n (\zeta_h - \phi_h^{n+1}) + \frac{\pi}{4} \int_{\Omega_L} (u_h^n)^2 (\zeta_h - \phi_h^{n+1}) \geq 0 \end{aligned}$$

$$\frac{\epsilon}{\Delta t} \langle u_h^{n+1} - u_h^n, \eta_h \rangle_{\Omega_L}^h + \int_{\Omega_L} D(\phi_h^n) \nabla u_h^{n+1} \cdot \nabla\eta_h + \alpha \langle D(\phi_h^n)(u_h^{n+1} - 1), \eta_h \rangle_{\partial\Omega_L}^h = 0$$

for all $\zeta_h \in K_h$ and all $\eta_h \in X_h$. Here, the discrete inner products are defined by

$$\langle f, g \rangle_{\Omega_L}^h = \int_{\Omega_L} I_h(fg), \qquad \langle f, g \rangle_{\partial\Omega_L}^h = \int_{\partial\Omega_L} I_h(fg).$$

If we assume in addition, that the triangulation is weakly acute, then the use of numerical integration in the second equation combined with the fact that $0 \leq u_0 \leq 1$ ensures that $0 \leq u_h^n \leq 1$ for all $0 \leq n \leq [\frac{T}{\Delta t}]$.

Figures 1 and 2 show examples of calculated solutions ϕ_h and u_h. The initial conditions were chosen as

$$\phi_0(x) := \begin{cases} -1, & x_1 \leq -\frac{\epsilon\pi}{2} \\ \sin(\frac{x_1}{\epsilon}), & -\frac{\epsilon\pi}{2} < x_1 < \frac{\epsilon\pi}{2} \\ +1, & x_1 \geq \frac{\epsilon\pi}{2} \end{cases}$$

and $u_0 \equiv 0$. The computations were carried out on a uniform grid with $H = 2$, $h = \frac{1}{100}$, $\Delta t = \frac{h^2}{100}$, $\epsilon = 20h$, $\rho = 0.8$ and the numerical solutions are shown at $t = 0.2$. While the function ϕ_h keeps its sinusoidal shape, the interfacial region has width $\approx 0.6 \approx \pi\epsilon$ and is moving in the positive x_1-direction. The concentration u_h, which initially was identically zero, now has non-zero values in the region through which the interface has passed. Once a point has been left behind the interfacial region, the values of u_h at this point do not change at later times. We remark that it is sufficient to carry out the computations in a small neighborhood of the discrete free boundary $|\phi_h| < 1$.

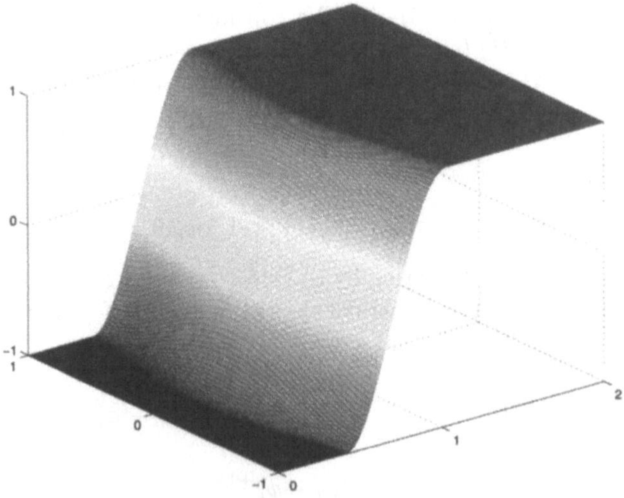

FIGURE 1. $\phi_h(\mathbf{x}, t)$

Apart from presenting calculations for the phase-field model, [4] also investigates the convergence as $\epsilon \to 0$ to the sharp interface limit (8) from a numerical point of view. Furthermore, convergence to traveling wave solutions as $t \to \infty$ is studied.

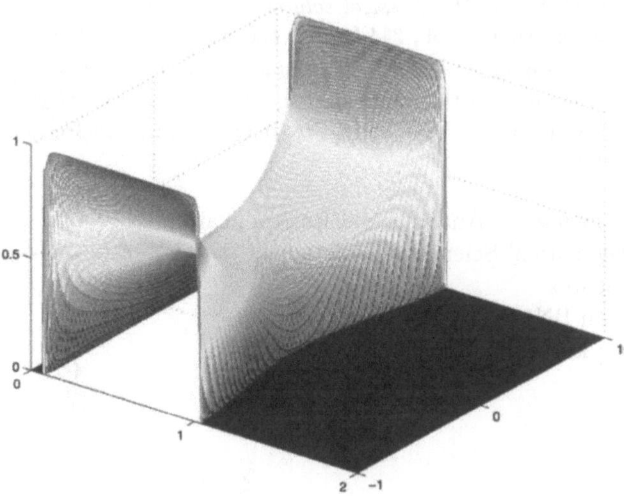

FIGURE 2. $u_h(\mathbf{x}, t)$

Acknowledgement

The author wants to thank V. Styles for computing the numerical example.

References

[1] J. F. Blowey, C. M. Elliott, *The Cahn-Hilliard gradient theory for phase separation with non-smooth free energy Part I: Mathematical Analysis*, Eur. J. Appl. Math., **2** (1991), 233–280.

[2] J. W. Cahn, P. C. Fife, O. Penrose, *A phase-field model for diffusion-induced grain boundary motion*, Acta Mater., **45** (1997), 4397–4413.

[3] K. Deckelnick, C. M. Elliott, *An existence and uniqueness result for a phase-field model of diffusion-induced grain-boundary motion*, Research Report No: 99–12, Centre for Mathematical Analysis and Its Applications, University of Sussex (1999), to appear in Proc. R. Soc. Edinb. A.

[4] K. Deckelnick, C. M. Elliott, V. Styles. *Numerical diffusion-induced grain boundary motion*, Research Report No: 00-09, Centre for Mathematical Analysis and Its Applications, University of Sussex (2000), to appear in Interfaces and Free Boundaries.

[5] P. C. Fife, J. W. Cahn, C. M. Elliott, *A free boundary model for diffusion induced grain boundary motion*, Research Report No: 99–10, Centre for Mathematical Analysis and Its Applications, University of Sussex (1999), to appear in Interfaces ind Free Boundaries.

[6] C. Handwerker, *Diffusion-induced grain boundary migration in thin films*, in Diffusion Phenomena in Thin Films and Microelectronic Materials, D. Gupta and P.S. Ho, eds., 245–322, Noyes Pubs. Park Ridge, N.J., 1988.

[7] U. F. Mayer, G. Simonett, *Classical solutions for diffusion-induced grain-boundary motion*, J. Math. Anal. Appl., **234** (1999), 660–674.

[8] U. F. Mayer, G. Simonett, *On diffusion-induced grain-boundary motion*, Chen, G.-Q. (ed.) et al., Nonlinear partial differential equations. Proceedings of the international conference, Evanston IL, USA, 1998. Providence, RI: AMS Contemp. Math. **238**, 231–240 (1999).

Centre for Mathematical Analysis and Its Applications
School of Mathematical Sciences
University of Sussex
Falmer, Brighton BN1 9QH, England
E-mail address: `k.p.deckelnick@sussex.ac.uk`

Evolution of a Closed Interface between Two Liquids of Different Types

Irina V. Denisova

Abstract. We study a free boundary problem governing the motion of two immiscible viscous capillary fluids. The fluids occupy the whole space $\mathbb{R}^3$ but one of them should have a finite volume. Every liquid may be of both types: compressible and incompressible.

Local (with respect to time) unique solvability of the problem is obtained in the Sobolev-Slobodetskiĭ spaces. After the passage to Lagrangian coordinates, one obtains a nonlinear, noncoercive initial boundary-value problem the proof of the existence theorem for which is based on the method of successive approximations and on an explicit solution of a model linear problem with a plane interface between the liquids.

Some restrictions to the fluid viscosities appear in the case when at least, one of the liquids is compressible.

1. Introduction

In this paper, we summarize the study of the solvability of problems governing the motion of two viscous liquids separated by an unknown closed interface. Every fluid may be of both types: compressible and incompressible. They occupy the whole space $\mathbb{R}^3$. On the interface, we take the capillary forces into account. All results remain valid for noncapillary fluids too.

The main result of this investigation is the unique solvability of the problems mentioned above in the Sobolev spaces in sufficient small time intervals. As auxiliary results, one can consider the proof of existence of unique solutions for linearized problems in any finite time interval and global unique solvability in the weighted Sobolev spaces of linear model problems with a plane interface between the liquids. We compare explicit solutions in the dual spaces of these linear problems. We make passage to the limit from the solution of the problem for two compressible fluids through the solution of the "mixed type" system to the solution of the problem for two incompressible liquids. We remark that there is no restriction to the fluid viscosities for the last problem whereas for two compressible liquids the stated results are proved only if the dynamic viscosities of the fluids are not different by more than two times. As for the "mixed type" problem, these results are valid only for fluids with low viscosity.

2. Statement of the Problems and Formulation of the Main Results

First, we formulate the most complicated problem, for the case of two liquids of different types. A study of this problem is made in [2].

Let, for definiteness, at the initial moment $t = 0$, the compressible fluid have a finite volume and be situated in a bounded domain $\Omega_0^+ \subset \mathbb{R}^3$ inside an incompressible one occupying the domain $\Omega_0^- \equiv \mathbb{R}^3 \setminus \overline{\Omega_0^+}$.

Let $\mu^+ > 0$, $\lambda^+ > 0$ be the dynamic viscosities of the compressible liquid. We denote the kinematic viscosity of the incompressible fluid by constant $\nu^- > 0$ and its density coefficient by $\rho^- > 0$. We consider that the compressible fluid is barotropic. We note that we could also suppose the compressible fluid to be exterior to the incompressible one.

For $t > 0$, it is necessary to find the free interface Γ_t between the liquids evolving in the domains Ω_t^- and Ω_t^+, the density function $\rho^+(x,t) > 0$ of the compressible fluid, the pressure function $p^-(x,t)$ of the incompressible fluid, as well as the velocity vector field of both liquids $\boldsymbol{v}(x,t) = (v_1, v_2, v_3)$ satisfying the initial-boundary value problem for the Navier-Stokes system:

$$\begin{aligned} \rho^\pm(\mathcal{D}_t\boldsymbol{v} + (\boldsymbol{v}\cdot\nabla)\boldsymbol{v}) - \nabla\mathbb{T} &= \rho^\pm \boldsymbol{f}, \ \ \mathcal{D}_t\rho^\pm + \nabla\cdot(\rho^\pm\boldsymbol{v}) = 0 \text{ in } \Omega_t^- \cup \Omega_t^+, \\ \boldsymbol{v}|_{t=0} &= \boldsymbol{v}_0 \ \text{ in } \ \Omega_0^- \cup \Omega_0^+, \end{aligned} \tag{1}$$

$$\rho^+|_{t=0} = \rho_0^+ \quad \text{in} \quad \Omega_0^+; \quad \boldsymbol{v} \xrightarrow[|x|\to\infty]{} 0, \quad p^- \xrightarrow[|x|\to\infty]{} 0; \tag{2}$$

$$[\boldsymbol{v}]\Big|_{\Gamma_t} \equiv \lim_{\substack{x\to x_0\in\Gamma_t,\\ x\in\Omega_t^+}} \boldsymbol{v}(x) - \lim_{\substack{x\to x_0\in\Gamma_t,\\ x\in\Omega_t^-}} \boldsymbol{v}(x) = 0, \quad [\mathbb{T}\boldsymbol{n}]\Big|_{\Gamma_t} = \sigma H \boldsymbol{n} \quad \text{on} \quad \Gamma_t. \tag{3}$$

Here $\mathcal{D}_t = \partial/\partial t$, $\nabla = (\partial/\partial x_1, \partial/\partial x_2, \partial/\partial x_3)$, function $\rho^\pm$ is equal to $\rho^+(x)$ in Ω_t^+ and to the constant ρ^- in Ω_t^-; the stress tensor is given by

$$\mathbb{T} = \begin{cases} (-p^+(\rho^+) + \lambda\nabla\cdot\boldsymbol{v})\,\mathbb{I} + \mu^+\mathbb{S}(\boldsymbol{v}) & \text{in } \Omega_t^+, \\ -p^-\mathbb{I} + \mu^-\mathbb{S}(\boldsymbol{v}) & \text{in } \Omega_t^-, \end{cases} \tag{4}$$

$(\mathbb{S}(\boldsymbol{v}))_{ik} = \partial v_i/\partial x_k + \partial v_k/\partial x_i$, $i,k = 1,2,3$; $\mathbb{I}$ is the unit matrix; $\mu^- = \nu^-\rho^-$; $p^+(\rho^+)$ is the pressure of the compressible fluid given by a smooth function of its density; $\boldsymbol{f}$ is the given vector field of mass forces; $\boldsymbol{v}_0$ is the initial value of the velocity vector field; ρ_0^+ is the initial density distribution of the compressible fluid; $\sigma \geqslant 0$ is the surface tension coefficient, $\boldsymbol{n}$ is the outward normal vector to Ω_t^+, $H(x,t)$ is twice the mean curvature of Γ_t ($H < 0$ at the points where Γ_t is convex towards Ω_t^-); $\nabla\mathbb{T}$ means the vector with the components $(\nabla\mathbb{T})_j = \frac{\partial T_{ij}}{\partial x_i}$, $T_{ij} = (\mathbb{T})_{ij}$, $j = 1,2,3$. We imply the summation from 1 to 3 with respect to repeated indices. A Cartesian coordinate system $\{x\}$ is introduced in $\mathbb{R}^3$. The central dot denotes the scalar product. We identify the vectors and the vector spaces by boldface letters.

Since we suppose the liquids to be immiscible it is natural to impose on Γ_t a condition excluding the mass transportation through this surface. Mathematically, this condition means that Γ_t consists of the points $x(\xi,t)$ whose radius vector $\boldsymbol{x}(\xi,t)$ is a solution of the Cauchy problem

$$\mathcal{D}_t\boldsymbol{x} = \boldsymbol{v}(x(\xi,t),t)\,, \quad \boldsymbol{x}(\xi,0) = \boldsymbol{\xi}\,, \quad \xi \in \Gamma\,, \quad t > 0\,, \tag{5}$$

where $\Gamma \equiv \Gamma_0 = \partial\Omega_0^+$ is a surface given at the initial moment. Hence, $\Omega_t^\pm = \{x = x(\xi,t) | \xi \in \Omega_0^\pm\}$.

Condition (5) completes system (1)–(3).

For two compressible fluids, the problem formulation differs from (1)–(5) by conditions (2) that look as follows

$$\rho^-|_{t=0} = \rho_0^- \quad \text{in} \quad \Omega_0^-\,; \qquad \rho^+|_{t=0} = \rho_0^+ \quad \text{in} \quad \Omega_0^+\,; \quad \boldsymbol{v} \xrightarrow[|x|\to\infty]{} 0\,. \tag{6}$$

The stress tensor in this case is

$$\mathbb{T} = \begin{cases} (-p^+(\rho^+) + \lambda\nabla\cdot\boldsymbol{v})\,\mathbb{I} + \mu^+\mathbb{S}(\boldsymbol{v}) & \text{in } \Omega_t^+\,, \\ (-p^-(\rho^- - +\lambda\nabla\cdot\boldsymbol{v})\,\mathbb{I} + \mu^-\mathbb{S}(\boldsymbol{v}) & \text{in } \Omega_t^-\,. \end{cases} \tag{7}$$

This problem has been studied in [3].

In the case of two incompressible liquids, we would have (2) in the form

$$\boldsymbol{v} \xrightarrow[|x|\to\infty]{} 0, \quad p \xrightarrow[|x|\to\infty]{} 0\,. \tag{8}$$

The stress tensor would be given by

$$\mathbb{T} = \begin{cases} -p\mathbb{I} + \mu^+\mathbb{S}(\boldsymbol{v}) & \text{in } \Omega_t^+\,, \\ -p\mathbb{I} + \mu^-\mathbb{S}(\boldsymbol{v}) & \text{in } \Omega_t^-\,. \end{cases} \tag{9}$$

We have analysed the latter problem in [1, 5, 4].

We present an investigation scheme common for all three problems considering the example of (1)–(5). This technique was proposed by V. A. Solonnikov in [6, 7] for the study of the drop evolution in vacuum and it was modified by him and A. Tani in [8] for the case of the bubble motion in vacuum.

We transform the Eulerian coordinates $\{x\}$ into the Lagrangian ones $\{\xi\}$ by the formula

$$\boldsymbol{x}(\xi,t) = \boldsymbol{\xi} + \int_0^t \boldsymbol{u}(\xi,\tau)\mathrm{d}\tau \equiv \boldsymbol{X}_{\boldsymbol{u}}(\xi,t) \tag{10}$$

where $\boldsymbol{u}(\xi,t)$ is the velocity vector field in the Lagrangian coordinates.

The Jacobian of transformation (10) $\mathcal{J}_{\boldsymbol{u}}(\xi,t) = \det\{a_{ij}\}_{i,j=1}^3$, $a_{ij}(\xi,t) = \delta_j^i + \int_0^t \frac{\partial u_i}{\partial \xi_j}\,\mathrm{d}\tau$, being a solution of the Cauchy problem

$$\mathcal{D}_t\mathcal{J}_{\boldsymbol{u}}(\xi,t) = A_{ij}\frac{\partial u_i}{\partial \xi_j} \equiv \mathcal{J}_{\boldsymbol{u}}(\xi,t)(\nabla\cdot\boldsymbol{v}|_{x=X_{\boldsymbol{v}}})\,, \qquad \mathcal{J}_{\boldsymbol{u}}(\xi,0) = 1\,,$$

is expressed by the formula

$$\mathcal{J}_{\boldsymbol{u}}(\xi,t)=\exp\left(\int_0^t \nabla\cdot\boldsymbol{v}|_{x=X_{\boldsymbol{u}}}\mathrm{d}\tau\right)\equiv\exp\left(\int_0^t \nabla_{\boldsymbol{u}}\cdot\boldsymbol{u}\,\mathrm{d}\tau\right). \tag{11}$$

Here we use the standard notation $\{\delta^i_j\}^3_{i,j=1}$ for the Kronecker symbols and also $\nabla_{\boldsymbol{u}}\equiv\left\{\frac{\partial\xi_i}{\partial x_k}\frac{\partial}{\partial\xi_i}\right\}^3_{k=1}=\mathcal{J}_{\boldsymbol{u}}^{-1}\mathbb{A}\nabla$; $\mathbb{A}\equiv\{A_{ij}\}^3_{i,j=1}$ is the cofactor matrix of the Jacobi matrix $\{a_{ij}\}$ in (10). We note that $\mathcal{J}_{\boldsymbol{u}}(\xi,t)\equiv 1$ in the domains with incompressible fluid.

After tranformation (10), we can integrate the second equation in (1) for the compressible liquid. Then we obtain the following expression for the density ρ^+ in the Lagrangian coordinates:

$$\widehat{\rho^+}(\xi,t)=\rho^+_0(\xi)\exp\left(-\int_0^t\nabla_{\boldsymbol{u}}\cdot\boldsymbol{u}\,\mathrm{d}\tau\right)=\rho^+_0(\xi)\mathcal{J}_{\boldsymbol{u}}^{-1}(\xi,t).$$

Next, we use the well-known formula for twice the surface mean curvature:

$$H\boldsymbol{n}=\Delta(t)\boldsymbol{x}=\Delta(t)\boldsymbol{X}_{\boldsymbol{u}}$$

where $\Delta(t)$ is the Beltrami-Laplace operator on Γ_t. Moreover, we separate the last boundary condition in (3) on the tangential and normal components. To this end, we project it first onto the tangent plane of Γ_t and then onto that of Γ by means of projectors Π and Π_0, respectively.

Let $\boldsymbol{n}_0$ be the outward normal to Γ. It is connected with $\boldsymbol{n}$ by the relation $\boldsymbol{n}=\frac{\mathcal{J}_{\boldsymbol{u}}^{-1}\mathbb{A}\boldsymbol{n}_0}{|\mathcal{J}_{\boldsymbol{u}}^{-1}\mathbb{A}\boldsymbol{n}_0|}=\frac{\mathbb{A}\boldsymbol{n}_0}{|\mathbb{A}\boldsymbol{n}_0|}$.

As a result of the above transformation, we obtain the system:

$$\mathcal{D}_t\boldsymbol{u}-\frac{1}{\rho^+_0(\xi)}\mathbb{A}\nabla\mathbb{T}'_{\boldsymbol{u}}(\boldsymbol{u})=\boldsymbol{f}(X_{\boldsymbol{u}},t)-\frac{1}{\rho^+_0(\xi)}\mathbb{A}\nabla p^+(\rho^+_0\mathcal{J}_{\boldsymbol{u}}^{-1})\text{ in }Q^+_T\equiv\Omega^+_0\times(0,T),$$

$$\mathcal{D}_t\boldsymbol{u}-\nu^-\nabla^2_{\boldsymbol{u}}\boldsymbol{u}+\frac{1}{\rho^-}\nabla_{\boldsymbol{u}}q=\boldsymbol{f}(X_{\boldsymbol{u}},t),\quad \nabla_{\boldsymbol{u}}\cdot\boldsymbol{u}=0\quad\text{in}\quad Q^-_T\equiv\Omega^-_0\times(0,T),$$

$$\boldsymbol{u}\Big|_{t=0}=\boldsymbol{v}_0\quad\text{in}\quad\Omega^-_0\cup\Omega^+_0,\quad \boldsymbol{u}\xrightarrow[|\xi|\to\infty]{}0,\quad q\xrightarrow[|\xi|\to\infty]{}0, \tag{12}$$

$$[\boldsymbol{u}]\Big|_{G_T}=0,\quad[\mu^\pm\Pi_0\Pi\mathbb{S}_{\boldsymbol{u}}(\boldsymbol{u})\boldsymbol{n}]\Big|_{G_T}=0\quad(G_T\equiv\Gamma\times(0,T)),$$

$$[\boldsymbol{n}_0\cdot\mathbb{T}'_{\boldsymbol{u}}(\boldsymbol{u},q)\boldsymbol{n}]\Big|_{G_T}-\sigma\boldsymbol{n}_0\cdot\Delta(t)\boldsymbol{X}_{\boldsymbol{u}}\Big|_{G_T}=(\boldsymbol{n}_0\cdot\boldsymbol{n})p^+(\rho^+_0\mathcal{J}_{\boldsymbol{u}}^{-1})\Big|_{G_T},$$

which is equivalent to (1)–(5) provided that $\boldsymbol{n} \cdot \boldsymbol{n}_0 > 0$. In (12) we used the notation: $q(\xi, t)$ was the pressure function in the Lagrangian coordinates;

$$(\mathbb{T}'_{\boldsymbol{u}}(\boldsymbol{w}, q))_{i,j} = \begin{cases} (\lambda^+ \nabla_{\boldsymbol{u}} \cdot \boldsymbol{w})\delta^i_j + \mu^+(\mathbb{S}_{\boldsymbol{u}}(\boldsymbol{w}))_{ij} & \text{in } Q^+_T, \\ -\delta^i_j q + \mu^-(\mathbb{S}_{\boldsymbol{u}}(\boldsymbol{w}))_{ij} & \text{in } Q^-_T; \end{cases}$$

$$(\mathbb{S}_{\boldsymbol{u}}(\boldsymbol{w}))_{ij} = \mathcal{J}^{-1}_{\boldsymbol{u}} \left(A_{ik} \frac{\partial w_j}{\partial \xi_k} + A_{jk} \frac{\partial w_i}{\partial \xi_k} \right);$$

$$\Pi_0 \boldsymbol{\omega} = \boldsymbol{\omega} - (\boldsymbol{n}_0 \cdot \boldsymbol{\omega})\boldsymbol{n}_0, \quad \Pi \boldsymbol{\omega} = \boldsymbol{\omega} - (\boldsymbol{n} \cdot \boldsymbol{\omega})\boldsymbol{n}.$$

We shall use the standard normalization for the Sobolev-Slobodetskiĭ spaces $W^m_2(\Omega)$ for $m > 0$, Ω being a domain in $\mathbb{R}^n$, $n \in \mathbb{N}$. $\|\cdot\|_\Omega$ is the norm of $L_2(\Omega)$.

The anisotropic space $W^{m,m/2}_2(Q_T)$ consists of functions defined in the cylinder $Q_T = \Omega \times (0, T)$, $0 < T \leqslant \infty$, and having finite norm

$$\|u\|_{W^{m,m/2}_2(Q_T)} = \left(\int_0^T \|u\|^2_{W^m_2(\Omega)} \mathrm{d}t + \int_\Omega \|u\|^2_{W^{m/2}_2(0,T)} \mathrm{d}x \right)^{1/2}.$$

Now we define three norms necessary for formulating the main result of this paper. The first of them is

$$\|u\|^{(m,m/2)}_{Q^-_T \cup Q^+_T} = \left(\|u\|^2_{\bigcup\limits_{i=-,+} W^{m,m/2}_2(Q^i_T)} + T^{-m} \|u\|^2_{\mathbb{R}^3_T} \right)^{1/2}.$$

It is equivalent to $\|u\|^2_{\bigcup\limits_i W^{m,m/2}_2(Q^i_T)}$ for $\forall T < \infty$. The square of the second one is determined by the formula

$$\left(\|u\|^{(2+l,1+l/2)}_{Q^-_T \cup Q^+_T} \right)^2 = \|u\|^2_{\bigcup\limits_{i=-,+} W^{2+l,1+l/2}_2(Q^i_T)} + T^{-l} \Big\{ \|\mathcal{D}_t u\|^2_{Q^-_T \cup Q^+_T}$$
$$+ \sum_{|\boldsymbol{\alpha}|=2} \|\mathcal{D}^{\boldsymbol{\alpha}}_x u\|^2_{Q^-_T \cup Q^+_T} \Big\} + \sup_{t \leq T} \|u(\cdot, t)\|^2_{\bigcup\limits_i W^{1+l}_2(\Omega^i_0)}.$$

For $\beta \in (0, 1)$ we will consider the following Hölder norm of u in $\mathbb{R}^3_T \equiv \mathbb{R}^3 \times (0, T)$

$$|||u|||_{\mathbb{R}^3_T} = \sup_{\mathbb{R}^3_T} |u| + \max_k \sup_{(x,t) \in \mathbb{R}^3_T} |\mathcal{D}_{x_k} u(x, t)| + \sup_{(x,t), \tau \leq T} \frac{|u(x, t) - u(x, \tau)|}{|\tau - t|^\beta}.$$

Let B_d be the ball $\{x : |x| < d\}$. We choose a coordinate system $\{x\}$ so that Ω^+_0 is contained in the ball B_d, $d < \infty$, and we set $B^-_{dT} \equiv (B_d \setminus \overline{\Omega^+_0}) \times (0, T)$.

Theorem 2.1. *Assume that for some* $l \in (1/2, 1)$ *we have* $\Gamma \in W^{5/2+l}_2$, $\rho^+_0 \in W^{1+l}_2(\Omega^+_0)$, $0 < R_0 \leqslant \rho^+_0(\xi) \leqslant R_\infty < \infty$, $\xi \in \Omega^+_0$, $p^+ \in C^3(\mathbb{R}_+)$, $\boldsymbol{f} \in \boldsymbol{W}^{l,l/2}_2(\mathbb{R}^3_T)$, $0 < T < \infty$, $\boldsymbol{f}(\cdot, t) \in \boldsymbol{C}^2(\mathbb{R}^3)$ *for* $\forall t \in [0, T]$, $\boldsymbol{f}(\xi, \cdot)$, $\nabla \boldsymbol{f}(\xi, \cdot) \in \boldsymbol{C}^\beta(0, T)$ *for*

$\forall\xi\in\mathbb{R}^3$ with some $\beta\in(1/2,1)$. In addition, let the initial velocity vector $\boldsymbol{v}_0\in\bigcup\limits_{i=-,+}\boldsymbol{W}_2^{1+l}(\Omega_0^i)$ satisfy the compatibility conditions

$$\nabla\cdot\boldsymbol{v}_0=0\quad\text{in}\quad\Omega_0^-,\qquad[\boldsymbol{v}_0]\Big|_\Gamma=0,\quad[\Pi_0\mathbb{S}(\boldsymbol{v}_0)\boldsymbol{n}_0]\Big|_\Gamma=0\,,$$

and for the viscosities of the liquids, the inequalities

$$\mu^->\mu^+,\quad\nu^-<\mu^+/R_\infty\tag{13}$$

hold.

Under these hypotheses, there exists a constant $T_0\in(0,T]$ such that problem (12) is uniquely solvable on the interval $(0,T_0)$ and its solution $(\boldsymbol{u},q)$ has the properties: $\boldsymbol{u}\in\bigcup\limits_{i=-,+}\boldsymbol{W}_2^{2+l,1+l/2}(Q_{T_0}^i)$, $q\in W_{2,loc}^{l,l/2}(Q_{T_0}^-)$, $\nabla q\in\boldsymbol{W}_2^{l,l/2}(Q_{T_0}^-)$, $q|_{G_{T_0}}\in W_2^{l+1/2,l/2+1/4}(G_{T_0})$ and

$$\begin{aligned}&\|\boldsymbol{u}\|^{(2+l,1+l/2)}_{Q_{T_0}^-\cup Q_{T_0}^+}+\|\nabla q\|^{(l,l/2)}_{Q_{T_0}^-}+\|q\|^{(l,l/2)}_{B_{dT_0}}+\|q\|_{W_2^{l+1/2,l/2+1/4}(G_{T_0})}\\&\quad\leqslant c_1\big(c_2+c_3T_0^{\frac{1-l}{2}}\|\boldsymbol{v}_0\|_{\bigcup\limits_i\boldsymbol{W}_2^{1+l}(\Omega_0^i)}\big)\Big\{|||\boldsymbol{f}|||_{\mathbb{R}^3_{T_0}}+\|\boldsymbol{v}_0\|_{\bigcup\limits_i\boldsymbol{W}_2^{1+l}(\Omega_0^i)}\\&\qquad+\sigma\|H_0\|_{W_2^{l+1/2}(\Gamma)}+\|\frac{1}{\rho_0^+}\nabla p^+(\rho_0^+)\|_{W_2^l(\Omega_0^+)}+\|p^+(\rho_0^+)\|_{W_2^{1+l}(\Omega_0^+)}\Big\}\,.\end{aligned}$$

The value T_0 depends on the norms of $\boldsymbol{f}$, $\boldsymbol{v}_0$, ρ_0, p^+ and on the curvature magnitude of Γ.

This theorem is proved by successive approximations in the same way as the analogous theorems for the case of a single incompressible fluid [7] or for the case of a single compressible one [8]. The role of the successive approximations is played by the solutions of the following linearized problems:

$$\mathcal{D}_t\boldsymbol{w}-\frac{1}{\rho_0^+(\xi)}\mathbb{A}\nabla\mathbb{T}'_{\boldsymbol{u}}(\boldsymbol{w})=\boldsymbol{f}\quad\text{in}\quad Q_T^+\,,$$

$$\mathcal{D}_t\boldsymbol{w}-\nu^-\nabla^2_{\boldsymbol{u}}\boldsymbol{w}+\frac{1}{\rho_0^-}\nabla_{\boldsymbol{u}}s=\boldsymbol{f},\quad\nabla_{\boldsymbol{u}}\cdot\boldsymbol{w}=r\quad\text{in}\quad Q_T^-\,,$$

$$\boldsymbol{w}\Big|_{t=0}=\boldsymbol{w}_0\quad\text{in}\quad\Omega_0^-\cup\Omega_0^+,\quad\boldsymbol{w}\xrightarrow[|\xi|\to\infty]{}0,\quad s\xrightarrow[|\xi|\to\infty]{}0\,,\tag{14}$$

$$[\boldsymbol{w}]\Big|_{G_T}=0,\quad[\mu^\pm\Pi_0\Pi\mathbb{S}_{\boldsymbol{u}}(\boldsymbol{w})\boldsymbol{n}]\Big|_{G_T}=\Pi_0\boldsymbol{a}\,,$$

$$[\boldsymbol{n}_0\cdot\mathbb{T}'_{\boldsymbol{u}}(\boldsymbol{w},s)\boldsymbol{n}]\Big|_\Gamma-\sigma\boldsymbol{n}_0\cdot\Delta(t)\int\limits_0^t\boldsymbol{w}\Big|_\Gamma d\tau=b+\sigma\int\limits_0^t B\,d\tau,\quad t\in(0,T)\,.$$

The proof of the existence theorem for problem (14) is also based on the successive approximation method, the solution of system (14) with $\boldsymbol{u} = 0$ being taken as the first approximation.

Existence of a unique solution of the latter problem and its smoothness may be proved by constructing a regularizer [7, 8] or by means of generalized solution [5]. Both these methods are based on the Schauder estimates of the solution and considering a model problem with plane interface Γ.

For problem (14), a theorem on the unique solvability takes place for an arbitrary finite time interval $(0,T)$ and for a vector $\boldsymbol{u} \in \bigcup\limits_{i=-,+} \boldsymbol{W}_2^{2+l,1+l/2}(Q_T^i)$ being continuous across the boundary Γ, such that the inequality

$$T^{1/2}\|\boldsymbol{u}\|^{(2+l,1+l/2)}_{Q_T^-\cup Q_T^+} \leqslant \delta$$

holds with a sufficiently small number δ.

For one-type liquid problems, similar results take place too. We note that in the case of two incompressible fluids, they are formulated without any restrictions to the liquid viscosities [4]. As for two compressible fluids, we have the inequalities

$$\frac{\mu^-}{2} \leqslant \mu^+ \leqslant 2\mu^-, \ 0 < \lambda^\pm \leqslant \mu^\pm \tag{15}$$

instead of (13). A theorem similar to Theorem 2.1 can be found in [3].

3. The Model Problem with a Plane Interface between the Fluids

In this section, we consider the problem

$$\begin{gathered}
\mathcal{D}_t\boldsymbol{v} - \nu^+\nabla^2\boldsymbol{v} + \frac{1}{\rho_0^+}\nabla p = 0, \quad \nabla\cdot\boldsymbol{v} = 0, \quad \text{in} \quad D_\infty^+ = \mathbb{R}_+^3 \times (0,\infty)\,, \\
\mathcal{D}_t\boldsymbol{v} - \nu^-\nabla^2\boldsymbol{v} - (\nu^- + \kappa^-)\nabla(\nabla\cdot\boldsymbol{v}) = 0 \quad \text{in} \quad D_\infty^- = \mathbb{R}_-^3 \times (0,\infty)\,, \\
\boldsymbol{v}\Big|_{t=0} = 0 \quad \text{on} \quad \mathbb{R}_-^3 \cup \mathbb{R}_+^3, \quad \boldsymbol{v} \xrightarrow[|x|\to\infty]{} 0, \quad p \xrightarrow[|x|\to\infty]{} 0\,, \\
[\boldsymbol{v}]\Big|_{x_3=0} = 0, \quad -\left[\mu^\pm\left(\frac{\partial v_\alpha}{\partial x_3} + \frac{\partial v_3}{\partial x_\alpha}\right)\right]\Bigg|_{x_3=0} = b_\alpha(x',t), \quad \alpha = 1,2\,; \\
-p - \lambda^-\nabla\cdot\boldsymbol{v} + \left[2\mu^\pm\frac{\partial v_3}{\partial x_3}\right]\Bigg|_{x_3=0} + \sigma\Delta'\int_0^t v_3 \mathrm{d}\tau\Bigg|_{x_3=0} = b_3 + \sigma\int_0^t B\,\mathrm{d}\tau \equiv b_3' \text{ on } \mathbb{R}_\infty^2\,.
\end{gathered} \tag{16}$$

Here we have used the notation: $\mathbb{R}_\pm^3 = \{\pm x_3 > 0\}$, $\mathbb{R}_\infty^2 = \mathbb{R}^2\times(0,\infty)$, $\kappa^- = \lambda^-/\rho_0^-$, $\rho_0^- = \text{constant} > 0$, $\nu^- = \mu^-/\rho_0^-$, $\Delta' = \partial^2/\partial x_1^2 + \partial^2/\partial x_2^2$, $x' = (x_1,x_2)$.

We take the Fourier transform on the tangent space variables $(x_1,x_2) = x'$ and the Laplace transform on t. We denote the dual variables by $\xi = (\xi_1,\xi_2)$ and s, respectively. The problem (16) is then transformed into the system of ordinary differential equations with respect to unknown functions $\widetilde{\boldsymbol{v}}$, $\widetilde{p}$. We solve this system and we write a solution in the form convenient for the following estimates:

$$\begin{aligned} \widetilde{\boldsymbol{v}} &= \boldsymbol{W} e_0^{\pm} + \boldsymbol{V}^{\pm} e_1^{\pm}, \quad \pm x_3 > 0, \\ \widetilde{p} &= -C_3^+ \rho_0^+ s e^{-|\xi| x_3} = -\mu^+ C_3^+ (r^+ - |\xi|)(r^+ + |\xi|) e^{-|\xi| x_3}, \quad x_3 > 0, \end{aligned} \tag{17}$$

where $r^{\pm} = \sqrt{\frac{s}{\nu^{\pm}} + \xi^2}$, $r_1^- = \sqrt{\frac{s}{(2+\beta^-)\nu^-} + \xi^2}$, $|\xi| = \sqrt{\xi_1^2 + \xi_2^2}$, $|\arg \sqrt{z}| < \pi/2$ for $\forall z \in \mathbb{C}$, $\beta^- = \kappa^-/\nu^-$; $e_0^{\pm} = e^{\mp r^{\pm} x_3}$, $e_1^+ = \frac{e^{-r^+ x_3} - e^{-|\xi| x_3}}{r^+ - |\xi|}$, $e_1^- = \frac{e^{r^- x_3} - e^{r_1^- x_3}}{r^- - r_1^-}$,

$$\boldsymbol{W} = \begin{pmatrix} \omega_1 \\ \omega_2 \\ \omega_3 \end{pmatrix}, \ \boldsymbol{V}^+ = -C_3^+ (r^+ - |\xi|) \begin{pmatrix} i\xi_1 \\ i\xi_2 \\ -|\xi| \end{pmatrix}, \ \boldsymbol{V}^- = -C_3^- (r^- - r_1^-) \begin{pmatrix} i\xi_1 \\ i\xi_2 \\ r_1^- \end{pmatrix},$$

$$\begin{aligned} C_3^+ &= -\frac{A\{[2\mu^+ r^+ + \frac{\sigma}{s}\xi^2](r^- r_1^- - \xi^2) + \mu^- r^+ (r^{-2} - 2r^- r_1^- + \xi^2) + \rho_0^- s r^-\}}{(r^+ - |\xi|)P} \\ &\quad - \frac{\widetilde{b}_3'\{\mu^+ (r^- r_1^- - \xi^2)(r^{+2} + \xi^2) + \mu^- \xi^2 (r^{-2} + \xi^2 - 2r^- r_1^-) + \rho_0^- s r^+ r_1^-\}}{(r^+ - |\xi|)P}, \\ C_3^- &= -\frac{A}{P}\Big\{\mu^+ \left[r^+(r^+ + |\xi|) + r^-(r^+ - |\xi|)\right] + 2\mu^- r^- |\xi| + \frac{\sigma}{s}|\xi|^3\Big\} \\ &\quad + \frac{\widetilde{b}_3'}{P}\Big\{\mu^+ \left[r^+ |\xi| (r^- + |\xi|) + (r^- - |\xi|)\xi^2\right] + \mu^- (r^{-2} + \xi^2)|\xi|\Big\}, \\ \omega_\alpha &= \frac{\{\widetilde{b}_\alpha + i\xi_\alpha [\mu^+ (r^+ - |\xi|) C_3^+ + \mu^- (r^- - r_1^-) C_3^- + (\mu^+ - \mu^-)\omega_3]\}}{\mu^+ r^+ + \mu^- r^-}, \\ &\qquad \alpha = 1, 2, \\ \omega_3 &= \frac{-(r^+ - |\xi|)|\xi| C_3^+ + (r^- r_1^- - \xi^2) C_3^-}{r^+ + r^-}. \end{aligned} \tag{18}$$

In formulas (18) we have used the notation: $A = i\xi_1 \widetilde{b}_1 + i\xi_2 \widetilde{b}_2$, $\widetilde{b}_3' = \widetilde{b}_3 + \frac{\sigma}{s}\widetilde{B}$,

$$\begin{aligned} P &= \mu^+ s (r^+ + |\xi|)\{\rho_0^+ (r^- r_1^- - \xi^2) + \rho_0^- (r^- |\xi| + r^+ r_1^- + 2\xi^2)\} + \rho_0^{-2} s^2 |\xi| \\ &\quad + 4(\mu^+ - \mu^-)\xi^2 \{\mu^+ r^+ (r^- r_1^- - \xi^2) - \mu^- r^- (r^- - r_1^-)|\xi|\} \\ &\quad + \frac{\sigma |\xi|^3}{s}\{\mu^+ (r^+ + |\xi|)(r^- r_1^- - \xi^2) + \rho_0^- r_1^- s\}. \end{aligned}$$

We observe that solution (17), (18) may be obtained by the passage to the limit $r_1^+ \to |\xi|$ from a solution of the model problem with a plane interface between two compressible fluids (see [3]). On the other hand, it goes over as $r_1^- \to |\xi|$ into a solution of the corresponding problem for two incompressible liquids [1]. This passage may also be demonstrated in the equations. Let us turn r_1^- to $|\xi|$ that corresponds to $\beta^- \to \infty$. In this case, the Navier-Stokes equations for a compressible fluid in the domain $\{x_3 < 0\}$ become the equations for the incompressible one.

The estimates of solution (17), (18) have been considered in detail in [2]. We remark only that C_3^+ is contained in all the expressions with multiplier $(r^+ - |\xi|)$, which is cancelled with the denominator of C_3^+. Hence, the estimate of solution (17), (18) depends only on the lower bound of $|P|$.

It should be noted also that the uniqueness of solution (17), (18) for $\gamma \geqslant \gamma_0 > 0$ is guaranteed by the fact that in this case $|P|$ is separated from zero.

Lemma 3.1. *Assume that for the viscosities of the fluids the inequalities*

$$\nu^+ < \nu^-, \quad \mu^+ > \mu^- \tag{19}$$

hold and that $\sigma \geqslant 0$. *Then for* $\forall \xi \in \mathbb{R}^2$, $\forall s \in \mathbb{C}$, $\operatorname{Re} s = \gamma > 0$,

$$|P| \geqslant c\left(|s|^2 + |s|^{3/2}|\xi| + |s|\xi^2 + \sigma|\xi|^3\right)\left(|s|^{1/2} + |\xi|\right).$$

The main difficulty in the evaluation of $|P|$ has been to find a universal multiplier q for P in the following sense: the quotient of the division of P by q has a positive real part. There are many ways to write down P. We have needed to find such an expression for $P = \sum\limits_{j=1}^{m} I_j$ that $\operatorname{Re}\frac{I_j}{q} \geqslant 0$, $j = 1, \dots, m$, and we could apply the inequality

$$|P| \geq |q| \left|\operatorname{Re}\left(\sum_{j=1}^{m} \frac{I_j}{q}\right)\right| \geqslant |q| \sum_{j=1}^{m} \operatorname{Re}\frac{I_j}{q} > 0.$$

We observe that it is the polynomial multiplier of the term with $\sigma|\xi|^3/s$ that plays the role of such a multiplier q in all three cases. For example, in the case of two incompressible fluids [1] $q = s[\mu^+(r^+ + |\xi|) + \mu^-(r^- + |\xi|)]$.

Remark 3.1. *We note that inequalities* (13) *follow from* (19) *so as for the case of two compressible fluids: restrictions* (15) *appear in the estimate process of the corresponding denominator* P *(see* [3]*).*

The author is grateful to the organizers and personally to Professor J.-F. Rodrigues for the possibility to attend so significant meeting and to give there a talk.

References

[1] I. V. Denisova, *A priori estimates of the solution of a linear time-dependent problem connected with the motion of a drop in a fluid medium,* Trudy Mat. Inst. Steklov **188** (1990), 3–21 (English transl. in Proc. Steklov Inst. Math. (1991), no. 3, 1–24).

[2] I. V. Denisova, *Evolution of compressible and incompressible fluids separated by a closed interface,* Interfaces and Free Boundaries **2** (2000), no. 3, 283–312.

[3] I. V. Denisova, *Problem of the motion of two compressible fluids separated by a closed free interface,* Zap. Nauchn. Semin. Petersburg. Otdel. Mat. Inst. Steklov. (POMI) **243** (1997), 61–86 (English transl. in J. Math. Sci. **99** (2000), no. 1, 837–853).

[4] I. V. Denisova, *Problem of the motion of two viscous incompressible fluids separated by a closed free interface,* Acta Appl. Math. **37** (1994), 31–40.

[5] I. V. Denisova and V. A. Solonnikov, *Solvability of the linearized problem on the motion of a drop in a liquid flow,* Zap. Nauchn. Semin. Leningrad. Otdel. Mat. Inst. Steklov. (LOMI) **171** (1989), 53–65 (English transl. in J. Soviet Math. **56** (1991), no. 2, 2309–2316).

[6] V. A. Solonnikov, *On an initial-boundary value problem for the Stokes systems arising in the study of a problem with a free boundary,* Trudy Mat. Inst. Steklov. **188** (1990), 150–188 (English transl. in Proc. Steklov Inst. Math. (1991), no. 3, 191–239).

[7] V. A. Solonnikov, *Solvability of the problem of evolution of a viscous incompressible fluid bounded by a free surface on a finite time interval,* Algebra i Analiz **3** (1991), no. 1, 222–257 (English transl. in St.Petersburg Math. J. **3** (1992), no. 1, 189–220).

[8] V. A. Solonnikov and A. Tani, *Free boundary problem for a viscous compressible flow with surface tension,* in: Constantin Carathéodory: An International Tribute, World Scientific (1991), 1270–1303.

Institute of Mechanical Engineering Problems
Russian Academy of Sciences
V. O., Bolshoy pr.,61
St. Petersburg, 199178 Russia
E-mail address: `ira@wave.ipme.ru`

Applications of a Local Energy Method to Systems of PDE's Involving Free Boundaries

Gonzalo Galiano

Abstract. We present a method of analysis for free boundary problems which is based on local energy estimates. This method allows us to deal with a great variety of problems formulated in a very general form, where *generality* stands for:

- No global information, like boundary conditions or boundedness of the domain, is needed.
- No monotonicity assumption on the nonlinearities is required, as the comparison principle is not invoked.
- Coefficients may depend on space and time variables and only a weak regularity is required.
- No restriction on the space dimension is assumed.

In this article we first show how the energy method applies to a simple example, proving the well known property of finite speed of propagation for the porous medium equation. We then give an outline of how the method works in more complicated situations: the occurrence of dead cores for the mangroves' problem and the finite speed of propagation along the characteristics for an evolution convection-diffusion equation.

1. Introduction

In this article we describe by means of three examples the so called *energy method for free boundary problems.* The method was introduced by Antontsev in [1], and was later extended and rendered to a mathematically rigorous form by Antontsev, Díaz and Shmarev [2] and Díaz and Verón [6]. Other contributions have been given by Bernis to PDE's of arbitrary order, see [3]. See also some applications of the method to systems of equations in [9, 5, 7, 8]. The idea of the method is to find a differential inequality which is satisfied by the natural energies associated to the problem and to deduce from its resolution some qualitative properties of the solutions of the original problem. Among these properties we find the finite speed of propagation and the formation of dead cores. Both are related to a degeneration introduced in the problem when the solution attains certain value, usually normalized to be the zero value. *Finite speed of propagation* means that solutions

---

Partially supported by DGE (Spain), project PB98/1564.

corresponding to compact supported initial datum remain with compact support at least for some time, meanwhile *dead core* means that even if the initial datum is strictly positive, a region where the zero level is attained may appear in finite time. Both phenomena are purely nonlinear.

2. The Porous Medium Equation and Finite Speed of Propagation

We consider the Cauchy problem for the porous medium equation,

$$v_t - (v^m)_{xx} = 0, \quad v(\cdot,0) = v_0\,, \tag{1}$$

with $m > 1$ and $v_0 \geq 0$ of compact support in $\mathbb{R}$. It is well known that the support of $v(\cdot,t)$ remains bounded for all $t > 0$. This is a peculiar behaviour of solutions of some type of parabolic problems which degenerate in some subset of the domain, and which is called the *finite speed of propagation* property. The aim of this section is to show how the energy method works for this simple example. Introducing the change of unknown $u := v^m$ and defining $p := 1/m$, (1) transforms into

$$(u^p)_t - u_{xx} = 0, \quad u(\cdot,0) = u_0 := v_0^m\,. \tag{2}$$

Let us denote by B_r the ball $\{x \in \Omega \colon |x - x_0| < r\}$, for any $r > 0$. Let $x_0 \in \mathbb{R}$ and $\rho_0 > 0$ be such that $u_0(x_0) = 0$ in B_{ρ_0}. Multiplying (2) by u and integrating in $B_\rho \times (0,t)$, with $\rho < \rho_0$ and $t > 0$, we obtain

$$\tfrac{p}{p+1}\int_{B_\rho} u(t)^{p+1} + \int_0^t\int_{B_\rho} |u_x|^2 = \int_0^t\int_{\partial B_\rho} u u_x n\,, \tag{3}$$

with n the outer normal to B_ρ. We define the *energy functions*

$$b(\rho,t) := \sup_{0\leq\tau\leq t}\int_{B_\rho} u(\tau)^{p+1} \quad \text{and} \quad E(\rho,t) := \int_0^t\int_{B_\rho} |u_x|^2\,,$$

which are clearly non-negative and non-decreasing with respect to ρ and t. Applying Hölder's inequality in (3), taking into account the increasing character of the different terms with respect to t and using that $E_\rho(\rho,t) = \int_0^t\int_{\partial B_\rho} |u_x|^2$, with E_ρ the partial derivative of E with respect to ρ, we obtain

$$\tfrac{p}{p+1} b(\rho,t) + E(\rho,t) \leq \left(\int_0^t\int_{\partial B_\rho} u^2\right)^{\frac{1}{2}} (E_\rho)^{\frac{1}{2}} =: I(E_\rho)^{\frac{1}{2}}\,. \tag{4}$$

To estimate I, which is a key step of the method, we apply a simple version of an interpolation-trace inequality introduced in a general setting in [6].

Lemma 2.1. *Let $\varphi \in H^1(x_0 - \rho, x_0 + \rho)$, for $x_0 \in \mathbb{R}$ and $\rho > 0$. Then*

$$|\varphi(x_0-\rho)| + |\varphi(x_0+\rho)| \leq L_0\left(\|\varphi_x\|_2 + \rho^{-\delta}\|\varphi\|_{p+1}\right)^\gamma \|\varphi\|_r^{1-\gamma}\,, \tag{5}$$

with $L_0 > 0$, $p > 0$, $\gamma := \frac{2}{2+r}$, $\delta := \frac{p+3}{2(p+1)}$ and $r \in [1,2]$.

Here we used the notation $\|\cdot\|_s := \|\cdot\|_{L^s(x_0-\rho,x_0+\rho)}$. The proof of this lemma is worked out by using some basic inequalities for L^p and Sobolev spaces. Using this lemma together with Young and Hölder inequalities we obtain

$$I \leq ct^{\frac{1-\gamma}{2}}(E+b)^{\beta}, \tag{6}$$

with $\beta = \gamma/2+(1-\gamma)/(p+1)$, and c a positive constant. It is important to observe here that $\beta > 1/2$ if and only if $p < 1$, since the argument we use now requires this condition. This is the moment in which the degenerate parabolicity of the porous medium equation comes in play. Applying Young's inequality to (6), we obtain

$$I(E_\rho)^{\frac{1}{2}} \leq \varepsilon(E+b) + c_\varepsilon t^{\frac{1-\gamma}{\kappa}}(E_\rho)^{\frac{1}{\kappa}}, \tag{7}$$

for some $\varepsilon > 0$ and $\kappa := 2(1-\beta)$. Combining (4) and (7) and using $b \geq 0$ we get

$$E \leq ct^{\frac{1-\gamma}{\kappa}}(E_\rho)^{\frac{1}{\kappa}},$$

for $0 < \rho < \rho_0$ and $0 < t < T$. Since $\beta > 1/2$, we have $\kappa < 1$ and a direct integration of this differential inequality in (ρ, ρ_0) gives

$$E^{1-\kappa}(\rho,t) \leq E^{1-\kappa}(\rho_0,t) - c(\rho_0-\rho)t^{1-\gamma}.$$

Therefore we deduce that choosing $\rho \leq c\rho_0 - E^{1-\kappa}(\rho_0,t)t^{\gamma-1} =: r(t)$, which is possible till certain instant $t^* > 0$ then $E(\rho,t) = 0$ for all $t \in (0,t^*)$ and ρ such that $\rho \leq r(t)$. And the result follows, i.e. $u(\cdot,t) = 0$ in $B_{r(t)}$ for all $t \in (0,t^*)$, or in other words, the support of $u(\cdot,t)$ remains bounded in $\mathbb{R}$ at least while $t < t^*$.

3. The Mangroves' Problem and Formation of Dead Cores

Mangroves grow on saturated soils or muds which are subject to regular inundation by tidal water with salt concentration c_w close to that of sea water. The mangrove roots take up fresh water from the saline soil and leave behind most of the salt, resulting in a net flow of water downward from the soil surface, which carries salt with it. In the absence of lateral flow, the steady state salinity profile in the root zone must be such that the salinity around the roots is higher than c_w, and that the concentration gradient is large enough so that the advective downward flow of salt is balanced by the diffusive flow of salt back up to the surface. In [7] we studied the following dimensionless model for the mangroves' problem

$$\begin{cases} v_t + (vq)_x - v_{xx} + f(x,v) = 0 \\ q_x + f(x,v) = 0 \end{cases} \quad \text{in } Q_T := (0,d)\times(0,T), \tag{8}$$

with

$$v(0,\cdot) = v_D, \quad v_x(d,\cdot) = q(d,\cdot) = 0 \quad \text{in } (0,T), \quad \text{and } v(\cdot,0) = v_0 \quad \text{in } (0,d). \tag{9}$$

Here $v := 1-u$ and $0 \leq u \leq 1$ denotes the water salt concentration, q the water discharge and $f(x,v)$ a function modeling the fresh water uptake by the roots of mangroves, typically given by $f(x,v) := k(x)v^p$, with $k \geq 0$ the root distribution (a bounded function) and p a positive constant. The variable $x \in \Omega$ denotes depth. The equations of (8) state the conservation of mass of salt and water.

The question is whether or not the water surrounding the roots of mangroves may reach a threshold level of salinization, $v = 0$, in finite time. We know that if $p = 1$, i.e. f is a linear function on the second variable, then $v = 0$ is not possible in finite time. This is the same situation as that of the heat equation with a linear source term. However, if $p < 1$, then f is non-Lipschitz continuous and a different behaviour of v may appear. In fact, we prove, see Theorem 3.1, that even when v_0 is positive, a *dead core* (region with $v = 0$) may appear in finite time. To apply the energy method we consider the set

$$\mathcal{P}(t) := \{(x,\tau)\colon |x - x_0| < R(\tau;t), \quad \tau \in (t,T)\}, \quad \text{for } t \in (0,T),$$

with $R(\tau;t) := (\tau - t)^\nu$, $0 < \nu < 1$ to be fixed, $\alpha > 0$ and $x_0 \in (0,d)$ such that $R(T;0) < x_0 < d - R(T;0)$, implying $\mathcal{P}(t) \subset Q_T$ for all $t \in (0,T)$. For brevity, we shall write $\mathcal{P}$ instead of $\mathcal{P}(t)$. We decompose the boundary of $\mathcal{P}$ as

$$\partial\mathcal{P}(t) := \partial_f\mathcal{P}(t) \cup \partial_l\mathcal{P}(t),$$

with $\partial_f\mathcal{P}(t) := \{(x,T) \in \partial\mathcal{P}\}$ and $\partial_l\mathcal{P}(t) := \{(x,\tau) \in \partial\mathcal{P}\colon t < \tau < T\}$. Finally, we define the *local energy functions*

$$E(t) := \int_{\mathcal{P}(t)} |v_x|^2 \, dx\, d\tau \quad \text{and} \quad C(t) := \int_{\mathcal{P}(t)} v^{p+1} \, dx\, d\tau. \tag{10}$$

Observe that E and C are non-increasing functions with respect to t. We make the following assumption on f: there exist positive constants k_0, k_1 such that

$$0 < k_0 s^{p+1} \le s f(\cdot, s) \le k_1 s^{p+1} \quad \text{for } s \in [0,1] \tag{11}$$

in $\mathcal{P}(t)$ for a.e. $t \in (0,T)$, with $p \in (0,1)$ and $k_0 > k_1/2$.

Theorem 3.1. *Let (v,q) be a solution of (8)–(9) and assume (11). Then there exists a constant $M > 0$ such that if $E(0) + C(0) \le M$ then $v \equiv 0$ in $\mathcal{P}(t^*)$, for some $t^* \in (0,T)$.*

Let us observe that testing the first equation of (8) with v and using the second equation of (8) leads to the estimate

$$E(0) + C(0) \le \int_\Omega v_0^2(x)dx + \int_0^T v_D(t) v_x(0,t)dt.$$

In some situations, for instance when $v_D \ge v_0$ the maximum principle implies $v_x(0,\cdot) \le 0$, allowing us to obtain an estimate of $E(0) + C(0)$ only in terms of v_0.

Proof of Theorem 3.1. The proof consists of three steps.
Step 1. Multiplying the first equation of (8) by v and integrating in $\mathcal{P}$ gives

$$\int_{\mathcal{P}} \{\tfrac{1}{2}(v^2)_t + \tfrac{1}{2}((v^2 q)_x + v^2 q_x) + (|v_x|^2 - (v v_x)_x) + v f(x, 1-v)\}\, dx d\tau = 0.$$

Using the divergence theorem, the second equation of (8) and (11) we find

$$\int_{\mathcal{P}} |v_x|^2\, dx d\tau + k_0 \int_{\mathcal{P}} v^{p+1}\, dx d\tau \le \int_{\partial_l \mathcal{P}} v v_x n_x\, dx d\tau$$

$$- \tfrac{1}{2} \int_{\partial_l \mathcal{P}} v^2 (n_\tau + q n_x)\, dx d\tau + \tfrac{k_1}{2} \int_{\mathcal{P}} v^{p+2},$$

with (n_x, n_τ) the unitary outward normal vector to $\mathcal{P}$. Using $v \le 1$, $q \le dk_1$,

$$E(t) + (k_0 - \tfrac{k_1}{2}) C(t) \le \tfrac{1+dk_1}{2} \int_t^T [v^2] d\tau + \int_{\partial_l \mathcal{P}} |v||v_x|\, dx d\tau\,, \tag{12}$$

where we introduced the notation $[v] := |v(x_0 + R(\tau;t),\tau)| + |v(x_0 - R(\tau;t),\tau)|$.
Step 2. Our aim is to estimate the right-hand side of (12) by means of E, C and their derivatives. First notice that $E_t(t) = \int_t^T [|v_x|^2] R_t(\tau;t) d\tau$, with the subindex t denoting derivative with respect to t. We use Hölder's inequality to get

$$\int_{\partial \mathcal{P}} |v||v_x|\, dx d\tau \le \left(\int_t^T -R_t [|v_x|^2]\, d\tau \right)^{1/2} \left(\int_t^T (-R_t)^{-1} [v^2]\, d\tau \right)^{1/2}$$

$$=: (-E_t)^{1/2} I_1 \le (-(E+C)_t)^{1/2} I_1\,. \tag{13}$$

To handle $I_1(t)$ and $I_2(t) := \int_t^T [v^2]$ of (12) we apply the interpolation-trace inequality stated in Lemma 2.1 to the function $\varphi := v(\cdot,t)$. We take $r < 2$ and find, by applying Hölder's inequality with exponent $\theta := \frac{1-p}{2-r}$

$$\|v(\cdot,t)\|_r \le \|v(\cdot,t)\|_2^{\frac{2}{r\theta'}} \|v(\cdot,t)\|_{p+1}^{\frac{p+1}{r\theta}}\,. \tag{14}$$

Combining (5) and (14) and using $v \le 1$ we get

$$[v^2] \le [v]^2 \le L_0^2 m(R) (\|v_x\|_2^2 + \|v\|_{p+1}^{p+1})^\gamma |Q_T|^{\frac{2(1-\gamma)}{r\theta'}} \|v\|_{p+1}^{\frac{2(1-\gamma)(p+1)}{r\theta}} \tag{15}$$

with $m(R) := \max\{1, R^{-2\delta\gamma}\}$. We then deduce from (15)

$$I_1 \le L_0 |Q_T|^{\frac{1-\gamma}{r\theta'}} \left(\int_t^T m(R)(-R_t)^{-1} (\|v_x\|_2^2 + \|v\|_{p+1}^{p+1})^{\gamma + \frac{2(1-\gamma)}{r\theta}} d\tau \right)^{1/2}. \tag{16}$$

Due to the crucial assumption $p < 1$, it is compatible to choose $r < 2$ and $r \ge 4/(3-p)$. Then we obtain that μ given by $\mu^{-1} := \gamma + (2(1-\gamma)/r\theta)$ satisfies $\mu \ge 1$. Using Hölder's inequality with exponent μ and substituting R we obtain from (16)

$$I_1 \le \Lambda (E+C)^{\frac{\gamma}{2} + \frac{1-\gamma}{r\theta}}\,, \tag{17}$$

with $\Lambda(t) := L_0 |Q_T|^{\frac{1-\gamma}{r\theta'}} \nu^{-1/2} \left(\int_t^T (\tau - t)^{\mu'(1-\nu-2\delta\nu\gamma)} d\tau \right)^{1/2\mu'}$. Function Λ is finite whenever we choose $\nu < (\mu+1)/(\mu(1+2\delta))$ which is always possible since the only restriction assumed on ν is $0 < \nu < 1$. Combining (13) and (17) we get

$$\int_{\partial \mathcal{P}} |v||v_x|\, dx d\tau \le \Lambda(t)(-(E+C)_t)^{1/2} (E+C)^{\frac{\gamma}{2} + \frac{1-\gamma}{r}}\,. \tag{18}$$

In a similar way, but choosing $r = 2$ in (5), we get the estimate

$$I_2 \leq L_0 \Gamma(E + C), \tag{19}$$

with $\Gamma^2(t) := \int_t^T (\tau - t)^{-\delta\nu} d\tau < \infty$ if $\nu < 1/\delta$.

Step 3. From (12), (18) and (19) we deduce

$$c_0(E + C) \leq \Lambda(-(E + C)_t)^{1/2}(E + C)^{\frac{\gamma}{2} + \frac{1-\gamma}{r}},$$

with $c_0 \leq k_0 - \frac{k_1}{2} - \frac{1+dk_1}{2} L_0 \Gamma(t)$. Notice that making $T - t$ small enough, say $T - t \leq \varepsilon$, we can ensure $c_0 > 0$. Making the assumption, to force a contradiction, that $E(t) + C(t) > 0$ for all $t \in [0, T]$, we arrive at the inequality

$$c_0^2 \left(E(t) + C(t)\right)^{2(1 - \frac{\gamma}{2} - \frac{1-\gamma}{r})} \leq -\Lambda(t)^2 (E + C)_t(t). \tag{20}$$

Due again to $p < 1$ we find $\sigma := 2\left(1 - \frac{\gamma}{2} - \frac{1-\gamma}{r}\right) < 1$. We assume $T > \varepsilon$ and restrict t to take values on $(T - \varepsilon, T)$ (so $T - t \leq \varepsilon$ is fulfilled). Integrating (20) in $t \in (T - \varepsilon, t^*)$ with $t^* \in (T - \varepsilon, T)$ we obtain

$$(E + C)^{1-\sigma}(t^*) \leq (E + C)^{1-\sigma}(T - \varepsilon) - (1 - \sigma)c_0^2 \int_{T-\varepsilon}^{t^*} \Lambda(t)^{-2} dt.$$

Therefore, since $E + C$ is non-increasing we have that if the initial energy satisfies

$$(E + C)^{1-\sigma}(0) \leq (1 - \sigma)c_0^2 \int_{T-\varepsilon}^{t^*} \Lambda(t)^{-2} dt =: M^{1-\sigma}$$

then $E(t^*) + C(t^*) = 0$ and therefore $v = 0$ in $\mathcal{P}(t^*)$.

4. The Boussinesq System and Finite Speed of Propagation Along the Characteristics

The Boussinesq system of hydrodynamics equations arises from a zero order approximation to the coupling between the Navier-Stokes equations and the thermodynamic equation. The presence of density gradients in a fluid means that gravitational potential energy can be converted into motion through the action of bouyant forces. Density differences are induced, for instance, by gradients of temperature arising by heating non-uniformly the fluid. In the Boussinesq approximation of a large class of flows problems, thermodynamical coefficients, such as viscosity, specific heat and thermal conductivity, can be assumed as constants leading to a coupled system with linear second order operators in the Navier-Stokes equations and in the heat conduction equation. However, there are some fluids, such as lubricants or some plasma flows, for which this is no longer an accurate assumption. In this case the following problem must be considered, see [4]

$$\begin{cases} \mathbf{u}_t + (\mathbf{u} \cdot \nabla)\mathbf{u} - \operatorname{div}(\mu(\theta) D(\mathbf{u})) = \mathbf{F}(\theta) \\ \theta_t + \mathbf{u} \cdot \nabla\theta - \Delta\varphi(\theta) = 0 \end{cases} \quad \text{in } Q_T, \tag{21}$$

with $\operatorname{div} \mathbf{u} = 0$ in Q_T, and

$$\begin{aligned} \mathbf{u} &= \mathbf{0}, & \varphi(\theta) &= 0 & &\text{on } \Sigma_T, \\ \mathbf{u}(\cdot,0) &= \mathbf{u}_0, & \theta(\cdot,0) &= \theta_0 & &\text{on } \Omega. \end{aligned} \tag{22}$$

We shall study the finite speed of propagation of the temperature variable, θ, along the characteristics defined by $\mathbf{u}$. We suppose that φ is given by $\varphi(s) := s^m$ with $m > 1$, although more general functions may be considered, see [8]. Assume that the velocity component of a solution of (21)–(22) is such that $\mathbf{u} \in C([0,T]; C^1_\sigma(\Omega))$, where σ denotes free divergence. Since $\mathbf{u} = 0$ on Σ_T, we may consider the prolongation of $\mathbf{u}$ by zero to the whole $\mathbb{R}^N$, and formulate the problem

$$\begin{cases} \dfrac{\partial \mathbf{X}}{\partial t}(\mathbf{x},t) = \mathbf{u}(\mathbf{X}(\mathbf{x},t),t) & \text{in } \mathbb{R}^N \times (0,T), \\ \mathbf{X}(\mathbf{x},0) = \mathbf{x} & \text{in } \mathbb{R}^N, \end{cases}$$

which admits a unique solution $\mathbf{X} \in C^1(\mathbb{R}^N \times [0,T])$. In the following we shall suppose that θ_0 vanishes in some ball B_{ρ_0} centered in $\mathbf{x}_0$ and compactly embedded in Ω. The following property is a consequence of the continuity of $\mathbf{X}$:

$$\begin{cases} \text{there exist } \hat{t} > 0 \text{ and } \rho_1 > \rho_0 \text{ such that if } t < \hat{t} \text{ and } \rho < \rho_1 \\ \text{then } \mathbf{X}(B_\rho, t) \in \Omega. \end{cases} \tag{23}$$

Theorem 4.1. *Let $(\mathbf{u},\theta)$ be any solution of (21)–(22) and suppose that $\mathbf{u}$ is locally Lipschitz continuous in Q_T. Then there exists $t^* \in (0,\hat{t})$ and a continuous function $r\colon [0,t^*] \to \mathbb{R}_+$, with $r(0) = \rho_0$ such that*

$$\theta(\mathbf{x},t) \equiv 0 \quad a.e.\ in \quad \{(\mathbf{x},t)\colon \mathbf{x} \in \mathbf{X}(B_{\rho_0},t), \quad t \in (0,t_*)\}\,.$$

Proof. We introduce the change of unknown $v := \theta^m$. Then v satisfies:

$$(v^p)_t + \mathbf{u}\cdot\nabla(v^p) - \Delta v = 0\,,$$

for $p := 1/m$. Multiplying by $v(\cdot,t)$, for $t > 0$ fixed and integrating in $\mathbf{X}(B_\rho,t)$

$$\tfrac{p}{p+1}\int_{\mathbf{X}(B_\rho,t)} \left((v^{p+1})_t + \mathbf{u}\cdot\nabla v^{p+1}\right) + \int_{\mathbf{X}(B_\rho,t)} |\nabla v|^2 = \int_{\partial\mathbf{X}(B_\rho,t)} v\nabla v\cdot\mathbf{n}\,. \tag{24}$$

The Reynolds Transport Lemma asserts that for any regular ψ,

$$\int_{\mathbf{X}(B_\rho,t)} \frac{\partial}{\partial t}\psi(\mathbf{y},t)\,\mathbf{dy} = \frac{d}{dt}\int_{\mathbf{X}(B_\rho,t)} \psi(\mathbf{y},t)\,\mathbf{dy} - \int_{\mathbf{X}(B_\rho,t)} \mathbf{u}(\mathbf{y},t)\cdot\nabla\psi(\mathbf{y},t)\mathbf{dy}\,. \tag{25}$$

Thus, integrating (24) in $(0,t)$ and using (25) and $v(\cdot,0) = 0$ in B_{ρ_0} we obtain

$$\int_{\mathbf{X}(B_\rho,t)} v^{p+1}(t) + \int_0^t\int_{\mathbf{X}(B_\rho,t)} |\nabla v|^2 \le c\int_0^t\int_{\partial\mathbf{X}(B_\rho,t)} v\nabla v\cdot\mathbf{n}\,.$$

Note that this expression is similar to (3) but integrated in a ball transformed along the characteristics defined by $\mathbf{u}$. As usual, we define the energies

$$b(\rho,t) := \sup_{0\le\tau\le t}\int_{\mathbf{X}(B_\rho,t)} v(\tau)^{p+1} \quad \text{and} \quad E(\rho,t) := \int_0^t\int_{\mathbf{X}(B_\rho,t)} |\nabla v|^2\,.$$

Due to the regularity of $\mathbf{X}$, we have $\partial(\mathbf{X}(B_\rho, t)) \equiv \mathbf{X}(\partial B_\rho, t)$ and therefore,

$$E_\rho(\rho, t) = \int_0^t \int_{\partial \mathbf{X}(B_\rho, t)} |\nabla v|^2 \quad \text{for a.e. } \rho > 0 . \tag{26}$$

To finish the proof we follow the steps given in Section 2 using now a version of the interpolation-trace trace inequality for sets transformed along the characteristics, see [8]. We deduce $b + E \leq c(E_\rho)^{1/2}(b + E)^{1/\kappa}$ where $\kappa \in (0,1)$ due to $p < 1$. Integrating this inequality the assertion follows. □

References

[1] S. N. Antontsev, *On the localization of solutions of nonlinear degenerate elliptic and parabolic equations,* Soviet Math. Dokl., **24**,2 (1981), 420–424.

[2] S. N. Antontsev, J. I. Díaz and S. I. Shmarev. *Energy methods for free boundary problems in continuum mechanics.* To appear in Birkhäuser, 2000.

[3] F. Bernis, *Compactness of the support for some nonlinear elliptic problems of arbitrary order in dimension n,* Comm. P. D. E., **9** (1984), 271–312.

[4] J. I. Díaz and G. Galiano, *Existence and uniqueness of solutions of the Boussinesq system with nonlinear thermal diffusion,* Top. Meth. in Nonlinear Analysis, **11**, 1, (1998), 59–82.

[5] J. I. Díaz, G. Galiano and A. Jüngel, *On a quasilinear degenerate system arising in semiconductors theory. Part II: Localization of vacuum solutions,* Nonlinear Analysis, **36**, 5 (1998), 569–594.

[6] J. I. Díaz and L. Véron, *Local vanishing properties of solutions of elliptic and parabolic quasilinear equations,* Trans. AMS, **290** (1985), 787–814.

[7] C. J. van Duijn, G. Galiano and M. A. Peletier, *A diffusion-convection problem with drainage arising in the ecology of mangroves.* Interfaces and Free Boundaries, Vol. 3, 15–44, 2001.

[8] G. Galiano, *Spatial and time localization of solutions of the Boussinesq system with nonlinear thermal diffusion.* Nonlinear Analysis, Vol. 42, N° 3, 423–438, 2000.

[9] G. Galiano and M. A. Peletier, *Spatial localization for a general reaction-diffusion system,* Annales de la Faculté des Sciences de Toulouse, **VII**, 3 (1998), 419–441.

[10] S. I. Shmarev, *Local properties of solutions of nonlinear equation of heat conduction with absorption,* Dinamika Sploshnoi Sredy, **95** (1990), 28–42.

Departamento de Matemáticas
Universidad de Oviedo
Avda. Calvo Sotelo s/n
33007 Oviedo, Spain
E-mail address: `galiano@orion.ciencias.uniovi.es`

Phase Boundaries in Alloys with Elastic Misfit

Harald Garcke

Abstract. We study a mathematical model describing phase transformations in alloys with kinetics driven by mass transport and stress. To describe the dynamics, a Cahn-Hilliard system taking elastic effects into account is studied. Existence and uniqueness results for the resulting singular elliptic-parabolic system are given.

In the Cahn-Hilliard model, phase boundaries are described by a diffuse interface with small positive thickness. In the stationary case we identify the sharp interface free boundary problem that arises when the interfacial thickness tends to zero. In particular, we obtain a geometric partition problem generalizing variants of isoperimetric problems to situations where elastic interactions cannot be neglected.

1. Introduction

Many problems involving free boundaries arise in the theory of phase transitions. In this context the interfaces separating two or more phases are the boundaries that have to be determined as part of the problem. We will consider phase transformations in alloys with kinetics driven by mass transport and stress. Different phases in such systems are characterized by certain distinct mixtures of the alloy components. For example if the alloy consists of just two types of atoms, A and B, and if two phases are possible, then one phase, which is called the α phase, has a higher concentration of A atoms then the other which is called the β phase. The interest now lies in a mathematical description of interfaces between different phases. We will first mathematically study a theory describing the evolution in such systems and then we will analyze stationary solutions that arise as absolute minimizer of an underlying energy.

Classical mathematical theories for phase transformations describe interfaces as hypersurfaces across which certain quantities, e.g. the mass density or certain components of the stress tensor, suffer jump discontinuities. These models lead to free boundary problems. The domains occupied by the individual phases are a priori unknown, i.e. the boundaries of the phase regions are free boundaries that have to be determined. In the interior of the phase domains, partial differential equations have to hold and on the boundaries certain conditions, e.g. jump conditions ensuring mass or energy conservation, are prescribed. Through the boundary conditions the solutions in the individual phases are coupled. Models of this type are called sharp interface models and a classical example is the Stefan problem describing melting and solidification (see e.g. the book of Meirmanov [11]).

In another approach phase boundaries are modeled by transition layers with a small positive thickness. In these theories usually all quantities describing the physical system are continuous. Some of the quantities involved might have steep gradients in the interfacial layers, those quantities that would jump in a sharp interface model. Models of this type are called diffuse interface models; examples are the phase field models that are the diffuse interface counterpart of Stefan problems (see e.g. Caginalp [1]).

To study sharp interface models analytically or computationally one has to somehow resolve the hypersurfaces making up the free boundary. This can lead to serious difficulties when the phase boundaries develop singularities and/or undergo topological changes during the evolution. Diffuse interface models instead use quantities that are smooth in the whole domain and at least conceptionally they handle topological changes without difficulties.

Our goal in the first part of this contribution is to mathematically analyse a diffuse interface model describing the motion of interfaces in alloys in which stresses caused by an elastic misfit have a pronounced effect on the evolution (see e.g. Leo, Lowengrub and Jou [10] and Garcke, Rumpf and Weikard [6] for numerical simulations on the basis of such models). The evolution will be governed by a fourth order parabolic system resulting from mass balance laws for the alloy components coupled to an elliptic system describing mechanical equilibrium. In a second part we will show that the energies related to the diffuse interface model converge in the sense of Γ-limits to the energy related to the sharp interface model when the interfacial thickness tends to zero. Moreover, we can show that minimizers converge (along subsequences) and it is also possible to pass to the limit in the first variation formula. The results on this limit are the first rigorous results relating a diffuse interface model to a sharp interface model in the case of alloys with elastic misfit.

This contribution is a review of results obtained by the author in his habilitation thesis [5]. Readers who are interested in more details may contact the author directly for a copy of the thesis.

2. The Diffuse Interface Model

Let us consider an alloy that consists of N components. We denote by c_k ($k = 1, \ldots, N$) the concentration of component k. Therefore the vector $\mathbf{c} = (c_k)_{k=1,\ldots,N}$ has to fulfill the constraint $\sum_{k=1}^N c_k = 1$, i.e. $\mathbf{c}$ lies in the affine hyperplane

$$\Sigma := \{\mathbf{c}' = (c'_k)_{k=1,\ldots,N} \in \mathbb{R}^N \mid \sum_{k=1}^N c'_k = 1\}.$$

To describe elastic effects we define the displacement field $\mathbf{u}(\mathbf{x})$, i.e. a material point $\mathbf{x}$ in the undeformed body will be at the point $\mathbf{x} + \mathbf{u}(\mathbf{x})$ after deformation. Since in phase separation processes the displacement gradient usually is small, we consider an approximative theory based on the linearized strain tensor

$$\mathcal{E}(\mathbf{u}) = \frac{1}{2}\left(\nabla \mathbf{u} + (\nabla \mathbf{u})^t\right) .$$

A free energy of alloy systems taking elastic interactions into account is of the form

$$E(\mathbf{c}, \mathbf{u}) = \int_\Omega \left\{ \frac{\gamma}{2} |\nabla \mathbf{c}|^2 + \Psi(\mathbf{c}) + W(\mathbf{c}, \mathcal{E}(\mathbf{u})) \right\} \tag{1}$$

where $\Omega \subset \mathbb{R}^n$, $n \in \mathbb{N}$, is a bounded domain with Lipschitz boundary and $\gamma > 0$ is a small positive constant. The first term in the energy is the gradient part penalizing rapid spatial variations in the concentrations. The second summand Ψ is the chemical energy of the system and in the case of phase separation Ψ is a non-convex function of $\mathbf{c}$. The non-convexity in Ψ gives rise to the appearance of different phases. Roughly speaking one can say that local minima of Ψ correspond to different phases in the system. The last term in the free energy takes elastic effects into account. The dependence on $\mathbf{c}$ reflects the fact that the stress free strain depends on the concentration. As the concentration is typically inhomogeneous in space, the stress free strain will vary in space.

To describe evolution phenomena, we consider mass diffusion for the individual components leading to diffusion equations for the concentrations. Mechanical equilibrium is attained on a much faster time scale than diffusion takes place. Therefore, we will assume a quasi-static equilibrium for $\mathbf{u}$, i.e. for all times

$$\nabla \cdot \mathcal{S} = 0 ,$$

where

$$\mathcal{S} = W_{,\mathcal{E}}(\mathbf{c}, \mathcal{E}(\mathbf{u}))$$

is the stress tensor. Here and below we will denote by $W_{,\mathcal{E}}$, $W_{,\mathbf{c}}$, $\Psi_{,\mathbf{c}}, \ldots$ the derivative with respect to the variable appearing as a subscript. We point out that $\mathbf{c}$ in general is time dependent, and hence the solution of the elastic system will also depend on time.

The diffusion equations for the concentrations c_k $(k = 1, \ldots, N)$ are based on mass balances for the individual components. To define the evolution equations for the concentrations, we need to introduce the vector of generalized chemical potential differences $\mathbf{w}$, which is given by the first variation of the energy E with respect to $\mathbf{c}$. Since values of $\mathbf{c}$ are restricted to lie on the affine hyperplane Σ, we need to take this constraint into account when computing the first variation. One obtains

$$\mathbf{w} = \mathbf{P}\left(-\gamma \Delta \mathbf{c} + \Psi_{,\mathbf{c}}(\mathbf{c}) + W_{,\mathbf{c}}(\mathbf{c}, \mathcal{E}(\mathbf{u}))\right)$$

where $\mathbf{P}$ is the euclidian projection of $\mathbb{R}^N$ onto

$$T\Sigma = \{\mathbf{d}' = (d'_k)_{k=1,\dots,N} \in \mathbb{R}^N \mid \sum_{k=1}^{N} d'_k = 0\}$$

which is the tangent space to Σ. Using the projection $\mathbf{P}$ is one way of taking the Lagrange multiplier associated to the constraint $\sum_{k=1}^N c_k = 1$ into account. Since $\mathbf{w}$ serves as a diffusional potential, the balance law for mass becomes $\partial_t \mathbf{c} = \Delta \mathbf{w}$. Altogether we obtain the system of equations

$$\partial_t \mathbf{c} = \Delta \mathbf{w}\,, \tag{2}$$

$$\mathbf{w} = \mathbf{P}\left(-\gamma \Delta \mathbf{c} + \Psi_{,\mathbf{c}}(\mathbf{c}) + W_{,\mathbf{c}}(\mathbf{c}, \mathcal{E}(\mathbf{u}))\right)\,, \tag{3}$$

$$\nabla \cdot W_{,\mathcal{E}}(\mathbf{c}, \mathcal{E}(\mathbf{u})) = 0\,. \tag{4}$$

In the case of two components, the equations (2)–(4) were first stated by Larché and Cahn [9] for $\gamma = 0$ and for nonzero γ by Onuki [12].

We will consider the system (2)–(4) together with initial conditions for $\mathbf{c}$

$$\mathbf{c}(\mathbf{x}, 0) = \mathbf{c}^0(\mathbf{x}) \tag{5}$$

and boundary conditions

$$\nabla \mathbf{w} \cdot \mathbf{n} = 0\,, \tag{6}$$

$$\nabla \mathbf{c} \cdot \mathbf{n} = 0\,, \tag{7}$$

$$\mathcal{S} \cdot \mathbf{n} = 0 \tag{8}$$

where $\mathbf{n}$ is the outer unit normal to Ω.

Two simple properties of the above system are:

i) *mass conservation:* i.e., for all $t > 0$,

$$\int_\Omega \mathbf{c}(\mathbf{x}, t) d\mathbf{x} = \int_\Omega \mathbf{c}^0(\mathbf{x}) d\mathbf{x}\,,$$

and

ii) *decay of the free energy:* for all $t_2 > t_1 \geq 0$,

$$E\left((\mathbf{c}, \mathbf{u})(t_2)\right) + \int_{t_1}^{t_2} |\nabla \mathbf{w}|^2 = E\left((\mathbf{c}, \mathbf{u})(t_1)\right)\,.$$

3. Existence for the Diffuse Interface System

To keep the presentation relatively simple, we present the existence result for a specific case which is of physical interest and still contains all mathematical difficulties. A chemical energy Ψ derived from a mean-field theory is the sum of a logarithmic entropy term and a pairwise interaction term and has the form

$$\Psi(\mathbf{c}) = \sum_{k=1}^{N} c_k \ln c_k + \tfrac{1}{2} \mathbf{c} \cdot \mathbf{A}\mathbf{c}\,, \tag{9}$$

where the entries of the matrix $\mathbf{A} = (A_{kl})_{k,l=1,\dots,N}$ are the constant parameters that describe pairwise interactions between the components. A typical form for the free energy density is

$$W(\mathbf{c}, \mathcal{E}) = \frac{1}{2} (\mathcal{E} - \mathcal{E}^\star(\mathbf{c})) : \mathcal{C}(\mathbf{c}) (\mathcal{E} - \mathcal{E}^\star(\mathbf{c})) . \tag{10}$$

Here, $\mathcal{C}(\mathbf{c})$ is the concentration dependent elasticity tensor mapping symmetric tensors in $\mathbb{R}^{n\times n}$ into itself. We require $\mathcal{C}(\mathbf{c})$ to be symmetric, positive definite and bounded. The quantity $\mathcal{E}^\star(\mathbf{c})$ is the symmetric stress free strain (or eigenstrain) at concentration $\mathbf{c}$. This is the value the strain tensor attains if the material were uniform with concentration $\mathbf{c}$ and unstressed. For simplicity, we assume that $\mathcal{E}^\star$ grows at most linearly in $\mathbf{c}$. Finally, for matrices $\mathbf{A}$ and $\mathbf{B}$ we use the notation $A : B = \operatorname{tr}(A^t B)$.

The main purpose of this section is to present a result on existence of weak solutions to the problem (2)–(8). The main difficulties in proving such a result are the facts that $\Psi_{,\mathbf{c}}$, a term entering the equation for the chemical potential differences (see (3)), is singular if c_k tends to zero and that $W_{,\mathbf{c}}$ growths quadratically in $\nabla\mathbf{u}$. We remark that the boundary condition (8) for the elasticity system only specifies the displacement $\mathbf{u}$ up to infinitesimal rigid displacements (i.e. translations and infinitesimal rotations). We note that the strain $\mathcal{E}(\mathbf{u})$ is uniquely determined and hence in the evolution equation for $\mathbf{c}$ the non-uniqueness of $\mathbf{u}$ will not play any role. To fix the freedom for the displacement we introduce the space

$$X_2 := \{\mathbf{u} \in H^1(\Omega, \mathbb{R}^n) \mid (\mathbf{u}, \mathbf{v})_{H^1} = 0 \text{ for all } \mathbf{v} \in X_{\text{ird}}\} = X_{\text{ird}}^\perp ,$$

where X_{ird} is the space of all infinitesimal rigid displacements (see e.g. [3]).

Let us state the definition of weak solutions we will use.

Definition 3.1. (Weak solution) *A triple*

$$(\mathbf{c}, \mathbf{w}, \mathbf{u}) \in L^2(0,T; H^1(\Omega, \mathbb{R}^N)) \times L^2(0,T; H^1(\Omega, \mathbb{R}^N)) \times L^2(0,T; X_2)$$

with $\mathbf{P}\Psi_{,\mathbf{c}}(\mathbf{c}) \in L^1(\Omega_T)$ *is called a weak solution of (2)–(8) if and only if*

(i)

$$-\int_{\Omega_T} \partial_t \boldsymbol{\xi} \cdot (\mathbf{c} - \mathbf{c}^0) + \int_{\Omega_T} \nabla\mathbf{w} : \nabla\boldsymbol{\xi} = 0 \tag{11}$$

for all $\boldsymbol{\xi} \in L^2(0,T; H^1(\Omega, \mathbb{R}^N))$ *with* $\partial_t\boldsymbol{\xi} \in L^2(\Omega_T)$ *and* $\boldsymbol{\xi}(T) = 0$,

(ii)

$$\int_{\Omega_T} \mathbf{w} \cdot \boldsymbol{\zeta} = \int_{\Omega_T} \{\nabla\mathbf{c} \colon \nabla\boldsymbol{\zeta} + \mathbf{P}\Psi_{,\mathbf{c}}(\mathbf{c}) \cdot \boldsymbol{\zeta} + \mathbf{P}W_{,\mathbf{c}}(\mathbf{c}, \mathcal{E}(\mathbf{u})) \cdot \boldsymbol{\zeta}\} \tag{12}$$

for all $\boldsymbol{\zeta} \in L^2(0,T; H^1(\Omega, \mathbb{R}^N)) \cap L^\infty(\Omega_T, \mathbb{R}^N)$, *and*

(iii)

$$\int_{\Omega_T} W_{,\mathcal{E}}(\mathbf{c}, \mathcal{E}(\mathbf{u})) \colon \nabla\boldsymbol{\eta} = 0 \tag{13}$$

for all $\boldsymbol{\eta} \in L^2(0,T; H^1(\Omega, \mathbb{R}^n))$.

In the following theorem the existence result is stated.

Theorem 3.2. *Let Ψ and W be of the form (9) and (10) respectively, with smooth C and $\mathcal{E}^\star$. In addition, assume that $\mathbf{c}^0 \in H^1(\Omega, \Sigma)$ fulfill $c_k^0 \geq 0$ and $⨍_\Omega c_k^0 > 0$ for $k = 1, \ldots, N$.*

Then there exists a weak solution of the elastic Cahn-Hilliard system which has the following properties:

(i) $\mathbf{c} \in C^{0,\frac{1}{4}}([0,T]; L^2(\Omega))$,
(ii) $\partial_t \mathbf{c} \in L^2\left(0,T; \left(H^1(\Omega)\right)^*\right)$,
(iii) *there exists a $p > 2$ such that $\mathbf{u} \in L^\infty\left(0,T; W^{1,p}(\Omega, \mathbb{R}^n)\right)$,*
(iv) *there exists a $q > 1$ such that for $k \in \{1, \ldots, N\}$*

$$\ln c_k \in L^q(\Omega_T)\,.$$

In particular, $c_k > 0$ almost everywhere.

We remark that the assumptions on $\mathbf{c}^0$ are physical. As $\mathbf{c}^0$ represents concentrations, only non-negative values for c_k^0 are physically meaningful. Furthermore, if $⨍_\Omega c_k^0 = 0$, the component k would not be present at all and the system could be reduced to $N-1$ components.

Idea of the proof. The proof is rather technical and we only sketch the methods. A full proof can be found in [5]. First of all one has to regularize the logarithmic part of Ψ by approximating smooth functions. Then one obtains time discrete solutions to the regularized problems by the method of implicit time discretisation. To solve the time discrete problems, one can use that the system is the H^{-1}-gradient flow with respect to the variable $\mathbf{c}$. This observation allows us to use variational methods to show existence of time discrete solutions. The free energy decay property stated in Section 2 has a discrete counterpart that first gives important a priori estimates. Anyhow, these are not enough to use compactness arguments to pass to the limit. To proceed, one has to show a higher integrability property for $\nabla \mathbf{u}$ (see (iii) in Theorem 3.2). This is necessary since $\nabla \mathbf{u}$ appears quadratically on the right-hand side in the equation for the chemical potential (see equation (3)). This can be achieved by using a perturbation technique introduced by Giaquinta and Modica (see [7]). Finally, one has to use that the non-smooth term appearing in Ψ is convex to obtain higher integrability of $\ln c_k$ (see (iv) in Theorem 3.2). The properties (iii) and (iv) together with more or less standard compactness arguments are then enough to pass to the limit in the time discrete version of the regularized problem with both the time step and the regularizing parameter.

Remark 3.3.

a) *A uniqueness result can be shown by an energy method in the case that C does not depend on $\mathbf{c}$ (see [5]).*
b) *If $N = 2$, if C does not depend on $\mathbf{c}$ and if Ψ is assumed to have polynomial growth with a strictly positive leading term, an existence and uniqueness result was shown independently by Carrive, Miranville and Piétrus [2]. However, the*

main difficulties in our study arise due to the facts that C is concentration dependent and that the chemical energy Ψ contains non-smooth logarithmic terms and these cases are not covered by the analysis of [2].

4. The Sharp Interface Limit

In this section we study solutions of the variational problems.

($\mathbf{P}^\varepsilon$) Find a minimizer $(\mathbf{c},\mathbf{u}) \in X_1 \times X_2$ of

$$E^\varepsilon(\mathbf{c},\mathbf{u}) := \int_\Omega \left(\varepsilon|\nabla \mathbf{c}|^2 + \frac{1}{\varepsilon}\Psi(\mathbf{c}) + W(\mathbf{c},\mathcal{E}(\mathbf{u}))\right), \quad \varepsilon > 0,$$

subject to the constraint $\fint_\Omega \mathbf{c} = \mathbf{m}$.

Here $\mathbf{m} \in \Sigma$ is a fixed constant. Under some natural assumptions it turns out that minimizers of the above variational problem are, roughly speaking, of the following form. In most of Ω the solution is close to values that minimize Ψ and the regions where the solution is close to minimizers of Ψ are separated by transition layers which are of a thickness proportional to ε. It is the goal of this section to study the limiting behaviour of E^ε and its minimizers as ε tends to zero. The scaling in ε (i.e. $\gamma \to \varepsilon^2$, $W \to \varepsilon W$, $E \to \frac{1}{\varepsilon}E$) is motivated by former studies for the case when no elastic contributions are present and by formally matched asymptotic expansions by Leo, Lowengrub and Jou [10].

For the rest of this section we assume that the chemical energy Ψ is such that

$$\Psi \geq 0 \quad \text{and} \quad \Psi(\mathbf{c}') = 0 \quad \Leftrightarrow \quad \mathbf{c}' \in \{\mathbf{p}_1, \dots, \mathbf{p}_M\}, \tag{14}$$

where $\mathbf{p}_1, \dots, \mathbf{p}_M \in \Sigma$ are mutually different and $M \geq 2$. Under the assumption (14) it is energetically favourable for $\mathbf{c}$ to attain the values $\mathbf{p}_1, \dots, \mathbf{p}_M$. And in fact minimizers of E^ε in most of the domain Ω have values close to $\mathbf{p}_1, \dots, \mathbf{p}_M$. The values $\mathbf{p}_1, \dots, \mathbf{p}_M$ correspond to M different phases and if $\mathbf{c}$ is close to a $\mathbf{p}_k$ we say that $\mathbf{c}$ is in phase k. It turns out that minimizers of E^ε converge (along subsequences) to minimizers of the functional

$$E^0\colon L^1(\Omega,\Sigma) \times X_2 \to \mathbb{R} \cup \{\infty\}$$

with

$$E^0(\mathbf{c},\mathbf{u}) = \begin{cases} \displaystyle\sum_{\substack{k,l=1 \\ k<l}}^{M} \sigma_{kl}\mathcal{H}^{n-1}\left(\partial^*\{\mathbf{c}=\mathbf{p}_k\} \cap \partial^*\{\mathbf{c}=\mathbf{p}_l\}\right) + \int_\Omega W(\mathbf{c},\mathcal{E}(\mathbf{u})), & \\ \qquad\qquad \text{if} \quad \mathbf{c} \in BV(\Omega),\ \Psi(\mathbf{c}) = 0 \text{ a.e.},\ \displaystyle\fint_\Omega \mathbf{c} = \mathbf{m}, & \\ \infty \qquad\qquad \text{otherwise}. & \end{cases}$$

Here, $\{\mathbf{c} = \mathbf{p}_k\} := \{\mathbf{x} \in \Omega \mid \mathbf{c}(\mathbf{x}) = \mathbf{p}_k\}$, $BV(\Omega)$ is the space of functions with bounded variation and $\partial^* D$ is the reduced boundary of a set D (for more details on these concepts we refer to Giusti [8]). The real numbers σ_{kl} are the surface tensions of the interfaces between phases k and l and they are defined via the chemical energy Ψ (see [5] for details).

Theorem 4.1. *Let W be of the form (10) with smooth $\mathcal{C}$ and $\mathcal{E}^\star$ and assume that $\Psi \in C^1(\mathbb{R}^N, \mathbb{R})$ growths at least quadratically for large* $\mathbf{c}$. *Then the variational problems* $(\mathbf{P}^\varepsilon)$ *possess minimizers $(\mathbf{c}^\varepsilon, \mathbf{u}^\varepsilon) \in H^1(\Omega, \Sigma) \times X_2$ provided that ε is small enough. Furthermore, there exists a sequence $\{\varepsilon^\kappa\}_{\kappa \in \mathbb{N}} \subset \mathbb{R}^+$ with $\lim_{\kappa\to\infty} \varepsilon^\kappa = 0$ and a $(\mathbf{c}, \mathbf{u}) \in L^2(\Omega, \Sigma) \times H^1(\Omega, \mathbb{R}^n)$ such that*

i)

$$\begin{aligned} \mathbf{c}^{\varepsilon^\kappa} &\to \mathbf{c} \quad \textit{in} \quad L^2(\Omega, \Sigma)\,, \\ \mathbf{u}^{\varepsilon^\kappa} &\to \mathbf{u} \quad \textit{in} \quad H^1(\Omega, \mathbb{R}^n)\ ; \end{aligned}$$

ii) $(\mathbf{c}, \mathbf{u})$ *is a global minimizer of E^0.*

Remark 4.2.

i) *To obtain the convergence result above, one shows that the energy E^ε converges in the sense of Γ-limits to the energy E^0. In a second step one has to show compactness for the sequence of minimizers $(\mathbf{c}^\varepsilon, \mathbf{u}^\varepsilon)_{\varepsilon>0}$. With respect to the concentration* $\mathbf{c}$ *one uses the L^2-topology for the Γ-limit and for the compactness argument.*

ii) *In the binary case, i.e. $N = 2$, it is also possible to analyze the limiting behaviour of the Euler-Lagrange equation for minimizers of the elastic Ginzburg-Landau energies as the interfacial thickness tends to zero. In [5] it is shown that the Lagrange multiplier entering the Euler-Lagrange equations converge and in the sharp interface limit we obtain a weak formulation of the Gibbs-Thomson relation, which is one of the conditions that has to hold on the free boundary.*

References

[1] G. Caginalp, *An analysis of a phase field model of a free boundary,* Arch Rat. Mech. Anal. **92** (1986), 205–245.

[2] M. Carrive, A. Miranville and A. Piétrus, *The Cahn-Hilliard equation for deformable elastic media,* Adv. Math. Sci. Appl. (to appear).

[3] P. G. Ciarlet, *Mathematical Elasticity,* North Holland, Amsterdam, New York, Oxford, **1988**.

[4] C. M. Elliott and S. Luckhaus, *A generalised diffusion equation for phase separation of a multi-component mixture with interfacial free energy,* SFB256 University Bonn, Preprint 195 (1991).

[5] H. Garcke, *On mathematical models for phase separation in elastically stressed solids,* habilitation thesis, University Bonn, **2000**.

[6] H. Garcke, M. Rumpf and U. Weikard, *The Cahn-Hilliard equation with elasticity: Finite element approximation and qualitative studies,* Interfaces and Free Boundaries **3** (2001), 101–118.

[7] M. Giaquinta, *Multiple integrals in the calculus of variations and nonlinear elliptic systems,* Annals of Mathematics Studies, Princeton University Press, **1983**.

[8] E. Giusti, *Minimal Surfaces and Functions of Bounded Variation*, Birkhäuser, Basel, Boston, Stuttgart, **1984**.

[9] F. C. Larché and J. W. Cahn, *The effect of self-stress on diffusion in solids,* Acta Metall., 30 (1982), 1835–1845.

[10] P. H. Leo, J. S. Lowengrub and H. J. Jou, *A diffuse interface model for microstructural evolution in elastically stressed solids*, Acta Mater., 46 (1998), 2113–2130.

[11] A. M. Meirmanov, *The Stefan Problem,* De Gruyter, Berlin, **1992**.

[12] A. Onuki, *Ginzburg-Landau approach to elastic effects in the phase separation of solids*, J. Phys. Soc. Jpn., 58 (1989), 3065–3068.

Institut für Angewandte Mathematik
Universität Bonn
Wegelerstraße 6
D-53115 Bonn, Germany
E-mail address: `harald@iam.uni-bonn.de`

Some Aspects of the Thin Film Equation

Josephus Hulshof

Abstract. I discuss some aspects of the mathematical theory for the Thin Film Equation in comparison to the Porous Medium Equation. The starting point of this overview is that of self-similar solutions.

1. Introduction

In this lecture I will discuss some aspects of nonnegative solutions $h(x,t)$ of the Thin Film Equation

$$h_t + (h^n h_{xxx})_x = 0\,. \tag{TFE}$$

Here x is the (one-dimensional) space variable, t is time and $n > 0$ is a parameter. The emphasis will be on self-similar solutions and dynamical systems methods for the resulting ordinary differential equations (ODEs).

Equation (TFE) arises in numerous applications of which I mention the spreading of a liquid film along a solid surface ($n = 3$) and Hele-Shaw flows ($n = 1$), see e.g., [21, 13, 20, 49, 5, 28, 30, 34, 52]. The most common derivation of (TFE) is as a lubrication approximation of the Navier–Stokes equations for thin film viscous flows, $h(x,t)$ being the film height, driven by surface tension and occuring with various values of $1 \leq n \leq 3$.

From a mathematical point of view it is natural to consider (TFE) with n as a parameter. At first sight it looks like a nonlinear degenerate version of the linear parabolic fourth order equation $h_t + h_{xxxx} = 0$. Another perspective is to see (TFE) as a fourth order version of the classical Porous Medium Equation (see e.g., [6])

$$h_t = (h^n h_x)_x\,, \tag{PME}$$

which may be considered as *the* prototype second order nonlinear degenerate diffusion equation. It derives its name from applications to flows in porous media and goes back as far as [48]. In applications, $h(x,t)$ is typically a density but it also occurs as the height of a groundwater mound, see [9].

Due to their homogeneous nature, both (TFE) and (PME) allow self-similar solutions of the form

$$h(x,t) = t^{-\alpha} f(\eta), \quad \eta = xt^{-\beta}\,, \tag{1}$$

where the similarity exponents respectively have to satisfy $n\alpha+4\beta=1$ (TFE) and $n\alpha+2\beta=1$ (PME). Such self-similar solutions are the guideline for the general PDE theory of the full partial differential equation (PDE). I will discuss some of these self-similar solutions of (TFE), taking (PME) as a, as we shall see, somewhat deceptive perspective.

2. Source-Type Solutions

Let us first consider (PME) and (TFE) on the whole line. Both equations can be written as the conservation law $h_t + q_x = 0$. Assuming the flux q to tend to zero as x approaches the boundary of the support of h, one expects conservation of mass: the spatial integral $\int h(x,t)dx$ (the mass) is constant in time. For (1) this implies $\alpha = \beta$. For (PME) this gives rise to the famous Zel'dovich–Kompaneetz–Barenblatt (ZKB) solutions given by

$$\alpha = \beta = \frac{1}{n+2}, \quad f(\eta) = \{C - \frac{n}{2(n+2)}\eta^2\}_+^{\frac{1}{n}}, \tag{2}$$

which feature an important property of (PME). The ZKB-solutions are nonclassical weak solutions with a propagating compact support. As $t \downarrow 0$, they converge, in the sense of distributions, to a multiple $M\delta$ of the Dirac measure δ, where $M = \int f(\eta)d\eta$. Any compactly supported nonnegative weak solution has intermediate asymptotics described by a ZKB-solution: transforming to self-similar variables the ZKB-solutions become equilibria attracting all nonnegative solutions with the same mass. In establishing this behaviour the comparison principle for second order diffusion equations, which says that a pointwise ordering of two initial data is carried over to the corresponding solutions of the initial value problem, is of great help. It allows us to control the solutions and their supports using the ZKB-solutions as comparison functions.

The boundary of the support, usually called the interface, has been shown to have interesting properties. It may be stagnant for a while, a phenomenon observed in [47] from an analysis of self-similar solutions, but then, after a finite waiting time [7] it starts to move with a positive speed and is real analytic in time [1].

This discussion of (PME) in a nutshell applies to all values of $n > 0$. Note that as $n \to 0$, the ZKB-solutions converge to multiples of the heat kernel, i.e., the fundamental solution of the linear heat equation $h_t = h_{xx}$. However, if (PME) is written in terms of the (pressure) variable $v = \frac{1}{n}h^n$ as $v_t = nvv_{xx} + v_x^2$, a very different limiting behaviour occurs, see again [6].

Looking at source-type solutions of (TFE), one arrives at

$$\alpha = \beta = \frac{1}{n+4}, \quad f^n f''' = \frac{\eta f}{n+4}, \tag{3}$$

which is to be compared with

$$\alpha = \beta = \frac{1}{n+2}, \quad f^n f' + \frac{\eta f}{n+2} = 0,$$

leading to (2) for (PME). In other words, where the source-type solutions of (PME) appear as solutions of a simple first order ordinary differential equation (ODE), source-type solutions have to be found from a much more difficult third order ODE. In view of the physical background the solution f has to be nonnegative and the analogy with (PME) suggests that f should have compact support. At the free boundary of the support obviously $f = 0$. A second physically appropriate condition is that also $f' = 0$ (zero contact angle).

This problem may be solved using Green's functions and a regularisation, see [32], but it was first studied in [17] for $n > 0$ using a shooting argument. The third order ODE in (3) is solved with initial data

$$f(0) = 1, \quad f'(0) = 0, \quad f''(0) = -\mu < 0. \tag{4}$$

Varying μ one expects to find a solution f touching zero (i.e., $f = f' = 0$) at some $\eta = a > 0$ as a borderline case between solutions that have a positive minimum before they can reach zero and solutions that hit zero with a negative slope. Having found this borderline case one then scales the solution f and obtains a source-type solution for any given mass $M > 0$. Surprisingly this method fails for $n \geq 3$ because the solutions with (4) always turn around and never hit zero. Thus source-type solutions can exist only for $n < 3$ and their existence is established along the lines indicated: there is one unique μ for which $f(\eta)$ decreases to touch zero.

Thus $n = 3$ is a critical value. One can show [33] that as $n \to 3$ from below, the source-type solutions with fixed mass M, as functions of x and t, converge to a Dirac mass which is just sitting there in $x = 0$. Apparently, (TFE) does not have moving interfaces and does not behave at all like (PME) when $n \geq 3$. Since $n = 3$ is also the parameter value most common in physical situations, there have been efforts to modify the derivation of the corresponding mathematical models, see [23] and [10].

For $0 < n < 3$ one may still have some validity of the analogy between (TFE) and (PME) but in fact this is only so for $\frac{3}{2} < n < 3$ as the local behaviour of the source-type profile f near the interface indicates. As $\eta \to a$ one has that $f(\eta) \sim A(a - \eta)^b$ with $b = 2$ for $0 < n < \frac{3}{2}$ and $b = \frac{3}{n}$ for $\frac{3}{2} < n < 3$. Note that (2) implies $b = \frac{1}{n}$ for (PME) and there is no way to see $0 < n < \frac{3}{2}$ for (TFE) as being reminiscent of that. I will come back to this in the discussion of self-similar solutions with general α and β.

Another surprise is that, as n goes to and crosses zero, there is no real change as far as source-type solutions are concerned. Source-type solution with compact support and quadratic behaviour at the interface exist throughout the range $-4 < n < \frac{3}{2}$ [19], in sharp contrast to the PME case, where a sharp transition from compact support to algebraic decay occurs as n crosses zero.

3. Strong Solutions

The discussion in Section 2 calls for some comments concerning the PDE theory for (TFE). Source-type solutions of (TFE) cannot converge to source-type solutions of the linear equation $h_t + h_{xxxx} = 0$ as $n \to 0$. This linear fourth order parabolic equation has a fundamental solution of the form

$$E(x,t) = t^{-\frac{1}{4}} f(xt^{-\frac{1}{4}}),$$

where f is the Fourier inverse of $\exp(-k^4)$ and has oscillating tails. Thus the source-type solutions of (TFE) cannot be obtained by perturbing from $n = 0$ and the linear equation and therefore one cannot expect a PDE theory for (TFE) which starts from the theory for the linear equation. In this context we mention the equation $h_t + (|h|^n h_x)_{xxx} = 0$, see [16], which has compactly supported source-type solutions with infinitely many oscillations near the free boundary. This equation is in fact a natural nonlinear version of $h_t + h_{xxxx} = 0$ but is much less relevant for applications.

So what kind of PDE theory can one expect for (TFE)? The first results have been obtained in [15] and [11]. Since the equation is degenerate at $h = 0$ a weak solution approach is necessary. Multiplying by a testfunction, one can integrate by parts once, but a second integration by parts runs into complications because the third order derivatives are on the inside and not on the outside of the expression to be dealt with. Therefore the commonly used definition uses only one integration, putting only one space derivative on the testfunction, and restricts this integral to the positivity set. As a consequence, this concept of solution is too weak to guarantee uniqueness of solutions with prescribed sufficiently smooth nonnegative initial data. Moreover, it does not imply a weak form of the zero-contact angle condition on the boundary of the support. Zero-contact angle weak solutions are called strong solutions in most of the papers on (TFE) but there is no natural PDE formulation of this concept.

For $n = 0$ the source-type solutions in Section 2 are positive on the interior of their support and satisfy $u = u_x = u_{xxx} = 0$ at the interface. This strongly suggests that strong solutions of (TFE) converge to solutions of a free boundary problem for the linear equation with these three conditions at the free boundary. This limit free boundary problem is discussed in [19] and remains to be solved in a PDE setting. It also suggests that strong solutions of (TFE) have to be considered as solutions to a free boundary problem for (TFE) with free boundary conditions $u = u_x = u^n u_{xxx} = 0$.

It is hard to obtain strong nonnegative solutions. For $0 < n < 3$ they are constructed as limits of solutions of approximating problems with very carefully modified initial data and h^n replaced by $D(h)$ with $D(h) \sim h^4$ as $h \to 0$. The approximating solutions in turn are constructed using a regularisation of the equation $h_t + (D(h)h_{xxx})_x = 0$. The nonnegativity of the approximating solutions follows from a nonnegativity property for solutions of (TFE) established in [15] for $n \geq 4$. Precise information about interface behaviour of solutions obtained through different regularisations of (TFE) has also been obtained using formal asymptotics in [25].

There is no comparison principle for strong solutions so it is not possible to use the source-type solutions as comparison functions and thereby control the solution and its free boundary. Nevertheless, some statements derived through such incorrect comparison arguments are essentially correct for strong solutions obtained through some approximation. For instance, as a first step towards a decription of the intermediate asymptotics, the support of a strong solution obtained in a suitable manner ($0 < n < 3$) is indeed controlled by $C(1+t)^{\frac{1}{n+4}}$ where C is a constant depending only on the mass of the initial data, see [14] and [43]. These are a posteriori results valid for the constructed strong solutions only. Unfortunately uniqueness of strong solutions is not known.

4. Dipole Solutions for $n = 1$

Equation (PME) has a second class of explicit self-similar solutions, given by

$$\alpha = 2\beta = \frac{1}{n+1}, \quad f(\eta) = \eta^{\frac{1}{n+1}} \{C - \frac{n}{2(n+2)}\eta^{\frac{n+2}{n+1}}\}_+^{\frac{1}{n}}. \tag{5}$$

These are the dipole profiles. The corresponding dipole similarity solutions of (PME) are nonnegative on the half-line $\{x > 0\}$, zero on the lateral boundary $x = 0$ and conserve the first moment. These dipole solutions describe the intermediate asymptotics of general compactly supported solutions with zero boundary values in $x = 0$ in exactly the same fashion as the ZKB-solutions describe the intermediate asymptotics of compactly supported solutions living on the whole line. In the afterword of his book [9], Barenblatt poses the question as to whether such dipole solutions exist for (TFE). If dipoles are understood to be of fixed first moment ($\alpha = 2\beta$), it follows that $n = 1$ because only then is there the appropriate conservation law, reading

$$(xh)_t = -(xhh_{xxx})_x + (hh_{xx})_x - \frac{1}{2}(h_x^2)_x. \tag{6}$$

Thus

$$n = 1, \quad \alpha = \frac{1}{3}, \quad \beta = \frac{1}{6}, \tag{7}$$

and the equation for the profile f comes out as

$$f''' = \frac{f''}{\eta} - \frac{f'^2}{2\eta f} + \frac{\eta}{6}. \tag{8}$$

The ODE (8) looks more difficult to analyse than the ODE in (3). Also it is not clear a priori what the boundary conditions at $\eta = 0$ should be, but as we shall see (8) there is a natural choice which follows from the analysis sketched below, see [18] for details. Because of its scaling invariance, the third order nonautonomous ODE may be reduced to a third order autonomous system. Introducing

$$x = \frac{\eta f'}{f}; \quad y = \frac{\eta^2 f''}{f}; \quad z = \frac{\eta^4}{6f}; \quad s = \log \eta, \tag{9}$$

the new unknowns x, y, z must solve

$$\dot{x} = x(1-x) + y; \quad \dot{y} = y(3-x) - \frac{x^2}{2} + z; \quad \dot{z} = z(4-x), \tag{10}$$

where dots denote differentiation with respect to s. The dipole profile will now be identified as corresponding to a unique intersection of a two-dimensional unstable manifold of a finite critical point and the two-dimensional stable manifold of a curve of critical points at infinity.

There are four finite critical points. The dimensions of their unstable manifolds and the characterisation of solutions $f(\eta)$ near $\eta = 0$ corresponding to orbits in these manifolds are listed below. Note that the x-value of the critical point corresponds to the exponent of $f(\eta)$ and $s \to -\infty$ to $\eta = 0$.

$$\dim W^u(0,0,0) = 3 \ (f(0) > 0); \qquad \dim W^u(\tfrac{3}{2}, \tfrac{3}{4}, 0) = 2 \quad (f(\eta) \sim A\eta^{\frac{3}{2}}); \tag{11}$$

$$\dim W^u(2,2,0) = 1 \ (f(\eta) \sim A\eta^2); \quad \dim W^u(4,12,20) = 0\,. \tag{12}$$

Since the dipole solution is expected to have compact support, its corresponding orbit does not exist globally in forward s-time so it must escape to infinity where the flow may be analysed using a suitable version of the classical Poincaré transformation, see also [12]. The appropriate transformation reads

$$x = \frac{X}{V}, \quad y = \frac{Y}{V^2}, \quad z = \frac{Z}{V^2}; \quad X^2 + Y^2 + 2Z = 1; \quad Z \geq 0\,, \tag{13}$$

$$(X^2 + 2Y^2 + 2Z)V\frac{d}{ds} = (1 + Y^2)V\frac{d}{ds} = \frac{d}{d\tau}\,. \tag{14}$$

Thus $\{(x, y, z) : z \geq 0\}$ corresponds to $\{(X, Y, V) : X^2 + Y^2 \leq 1, V > 0\}$ where the flow is given by

$$\frac{dX}{d\tau} = \frac{(X^2 - 2Y)(X^2 - Y^2 - 1)}{2} + X(X^2 - 1 + \frac{Y}{2}(2X^2 + Y^2 - 1))V\,; \tag{15}$$

$$\frac{dY}{d\tau} = XY(X^2 - 2Y) + \frac{(Y^2 + 2Y - 1)(2X^2 + Y^2 - 1)V}{2}\,; \tag{16}$$

$$\frac{dV}{d\tau} = \frac{X(X^2 + (Y-1)^2)V}{2} + (X^2 - Y^2 - 2 + Y(X^2 + \frac{Y^2 - 1}{2})V^2\,. \tag{17}$$

The critical points at infinity ($V = 0$) are

$$(X, Y) = (\pm 1, 0) \text{ and the parabola } X^2 = 2Y\,, \tag{18}$$

while in $V = 0$ the system is integrable with orbits

$$X^2 + (Y - C)^2 = C^2 + 1\,. \tag{19}$$

Dimensions of stable manifolds contained in $\{V > 0\}$ with corresponding local behaviour of $f(\eta)$ near the boundary $\eta = a > 0$ of the support of f are

$$\dim W^s(X = -1, Y = V = 0) = 3: \quad f(\eta) \sim A(a - \eta)\,; \tag{20}$$

$$\dim W^s(X < 0, X^2 = 2Y) = 2: \qquad f(\eta) \sim A(a - \eta)^2\,. \tag{21}$$

One can show that

$$W^u(\frac{3}{2},\frac{3}{4},0)\cap W^s(X<0,X^2=2Y)=\{\text{one single orbit}\}\,, \tag{22}$$

and this gives the only possible dipole profile, which turns out to satisfy

$$f(0)=f'(0)=f(a)=f'(a)=(ff''')(a)=0\,. \tag{23}$$

This means zero contact angle on both sides and zero flux at $\eta=0$. The corresponding dipole similarity solutions lose mass in $x=0$. Translations in s correspond to scaling a and the first moment of f.

5. General Self-Similar Solutions

In the light of the outcome of the analysis for $n=1$ it seems natural to expect (for other values of n) dipole-type self-similar solutions with zero contact angle in $x=0$ and zero contact angle, zero flux at the free boundary to be characteristic for solutions of (TFE) on the half-line. Since there is no conservation law for the first momentum if $n\neq 1$, the values of α and β are unknown a priori and have to be found from the analysis of the full fourth order ODE

$$(f^n f''')'=\beta\eta f'+\alpha f\,. \tag{24}$$

Such exponents are called anomalous and the corresponding solutions are called similarity solutions of the second kind, see again [9], but also [44] and [8].

Equation (24) is the fourth order version of a similar second order ODE for (PME), which, together with its N-dimensional version has been completely analysed in [40] and [42] for all $n>-1$. In particular [42] contains all the relevant forwards and backwards self-similar solutions of the equation $h_t=\nabla\cdot(|h|^n\nabla h)$. The philosophy in the analysis is basically to find similarity exponents for which two one-dimension manifolds (separatrices) of a two-dimensional system coincide. For instance, the ZKB-solutions are contained in an orbit which is the one-dimensional fast unstable manifold of a finite critical point and coincides with the one-dimensional stable manifold of a critical point at infinity. The following transformation mimics the reduction of the ODE for self-similar solutions of (PME) and the reduction of (8) to (10). Let

$$x=\frac{\eta f'}{f};\quad y=\frac{\eta^2 f''}{f};\quad z=\frac{\eta^3 f'''}{f};\quad u=\eta^4 f^{-n};\quad t=\log\eta\,. \tag{25}$$

Then (24) transforms to

$$\begin{aligned}\dot{x}&=x(1-x)+y; & \dot{z}&=z(3-(n+1)x)+u(\alpha+\beta x);\\ \dot{y}&=y(2-x)+z; & \dot{u}&=u(4-nx)\,,\end{aligned}$$

and in this system all (forwards and backwards) self-similar solutions can be found as intersections of two-dimensional unstable manifolds of finite critical points and two-dimensional stable manifolds of critical points at infinity. This is a fairly extensive programme because even the local analysis is involved. Besides five finite critical points there are also five critical points at infinity, as the transformation

$$x = \frac{X}{V};\quad y = \frac{Y}{V^2};\quad z = \frac{Z}{V^3};\quad u = \frac{U}{V^3};\quad X^2 + Y^2 + Z^2 + 2U = C\,,$$

see [27] reveals. This transformation gives a much clearer picture than the one used in [41]. There is no space here to list all the relevant dimensions of stable and unstable manifolds but I will briefly mention some results.

Dipole-type solutions with zero-contact angle in $x = 0$ can exist only for $n < 2$ because, as n crosses $n = 2$, the dimension of the unstable manifold of the relevant finite critical point drops. In [27] these dipole-type solutions are found using a numerical shooting technique. The results are consistent with a formal perturbation argument giving the derivatives of α and β with respect to n in $n = 1$. As $n \to 2$ the ratio $\frac{\alpha}{\beta} \to 1$, which means conservation of mass. For all $0 < n < 3$ the ODE in (2) allows solutions supported on $[0, a]$ with zero flux in both 0 and a but zero contact-angle only in $x = a$. As $n \uparrow 2$, the dipole-type solutions merge with these fixed mass solutions. A formal analysis of the boundary behaviour of solutions of (TFE) confirms that a zero-contact angle at $x = 0$ may be imposed only if $n < 2$.

The jump from exponent 2 to $\frac{3}{n}$ as n crosses $n = \frac{3}{2}$ is due to an exchange of dimensions of stable manifolds of two critical points at infinity, the relevant critical point for exponent $\frac{3}{n}$ disappearing into $\{u < 0\}$ as n is taken outside the range $\frac{3}{2} < n < 3$ ($\beta > 0$). Moreover, when n drops below $\frac{1}{2}$ the two-dimensional stable manifold of another critical point at infinity, giving profiles f with exponent $\frac{3}{n+1}$, becomes two-dimensional. Such solutions are flatter at the free boundary than the zero flux, zero contact-angle solutions and, taking the limit $n \to 0$, one has $f = f' = f'' = 0$ in $\eta = a > 0$. This suggests that there is another free boundary formulation for (TFE) for $n < \frac{1}{2}$ which imposes optimal flatness at the free boundary and leads to the variational obstacle problem for the linear equation in the limit $n \to 0$.

Other relevant second kind self-similar solutions to be found are solutions which describe rupture (i.e., the initially positive solution develops an interior zero and possibly a dead core. This is of course also and even more relevant in dimensions $N = 2$ and $N = 3$, just as the question of how dry spots disappear. This last issue has been extensively studied for (PME) in which case there is a radial family of second kind self-similar solutions (the Aronson–Graveleau solutions) which describe this phenomenon for the radial solutions, see [2]. Recently [3] however symmetry breaking results have been obtained for this class of solutions using n as parameter. Moreover the linearised system obtained in [3] after an ingenious linearisation procedure may be analysed using transformations of the same nature as the one sketched in this lecture. In particular the two-dimensional system in which the Aronson–Graveleau solutions can be found has to be appended with only one first order equation. In [4] we use this to prove that the Aronson–Graveleau profiles are unstable for all $n > 0$. Obviously, the same problem for (TFE) poses a formidable task.

Acknowledgement

My participation in 3ECM2000 and this research were supported by the TMR network Nonlinear Parabolic Partial Differential Equations: Methods And Applications, ERBFMRXCT980201.

References

[1] Angenent S B 1988 Analyticity of the Interfaces of the Porous Media Equation after the Waiting Time *Proc. A.M.S.* **102** 329–336.

[2] Angenent S B and Aronson D G 1996 The focusing problem for the radially symmetric porous medium equation *Comm. P.D.E.* **20** 1217–1240.

[3] Angenent S B, Aronson D G and Betelú S I 2000 Renormalization study of two-dimensional convergent solutions of the porous medium equation *Phys. D* **138** 344–359.

[4] Aronson D G and Hulshof J (in preparation, `http://www.cs.vu.nl/~jhulshof`).

[5] Almgren R 1996 Singularity formation in Hele-Shaw bubbles *Phys. Fluids* **8** 344–352.

[6] Aronson D G 1986 Some Problems in Nonlinear Diffusion, eds. A. Fasano & M. Primicerio, Lecture Notes in Math., CIME Foundation Series, Springer Verlag No. 1224.

[7] Aronson D G, Caffarelli L A and Kamin S 1983 How an initially stationary interface begins to move in porous medium flow *SIAM J. Math. Anal.* **14** 639–658.

[8] Aronson D G and Graveleau J 1993 A selfsimilar solution for the focussing problem for the porous medium equation *Eur. J. Appl. Math.* **4** 65–81.

[9] Barenblatt G I 1996 *Scaling, Self-Similarity and Intermediate Asymptotics* (Cambridge: Cambridge University Press).

[10] Barenblatt G I, Beretta E and Bertsch M 1997 The problem of the spreading of a liquid film along a solid surface: A new mathematical formulation *Proc. Nat. Acad. Sci. U.S.A.* **94** 10024–10030.

[11] Beretta E, Bertsch M and Dal Passo R 1995 Nonnegative solutions of a fourth order nonlinear degenerate parabolic equation *Arch. Rational Mech. Anal.* **129** 175–200.

[12] van den Berg G J B, Hulshof J and van der Vorst R C A M 1999 Travelling waves for fourth-order semilinear parabolic equations (Leiden preprint MI 16–99).

[13] Bernis F 1995 Viscous flows, fourth order nonlinear degenerate parabolic equations and singular elliptic problems *Free boundary problems: theory and applications (Pitman Research Notes in Mathematics 323)* ed J I Diaz *et al* (Harlow: Longman) pp 40–56.

[14] Bernis F 1996 Finite speed of propagation and continuity of the interface for thin viscous flows *Adv. Differential Equations* **1** 337–368.

[15] Bernis F and Friedman A 1990 Higher order nonlinear degenerate parabolic equations *J. Differential Equations* **83** 179–206.

[16] Bernis F and McLeod J B 1991 Similarity solutions of a higher order nonlinear diffusion equation. *Nonlinear Anal.* **17** 1039–1068.

[17] Bernis F, Peletier L A and Williams S M 1992 Source type solutions of a fourth order nonlinear degenerate parabolic equation *Nonlinear Anal. TMA* **18** 217–234.

[18] Bernis F, Hulshof J and King J R 2000 Dipoles and similarity solutions of the thin film equation in the half-line *Nonlinearity* **13** 1–27.

[19] Bernis F, Hulshof J and Quirós F 1999 The "linear" limit of thin film flows as an obstacle-free boundary problem Leiden preprint MI 24–99, to appear in EJAM.

[20] Bertozzi A L 1998 The mathematics of moving contact lines in thin liquid films *Notices Amer. Math. Soc.* **45** 689–697.

[21] Bertozzi A L, Brenner M P, Dupont T F and Kadanoff L P 1994 Singularities and similarities in interface flows *Trends and Perspectives in Applied Mathematics (Applied Mathematical Sciences 100)* ed L Sirovich (Berlin: Springer-Verlag) pp 155–208.

[22] Bertozzi A L and Pugh M 1996 The lubrication approximation for thin viscous films: regularity and long time behaviour of weak solutions *Comm. Pure Appl. Math.* **49** 85–123.

[23] Bertsch M, Dal Passo R, Davis S H and Giacomelli L 1999 Effective and microscopic contact angles in thin film dynamics *European J. Appl. Math.* **11** (2000), no. 2, 181–201.

[24] Bertsch M, Dal Passo R, Garcke H and Grün G 1998 The thin viscous flow equation in higher space dimensions *Adv. Differential Equations* **3** 417–440.

[25] Bowen M and King J R 1999 Moving boundary problems and non-uniqueness for the thin film equation Leiden preprint MI 28–99.

[26] Bowen M and King J R 1999 Asymptotic behaviour of the thin film equation in bounded domains Leiden preprint MI 29–99.

[27] Bowen M, Hulshof J and King J R 2000 Anomalous exponents and dipole solutions for the thin film equation Leiden preprint MI 01–2000.

[28] Constantin P, Dupont T F, Goldstein R E, Kadanoff L P, Shelley M J and Su-Min Zhou 1993 Droplet breakup in a model of the Hele-Shaw cell *Phys. Rev. E* **47** 4169–4181.

[29] Dal Passo R and Garcke H 1999 Solutions of a fourth order degenerate parabolic equation with weak initial trace *Ann. Scuola Norm. Sup. Pisa Cl. Sci.* **28** 153–181.

[30] Dupont T F, Goldstein R E, Kadanoff L P and Su-Min Zhou 1993 Finite-time singularity formation in Hele-Shaw systems *Phys. Rev. E* **47** 4182–4196.

[31] Ehrhard P and Davis S H 1991 Non-isothermal spreading of liquid drops on horizontal plates *J. Fluid Mech.* **229** 365–388.

[32] Ferreira R and Bernis F 1997 Source-type solutions to thin-film equations in higher dimensions *European J. Appl. Math.* **8** 507–524.

[33] Ferreira R and Bernis F 1999 Source-type solutions to thin-film equations: the critical case, *Applied Math. Lett.* **12** 45–50.

[34] Goldstein R E, Pesci A I and Shelley M J 1998 Instabilities and singularities in Hele-Shaw flow *Phys. Fluids* **10** 2701–2723.

[35] Greenspan H P 1978 On the motion of a small viscous droplet that wets a surface *J. Fluid Mech.* **84** 125–143.

[36] Greenspan H P and McCay B M 1981 On the wetting of a surface by a very viscous fluid *Stud. Appl. Math.* **64** 95–112.

[37] Hocking L M 1981 Sliding and spreading of thin two-dimensional drops *Quart. J. Mech. Appl. Math.* **34** 37–55.

[38] Hocking L M 1983 The spreading of a thin drop by gravity and capillarity *Quart. J. Mech. Appl. Math.* **36** 55–69.

[39] Hocking L M 1992 Rival contact-angle models and the spreading of drops *J. Fluid Mech.* **239** 671–681.

[40] Hulshof J 1991 Similarity solutions of the porous medium equation with sign changes *J. Math. Anal. Appl.* **157** 75–111.

[41] Hulshof J 1996 A local analysis of similarity solutions of the thin film equation, *Nonlinear analysis and applications Warsaw 1994 (GAKUTO Internat. Ser. Math. Sci. Appl. 7)* (Tokyo: Gakkotosho) pp 179–192.

[42] Hulshof J, King J R and Bowen M 1998 Intermediate asymptotics of the porous medium equation with sign changes Leiden preprint W98–20, to appear in Adv. Diff. Equ.

[43] Hulshof J and Shishkov A E 1998 The thin film equation with $2 \leq n < 3$: finite speed of propagation in terms of the L^1-norm *Adv. Differential Equations* **3** 625–642.

[44] Hulshof J and Vazquez J L 1994 Selfsimilar solutions of the second kind for the modified porous medium equation, *Eur. J. Appl. Math* **5** 391–403.

[45] Kamin S and Vazquez J L 1991 Asymptotic behaviour of solutions of the porous medium equation with changing sign *SIAM J. Math. Anal.* **22** 34–45.

[46] Lacey A A 1982 The motion with slip of a thin viscous droplet over a solid surface *Stud. Appl. Math.* **67** 217–230.

[47] Lacey A A, Ockendon J R and Tayler A B 1982 'Waiting time' solutions of a nonlinear diffusion equation *Siam J. Appl. Math.* **42** 1252–1264.

[48] Muskat M 1937 The Flows of Homogeneous Fluids through Porous Media McGraw-Hill, New York.

[49] Myers T G 1998 Thin films with high surface tension *SIAM Rev.* **40** 441–462.

[50] Neogi P and Miller C A 1983 Spreading kinetics of a drop on a rough solid surface *J. Colloid Interface Sci.* **92** 338–349.

[51] Otto F 1998 Lubrication approximation with prescribed nonzero contact angle *Comm. Partial Differential Equations* **23** 2077–2164.

[52] Pesci A I, Goldstein R E and Shelley M J 1999 Domain of convergence of perturbative solutions for Hele-Shaw flow near interface collapse *Phys. Fluids* **11** 2809–2811.

[53] Vazquez J L 1992 An Introduction to the Mathematical Theory of the Porous Medium Equation, in: Shape Optimization and Free Boundaries, ed. M. C. Delfour, Mathematical and Physical Sciences, Series C, 380: 347–389, Kluwer, Dordrecht-Boston-Leiden.

Faculty of Sciences
Mathematics and Computer Science division
Free University Amsterdam
De Boelelaan 1081
1081 HV Amsterdam, The Netherlands
E-mail address: jhulshof@cs.vu.nl

A Brief Overview on The Obstacle Problem

Regis Monneau

Abstract. We present a short survey on the obstacle problem including the theory developed by L. A. Caffarelli and the theory developed independently by G. S. Weiss. We also present some other recent results on the regularity of the free boundary.

1. Introduction

In this note we will present a short survey on the obstacle problem. This presentation is obviously far from being exhaustive. The author has decided to present some new and old important results chosen from a wide literature. In particular the references of this note are deliberately limited.

A number of books deal with the obstacle problem, e.g., the 1980 book of D. Kinderlehrer and G. Stampacchia [17] on variational inequalities, and the book of A. Friedman [12], which covers some results on the obstacle problem through 1982, as well as many other free boundary problems. For examples of applications of the obstacle problem to Mathematical Physics and many references until 1987 look at the book of J. F. Rodrigues [20]. Its Chapter 6 gives in particular a good overview and detailed results on the obstacle problem.

1.1. A physical example

Let us consider a horizontal wire and a membrane hanging on this wire. We assume that this membrane is above a plate. We push up the plate on the membrane. We get a contact area between the membrane and the obstacle which is the plate. This contact area is called the coincidence set (see Figure 1). The boundary of the coincidence set is called the free boundary for the obstacle problem. The goal of this note is to throw some light on the properties of this free boundary.

1.2. Mathematical formulation

We now give a simple mathematical model for this problem. We assume that the membrane is given by the graph of a function

$$u\colon \Omega \subset \mathbf{R}^n \longrightarrow \mathbf{R}$$

1991 *Mathematics Subject Classification.* 35R35.

Key words and phrases. Obstacle problem, Coincidence set, Free boundary, Blowup, Liouville result, Monotonicity formula, Stability, Convexity, Generic regularity.

where Ω is a smooth bounded domain.

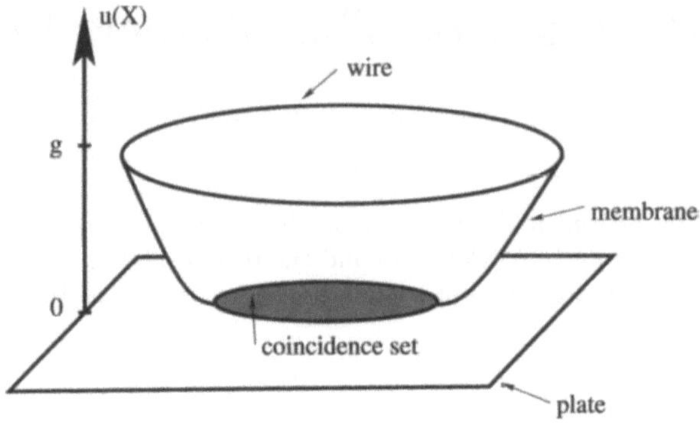

FIGURE 1. Membrane over a plate obstacle

We assume that the wire is given by the value of u on the fixed boundary $\partial\Omega$ which is assumed to be a constant

$$u = \text{constant} = g \quad \text{on} \quad \partial\Omega\,. \tag{1}$$

We assume that the plate is at the 0-level, and we assume that the membrane stays above the plate, i.e., the obstacle

$$u \geq 0 \quad \text{on} \quad \Omega\,.$$

We assume that u minimizes an energy on Ω which is given by

$$\int_\Omega |\nabla u|^2 + 2u \tag{2}$$

where the minimization is mathematically taken on the Sobolev space of functions

$$K = \left\{u \in H^1(\Omega), \quad u = g \text{ on } \partial\Omega, \quad u \geq 0 \text{ on } \Omega\right\}\,. \tag{3}$$

The first term in the energy (2) is a kind of elastic energy and the second term is a kind of gravity energy for a heavy membrane.

The well-posedness and good properties of the obstacle problem are guaranteed by the following result for $g > 0$:

Theorem 1.1. (J. Frehse [11], $C^{1,1}$ regularity of the minimizer) *There exists a unique solution u minimizing energy* (2) *on the convex set K defined in* (3). *Moreover this solution satisfies*

$$\begin{cases} \Delta u = 1 & \textit{on} \quad \{u > 0\} \cap \Omega \\ u \geq 0 & \textit{on} \quad \Omega \\ u \in C^{1,1}(\overline{\Omega}) \end{cases}\,. \tag{4}$$

The coincidence set is

$$\{u = 0\}$$

and the free boundary is $\partial\{u = 0\}$. Moreover it is proved that the free boundary has a finite (hyper)area:

Theorem 1.2. (H. Brezis, D. Kinderlehrer [3], L. A. Caffarelli [6], Finite Hausdorff measure)

$$\mathcal{H}^{n-1}(\partial\{u = 0\}) \leq C\,.$$

1.3. Examples of singularities

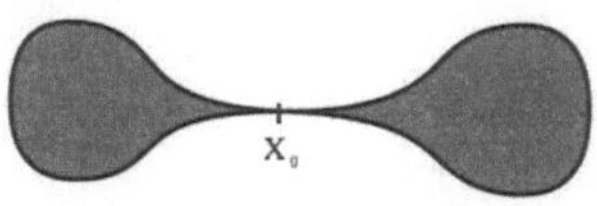

FIGURE 2. Contact point

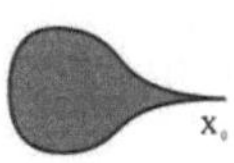

FIGURE 3. Cusp

Let us now give some examples of singularities presented by D. G. Schaeffer in [23]. In two dimensions, the free boundary can have a contact point (see Figure 2) or a cusp point (see Figure 3). Following Schaeffer [23], the intuitive origin of these singularities can be obtained by "continuous deformation" with varying obstacles with different components, the case of Figure 2 being the "product" of the joining of two components, and the case of Figure 3 a further consequence of the vanishing of the diameter in the right part of Figure 2. Notice that the double point of $\partial\{u = 0\}$ consists of two tangent curves and not two curves intersecting at a non-zero angle.

2. The Blowup Method

To prove regularity results on the free boundary, the main tool (first introduced for the obstacle problem by L. A. Caffarelli in [4]) is the notion of blowup.

Let us consider a solution u of (4) and assume that X_0 is a point on the free boundary $\partial\{u = 0\}$. Let us consider the sequence of functions

$$u^\epsilon(X) = \frac{u(X_0 + \epsilon X)}{\epsilon^2}\,.$$

By Theorem 1.1, $u^\epsilon(0) = \nabla u^\epsilon(0) = 0$ and the second derivatives $|D^2 u^\epsilon|$ are bounded by a constant independent of $\epsilon > 0$. Up to extraction of a subsequence we get a limit as $\epsilon \to 0$,

$$u^0(X) \quad \text{defined on} \quad \mathbf{R}^n .$$

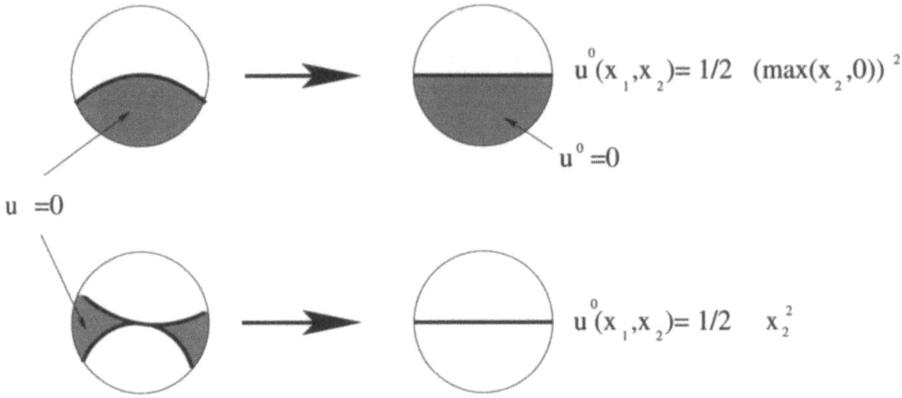

FIGURE 4. Blowup limits

In Figure 4 we have represented some possible limits in two dimensions. In any dimensions, the main result for the regularity of the free boundary is the following

Theorem 2.1. (L. A. Caffarelli [5, 7], Characterization of the blowup limit) *The blowup limit is unique and depends only on the point X_0 on the free boundary. Moreover either there exists a unit vector $\nu_{X_0} \in \mathbf{S}^{n-1}$ such that*

$$u^0(X) = \frac{1}{2}(\max(\langle X, \nu_{X_0}\rangle, 0)^2$$

and the point X_0 is called regular,

or u^0 is a quadratic form, i.e.,

$$u^0(X) = \frac{1}{2} \quad {}^t X \cdot Q_{X_0} \cdot X \geq 0$$

where Q_{X_0} is a symmetric matrix $n \times n$ such that $tr \sim Q_{X_0} = 1$. Such a point X_0 is called singular.

This theorem says two things:

i) A Liouville result which classifies all possible blowup limits on $\mathbf{R}^n$.
ii) The uniqueness of the blowup limit for each point X_0.

The Liouville result was obtained by L. A. Caffarelli in [5] and essentially uses the maximum principle. We can also cite the work of G. S. Weiss [25] which helps to classify all solutions on $\mathbf{R}^n$ invariant by some dilatations.

The uniqueness of the blowup limit is difficult to prove in the case of regular points and will be discussed in Section 3. On the contrary, in the case of singular points, this uniqueness is quite easy to prove. In [7] L. A. Caffarelli uses the monotonicity formula of H. W. Alt, L. A. Caffarelli and A. Friedman [1] applied to the first derivatives of the solution to prove it. A more elementary fact which avoids deriving the solution u is the following:

Theorem 2.2. (R. Monneau [19], Monotonicity formula for singular points) *Let $v(X) = \frac{1}{2}\ {}^tX \cdot Q \cdot X \geq 0$ where Q is a symmetric matrix $n \times n$ such that $tr \sim Q = 1$. If u is a solution to (4) and the origin 0 is a singular point, then*

$$\frac{1}{r^{n+3}} \int_{\partial B_r(0)} (u-v)^2 \quad \text{is nondecreasing in} \quad r\,.$$

This result is a simple corollary of the following monotonicity formula discovered by G. S. Weiss:

Theorem 2.3. (G. S. Weiss [25], Monotonicity formula for the obstacle problem) *For each point $X_0 \in \Omega$ such that the ball $B_r(X_0) \subset \Omega$ we have*

$$\Phi_{X_0,u}(r) := \left(\frac{1}{r^{n+2}} \int_{B_r(X_0)} |\nabla u|^2 + 2u - \frac{1}{r^{n+3}} \int_{\partial B_r(X_0)} 2u^2 \right)$$
$$\text{is nondecreasing in} \quad r\,.$$

In particular with these monotonicity formulas it is possible to prove

Theorem 2.4. (L. A. Caffarelli [7], Continuity on the set of singular points) *The map*

$$X_0 \longmapsto Q_{X_0}$$

is continuous on the set of singular points of the free boundary. Moreover the set of singular points is a closed set included in a C^1 $(n-1)$-dimensional submanifold.

More precise results can be found in [7].

Remark 2.5. *We have*

$$\Phi_{0,u^\epsilon}(r) = \Phi_{X_0,u}(\epsilon r)$$

and taking the limit we get

$$\Phi_{0,u^0}(r) = constant = \Phi_{X_0,u}(0^+)\,.$$

Using Theorem 2.1, it is possible to compute for a point X_0 of the free boundary

$$\Phi_{X_0,u}(0^+) = \begin{cases} \alpha_n & \text{for a regular point} \quad X_0 \\ 2\alpha_n & \text{for a singular point} \quad X_0 \end{cases} \tag{5}$$

where α_n is a constant which depends only on the dimension n.

3. Regular Points

The expected property for regular points is the following

Theorem 3.1. (L. A. Caffarelli [4, 5], **G. S. Weiss** [25], **Smoothness of the free boundary near a regular point)** *If X_0 is a regular point, then in some appropriate coordinates, the coincidence set is a subgraph $\{x_n \leq h(x_1, \dots, x_{n-1})\}$ in a neighborhood of X_0, where the function h is of class C^1.*

This result is completed by

Theorem 3.2. (D. Kinderlehrer, L. Nirenberg [16], **V. Isakov** [14], **Higher regularity of the free boundary)** *If for a solution of* (4), *the free boundary is locally C^1, this free boundary is analytic.*

3.1. Criteria for regular points

We will present an energetic criterion first and then a geometric criterion.

Using the monotonicity formula (Theorem 2.3) and the property (5), we get

Theorem 3.3. (G. S. Weiss [25], **Energetic criterion)** *If*

$$\Phi_{X_0,u}(r) < 2\alpha_n$$

for some r, then the point X_0 is a regular point.

To introduce the geometric criterion found by L. A. Caffarelli, we need the following

Definition 3.4. (Thickness of the coincidence set) *We define the thickness of the coincidence set $\{u = 0\}$ in a ball $B_r(X_0)$ by*

$$\delta_r(X_0) = \frac{1}{r} m.d. \left(\{u = 0\} \cap B_r(X_0)\right)$$

where the minimum diameter (m.d.) of $\{u = 0\} \cap B_r(X_0)$ is the infimum of the distances between pairs of parallel hyperplanes whose strip determined by them contains it.

Theorem 3.5. (L. A. Caffarelli [7], **Geometric criterion)** *For each $r > 0$, there exists a critical thickness $\sigma(r)$ with $\sigma(r) \to 0$ as $r \to 0$, such that if*

$$\delta_r(X_0) > \sigma(r)$$

for some point X_0 of the free boundary and for one radius $r > 0$, then the point X_0 is regular.

In fact the uniqueness of the blowup limit for regular points is a consequence of Theorem 3.1. The main difficulty is to avoid the free boundary rotating (slowly and slowly) around X_0. In the approach of G. S. Weiss, the proof of Theorem 3.1 is based on an energy power decay estimate on $\Phi_{X_0,u}(r) - \Phi_{X_0,u}(0^+) \geq 0$ which is enough to control the axis of the blowup limit.

The approach of L. A. Caffarelli to prove Theorem 3.1 is based on classical tools like the Maximum Principle and Harnack Inequality, used very cleverly to control the rotation of the axis.

3.2. A counterexample

Let us consider the slightly more general obstacle problem

$$\begin{cases} \Delta u = f \geq 1 & \text{on} \quad \{u > 0\} \cap \Omega \\ u \geq 0 & \text{on} \quad \Omega \\ u \in C^{1,1}(\overline{\Omega}) \end{cases} \tag{6}$$

where $f \in C^{\alpha}(\overline{\Omega})$; then Theorems 2.1 and 3.1 are still true. But we cannot go up to the regularity $f \in C^0$ as shown by the following counterexample

Theorem 3.6. (I. Blank [2], Example of non-uniqueness of the blowup limit) *In* $\mathbf{R}^2$, *there exists a continuous function* f, *a solution* u *to* (6) *and a point* X_0 *of the free boundary such that the blowup limit in* X_0 *is not unique. More precisely for each* $\nu \in \mathbf{S}^1$, *there exists a sequence* $\epsilon_k \to 0$ *such that*

$$u^{\epsilon_k}(X) \longrightarrow \frac{1}{2}(\max(\langle X, \nu\rangle, 0))^2 .$$

4. More Results in Two Dimensions

Let us recall

Theorem 4.1. (L. A. Caffarelli, N. M. Rivière [8], Regularity of the free boundary in two dimensions) *If* C *is a connected component of the interior of* $\{u = 0\}$ *in* $\mathbf{R}^2$, *then* ∂C *is analytic except in a finite number of points.*

Moreover these authors give a precise behaviour of the free boundary near a singularity in $\mathbf{R}^2$ in [9]. See also [24]. We now give more information on the number of singularities:

Theorem 4.2. (R. Monneau [18], Small components have at most two singular points) *If* C *is the connected component of the interior of* $\{u = 0\}$ *in* $\mathbf{R}^2$, *then there exists*

$$\rho = \rho(|D^2 u|_{L^\infty(\Omega)}, d(C, \partial\Omega)) > 0$$

such that

i) *If* $\text{diam}(C) < \rho$, *then* ∂C *has at most two singular points.*
ii) *If* $\text{diam}(C) \geq \rho$ *and* $X_1, X_2 \in \partial C$ *are singular points, then* $|X_2 - X_1| > \frac{\rho}{2}$.

5. Stability and Genericity

In this section we consider solutions u to the obstacle problem (4) such that $u =$ constant $= g$ on $\partial\Omega$. We will make perturbations varying the parameter $g > 0$. To this end we denote u by u_g to make explicit the dependence of the solution on g.

Theorem 5.1. (Stability of the free boundary) *If the free boundary* $\partial\{u_{g_0} = 0\}$ *is analytic for a particular value* g_0, *then* $\partial\{u_g = 0\}$ *is analytic for* g *in a neighborhood of* g_0.

This result was first proved by D. G. Schaeffer in [22] when $\partial\Omega \in C^\infty$, using the Nash–Moser inverse function theorem. Using the geometric criterion (Theorem 3.5) of L. A. Caffarelli, it is possible to see that Theorem 5.1 is a consequence of

Theorem 5.2. (L. A. Caffarelli [6], Measure stability) *If u_1, u_2 satisfy the same obstacle problem (4), then we can estimate the measure of the symmetric difference of the coincidence sets by*

$$|\{u_2 = 0\}\Delta\{u_1 = 0\}| \leq C|u_2 - u_1|^{\frac{1}{2}}_{L^\infty(\Omega)}$$

where $C = C(|D^2u_1|_{L^\infty(\Omega)}, |D^2u_2|_{L^\infty(\Omega)}, \Omega, n)$.

See also the book of J. F. Rodrigues [20] for detailed results on more general operators.

Let us recall

Conjecture 5.3. (D. G. Schaeffer [21] (1974)) *We conjecture that generically the weak solution of the obstacle problem that one obtains variationally is also a strong solution, by which we mean that the free boundary is a C^∞ manifold.*

The following result is a first answer in a simple case:

Theorem 5.4. (R. Monneau [18], Generic regularity of the free boundary) *In $\mathbf{R}^2$, for almost every constant $g > 0$, the free boundary $\partial\{u_g = 0\}$ is analytic.*

The question remains open in higher dimensions.

6. Convexity Property of The Coincidence Set

Theorem 6.1. (B. Kawohl [15], Convexity of the coincidence set) *Let $\Omega \subset \mathbf{R}^n$ be a smooth bounded domain. If Ω is convex, then for each constant $g > 0$, the coincidence set $\{u_g = 0\}$ is convex (and analytic).*

This result was first proved in the particular case $n = 2$ by A. Friedman and D. Phillips in [13]. Let us give an example of a generalization of this result in two dimensions

Theorem 6.2. (J. Dolbeault, R. Monneau [10], Convexity for nonlinear obstacle problems) *Let $\Omega \subset \mathbf{R}^2$ a smooth bounded domain. Let $u_g \in H^1(\Omega)$ minimizing*

$$\int_\Omega F(|\nabla u|^2) + G(u)$$

under the constraints $u \geq 0$ on Ω and $u =$ constant $= g$ on $\partial\Omega$.

We assume that $F, G \in C^2$ are nondecreasing convex functions, with $F'(0) > 0$, $G'(0) > 0$. If Ω is convex, then the coincidence set $\{u_g = 0\}$ is convex (and C^1).

Acknowledgements

I would like to thank Jose Francisco Rodrigues who invited me to present this note at the third European Congress of Mathematics (2000), and for helpful comments on the redaction of this note. Part of this work was done with the help of a NATO grant for a postdoctoral position at MIT (Cambridge, USA) and I would like to thank these two institutions. In particular I would like to thank David Jerison for stimulating discussions and his help for the references.

Most of all I would like to express my deepest gratitude to my advisors Alexis Bonnet and Henri Berestycki.

References

[1] H. W. ALT, L. A. CAFFARELLI AND A. FRIEDMAN, *Variational Problems with two phases and their free boundaries*, Trans. Amer. Math. Soc. **282**(2), (1984), 431–461.

[2] I. BLANK, *Sharp Results for the Regularity and Stability of the Free Boundary in the Obstacle Problem*, to appear in Indiana Univ. Math. J.

[3] H. BRÉZIS AND D. KINDERLEHRER, *The Smoothness of Solutions to Nonlinear Variational Inequalities*, Indiana Univ. Math. J. **23**(9), (1974), 831–844.

[4] L. A. CAFFARELLI, *Free boundary problem in highter dimensions*, Acta Math. **139**, (1977), 155–184.

[5] L. A. CAFFARELLI, *Compactness Methods in Free Boundary Problems*, Comm. Partial Differential Equations **5**(4), (1980), 427–448.

[6] L. A. CAFFARELLI, *A remark on the Hausdorff measure of a free boundary, and the convergence of coincidence sets*, Boll. Un. Mat. Ital. A **18**(5), (1981), 109–113.

[7] L. A. CAFFARELLI, *The Obstacle Problem revisited*, J. Fourier Anal. Appl. **4**, (1998), 383–402.

[8] L. A. CAFFARELLI AND N. M. RIVIÈRE, *Smoothness and Analyticity of Free Boundaries in Variational Inequalities*, Ann. Scuola Norm. Sup. Pisa, serie IV **3**, (1975), 289–310.

[9] L. A. CAFFARELLI AND N. M. RIVIÈRE, *Asymptotic behaviour of free boundaries at their singular points*, Ann. of Math. **106**, (1977), 309–317.

[10] J. DOLBEAULT AND R. MONNEAU, *Convexity Properties of the Free Boundary and Gradient Estimates for Quasi-linear Elliptic Equations*, to appear.

[11] J. FREHSE, *On the Regularity of the Solution of a Second Order Variational Inequality*, Boll. Un. Mat. Ital. B (7) **6**(4), (1972), 312–315.

[12] A. FRIEDMAN, *Variational Principles and Free Boundary Problems*, Pure and applied mathematics, ISSN 0079–8185, a Wiley-Interscience publication, (1982).

[13] A. FRIEDMAN AND D. PHILLIPS, *The free boundary of a semilinear elliptic equation*, Trans. Amer. Math. Soc. **282**, (1984), 153–182.

[14] V. ISAKOV, *Inverse theorems on the smoothness of potentials*, Differential Equations **11**, (1976), 50–57.

[15] B. KAWOHL, *When are solutions to nonlinear elliptic boundary value problems convex?*, Comm. Partial Differential Equations **10**, (1985), 1213–1225.

[16] D. KINDERLEHRER AND L. NIRENBERG, *Regularity in free boundary problems*, Ann. Scuola Norm. Sup. Pisa **4**, (1977), 373–391.

[17] D. KINDERLEHRER AND G. STAMPACCHIA, *An Introduction to Variational Inequalities and Their Applications*, Academic Press, New York, (1980).

[18] R. MONNEAU, *Problèmes de frontières libres, EDP elliptiques non linéaires et applications en combustion, supraconductivité et élasticité*, Doctoral Dissertation, Université Pierre et Marie Curie, Paris (1999).

[19] R. MONNEAU, *On the Number of Singularities for the Obstacle Problem in Two Dimensions*, to appear.

[20] J. F. RODRIGUES, *Obstacle Problems in Mathematical Physics*, North-Holland, (1987).

[21] D. G. SCHAEFFER, *An Example of Generic Regularity for a Non-Linear Elliptic Equation*, Arch. Rat. Mach. Anal. **57**, (1974), 134–141.

[22] D. G. SCHAEFFER, *A Stability Theorem for the Obstacle problem*, Advances in Math. **16**, (1975), 34–47.

[23] D. G. SCHAEFFER, *Some Examples of Singularities in a Free Boundary*, Ann. Scuola Norm. Sup. Pisa **4**(4), (1976), 131–144.

[24] D. G. SCHAEFFER, *One-sided Estimates for the Curvature of the Free Boundary in the Obstacle Problem*, Advances in Math. **24**, (1977), 78–98.

[25] G. S. WEISS, *A homogeneity improvement approach to the obstacle problem*, Invent. Math. **138**, (1999), 23–50.

Ecole Nationale des Ponts et Chaussées
CERMICS
6 et 8 avenue Blaise Pascal
Cité Descartes Champs-sur-Marne
77455 Marne-la-Vallée Cedex 2, France
E-mail address: monneau@cermics.enpc.fr

The Impact of Monotonicity Formulas in Regularity of Free Boundaries

Henrik Shahgholian

Abstract. In this note we give a survey on recent developments in the regularity of free boundaries of obstacle type in absence of the obstacle, giving rise to solutions that may change sign. The focus is on two techniques, the monotonicity formulas and global versus local analysis.

1. Introduction

For a bounded domain Ω in $\mathbb{R}^n$ $(n \geq 2)$ we assume that there exists a function u such that locally

$$(\Delta u - f)u = 0 \quad \text{in } B(x^0, r_0) \setminus \Omega, \quad x^0 \in \partial\Omega, \tag{1}$$

where $f > 0$ is Lipschitz, and the equation is satisfied in the sense of distributions.

Two main questions that come up immediately are the following:

Q1) How smooth is u across $\partial\Omega$?

Q2) How smooth is $\partial\Omega$ in a neighborhood of x^0?

The problem described above has its origin in inverse problems of potential theory, also known as harmonic continuations of potentials from the free space into the domain of integration. To explain this in more detail let U denote the Newtonian potential of Ω (bounded set) with constant density (i.e., the convolution of the fundamental solution with χ_Ω) and with $x^0 \in \partial\Omega$. Then suppose (this is not necessarily the case in general) there exists a harmonic function w in $B(x^0, r)$ (r small) such that $w = U$ in Ω^c (the complement of Ω); observe that U is harmonic in Ω^c. This property is referred to as harmonic continuation of potentials.

The reader familiar with elliptic theory can immediately see that Q1) can be partially answered. Indeed, by elliptic estimate u is $C^{1,\alpha}$ in $B(x^0, r_0/2)$ for $\alpha < 1$. A more elaborate estimate may also be shown using the potential representation. This yields the estimate

$$|\nabla u(x) - \nabla u(x^0)| \leq C|x - x^0| \log|x - x^0| \, .$$

The problem here is to get rid of the log-term.

Partially supported by the Swedish Natural Science Research Council.

In the case of obstacle problem, i.e., when $u \geq 0$ one may easily obtain that $u \in C^{1,1}(B(x^0, r_0/2))$. This can be done using Harnack's inequality or simple estimate based on the so-called "Schwarz potential" (the difference between the paraboloid $|x - x^0|^2/2n$ and the fundamental solution with source at x^0). We refer to [9] for some details in the latter case. For the application of the Harnack inequality we refer to [2].

Now the "no-sign-assumption" in problem (1) introduces a new and actually a very peculiar difficulty. To overcome this difficulty the authors in [8] have used the monotonicity formula developed by [1]. Unfortunately the technique works(ed) only for the case when $f \equiv 1$, since then $D_e u$, where D_e denotes partial derivative in direction e, will be a harmonic function and consequently $(D_e u)^{\pm}$ will be subharmonic functions with disjoint support. These are the exact conditions in the monotonicity formula.

2. The Use of the Monotonicity Formula

To discuss the approach of [8] in answering Q1)–Q2) we state the following monotonicity lemma.

Lemma 2.1. (See [1, Lemma 5.1]**)** *Let h_1, h_2 be two non-negative continuous subsolutions of $\Delta u = 0$ in $B(x^0, R)$ $(R > 0)$. Assume further that $h_1 h_2 = 0$ and that $h_1(x^0) = h_2(x^0) = 0$. Then the following function is monotone in r $(0 < r < R)$*

$$\varphi(r) = \frac{1}{r^4}\left(\int_{B(x^0,r)} \frac{|\nabla h_1|^2}{|x - x^0|^{n-2}}\right)\left(\int_{B(x^0,r)} \frac{|\nabla h_2|^2}{|x - x^0|^{n-2}}\right).$$

Now in the above lemma one replaces h_i with $(D_e u)^{\pm}$. Since, heuristically $\varphi(0, D_e u) \approx D_{ij}u(0)$ one expects to obtain a uniform bound. Let us briefly explain this, for more detail we refer to [8, Theorem I]. First let us define

$$S_r = \sup_{B_r} |u(x)|,$$

where B_r denotes the ball of radius r, centered at the origin. Here we assume that $x^0 \in \partial\Omega$ is the origin. Since the Laplacian is rotation and translation invariant we can make the same argument for any point of $\partial\Omega$ other than the origin. Next we expect to have

$$S_r \leq Cr^2, \quad \forall\, r < 1/2,$$

and for some constant C depending on the sup-norm of u and the space dimension only. Therefore we claim that there exists C such that

$$S_{r/2} \leq \max\left(\frac{S_r}{4}, Cr^2\right) \quad \forall\, r < 1/2. \tag{2}$$

If this is true then we are done using iteration. The idea of introducing (2) (rather than the more difficult approach in [8]) is due to N. Uraltseva; see [11]). Now we suppose, towards a contradiction, that (2) fails. Then there exists a sequence $\{r_j\}$

(and if we want to have uniformity for a class of solutions, also a sequence of solutions u_j to (1), but here for simplicity we do this for just one function) such that (2) fails. More exactly we have

$$S_{r_j/2} \geq \max\left(\frac{S_{r_j}}{4}, jr_j^2\right), \quad j = 1, 2, \dots . \tag{3}$$

Now the reader may verify that (3) along with the properties of u implies that

$$u_j(x) = \frac{u(r_j x)}{S_{r_j}}$$

is uniformly bounded on B_1 and

$$|\Delta u_j| \leq \frac{4}{j}.$$

Consequently a subsequence of $\{u_j\}$ converges uniformly to a harmonic function u_0 in the unit ball and $u_0(0) = |\nabla u_0|(0) = 0$. Moreover, from (3), it follows that

$$\max_{B_{1/2}} |u_0| = 1. \tag{4}$$

Next we use the monotonicity formula to obtain that

$$\varphi(1, D_e u_j) = \left(\frac{r_j^2}{S_{r_j}}\right)^4 \varphi(r_j, D_e u) \leq C' \left(\frac{r_j^2}{S_{r_j}}\right)^4 \to 0,$$

by (3).

In particular this implies that

$$\varphi(1, D_e u_0) = 0,$$

and consequently either of $(D_e u_0)^{\pm}$ is to be zero. In particular $D_e u_0$ is a harmonic function that does not change sign. Since also $|\nabla u_0(0)| = 0$ we must have $D_e u_0(0) = 0$ for all directions e and hence u_0 is constant. The constant is zero since $u_0(0) = 0$. This contradicts (4) and therefore (2) must be true. Now from the estimate in (2) one can further show that u is $C^{1,1}$ using classical elliptic regularity.

One thing that we have not been careful with is the convergence in $W^{2,2}$ in the monotonicity formula as $r_j \to 0$. However, the proof of this with all the details is given in [8, Theorem I]. For the general Lipschitz right-hand sides the above monotonicity formula is out of use. However, newly developed monotonicity formulas by L. Caffarelli, D. Jerison, and C. Kenig [7] will still help us to obtain the $C^{1,1}$ estimate for solutions of (1). Here we formulate their result.

Lemma 2.2. (See [7]) *Recall the assumptions in Lemma 2.1, and replace the subharmonicity assumption by the boundedness of the Laplacian of h_i, i.e., assume $|\Delta h_i| \leq 1$. Suppose moreover $|h_i(x)| \leq C|x|^\epsilon$ for some $\epsilon > 0$. Then, for $0 < r_1 \leq r_2 \leq R_0$,*

$$\varphi(r_1) \leq (1 + r_2^\epsilon)\varphi(r_2) + Cr_2^\epsilon .$$

The question that raises itself is: What is the best result in this direction?. Can we relax the Lipschitz condition on f and still obtain such a result. It is known that the regularity of f can be relaxed but has to be compensated somehow, e.g., it is known that if the complement of Ω is thick enough (in the sense of capacitary density) near the origin then we still obtain a quadratic growth for the function u; see [10]. Of course not a $C^{1,1}$ estimate, as is obvious. However, this is enough to blow up solutions and work with global solutions, a notion that is almost inevitable in the context of regularity of free boundaries. Let us explain this in more detail.

A global solution to problem (1) is a solution u of

$$(\Delta u - 1)u = 0 \quad \text{in } \mathbb{R}^n \setminus \Omega ,$$

with the extra assumption that

$$|u(x)| \leq (|x| + 1)^2 .$$

In other words a solution in the entire space. In [8, Theorem II] the authors prove that global solutions are either quadratic polynomials or non-negative convex functions. In particular if the complement of the set Ω has non-empty interior then global solutions are convex, and consequently $\mathbb{R}^n \setminus \Omega$ is convex.

Now this classification is the core of the analysis of the regularity of free boundaries. Indeed, if one assumes an a priori thickness condition on the complement of Ω near a free boundary point x^0, then a blow up of u at x^0 will result in a global convex solution u_0, with x^0 on the free boundary $\partial\Omega(u_0)$. Here we choose to explain the technique for just one function and not a whole class of functions. The technique for a class is very much similar but becomes more technical. We avoid this here.

The best way is to think of Ω as having a truncated cone outside it with vertex at x^0. Then one immediately verifies that the same property is inherited by the blow-ups of u, whatever the sequence of blow-up may be. The reader should notice that there might well be examples such that blow-up w.r.t. two different sequences $\{r_j\}$ and $\{t_j\}$ are different functions. One of the main difficulties in the theory is to prove that this is not the case, provided the free boundary does not develop cusp singularities.

Next having a convex global solution one needs to prove that these solutions have locally C^1 (actually analytic) boundary. This is done more easily, since convexity of the complement implies Lipschitz regularity of $\partial\Omega$. One may also easily see that the boundary is C^1 by just blowing up u at a boundary point x^0. Now a blow-up at a boundary point will give a homogeneous solution. Indeed we have

$$\varphi(s, D_e u_0) = \lim_{r\to 0} \varphi(s, D_e u_r) = \lim_{r\to 0} \varphi(sr, D_e u) = C_e , \tag{5}$$

for all $s > 0$.

The problem in (5) is that we cannot use a different sequence of functions u_j since then the limit C_e is not necessarily unique, and the problem becomes very involved in the case we consider a whole class of functions.

Now, going back to our simple situation of just one function, we have that global solutions with non-void $\mathbb{R}^n \setminus \overline{\Omega}$ are convex, and locally the free boundary is C^1. Now we use this idea for local solutions, by scaling the function u with u_{r_j} and assume that

$$B_{r_j} \setminus \Omega(u) \quad \text{is thick enough}. \tag{6}$$

Now u_{r_j} is a solution in B_{1/r_j} (almost a global solution, or which is known as approximate global solution) and by (6)

$$B_1 \setminus \Omega(u_{r_j}) \quad \text{is thick enough}; \tag{7}$$

the thickness property introduced by L. Caffarelli [4] is stable under scaling. Now (7), in conjunction with the classification of global solutions, imply that $\partial\Omega(u_{r_j})$ near the origin is almost C^1 (or flat). Now from here one wants to go further to prove that the boundary is actually C^1. One way of doing this, and this is the technique, developed by L. Caffarelli, is to apply the maximum principle to the function $w_j^e = Cr_jD_eu - u$ in $\Omega(u) \cap B_{r_j}$ to show that, for some e and small r_j, this function is non-negative in B_{r_j}, provided u_{r_j} is close enough to a global solution with $\partial\Omega(u_{r_j})$ flat enough in the unit ball; see [5] and [8] for the simple proof.

Once it is shown that w_j^e is nonnegative, we can integrate by parts to obtain $u \geq 0$ in $B_{r_j/4}$. Actually one can show that $w_j^e \geq 0$ for a whole set of vectors e, i.e., in B_{r_j}

$$w_j^e \geq 0 \quad \forall \quad e \in K_j := \{\nu \colon \nu \cdot e_0 \geq \frac{1}{j}\},$$

where e_0 depends on u. From here one obtains the Lipschitz regularity of the free boundary without much effort. The next step is to show that the Lipschitz norm gets better, if we choose smaller balls. One shows that for each $j > 0$ there exists r_j such that $w_j^e \geq 0$ in B_{r_j} for all $e \in K_j$. This gives the C^1-regularity of $\partial\Omega$.

Finally, to illuminate the importance of the monotonicity formula, we give an application of it in classification of global solutions, see [8, Theorem II]. Cf. also [11, Theorem B]. So let us assume u is a global solution, i.e., a solution to (1) in entire $\mathbb{R}^n$. To give a simplified example, suppose also that there exists a sequence $R_j \to \infty$ such that

$$\frac{\operatorname{vol}(B_{R_j} \setminus \Omega)}{\operatorname{vol}(B_{R_j})} \geq \epsilon_0 > 0, \quad \forall\, j\,.$$

The latter means that the set $\mathbb{R}^n \setminus \Omega$ has positive upper Lebesgue density at the infinity point. Now define the blow-up sequence u_{R_j} and apply the monotonicity formula as in (5), to obtain

$$\varphi(s, D_eu_0) = \lim_{j\to 0} \varphi(s, D_eu_{R_j}) = \lim_{j\to 0} \varphi(sR_j, D_eu) = C_e\,, \tag{8}$$

for all $s > 0$. Now according to the monotonicity formula (actually a stronger version of it; see [6]) $\varphi(s, D_eu_0)$ is either identically zero or strictly increasing, or the sets $\{(D_eu)^{\pm} > 0\} \cap B(0, s)$ are half spherical caps up to zero area. Now the latter case is not possible due to condition (8), and therefore the only possibility is that $\varphi(s, D_e u_0) \equiv 0$. Now this in turn implies that $\varphi(r, D_e u) \leq \varphi(\infty, D_e u) = \varphi(1, D_e u_0) = C_e = 0$. We thus arrive at the fact that at least one of the functions $(D_e u)^{\pm}$ is identically zero. Hence $D_e u > 0$ (say). The strict inequality is due to the maximum principle, since $D_e u$ is harmonic (we look at the case $f \equiv 1$ for global solutions). From here it is not hard to show that u is one-dimensional and consequently $u = (\max(x_1, 0))^2/2$ in some rotated system.

References

[1] H. W. Alt, L. A. Caffarelli and A. Friedman, *Variational problems with two phases and their free boundaries,* Trans. AMS, **282** (1984), 431–461.

[2] I. Blank, *Sharp results for the regularity and stability of the free boundary in the obstacle problem,* Manuscript.

[3] L. A. Caffarelli, *The regularity of free boundaries in higher dimension,* Acta Math., **139** (1977), 155–184.

[4] L. A. Caffarelli, *Compactness methods in free boundary problems,* Comm. P.D.E., **5** (1980), 427–448.

[5] L. A. Caffarelli, *The obstacle problem revisited,* Journal of Harmonic Analysis and Applications, J. Fourier Anal. Appl. **4** (1998), no. 4–5, 383–402.

[6] L. A. Caffarelli and C. E. Kenig, *Gradient estimates for variable coefficient parabolic equations and singular perturbation problems,* Amer. J. Math., **120(2)** (1998), 391–439.

[7] L. A. Caffarelli, D. Jerison and C. E. Kenig, *Manuscript.*

[8] L. A. Caffarelli, L. Karp and H. Shahgholian, *Regularity of a free boundary with application to the Pompeiu problem,* Ann. Math., **152** (2000), 269–292.

[9] L. Karp and H. Shahgholian, *Regularity of a free boundary problem,* J. Geometric Analysis **9** (1999), no. 4, 653–669.

[10] L. Karp and H. Shahgholian, *On the optimal growth of functions with bounded Laplacian,* Electron. J. Differential Equations, **2000(3)** (2000), 9 pp. (electronic).

[11] H. Shahgholian and N. Uraltseva, *Regularity properties of a free boundary near contact points with the fixed boundary,* Manuscript.

Department of Mathematics
Royal Institute of Technology
100 44 Stockholm, Sweden
E-mail address: henriks@math.kth.se

A Free Boundary Problem: Contributions from Modern Analysis

José Miguel Urbano

Abstract. We exemplify the role of Free Boundary Problems as an important source of ideas in modern analysis. With the help of a model problem we illustrate the use of analytical, algebraic and geometrical techniques obtaining uniqueness of weak solutions via the use of entropy inequalities, existence through nonlinear semigroup theory, and regularity using a method, called intrinsic scaling, based on interpreting a partial differential equation in a geometry dictated by its own structure.

1. Introduction

In this contribution to the mini-symposium on Free Boundary Problems (FBPs) we use a model problem to support the idea that this is an important topic in modern analysis, both because of the mathematical questions it raises and of the variety of techniques it employs to produce interesting answers.

In a brief definition we can say that an FBP is a boundary value problem defined in a domain that is not given *a priori*, thus being part of the unknown. This models a feature that is common to many physical phenomena, and it comes as no surprise that the main motivation to study FBPs lies in absolutely practical matters. But our concern here is different and somehow nonstandard. We will not use the physical motivations and the successful practical achievements to justify the study of FBPs; we will try to illustrate their strength and beauty as a modern topic in mathematical analysis. Although their origin can be traced back to the 19th century, FBPs flourished as mathematical problems only in the late 1960s and 1970s, mainly due to the systematic approach to existence provided by the theory of variational inequalities (cf. [3]). Since then the joint effort of an enormous number of mathematicians shed light into many difficult questions and opened up several new directions of research in analysis. We present here, in a descriptive form, a model problem that shares some of the aforementioned characteristics and address three main issues to illustrate it (the existence of a weak solution, its uniqueness and regularity) using three different methods in analysis. The existence is obtained from nonlinear semigroup theory, one of the most successful modern attempts of *algebrizing* analysis. The uniqueness is based on a typically *analytical* approach to PDEs, consisting of the establishment of integral inequalities related to the concept of entropy solution and using them as tools to prove a contraction property. The regularity, namely the continuity of the weak solution, follows from a technique, called intrinsic scaling, that provides an adequate setting to understand what we can heuristically summarize in the sentence (cf. [7]): "the equation behaves in its own *geometry* like the heat equation". The other basic question commonly attached to an FBP, the regularity of the free boundary, remains in this case a challenging open problem.

1.1. The model problem

The model problem we have in mind describes a phase transition at fixed temperature for a homogeneous material that diffuses in a nonlinear way. The free boundary is the interface between the two phases and we assume it moves according to a Stefan type condition (cf. [11]). Consider the maximal monotone graph γ defined by

$$\gamma(s) = s + \lambda H(s) \ , \qquad H(s) = \begin{cases} 0 & \text{if} \quad s < 0 \\ [0,1] & \text{if} \quad s = 0 \\ 1 & \text{if} \quad s > 0 \end{cases} ,$$

where λ stands for the latent heat of the phase transition, i.e. the amount of energy it requires to take place, and H is the Heaviside graph. The two main physical features of the phenomenon, the phase transition and the nonlinear diffusion, are captured by the PDE

$$\partial_t v - \Delta_p u - \nabla \cdot V(u) = f \ , \quad v \in \gamma(u) \tag{1}$$

through the maximal monotone graph γ and the nonlinear degenerate operator $\Delta_p u := \nabla \cdot \left(|\nabla u|^{p-2} \nabla u\right)$, the so-called p-Laplacian. We denote with u the temperature and v the enthalpy, with $V(u)$ a temperature dependent velocity field and with f a reaction term. For a derivation of (1) from the principles of continuum mechanics under suitable constitutive assumptions see, e.g., [12]. We attach to the equation a homogeneous Dirichlet boundary condition at the fixed boundary of the space-time domain $Q = \Omega \times (0,T)$ and an initial condition

$$u = 0 \ \text{ on } \ \partial\Omega \times (0,T) \qquad \text{and} \qquad v(0) = v_0 \ \text{ in } \ \Omega \tag{2}$$

and formulate the problem weakly performing the usual multiplication by test functions and formal integration by parts. This leads to the following

Definition 1.1. *A weak solution of* (1)–(2) *is a pair of functions*

$$(u,v) \in L^p(0,T; W_0^{1,p}(\Omega)) \times L^\infty(Q)$$

such that $v \in \gamma(u)$, a.e. in Q, and, for all testing functions φ,

$$-\iint_Q v \, \partial_t \varphi + \iint_Q \left(|\nabla u|^{p-2} \nabla u + V(u)\right) \cdot \nabla \varphi = \iint_Q f \varphi + \int_\Omega v_0 \, \varphi(0) .$$

Remark 1.2. *Observe that since $u = \gamma^{-1}(v)$ and γ^{-1} is continuous, we have,* a fortiori, *$u \in L^\infty(Q)$.*

Remark 1.3. *Note that in this weak or variational formulation the free boundary is (hidden or) implicit in the nonlinearity* $\gamma(u)$. *The price to be paid is the presence of the singularity "* $\gamma'(0) = \infty$ *".*

We introduce some notation and the main assumptions concerning the data of the problem. The set $\Omega \in \mathbf{R}^n$ is a domain with a smooth boundary and $T > 0$ is a real number. The inverse of γ, which is a non-decreasing continuous function in $\mathbf{R}$, is denoted by $\varphi \equiv \gamma^{-1}$; we have $\varphi(0) = 0$. We denote with Sign_0^+ the discontinuous function corresponding to the choice $\mathrm{Sign}_0^+(0) = 0$. Concerning the convective term, that we took independent of the jumps of γ, i.e., depending only on the temperature, we assume that $V : \mathbf{R} \longrightarrow \mathbf{R}^N$ is Lipschitz, with $V(0) = \mathbf{0}$. The initial data is taken as $\gamma(u_0) \ni v_0 \in L^\infty(\Omega)$ and such that

$$\exists\, M > 0 \; : \; \|u_0\|_{L^\infty(\Omega)} \leq M\,. \tag{3}$$

2. Uniqueness Via Entropy Inequalities

The basic idea in the proof of the uniqueness is to show that weak solutions satisfy certain integral inequalities, called *entropy inequalities* due to the fact that they are inspired in the notion of entropy solution introduced by S. Kruzkov in the context of quasilinear first-order equations. The word entropy had its roots in gas dynamics since the inequalities model, in that setting, the requirement of increasing entropy for shock waves. With the entropy inequalities at hand it is relatively easy to obtain a contraction property in L^1 and the uniqueness as an obvious corollary. This section is based on the article [9]. We start with

Definition 2.1. *An entropy solution of* (1)–(2) *is a weak solution that additionally satisfies the (entropy) inequalities*

$$\iint_Q \mathrm{Sign}_0^+(v-k)\Big\{(v-k)\,\xi_t - \big(|\nabla u|^{p-2}\nabla u + V(u)\big)\cdot\nabla\xi\Big\}$$

$$\geq -\int_\Omega (v_0-k)^+\,\xi(0) - \iint_Q \mathrm{Sign}_0^+(v-k)\, f\,\xi \tag{4}$$

for any $(k,\xi) \in \mathbf{R} \times \Big[L^p(0,T;W_0^{1,p}(\Omega)) \cap W^{1,1}(0,T;L^\infty(\Omega))\Big]$ *such that* $\xi \geq 0$, $\xi(T) = 0$; *and any* $(k,\xi) \in \mathbf{R}_0^+ \times \Big[L^p(0,T;W^{1,p}(\Omega)) \cap W^{1,1}(0,T;L^\infty(\Omega))\Big]$ *such that* $\xi \geq 0$, $\xi(T) = 0$.

The main step in the proof of the uniqueness is

Proposition 2.2. *Every weak solution of* (1)–(2) *is an entropy solution.*

Remark 2.3. *The entropy inequalities can basically be obtained by choosing as a test function in the definition of a weak solution the Steklov average of (an extension of) the function*

$$\Phi = H_\epsilon\Big(\varphi(v) - \varphi(k)\Big)\,\xi\,, \tag{5}$$

where H_ϵ is the approximation of the Heaviside graph defined by

$$H_\epsilon(r) = \min\left(\frac{r^+}{\varepsilon}, 1\right).$$

The main issue concerning this choice is whether Φ vanishes on the boundary of the domain or not. It is clear that if ξ vanishes on the boundary, the same happens with Φ, but if that is not the case, then we must have $H_\epsilon(-\varphi(k)) = 0$, i.e. $\varphi(k) \geq 0$, which follows if and only if $k \geq 0$. This explains a posteriori *the reason for choosing k and ξ in the above classes.*

We give now a flavor of how the proof of **Proposition 2.2** carries through. We first derive the entropy inequalities for constants k such that $\varphi(k) \notin [0, \lambda]$, which is, say, the favorable case. Then we establish a "filling" lemma which says that if the entropy inequalities are valid for $k = 0$ and $k = \lambda$, then they are also valid for any $k \in [0, \lambda]$, and we are left to prove the inequalities at the extreme points of this interval. These are relatively easy to handle if we can "approximate" the extreme points with k's for which the inequalities already hold; this is clearly possible in the case $(k, \xi) \in \mathbf{R} \times \left[L^p(0, T; W_0^{1,p}(\Omega)) \cap W^{1,1}(0, T; L^\infty(\Omega))\right]$. But when ξ does not vanish on the boundary we must choose $k \geq 0$ and we have problems since we cannot approximate $k = 0$ from the left; this also prevents us from using the filling lemma to cover the case $k \in (0, 1)$ since it requires the entropy inequality at $k = 0$. The key idea is then to deal with functions z (instead of constants k) such that $\varphi(z)$ vanish on the boundary $\partial\Omega \times (0, T)$, and to establish more general entropy inequalities for this case. Then, with the help of a strong maximum principle, we are able to choose a convenient sequence z_n such that $z_n < 0$ and $z_n \to 0$, thus approximating $k = 0$ in this way. Now z_n may be negative since it is the fact that $\varphi(z_n)$ vanishes on the boundary that guarantees that (5) is a good test function. From **Proposition 2.2** we obtain

Theorem 2.4. *Let $f_1, f_2 \in L^1(Q)$ and $v_{01}, v_{02} \in L^1(\Omega)$. If v_i $(i = 1, 2)$ are the corresponding weak solutions of* (1)–(2), *then, for a.e. $t \in (0, T)$,*

$$\int_\Omega \Big(v_1(t) - v_2(t)\Big)^+ \leq \int_\Omega (v_{01} - v_{02})^+ + \int_0^t \int_\Omega (f_1 - f_2)^+.$$

The proof is outlined in [9] and is a simple modification to the nonlinear case of a result in [4]. It makes essential use of the entropy inequalities to derive the contraction property. The uniqueness is an obvious corollary.

Remark 2.5. *It is very important for the uniqueness result that the convective term is independent of the jumps of γ, i.e., that it depends only on the temperature and not on the enthalpy. We now present a counterexample to the uniqueness in this more general situation. Consider the one-dimensional problem in $(t, x) \in Q = (0, 1) \times \Omega$, $\Omega = (0, 1)$*

$$v_t = u_{xx} + v_x \qquad u_{|\partial\Omega} = 0 \qquad v(0) = 1 .$$

A weak solution is a pair $v \in L^\infty(Q)$, $u \in L^2(0,1;H_0^1(\Omega))$, *such that* $v \in \gamma(u)$, *and solving the equation in the sense of distributions:*

$$-\int_0^1\int_0^1 v\,\xi_t + \int_0^1\int_0^1 u_x\,\xi_x + \int_0^1\int_0^1 v\,\xi_x = \int_0^1 \xi(0)\ , \quad \forall \xi \in \mathcal{T}\,.$$

It is obvious that the pair $(v_1,u_1) \equiv \big(1,\gamma^{-1}(1)\big)$ *is a weak solution. Let us construct a different one; take the* C^1 *function* F *defined in* $[0,2]$ *by*

$$F(s) = \begin{cases} 1 & \text{if } \ 0 \le s \le 1 \\ 1-(s-1)^2 & \text{if } \ 1 < s \le 2 \end{cases}$$

and put $v_2(t,x) = F(t+x)$. *We have* $(v_2)_t(t,x) = F'(t+x) = (v_2)_x(t,x)$ *and* $v_2(0,x) = F(x) = 1$ *because* $x \in (0,1)$. *Since* $0 \le F \le 1$, *we conclude that* $\big(v_2,\gamma^{-1}(v_2)\big)$ *is also a weak solution and it is of course different from* (v_1,u_1). *So there is no uniqueness for this problem.*

3. Existence Via Nonlinear Semigroup Theory

In this section we denote the temperature with $\varphi(v)$ instead of u. Consider the stationary problem

$$\begin{cases} v \in L^\infty(\Omega), \quad \varphi(v) \in W_0^{1,p}(\Omega) \\ v - \Delta_p\,\varphi(v) - \nabla\cdot V(\varphi(v)) = h \quad \text{in} \quad \mathcal{D}'(\Omega)\ . \end{cases} \tag{6}$$

In [9] we prove

Proposition 3.1. *Let* $h \in L^\infty(\Omega)$. *Then there exists a unique weak solution* v *of* (6). *Moreover* $\|v\|_\infty \le \|h\|_\infty$ *and, for any* $h_i \in L^\infty(\Omega)$, $i = 1,2$, *and* v_i *the corresponding weak solutions,*

$$\left\|(v_1-v_2)^+\right\|_1 \le \left\|(h_1-h_2)^+\right\|_1\ .$$

Now define in $L^1(\Omega)$ an operator A by

$$Az = -\Delta_p\,\varphi(z) - \nabla\cdot V(\varphi(z)) \ \text{ in } \mathcal{D}'(\Omega)$$

with domain

$$D(A) = \left\{ z \in L^\infty(\Omega) \ : \ \varphi(z) \in W_0^{1,p}(\Omega) \ \text{ and } \ Az \in L^1(\Omega) \right\}.$$

It follows from **Proposition 3.1** that A is T-accretive and that

$$R(I+\lambda A) \supseteq L^\infty(\Omega)\ , \quad \text{for all } \lambda > 0\,.$$

Moreover, the domain $D(A)$ of the operator A can be shown to be dense in $L^1(\Omega)$ (cf. [2]), i.e.,

$$\overline{D(A)}^{L^1(\Omega)} = L^1(\Omega)\,.$$

We can now use the general theory of evolution equations (cf. [5]) to obtain, for any $v_0 \in L^1(\Omega)$ and $f \in L^1(Q)$, a unique mild solution of

$$v \in C([0,T);L^1(\Omega)) \quad : \quad \frac{dv}{dt} + Av = f \quad \text{on} \quad (0,T)\,, \qquad v(0) = v_0\,.$$

Finally, for data in L^∞, we can show (cf. [9, Prop. 4]) that this mild solution is also a weak solution of (1)–(2), thus obtaining

Theorem 3.2. *Given $v_0 \in L^\infty(\Omega)$ and $f \in L^\infty(Q)$ there exists at least one weak solution $(\varphi(v), v)$ of (1)–(2).*

Remark 3.3. *For data in L^1 we are not able to prove that the mild solution is a weak solution and so there is no existence result in this case. It remains the possibility that the mild solution coincides with another type of solution, maybe the solution in the renormalized sense. This topic is to be investigated in the future.*

Remark 3.4. *The existence was treated in [12], in the more general setting of a convective term depending on the enthalpy v, using a regularization method and a* priori *estimates.*

4. Regularity Via Intrinsic Scaling

We consider finally the problem of the regularity of the weak solution, showing that the temperature u is in fact a continuous function. We only mention here the interior continuity obtained in [13], although the results hold up to the Dirichlet boundary (cf. [14]). Also, for the sake of simplicity of the arguments, the analysis is performed with the restrictions $V \equiv 0$ and $f \equiv 0$ but the techniques employed can easily be adapted to cover the general case. A more crucial restriction is the assumption, in this section, that $p > 2$.

The proof consists in showing that a sequence of uniformly bounded approximate solutions is equicontinuous, i.e., in deriving, at least implicitly, a modulus of continuity that is independent of the approximation. The approximated problem is obtained by regularization of the maximal monotone graph γ. Let $0 < \epsilon \ll 1$ and consider the function

$$\gamma_\epsilon(s) = s + \lambda H_\epsilon(s)\,,$$

where H_ϵ is a C^∞-approximation of the Heaviside function, such that

$$H_\epsilon(s) = 0 \quad \text{if} \quad s \le 0\;, \qquad H_\epsilon(s) = 1 \quad \text{if} \quad s \ge \epsilon\,,$$

$H'_\epsilon \ge 0$ and $H_\epsilon \to \mathrm{Sign}_0^+$ uniformly in the compact subsets of $\mathbf{R} \setminus \{0\}$, as $\epsilon \to 0$. The function γ_ϵ is bi-Lipschitz and satisfies

$$1 \le \gamma'_\epsilon(s) \le 1 + \lambda L_\epsilon\;, \quad s \in \mathbf{R}\,, \tag{7}$$

with $L_\epsilon \equiv \mathcal{O}(\frac{1}{\epsilon})$ being the Lipschitz constant of H_ϵ. Taking also a uniformly bounded sequence of functions $u_{0\epsilon}$ that appropriately approximates the initial data, the approximated problem is defined as follows.

Definition 4.1. *An approximate solution of* (1)–(2) *is a function*

$$u_\epsilon \in H^1(0,T;L^2(\Omega)) \cap L^\infty(0,T;W_0^{1,p}(\Omega)) \cap L^\infty(Q)$$

such that, for all testing functions φ*,*

$$-\iint_Q \gamma_\epsilon(u_\epsilon)\,\partial_t\varphi + \iint_Q |\nabla u_\epsilon|^{p-2}\nabla u_\epsilon \cdot \nabla\varphi = \int_\Omega \gamma_\epsilon(u_{0\epsilon})\,\varphi(0)\,. \tag{8}$$

For each $\epsilon > 0$ this problem has a unique solution u_ϵ that satisfies a uniform estimate in L^∞ ($\|u_\epsilon\|_\infty \le M$; cf. (3)) and the sequence $(u_\epsilon, \gamma_\epsilon(u_\epsilon))$ converges to the solution of the original problem as $\epsilon \to 0$ (cf. [12]). It is also clear, from the available theory (cf. [7]), that the solution u_ϵ of the approximated problem is Hölder continuous.

The proof of the equicontinuity of (u_ϵ) is based on certain uniform local estimates of energy and logarithmic type. We will mention here only the energy estimates since that is enough to illustrate the main difficulties involved and to give a clear idea of the essential parts of the proof. Consider a cylinder in Q

$$(x_0,t_0) + Q(\tau,\rho) := K_\rho(x_0) \times (t_0 - \tau, t_0)$$

and let $0 \le \zeta \le 1$ be a piecewise smooth cutoff function in $(x_0,t_0) + Q(\tau,\rho)$ such that

$$|\nabla\zeta| < \infty \quad \text{and} \quad \zeta(x,t) = 0\,,\ \ x \notin K_\rho(x_0)\,. \tag{9}$$

For the sake of simplicity and without loss of generality, we will be restricted to cylinders that are centered at the origin $(0,0)$, the changes being obvious in the case the center is a point (x_0,t_0). The energy estimates are obtained for the truncated functions $(u_\epsilon - k)_\pm$ choosing $\varphi = \pm(u_\epsilon - k)_\pm\zeta^p$ in (8). We mention only the negative case.

Proposition 4.2. *Let* u_ϵ *be a solution of the approximated problem and* $k < M$*. There exists a constant* $C > 0$*, independent of* ϵ*, such that for every cylinder* $Q(\tau,\rho) \subset Q$*,*

$$\sup_{-\tau<t<0} \int_{K_\rho\times\{t\}} (u_\epsilon - k)_-^2\zeta^p + \int_{-\tau}^0\int_{K_\rho} |\nabla(u_\epsilon - k)_-\zeta|^p$$

$$\le C\int_{-\tau}^0\int_{K_\rho}(u_\epsilon - k)_-^p|\nabla\zeta|^p + C\int_{K_\rho\times\{-\tau\}}(u_\epsilon - k)_-\zeta^p + C\int_{-\tau}^0\int_{K_\rho}(u_\epsilon - k)_-\zeta^{p-1}\partial_t\zeta\,.$$

Once these estimates are obtained, the problem becomes a problem in analysis and we can forget the PDE; they are essential to set forward an iterative argument consisting of showing that, for every point $(x_0,t_0) \in Q$, we can find a sequence of nested and shrinking cylinders $(x_0,t_0) + Q(\tau_n,\rho_n)$, such that, as the cylinders shrink to the point, the essential oscillation of each function θ_ϵ in the cylinders converges to zero; and this in a way that is qualitatively independent of ϵ. The iterative argument was introduced for strongly elliptic equations by DeGiorgi in [6] and later adapted by the Russian school to the parabolic case (cf. [10]). But it was essential in the argument that the equation was nondegenerate so that the integral norms appearing in the energy estimates were homogeneous.

This is not the case in **Proposition 4.2**: the presence of the powers p and 1 jeopardizes the homogeneity in the energy estimates and the recursive process (leading to the conclusion sought) itself. The power p clearly comes from the p-Laplacian and we can say that the nonlinear diffusion at the physical level produces a degeneracy at the PDE level which in turn leaves its trace at the analytical level in the form of such a power in the integral norms. The power 1 is, in the same spirit, the trace of the phase transition and the singularity in the PDE. Since this is not so obvious let us reproduce the part of the estimate responsible for the appearance of the power 1, which is the one that involves the time derivative. The crux of the matter is to estimate uniformly the regularization of the maximal monotone graph:

$$-\int_{-\tau}^{t}\int_{K_\rho} \partial_t\left[\gamma_\epsilon(u_\epsilon)\right]\left((u_\epsilon-k)_-\zeta^p\right) = \int_{K_\rho}\int_{-\tau}^{t}\partial_t\left(\int_0^{(u_\epsilon-k)_-}\gamma_\epsilon'(k-s)s\,\mathrm{d}s\right)\zeta^p$$

$$\geq \frac{1}{2}\int_{K_\rho\times\{t\}}(u_\epsilon-k)_-^2\zeta^p - C\int_{K_\rho\times\{-\tau\}}(u_\epsilon-k)_-\zeta^p - C\int_{-\tau}^{t}\int_{K_\rho}(u_\epsilon-k)_-\zeta^{p-1}\partial_t\zeta\,,$$

where C is a constant depending only on p, λ and M, the uniform bound in L^∞ for u_ϵ. The inequality is justified, recalling (7), by

$$\int_0^{(u_\epsilon-k)_-}\gamma_\epsilon'(k-s)s\,\mathrm{d}s \geq \int_0^{(u_\epsilon-k)_-} s\,\mathrm{d}s = \frac{1}{2}(u_\epsilon-k)_-^2$$

and

$$\begin{aligned}\int_0^{(u_\epsilon-k)_-}\gamma_\epsilon'(k-s)s\,\mathrm{d}s &\leq (u_\epsilon-k)_-\int_0^{(u_\epsilon-k)_-}\gamma_\epsilon'(k-s)\,\mathrm{d}s\\ &= (u_\epsilon-k)_-\Big[\gamma_\epsilon(k)-\gamma_\epsilon(u_\epsilon)\Big] \leq 2(M+\lambda)\,(u_\epsilon-k)_-\,.\end{aligned}$$

The key idea to overcome the difficulty presented by the inhomogeneity was introduced by DiDenedetto (cf. [7] and [8] for an account of the theory) in the nonsingular case ($\gamma(s)\equiv s$) and consists essentially in looking at the equation in its own geometry, i.e., in a geometry dictated by its degenerate structure. This amounts to rescaling the standard parabolic cylinders by a factor depending on the oscillation of the solution. This procedure, which can be called accommodation of the degeneracy, allows one to recover the homogeneity in the energy estimates written over these rescaled cylinders and to carry on with the proof. We can say heuristically that the equation behaves in its own geometry like the heat equation. In the present singular-degenerate case, no rescaling permits the compatibility of the three powers involved so we use the geometry of the nonsingular case to deal with the degeneracy and pay the price of a dependence on the oscillation in the various constants that are determined during the proof. Owing to this fact we are no longer able to exhibit a modulus of continuity but only to define it implicitly independently of the regularization. This is enough to obtain the equicontinuity of the approximations and a continuous solution for the original problem; but the Hölder continuity, that holds in the nonsingular case, is lost. Let us briefly describe the procedure. From now on we will drop the ϵ in u_ϵ. Consider a point $(x_0, t_0) \in Q$ and, by translation and to simplify, assume $(x_0, t_0) = (0, 0)$. Consider $R > 0$ such that $Q(R^{p-1}, 2R) \subset Q$, define

$$\omega := \text{ess osc}_{Q(R^{p-1},2R)} \ u$$

and construct the cylinder

$$Q(a_0 R^p, R) , \qquad \text{with} \quad a_0 = \left(\frac{\omega}{A}\right)^{2-p}$$

where the number A is to be chosen of the form

$$A = 2^{s_3} , \qquad \text{with} \qquad s_3 > C\,\omega^{-\alpha} , \quad \alpha = \frac{2(p+1)(N+p)}{p} . \tag{10}$$

Note that for $p = 2$, i.e. in the nondegenerate case, $a_0 = 1$ and these are the standard parabolic cylinders. We will assume, without loss of generality, that $\omega < 1$ and also that

$$\frac{1}{a_0} = \left(\frac{\omega}{A}\right)^{p-2} > R$$

which implies that $Q(a_0 R^p, R) \subset Q(R^{p-1}, 2R)$ and the relation

$$\text{ess osc}_{Q(a_0 R^p,R)} \ u \leq \omega \tag{11}$$

which will be the starting point of the iteration process. We now consider sub-cylinders of $Q(a_0 R^p, R)$ of the form

$$(0, t^*) + Q(dR^p, R) , \qquad \text{with} \quad d = \left(\frac{\omega}{2}\right)^{2-p}$$

that are contained in $Q(a_0 R^p, R)$, since $A > 2$ and if

$$\left(2^{p-2} - A^{p-2}\right) \frac{R^p}{\omega^{p-2}} < t^* < 0 .$$

The proof follows from the analysis of two complementary cases and the achievement of the same type of conclusion for both. We can briefly describe them in the following way: in the first case, we assume that there is a cylinder of the type $(0, t^*) + Q(dR^p, R)$ where u is essentially away from its infimum. We show that going down to a smaller cylinder the oscillation decreases by a small factor that we can exhibit and that depends on the oscillation. If that cylinder cannot be found, then u is essentially away from its supremum in all cylinders of that type, and we can add up this information to reach the same conclusion as in the previous case. We summarize:

Lemma 4.3. *There exists a constant $\sigma = \sigma(\omega) \in (0, 1)$, that depends only on the data and ω, such that*

$$\text{ess osc}_{Q\left(d(\frac{R}{8})^p, \frac{R}{8}\right)} \ u \ \leq \ \sigma(\omega)\,\omega .$$

Note the dependence of σ on the oscillation which is responsible for the loss of the Hölder continuity. Still, with this result at hand, we can define recursively two sequences of positive real numbers $(\omega_n)_n$ and $(R_n)_n$ and obtain

Proposition 4.4. *The sequences $(\omega_n)_n$ and $(R_n)_n$ are decreasing sequences that converge to zero. Moreover, for all $n = 0, 1, 2, \ldots$,*

$$Q_{n+1} \subset Q_n \qquad \text{and} \qquad \operatorname{ess\,osc}_{Q_n} u \leq \omega_n\,.$$

An immediate consequence is that we can choose a continuous representative for each u_ϵ out of its equivalence class and implicitly obtain an interior modulus of continuity, i.e., for each $K \subset Q$, a continuous and nondecreasing function $F_K \colon \mathbf{R}^+ \to \mathbf{R}^+$, depending only on the data and K, such that

$$\left| \, u_\epsilon(x,t) - u_\epsilon(x',t') \, \right| \leq F_K\Big(|x - x'| + |t - t'|^{\frac{1}{p}} \Big)\,.$$

Since this modulus of continuity is independent of ϵ, we find that u is locally continuous as a consequence of Ascoli's theorem.

Theorem 4.5. *The function u in the definition of the weak solution for* (1)–(2) *is locally continuous.*

Acknowledgements

Research partially supported by CMUC/FCT.

References

[1] H. W. Alt and S. Luckhaus, *Quasilinear elliptic-parabolic differential equations*, Math. Z., **183** (1983), 311–341.

[2] Ph. Bénilan and P. Wittbold, *On mild and weak solutions of elliptic-parabolic problems*, Adv. Diff. Eqs., **1** (1996), 1053–1073.

[3] H. Brézis and F. Browder, *Partial differential equations in the 20th century*, Adv. Math., **135** (1998), no. 1, 76–144.

[4] J. Carrillo, *Entropy solutions for nonlinear degenerate problems*, Arch. Rat. Mech. Anal., **147** (1999), 269–361.

[5] M. Crandall and T. Liggett, *Generation of semi-groups of nonlinear transformations in general Banach spaces*, Amer. J. Math., **93** (1971), 265–298.

[6] E. DeGiorgi, *Sulla differenziabilitá e l'analiticitá delle estremali degli integrali multipli regolari*, Mem. Accad. Sci. Torino, Cl. Sci. Fis. Mat. Nat., Ser. 3, **3** (1957), 25–43.

[7] E. DiBenedetto, *Degenerate Parabolic Equations*, Springer-Verlag, New York, 1993.

[8] E. DiBenedetto, *Parabolic equations with multiple singularities*, in: Proceedings of Equadiff 9, Brno, 1997.

[9] N. Igbida and J. M. Urbano, *Uniqueness for nonlinear degenerate problems,* Preprint, (2000).

[10] O. Ladyzenskaja, V. Solonnikov and N. Ural'ceva, *Linear and Quasi-linear Equations of Parabolic Type*, A.M.S. Transl. Monog. **23**, Providence, R.I., 1968.

[11] A. Meirmanov, *The Stefan Problem,* Walter de Gruyter, Berlin, 1992.

[12] J. M. Urbano, *A free boundary problem with convection for the p-Laplacian,* Rend. Mat. Appl. (VII), **17** (1997), 1–19.

[13] J. M. Urbano, *Continuous solutions for a degenerate free boundary problem,* Ann. Mat. Pura Appl., **178** (2000), 195–224.

[14] J. M. Urbano, *A singular-degenerate parabolic problem: Regularity up to the Dirichlet boundary,* in: Free Boundary Problems: Theory and Applications I, GAKUTO International Series Mathematical Sciences and Applications, **13** (2000), 399–410.

Departamento de Matemática
Universidade de Coimbra
Apartado 3008
3001–454 Coimbra, Portugal
E-mail address: `jmurb@mat.uc.pt`

Risk Sensitive Control with Applications to Fixed Income Portfolio Management

Tomasz R. Bielecki and Stanley R. Pliska

Abstract. This paper presents an application of risk sensitive control theory in financial decision making. Specifically, we develop optimal, risk-sensitive investment strategies for a long-term investor who is interested in optimal allocation of her/his capital between cash, equities and fixed income instruments. The long-term fixed income instruments used are so-called rolling-horizon bonds. In order to construct the optimal risk-sensitive control policies relevant for the present application we advance the risk sensitive control theory developed in our previous papers.

1. Introduction

Beginning with the pioneering work by Merton [19, 20, 21] and continuing through the recent books by Karatzas and Shreve [16] and Korn [18], some very sophisticated stochastic control methods have been applied to portfolio management. But most of these applications have been concerned with, at least implicitly, the management of equities. In spite of an abundance of well-known mathematical interest rate models, exemplified by the classical models of Vasicek [28] and Heath, Jarrow, and Morton [13], one can find very few applications of modern control theory to fixed income management.

There are at least two possible explanations for this deficiency in the literature. First, most fixed income assets have finite lives, so they cannot be modeled by simple stochastic processes such as the infinitely-lived geometric Brownian motions that are commonly used for equities. Moreover, the maturing of bonds before the planning horizon causes modeling difficulties by forcing discontinuities in the trading strategy. Another possible explanation for the limited number of control

---

JEL Classification. C61, C63, G11, E21.

1991 *Mathematics Subject Classification.* 60J20, 90A09, 90C40, 93E20.

Key words and phrases. risk sensitive control, portfolio management, fixed income management, intertemporal capital asset pricing model.

The research of the authors was partially supported by NSF Grant DMS-9971307, and NSF Grant DMS-9971424, respectively. We appreciate helpful comments by Marek Musiela and Marek Rutkowski as well as by participants at the December 1998 control theory workshop at the University of Southern California.

theory applications to fixed income management is that, unlike equities, the risk and return characteristics of fixed income assets explicitly depend upon underlying exogenous factors, namely, levels of interest rates. Consequently, sensible trading strategies will need to be explicit functions of these interest rates, and so these interest rate factors need to be explicitly incorporated in any mathematical model.

Of course a stochastic control model having a general form suitable for fixed income management was formulated more than 25 years ago by Merton [20]. His famous intertemporal capital asset pricing model (ICAPM) features both assets and the exogenous factors which affect them, all of which are modeled as diffusion processes. For the problem of maximizing expected utility of consumption and/or terminal wealth, he then derived and provided economic interpretations of the corresponding Hamilton–Jacobi–Bellman equation. It is straightforward, in principle, to adapt his model to fixed income management, namely, by letting the factors be interest rates in accordance with a desired interest rate model and then letting the assets be securities like zero coupon, discount bonds. Unfortunately, this approach is unlikely to be successful because, except for a very few special cases, the ICAPM is intractable.

In one successful special case Kim and Omberg [17] derived explicit solutions of the ICAPM where the interest rate is constant and the Sharpe ratio of the only available risky asset follows an Ornstein–Uhlenbeck process. Canestrelli [8] and Canestrelli and Pontini [9] obtained explicit solutions where the short interest rate is described by an Ornstein–Uhlenbeck process and there is also a single risky asset. In all these cases the only fixed income asset is the bank account which earns interest at the short, locally riskless rate.

There have also been a few studies of the ICAPM where one or more of the risky assets are taken to be fixed income securities. Brennan and Schwartz [7] and Brennan, Schwartz, and Lagnado [6] used numerical methods to solve the Hamilton–Jacobi–Bellman partial differential equation for the optimal trading strategy, but they were hard pressed to derive a solution even though there were only three factors and a similar number of assets. Explicit results on the application of the ICAPM to fixed income management were recently obtained by Bajeux-Besnainou, Jordan, and Portait [1] and Sørensen [27]. Apparently working independently, they studied the same special case. They both assumed there is one factor, an Ornstein–Uhlenbeck process for the short, locally riskless interest rate, and there are three assets. The first is a bank account where money grows according to the short rate. The second is the zero coupon bond which matures at a fixed planning horizon T. With the Brownian motion driving the short rate being the sole Brownian motion affecting the term structure of interest rates, the zero coupon bond thus evolves according to the well-known model due to Vasicek [28]. The third asset is a stock which evolves according to a process that is a geometric Brownian motion except that the appreciation rate equals the short interest rate plus a constant; the Brownian motion driving this process is correlated with the one driving the interest rate term structure.

Both studies focus on the problem of maximizing expected utility of wealth at a finite planning horizon T, where the utility function is of the form $u(w) = w^{1-\gamma}/(1-\gamma)$ and $\gamma > 0$ is the parameter of constant relative risk aversion. For $\gamma = 1$ one actually has as a special limiting case $u(w) = \ln(w)$, giving rise to what is sometimes called the growth optimal or numeraire portfolio. Note that increasing values of the parameter γ correspond to increasing levels of risk aversion for the investor.

Using the risk neutral computational approach introduced by Pliska [24, 25], both studies derived the same general form for the optimal trading strategy: at every point in time hold the fraction $1/\gamma$ of one's wealth in the growth optimal portfolio and invest the rest in the zero coupon bond. This makes some intuitive sense, because the zero coupon bond is the riskless asset for this problem, and the bigger the value of γ, the bigger the proportion in this asset. Unfortunately, however, their solutions leave unresolved a troublesome issue. Although their proportions in the stock are constants, their proportions in the zero coupon bond are deterministic functions of time which are unbounded in every neighborhood of the planning horizon T. This difficulty is associated with the fact that the zero coupon bond's volatility converges to zero as time approaches maturity.

Research in mathematical control theory has progressed to the point where it is likely that soon there will be explicit solutions of much more general versions of Merton's ICAPM. This will be very important, both for fixed income management as well as many other kinds of consumption/investment problems. However, in this paper we pursue a variation of the ICAPM where explicit, general solutions can be obtained right now. Retaining the same diffusion process model for assets and factors, our approach calls for changing from Merton's finite horizon expected utility objective to the infinite horizon objective of maximizing the risk adjusted growth rate. As explained fully in our earlier work [2, 3, 4, 5], this criterion can be viewed as being analogous to the classical Markowitz single-period model except that instead of trading off single-period criteria we are trading off the long run growth rate (which by itself is maximized by the growth optimal portfolio) versus the asymptotic variance of the portfolio. Our risk adjusted growth rate objective emerges naturally from the application of recent mathematical results on risk sensitive control theory. The principal benefit of this objective is that it is an infinite horizon criterion and therefore, as with most control problems in general, the ICAPM is more tractable than if a finite horizon criterion is used.

Our initial work in this direction developed a model with Gaussian factors and with assets having constant volatilities and appreciation rates that are affine functions of the factor levels. The theoretical foundations of this model are found in [2], while its applications to asset allocation problems are explored in [5]. But while this model fully allows for there to be correlations between asset returns and movements of the factor levels, it has a critical shortcoming: the partial correlations between asset returns and movements of the factor levels must be zero, that is, the residuals of the asset returns must be independent of the residuals of the factors. While this assumption is reasonable for applications where the only interest rate asset is the bank account, it is unacceptable for models which include both interest rate factors and additional fixed income assets.

In this paper we develop a rather general model featuring our risk sensitive criterion and, as in the ICAPM, exogenous factors which explicitly affect the diffusion process dynamics of the assets. The details and main results are provided in the following section. A special case of our model was recently studied by Fleming and Sheu [12]. They obtained sub-optimal investment strategies for small values of the risk sensitivity parameter.

In Section 3 we turn to the special case of fixed income management. By utilizing the concept of rolling-horizon bonds, a concept introduced by Rutkowski [26], we are able to incorporate fixed income assets having infinite lives. These assets can be viewed, roughly, as mutual funds of zero coupon bonds, all of which mature at about the same fixed distance in the future; these bonds are rolled over in a self financing manner so this same fixed distance is preserved through the course of time. We provide explicit theoretical results for a model having three kinds of assets: the bank account, rolling horizon bonds, and (possibly) equities; all the factors are various interest rates. Proofs and additional results will be provided in a future, extended version of this paper.

2. Formulation of the Optimal Risk Sensitive Asset Management Problem and the Main Results

In this section we formulate a general optimal dynamic asset management problem and we provide the characterization of optimal investment strategies. We consider a market consisting of $n \geq 2$ securities and $m+1$, $m \geq 0$, economic factors. The set of securities may include stocks, certain fixed income assets, cash and derivative securities, as in Brennan, Schwartz and Lagnado [6] and Bajeux-Besnainou, Jordan, and Portait [1] for example. The set of factors may include dividend yields, price-earning ratios, short-term interest rate, yields on various bonds, the rate of inflation, etc., as in Pesaran and Timmermann [22, 23] for example.

Let $(\Omega, \{\mathcal{F}_t\}_{t\geq 0}, \mathcal{F}, \mathbf{P})$ be the underlying probability space. Denoting by $S_i(t)$ the price of the i-th security and by $X_j(t)$ the level of the j-th factor at time t, we consider the following market model for the dynamics of the security prices and factors:

$$\frac{dS_i(t)}{S_i(t)} = (a + AX(t))_i dt + \sum_{k=0}^{N} \sigma_{ik} dW_k(t), \quad S_i(0) = s_i, \quad i = 1, 2, \cdots, n \tag{1}$$

$$dX(t) = (b + BX(t))dt + \Lambda dW(t), \quad X(0) = x\,, \tag{2}$$

where $W(t) = (W_0(t), W_1(t), \ldots, W_N)'$ is an R^{N+1} valued standard Brownian motion process, $X(t) = (X_0(t), X_1(t), \ldots, X_m)'$ is the R^{m+1} valued factor process, the market parameters a, A, $\Sigma := [\sigma_{ij}]$, b, B, $\Lambda := [\Lambda_{ij}]$ are matrices of appropriate dimensions, and $(a+Ax)_i$ denotes the i-th component of the vector $a+Ax$.

It is well known that a unique, strong solution exists for (1), (2), and that the processes $S_i(t)$ are positive with probability 1 (see e.g., [15], chapter 5).

Let $\mathcal{G}_t := \sigma((S(s), X(s)), 0 \le s \le t)$, where $S(t) = (S_1(t), S_2(t), \dots, S_n(t))$ is the security price process. Let $h(t)$ denote an R^n valued investment process (or strategy) whose components $h_i(t)$, $i = 1, 2, \dots, n$, represent the time-t proportions of wealth in the corresponding assets.

Definition 2.1. *An investment process $h(t)$ is* **admissible** *if the following conditions are satisfied:*

(i) $\sum_{i=1}^n h_i(t) = 1$,

(ii) *$h(t)$ is measurable and $\mathcal{G}_t$-adapted,*

(iii) *for every $\theta > 0$ there exists a probability measure $\mathbf{P}^{h,\theta}$ on $(\Omega, \mathcal{F})$ so that*

$$\frac{d\mathbf{P}^{h(\cdot),\theta}}{dP}\bigg|_{\mathcal{F}_t} = \eta_t(h(\cdot), \theta),$$

where

$$\eta_t(h(\cdot), \theta) = \exp\{-\frac{\theta^2}{8}\int_0^t \|h'(s)\Sigma\|^2 ds - \frac{\theta}{2}\int_0^t h'(s)\Sigma dW_s\},$$

(iv) $\limsup_{t\to\infty} t^{-1} \ln \mathbf{E}^{h(\cdot),\theta} \exp\{(\frac{\theta}{2})(X_t' K_1(\theta) X_t + K_2(\theta) X_t\} \le 0$, *where $K_1(\theta)$ and $K_2(\theta)$ are matrices defined later in the text, and $\mathbf{E}^{h(\cdot),\theta}$ denotes expectation under $\mathbf{P}^{h(\cdot),\theta}$.*

The class of admissible investment strategies will be denoted by $\mathcal{H}$. □

Let now $h(t)$ be an admissible investment process. Then there exists a unique, strong, and almost surely positive solution $V(t)$ to the following equation:

$$dV(t) = \sum_{i=1}^m h_i(t)V(t)\left[\mu_i(X(t))dt + \sum_{k=0}^N \sigma_{ik} dW_k(t)\right], \quad V(0) = v > 0, \qquad (3)$$

where $\mu_i(x)$ is the i-th coordinate of the vector $a + Ax$ for $x \in R^{m+1}$. The process $V(t)$ represents the investor's capital at time t, and $h_i(t)$ represents the proportion of capital that is invested in security i, so that $h_i(t)V(t)/S_i(t)$ represents the number of shares invested in security i, just as in, for example, Section 3 of [14].

In this paper we consider the following family of risk sensitized optimal investment problems, labeled as $\mathcal{P}_\theta$:

for $\theta \in (0, \infty)$, maximize the risk sensitized expected growth rate

$$J_\theta(v, x; h(\cdot)) := \liminf_{t\to\infty}(-2/\theta)t^{-1} ln\ \mathbf{E}^{h(\cdot)}\ [e^{-(\theta/2) ln V(t)} | V(0) = v, X(0) = x] \quad (4)$$

over the class of all admissible investment processes $h(\cdot)$, subject to (1) and (2),

where $\mathbf{E}^{h(\cdot)}$ is the expectation with respect to $\mathbf{P}$. The notation $\mathbf{E}^{h(\cdot)}$ emphasizes that the expectation is evaluated for process $V(t)$ generated by (3) under the investment strategy $h(t)$.

Remark 2.2. *The positive value of the risk sensitivity parameter θ corresponds to a risk averse investor. The techniques used in this paper can also be used to study problems $\mathcal{P}_\theta$ for negative values of θ, corresponding to risk seeking investors. The risk null case, for $\theta = 0$, can be studied independently or as the limit of the risk averse situation when the risk sensitivity parameter θ goes to zero. However, we shall not consider the case of $\theta \leq 0$ here.*

We make the following standing assumptions:

Assumption (A1) *The spectrum of the matrix B is contained in the left half plane [i.e., the matrix B is stable].*

Assumption (A2) *(a) The matrix $\Sigma\Sigma'$ is positive definite; (b) The matrix $\Lambda\Lambda'$ is positive definite.*

Remark 2.3. *It is worth emphasizing at this point that we do not assume independence of random perturbations driving the price and the factor dynamics. That is, we do not assume that $\Sigma\Lambda' = 0$. Such an assumption was made in* [2].

Let us consider now the following continuous algebraic Riccati equation (CARE)

$$K'R_1(\theta)K + K'R_2(\theta) + R_2(\theta)'K + R_3(\theta) = 0\,, \tag{5}$$

where

$$R_1(\theta) := \frac{\theta^2}{2}\big(\frac{\theta}{2}+1\big)^{-1}\Lambda\Sigma'\Psi\Sigma\Lambda' - \theta\Lambda\Lambda'\,,$$
$$R_2(\theta) := -\frac{\theta}{2}\big(\frac{\theta}{2}+1\big)^{-1}\Lambda\Sigma'\Psi A + B\,,$$
$$R_3(\theta) := (1/2)\big(\frac{\theta}{2}+1\big)^{-1}A'\Psi A\,,$$

and

$$\Psi := (\Sigma\Sigma')^{-1} - \Gamma\,,$$
$$\Gamma := \frac{(\Sigma\Sigma')^{-1}\mathbf{1}\mathbf{1}'(\Sigma\Sigma')^{-1}}{\mathbf{1}'(\Sigma\Sigma')^{-1}\mathbf{1}}\,,$$
$$\mathbf{1} = (1,1,\ldots,1)'\,.$$

In order to formulate our main results we need one more assumption:

Assumption (A3) *For every $\theta > 0$ the equation (5) admits a unique, positive semi-definitive solution, say $K_1(\theta)$. Moreover, the matrix $G(\theta)$ defined below is stable:*

$$G(\theta) := R_2(\theta) + R_1(\theta)K_1(\theta)\,. \tag{6}$$

Remark 2.4. *Standard controllability and observability conditions imposed on the parameters of problems $\mathcal{P}_\theta$ will guarantee that Assumption (A3) is satisfied. See Wonham* [29], *for example.*

In view of the Assumption (A3), for every $\theta > 0$ we may define a matrix $K_2(\theta)$ as follows:

$$K_2(\theta) := -\Big(G(\theta)'\Big)^{-1}\Big\{\Big(A - \theta\Sigma\Lambda' K_1(\theta)\Big)'\Big[\frac{(\Sigma\Sigma')^{-1}\mathbf{1}}{\mathbf{1}'(\Sigma\Sigma')^{-1}\mathbf{1}} + (\frac{\theta}{2}+1)^{-1}\Psi a\Big] - 2K_1(\theta)b'\Big\}. \tag{7}$$

Theorems 2.5 and 2.6 below contain the main results of this paper. Theorem 2.5 characterizes optimal, risk-sensitive investment strategies. Theorem 2.6 characterizes the optimal value of the objective criterion.

Theorem 2.5. *Assume (A1), (A2) and (A3). Fix $\theta > 0$ and consider the process $h^\theta(t)$ defined as*

$$h^\theta(t) := (\frac{\theta}{2}+1)^{-1}(\Sigma\Sigma')^{-1}\Big[\beta(X_t) + \lambda(X_t)\mathbf{1}\Big], \tag{8}$$

where

$$\beta(x) := \big(A - \theta\Sigma\Lambda' K_1(\theta)\big)x + a - \frac{\theta}{2}\Sigma\Lambda' K_2(\theta), \tag{9}$$

$$\lambda(x) := \frac{(\frac{\theta}{2}+1) - \mathbf{1}'(\Sigma\Sigma')^{-1}\beta(x)}{\mathbf{1}'(\Sigma\Sigma')^{-1}\mathbf{1}}, \tag{10}$$

the matrix $K_1(\theta)$ is as in Assumption (A3), and the corresponding matrix $K_2(\theta)$ is defined in (7). Then for θ sufficiently small the investment process $h_\theta(t)$ is optimal for problem (P_θ), that is,

$$J_\theta(v,x;h(\cdot)) \le J_\theta(v,x;h_\theta(\cdot))$$

holds for all $h(\cdot) \in \mathcal{H}$, $v > 0$, $x \in R^{m+1}$.

Theorem 2.6. *Assume (A1), (A2) and (A3), fix $\theta > 0$ sufficiently small, and consider the problem P_θ. Let $h_\theta(t)$ be as in Theorem 2.5. Then*

(a) *For all $v > 0$ and $x \in R^{m+1}$ we have*

$$J_\theta(v,x;h_\theta(\cdot)) = \lim_{t\to\infty}(-2/\theta)t^{-1}ln\ \mathbf{E}^{h_\theta(\cdot)}\ [e^{-(\theta/2)lnV(t)}|V(0)=v, X(0)=x] =: \rho(\theta).$$

(b) *The constant $\rho(\theta)$ in (a) is given by the formula*

$$\begin{aligned}\rho(\theta) =& b'K_2(\theta) + (1/2)\Big[(-\theta/2)K_2(\theta)'\Lambda\Lambda' K_2(\theta) + 2tr\Lambda' K_1(\theta)\Lambda\Big]\\ &+ (1/2)(\frac{\theta}{2}+1)^{-1}\Big(a - (\theta/2)\Sigma\Lambda' K_2(\theta)\Big)'\Psi\Big(a - (\theta/2)\Sigma\Lambda' K_2(\theta)\Big)\\ &- \frac{\mathbf{1}'(\Sigma\Sigma')^{-1}}{\mathbf{1}'(\Sigma\Sigma')^{-1}\mathbf{1}}(a - (\theta/2)\Sigma\Lambda' K_2(\theta)) + \frac{(1/2)(\frac{\theta}{2}+1)}{\mathbf{1}'(\Sigma\Sigma')^{-1}\mathbf{1}},\end{aligned} \tag{11}$$

where the matrix $K_1(\theta)$ is as in Assumption (A3) and the corresponding matrix $K_2(\theta)$ is defined in (7).

Remark 2.7. *The key point of the first equality in* (*a*) *is, of course, that the optimal objective value is given by an ordinary* lim *rather than the* lim inf *as in (4). The key point of the second equality in* (*a*) *is that the optimal objective value does not depend on either the initial amount of the investor's capital* (v) *or on the initial values of the underlying economic factors* (x), *although it depends, of course, on the investor's attitude towards risk (encoded in the value of* θ*).*

Remark 2.8. *It appears that the results of Theorems 2.5 and 2.6 are stronger than the results of Section 7 in Fleming and Sheu* [12], *where only suboptimal investment strategies are discussed.*

Remark 2.9. *Even though we require in Theorems 2.5 and 2.6 that the parameter* θ *is sufficiently small [the case that we can prove at the moment], we conjecture that the theorems remain true for all* θ *positive.*

Remark 2.10. *Theorems 2.5 and 2.6 are demonstrated in the extended version of this paper.*

2.1. The case where one of the assets is "risk free"

Let us consider the market (1) and (2) provided with an additional instantaneously risk-free asset, say $S_0(t)$, whose dynamics are given as

$$\frac{dS_0(t)}{S_0(t)} = (a_0 + A_0 X(t))dt, \quad S_0(0) = s_0 \,. \tag{12}$$

Let $h(t)$ denote an R^{n+1}-valued investment process or strategy whose components are $h_i(t)$, $i = 0, 1, 2, \ldots, n$, and which satisfies conditions analogous to the ones specified in Definition 2.1. It is convenient to partition $h(t)$ as

$$h(t) = (h_0(t), \tilde{h}(t)) \,.$$

Define

$$\tilde{A} := A - \mathbf{1}A_0, \quad \tilde{a} := a - \mathbf{1}a_0, \quad \tilde{\Psi} := (\Sigma\Sigma')^{-1} \,.$$

In order to characterize the optimal investment strategy $h^\theta(t)$ in this case we need to consider the CARE (5) and the formulas (6) and (7) with A, a, and Ψ replaced with $\tilde{A}$, $\tilde{a}$ and $\tilde{\Psi}$, respectively. Then we have that an optimal investment strategy is given as

$$h^\theta(t) = (h_0^\theta(t), \tilde{h}^\theta(t)) \,, \tag{13}$$

where

$$\tilde{h}^\theta(t) = (\theta/2 + 1)^{-1}\tilde{\Psi}\big[(\tilde{A} - \theta\Sigma\Lambda'\tilde{K}_1(\theta))X_t + \tilde{a} - (\theta/2)\Sigma\Lambda'\tilde{K}_2(\theta)\big] \,, \tag{14}$$

$$h_0^\theta(t) = 1 - \mathbf{1}'\tilde{h}^\theta(t) \,, \tag{15}$$

$\tilde{K}_1(\theta)$ is the unique solution to (5) with A and Ψ replaced with $\tilde{A}$ and $\tilde{\Psi}$, respectively, and

$$\tilde{K}_2(\theta) = -\left(\tilde{G}(\theta)'\right)^{-1}\left((\theta/2+1)^{-1}[\tilde{A} - \theta\Sigma\Lambda'\tilde{K}_1(\theta)]'\tilde{\Psi}\tilde{a} + A_0' + 2\tilde{K}_1(\theta)'b\right). \tag{16}$$

In the above formula the matrix $\tilde{G}(\theta)$ is defined just as $G(\theta)$ with A and Ψ replaced with $\tilde{A}$ and $\tilde{\Psi}$, respectively.

Finally, in this case we have

$$\begin{aligned}\rho(\theta) =&(1/2)(\theta/2+1)^{-1}\left(\tilde{a} - (\theta/2)\Sigma\Lambda'\tilde{K}_2(\theta)\right)'\tilde{\Psi}\left(\tilde{a} - (\theta/2)\Sigma\Lambda'\tilde{K}_2(\theta)\right)\\ &+ (1/2)\left[-(\theta/2)\tilde{K}_2(\theta)'\Lambda\Lambda'\tilde{K}_2(\theta) + 2\mathrm{tr}\Lambda'\tilde{K}_1(\theta)\Lambda\right] + b'\tilde{K}_2(\theta) + a_0\,.\end{aligned} \tag{17}$$

In Section 4 we shall apply the general results in this section to a fixed income management problem where there are a money market account, some fixed income assets, and possibly some equities.

3. Rolling-Horizon Bonds and Gaussian Yield-Factor Models

In the finance literature one can find studies where the authors assume infinitely-lived geometric Brownian motion models of fixed income securities, but this choice of model is never justified. In particular, no arguments can be found which start with a specific interest rate model together with a specific fixed income asset and then derive the stochastic dynamics of this asset's price process, showing this process to be a geometric Brownian motion. A purpose of this section is to make such an argument.

The appropriate asset is called a *rolling-horizon bond*, a concept introduced by Rutkowski [26]. To understand this it is convenient to assume the so-called Heath–Jarrow–Morton framework for interest rate models, namely, that for any maturity date T the instantaneous forward rate $f(\cdot, T)$ satisfies the stochastic differential equation

$$df(t,T) = \alpha(t,T)dt + \nu(t,T)\cdot dW_t\,, \tag{18}$$

where $\alpha(\cdot,T)$ and $\nu(\cdot,T)$ are adapted stochastic processes, W is a standard (possibly vector valued) Brownian motion defined on a filtered probability space $(\Omega, \{\mathcal{F}_t\}_{t\geq 0}, \mathcal{F}, \mathbf{P})$, and $\mathbf{P}$ is the actual, real-world probability measure. Letting $P(t,T)$ denote the time-t price of a zero-coupon (i.e., pure discount) bond that pays one unit of cash at time T, it is well known (see Heath *et al.* [13]) that under a martingale measure $\mathbf{P}^*$ the dynamics of P are given by

$$dP(t,T) = P(t,T)(r_t dt + b(t,T)\cdot dW_t^*)\,, \tag{19}$$

where W^* is a standard (possibly vector valued) Brownian motion under $\mathbf{P}^*$, $r_t := f(t,t)$ is the short-term, locally riskless interest rate, and $b(t,T) := -\int_t^T \nu(t,u)du$ is the bond's volatility.

Now consider a self-financing trading strategy in which the total wealth is continuously reinvested in zero-coupon bonds which mature exactly T time units in the future. This strategy is a bit abstract, but it can be viewed as the limit as $\delta \to 0$ of the self-financing strategy which every δ time units rolls over all the funds from the bonds that mature in $T-\delta$ time units into bonds that mature in T time units. The wealth process denoted $U(\cdot,T)$ of such a strategy is called a rolling horizon bond. As reported by Rutkowski [26], the dynamics of $U(\cdot,T)$ under the martingale measure $\mathbf{P}^*$ are

$$dU(t,T) = U(t,T)(r_t dt + b(t,t+T)\cdot dW_t^*)\,. \tag{20}$$

Meanwhile, under the actual probability measure $\mathbf{P}$ the rolling-horizon bond process will have the same volatility $b(t,t+T)$, while its drift will depend in a simple way upon the market price of risk (i.e., the Girsanov transformation relating $\mathbf{P}$ and $\mathbf{P}^*$). In particular, and this is the case that will be studied in this paper, for the Gaussian HJM model, where the diffusion coefficient ν in the forward rate SDE is nonrandom, the volatility $b(t,t+T)$ of the rolling-horizon bond is also a deterministic function of time. Hence the rolling-horizon bond can be a geometric Brownian motion whose appreciation rate depends upon the market price of risk.

We now turn to the interest rate model that is the subject of this paper. We fix k times $T_1 < T_2 < \cdots < T_k$, where $T_1 > 0$, and consider the yields $Y_t^1, Y_t^2, \ldots, Y_t^k$ of the corresponding *sliding bonds* (this is Rutkowski's [26] terminology) $P(t,t+T_i), i = 1,2,\ldots,k$. In other words,

$$Y_t^i = -\frac{1}{T_i}\ln P(t,t+T_i), i = 1,2,\ldots,k\,. \tag{21}$$

There is also the usual short-term, locally riskless interest rate r_t, the yield corresponding to a zero coupon bond having a maturity of zero, if you will. These $k+1$ interest rates will be the factors affecting and thus characterizing our interest rate model. We assume they satisfy, under the actual, real-world probability measure $\mathbf{P}$, the system of stochastic differential equations:

$$dr_t = (b_0 + B_{00}r_t + B_{01}Y_t^1 + \cdots + B_{0k}Y_t^k)dt + \sum_{j=0}^N \lambda_{0j}dW_j(t)\,, \tag{22}$$

$$dY_t^i = (b_i + B_{i0}r_t + B_{i1}Y_t^1 + \cdots + B_{ik}Y_t^k)dt + \sum_{j=0}^N \lambda_{ij}dW_j(t), \quad i = 1,2,\ldots,k\,. \tag{23}$$

Here b_i, B_{ij}, and λ_{il} are all fixed constants for all $i,j = 0,1,\ldots,k$, $l = 0,1,\ldots,N$ and $W = (W_0, W_1, \ldots, W_N)$ is a standard $(N+1)$-dimensional Brownian motion under $\mathbf{P}$, as before. Note that the $(k+1)$-dimensional factor process $(r, Y^1, \ldots, Y^k)$

is Markovian with drift coefficients that are affine functions of the factor levels and with constant diffusion coefficients.

We now make an important

Assumption (A4). *The market price of risk is an affine function of the* $k+1$ *interest rate factors.*

The market price of risk, of course, provides the connection via Girsanov between the real-world probability measure $\mathbf{P}$ and the martingale measure $\mathbf{P}^*$. Therefore, as a consequence of our assumption (or "equivalently", if you prefer), under $\mathbf{P}^*$ the dynamics of the $k+1$ interest rate factor will have exactly the same form as under $\mathbf{P}$. In other words, under $\mathbf{P}^*$ the SDE's describing the interest rate dynamics will have drift coefficients that are affine functions of the interest rate levels and diffusion coefficients that are constants.

We thus have specified what is a special case of the affine yield-factor model developed by Duffie and Kan [10]. Accordingly, there exist deterministic, real-valued functions $\alpha(\cdot), \beta_0(\cdot), \ldots, \beta_k(\cdot)$ on $[0, \infty)$ such that the price of any sliding bond is given by

$$P(t, t+T) = \exp[\alpha(T) + \beta_0(T) r_t + \beta_1(T) Y_t^1 + \cdots + \beta_k(T) Y_t^k], \quad \forall t \geq 0. \tag{24}$$

These $k+1$ deterministic functions can be computed from the parameters in equations (22) and (23) together with the market price of risk parameters, but doing so will not be necessary for our purposes. Instead, it suffices to observe that by taking $T = T_i$ for $i = 1, 2, \ldots, k$ and using equation (21) we must have (see also (5.2) of Duffie and Kan [10])

$$\alpha(T_i) = \beta_j(T_i) = 0,\ j \neq i;\ \beta_i(T_i) = -T_i\,, \quad i, j = 1, \ldots, k\,. \tag{25}$$

We shall use this to derive the real-world dynamics of the rolling-horizon bonds. Applying Ito's formula to the above expression for the sliding bond we obtain the following real-world dynamics:

$$dP(t, t+T)/P(t, t+T) = (\ldots)dt + \beta(T)\Lambda \cdot dW(t)\,, \tag{26}$$

where $\beta(T) := (\beta_0(T), \beta_1(T), \ldots, \beta_k(T))$ is a row vector and $\Lambda := (\lambda_{ij})$ is the $(k+1) \times (N+1)$ matrix of diffusion coefficients (as in (22), (23) as well as the corresponding risk neutral SDE's for the interest rates); the drift coefficient here does not matter, so we omit the details.

In particular, when $T = T_i$ for any $i = 1, \ldots, k$ we can apply (25) and thus obtain

$$dP(t, t+T_i)/P(t, t+T_i) = (\ldots)dt - T_i \lambda_{i0} dW_0(t) - \cdots - T_i \lambda_{iN} dW_N(t). \tag{27}$$

As explained by Rutkowski [26], it thus follows that our bond price volatilities $b(t, T)$ satisfy

$$b(t, t+T_i) = -T_i(\lambda_{i0}, \lambda_{i1}, \ldots, \lambda_{iN}), \quad i = 1, \ldots, k\,. \tag{28}$$

Hence substituting this in (20) and using again our assumption that the market price of risk is an affine function of the $k+1$ interest rate factors, we obtain the following characterization of the real-world dynamics of the k rolling-horizon bonds corresponding to the k specified maturities and interest rate factors:

$$dU(t,T_i)/U(t,T_i) = (\alpha_i + \mathcal{A}_{i0}r_t + \mathcal{A}_{i1}Y_t^1 + \cdots + \mathcal{A}_{ik}Y_t^k)dt \\ - T_i \sum_{j=0}^{N} \lambda_{ij} dW_t^j \,, \qquad i = 1\ldots,k. \quad (29)$$

Here α_i and $\mathcal{A}_{ij}$ for $i = 1,\ldots,k$ and $j = 0,1,\ldots,k$ are all fixed constants that can be computed from the market price of risk and other parameters (recall that each $U(t,T_i)$ is an asset price which under the martingale measure has drift coefficient $r_tU(t,T_i)$ and the same diffusion coefficient as under the real-world probability measure). Alternatively, and this is more convenient for implementation purposes, one can estimate these drift parameters directly, ignoring the market price of risk parameters.

In the following section we incorporate this Gaussian yield-factor model and the corresponding rolling-horizon bonds as part of the general risk sensitive asset management model that was developed in Section 2.

4. Optimal Dynamic Risk-Sensitive Management of Cash, Equities and Bonds

In this section we consider a specification of the general model of Section 2.1. Specifically, we consider the case where [in the notation of Sections 2 and 3]:

$$X(t) = (r_t, Y_t^1, \ldots, Y_t^k)' . \quad (30)$$

That is to say, the economic factors considered are the short rate r_t and the yields Y_t^i. As the tradable assets we take the bank account $S_0(t)$ whose dynamics are given as

$$\frac{dS_0(t)}{S_0(t)} = r_t dt, \quad S_0(0) = s_0 \,, \quad (31)$$

$\bar{m}$ stocks and stock indices $S_i(t)$, $i = 1,\ldots,\bar{n}$ whose dynamics are given by (1), and k rolling horizon bonds $S_{\bar{n}+j}(t) := U(t,T_j)$, $j = 1,\ldots k$ whose dynamics are given by (29).

Thus, we are considering the model of Section 2.1 with the following parameterization:

$$n = \bar{n} + k \,, \qquad m = k \,, \quad (32)$$

$$a_0 = 0, \qquad A_0 = (1,0,\ldots,0)_{1\times(1+k)} \,, \quad (33)$$

$$a_{\bar{n}+j} = \alpha_j \,, \qquad A_{\bar{n}+j,l} = \mathcal{A}_{j,l} \,, \quad j,l = 1,\ldots,k \,, \quad (34)$$

$$\Sigma_{i,l} = \sigma_{i,l}\,, \qquad i = 1, \ldots, \bar{n}\,, \qquad l = 0, 1, \ldots, N\,, \tag{35}$$

$$\Sigma_{i,l} = -T_{i-\bar{n}}\lambda_{i-\bar{n},l}\,, \qquad i = \bar{n}+1, \ldots, \bar{n}+k\,, \qquad l = 0, 1, \ldots, N\,, \tag{36}$$

$$\Lambda_{i,l} = \lambda_{i,l}\,, \qquad i = 0, 1, \ldots, k\,, \qquad l = 0, 1, \ldots, N\,, \tag{37}$$

Remark 4.1. *For example, letting $\bar{n} = k = 1$ and $N = 3$ implies*

$$n = 2, \quad m = 1\,, \tag{38}$$

$$W(t) = (W_0(t), W_1(t), W_2(t), W_3(t))'\,, \tag{39}$$

$$S_2(t) = U(t, T_1), \quad X_0(t) = r_t, \quad X_1(t) = Y_t^1\,, \tag{40}$$

$$A_0 = (1, 0)\,, \quad a = \begin{pmatrix} a_1 \\ \alpha_1 \end{pmatrix}, \quad A = \begin{pmatrix} A_{1,0} & A_{1,1} \\ \mathcal{A}_{1,0} & \mathcal{A}_{1,1} \end{pmatrix}, \tag{41}$$

$$\Sigma = \begin{pmatrix} \sigma_{1,0} & \sigma_{1,1} & \sigma_{1,2} & \sigma_{1,3} \\ -T_1\lambda_{1,0} & -T_1\lambda_{1,1} & -T_1\lambda_{1,2} & -T_1\lambda_{1,3} \end{pmatrix}, \tag{42}$$

$$b = \begin{pmatrix} b_0 \\ b_1 \end{pmatrix}, \quad B = \begin{pmatrix} B_{00} & B_{01} \\ B_{10} & B_{11} \end{pmatrix}, \tag{43}$$

$$\Lambda = \begin{pmatrix} \lambda_{0,0} & \bar{\lambda}_{0,1} & \lambda_{0,2} & \lambda_{0,3} \\ \lambda_{1,0} & \lambda_{1,1} & \lambda_{1,2} & \lambda_{1,3} \end{pmatrix}. \tag{44}$$

On the other hand, letting $\bar{n} = 1$, $k = 2$ and $N = 2$ implies

$$n = 3, \quad m = 2\,, \tag{45}$$

$$W(t) = (W_0(t), W_1(t), W_2(t))'\,, \tag{46}$$

$$S_2(t) = U(t, T_1)\,, \; S_3(t) = U(t, T_2)\,, \; X_0(t) = r_t\,, \; X_1(t) = Y_t^1\,, \; X_2(t) = Y_t^2\,, \tag{47}$$

$$A_0 = (1, 0, 0)\,, \quad a = \begin{pmatrix} a_1 \\ \alpha_1 \\ \alpha_2 \end{pmatrix}, \quad A = \begin{pmatrix} A_{1,0} & A_{1,1} & A_{1,2} \\ \mathcal{A}_{1,0} & \mathcal{A}_{1,1} & \mathcal{A}_{1,2} \\ \mathcal{A}_{2,0} & \mathcal{A}_{2,1} & \mathcal{A}_{2,2} \end{pmatrix}, \tag{48}$$

$$\Sigma = \begin{pmatrix} \sigma_{1,0} & \sigma_{1,1} & \sigma_{1,2} \\ -T_1\lambda_{1,0} & -T_1\lambda_{1,1} & -T_1\lambda_{1,2} \\ -T_2\lambda_{2,0} & -T_2\lambda_{2,1} & -T_2\lambda_{2,2} \end{pmatrix}, \tag{49}$$

$$b = \begin{pmatrix} b_0 \\ b_1 \\ b_2 \end{pmatrix}, \quad B = \begin{pmatrix} B_{00} & B_{01} & B_{02} \\ B_{10} & B_{11} & B_{12} \\ B_{20} & B_{21} & B_{22} \end{pmatrix}, \tag{50}$$

$$\Lambda = \begin{pmatrix} \lambda_{0,0} & \lambda_{0,1} & \lambda_{0,2} \\ \lambda_{1,0} & \lambda_{1,1} & \lambda_{1,2} \\ \lambda_{2,0} & \lambda_{2,1} & \lambda_{2,2} \end{pmatrix}. \tag{51}$$

Formulas of Section 2.1 may now be applied to compute optimal investment strategies and the optimal value of the objective criterion for the problem (4) specified to the present setting. Economic interpretations of illustrative numerical examples based on historical data will be provided in the extended version of this paper.

References

[1] I. Bajeux-Besnainou, J. V. Jordan and R. Portait, (1999), "Dynamic Allocation of Stocks, Bonds and Cash for Long Horizons," *working paper.*

[2] T. R. Bielecki and S. R. Pliska, (1999), "Risk Sensitive Dynamic Asset Management," *Appl. Math. Optim.*, vol. 39, pp. 337–360.

[3] T. R. Bielecki, D. Hernandez-Hernandez and S. R. Pliska, (1999), "Risk sensitive control of finite state Markov chains in discrete time, with applications to portfolio management," *Math. Meth. Oper. Res.*, vol. 50, pp. 167–188.

[4] T. R. Bielecki and S. R. Pliska, (2000), "Risk-sensitive dynamic asset management in the presence of transaction costs," *Finance and Stochastics*, vol. 4, pp. 1–33.

[5] T. R. Bielecki, S. R. Pliska and M. Sherris (1998), "Risk Sensitive Asset Allocation," *Journal of Economic Dynamics and Control*, to appear.

[6] M. J. Brennan, E. S. Schwartz and R. Lagnado, (1997) "Strategic asset allocation," *J. Economic Dynamics and Control*, vol. 21, pp. 1377–1403.

[7] M. J. Brennan and E. S. Schwartz, (1996), "The use of treasury bill futures in strategic asset allocation programs," In *World-Wide Asset and Liability Modeling*, J. M. Mulvey and W. T. Ziemba, eds. Cambridge University Press.

[8] E. Canestrelli, (1998), "Applications of Multidimensional Stochastic Processes to Financial Investments," *preprint*, Dipartimento di Matematica Applicata ed Informatica, Universita "Ca' Foscari" di Venezia.

[9] E. Canestrelli and S. Pontini, (1998), "Inquires on the Applications of Multidimensional Stochastic Processes to Financial Investments," *preprint*, Dipartimento di Matematica Applicata ed Informatica, Universita "Ca' Foscari" di Venezia.

[10] D. Duffie and R. Kan, (1996), "A yield-factor model of interest rates," *Math. Finance*, vol. 6, pp. 379–406.

[11] W. H. Fleming and S. J. Sheu, (1998), "Optimal long term growth rate of expected utility of wealth," preprint, Brown University.

[12] W. H. Fleming and S. J. Sheu, (2000), "Risk Sensitive Control and an Optimal Investment Model," *Mathematical Finance*, forthcoming.

[13] D. Heath, R. Jarrow and A. Morton, (1992), "Bond pricing and the term structure of interest rates: a new methodology for contingent claims valuation," *Econometrica*, vol. 60, pp. 77–105.

[14] I. Karatzas and S. G. Kou, (1996), "On the pricing of contingent claims under constraints," *Annals of Applied Probability*, vol. 6, pp. 321–369.

[15] I. Karatzas and S. E. Shreve, (1988), "Brownian Motion and Stochastic Calculus," Springer-Verlag, New York.

[16] I. Karatzas and S. E. Shreve, (1998), "Methods of Mathematical Finance," Springer-Verlag, New York.

[17] T. S. Kim and E. Omberg, (1996), "Dynamic Non-myopic Portfolio Behavior," *The Review of Financial Studies*, vol. 9, pp. 141–161.

[18] R. Korn, (1997), "Optimal Portfolios. Stochastic Models for Optimal Investment and Risk Management in Continuous Time," World Scientific, Singapore.

[19] R. C. Merton, (1971), "Optimum consumption and portfolio rules in a continuous time model," *J. Econ. Th.* vol. 3, pp. 373–413.

[20] R. C. Merton, (1973), "An intertemporal capital asset pricing model," *Econometrica* vol. 41, pp. 866–887.

[21] R. C. Merton, (1990), "Continuous-Time Finance," Basil Blackwell, Cambridge.

[22] M. H. Pesaran and A. Timmermann, (1995), "Predictability of stock returns: robustness and economic significance," *J. Finance*, vol. 50, pp. 1201–1228.

[23] M. H. Pesaran and A. Timmermann, (1998), "A recursive modelling approach to predicting UK stock returns," *working paper*, Trinity College, Cambridge.

[24] S. R. Pliska, (1982), "A discrete time stochastic decision model," in *Advances in Filtering and Optimal Stochastic Control*, eds. W. H. Fleming and L. G. Gorostiza, Lecture Notes in Control and Information Sciences 42, Springer Verlag, New York, pp. 290–304.

[25] S. R. Pliska, (1986), "A stochastic calculus model of continuous trading: optimal portfolios," *Math. Operations Research*, vol. 11, pp. 371–384.

[26] M. Rutkowski, (1997), "Self-financing Trading Strategies for Sliding, Rolling-horizon, and Consol Bonds," *Mathematical Finance* vol. 9, pp. 361–385.

[27] C. Sørensen, (1999), "Dynamic Asset Allocation and Fixed Income Manegement," *Journal of Financial and Quantitative Analysis*, forthcoming.

[28] O. A. Vasicek, (1977), "An Equilibrium Characterization of the Term Structure," *Journal of Financial Economics* vol. 5, pp. 177–188.

[29] W. M. Wonham, (1979), "Linear Multivariable Control: a Geometric Approach," Springer-Verlag, New York.

T. R. Bielecki
Department of Mathematics,
The Northeastern Illinois University,
5500 North St. Louis Avenue,
Chicago, IL 60625-4699 USA
E-mail address: T-Bielecki@neiu.edu

S. R. Pliska
Department of Finance,
University of Illinois at Chicago,
601 S. Morgan Street,
Chicago, IL 60607-7124 USA
E-mail address: srpliska@uic.edu

Wavelet Based PDE Valuation of Derivatives

M. A. H. Dempster and A. Eswaran

Abstract. We investigate the application of a wavelet method of lines solution to financial PDEs and demonstrate the suitability of a numerical scheme based on biorthogonal interpolating wavelets to problems where there are discontinuities or regions of sharp transitions in the solution. The examples treated are the Black–Scholes PDE with discontinuous payoffs and a 3-dimensional cross currency swap PDE for which a speedup over standard finite difference methods of two orders of magnitude is reported. We expect that a thresholded version of the algorithm currently being developed will improve speedup by at least a further order of magnitude.

1. Introduction

Wavelets are nonlinear functions which can be scaled and translated to form a basis for the Hilbert space $L^2(\mathbb{R})$ of square integrable functions. Wavelets thus generalize the trigonometric functions given by e^{ist} ($s \in \mathbb{R}$) which generate the classical Fourier basis for L^2, and wavelet and fast wavelet transforms exist which generalize the time-to-frequency map of the Fourier transform to pick up *both* the space and time behaviour of a function [11]. Wavelets have been used in the fields of image compression and image analysis for quite some time to compute compact representations of functions and data sets based on exploiting structure in the underlying functions. In the solution of PDEs using wavelets [1, 4, 5, 15, 18, 19, 21], functions and operators are expanded in a wavelet basis to allow a combination of the desirable features of finite-difference methods, spectral methods and front-tracking or adaptive grid approaches. In solving PDEs that occur in finance, wavelets exploit the fact that large classes of operators and functions which occur in this area are sparse, or sparse to some high degree of accuracy, when transformed into the wavelet domain. Wavelets are also suitable for problems with multiple spatial scales which arise frequently in finance, since they give an accurate representation of the value function solution in regions of sharp transitions and combine the advantages of both spectral and finite-difference methods.

In this paper we implement a wavelet method of lines scheme using biorthogonal interpolating wavelets to solve the Black–Scholes PDE for option values with discontinuous payoff structures and a 3-dimensional cross currency swap PDE based on extended Vasicek interest rate models. The numerical advantages of using a wavelet based PDE method to solve these kinds of problem are demonstrated. In Section 2 we give a brief introduction to wavelet theory. In Sections 3 and 4 we introduce wavelet based PDE methods and explain the biorthogonal wavelet approach in more detail. Sections 5 and 6 contain respectively the problems and numerical results for the Black–Scholes and cross currency swap PDEs and Section 7 concludes and describes current research.

2. Basic Wavelet Theory

This section contains a brief introduction to wavelets for real valued functions of a real argument [6]; the extension to several variables is treated in Sections 3 and 4.

2.1. Daubechies based wavelets

Consider the following functions. The *scaling function* is the solution of a *dilation equation*

$$\phi(x) = \sqrt{2}\sum_{k=0}^{\infty} h_k \phi(2x-k)\,, \tag{1}$$

where ϕ is normalised so that $\int_{-\infty}^{\infty}\phi(x)dx = 1$, and the *wavelet function* is defined in terms of the scaling function as

$$\psi(x) = \sqrt{2}\sum_{k=0}^{\infty} g_k \phi(2x-k)\,. \tag{2}$$

We can build up an *orthonormal basis* for the *Hilbert space* $L^2(\mathbb{R})$ of (equivalence classes of) square integrable functions from the functions ϕ and ψ by dilating and translating them to obtain the *basis functions*:

$$\phi_{j,k}(x) = 2^{-j/2}\phi(2^{-j}x-k) = 2^{-j/2}\phi\left(\frac{x-2^j k}{2^j}\right) \tag{3}$$

$$\psi_{j,k}(x) = 2^{-j/2}\psi(2^{-j}x-k) = 2^{-j/2}\psi\left(\frac{x-2^j k}{2^j}\right)\,. \tag{4}$$

In (3) and (4) j is the *dilation* or *scaling parameter* and k is the *translation parameter*. All wavelet properties are specified through the *filter* coefficients $H := \{h_k\}_{k=0}^{\infty}$ and $G := \{g_k\}_{k=0}^{\infty}$ which are chosen so that dilations and translations of the wavelet ψ form an orthonormal basis of $L^2(\mathbb{R})$, i.e.,

$$\int_{-\infty}^{\infty}\psi_{j,k}(x)\psi_{l,m}(x)dx = \delta_{jl}\delta_{km} \quad j,k,l,m \in \mathbb{Z}_+\,, \tag{5}$$

where $\mathbb{Z}_+ := \{0,1,2,\dots\}$ and δ_{jl} is the Kronecker delta function.

Under these conditions for any function $f \in L^2(\mathbf{R})$ there exists a set $\{d_{jk}\}$ such that

$$f(x) = \sum_{j\in\mathbb{Z}_+} \sum_{k\in\mathbb{Z}_+} d_{jk}\psi_{j,k}(x)\,, \tag{6}$$

$$d_{jk} := \int_{-\infty}^{\infty} f(x)\psi_{j,k}(x)dx\,. \tag{7}$$

It is usual to denote the spaces spanned by $\phi_{j,k}$ and $\psi_{j,k}$ over the location parameter k, with scale parameter j fixed, by

$$\mathbf{V}_j := \mathrm{span}_{k\in\mathbb{Z}_+}\,\phi_{j,k} \quad \mathbf{W}_j := \mathrm{span}_{k\in\mathbb{Z}_+}\,\psi_{j,k}\,. \tag{8}$$

In the expansion (6) functions with arbitrarily small scales can be represented, however in practice there is a limit on how small the smallest structure can be, which could be dependent on a required grid size in a numerical computation as we shall see below, *cf.* Section 3. To implement wavelet analysis on a computer, we also need to have a bounded range and domain to generate approximations to functions $f \in L^2(\mathbb{R})$ and thus must limit the filters H and G to *finite* sets. Approximation *accuracy* is specified by requiring that the wavelet function ψ satisfies

$$\int_{-\infty}^{\infty} \psi(x)x^m dx = 0 \tag{9}$$

for $m = 0, \ldots, M-1$, which implies exact approximation for polynomials of degree $M-1$. For *Daubechies wavelets* [6] the number of coefficients or *length* L of the filters H and G is related to the number M of vanishing moments in (9) by $2M = L$ and their elements are related by $g_k = (-1)^k h_{L-k}$ for $k = 0, \ldots, L-1$. In the signal processing literature such finite filters H and G are known as *quadrature mirror* filters. The coefficients H needed to define compactly supported wavelets with high degrees of regularity can be derived, see e.g., [6]. Therefore on a computer an approximation subspace expansion would be in the form of a finite direct sum of finite dimensional vector spaces as

$$\mathbf{V}_J = \mathbf{W}_0 \oplus \mathbf{W}_1 \oplus \mathbf{W}_2 \oplus \cdots \oplus \mathbf{W}_{J-1} \oplus \mathbf{V}_0\,,$$

where the spaces $\mathbf{W}_j$ and $\mathbf{V}_j$ are termed *scaling function* and *approximation subspaces* respectively. The corresponding orthogonal wavelet series *approximation* to a continuous function f on a compact domain is given by

$$f(x) \approx \sum_k d_{0,k}\psi_{0,k}(x) + \cdots + \sum_k d_{J-1,k}\psi_{J-1,k}(x) + \sum_k s_{0,k}\phi_{0,k}\,, \tag{10}$$

where J is the number of *multiresolution components* (or *scales*) and k ranges from 0 to the number of coefficients in the specified component $j \in \{0, \ldots, J-1\}$. The coefficients $d_{0,k}, \ldots, d_{J-1,k}, s_{0,k}$ are termed the *wavelet transform coefficients* and the functions $\phi_{j,k}$ and $\psi_{j,k}$ are the *approximating wavelet functions.* Some examples of basic wavelets are the *Haar wavelet* which is just a square wave (the indicator function of the unit interval), the *Daubechies wavelet* [6] and *the Coiflet wavelet* [2].

2.2. Biorthogonal wavelets

Biorthogonal wavelets are a generalization of orthogonal wavelets first introduced by Cohen, Daubechies and Feauveau [3] which are symmetric and do not introduce phase shifts in the coefficients. Biorthogonal wavelet analysis uses four basic function types: ϕ and ψ termed *mother* and *father* wavelets and $\tilde{\phi}$ and $\tilde{\psi}$ termed *dual wavelets*. The father and mother wavelets are used to compute the wavelet coefficients as in the orthogonal case, but now the *biorthogonal wavelet approximation* of a continuous function on a compact domain is expressed in terms of the dual wavelet functions as

$$f(x) \approx \sum_k d_{0,k}\tilde{\psi}_{0,k}(x) + \cdots + \sum_k d_{J-1,k}\tilde{\psi}_{J-1,k}(x) + \sum_k s_{0,k}\tilde{\phi}_{0,k}\,, \tag{11}$$

where k is summed over an appropriate subset of $\{1,\ldots,2^J\}$ determined by the resolution components $j \in \{0,\ldots,J-1\}$. In general biorthogonal wavelets are not mutually orthogonal, but rather satisfy *biorthogonal* relationships of the form

$$\begin{aligned}
&\int \phi_{j,k}\tilde{\phi}_{j',k'}(x)dx = \delta_{j,j'}\delta_{k,k'}\\
&\int \phi_{j,k}\tilde{\phi}_{j',k}(x)dx = 0\\
&\int \psi_{j,k}\tilde{\phi}_{j,k'}(x)dx = 0\\
&\int \psi_{j,k}\tilde{\psi}_{j',k'}(x)dx = \delta_{j,j'}\delta_{k,k'}\,,
\end{aligned} \tag{12}$$

where j, j', k, k' range over appropriate subsets of $\{1,\ldots,2^J\}$ as above.

3. Wavelets and PDEs

Wavelet based approaches to the solution of PDEs have been presented by Xu and Shann [21], Beylkin [1], Vasilyev *et al.* [18, 19], Prosser and Cant [15], Dahmen *et al.* [5] and Cohen *et al.* [4]. In order to describe briefly the two main approaches to the numerical solution of PDEs using wavelets, consider the most general first order form for a system of parabolic PDEs of any order given by

$$\begin{aligned}
&\frac{\partial u}{\partial t} = F(x,t,u,\nabla u)\\
&\Phi(x,t,u,\nabla u) = 0\,,
\end{aligned} \tag{13}$$

which describe the time evolution of a vector valued function u and boundary conditions which are algebraic or differential constraints. The *wavelet-Galerkin* method [1, 21] assumes that the wavelet coefficients are functions of time. An appropriate wavelet decomposition for each component of the solution is substituted into (13) and a Galerkin projection is used to derive a nonlinear system of ordinary differential-algebraic equations which describe the time evolution of the wavelet coefficients. In a *wavelet-collocation* method [18, 19] the system (13) is evaluated at collocation points of the domain of the solution u and a system of nonlinear ordinary differential-algebraic equations describing the evolution of u at these collocation points is obtained.

If we want the numerical algorithm to be able to resolve all structures appearing in the solution and also to be efficient in terms of minimising the number of unknowns, the *basis elements* corresponding to active wavelets – and thus the *computational grid points* – for the wavelet-collocation algorithm should adapt dynamically in time to reflect local changes in the solution based on an analysis of currently significant wavelet coefficients. The contribution of a particular wavelet to the approximation is *significant* if and only if the nearby structures of the solution have a size comparable with that wavelet's scale. Thus by using a *thresholding* technique a large number of the fine scale wavelets may be dropped in regions where the solution is smooth. In a wavelet-collocation method every wavelet is uniquely associated with a collocation point and hence a point can be omitted from the grid if the associated wavelet is omitted from the approximation. This property of the *multiresolution scale* wavelet approximation allows local grid refinement up to a prescribed small scale without a drastic increase in the number of collocation points maintained at each time step. A fast adaptive wavelet collocation algorithm for two dimensional PDEs is presented in [18] and a spatial discretization scheme using bi-orthogonal interpolating wavelets is implemented in [12]–[14] to solve the *reactive Navier–Stokes* equations in three dimensions. The main advantage of the latter approach is that for the solution computed in wavelet space it is possible to exploit sparsity in order to reduce storage costs and speed up solution times. We next explain the wavelet-collocation method in greater detail.

3.1. The biorthogonal wavelet approach

The main difference in using biorthogonal systems is that we have both *primal* and *dual* basis functions derived from primal and dual scaling and wavelet functions. Biorthogonal wavelet systems are derived from a *paired* hierarchy of scaling function subspaces

$$\cdots \mathbf{V}_{j-1} \subset \mathbf{V}_j \subset \mathbf{V}_{j+1} \cdots$$
$$\cdots \tilde{\mathbf{V}}_{j-1} \subset \tilde{\mathbf{V}}_j \subset \tilde{\mathbf{V}}_{j+1} \cdots .$$

(Note that here increasing $j = 1, 2, \ldots$ denotes refinement of the grid.) For periodic discretizations, $\dim(\mathbf{V}_j) = 2^j$. The basis functions for these spaces are generated from the *primal* scaling function ϕ and the *dual* scaling function $\tilde{\phi}$. Define two *innovation wavelet spaces* $\mathbf{W}_j$ and $\tilde{\mathbf{W}}_j$ such that

$$\begin{aligned} \mathbf{V}_{j+1} &:= \mathbf{V}_j \oplus \mathbf{W}_j \\ \tilde{\mathbf{V}}_{j+1} &:= \tilde{\mathbf{V}}_j \oplus \tilde{\mathbf{W}}_j \end{aligned} \tag{14}$$

where $\tilde{\mathbf{V}}_j \perp \mathbf{W}_j$, $\mathbf{V}_j \perp \tilde{\mathbf{W}}_j$ and $j \in \mathbb{Z}_+$. The innovation spaces so defined satisfy

$$\bigoplus_{j=0}^{\infty} \mathbf{W}_j = \mathbf{L}^2(\mathbb{R}) = \bigoplus_{j=0}^{\infty} \tilde{\mathbf{W}}_j \tag{15}$$

and the innovation space basis functions are generated from ψ and $\tilde{\psi}$.

3.2. Biorthogonal interpolating wavelet transform

The *biorthogonal interpolating wavelet transform* [8, 9] has a finite number of basis functions given by

$$\begin{aligned} \phi_{j,k}(x) &= \phi(2^j x - k) \\ \psi_{j,k}(x) &= \phi(2^{j+1} x - 2k - 1) \\ \tilde{\phi}_{j,k}(x) &= \delta(x - x_{j,k}) \, , \end{aligned} \tag{16}$$

where δ is the Dirac delta function and $x_{j,k}$ is a grid point in the spatial domain at resolution level j. An explicit form for $\tilde{\psi}$ is unknown but the corresponding wavelet coefficients can be derived using the biorthogonality conditions (14). These biorthogonal wavelets are said to be *interpolating* because the primal scaling function ϕ satisfies

$$\phi(k) = \begin{cases} 1 & k = 0\, , \\ 0 & k \neq 0\, . \end{cases}$$

Fast transform methods for evaluation of the wavelet and scaling function coefficients are given in [15, 17]. The *projection* of a function f onto a *finite dimensional* space of scaling functions $\mathbf{V}_j$, $j = 1, \ldots, J$, is given (assuming the spatial domain to be discretized is $[0, 1]$) by

$$P_{\mathbf{V}_j} f(x) = \sum_{k=0}^{2^j} s^f_{j,k} \phi^{\square}_{j,k}(x) \, , \tag{17}$$

where $s^f_{j,k}$ is defined as $\langle f, \tilde{\phi}_{j,k} \rangle$ in terms of a suitable inner product and $\phi^{\square}$ is used to denote a *boundary* or *internal wavelet* given by

$$\phi^{\square}_{j,k}(x) = \begin{cases} \phi^L_{j,k}(x) & k = 0, \ldots, M-1 \\ \phi_{j,k}(x) & k = M, \ldots, 2^j - M \\ \phi^R_{j,k}(x) & k = 2^j - M + 1, \ldots, 2^j \, . \end{cases} \tag{18}$$

3.3. Fast biorthogonal interpolating wavelet transform algorithm

The projection of a function f onto a scaling function space $\mathbf{V}_j$ is given as above by

$$\begin{aligned} P_{\mathbf{V}_j} f(x) &= \sum_k \langle f(u), \tilde{\phi}_{j,k}(u)\rangle \phi_{j,k}(x) \\ &= \sum_k f(k/2^j)\phi_{j,k}(x) \\ &= \sum_k s^f_{j,k}\phi_{j,k}(x)\,, \end{aligned} \tag{19}$$

where $s^f_{j,k} = f(k/2^j)$. The coefficients at resolution level j must be derived using

$$\begin{aligned} P_{\mathbf{W}_j} f(x) &= P_{\mathbf{V}_{j+1}} f(x) - P_{\mathbf{V}_j} f(x) \\ \sum_l d^f_{j,l}\psi_{j,l}(x) &= \sum_m s^f_{j+1,m}\phi_{j+1,m}(x) - \sum_n s^f_{j,n}\phi_{j,n}(x)\,. \end{aligned} \tag{20}$$

An arbitrary *wavelet coefficient* $d^f_{j,m}$ can be calculated from

$$d^f_{j,m} = s^f_{j+1,2m+1} - \sum_n s^f_{j,n}\phi(m-n+1/2) = s^f_{j+1,2m+1} - \sum_n \Gamma_{mn} s^f_{j,n}\,, \tag{21}$$

where Γ is a square matrix of size $2^j \times 2^j$ for periodic discretizations defined by

$$\Gamma_{mn} := \phi(m-n+1/2)\,.$$

Because of the compact support of the interpolating primal scaling function this matrix has a band diagonal structure. A general primal scaling function ϕ with compact support can be defined through the use of the *two scale* relation

$$\phi(x) = \sum_{\xi\in\Xi} \phi(\xi/2)\phi(2x-\xi)\,, \tag{22}$$

where Ξ is a suitable subset of $\mathbb{Z}_+$. The smoothness of ϕ is dictated by its $(M-1)^{st}$ degree polynomial span which in turn depends on the $M+1$ non-zero values of $\phi(\xi/2)$. The values $\phi(\xi/2)$ of the *interpolating* primal scaling function can be calculated explicitly [15]. The transform vector that arises from the fast wavelet transform will in this case have a structure of the following form:

$$\begin{array}{c}
\{s^f_{J,0}, s^f_{J,1}, s^f_{J,2} \cdots s^f_{J,2^J-1}\}^T \\
\downarrow \\
\{d^f_{J-1,0}, \dots, d^f_{J-1,2^{J-1}-1} \mid s^f_{J-1,0}, s^f_{J-1,1} \cdots s^f_{J-1,2^{J-1}-1}\}^T \\
\downarrow \\
\{d^f_{J-1,0}, \dots, d^f_{J-1,2^{J-1}-1}, \mid d^f_{J-2,0}, \cdots d^f_{J-2,2^{J-2}-1} \mid s^f_{J-2,0} \cdots s^f_{J-2,2^{J-2}-1}\}^T \\
\downarrow \\
\{d^f_{J-1,0}, \dots, \mid \cdots d^f_{J-P-1,0}, \cdots d^f_{J-P-1,2^{J-P-1}-1} \mid s^f_{J-P-1,0} \cdots s^f_{J-P-1,2^{J-P-1}-1}\}^T \\
\mathbf{W}_{J-1} \oplus \mathbf{W}_{J-2} \oplus \mathbf{W}_{J-3} \oplus \cdots \oplus \mathbf{V}_{J-P}
\end{array} \tag{23}$$

involving P resolution levels and a finest spatial discretization of 2^J.

3.4. Algorithm complexity

The number of floating point operations required for the fast biorthogonal interpolating wavelet transform algorithm for P *resolution* levels is $2M\sum_{j=J-P}^{J} 2^j = 2^{J-P}M\{2^{P+1}-1\}$. This comes from the fact that we require $2M$ filter coefficients to define the primal scaling function ϕ which spans the space of polynomials of degree less than $M-1$. The calculation of the wavelet coefficients $d^f_{j,k}$ for a given resolution j can be accomplished in $2(M-1)+1$ floating point operations. The sub-sampling process for the scaling function coefficients $s^f_{j,k}$ requires a further 2^j operations and a total of $2^{j+1}M$ operations are required per resolution j. Thus for fixed J and P the complexity of the fast interpolating wavelet transform algorithm is $O(M)$ [15]. Since the finest resolution in a PDE spatial grid of N points is $J = \log_2 N$, for fixed M and P the complexity of the transform is $O(N)$.

3.5. Decomposition of differential operators

Defining

$$\partial_J^{(n)} f(x) := P_{V_J} \frac{d^n}{dx^n} P_{V_J} f(x)\,, \tag{24}$$

repeated application of the approximation subspace decomposition gives us

$$\partial_J^{(n)} f(x) := \left(P_{\mathbf{V}_{J-P}} + \sum_{j=J-P}^{J-1} P_{\mathbf{W}_j} \right) \frac{d^n}{dx^n} \left(P_{\mathbf{V}_{J-P}} + \sum_{j=J-P}^{J-1} P_{\mathbf{W}_j} \right) f(x)\,. \tag{25}$$

For example, the decomposition of the first derivative operator $\frac{d}{dx}$ is given by

$$\partial_J = \left(P_{\mathbf{V}_{J-P}} + \sum_{j=J-P}^{J-1} P_{\mathbf{W}_i} \right) \frac{d}{dx} \left(P_{\mathbf{V}_{J-P}} + \sum_{j=J-P}^{J-1} P_{\mathbf{W}_j} \right), \tag{26}$$

where $\partial_J := W\partial_J^1 W^{-1}$ and W and W^{-1} are matrices denoting the forward and inverse transforms with

$$\partial_J^1 = P_{\mathbf{V}_J} \frac{d}{dx} P_{\mathbf{V}_J} . \tag{27}$$

We can analyse ∂_J^1 instead of ∂_J without loss of generality because the forward and inverse transforms are exact up to machine precision. The matrix ∂_J^1 has a band diagonal structure and can be treated as a finite difference scheme for analysis. The biorthogonal expansion for $\frac{df}{dx}$ requires information on the interaction between the differentiated and original scaling functions along with information about both the primal and dual basis functions. Using the sampling nature of the dual scaling function, ∂_J^1 can be written as

$$\partial_J^1 = 2^J \sum_{\alpha,k} s_{J,k}^f \frac{d\phi^\square}{dx} \mid_{x=\alpha-k} , \tag{28}$$

from which using equation (18) yields

$$\partial_J^1 = \begin{cases} 2^J \sum_{\alpha,k} s_{J,k}^f \frac{d\phi^L}{dx} \mid_{x=\alpha-k} \phi_{J,\alpha}^L(x) & k = 0, \dots, M-1 \\ 2^J \sum_{\alpha,k} s_{J,k}^f \frac{d\phi}{dx} \mid_{x=\alpha-k} \phi_{J,\alpha}(x) & k = M, \dots, 2^J - M \\ 2^J \sum_{\alpha,k} s_{J,k}^f \frac{d\phi^R}{dx} \mid_{x=\alpha-k} \phi_{J,\alpha}^R(x) & k = 2^J - M + 1, \dots, 2^J . \end{cases} \tag{29}$$

The entire operator ∂_J^1 can be determined provided values of $r_{\alpha-k}^{(1)} := \frac{d\phi}{dx} \mid_{x=\alpha-k}$ can be obtained. An approach to determining filter coefficients for higher order derivatives is given in [13].

3.6. Extension to multiple dimensions

The entire wavelet multiresolution framework presented so far can be extended to several spatial dimensions by taking straightforward tensor products of the appropriate 1D wavelet bases. The imposition of boundary conditions on nonlinearly bounded domains is nontrivial, but these are fortunately rare in derivative valuation PDE problems which are usually Cauchy initial value problems on a strip.

The fast biorthogonal interpolating wavelet transform used with wavelet collocation methods for problems posed over d-dimensional domains exhibits better complexity than its alternatives. Indeed, since one basis function is needed for each collocation point, using a spatial grid of $n := 2^J$ points in each dimension there are $N := n^d$ points in the spatial domain to result in transform complexity $O(n^d)$ – versus $O(n^d \log_2 n)$ for the *Fast Fourier Transform* (which is applicable only to time independent coefficient PDEs), $O(n^{2d})$ for an *explicit* finite difference scheme and $O(n^{3d})$ for a *Crank–Nicholson* or *implicit* scheme (which makes these methods impractical for $d > 2$, *cf.* [7]).

4. Wavelet Method of Lines

In a traditional finite difference scheme, partial derivatives are replaced with algebraic approximations at grid points and the resulting system of algebraic equations is solved to obtain the numerical solution of the PDE. In the *wavelet method of lines* we transform the PDE into a vector ODE by replacing the spatial derivatives with their wavelet transform approximations but retaining the time derivatives. We then solve this vector ODE using a suitable stiff ODE solver. We have implemented both a fourth order Runge Kutta method and a method based on the backward differentiation formula (LSODE) developed at the Lawrence Livermore Laboratories [16]. The fundamental complexity of this method is $O(\tau n^d)$ for space and time discretizations of size n and τ respectively over domains of dimension d (*cf.* Section 3, [16]).

4.1. An example

Consider a first order nonlinear hyperbolic *transport* PDE defined over an interval $\Omega = [x_l, x_r]$:

$$\begin{aligned} \frac{\partial u}{\partial t} &= \frac{\partial u}{\partial x} + S^{u/\rho} & x &\notin \partial\Omega \\ \frac{\partial u}{\partial t} &= -\chi^L(t) & x &= x_l \\ \frac{\partial u}{\partial t} &= -\chi^R(t) & x &= x_r \,. \end{aligned} \tag{30}$$

The numerical scheme is applied to the wavelet transformed counterpart of the above equations

$$\frac{\partial}{\partial t}\wp_{J-P}^{J-1} u = -\partial_J^{(1)} u + \wp_{J-P}^{J-1} S^{u/\rho} \quad x \notin \partial\Omega\,, \tag{31}$$

where $\wp_{J-P}^{J-1} := \left(P_{\mathbf{V}_{J-P}} + \sum_{j=J-P}^{J-1} P_{\mathbf{W}_j}\right)$ and $\partial_J^{(1)}$ is the standard decomposition of $\frac{d}{dx}$ defined as $\wp_{J-P}^{J-1}\frac{d}{dx}\wp_{J-P}^{J-1}$. In using the multiresolution strategy to discretize the problem we represent the domain $P+1$ times, where P is the number of different resolutions in the discretization, because of the P wavelet spaces and the coarse resolution scaling function space $\mathbf{V}_{J-P}$, $P \geq 1$. In the transform domain each representation of the solution defined at some resolution p should be supplemented by boundary conditions and [15] shows how to impose boundary conditions in both the scaling function spaces and the wavelet spaces.

5. Financial Derivative Valuation PDEs

In this section we introduce briefly the PDEs for financial derivative valuation and the products we have valued using the wavelet method of lines described above. More details may be found in [7, 20].

5.1. Black–Scholes products

We have applied the biorthogonal interpolating wavelet collocation method to solve the Black–Scholes PDE for a vanilla European call option and two binary options. The *Black–Scholes* quasilinear parabolic *PDE* is

$$\frac{\partial C}{\partial t} + \frac{1}{2}\sigma^2 S^2 \frac{\partial^2 C}{\partial S^2} + rS\frac{\partial C}{\partial S} - rC = 0\,, \tag{32}$$

where S is the *stock price*, σ is *volatility*, r is the *risk free rate* of interest. This becomes the *heat diffusion* equation

$$\frac{\partial u}{\partial \tau} = \frac{\partial^2 u}{\partial x^2} \quad \text{for} \quad -\infty < x < \infty,\, \tau > 0 \tag{33}$$

making the transformations

$$S := Ke^x, \quad t := T - \frac{2\tau}{\sigma^2}$$
$$C := e^{-1/2(k-1)x - 1/4(k+1)^2\tau} K u(x,\tau)\,,$$

where $k := 2r/\sigma^2$, K is the *exercise price* and T is the *time to maturity* of the option to be valued. The boundary conditions for the PDE depend on the specific type of option. For a vanilla *European call option* these are:

$$C(0,t) = 0, \quad C(S,t) \sim S \quad \text{as} \quad S \to \infty, \quad C(S,T) = \max(S-K,0)\,.$$

The boundary conditions for the transformed PDE are

$$\begin{aligned} u(x,\tau) &= 0 && \text{as} \quad x \to -\infty\,, \\ u(x,\tau) &= e^{1/2(k+1)x + 1/4(k+1)^2\tau} && \text{as} \quad x \to \infty\,, \\ u(x,0) &= \max(e^{1/2(k+1)x} - e^{1/2(k-1)x}, 0)\,. \end{aligned}$$

The first binary option we solve is the *cash-or-nothing call* option with payoff given by $B\mathcal{H}(S-K)$, where $\mathcal{H}$ is the Heaviside function, i.e., the payoff is B if at expiry the stock price $S > K$. The boundary conditions for this option in the transformed domain are

$$\begin{aligned} u(x,\tau) &= 0 && \text{as} \quad x \to -\infty\,, \\ u(x,\tau) &= \frac{B}{K} e^{\frac{1}{2}(k-1)x + \frac{1}{4}(k+1)^2\tau} && \text{as} \quad x \to \infty\,, \\ u(x,0) &= e^{\frac{1}{2}(k-1)x}\frac{B}{K}\mathcal{H}(Ke^x - K)\,. \end{aligned}$$

The second binary option is a *supershare call* [20] option that pays an amount $1/d$ if the stock price lies between K and $K+d$ at expiry. Its payoff is thus $(\mathcal{H}(S-K) - \mathcal{H}(S-K-d))/d$ which becomes a *delta function* in the limit $d \to 0$. The initial boundary condition for this option is

$$u(x,0) = \frac{1}{dK} e^{\frac{1}{2}(k-1)x}(\mathcal{H}(Ke^x - K) - \mathcal{H}(Ke^x - K - d))\,.$$

For all three options valued the solution is transformed back to the original variables using the transformation

$$C(S,t) = K^{\frac{1}{2}(k+1)} S^{\frac{1}{2}(1-k)} e^{\frac{1}{8}(k+1)^2\sigma^2(T-t)} u(\log(S/K), 1/2\sigma^2(T-t)),$$

where $k = 2r/\sigma^2$.

There are closed form solutions for all the above options (see for example [20]). The Black–Scholes solution for the vanilla European call option is

$$C(S,t) = SN(d_1) - Ke^{-r(T-t)}N(d_2)$$

$$\begin{aligned} d_1 &:= \frac{\log(S/K) + (r + \frac{1}{2}\sigma^2)(T-t)}{\sigma\sqrt{(T-t)}} \\ d_2 &:= \frac{\log(S/K) + (r - \frac{1}{2}\sigma^2)(T-t)}{\sigma\sqrt{(T-t)}} \, . \end{aligned} \tag{34}$$

The solution for a cash or nothing call is

$$C(S,t) = Be^{-r(T-t)}N(d_2) \, . \tag{35}$$

The solution for the supershare option is

$$C(S,t) = \frac{1}{d} e^{-r(T-t)}(N(d_2) - N(d_3))$$

$$d_3 := d_2 - \frac{\log(1 + \frac{d}{K})}{\sigma\sqrt{T-t}} \, . \tag{36}$$

5.2. Cross currency swap products

A *cross currency swap* is a derivative contract between two counterparties to exchange cash flows in their respective domestic currencies. Such contracts are an increasing share of the global swap markets and are individually structured products with many complex variations. With one domestic and one foreign economy there are different term structure processes and risk preferences in each and a rate of currency exchange between them. We model the interest rates in a single factor extended Vasicek framework [7].

To value any European-style derivative security whose payoff is a measurable function with respect to a filtration $\mathcal{F}_T$ we may derive a PDE for its value. The *domestic* and *foreign bond prices* and *exchange rate* are specified in terms of the Gaussian state variables X_d, X_f and X_S whose corresponding driftless processes $\mathbf{X}_d$, $\mathbf{X}_f$ and $\mathbf{X}_S$ are sufficient statistics for movements in the term structure dynamics. Let $V = V(X_d, X_f, X_S, t)$ be the *domestic value function* of a security with a terminal payoff measurable with respect to $\mathcal{F}_T$ and no intermediate payments, and assume that $V \in C^{2,1}\left(\mathbb{R}^3 \times [0,T)\right)$. Then the *normalised domestic* *value*, defined by $V^*(t) := V(t)/P_d(t,T)$, satisfies the quasilinear parabolic PDE with time dependent coefficients given by

$$-\frac{\partial V^*}{\partial t} = \frac{1}{2}\lambda_d^2 \frac{\partial^2 V^*}{\partial X_d^2} + \frac{1}{2}\lambda_f^2 \frac{\partial^2 V^*}{\partial X_f^2} + \frac{1}{2} H^{SS} \frac{\partial^2 V^*}{\partial X_S^2} + H^{df} \frac{\partial^2 V^*}{\partial X_d \partial X_f} + H^{dS} \frac{\partial^2 V^*}{\partial X_d \partial X_S} + H^{fS} \frac{\partial^2 V^*}{\partial X_f \partial X_S}, \quad (37)$$

on $\mathbb{R}^3 \times [0,T)$. Here the functions H^{SS}, H^{df}, H^{dS} and H^{fS} are defined by

$$\begin{aligned} H^{SS}(s) &:= G_d^2(s)\lambda_d^2(s) + G_f^2(s)\lambda_f^2(s) + \sigma_S^2(s) \\ &\quad - 2\rho_{df}(s)G_d(s)\lambda_d(s)G_f(s)\lambda_f(s) \\ &\quad + 2\rho_{dS}(s)G_d(s)\lambda_d(s)\sigma_S(s) - 2\rho_{fS}(s)G_f(s)\lambda_f(s)\sigma_S(s) \\ H^{df}(s) &:= \rho_{df}(s)\lambda_d(s)\lambda_f(s) \\ H^{dS}(s) &:= \lambda_d(s)\left[G_d(s)\lambda_d(s) - \rho_{df}(s)G_f(s)\lambda_f(s) + \rho_{dS}(s)\sigma_S(s)\right] \\ H^{fS}(s) &:= \lambda_f(s)\left[\rho_{df}(s)G_d(s)\lambda_d(s) - G_f(s)\lambda_f(s) + \rho_{fS}(s)\sigma_S(s)\right] \end{aligned} \quad (38)$$

and the volatility is of the form

$$\begin{aligned} \sigma_k(t,T) &= [G_k(T) - G_k(t)]\lambda_k(t) \quad k = d, f, \\ G_k(t) &:= \frac{1 - e^{-\xi_k t}}{\xi_k} \quad k = d, f, \end{aligned} \quad (39)$$

for some *mean reversion rates* ξ_d and ξ_f and

$$\lambda_k(t) = e^{\xi_k t}\kappa_k(t) \quad k = d, f, \quad (40)$$

where $\kappa_k(t)$ is the *prospective variability* of the short rate. For the derivation of the PDE and further details of the extended Vasicek model see [7]. We solve the PDE with boundary conditions appropriate to any standard European-style derivative security.

The most common type of cross-currency swap is the exchange of floating or fixed rate interest payments on *notional principals* Z_d and Z_f in the domestic and foreign currencies respectively. Now we describe precisely the deal that we are going to value which differs from that of [7], see also [10].

5.3. Fixed-for-fixed cross-currency swap with a Bermudan option to cancel

The cross-currency swap tenor $[0,T]$ is divided into N_{cpn} *coupon periods*. The start and end dates for these periods are given by $T_0, \ldots, T_{N_{cpn}}$ and cashflows are exchanged at coupon period end dates $T_1, \ldots, T_{N_{cpn}}$. Typically the swap cashflows consist of coupon payments at annualized rates on notional amounts Z_d for the first currency and Z_f for the second currency. In addition, notional amounts Z_d and Z_f may be exchanged at the swap start and/or end dates. The size of a coupon payment is given by: *coupon rate* $\times$ *notional amount* $\times$ *coupon period day count*. Both interest rates R_f and R_d are *fixed* at the outset of the contract, as opposed to those for a LIBOR swap where they are floating LIBOR rates for each currency [7]. The path-dependence in the payoffs is completely due to foreign exchange risk since payments in the respective currencies are fully determined at payment dates. *Payments* p_j made at the end of each period at time t_j^- are of size

$$p_j = \delta_j \left(S(t_j^-) R_f Z_f - m - Z_d R_d \right) ,$$

where m is the *margin* to the issuing counterparty. This is the terminal condition for the period $[t_{j-1}, t_j)$. The *value* of the deal is the sum of the expected present values of all payments.

When the contract has a *Bermudan option to cancel*, one of the counterparties is given an option to cancel all the future payments at times $t_1, \ldots, t_n$. Typically $t_1, \ldots, t_n$ are set a fixed number of calendar days before the start date of each period, i.e., $t_1 = T_{N_{cpn}-1-n} - \Delta, \ldots, T_{N_{cpn}-1} - \Delta$, where Δ is the *notification period.* We assume that net principal amounts $(Z_d - S(0)Z_f))$ are paid at time 0 and at time t_N if the option is *not* cancelled, or at time t_{k+1} if the option *is* cancelled. The *terminal condition* at t_N^- is given by

$$V(t_N^-) = \delta_N \left(S(t_N^-) Z_f (1 + R_f) - m - Z_d (R_d + 1) \right) . \tag{41}$$

When the option to cancel is exercised at t_k, coupon payments due on $T_{N_{cpn}-2-n+k}$ and notional amounts are exchanged. The holder of the option will terminate the deal if its current expected present value is less than the termination cost. Thus the decision at time $t_{k+1-\Delta}$ is to *continue* if

$$\begin{aligned} &P_d(t_{k+1} - \Delta, t_{k+1}^-) V(t_{k+1}) \\ &\qquad < S(t_{k+1} - \Delta) Z_f P_f(t_{k+1} - \Delta, t_{k+1}^-) - P_d(t_{k+1} - \Delta, t_{k+1}^-) Z_d \, . \end{aligned}$$

The deal is valued by dynamic programming backward recursion, solving the PDE for the last period using the terminal condition (41) and stepping backwards in time using the initial condition (in time remaining to payment date)

$$\begin{aligned} &V(t_{k+1} - \Delta) \\ &= \min \left\{ P_d(t_{k+1} - \Delta, t_{k+1}^-) V(t_{k+1}), S(t_{k+1} - \Delta) Z_f P_f(t_{k+1} - \Delta, t_{k+1}^-) - P_d(t_{k+1} - \Delta, t_{k+1}^-) Z_d \right\} \end{aligned}$$

for earlier periods. The exchange of principals is added to the computed value at time 0.

6. Numerical Results

The numerical results using a 1D and 3D implementation of the wavelet method of lines algorithm and the LSODE stiff vector ODE solver [16] computed on an RS6000/590 running AIX 4.3 are given below. In each case the numerical deal values are compared with a standard PDE solution technique and the known exact solution. Practical speed-up factors are reported which increase with both boundary condition discontinuities and spatial dimension.

6.1. European call option

Stock price: 10 *Strike price*: 10 *Interest rate*: 5% *Volatility*: 20% *Time to maturity*: 1 Year.

The exact value of this option is: **1.04505**.

Comparing tables 1 and 2 shows a speedup of **1.9**.

TABLE 1. Wavelet Method of Lines Solution

Space Steps	Time Steps	Value	Solution Time in Seconds
64	60	1.03515	.05
128	100	1.04220	.10
256	200	1.04502	.13
512	200	**1.04505**	**.30**
1024	200	1.04505	.90

TABLE 2. Crank–Nicolson Finite Difference Method

Space Steps	Time Steps	Value	Solution Time in Seconds
64	60	1.03184	.02
128	100	1.04184	.04
256	200	1.04426	.09
512	200	1.04486	.16
1024	200	1.04501	.30
2000	200	**1.04505**	**.57**

6.2. Cash-or-nothing call

The option with the same parameters as the European call.

The payoff is $B.\mathcal{H}(S-K)$, where $B := 3$ is the cash given, with a single discontinuity.

The exact value of this option is: **1.59297**.

Comparing Tables 3 and 4 shows a speedup of **2.5**.

TABLE 3. Wavelet Method of Lines Solution

Space Steps	Time Steps	Value	Solution Time in Seconds
128	100	1.49683	.10
256	200	1.54904	.13
512	200	1.59216	.30
1024	400	**1.59288**	**1.02**

TABLE 4. Crank–Nicolson Finite Difference Scheme

Space Steps	Time Steps	Value	Solution Time in Seconds
128	200	1.46296	.04
256	400	1.53061	.10
512	400	1.56391	.18
1024	400	1.58046	.31
2048	800	1.58872	1.35
4096	800	**1.59285**	**2.56**

6.3. Supershare call

Stock price: 10 *Strike price*: 10 *Parameter* d: 3 *Interest rate*: 5% *Volatility*: 20% *Time to maturity*: 1 Year.

The option pays an amount $1/d$ if the stock price lies between K and $K + D$ i.e., the option has a payoff $1/d.(\mathcal{H}(S-K)-\mathcal{H}(S-K-D))$ with two discontinuities.

The exact value of this option is: **0.13855**.

Comparing Tables 5 and 6 shows a speedup of **4.9**.

TABLE 5. Wavelet Method of Lines Solution

Space Steps	Time Steps	Value	Solution Time in Seconds
128	100	0.12796	.10
256	200	0.13310	.14
512	200	0.13808	.30
1024	400	**0.13848**	**1.04**

TABLE 6. Crank–Nicolson Finite Difference Scheme

Space Steps	Time Steps	Value	Solution Time in Seconds
128	200	0.12369	.04
256	400	0.13290	.09
512	400	0.13435	.16
1024	400	0.13666	.34
2048	800	0.13787	1.35
4096	800	0.13800	2.56
8000	800	**0.13835**	**5.11**

6.4. Cross currency swap

Domestic fixed rate: 10%, *Foreign fixed rate:* 10%.

The exact value of this swap is: **0.0**.

Comparing Tables 7 and 8 shows a speedup exceeding **81**.

TABLE 7. Wavelet Method of Lines Solution

Discretization	Value	Solution Time in Seconds
$20 \times 8 \times 8 \times 8$	−0.00082	1.2
$20 \times 16 \times 16 \times 16$	−0.00052	6.54
$20 \times 32 \times 32 \times 32$	−0.00047	40.40
$40 \times 64 \times 64 \times 64$	**−0.00034**	**410.10**
$100 \times 128 \times 128 \times 128$	−0.00028	4240.30
$160 \times 256 \times 256 \times 256$	−0.00025	53348.10

TABLE 8. Explicit Finite Difference Scheme

Discretization	Value	Solution Time in Seconds
$20 \times 8 \times 8 \times 8$	−0.00109	0.28
$20 \times 16 \times 16 \times 16$	−0.00101	1.70
$20 \times 32 \times 32 \times 32$	−0.00074	16.82
$40 \times 64 \times 64 \times 64$	−0.00058	188.10
$100 \times 128 \times 128 \times 128$	−0.00046	2421.6
$160 \times 256 \times 256 \times 256$	**−0.00038**	**33341.8**

7. Conclusions and Future Directions

The wavelet method of lines performs well on problems with one spatial dimension and discontinuities or spikes in the payoff. For example for the supershare option the wavelet method requires a lower discretization than the Crank–Nicolson finite difference scheme for equivalent accuracy (three decimal places), as the discontinuities in the payoff can be resolved better in wavelet space. We also see that for the cross currency swap PDE in three spatial dimensions the (prototype) wavelet method outperforms the (tuned) explicit finite difference scheme by approximately two orders of magnitude – a very promising result. One of the important things to note is that $O(N)$ wavelet based PDE methods generalize $O(N \log N)$ spectral methods without their drawbacks. This lower basic complexity feature of the wavelet PDE method makes it suitable to solve higher dimensional PDEs. Further, to improve basic efficiency of the method we are currently implementing an adaptive wavelet technique in which the wavelet coefficients are thresholded at each time step (*cf.* Section 3). This should result in an improvement in both speed and memory usage because of sparse wavelet representation. Such a technique has resulted in a further order of magnitude speedup in other applications [18]. Future work will thus involve applying the thresholded wavelet collocation technique to solve cross currency swap problems with two and three factor interest rate models for each currency to result in solving respectively 5 and 7 spatial dimension parabolic PDEs.

8. Acknowledgements

We would like to acknowledge the partial support if Citicorp and of Piotr Karasinski and Jan Coatelem of Schroder Salomon Smith Barney and also thank Darren Richards and particularly Robert Prosser for their help with this research.

References

[1] G. Beylkin (1993). Wavelets and fast numerical algorithms. Lecture Notes for Short Course, AMS 93.

[2] G. Beylkin, R. Coifman and V. Rokhlin (1991). Fast wavelet transforms and numerical algorithms. *Comm. Pure and Appl. Math* **44** 141–183.

[3] A. Cohen, I. Daubechies and J. Feauveau (1992). Biorthogonal bases of compactly supported wavelets. *Comm. Pure and Appl. Math.* **45** 485–560.

[4] A. Cohen, S. Kaber, S. Muller and M. Postel (2000). Accurate adaptive multiresolution scheme for scalar conservation laws. Preprint, LAN University Paris.

[5] W. Dahmen, S. Muller and T. Schlinkmann (1999). On a robust adaptive multigrid solver for convection-dominated problems. Technical report, RWTH Aachen. IGPM Report No 171.

[6] I. Daubechies (1992). *Ten Lectures on Wavelets.* SIAM, Philadelphia.

[7] M. A. H. Dempster and J. P. Hutton (1997). Numerical valuation of cross-currency swaps and swaptions. In *Mathematics of Derivative Securities*, eds. M. A. H. Dempster and S. R. Pliska. Cambridge University Press, 473–503.

[8] G. Deslauriers and S. Dubuc (1989). Symmetric iterative interpolation processes. *Constr. Approx* **5** 49–68.

[9] D. Donoho (1992). Interpolating wavelet transforms. Presented at the NATO Advanced Study Institute conference, Ciocco, Italy.

[10] B. Magee (1999). Pay attention to interest. *Risk* (October) 67–71.

[11] R. Polikar The wavelet tutorial. `http://www.public.iastate.edu/~rpolikar/wavelet.html`.

[12] R. Prosser and R. Cant (1998). Evaluation of nonlinear terms using interpolating wavelets. Working paper, CFD laboratory, Department of Engineering, University of Cambridge.

[13] R. Prosser and R. Cant (1998). On the representation of derivatives using interpolating wavelets. Working paper, CFD laboratory, Department of Engineering, University of Cambridge.

[14] R. Prosser and R. Cant (1998). On the use of wavelets in computational combustion. Working paper, CFD laboratory, Department of Engineering, University of Cambridge.

[15] R. Prosser and R. Cant (1998). A wavelet-based method for the efficient simulation of combustion. *J. Comp. Phys.* **147** (2) 337–361.

[16] W. Schiesser (1991). *The Numerical Method of Lines.* Academic Press, Inc.

[17] W. SWELDENS (1996). Building your own wavelets at home. ACM SIGGRAPH Course Notes.

[18] O. VASILYEV, D. YUEN AND S. PAOLUCCI (1996). A fast adaptive wavelet collocation algorithm for multi-dimensional PDEs. *J. Comp. Phys.* **125** 498–512.

[19] O. VASILYEV, D. YUEN AND S. PAOLUCCI (1997). Wavelets: an alternative approach to solving PDEs. Research Report, Supercomputer Institute, University of Minnesota.

[20] P. WILMOTT, S. HOWISON AND J. DEWYNNE (1995). *The Mathematics of Financial Derivatives.* Cambridge University Press.

[21] J. XU AND W. SHANN (1992). Galerkin-wavelet methods for two-point boundary value problems. *Numerische Mathematik* **63** 123–144.

Centre for Financial Research
Judge Institute of Management
University of Cambridge
Cambridge CB2 1AG, United Kingdom
E-mail address: `mahd2@cam.ac.uk`
`www-cfr.jims.cam.ac.uk`
E-mail address: `ae206@cam.ac.uk`

Some Analytic Facts on the Generalized Hyperbolic Model

Ernst Eberlein and Sebastian Raible

Abstract. We study some properties of a new continuous-time model for financial time series which is driven by a class of Lévy processes instead of a Brownian motion. This model emerged from extensive empirical investigations. We discuss path properties of the driving process and aspects of the valuation of derivatives.

1. Introduction

Risk management based on quantitative methods has become a new challenge for the financial industry. On the one hand, heightened scrutiny on capital requirements forces financial institutions to implement risk management systems. On the other hand, management itself as well as shareholders have a vital interest in assuring that sound models are used to drive business and strategy decisions. The accuracy of risk measurement depends crucially on how key risk factors such as equity prices, interest rates, indices, foreign exchange rates, default risk, and volatilities are modeled. Most of the models used at present are based on Brownian motion as the driving process. Solutions of differential equations of the form

$$dS_t = \mu(t, S_t)dt + \sigma(t, S_t)dB_t \,,$$

where $(B_t)_{t\geq 0}$ denotes Brownian motion, are those models which are mathematically best understood. Usually, it is assumed that the normal distribution assumption underlying this class of models is not too far from reality. But empirical studies show that this is not the case. As an example, consider the empirical distribution of six hours USD/DEM returns shown in Figure 1. The deviation from normality is substantial.

This observation, which holds for most other risk factors as well, is motivation enough to study models that are driven by more general processes. *Lévy processes* are an excellent class to be considered here. They are generated by infinitely divisible distributions in the same way as Brownian motion is generated by the normal distribution. One of the first to use a non-normal Lévy process in finance was Mandelbrot [23]. He replaced Brownian motion by a symmetric α-stable Lévy motion with index $\alpha < 2$. More recently Madan and Seneta ([21, 22])

introduced the variance gamma model which is based on Lévy processes generated by variance gamma distributions.

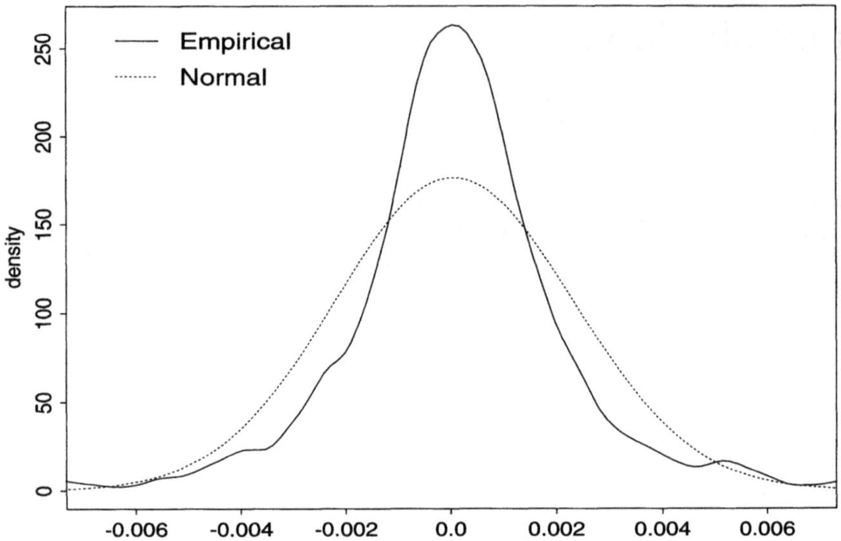

FIGURE 1. USD/DEM, six hours returns.

The *generalized hyperbolic model*, which was introduced in a series of papers ([10, 2, 11, 19, 12, 13]), emerged from extensive empirical investigations of financial time series. [8] is a survey of a number of aspects of this model. In the present article we add some analytic facts. Some of these properties are specific for the generalized hyperbolic model, while some properties are valid for models driven by Lévy processes in general.

2. The Generating Distribution

The basic building block of the generalized hyperbolic model is the *generalized hyperbolic distribution*, which generates the return process. It is given by its Lebesgue density

$$
\begin{aligned}
d_{GH}(x) = a(\lambda, \alpha, \beta, \delta) \left(\delta^2 + (x-\mu)^2\right)^{(\lambda - \frac{1}{2})/2} \\
\times K_{\lambda - \frac{1}{2}} \left(\alpha \sqrt{\delta^2 + (x-\mu)^2}\right) \exp(\beta(x-\mu))
\end{aligned}
\tag{1}
$$

where

$$
a(\lambda, \alpha, \beta, \delta) = \frac{\left(\alpha^2 - \beta^2\right)^{\lambda/2}}{\sqrt{2\pi}\, \alpha^{\lambda - \frac{1}{2}} \delta^\lambda K_\lambda\left(\delta\sqrt{\alpha^2 - \beta^2}\right)}
$$

and K_γ denotes the modified Bessel function of the third kind with index γ. This class of distributions was introduced in [1] and has been successfully used in the past in particular in turbulence studies. Its parameter space is $\lambda, \mu \in \mathbf{R}$, $\alpha > 0$, $-\alpha < \beta < \alpha$, and $\delta > 0$. The parameters have a clear meaning: λ is a class parameter that determines the shape of the distribution. For fixed λ, α determines the heaviness of the tails, β is a skewness parameter, and δ specifies the peakedness of the distribution in the center. μ is a location parameter. The class given by (1) contains a number of well-known distributions as special cases – either directly or as a limiting case. Among those are the generalized inverse Gaussian, the Student t, the Cauchy, the skewed Laplace, the exponential, the variance gamma, and, of course, the hyperbolic ($\lambda = 1$), and the normal inverse Gaussian ($\lambda = -\frac{1}{2}$) distributions. For details see the table in Chapter 1 of [24]. Some densities from the last mentioned subclass are plotted in Figure 2.

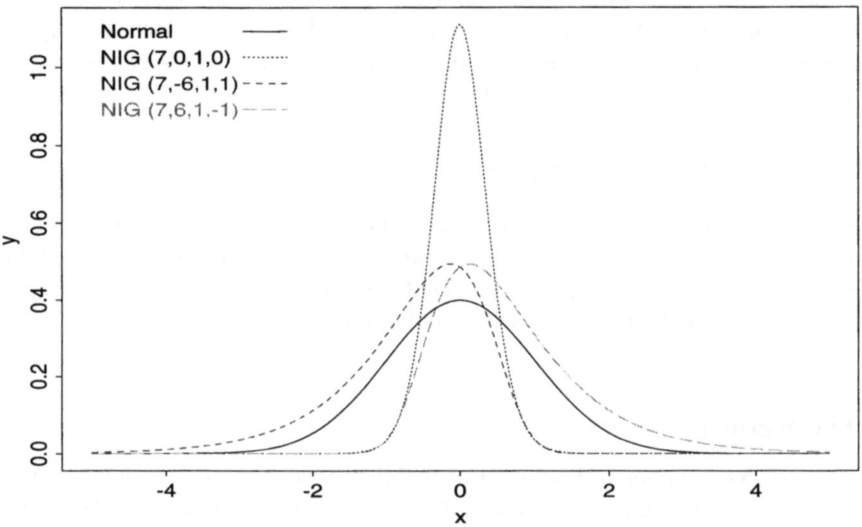

FIGURE 2. Densities of normal inverse Gaussian distributions with varying parameters $(\alpha, \beta, \delta, \mu)$

Looking at these graphs it is not surprising that the class of generalized hyperbolic distributions is flexible enough in order to allow an almost perfect fit to empirical return distributions as given in Figure 1.

The properties of the return process which we shall consider in the next section are completely determined by the Fourier transform Φ_{GH} of the distribution given by d_{GH}.

$$\Phi_{GH}(u) = e^{i\mu u} \frac{\left(\delta\sqrt{\alpha^2-\beta^2}\right)^\lambda}{K_\lambda\left(\delta\sqrt{\alpha^2-\beta^2}\right)} \cdot \frac{K_\lambda\left(\delta\sqrt{\alpha^2-(\beta+iu)^2}\right)}{\left(\delta\sqrt{\alpha^2-(\beta+iu)^2}\right)^\lambda}. \tag{2}$$

This formula can be derived easily by choosing the appropriate hyperbolic parameters under the integral $\int_{\mathbf{R}} e^{iux} d_{GH}(x)dx$ and using the fact that $\int_{\mathbf{R}} d_{GH}(x)dx = 1$. In its Lévy–Khintchine representation Φ_{GH} can be written as

$$\Phi_{GH}(u) = \exp\left(iub + \int_{-\infty}^{+\infty} \left(e^{iux} - 1 - iux\right) g(x)dx\right). \tag{3}$$

Note that there is no Gaussian component $-\frac{c}{2}u^2$ in the exponent. Since the expectation $E[GH] = \int_{\mathbf{R}} x d_{GH}(x)$ is finite, the drift coefficient b is exactly $E[GH]$. $g(x)$, the density of the Lévy measure F is explicitly given in [8], see equations (12) and (13) there, in an integral form.

For the qualitative path properties of the process generated by d_{GH} it is essential to understand the behavior of $g(x)$ near the origin. From the integral form mentioned above this cannot be derived easily, but the following expansion for $\varrho(x) = x^2 g(x)$ was obtained in [26]

$$\varrho(x) = \frac{\delta}{\pi} + \frac{\lambda + \frac{1}{2}}{2}|x| + \frac{\delta\beta}{\pi}x + o(|x|) \quad (x \to 0). \tag{4}$$

A direct consequence of this expansion is that $\int_D g(x)dx$ is not finite for any neighborhood D of the origin. In other words, the Lévy measure concentrates infinite mass around the origin. In the next section we shall state several deeper analytic results which are proved with the help of (4).

3. The Return Process

As exposed in a number of books on probability theory there is a one-to-one relationship between infinitely divisible distributions μ and Lévy processes $X = (X_t)_{t\geq 0}$, i.e., processes with stationary independent increments (see e.g., [4, Chapter 14.4]). The process X is constructed in such a way that μ is the law of X_1.

Generalized hyperbolic distributions are infinitely divisible as has been shown in [3]. Thus there is a Lévy process $(X_t)_{t\geq 0}$ uniquely defined by the fact that the law of X_1 has density d_{GH}. We call this process the *generalized hyperbolic Lévy motion* with parameters $(\lambda, \alpha, \beta, \delta, \mu)$. Figure 3 shows a simulation of a path of $(S_0 \exp(X_t))_{t\geq 0}$ for $S_0 = 100$ and $\lambda = -\frac{1}{2}$, i.e., for a generating normal inverse Gaussian distribution.

As we mentioned above, the exponent of Φ_{GH} in (3) has only a drift term and the integral describing the jumps, but no Gaussian term $-\frac{c}{2}u^2$. As a consequence the usual decomposition of Lévy processes (see e.g., [25, Theorem I.42])

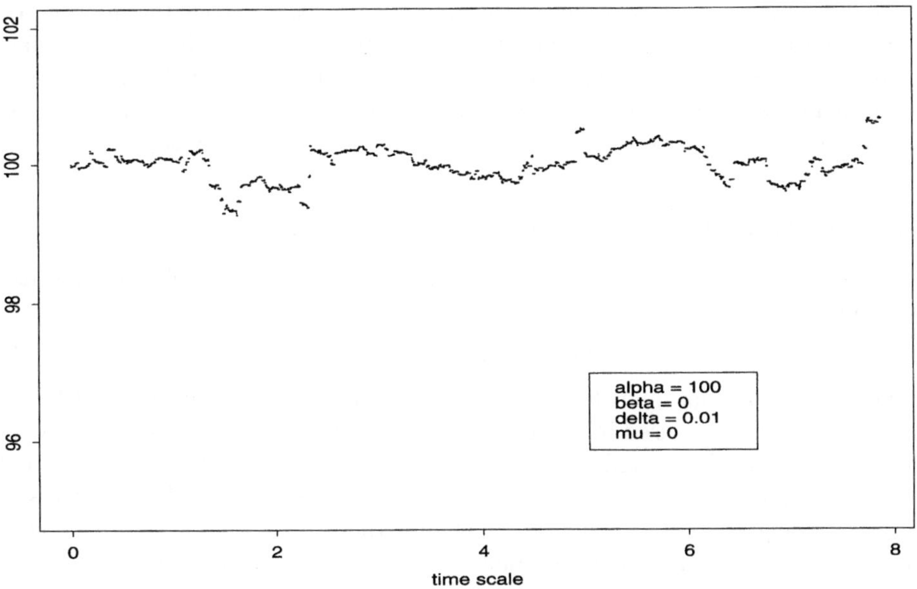

FIGURE 3. Simulation of a path $S_0 \exp(X_t)$.

into continuous martingale part, purely discontinuous martingale and drift is of the simple form

$$X_t = Z_t + \alpha t \tag{5}$$

where $(Z_t)_{t\geq 0}$ is a purely discontinuous martingale. This explains the qualitative behavior of paths as shown in Figure 3. The paths of $(X_t)_{t\geq 0}$ and thus of $(S_0 \exp(X_t))_{t\geq 0}$ change their values by jumps only. Since $\int_D g(x)dx$ is not finite as we pointed out in the previous section, there is an infinite number of small jumps in any interval of finite length. Do the paths of $(X_t)_{t\geq 0}$ contain also quantitative information? In particular, do the paths contain information on the parameters of the process? In [26, Chapter 2.6], the following answers are given.

Denote by $\Delta X_s = X_s - X_{s-}$ the jump of X at time s and by $\#A$ the number of elements of a finite set A. Let $(X_t)_{t\geq 0}$ be a generalized hyperbolic Lévy motion with parameters $(\lambda, \alpha, \beta, \delta, \mu)$; then for any $t > 0$ as $n \to \infty$,

$$N_n = \frac{1}{nt} \# \left\{ s \leq t \mid \Delta X_s \geq \frac{1}{n} \right\} \to \frac{\delta}{\pi} \text{ a.s.} \tag{6}$$

and

$$Y_n = X_t - \sum_{0 \leq s \leq t} \Delta X_s 1_{\{|\Delta X_s| \geq \frac{1}{n}\}} \to \mu t \text{ a.s.} \tag{7}$$

The first result follows from (4) and an application of the strong law of large numbers. For the second a convergence theorem for inverse martingales is needed.

The two convergence results mean that any typical path carries the information on the parameters δ and μ on intervals of arbitrary small length. Given this fact one would expect that these two parameters stay constant under the change to an equivalent martingale measure. In the next section we shall obtain a more precise statement: these two parameters and only these two do not change.

4. The Dynamic Asset Price Model

Let $X = (X_t)_{t \geq 0}$ be a Lévy process. The natural asset price model which we shall consider is given by

$$S_t = S_0 \exp(rt) \exp(X_t) , \tag{8}$$

where r is a constant interest rate. This is a slight modification of the model $S_t = S_0 \exp(X_t)$ which we used in previous papers ([12, 8]). The reason is that we shall usually consider discounted price processes $(\exp(-rt)S_t)_{t \geq 0}$ which are somewhat simpler in notation if we start with (8). Since $(rt + X_t)_{t \geq 0}$ is a Lévy process as well, choosing (8) only means that we change the drift of the log-return process. In case X is a generalized hyperbolic Lévy motion, (8) is called the *generalized hyperbolic model.* Formula (8) is a natural way to model asset prices. The reason is the following: If one looks at log returns $\log(S_t/S_{t-1})$ one gets, up to the constant r, the increments of length 1 of the process X in the exponent. By construction, these increments have a generalized hyperbolic distribution. Therefore, fitting the class of generalized hyperbolic distributions to the empirical distribution of daily price data of an asset yields a model where increments of length 1 of log-prices correspond exactly to what one observes in daily price changes. In this model this property is not restricted to time intervals of length 1. Once one has fitted a model on the basis of daily data, an interesting question is whether the distribution of increments along different time intervals – e.g., intervals which correspond to hourly or weekly price changes – are close to the corresponding empirical distributions. Empirical investigations confirm that for the generalized hyperbolic model this holds true. We call such a model *consistent.* This is a clear advantage of the general model (8) over the classical geometric Brownian motion, i.e., the special case where $(X_t)_{t \geq 0}$ is a Brownian motion with drift. In the classical model the log-increments are always normally distributed random variables, whatever time interval one considers. This property contradicts the observation (see Figure 4) that empirical return distributions move even further away from the normal distribution if the time span considered is less than one day.

Let us mention that (8) could also be introduced by a stochastic differential equation

$$dS_t = S_{t-} d\widetilde{X}_t \tag{9}$$

where $\widetilde{X} = (\widetilde{X}_t)_{t \geq 0}$ is a Lévy process whose jumps are strictly larger than -1. Conversely if $\widetilde{X}$ is a Lévy process with jumps strictly larger than -1, then the solution $S = (S_t)_{t \geq 0}$ of (9) which is the *stochastic* or *Doléans–Dade exponential* is

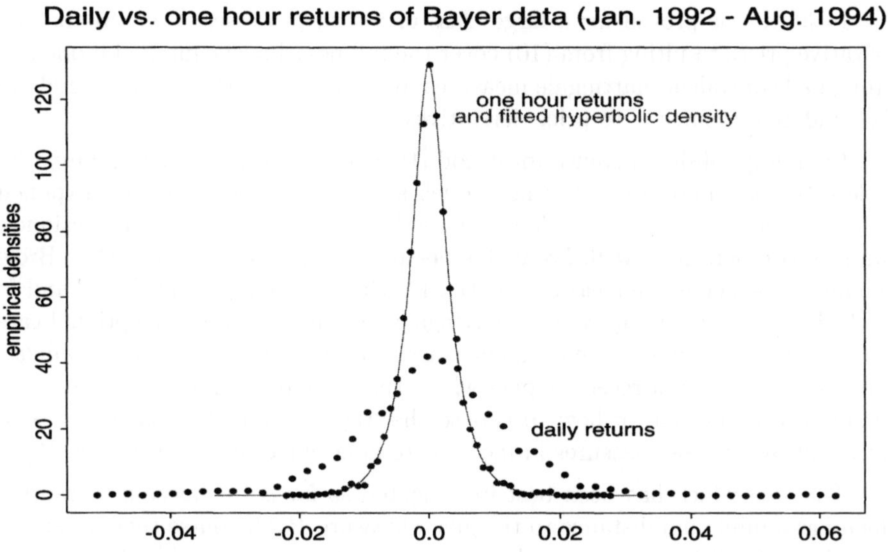

FIGURE 4. Comparison of return distributions along different time grids.

of the form (8) for some Lévy process X. The computation of $\widetilde{X}$ from X and vice versa is an easy exercise using the Lévy triplet which characterizes Lévy processes (see [15, Lemma 5.8]).

One of the most useful and most exciting applications of stochastic models in finance is the pricing of derivatives. Given a model $S = (S_t)_{t\geq 0}$ for the price fluctuations of the underlying instrument such as a stock or an index (DAX, S&P 500, etc.), let $w(S_T)$ denote the *payoff* of the derivative at maturity, i.e., at time $T > 0$. In the case of a European call option $w(S_T)$ would be $(S_T - K)^+$, where K is the strike price of the option. According to the fundamental theorem of asset pricing (see [7] for an advanced version) the correct, i.e., arbitrage-free price V_t of the derivative at time $t \in [0, T]$ is given by the conditional expected discounted payoff, where expectation is taken with respect to an equivalent martingale measure Q.

$$V_t = E^Q[\exp(-r(T-t))w(S_T) \mid \mathcal{F}_t]. \tag{10}$$

An *equivalent martingale measure* is a probability measure Q which is equivalent to the given (historical) probability measure P and under which the discounted price process $(e^{-rt}S_t)_{t\geq 0}$ is a martingale. Unfortunately for most models, in particular for those which are more realistic, the class of equivalent martingale measures is rather large.

In [9] it was proved for a large class of return processes X in (8) that the derivative prices resulting from (10) cover the full no-arbitrage interval if one goes through all equivalent martingale measures. In other words, there is an abundance of candidate measures Q to price derivatives.

In an unpublished manuscript Jacod [18] extended this result to many other models. Besides simple discrete-time models such as a one-period model, exponential Lévy processes (8) are considered where the return process X has a continuous Gaussian component as well. From this perspective the classical geometric Brownian motion which is the basis for the Black–Scholes option pricing theory and for which there is a unique equivalent martingale measure, is a very exceptional case. Models with more than one equivalent martingale measure are called *incomplete*. There are a number of recent papers where the authors obtain a more satisfying answer to the question of how to choose the 'right' candidate from the class of equivalent martingale measures in incomplete models (see e.g., [20, 16, 14, 19]).

The idea which has been best investigated so far is to choose a measure Q which has a minimum distance to the given measure P. Of course this measure Q depends on the distance one considers. Various distances such as L^1-, L^2-, entropy, or Hellinger distances have been studied in this context. A new aspect is that via duality theory the problem of choosing the minimum distance martingale measure can be translated into maximizing an expected utility and vice versa. The most general result in this context seems to be Goll and Rüschendorf [16] where the so-called f-divergence is used to measure the distance between probability measures. In this sense taking the equivalent martingale measure derived from the Esscher transform means that expected utility is maximized with respect to a power utility function $u(x) = \frac{x^p}{p}$. The Esscher transform measure has been used in a number of papers (see [10, 11, 12, 19, 24, 26]).

Now let Q be any equivalent martingale measure under which the return process $X = (X_t)_{t\geq 0}$ is again a Lévy process. Denote by (b, c, F) the Lévy–Khintchine triplet of X under Q (see [17, Chapter II.4c], for details) and denote by $\zeta_t = \ln\left(e^{r(T-t)}S_t\right)$ the log forward price process. Then V_t in (10) can be written as a function $f(\zeta_t, t)$ where

$$f(x,t) = \exp(-r(T-t))E^Q[w(\exp(x + X_{T-t}))]\,.$$

Write $\partial_i f$ for the partial derivative of f with respect to its ith argument and $\partial_{ii} f$ for the partial derivative of second order. Then the following analogue of the Black–Scholes partial differential equation holds.

Proposition 4.1. ([26, Proposition 1.12]) *Assume that $f(x,t)$ is of class $C^{(2,1)}$, that is, it is twice continuously differentiable in the first variable and continuously differentiable in the second variable. Assume that the law of X_t has support* $\mathbf{R}$.

Then $f(x,t)$ satisfies the following integrodifferential equation

$$-rf(x,t) + (\partial_2 f)(x,t) + (\partial_1 f)(x,t)b + \frac{1}{2}(\partial_{11} f)(x,t)c + \int \big(f(x+y,t) - f(x,t) - (\partial_1 f)(x,t)y\big)\, F(dy) = 0$$

$$f(x,T) = w(\exp(x)) \qquad (x \in \mathbf{R}, t \in (0,T)).$$

Although the interest rate r and the Lévy characteristics b, c enter linearly in this equation, it seems not to be easy to solve equations of this type. To actually compute option prices we use two different ways. In the case of the generalized hyperbolic model the risk-neutral densities are explicitly known and therefore from the general formula (10) one can derive a formula of Black–Scholes type for European calls

$$S_0 \int_\gamma^\infty d_{GH}^{*T}(x;\theta+1)dx - e^{-rT} K \int_\gamma^\infty d_{GH}^{*T}(x;\theta)dx$$

where $\gamma = \ln(K/S_0)$ and

$$d_{GH}^{*t}(x;\theta) = \frac{e^{\theta x} d_{GH}^{*t}(x)}{\int_{-\infty}^{+\infty} e^{\theta y} d_{GH}^{*t}(y)dy}$$

is the density of the law of X_t under the risk-neutral measure (see [12]).

The second approach, which was developed in [26], is much more powerful. It applies to general models (8) as soon as one knows the Fourier transform of the return distribution under the risk-neutral pricing measure Q. Explicit formulas of densities are not needed. The approach works for a number of exotic options as well, as long as the payoff depends on S_T only (e.g., power call and put options, self-quantos). It is based on the observation that the pricing formula can be represented as a convolution and that the bilateral Laplace transform of a convolution is the product of the corresponding transforms of the factors. The bilateral Laplace transform can be inverted very efficiently via Fast Fourier Transform. The method is more general than a recent approach of Carr and Madan [6] where these authors introduced dampened call values. The Fast Fourier Transform algorithm allows one to compute prices for all strikes of a series simultaneously. Figure 5 shows a typical profile of how generalized hyperbolic call option prices differ from Black–Scholes prices in the moneyness-time to maturity plane.

Understanding the class of equivalent martingale measures is the key to derivatives pricing. As has been shown in [9] it is sufficient to consider only those equivalent martingale measures Q under which the process $(X_t)_{t\geq 0}$ is again a Lévy process, since option values produced by the last mentioned subclass already cover the no-arbitrage interval. Let us consider now the generalized hyperbolic model in this context. Under the equivalent martingale measure Q the given process X will in general no longer be a generalized hyperbolic Lévy motion. One might ask if

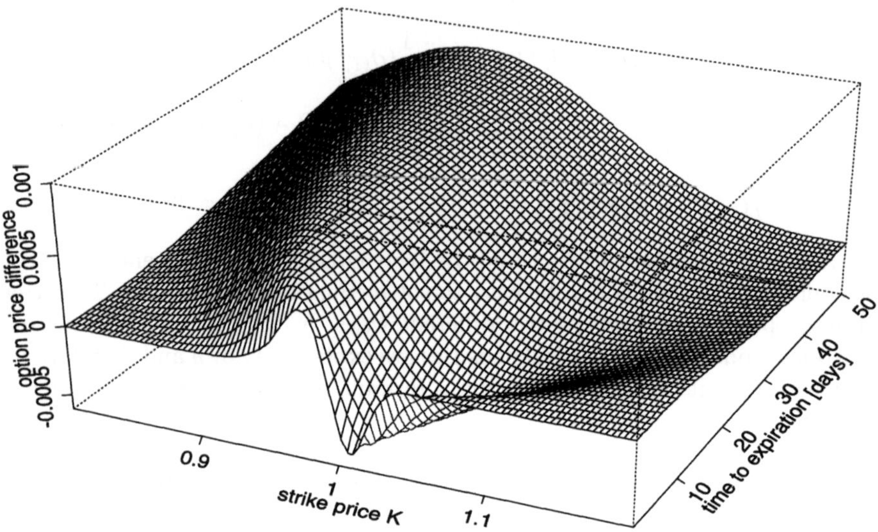

FIGURE 5. Difference of generalized hyperbolic call option prices and Black–Scholes call option prices as a function of strike price and time to expiration.

the range of option prices is narrowed if one considers only the subclass of equivalent martingale measures Q under which X is again a generalized hyperbolic Lévy process. It turns out that this is not the case ([26, Chapter 2.7]). As a consequence it is of great interest for option pricing to study equivalent martingale measures on the level of hyperbolic parameters. More precisely we shall ask which of the parameters $(\lambda, \alpha, \beta, \delta, \mu)$ can change if one replaces the given probability P by a (locally) equivalent probability Q. The answer is the following: X is again a generalized hyperbolic Lévy motion with parameters $(\lambda', \alpha', \beta', \delta', \mu')$ under a probability measure Q which is locally equivalent to P if and only if $\delta' = \delta$ and $\mu' = \mu$. For the proof of this result (see [26, Chapter 2.5]) one needs the expansion (4) and the general characterization of (locally) equivalent martingale measures in terms of Lévy–Khintchine triplets (b, c, F) as given in [17] (Theorem IV.4.39). A similar result is easy to obtain for classes of Lévy processes for which the Lévy density of the generating infinitely divisible distribution is of a simple structure. As an example we mention the class of CGMY Lévy processes introduced in [5].

References

[1] O. E. Barndorff-Nielsen, *Exponentially decreasing distributions for the logarithm of particle size*, Proceedings of the Royal Society London A, **353** (1977), 401–419.

[2] O. E. Barndorff-Nielsen, *Processes of normal inverse Gaussian type*, Finance and Stochastics, **2** (1998), 41–68.

[3] O. E. Barndorff-Nielsen and O. Halgreen, *Infinite divisibility of the hyperbolic and generalized inverse Gaussian distributions*, Zeitschrift für Wahrscheinlichkeitstheorie und verwandte Gebiete, **38** (1977), 309–312.

[4] L. Breiman, *Probability*, Addison-Wesley (1968).

[5] P. Carr, H. Geman, D. B. Madan and M. Yor, *The fine structure of asset returns: an empirical investigation*, Preprint (2000).

[6] P. Carr and D. B. Madan, *Option valuation using the fast Fourier transform*, J. of Computational Finance, **2** (1999), 61–73.

[7] F. Delbaen and W. Schachermayer, *A general version of the fundamental theorem of asset pricing*, Math. Ann., **300** (1994), 463–520.

[8] E. Eberlein, *Application of generalized hyperbolic Lévy motions to finance*, in: Lévy Processes - Theory and Applications, O. E. Barndorff-Nielsen, T. Mikosch, and S. I. Resnick (eds.), Birkhäuser Boston (2000).

[9] E. Eberlein and J. Jacod, *On the range of options prices*, Finance & Stochastics, **1** (1997), 131–140.

[10] E. Eberlein and U. Keller, *Hyperbolic distributions in finance*, Bernoulli, **1** (1995), 281–299.

[11] E. Eberlein, U. Keller and K. Prause, *New insights into smile, mispricing and value at risk: The hyperbolic model*, Journal of Business, **71** (1998), 371–406.

[12] E. Eberlein and K. Prause, *The generalized hyperbolic model: Financial derivatives and risk measures*, FDM-Preprint, **56** (1998).

[13] E. Eberlein and S. Raible, *Term structure models driven by general Lévy processes*, Mathematical Finance, **9**(1) (1999), 31–53.

[14] M. Frittelli, *The minimal entropy martingale measure and the valuation problem in incomplete markets*, Mathematical Finance, **10**(1) (2000), 39–52.

[15] Th. Goll and J. Kallsen, *Optimal portfolios for logarithmic utility*, Stochastic Processes and their Applications, **89** (2000), 31–48.

[16] Th. Goll and L. Rüschendorf, *Minimax and minimal distance martingale measures and their relationship to portfolio optimization*, Preprint Universität Freiburg (2000).

[17] J. Jacod and A. N. Shiryaev, *Limit Theorems for Stochastic Processes*, Springer-Verlag (1987).

[18] J. Jacod, unpublished manuscript (1997).

[19] U. Keller, *Realistic Modelling of Financial Derivatives*, Dissertation Universität Freiburg (1997).

[20] D. Kramkov and W. Schachermayer, *The asymptotic elasticity of utility functions and optimal investment in incomplete markets*, Ann. of Applied Probability, **9** (1999), 904–950.

[21] D. B. Madan and E. Seneta, *Chebyshev polynomial approximations and characteristic function estimation*, J. of the Royal Stat. Society Series B, **49**(2) (1987), 163–169.

[22] D. B. Madan and E. Seneta, *The variance gamma (V.G.) model for share market returns*, J. of Business, **63** (1990), 511–524.

[23] B. Mandelbrot, *The variation of certain speculative prices*, J. of Business, **36** (1963), 394–419.

[24] K. Prause, *The Generalized Hyperbolic Model: Estimation, Financial Derivatives, and Risk Measures*, Dissertation Universität Freiburg (1999).

[25] Ph. Protter, *Stochastic Integration and Differential Equations*, Springer-Verlag (1990).

[26] S. Raible, *Lévy Processes in Finance: Theory, Numerics, and Empirical Facts*, Dissertation Universität Freiburg (2000).

Institute for Mathematical Stochastics
University of Freiburg
Eckerstr. 1
D-79104 Freiburg, Germany
E-mail address: eberlein@stochastik.uni-freiburg.de
E-mail address: s.raible@insiders-fs.com

Functionals of Brownian Motion in Path-Dependent Option Valuation

Hélyette Geman

Abstract. Path-dependent options have become increasingly popular over the last few years, in particular in FX markets, because of the greater precision with which they allow investors to choose or avoid exposure to well-defined sources of risk. The goal of the paper is to exhibit the power of stochastic time changes and Laplace transform techniques in the evaluation and hedging of path-dependent options in the Black-Scholes-Merton setting. We illustrate these properties in the specific case of Asian options and continuously (de)activating double-barrier options and show that in both cases, the pricing, and as importantly, the hedging results are more accurate than the ones obtained through Monte Carlo simulations.

1. Introduction

Over the last ten years, the so-called "exotic" or path-dependent options have become increasingly popular in equity markets, and even more so in commodity and FX markets. As of today, ninety five per cent of options exchanged on oil and oil spreads are Asian. On the other hand, barrier options allow portfolio managers to hedge at a lower cost against extreme moves of stock or currency prices.

In order to overcome the technical difficulties associated with the valuation and hedging of path-dependent options even in the classical geometric Brownian motion setting of the Black-Scholes-Merton model, practitioners, taking advantage of the power of new computers and workstations, make with good reasons a great use of Monte Carlo simulations to price path-dependent options. However, our claim is that the results are not always extremely accurate: the most obvious example is the case of "continuously exploding" barrier options, heavily traded in the FX markets, and where the option is activated (or desactivated) at any point in the day where the underlying exchange rate hits a barrier. We recall that a barrier option provides the standard Black-Scholes payoff $\max(0, S_T - k)$ only if (or unless) the underlying asset S has reached a prespecified barrier L, smaller or greater than the strike price k, during the lifetime $[0, T]$ of the option. In the

Key words and phrases. Path-dependent options, Stochastic time changes, Laplace transform, Monte Carlo simulations, Hedging strategies.

equity markets, the classical situation for barrier contracts is that S_t is compared with L only at the end of each day (daily fixings). In contrast, in the FX markets, the comparison is made quasi-continuously and (de)activation may occur at any point in time. Obviously, the valuation of the option by Monte Carlo simulations, built piecewise by definition, may lead to fatal inaccuracies, in particular when the value of the underlying instrument is near the barrier close to maturity (entailing at the same time hedging difficulties well-known by option traders).

Along the same lines, when computing the Value at Risk (VaR) of a complex position or of a portfolio (for a given horizon T and a confidence level p, the value at risk is the loss in market value over the time horizon T that is exceeded with probability $(1-p)$), Monte Carlo simulations allow one to represent different scenarios on the state variables. But if the price of every exotic security in each scenario is in turn computed through Monte Carlo simulations, one has to face "Monte Carlo of Monte Carlo" and it becomes impossible, even with powerful computers, to calculate the VaR of the portfolio overnight. In an analytical methodology, since we obtain quasi-explicit solutions, the new values of the options can immediately be computed by incorporating in the pricing formulas the parameters corresponding to the different scenarios; hence, the problems mentioned above in estimating VaR are dramatically reduced.

The remainder of the article is organized as follows. Section 2 recalls the definition of stochastic time changes and shows why they are very useful to price (and hedge), via Laplace transforms, path-dependent options. Section 3 examines the specific case of Asian options and offers comparisons with Monte Carlo simulation prices. Section 4 addresses the case of barrier and double-barrier options and illuminates the hedging difficulties near maturity when the underlying asset price is close to a barrier. Section 5 contains some concluding remarks.

2. Stochastic Time Changes and Laplace Transforms

Representing the randomness of the economy by the probability space (Ω, F, F_t, P) where F_t represents the filtration of information available at time t and P the objective probability measure, we assume, as in the classical Black-Scholes-Merton setting, that the dynamics of the underlying asset price process $(S(t))_{t\geq 0}$ are driven by the stochastic differential equation

$$\frac{dS_t}{S_t} = \mu\, dt + \sigma\, d\hat{W}_t \tag{1}$$

where μ is a real number, σ is a strictly positive number and $(\hat{W}_t)_{t\geq 0}$ is a P-Brownian motion. Introducing the assumption of no arbitrage, we know from the seminal papers by Harrison-Kreps [12] and Harrison-Pliska [13], that there exists a so-called risk adjusted probability measure Q under which the dynamics of $S(t)$ become

$$\frac{dS_t}{S_t} = (r-y)\, dt + \sigma\, dW_t \tag{1'}$$

where $(W_t)_{t\geq 0}$ is a Q-Brownian motion and y denotes the continuous dividend rate of the underlying stock, supposed to be constant over the lifetime $[0,T]$ of this option. Equation (1) expresses the key mathematical assumption in the Black-Scholes model, namely that the underlying asset price is a geometric Brownian motion under the true probability measure. Since there is in this representation *one* source of risk, namely the Brownian motion $(\hat{W}_t)_{t\geq 0}$, it follows from central results of finance such as the Capital Asset Pricing Model or the Arbitrage Pricing Theory (S. Ross, 1976) that the expected return on a risky security should out perform the risk-free rate r over the period by *one* risk premium, i.e.,

$$E_p\left(\frac{dS_t}{S_t}\right) = dt\ (r + \textit{risk premium})$$

where r is supposed to be constant in the Black-Scholes model. The risk premium can be written as the positive constant σ times a quantity λ called the market price of equity risk. It is then possible to rewrite equation (1) as

$$E_p\left(\frac{dS_t}{S_t}\right) = r\,dt + \sigma(\lambda\,dt + d\hat{W}_t)$$

and Girsanov's theorem allows one to obtain equation (1′) (it also provides the expression of the Radon-Nykodim derivative $\frac{dQ}{dP}$ in terms of λ – constant or not – and $(\hat{W}_t)$).

From now on, we will be working under the probability measure Q in order to be allowed to write the price of an option as the expectation of the discounted terminal pay-off.

As explained for instance by Kemna and Vorst [15] who studied the valuation of average-rate options when these instruments started becoming very popular in the financial markets, the pricing difficulties are fundamentally conveyed by the fact that the representation of $(S_t)_{t\geq 0}$ by a geometric Brownian motion as described in equation (1) (and which is crucial for the simple proofs-through a partial differential equation or a probabilistic approach-of the Black-Scholes formula) is not transmitted to the average of S; hence, the idea of searching for a class of stochastic processes stable under additivity and related to the geometric Brownian motion. The so-called squared-Bessel processes, denoted hereafter BESQ(t), have the remarkable property of being Markov processes (which is the assumption common to nearly all models of option pricing and yield curve deformations) and of being stable by additivity. Moreover, a remarkable theorem due to Williams (1974), establishes that

$$S(t) = \text{BESQ}[X(t)] \tag{2}$$

where the processes X and BESQ are completely defined in terms of the parameters of the geometric Brownian motion $(S(t))_{t\geq 0}$. Equation (2) simply states that the value of S at time t is equal to the value of the squared-Bessel process BESQ at time $X(t)$. X defines a *stochastic time change* and formula (2) expresses that a geometric Brownian motion is a time-changed squared-Bessel process. In full generality, the only condition a process X has to satisfy in order to define a time change is to be (almost surely) increasing since time cannot go backwards. Other properties such as independent or identical increments may or may not be satisfied; when both are satisfied, the time change is called subordination (see [5]). Stochastic time changes are very useful for the pricing of exotic options, as this chapter will try to show. They have also become extremely popular when studying asset price dynamics: $X(t)$ may represent random sampling times in a financial time series as a function of calendar time t. The time change $X(t)$ may also account for differences in market activity at different hours in the day or because of new information release. [7] and [2] show that normality of asset returns which does not hold in calendar time can be recovered through a stochastic time change defined by the number of transactions.

Coming back to exotic option pricing, and assuming that the underlying asset return is a Brownian motion with drift as stated in equation (1), a very useful time change is obtained by choosing for $X(t)$ an exponential variable independent of the Brownian motion $(W(t))$. Indeed, when pricing barrier, double-barrier or corridor options, the quantities whose expectations (under the right probability measure) provide the option prices, involve functionals of Brownian motion such as the maximum M_T or the minimum I_T over the period $[0, T]$. The trivariate joint distribution of (M_T, I_T, S_T) has been known for some time (see [3]); however, its expression is complex. Pitman-Yor [19] show that this quantity becomes much simpler when the fixed time T (maturity of the option in our setting) is replaced by an exponential random variable τ independent of the Brownian motion contained in the dynamics of the process $(S_t)_{t>0}$. Remembering the expression of the density of an exponential variable, it is easy to see that in order to exploit the property above mentioned, we are naturally led to compute the Laplace transform of the option price with respect to its maturity, since $\int_0^\infty C(T)e^{-\lambda T}\,dT$ can be interpreted (up to the factor λ) as the expectation of $C(\tau)$ where τ is an exponential variable. Lastly, let us observe that the integral $\int_0^\infty S(s)\,ds$ is distributed as an inverse gamma variable. This interesting result was first proved by Dufresne (1990) in the analysis of perpetuities; another proof was given by Yor (1992). It must be noted however that the integral converges only if $r - q < \frac{\sigma^2}{2}$ (hence may not exist for non-dividend paying assets, since $\sigma = 0.02$ represents a fairly high volatility). Moreover, options traded in the financial markets have a finite maturity and the above described property cannot solve the Asian option valuation problem. The price $C(T)$ itself would have a much simpler expression for T infinite, which is obviously not the case for options traded on the markets.

3. The Case of Asian Options

As has been mentioned earlier and is substantiated by the continuously growing literature on the subject, Asian options have a number of attractive properties as financial instruments: for thinly traded assets and commodities (e.g., gold) or newly established exchanges, the averaging feature allows one to prevent possible manipulations on maturity day by investors or institutions holding large positions in the underlying asset. They are very popular among corporate treasurers who can hedge a series of cashflows denominated in a foreign currency by using an average-rate option as opposed to a portfolio of standard options; the hedge is obviously not identical but may be viewed as sufficient. Many domestic rates used in Europe and in the US as reference rates in floating-rate notes or interest rate swaps are defined as averages of spot rates; hence, caps and floors written on these rates are, by definition, Asian. To give an example very relevant in corporate finance, we can mention the so-called *contingent value rights*: a firm A wants to acquire a firm B. A is not willing to pay too high a price for the shares of company B but knows that this may lead to a failure of the takeover. Hence firm A will offer the shareholders of company B a share of the new firm AB accompanied by a contingent-value right on firm AB, maturing at time T (say two years later). This contingent-value right is nothing but an Asian put option. The put provides the classical protection of portfolio insurance; the Asian feature protects firm A for an exceptionally low market price of the share AB on day T, as well as the shareholders B in the case of a very high market price that day. These contingent-value rights were used when Dow Chemical acquired Marion Laboratory, when the French firm Rhône Poulenc acquired the American firm Rorer and more recently, when the insurance company Axa merged with Union des Assurances de Paris to form the second largest insurance company in the world (in the last case, the corresponding contingent value rights are still trading today). To give a last example of the usefulness of Asian options, we can mention that options written on oil or on oil spreads are mostly Asian since oil indices are generally defined as arithmetic averages. Many options written on gas have the same feature and the deregulation of the gas industry worldwide has entailed a significant growth of the gas derivatives market. Let us now turn to the valuation of these instruments.

3.1. The mathematical setting

We assume the asset price driven under the risk-adjusted probability measure Q by the dynamics described in (1′) where $y = 0$ for simplicity

$$dS(t) = rS(t)\,dt + \sigma S(t)\,dW(t)\,.$$

We also assume that the number of values whose average is computed is large enough to allow the representation of the average $A(T)$ over $[0, T]$ by the integral

$$A(T) = \frac{1}{T}\int_0^T S(u)\,du\,.$$

The value of an Asian call option at time t is expressed, by arbitrage arguments, as

$$C(t) = E_Q[e^{-r(T-t)}\max(A(T) - k, 0)/F_t] \tag{3}$$

where k is the strike price of the option and the discount factor may be pulled out of the expectation since we assumed constant interest rates. We know that the option has a unique price: there is only one source of randomness represented by the Brownian motion and a money-market instrument traded together with the risky security, which implies market completeness.

As mentioned earlier, the (important) mathematical difficulty in formula (3) stems from the fact that, denoting $A(t) = \frac{1}{T}\int_0^T S(u)\,du$, the process $(A(t))_{t\geq 0}$ is *not* a geometric Brownian motion. Many practitioners (see for instance Levy 1992) make this simplifying assumption and can then recover a Black-Scholes type pricing formula through the mere computation of the first two moments of $A(T)$. But to our knowledge, no upper bound of the error due to this approximation was ever provided.

Let us first observe that, when the option is traded at a date t posterior to date 0, the values of the underlying asset between 0 and t are fully known; the only randomness resides in the values to be taken by S between t and T. Hence, if the values observed between 0 and t are high enough, it may already be known at time t that the Asian call option will finish in the money and that we can make the simplification

$$\max(A(T) - k, 0) = A(T) - k,$$

since

$$A(T) > \frac{1}{T}\int_0^t S(u)\,du > k.$$

Writing

$$A(T) = \frac{1}{T}\int_0^t S(s)\,ds + \frac{1}{T}\int_t^T S(s)\,ds$$

and observing that the first term is fully known at date t, we obtain

$$E_Q[A(T) - k/F_t] = \frac{1}{T}\int_0^t S(s)\,ds - k + E_Q\left[\frac{1}{T}\int_0^T S(s)\,ds/F_t\right].$$

The linearity of the operators integral and expectation and the martingale property satisfied by the discounted price of S_t under Q allow one to compute explicitly the last term, namely

$$\begin{aligned}
E_Q\left[\frac{1}{T}\int_t^T S(s)\,ds/F_t\right] &= \frac{1}{T}\int_t^T E_Q[S(s)/F_t]\,ds \\
&= \frac{1}{T}\int_t^T S(t)e^{r(s-T)}\,ds \\
&= S(t)\frac{e^{r(t-T)-1}}{r(T-t)}.
\end{aligned}$$

We then obtain the Asian call price (when it is known at date T that the call is in the money) as

$$C(t) = S(t)\frac{1-e^{-r(T-t)}}{r(T-t)} + e^{-r(T-t)}\left[\frac{1}{T}\int_0^t S(s)\,ds - k\right]. \tag{4}$$

It is worth noting that the same type of considerations (Fubini theorem) allows one to compute fairly easily the exact moments of all orders of the arithmetic average, in contrast with the unnecessary approximations which are often offered in the literature.

Formula (4) has some striking resemblances with the Black-Scholes-Merton formula, the sign plus in the second term translating the moneyness of the Asian call in this situation.

Obviously, in most cases, this formula does not hold since at date t the quantity $\frac{1}{T}\int_0^t S(s)\,ds - k$ is likely to be non-positive. To address this difficult situation in an exact approach, a solution consists (see [9]) in

a) writing $S(t)$ as a time-changed squared Bessel process

b) choosing not to compute the option price itself but rather its Laplace transform with respect to maturity, namely the function $\varphi(\lambda) = \int_0^{+\infty} C(T)e^{-\lambda T}\,dT$.

Geman-Yor [9] give the details of the different mathematical steps which lead to the following expression for the call price

$$C(t) = \frac{4S(t)}{\sigma^2 T} e^{-r(T-t)} C^{(\nu)}(h,q) \tag{5}$$

where

$$\nu = \frac{2r}{\sigma^2} - 1; \quad h = \frac{\sigma^2}{4}(T-t); \quad q = \frac{\sigma^2}{4S(t)}\left\{kT - \int_0^t S(u)\,du\right\}$$

and the Laplace transform of the quantity $C^{(\nu)}$ with respect to h is given by

$$\int_0^\infty E^{-\lambda h} C^{(\nu)}(h,q)\,dh = \frac{\int_0^{1/2q} dx e^{-x} x^{\frac{\mu-\nu}{2}} (1-2qx)^{\frac{\mu-\nu}{2}+1}}{\lambda(\lambda - 2 - 2\nu)\Gamma\left(\frac{\mu-\nu}{2} - 1\right)} \tag{6}$$

where Γ denotes the gamma function and $\mu = \sqrt{2\lambda + \nu^2}$.

We can observe that when the underlying asset is a stock paying a continuous dividend y (y may also be the convenience yield of a commodity or the foreign interest rate in the case of a currency), the above results prevail exactly by replacing r by $r - y$ and ν by

$$\frac{2(r-y)}{\sigma^2} - 1.$$

The inversion of the Laplace transform in (6) provides not only the call price but also its delta through the same methodology. Indeed, the differentiation of formula (5) with respect to S gives

$$\Delta = \frac{\partial C_t}{\partial S_t} = \frac{C_t}{S_t} - \frac{e^{-r(T-t)}}{T}\frac{1}{S(t)}\left\{kT - \int_0^t S(u)\,du\right\}\frac{\partial C^\nu(h,q)}{\partial q} \qquad (7)$$

and we face an analogous problem of inversion of the Laplace transform.

Geman-Eydeland [8] on one hand, Fu-Madan-Wang [6] on the other hand apply different algorithms to invert the Geman-Yor formula but come up with results remarkably close (Fu-Madan-Wang use an algorithm developed by Abate and Whitt [1]; Geman-Eydeland use a method based on contour integration in the complex plane). The latter authors also provide comparisons with Monte Carlo simulations since this mathematically simple approach is very popular among practitioners and does not raise particular problems in the case of the smooth payoff of the average-rate option (in contrast to barrier options).

The following table gives some numerical results (the stock is assumed to pay no dividend over the period and the date of analysis t to be 0).

Interest rate r	Volatility σ	Maturity T	Strike price k	Initial value $S(0)$	$G-Y$	Monte Carlo
0.05	0.5	1	2	1.9	0.195	0.191
0.05	0.5	1	2	2	0.248	0.248
0.05	0.5	1	2	2.1	0.308	0.306
0.02	0.1	1	2	2	0.058	0.056
0.0125	0.25	2	2	2	0.1772	0.1771
0.05	0.5	2	2	2	0.351	0.347

The Monte Carlo values are obtained through a sample of 50 evaluations, each evaluation being performed on 500 Monte Carlo paths.

Turning to the computation of the delta of the option, for instance for $S(0) = 2$, we know that many practitioners use an elementary finite difference method with Monte Carlo values, which means in our example a delta equal to $\frac{0.306-0.191}{0.2} = 0.575$; by doing so, a much higher error appears in the delta than in the option price itself.

On the contrary, in the Laplace transform approach and thanks to the linearity of integration and derivation, the error does not deteriorate and the delta obtained in the above example is 0.56, a number significantly different than 0.575.

To end this section, let us observe that we have addressed the so-called fixed strike Asian option. A less popular type of Asian option has a floating strike, meaning that the pay-off at maturity is expressed as $\max(A_T - S_T, 0)$. Ingersoll in his book [14] conjectured that this case would be much simpler than the fixed-strike case. Indeed, taking the stock price as the numéraire (see Geman-El Karoui-Rochet 1995), one obtains a fairly simple partial differential equation satisfied by the Asian call option. The powerful change of numéraire technique, though still feasible, does not provide as simple a result for the fixed-strike Asian call options.

4. Barrier and Double-Barrier Options

Barrier options to which a vast body of literature is currently dedicated, represent the most common type of exotic options: they were traded in over-the counter markets in the United States many years before plain vanilla options were listed (see Snyder, 1969). The pricing of "single barrier" options is not very difficult in the standard Black-Scholes-Merton framework and closed-form solutions have been available for some time. The price of a down-and-out option was already in Merton's [18] seminal paper and in 1979, Goldman-Sosin-Gatto offered explicit solutions for all types of single barrier options.

We focus in this paragraph on double-barrier options which have become very popular recently. Not only, as mentioned earlier, do they provide a less expensive hedge which may be good enough in a number of situations. But they also allow investors with a specific view on the range of a stock price without any specific anticipation on the terminal value to take a position accordingly. We will be addressing the so-called "continuously deactivating" double-barrier options (meaning that the option vanishes at any time where the underlying asset price hits the upper barrier U or the lower barrier L), as opposed to comparing the daily fixings of a stock with the numbers U and L. This is the situation which prevails in the FX markets, where double-barrier options represent a significant fraction of options written every day. The methodology described below allows one to price, as a simpler case, the so-called corridor options which pay one at maturity if the underlying asset price has remained in the corridor during the lifetime of the option.

4.1. The mathematical setting

Assuming that the dynamics of the underlying asset are driven under Q by the same equation as before

$$\frac{dS_t}{S_t} = (r-y)\,dt + \sigma\,dW_t$$

and denoting by L the lower barrier and by U the upper barrier, we consider an option which vanishes as soon as either the upper or lower barrier is hit. The price of the call at time t is equal to

$$C(t) = E_Q\left[e^{-r(T-t)}\max(0, S(T)-k)1_{(\Sigma>T)}/F_t\right] \tag{8}$$

where $\Sigma = \{\inf t/S(t) \geq U \text{ or } S(t) \leq L\}$ is the first exit time of the process $(S(t))$ out of the interval $[L, U]$ and interest rates are supposed constant (as well as the possible dividend payment y). It is slightly easier to compute the quantity

$$D(t) = E_Q\left[e^{-r(T-t)}\max(0, S_t-k)1_{(\Sigma\leq T)}/F_t\right].$$

Obviously, the knowledge of $D(t)$ would give $C(t)$ since the two quantities add up to the Black-Scholes price.

Again, the expression whose expectation is computed (which is a functional of the brownian motion W_t through S_T and Σ) would be simpler if the fixed maturity date was replaced by an exponential time τ independent of (W_t). This leads us to compute the Laplace transform $\Psi(\lambda)$ of $D(t)$ with respect to maturity T. Geman-Yor [10] show that

$$\Psi(\lambda) = \frac{1}{\sigma^2}\zeta(\lambda/\sigma^2)$$

where

$$\zeta(\lambda) = \frac{sh(\mu b)}{sh[\mu(a+b)]}g_1(e^{-a}) + \frac{sh(\mu a)}{sh[\mu(a+b)]}g_2(e^b)$$

with

$$\alpha = \frac{1}{\sigma^2}\left(r - y - \frac{\sigma^2}{2}\right);\ L/S(t) = e^{-a};\ U/S(t) = e^b;\ h = k/S(t);\ \mu = \sqrt{2\lambda + \alpha^2};$$

$$g_1(e^{-a}) = \frac{h^{\alpha+1-\mu}e^{-\mu a}}{\mu(\mu-\alpha)(\mu-\alpha-1)};$$

$$g_2(e^b) = 2\left[\frac{e^{b(\alpha+1)}}{\mu^2 - (\alpha+1)^2}\right] + \frac{e^{-\mu b}h^{\alpha+1+\mu}}{\mu(\mu+\alpha)(\mu+\alpha+1)}.$$

Again, the numerical results obtained through the inversion of the Laplace transform are compared with Monte Carlo simulations. A first set of tests is performed with $t = 0$, $T = 1$ year

Parameters	$S(0) = 2$ $\sigma = 0.2$ $r = 0.02$ $k = 2$ $L = 1.5$ $u = 2.5$	$S(0) = 2$ $\sigma = 0.5$ $r = 0.05$ $k = 2$ $L = 1.5$ $U = 3$	$S(0) = 2$ $\sigma = 0.5$ $r = 0.05$ $k = 1.75$ $L = 1$ $U = 3$
$G - Y$ price	0.0411	0.0178	0.07615
Monte Carlo price (st. dev $= 0.003$)	0.0425	0.0191	0.0722

where the standard deviation is computed on a sample of 200 evaluations, each evaluation being performed on 5000 Monte Carlo paths with a step size of $1/365$ year.

In order to show the nonrobustness of Monte Carlo methods when we approach maturity while the price of the underlying asset is close to one of the barriers, we take the same parameters as in the first column of the above table except that $S(0)$ is supposed to be 2.4 and the time to maturity one month. The $G - Y$ method gives a call price equal to 0.17321 and there is no change in the accuracy nor in the computing time since the Laplace transform method is insensitive to the position of the underlying asset price with respect to the barrier. On the contrary, keeping the same step size of $1/365$ year gives a Monte Carlo standard deviation equal to 0.073 (which is clearly too high for practical purposes) and a Monte Carlo value for the call of 0.1930. By making the step four times smaller, the standard deviation is reduced to 0.008 and the price becomes 0.1739, which happens to be much closer to the $G - Y$ price and to be lower than 0.1930 (since in the first simulations, the option may have been overpriced through some trajectories "missing" the barrier while, in reality, the underlying asset had hit it, entailing the desactivation of the option).

5. Conclusion

The methodology involving stochastic time changes and Laplace transforms has been proved to be very efficient in the valuation and hedging of the most notoriously difficult European path-dependent options, namely the Asian and double-barrier options. The results have been obtained in the classical Black-Scholes-Merton model of a constant volatility. We can observe, however, that the introduction of a stochastic volatility in the underlying asset dynamics necessitates the use of a tree or of some numerical procedure (Monte Carlo or other). In all cases, the quasi-exact values obtained in the constant volatility case could be used as control variates to improve the accuracy of the numerical procedure.

References

[1] J. Abate and W. Whitt, *Numerical Inversion of Laplace Transforms of Probability Distributions,* ORSA Journal of Computing (1995).

[2] T. Ané and H. Geman, *Order Flow, Transaction Clock and Normality of Asset Returns,* the Journal of Finance, (2000).

[3] L. Bachelier, *Probabilités des Oscillations Maxima,* Note aux Comptes Rendus des Séances de l'Académie des Sciences (1941).

[4] F. Black and M. Scholes, *The Pricing of Options and Corporate Liabilities,* Journal of Political Economy **81** (1973), 637–654.

[5] P. Clark, *A subordinated Stochastic Process with Finite Variance for Speculative Price*, Econometrica **41**, no. 1 (1973), 135–156.

[6] M. Fu, D. Madan and T. Wang, *Pricing Continuous Time Asian Options: A Comparison of Analytical and Monte Carlo Methods*, Preprint, University of Maryland (1996).

[7] H. Geman and T. Ané, *Stochastic Subordination*, RISK, September (1996).

[8] H. Geman and A. Eydeland, *Domino Effect: Inverting the Laplace Transform*, RISK, April (1995).

[9] H. Geman and M. Yor, *Bessel Processes, Asian Options and Perpetuities*, Mathematical Finance **3**, no. 4 (1993), 349–375.

[10] H. Geman and M. Yor, *Pricing and Hedging Path-Dependent Options: A Probabilistic Approach*, Mathematical Finance **6**, no. 4 (1996), 365–378.

[11] M. Goldman, H. Sosin and M. Gatto, *Path Dependent Options: Buy at the Low, Sell at the High*, Journal of Finance **34** (1979), 111–127.

[12] J. M. Harrison and D. Kreps, *Martingales and Arbitrage in Multiperiod Securities Markets*, Journal of Economic Theory **20** (1979), 381–408.

[13] J. M. Harrison and S. R. Pliska, *Martingales and Stochastic Integrals in the Theory of Continuous Trading*, Stoch. Proc. Appl. **11** (1981), 215–260.

[14] J. Ingersoll, *Theory of Rational Decision Making, Rowman and Littlefield* (1987).

[15] A. Kemna and T. Vorst, *A Pricing Method for Options Based on Average Asset Values*, Journal of Banking and Finance **14** (1990), 113–129.

[16] N. Kunitomo and M. Ikeda, *Pricing Options with Curved Boundaries*, Mathematical Finance **2**, no. 4 (1992), 275–2.

[17] K. Itô and H. P. Mc Kean, *Diffusion Processes and Their Sample Paths*, Springer-Verlag (1965).

[18] R. C. Merton, *Theory of Rational Option Pricing*, Bell. J. Econ. Manag. Sci. **4** (1973), 141–183.

[19] J. Pitman and M. Yor, *The Laws of Homogeneous Functionals of Brownian Motion and Related Processes*, Preprint, University of Berkeley (1992).

Securities Markets, Commodities Markets and Risk Management
Université de Paris IX Dauphine
Place du Maréchal de Lattre de Tassigny
75775 Paris cedex 16, France
and
ESSEC Graduate Business School
E-mail address: `geman@essec.fr`

Optimal Portfolios under a Value at Risk Constraint

Ton Vorst

Abstract. Recently, financial institutions discovered that portfolios with a limited Value at Risk often showed returns that were close to the VaR and had large losses in the exceptional cases where losses exceeded VaR. In this paper we consider the construction of portfolios with options that maximize expected return with a restriction on the Value at Risk. These theoretically optimal portfolios indeed have the properties as experienced by financial institutions and illustrate that maximizing under a VaR-constraint is very dangerous. We also show that if one considers market prices of options there will be an even higher impetus to go for gambling portfolios.

1. Introduction

Over the last few years Value-at-Risk has become one of the standard instruments for measuring risk for banks and other financial institutions. Regulators such as the Bank for International Settlements recommend VaR-measures to determine capital adequacy requirements. Some institutions use the VaR-measure together with expected return to compare the attractiveness of different activities. Most research has focussed on methods to properly determine the VaR, where attention has been paid to the following issues: large portfolios, portfolios with derivatives, nonnormality of asset returns and description of the tails of distributions by extreme value theory. Furthermore, the concept of marginal VaR has been introduced to determine which parts of a portfolio have the highest impact on its VaR. However, there has only been very limited research on the determination of optimal portfolios in a VaR-framework. One of the few exceptions is a paper by Ahn, Boudoukh, Richardson and Whitelaw [1], who determine the portfolio with the lowest VaR that can be created by buying put options with a restriction on the option premium paid. Furthermore, Litterman [4, 5] describes assets that optimally reduce the VaR of a portfolio based on marginal VaR. These are called best hedges. In this paper we study a more general problem, where we try to determine the portfolio with the highest expected return under a restriction on the VaR. As stated before this combination of expected return and VaR is popular by some institutions to determine optimal portfolios. In most cases there are restrictions on the (number of) assets or positions that can be used in a portfolio. We will not invoke these kinds of restrictions and even allow trading in Arrow-Debreu securities. Within this general framework we will show that optimal portfolios are quite risky although they have a limited Value at Risk. They also bet on certain specific events to realize a high expected return. We are fully aware of the fact that Arrow-Debreu securities are not traded, but will argue that close substitutes can be created. Also, if only plain vanilla options are used to create an optimal portfolio our results will help in understanding the specific characteristics of these optimal portfolios. Our ideas are described with a simple 7-period Cox, Ross and Rubinstein [3] binomial model. We take a 7-period model since it is the simplest model within which we can demonstrate the intuition behind our results. Of course it can be extended to binomial models with more periods and to the continuous time Black-Scholes world as in Basak and Shapiro [2], but this only complicates the mathematics without providing much additional insight. Although our model is simple, according to Roth [7] it is believed to be able to explain some of the recent disasters in the financial world. Our results are based on theoretical option prices but we argue that if we consider market prices, which often include a "smirk" in implied option prices, our results will even be stronger.

2. The Optimization Model in Discrete Time

In this section we derive the basic result concerning portfolio strategies that maximize expected return under a value at risk constraint. If we could only select assets where returns are normally distributed, also all portfolios have normally distributed returns and a restriction on the Value at Risk would be equivalent to a restriction on the standard deviation. Now, maximizing expected return with a constraint on the standard deviation leads to the well-known Markowitz mean variance optimization. In this paper we will focus on the extra possibilities created in the market by options. These have definitely non-normal distributions and thus the mean variance framework does not apply. Hence, we focus on the case with only one basic asset and the complications come from the introduction of options. We assume that the investor wants to invest in the market index and uses (exotic) options to enhance the expected return of his portfolio and at the same time fulfill the VaR-constraint. The Value at Risk constraint is at the 99% level and the investor is not allowed to lose more than 10% of the initial portfolio value over the next ten days at this probability level.

Assume that the investor holds 1000 USD. We describe the movements of the index over the Value at Risk typical period by a non-recombining binomial tree with seven periods. Each period the index can go up by a factor u with probability 0.5 or go down by a factor d with the same probability. u and d should reflect the expected return and volatility of the index over a period in the binomial tree. For ease of exposition we assume that the riskless interest rate is equal to zero. This is not essential for our arguments. However, we also assume that investments in stocks come with a risk premium, i.e. the expected return of an investment in stocks is larger than zero. This implies that

$$0.5u + 0.5d - 1 > 0\,. \tag{1}$$

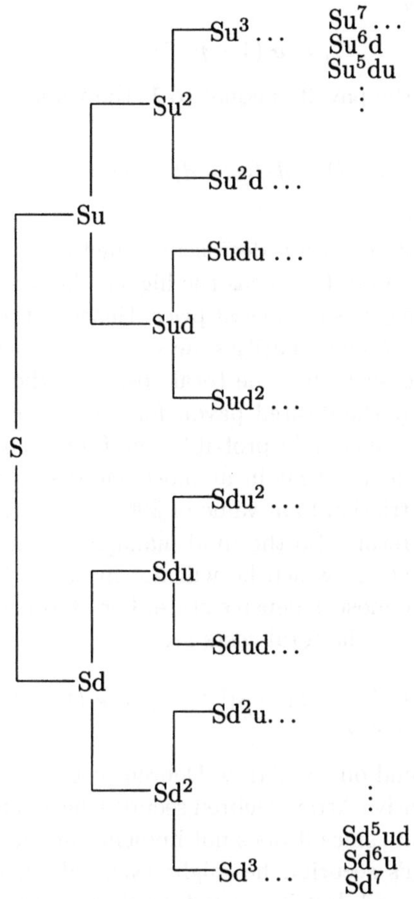

FIGURE 1. Non-recombining tree for price of stock index for seven periods.

This assumption is essential for our further arguments but is very natural in a world with risk averse investors. The left-hand side of equation (1) is known as the equity risk premium per period. If the initial index level is equal to S we get the binomial tree as given in Figure 1. In fact, we might also use a recombining binomial tree, but it is easier to illustrate our arguments with the non-recombining binomial tree. We assume one can trade in stocks and riskless bonds and hence it follows from the completeness of the market that for each final state of the binomial tree one can find a dynamic portfolio strategy that pays off 1 as this final state occurs and 0 in all other states. These are the payoffs of the well-known Arrow-Debreu securities. Since the payoffs can be constructed we assume that these Arrow-Debreu securities are also traded and the portfolio manager might use these securities to compose his optimal portfolio. It follows from the binomial Cox, Ross and Rubinstein option-pricing model that the cost of such an Arrow-Debreu security is given by

$$C_j = p^j(1-p)^{7-j} \tag{2}$$

if at the final node where the payoff is equal to 1, the value of the index is equal to Su^jd^{7-j}, where

$$p = (1-d)/(u-d) < 0.5 \tag{3}$$

since $0.5u + 0.5d > 1$.

Since $p < 0.5$ the costs C_j are a decreasing function of j. Hence, the state where the index goes down seven times has the highest Arrow-Debreu price, while the state with seven upswings has the lowest price. Hence, if the fund manager can spend his \$1000 on Arrow-Debreu securities, he simply specifies desired payoffs in all final states of the world such that the total costs, i.e. the sum of the Arrow-Debreu prices multiplied by the desired payoff for each state, is equal to \$1000. Since all states in the tree are equally probable, the fund manager has to specify the desired payoff in such a way that in at most one case, i.e. 1 out of 128, he does not meet the VaR-restriction final value of \$900. Furthermore, he would like to maximize his expected return. So the fund manager will select the state with seven downturns as the state in which he will not meet the \$900 level, since as remarked before, this is the most expensive state. For all other states he will buy 900 Arrow-Debreu securities. The total costs are

$$C = 900 \times \sum_{j=1}^{7} \binom{7}{j} p^j(1-p)^{7-j} = 900 \times (1-(1-p)^7)\,. \tag{4}$$

How much will he spend on the Arrow-Debreu security for the lowest state? Since this is the most expensive Arrow-Debreu security, he will not buy any Arrow-Debreu security for this state since it does not influence his Value at Risk. In fact, given the high price of this security, he might even take a short position. We assume that this is not allowed, but it is a potential danger in using the Value at Risk concept. What will the fund manager do with the remaining amount equal to $1000 - 900(1-(1-p)^7) = 100 + 900(1-p)^7$? Since, he has already secured his Value at Risk and all states are equally probable and have the same final payoffs, he will spend his money on the cheapest Arrow-Debreu securities, i.e. the state with seven upswings. This will give him the highest expected return. Hence, he will buy an extra

$$[100 + 900(1-p)^7]/p^7 \tag{5}$$

of these Arrow-Debreu securities. In Table 1 we give the total payoff in the seven upswings state for different values of the annualized volatility σ and the annualized equity risk premium μ of the index, where we set $u = e^{(\mu-\sigma^2/2)T/7+\sigma\sqrt{T/7}}$ and $d = e^{(\mu-\sigma^2/2)T/7-\sigma\sqrt{T/7}}$ and $T = 0.04$ for a ten trading days Value at Risk period.

	σ			
μ	0.15	0.2	0.25	0.3
0.02	15677.0	15398.7	15234.7	15126.5
0.04	16854.4	16252.5	15904.1	15677.0
0.06	18143.0	17165.9	16610.3	16252.5
0.08	19555.1	18142.9	17355.7	16854.3

Table 1. Payoffs of the optimal portfolio if the stock price goes up for seven consecutive days.

Hence, the portfolio that maximizes expected return under a Value at Risk constraint will end up with a probability of 126/128 at the level of 900, will end up worthless with a probability of $1/128 \leq 0.01$ and with the same probability will end up with the very high values given in Table 1. Given the Value at Risk constraint it is a gambling portfolio since it has a high probability of a loss and a very low probability of receiving 15 to 20 times the probable loss. It is not surprising that with a higher equity risk premium the payoff in the seven upswing state increases. If the underlying index is more volatile one has to offer some upside potential in order to remain within the Value at Risk limits. Hence the payoff in the extreme case decreases with an increasing volatility.

Our optimal portfolio is not path dependent, since the money invested in particular Arrow-Debreu securities depends only on the final value of the index. Hence, we can combine all Arrow-Debreu securities with the same final underlying stock price into one Arrow-Debreu security. Therefore only these securities have to be traded. We can also derive this result directly by considering a recombining tree instead of a non-recombining tree. In this case, there are only eight Arrow-Debreu securities, one for each $0 \leq j \leq 7$. Each state is specified by the number of upswings j. The probability of state j is equal to $\binom{n}{j}(0.5)^{n-j}/(0.5)^j$, while the Arrow-Debreu price is given by $\binom{n}{j}p^j(1-p)^{n-j}$. Hence, the price per unit of probability is given by

$$p^j(1-p)^{n-j}/(0.5)^n \tag{6}$$

which again is a decreasing function of j. Also in this setting the same portfolio will result that maximizes expected return given the Value at Risk constraint. Of course, one might argue that a seven period binomial tree is not very realistic and more periods or a continuous time model are needed to describe future portfolio values. If we would use more than seven periods the fund manager can do exactly the same, i.e. invest nothing for the lowest state, guarantee the Value at Risk level in all other states and use the remaining money to buy Arrow-Debreu securities, for the highest state. He might even select a few of the next lowest states to give a final payoff equal to 0 as long as the total probability of all states with 0 values does not exceed 1%. He uses the proceeds of this cost reduction to buy more

Arrow-Debreu securities in the highest states.The continuous geometric Brownian motion can be seen as a limiting case.

3. Relation with Markets in the Real World

One might argue that the Arrow-Debreu securities, as used in this paper, are not traded in financial markets. This is certainly true for the path dependent Arrow-Debreu securities. However, payoffs of non-path dependent Arrow-Debreu securities are offered by so-called digital options. These over-the-counter traded digital options pay off one unit of currency if the underlying value ends in a certain range $[X_2, X_3]$. Since, these digital options are only traded over the counter they might be quite expensive compared with their theoretical no-arbitrage price. However, the payoffs of these options can be approximated by combinations of standard options. For example, for exercise prices $X_1 < X_2 < X_3 < X_4$ with $X_2 - X_1 = X_4 - X_3 = 1/n$, one can buy n calls with exercise prices X_1 and n calls with exercise price X_4 and short the same number of call options with exercise prices X_2 and X_3.

Furthermore, our approach sheds light on the shape of optimal portfolios if one is allowed to buy and sell only standard options with a limited number of different exercise prices. The optimal portfolios have a strong tendency to have long positions in out-of-the money calls and short positions in out-of-the money puts. Especially, the tendency for out-of-the-money calls is also described in Oldenkamp [6].

In constructing the optimal portfolio we used theoretical Arrow-Debreu prices, which is equivalent to using theoretical option prices. One might wonder how the optimal portfolio would look if one used market prices, since it is well known that market prices deviate from theoretical prices. A way to measure these deviations is to consider implied volatilities, based on market prices, where options with high implied volatilities are relatively expensive compared with the theoretical prices and those with low implied volatilities are cheap. In equity derivatives markets one usually observes a "smirk" pattern, which means that out-of-the money calls have a low implied volatility and out-of-the money puts have a high implied volatility. This means that the Arrow-Debreu securities for the high final stock prices are priced cheaply in the market compared with the theoretical values and for low final stock prices, the Arrow-Debreu securities are expensive. Hence, based on market prices there will be an even stronger impetus to go for the gambling portfolios.

References

[1] D.-H. Ahn, J. Boudoukh, M. Richardson and R. F. Whitelaw, *Optimal Risk Management Using Options*, The Journal of Finance, **54(1)**, (1999), 359–375.

[2] S. Basak and A. Shapiro, *Value at Risk Based Risk Management: Optimal Policies and Asset Prices*, Working Paper Wharton School of the University of Pennsylvania (1999).

[3] J. C. Cox, S. A. Ross and M. Rubinstein, *Option Pricing: A Simplified Apprach*, Journal of Financial Economics, **7**, (1979), 229–264.

[4] R. Litterman, *Hot Spots and Hedges (I)*, in: Hedging with Trees, ed. Mark Broadie and P. Glasserman, 215–219, Risk Books, London (1998a).

[5] R. Litterman, *Hot Spots and Hedges (II)*, in: Hedging with Trees, ed. Mark Broadie and P. Glasserman, 221–228, Risk Books, London (1998b).

[6] B. Oldenkamp, *Derivatives in Portfolio Management*, Tinbergen Institute Research Series, Erasmus University Rotterdam (1999).

[7] B. Roth, *Risky Models for Traders*, Financial Times, April 8 (1999), 14.

Department of Finance
Erasmus University Rotterdam
P.O. Box 1738
NL-3000 DR Rotterdam, The Netherlands
E-mail address: `vorst@few.eur.nl`

Trace Formulas and Spectral Statistics of Diffractive Systems

Eugene Bogomolny

Abstract. Diffractive systems are quantum-mechanical models with point-like singularities where usual semiclassical approximation breaks down. An overview of recent investigations of such systems is presented. The following examples are considered in detail: (i) billiards (both integrable and chaotic) with small-size scatterers, (ii) pseudo-integrable polygonal plane billiards, and (iii) billiards with the Bohr-Aharonov flux lines. First, the diffractive trace formulas are discussed with particular emphasis on models where the diffractive coefficient diverges in certain directions. Second, it is demonstrated that the spectral statistics of diffractive models are different from the statistics of both integrable and chaotic systems. The main part of the lecture is devoted to analytical calculations of spectral statistics for certain diffractive models.

1. Introduction

Quantum chaos (or chaology) is a part of theoretical physics whose aim is the investigation of (multi-dimensional) quantum problems in the semiclassical limit $\hbar \to 0$. The modern development of this field is based on the semiclassical representation of the Green function, $G(\vec{x}, \vec{x}\,')$, through the sum over classical trajectories connecting points $\vec{x}$ and $\vec{x}\,'$ (see e.g. [16])

$$G(\vec{x}, \vec{x}\,') = \sum_{cl.\ tr.} A_{tr}(\vec{x}, \vec{x}\,') \exp(\frac{i}{\hbar} S_{cl}(\vec{x}, \vec{x}\,')) \qquad (1)$$

where the action, $S_{cl}(\vec{x}, \vec{x}\,')$, and the pre-factor, $A_{tr}(\vec{x}, \vec{x}\,')$ are computed from pure classical mechanics. In such an approach one implicitly assumes that the classical limit of quantum mechanical problem does exist which is not always the case. For a long time (see e.g. [17]) it was known that there are quantum systems without 'good' classical limit.

This lecture is devoted to the investigation of a particular type of such models, namely, diffractive systems whose characteristic property is the presence of point-like singularities (real or effective) in the quantum Hamiltonian. The existence of singularities in classical mechanics leads to the impossibility of continuing classical trajectories which hit these singularities. In quantum mechanics the situation is less dramatic [17]–[23]. Each point-like singularity can be described by a diffraction coefficient, $D(\vec{n},\vec{n}')$, which determines the scattering amplitude on the singularity. We normalize it in such a way that the semiclassical expansion of the Green function in the whole space in the presence of a singularity at point $\vec{x}_0$ has the following form

$$G(\vec{x},\vec{x}\,') = G_0(\vec{x},\vec{x}\,') + \frac{\hbar^2}{2m}\sum_{\vec{n},\vec{n}'} G_0(\vec{x},(\vec{x}_0,\vec{n}))D(\vec{n},\vec{n}')G_0((\vec{x}_0,\vec{n}'),\vec{x}\,') \quad (2)$$

where $G_0(\vec{x},\vec{x}\,')$ is the Green function (1) in the absence of singularity and $G_0(\vec{x},(\vec{x}_0,\vec{n}))$ is the term in (1) corresponding to a classical trajectory starting from point $\vec{x}$ and ending at point $\vec{x}_0$ with momentum in the direction $\vec{n}$.

The knowledge of the Green function permits us to write down the trace formula for diffractive systems [27, 19]. In the simplest case, when only one singularity is present, the density of states can be written as a sum of three terms

$$d(E) = \bar{d}(E) + d_p(E) + d_d(E) \quad (3)$$

where $\bar{d}(E)$ is the smooth part of the level density given by the Thomas-Fermi term (plus corrections if necessary) [16], $d_p(E)$ is the contribution from classical periodic orbits, and $d_d(E)$ is the diffractive contribution.

For integrable billiards [4]

$$d_p(E) = \frac{A}{2\pi}\sum_{ppo}\sum_{n=1}^{\infty}\frac{1}{\sqrt{2\pi k n l_p}}\cos(knl_p - \frac{\pi}{4} - \frac{\pi}{2}n\nu_p)\,. \quad (4)$$

For chaotic billiard systems [16]

$$d_p(E) = \sum_{ppo}\frac{l_p}{\pi k}\sum_{n=1}^{\infty}\frac{1}{|\det(M_p^n - 1)|^{1/2}}\cos(knl_p - \frac{\pi}{2}n\nu_p)\,. \quad (5)$$

In these formulas the summation is performed over all primitive periodic orbits (ppo) and their repetitions. l_p is the length of the ppo, ν_p is its Maslov index, and M_p is the monodromy matrix of the ppo.

In eq. (3) $d_d(E)$ is the contribution from classical orbits (called later diffractive orbits) which start from the singularity and end at it.

$$\begin{aligned} d_d(E) &= \sum_{m=1}^{\infty}\frac{1}{\pi m}\frac{\partial}{\partial E}\,\mathrm{Im}\sum_{\vec{n},\vec{n}'} G(\vec{n}_1,\vec{n}_1')D(\vec{n}_1',\vec{n}_2)G(\vec{n}_2,\vec{n}_2')D(\vec{n}_2',\vec{n}_3) \\ &\quad \cdots G(\vec{n}_m,\vec{n}_m')D(\vec{n}_m',\vec{n}_1)\,. \end{aligned} \quad (6)$$

Here $G(\vec{n},\vec{n}')$ is the contribution to the Green function (1) from a diffractive orbit which starts from the singular point with momentum in direction $\vec{n}$ and ends at this point with momentum in direction $\vec{n}'$. The sums in (6) can be transformed to

$$d_d(E) = -\frac{1}{\pi}\,\mathrm{Im}\,\frac{\partial}{\partial E}\ln\det(1-\hat{K}) \quad (7)$$

where the operator $\hat{K}$ is defined as

$$(\hat{K}\phi)(\vec{n}) = \sum_{\vec{n}',\vec{n}''} G(\vec{n},\vec{n}')D(\vec{n}',\vec{n}'')\phi(\vec{n}'') . \qquad (8)$$

Eqs. (7)–(8) mean that the formal 'quantization' condition of diffractive systems (from which energy levels can be computed) is

$$\det(1 - \hat{K}) = 0 . \qquad (9)$$

The main purpose of this paper is to discuss statistical properties of energy levels (i.e. the spectral statistics) of diffractive systems. The plan of the paper is the following. In Section 2 integrable systems with a diffractive center are considered and it is demonstrated that the spectral statistics of these systems can be computed analytically. In Section 3 the pseudo-integrable polygonal billiards and in Section 4 the billiards with an Aharonov-Bohm flux line are investigated. These models are examples of diffractive systems where the diffractive coefficient diverges in certain directions which considerably complicates all calculations. First, the trace formulas for these systems are derived and it is demonstrated that they differ from eq. (6). Second, by combining numerical and analytical arguments we investigate their spectral statistics. Finally, in Section 5 it is proved that the addition of a diffractive center to chaotic models does not change its spectral statistics.

2. Integrable Models with Diffractive Center

The simplest example of diffractive systems with a constant diffractive coefficient has been proposed in [21] and it consists of an integrable model (below we shall consider a 2-dimensional rectangular billiard) with a δ-function potential

$$V(\vec{x}) = \lambda\delta(\vec{x} - \vec{x}_0) . \qquad (10)$$

The quantization condition (9) in this case takes a particular simple form

$$\lambda G(\vec{x}_0, \vec{x}_0) = 1 \qquad (11)$$

or (ignoring renormalization problems [2])

$$\lambda \sum_{n=1}^{N} \frac{|\psi_n(\vec{x}_0)|^2}{E - e_n} = 1 \qquad (12)$$

where e_n and $\psi_n(\vec{x})$ are eigenvalues and eigenfunctions of the unperturbed system (i.e. without the δ-function potential). The natural question arises: what is the distribution of the new eigenvalues, E_n, provided the distributions of e_n and $\psi_n(\vec{x}_0)$ are known.

Let us for simplicity assume that

(i) e_n are independent random variables (as for integrable systems) with a step-like common distribution

$$d\mu(e) = \begin{cases} \frac{1}{2W} de\,, & \text{if } |e| \leq W \\ 0\,, & \text{otherwise} \end{cases} \tag{13}$$

and

(ii) $|\psi(\vec{x}_0)|^2 = 1$ (as for a rectangular billiard with periodic boundary conditions).

Under these assumptions eq. (12) takes the form

$$\sum_{j=1}^{N} \frac{1}{E - e_j} = \frac{1}{\lambda} \tag{14}$$

where each term in the sum is an independent random variable.

The exact density of solutions of this equation can be written in the form

$$\rho(E) = \delta\Big(\sum_{j=1}^{N} \frac{1}{E - e_j} - \frac{1}{\lambda}\Big) \sum_{k=1}^{N} \frac{1}{(E - e_k)^2}\,. \tag{15}$$

Representing the δ-function as a Fourier integral one can express the density of states through the characteristic function of the left-hand side of eq. (14)

$$\rho(E) = \int_{-\infty}^{\infty} \frac{d\alpha}{2\pi} \exp\Big(i\alpha\Big(\sum_{j=1}^{N} \frac{1}{E - e_j} - \frac{1}{\lambda}\Big)\Big) \sum_{k=1}^{N} \frac{1}{(E - e_k)^2}\,. \tag{16}$$

Because all e_j are assumed to be independent random variables, the correlation functions of E can, in principle, be computed straightforwardly.

In particular, the 2-point correlation function

$$R_2(E_1, E_2) = \langle \rho(E_1) \rho(E_2) \rangle \tag{17}$$

can be expressed in the form

$$\begin{aligned} R_2(E_1, E_2) &= \int \frac{d\alpha_1 d\alpha_2}{4\pi^2} [N (f(\alpha_1, \alpha_2))^{N-1} g(\alpha_1, \alpha_2) \\ &+ N(N-1)(f(\alpha_1, \alpha_2))^{N-2} \Psi_1(\alpha_1, \alpha_2) \Psi_2(\alpha_1, \alpha_2)] e^{-i(\alpha_1 + \alpha_2)/\lambda}\,, \end{aligned} \tag{18}$$

where

$$\begin{aligned} f(\alpha_1, \alpha_2) &= \int d\mu(e) \exp\Big(i\frac{\alpha_1}{E_1 - e} + i\frac{\alpha_2}{E_2 - e}\Big)\,, \\ g(\alpha_1, \alpha_2) &= \int d\mu(e) \exp\Big(i\frac{\alpha_1}{E_1 - e} + i\frac{\alpha_2}{E_2 - e}\Big) \frac{1}{(E_1 - e)^2 (E_2 - e)^2}\,, \\ \Psi_i(\alpha_1, \alpha_2) &= \int d\mu(e) \exp\Big(i\frac{\alpha_1}{E_1 - e} + i\frac{\alpha_2}{E_2 - e}\Big) \frac{1}{(E_i - e)^2}\,. \end{aligned}$$

When $N \to \infty$ the direct (but tedious) calculation of these integrals gives [6]

$$R_2(E_1, E_2) = \bar{\rho}^2 r_2(\epsilon)\,, \tag{19}$$

where

$$r_2(\epsilon) = 1 \qquad (20)$$
$$+\int_0^\infty d\alpha_1 \int_0^\infty d\alpha_2 (J_0^2(2\sqrt{\alpha_1\alpha_2}\,) + J_1^2(2\sqrt{\alpha_1\alpha_2}\,))e^{-2\pi\epsilon J(\alpha_1,\alpha_2)+2i(\alpha_1+\alpha_2)},$$

and

$$J(\alpha_1,\alpha_2) = \frac{1}{2}(\alpha_1+\alpha_2) + \frac{i}{2\pi\lambda_r}(\alpha_1-\alpha_2)$$
$$+ie^{i(\alpha_1+\alpha_2)}[(\alpha_1-\alpha_2)G(\alpha_1,\alpha_2) - i\alpha_1 J_0(2\sqrt{\alpha_1\alpha_2}\,) - \sqrt{\alpha_1\alpha_2}J_1(2\sqrt{\alpha_1\alpha_2}\,)],$$
$$G(\alpha_1,\alpha_2) = e^{i\alpha_2}\int_{\alpha_1}^\infty J_0(2\sqrt{\alpha_2 t}\,)e^{it}dt\,,$$
$$\frac{1}{\lambda_r} = \frac{1}{\bar\rho\lambda} + \ln\frac{W-E}{W+E},\ \epsilon = \bar\rho(E_1-E_2),\ \bar\rho = \frac{N}{2W}.$$

This exact formula permits us, in particular, to find the limiting behavior of the 2-point correlation function.

When $\epsilon \to 0$

$$r_2(\epsilon) \to A\epsilon \qquad (21)$$

where

$$A = 3\pi^2\int_0^\infty xJ_0((3+\delta)x)J_0^3(x)dx = \frac{\pi\sqrt{3}}{2} \approx 2.72 \qquad (22)$$

is independent of the coupling constant λ, and differs from the GOE prediction ($r_2(\epsilon) \to \pi^2/6\epsilon \approx 1.64\epsilon$).

When $\epsilon \to \infty$,

$$r_2(\epsilon) \to 1 + \frac{2}{\epsilon^2(\pi^2+1/\lambda_r^2)}\,. \qquad (23)$$

For GOE $r_2(\epsilon) \to 1 - 1/(\pi\epsilon)^2$.

These calculations give rise to different generalizations.

(i) The spectral statistics can also be computed for models where the residues, $|\psi(\vec{x}_0)|^2$ are either fixed quantity (different from 1) or independent random variables. In particular, for the rectangular billiard with the Dirichlet boundary conditions the introduction of the potential (10) (when coordinates of the scatterer are non-commensurable with the sides) leads to the following behavior of the 2-point correlation function at small ϵ,

$$r_2(\epsilon) \to \frac{\epsilon}{8\pi^3}\ln^4(\epsilon)\,. \qquad (24)$$

(ii) The spectral statistics of the Bohr-Mottelson model [14] which describes the interaction of one level with all others can be calculated by a similar method.

(iii) The asymptotic behavior of the 2-point correlation function (21)–(23) can be derived without the knowledge of the exact solution (20) even for more general cases.

(iv) In [9] it was demonstrated that, in the summation over periodic orbits of integrable systems, there exist hidden saddle points whose contributions give correct off-diagonal terms. This result permits us to construct a specific perturbation theory by which one can compute successive terms of the expansion at large ϵ of the 2-point correlation function for integrable systems with finite diffraction coefficient by semiclassical methods.

(v) Using the results of [9] it is possible to obtain an analog of trace formulas for composite operators (built from a product of the Green functions (1)), e.g. for conductance fluctuations [10].

3. Pseudo-Integrable Billiards

An interesting and important class of diffractive systems is 2-dimensional polygonal billiards with angles, α_i, equal rational multiples of π,

$$\alpha_i = \frac{m_i}{n_i}\pi \tag{25}$$

where m_i and n_i are coprime integers. These models (called pseudo-integrable billiards [20, 22]) have the characteristic property that their classical trajectories belong to a 2-dimensional surface with the genus

$$g = 1 + \frac{N}{2}\sum_i \frac{m_i - 1}{n_i} \tag{26}$$

where N is the least common multiplier of n_i.

Classical trajectories in these billiards are not defined after they hit a corner (25) with $m_i \neq 1$. These singular angles play the role of diffraction centers and the quantum diffraction coefficient can be read off from the exact Sommerfeld solution near a wedge with angle α [26],

$$D(\theta_f, \theta_i) = \frac{2}{\gamma}\sin\frac{\pi}{\gamma}\left[\frac{1}{\cos\frac{\pi}{\gamma} - \cos\frac{\theta_f+\theta_i}{\gamma}} - \frac{1}{\cos\frac{\pi}{\gamma} - \cos\frac{\theta_f-\theta_i}{\gamma}}\right] \tag{27}$$

where $\gamma = \alpha/\pi$.

An important difference between these models and the ones considered in the previous section is the fact that the diffraction coefficient (27) diverges in certain directions (called optical boundaries) to compensate the discontinuous behavior of classical trajectories. Consequently, diffractive orbits lying on optical boundaries cannot be described by eq. (6) and, first of all, a modification of the trace formulas is required [11] which can conveniently be done by using the Kirchoff approximation [26] valid in a vicinity of optical boundaries [11]. The basis of this approximation is the representation of the free (2-dimensional) Green function,

$G(\vec{x},\vec{x}\,')$ as an integral over a line separating points $\vec{x}$ and $\vec{x}\,'$

$$G(\vec{x},\vec{x}\,') = \int [G(\vec{q},\vec{x}\,')\frac{\partial}{\partial \vec{n}_q}G(\vec{x},\vec{q}\,) - G(\vec{x},\vec{q}\,)\frac{\partial}{\partial \vec{n}_q}G(\vec{q},\vec{x}\,')]d\vec{q} \tag{28}$$

where $\vec{n}_q$ is the normal to the line (parameterized by $\vec{q}\,$).

This formula is exact provided the integration is performed over the whole (infinite) line. The existence of the wedge reduces the integration to a semi-infinite region thus producing the Kirchoff approximation to the diffraction problem. The detailed calculations of many particular types of diffractive orbits in pseudo-integrable billiards have been performed in ref. [11]. Here we present only the contribution to the density of states from diffractive orbits lying on the boundary of multiple repetitions of a primitive periodic orbit of length l_0

$$d_{l_0}(E) = -\frac{l_0}{8\pi k}\left(\sum_{q=1}^{\infty}\frac{1}{q^{1/2}}e^{iqkl_0}\right)^2 + c.c. \tag{29}$$

This contribution with fixed l_0 and $k \to \infty$ is smaller than the periodic orbit contribution (4) but bigger than the contribution of diffractive orbits with finite diffraction coefficient (6).

The diffraction coefficient (27) can be written near an optical boundary as a pole term plus a finite part, D^{reg}. Eq. (29) corresponds to the contribution from the pole term. It is also possible to find analytically contributions from the interference of D^{reg} with the pole term

$$d'_{l_0}(E) = -\frac{1}{2\pi i}\frac{\partial}{\partial E}\ln\left(1 - \frac{D^{reg}}{\sqrt{8\pi k l_0}}\sum_{q=1}^{\infty}\frac{1}{q^{3/2}}e^{iqkl_0 - 3\pi i/4 - i\pi\nu_d/2}\right) + c.c. \tag{30}$$

To derive these results it was necessary to compute analytically certain multi-dimensional integrals. In particular, it has been shown in [11] that

$$\begin{aligned}&\int_0^{\infty}\cdots\int_0^{\infty} dx_1\cdots dx_n e^{-\Phi(\vec{x})}\\ &= \frac{1}{n+1}\int_{-\infty}^{\infty}\cdots\int_{-\infty}^{\infty} dx_1\cdots dx_n e^{-\Phi(\vec{x})} = \frac{\pi^{n/2}}{(n+1)^{3/2}}\end{aligned} \tag{31}$$

where $\Phi(\vec{x}) = x_1^2 + (x_1 - x_2)^2 + \cdots + (x_{n-1} - x_n)^2 + x_n^2$, and

$$\begin{aligned}&\int_0^{\infty} dx[\int_{-\infty}^{\infty}\cdots\int_{-\infty}^{\infty} dy_1\cdots dy_n e^{-\Psi(x,\vec{y})}\\ &-\int_0^{\infty}\cdots\int_0^{\infty} dy_1\cdots dy_n e^{-\Psi(x,\vec{y})}] = \frac{\pi^{(n-1)/2}}{4}\sum_{q=1}^{n-1}\frac{1}{\sqrt{q(n-q)}}\end{aligned} \tag{32}$$

where $\Psi(x,\vec{y}) = (x - y_1)^2 + (y_1 - y_2)^2 + \cdots + (y_{n-1} - y_n)^2 + (y_n - x)^2$. The calculation of these integrals is based on the invariance of the quadratic form $\Phi(\vec{x})$ under the action of a finite group generated by reflections.

Spectral statistics of pseudo-integrable billiards in the shape of the right triangle with one angle equals π/n for different n has been investigated numerically in ref. [8] and it was demonstrated that the energy level distribution for all n (except the integrable cases $n = 3, 4, 6$) differs from both the Poisson distribution typical for integrable models and the random matrix distribution [18] typical for chaotic systems. It was observed numerically that the spectral statistics of these billiards has the following characteristic properties:

1. The 2-point correlation function $R_2(\epsilon) \to A\epsilon$ when $\epsilon \to 0$, i.e. there exist linear level repulsion as for the Gaussian orthogonal ensemble of random matrices.
2. The nearest-neighbor distribution $p(s) \to e^{-\Lambda s}$ when $s \to \infty$ as for the Poisson distribution.
3. The number variance $\Sigma^2(L) \to \chi L$, when $L \to \infty$.
4. The distribution of energy levels does not change with increasing of energy (up to 30000 levels).
5. With increasing n, distributions stabilize quite far from random matrix predictions.
6. For $n = 5, 8, 10, 12$ the spectral statistics is close (better than 10^{-2}) to the so-called semi-Poisson distribution [7] which has the correlation functions

$$R_2(\epsilon) = 1 - e^{-4\epsilon}\,, \quad p(s) = 4se^{-2s}\,, \quad \chi = \frac{1}{2}\,. \tag{33}$$

Even the next-to-nearest distribution for these billiards agrees quite well with the prediction of the semi-Poisson model

$$p(2, s) = \frac{8}{3}s^3e^{-2s}\,. \tag{34}$$

The first step of analytical investigation of spectral statistics for these models is the calculation of the 2-point correlation form-factor in the diagonal approximation [5]. It is possible to prove that at small τ it is necessary to take into account only the contribution from periodic orbits which in these cases is the same as for integrable billiards [5]

$$K(\tau) = \frac{1}{2\pi^2\bar{d}}\sum_p \frac{A_p}{l_p}\delta(l - 4\pi k\bar{d}\tau)\,. \tag{35}$$

The summation here is performed over all periodic orbits. l_p is the periodic orbit length, A_p is the area occupied by the primitive periodic orbit family.

The billiard in the shape of the right triangle with one angle equals π/n belongs to the so-called Veech polygons [28] for which there exists a hidden group structure which permits the explicit calculation of the number of periodic orbits and surfaces occupied by them. By generalizing arguments of [28] it is possible to prove [12] that for these triangular billiards

$$K(0) = \frac{n + \epsilon(n)}{3(n-2)} \tag{36}$$

where $\epsilon(n) = 0$ for n odd, $\epsilon(n) = 2$ for n even but not divisible by 3, and $\epsilon(n) = 6$ for n even and divisible by 3. (Note that $K(0) > 1/3$.) These values of $K(0)$ are different from any known distributions. The existence of non-zero $K(0)$ leads in particular to the linear growth of the number variance, $\Sigma^2(L) \to K(0)L$ when $L \to \infty$.

The contributions from diffractive and non-diagonal terms are more difficult to find. The main problem is the divergent character of the diffraction coefficient which leads to the existence of terms as in eq. (29). Such terms grow too quickly with l and cannot be treated in the diagonal approximation. These terms should be cancelled by other terms and this cancellation is a delicate procedure.

4. Billiards with Aharonov-Bohm Flux Line

Another interesting diffractive model close to the above polygonal billiards is a rectangular billiard with the Aharonov-Bohm flux line [1]. This model is defined by the Schrödinger equation

$$[E + (\partial_\mu - iA_\mu)^2]\Psi = 0 \tag{37}$$

with the vector potential of a flux line $A_\phi = \alpha/r$ and, say, the Dirichlet boundary conditions on the boundary of the billiard.

The introduction of the flux line does not change classical trajectories. The only difference is that any time the trajectory encircles the flux line it is necessary to add the phase $2\pi\alpha$ to the semiclassical expansion (1).

The exact quantum-mechanical solution of the flux line in the whole space gives the value of the diffraction coefficient for the scattering on the flux line [1, 24]

$$D(\theta_f, \theta_i) = \frac{2\sin \pi\alpha}{\cos(\frac{\theta_f - \theta_i}{2})} e^{i(\theta_f - \theta_i)/2} . \tag{38}$$

This diffraction coefficient diverges in the forward direction and, as for polygonal billiards, the trace formula requires the careful study of multiple forward diffraction. This can be done, as in Section 3, by using the (generalized) Kirchoff approximation [11]. For example, the contribution to the trace formula from the simplest diffractive orbit parallel to the base of the rectangle (with 2-points of forward diffraction) takes the form

$$d(E) = -\frac{2\sqrt{l_0(a - l_0)}}{\pi^2 k} \sin^2 \pi\alpha \cos(2ka) \tag{39}$$

where a is the base of the rectangle and l_0 is the distance of the flux line from the rectangle side. Note that this expression has the same dependence on momentum and periodic orbit length as eq. (29).

For rectangular billiards with the flux line it is also possible to compute the value of the 2-point correlation function at small τ [12] When the ratios x_0/a

and y_0/b of coordinates of the flux line to the corresponding rectangle sides are non-commensurable irrational numbers

$$K(0) = 1 - 3\bar{\alpha} + 4\bar{\alpha}^2 \tag{40}$$

where $\bar{\alpha}$ is the fractional part of α, $0 \le \bar{\alpha} \le 1/2$ and it is symmetric with respect to $\bar{\alpha} = 1/2$ when $1/2 \le \bar{\alpha} \le 1$.

The numerically computed spectral statistics for this model shows considerable deviations from any known distributions.

5. Chaotic Systems with Diffractive Center

Consider now a chaotic system perturbed by a point-like scatterer. New energy levels, as before, should be computed from eq. (12). The only difference with the case considered in Section 2 is that here we shall assume that old (non-perturbed) energy levels, e_n, are distributed as eigenvalues of one of the standard random matrix ensembles [18]

$$P(\{e_k\}) \propto \prod_{i<j} |e_i - e_j|^{\beta} \tag{41}$$

where $\beta = 1, 2, 4$ for, respectively, GOE, GUE, and GSE cases. (We ignore here the one-body potential needed for the confinement of eigenvalues. One can e.g. assume that the eigenvalues are lying on a large radius circle.)

Within the random matrix theory the distribution of $v_k = |\psi(\vec{x_0})|$ is independent of the eigenvalues and is given by [18]

$$P(\{v_k\}) = \prod_i (\frac{\beta N}{2\pi})^{1-\beta/2} \exp(-\frac{\beta}{2} N v_i^2)\,. \tag{42}$$

The knowledge of the statistical distributions of the poles and the residues of eq. (12) permits the calculation of the distribution of new energy levels. The first step has been done in ref. [3] where the joint distribution of the new, E_j, and the old, e_i, levels has been computed,

$$P(\{E_j\},\{e_k\}) \propto \frac{\prod_{i<j}(e_i - e_j)(E_i - E_j)}{\prod_{i,j}|e_i - E_j|^{1-\beta/2}} \exp(-\rho \sum_i (E_i - e_i))\,, \tag{43}$$

where $\rho = \beta/(2\lambda N)$ and, due to the positivity of $|v_k|^2$, $e_i \le E_i \le e_{i+1}$. Here we assume that $\rho > 0$ and energy levels are ordered, $e_1 \le e_2 \le \cdots \le e_N$.

The resulting distribution of the new eigenvalues is defined by the expression

$$P(\{E_j\}) = \int_{-\infty}^{E_1} de_1 \int_{E_1}^{E_2} de_2 \cdots \int_{E_{N-1}}^{E_N} de_N P(\{E_j\},\{e_k\})\,. \tag{44}$$

In ref. [13] it was demonstrated that these integrals can be computed and the distribution $P(\{E_j\})$ has exactly the same form as the distribution of non-perturbed eigenvalues given by eq. (34)

$$P(\{E_j\}) \propto \prod_{i<j} |E_i - E_j|^{\beta} . \tag{45}$$

This result is not surprising. The random matrix theory distribution is often considered as the result of action of a large number of small-size scatterers. Consequently, the addition of one more center of scattering should not change the spectral statistics.

In ref. [25] the contribution to the 2-point correlation form-factor from diffractive orbits in the diagonal approximation has been computed for f-dimensional chaotic systems

$$K_d(\tau) = \frac{\tau^2}{8\beta\pi^2}\left(\frac{k}{2\pi}\right)^{2f-4} \int |D(\vec{n},\vec{n}')|^2 dO_{\vec{n}} dO_{\vec{n}'} . \tag{46}$$

According to the above arguments this additional contribution should be removed by other (non-diagonal) terms. In ref. [13] it has been demonstrated that it is the interference between diffractive orbits in the forward direction and periodic orbits close to the diffraction center that exactly cancel this term. In the derivation of this result two main ingredients were important. First, the uniformity principle for periodic orbits of chaotic systems, in particular

$$\sum_p \frac{\chi(\vec{q}_p,\vec{p}_p)}{|\det(M_p-1)|}\delta(T-T_p) = \frac{1}{\Sigma}\int \chi(\vec{q},\vec{p}\,) d^{f-1}q d^{f-1}p \tag{47}$$

where $\chi(\vec{q},\vec{p}\,)$ is a test function defined on a Poincaré surface of section $(\vec{q},\vec{p}\,)$. $(\vec{q}_p,\vec{p}_p)$ are coordinates of points of intersections of a periodic orbit with the surface of section. T_p is the periodic orbit period and $\Sigma = \int d^f q d^f p \delta(E - H(q,p))$ is the phase-space volume of the constant energy surface.

The second important point is the optical theorem for the diffractive coefficient which is a consequence of the unitarity of the scattering S-matrix [13]

$$\text{Im } D(\vec{n},\vec{n}) = -\frac{1}{8\pi}\left(\frac{k}{2\pi}\right)^{f-2} \int |D(\vec{n},\vec{n}')|^2 dO_{\vec{n}'} . \tag{48}$$

Using these relations one recovers [13] the invariance of random matrix results under the addition of short-range scatterers from the semiclassical methods. One can easily check that ignoring the optical theorem can change the spectral statistics.

6. Conclusion

In this paper a few typical examples of diffractive models have been considered. An integrable model with a small-size scatterer with a constant diffraction coefficient is the simplest and the most investigated case. Under the assumption that the unperturbed system obeys the Poisson statistics, it is possible to rigorously compute the spectral statistics of this model. The main result is that adding a diffractive center changes completely the spectral statistics. The resulting statistics is characterized by level repulsion and depends on the value of the diffractive coefficient. Models with finite diffraction coefficient permit the construction of a specific perturbation theory which allows the term-by term computation of expansion of 2-point correlation function $R_2(\epsilon)$ (and other correlation functions as well) in powers of $1/\epsilon$. The models such as pseudo-integrable polygonal billiards and rectangular billiards with a flux line are another type of diffractive model characterized by divergence of the diffraction coefficient. In this case the semiclassical contribution of classical periodic orbits to the trace formula differs from the unperturbed case and the spectral statistics of these models is characterized, first of all, by non-standard values of the 2-point correlation form factor at small τ, $K(0) < 1$, and consequently the linear growths of the number variance. For higher order terms of the expansion of correlation functions, the cancellation of rapidly growing terms (as in (29)) is not yet fully understood, and calculations are in progress. A simple example of such a cancellation is provided by chaotic systems with a point-like scatterer. Though it is evident (and can rigorously be proved) that, if the spectral statistics of an unperturbed chaotic system is described by one of standard random matrix ensembles, the addition of a diffraction center cannot change the statistics; in the semiclassical approach this invariance requires a compensation between diagonal and off-diagonal terms. The important point is that this strong cancellation is a general phenomenon connected mostly with the unitarity of quantum mechanical scattering.

Diffractive systems are an interesting and promising class of quantum-mechanical models. Their properties, in general, differ from standard expectations but often are accessible to analytical calculations, and they open new perspectives in the investigation of the semiclassical limit in quantum mechanics.

Acknowledgments

The author is very grateful to C. Schmit, N. Pavloff, P. Leboeuf, U. Gerland, and O. Giraud, whose collaboration permitted us to obtain most of the results of this paper.

References

[1] Y. Aharonov and D. Bohm, Phys. Rev. **115**, 485 (1959).

[2] S. Albeveris et al. *Solvable models in quantum mechanics* (Springer-Verlag, New York, 1988).

[3] I. L. Aleiner and K. A. Matveev, Phys. Rev. Lett. **80**, 814 (1998).

[4] M. V. Berry and M. Tabor, J. Phys. A: Math. Gen. **10**, 371 (1977).

[5] M. V. Berry, Proc. R. Soc. Lond. A **400**, 229 (1985).

[6] E. Bogomolny, U. Gerland, and C. Schmit, Phys. Rev. E **63**, 036206 (2001).

[7] E. Bogomolny, U. Gerland, and C. Schmit, Eor. Phys. J. **19**, 121 (2001).

[8] E. Bogomolny, U. Gerland, and C. Schmit, Phys. Rev. E **59**, R1315 (1999).
[9] E. Bogomolny, Nonlinearity **13**, 947 (2000).
[10] E. Bogomolny, O. Giraud, and C. Schmit, *Trace formulas for conductance fluctuations* (2000), unpublished.
[11] E. Bogomolny, N. Pavloff, and C. Schmit, Phys. Rev. E **61**, 3689 (2000).
[12] E. Bogomolny and O. Giraud, (2000) unpublished.
[13] E. Bogomolny, P. Leboeuf, and C. Schmit, Phys. Rev. Lett. **85**, 2486 (2000).
[14] A. Bohr and B. R. Mottelson, *Nuclear Structure* Vol. I, 302 (W.A. Benjamin, New York, 1969),
[15] H. Bruus and N. D. Whelan, Nonlinearity **9**, 1023 (1996).
[16] M. C. Gutzwiller, *Chaos in Classical and Quantum Mechanics* (Springer-Verlag, New York, 1990).
[17] J. B. Keller, J. Opt. Soc. Am. **52**, 116 (1962).
[18] M. L. Mehta, *Random Matrix Theory* (Springer-Verlag, New York, 1990).
[19] N. Pavloff and C. Schmit, Phys. Rev. Lett. **75**, 61 (1995); **75**, 3779(E) (1995).
[20] P. J. Richens and M.V. Berry, Physica D **2**, 485 (1981).
[21] P. Seba, Phys. Rev. Lett. **64**, 1855 (1990).
[22] A. Shudo and Y. Shimizu, Phys. Rev. E **47**, 54 (1993).
[23] M. Sieber, N. Pavloff, and C. Schmit, Phys. Rev. E **55**, 2279 (1997).
[24] M. Sieber, Phys. Rev. E **60**, 3982 (1999).
[25] M. Sieber, J. Phys. A. **32**, 7679 (1999).
[26] A. Sommerfeld, *Optics* (Academic Press, New York, 1954).
[27] G. Vattay, A. Wirzba, and P. E. Rosenqvist, Phys. Rev. Lett. **73**, 2304 (1994).
[28] W. A. Veech, Invent. Math. **97**, 553 (1989); Geom. Funt. Anal. **2**, 341 (1992).

Laboratoire de Physique Théorique et
Modelès Statistiques
Université de Paris XI, Bât. 100
F–91405 Orsay Cedex, France
E-mail address: `bogomol@ipno.in2p3.fr`

Semiclassical Results in the Linear Response Theory

Monique Combescure and Didier Robert

Abstract. We consider a quantum system of non-interacting fermions at temperature T, in the framework of linear-response theory. We show that semiclassical theory is an appropriate framework for describing some of their thermodynamic properties, in particular through exact expansions in $\hbar$ (Planck constant) of their dynamical susceptibilities. We show how the orbits of the classical motion in phase space manifest themselves in these expansions, in the regime where T is of order $\hbar$.

Consider a system of N non-interacting fermions confined by an external potential, and in contact with an exterior reservoir at temperature T. Assume that a time-varying external perturbation drives the system out of, but near to, its equilibrium state. The response of this quantum system to the external time-dependent perturbation is a subject of high physical interest, which can be investigated experimentally, in particular the so-called dynamical susceptibility. A complete rigorous analysis of this problem is still lacking, although recent progress is being made in the understanding of non-equilibrium statistical mechanics, and its link with the underlying chaotic dynamics [10, 11, 17, 18, 19].

A semi-empirical route which has been proposed and followed (see classical textbooks [14, 13]) consists, for small perturbations, of investigating the response function "to first order in the perturbation", i.e. the so-called "linear response theory". This semi-empirical route is being given a firmer foundation, in classical as well as quantum statistical mechanics, in terms of hyperbolicity properties of the dynamics, and the so-called KMS states [18, 19].

Here we are not following this line of research and do not address the question of validity of the linear response theory. We rederive, formally, the first order response function for the quantum fermionic system under study, i.e. the so-called "generalized Kubo formula" (see also [2]) and investigate semiclassical expansions for it, assuming suitable "chaoticity assumptions" on the one-body classical underlying dynamics. These semiclassical expansions are developed in a similar spirit as previous studies on the "semiclassical magnetic response for non-interacting electrons" [16, 1, 4, 9, 7, 12, 15], i.e. we exhibit a low temperature regime where the closed classical orbits of one-particle motion manifest themselves as oscillating corrections to the response function.

Consider a system of N non-interacting fermions, living in $\mathbb{R}^n$, subject to a one-body Hamiltonian $\widehat{H}$, which is the Weyl quantization of a classical Hamiltonian $H(q,p)$ of the form

$$H(q,p) = \frac{p^2}{2m} + V(q) \tag{H.1}$$

with $V \in C^\infty(\mathbb{R}^n)$ such that

$$V(q) \geq c_0(1+|q|^2)^{s/2} \quad \text{some} \quad s, c_0 > 0\,. \tag{H.2}$$

Under these assumptions, $\widehat{H}$ is self-adjoint in $L^2(\mathbb{R}^n)$, and its spectrum is pure point and contained in $[0,\infty)$. Furthermore, if μ is the Fermi level at temperature T, and f the Fermi-Dirac distribution

$$f(x) \equiv f_\beta(x-\mu) = \left(1 + e^{\beta(x-\mu)}\right)^{-1} \tag{1}$$

where

$$\beta = 1/kT \quad (k \text{ being the Boltzmann constant}), \tag{2}$$

$$\rho_{eq} = f(\widehat{H}) \tag{3}$$

is a trace-class operator in $L^2(\mathbb{R}^n)$, which describes the fermionic equilibrium state, and

$$N = \operatorname{Tr} f(\widehat{H})\,. \tag{4}$$

Now assume that a one-body perturbation $\widehat{A}$ is switched adiabatically. The one-body perturbed Hamiltonian takes the form

$$\widehat{H}(t) = \widehat{H} + \widehat{A}F(t) \tag{5}$$

where $\widehat{A}$ is self-adjoint in $L^2(\mathbb{R}^n)$ (being the Weyl quantization of a symbol $A(q,p)$ that we shall make precise later), and

$$F(t) = \begin{cases} e^{\eta t} & t < 0 \\ 1 & t \geq 0\,. \end{cases}$$

(In the usual Kubo formula for conductivity, $\widehat{A}$ is simply $\widehat{x} \cdot E$ where $\widehat{x}$ is the position operator in $\mathbb{R}^n$, and E an exterior electric field.) In the "linear response theory" we try to solve the following problem: find a "density matrix" $\rho(t)$ (i.e. a trace-class operator in $L^2(\mathbb{R}^n)$), which, to first order in the perturbation solves:

$$i\hbar \frac{\partial \rho}{\partial t} = [\widehat{H}(t), \rho] \tag{6}$$

with "initial condition" at $t = -\infty$ being

$$\lim_{t \to -\infty} \rho(t) = \rho_{eq}\,.$$

Let $V(t,t_0)$ be the unitary evolution operator induced by $\widehat{H}(t)$ (with $V(t,t_0) =$ Identity), and

$$U(t) =: e^{-it\widehat{H}/\hbar}. \tag{7}$$

We can show that $\rho(t)$ solution of (6) is

$$\rho(t) = \lim_{t_0 \to -\infty} V(t,t_0)\rho_{eq}\, V(t_0,t). \tag{8}$$

From (5) it follows that:

$$V(t,t_0) = U(t-t_0) + \frac{1}{i\hbar}\int_{t_0}^{t} dt'\; U(t-t')\; F(t')\; \widehat{A}\; V(t',t_0) \tag{9}$$

and the linear response density matrix $\rho_L(t)$ is obtained from $V(t,t_0)\rho_{eq}\, V(t_0,t)$ by

- retaining only the lowest order contributions with respect to perturbation $\widehat{A}$,
- letting $t_0 \to -\infty$.

But using (9) we easily see that, up to highest orders in $\widehat{A}$:

$$V(t,t_0)\rho_{eq}\, V(t_0,t) = \rho_{eq} + \frac{1}{i\hbar}\int_{t_0}^{t} dt'\; U(t-t')\; F(t')[\widehat{A},\rho_{eq}]U(t'-t) \tag{10}$$

so that, by the above prescription:

$$\rho_L(t) = \rho_{eq} + \frac{1}{i\hbar}\int_{-\infty}^{t} dt'\; F(t')\; U(t-t')[\widehat{A},\rho_{eq}]U(t'-t). \tag{11}$$

Let $\widehat{B}$ be a suitable self-adjoint operator that we want to "measure" in the state $\rho_L(t)$, as compared to its mean-value in the stationary state ρ. Thus we consider

$$J(t) = \mathrm{Tr}\left[\widehat{B}(\rho_L(t) - \rho_{eq})\right] \tag{12}$$

which, according to (11) can be rewritten as

$$J(t) = -\frac{1}{i\hbar}\int_{-\infty}^{t} dt'\; F(t')\; \mathrm{Tr}\left\{\widehat{B}U(t-t')[\widehat{A},\rho_{eq}]U(t'-t)\right\} \tag{13}$$

(the norm convergence of the integral at $t = -\infty$ is ensured by the function $F(t')$, which allows us to insert the trace operation inside the integral). If $\widehat{B}$ is the velocity operator $i[\widehat{H},\widehat{x}]$, and $\widehat{A} = \widehat{x}\cdot E$, $J(t)$ is the quantum current at time t (in the linear response framework).

(13) is therefore of the form

$$J(t) = \int_{-\infty}^{t} dt'\; F(t')\; \Phi(t-t') \tag{14}$$

where $\Phi(t)$ is by definition the "dynamical susceptibility" in the linear response framework. It is given by

$$\begin{aligned}\Phi(t) &= \frac{1}{i\hbar}\ \mathrm{Tr}\left\{\widehat{B}(t)[\rho_{eq},\widehat{A}]\right\}\\ &= \frac{1}{i\hbar}\,\mathrm{Tr}\left\{\rho_{eq}[\widehat{B}(t),\widehat{A}]\right\}\end{aligned} \tag{15}$$

where we have used the commutativity of the Trace, and we employ the usual notation for Heisenberg observables at time t evolved by the (unperturbed) Hamiltonian $\widehat{H}$:

$$\widehat{B}(t) = U(-t)\widehat{B}U(t) = e^{it\widehat{H}/\hbar}\ \widehat{B}e^{-it\widehat{H}/\hbar}\,. \tag{16}$$

Let us now take the Fourier transform, in the distributional sense, of $\Phi(t)$: this gives the so-called generalized susceptibility:

$$\chi(\omega) = \int_{-\infty}^{+\infty} \phi(t)\ e^{i\omega t}\ dt \tag{17}$$

which is the quantity we shall study now, in the particular case where $\widehat{A} = \widehat{B}$.

Let $(E_n)_{n\in\mathbb{N}}$ and $(\varphi_n)_{n\in\mathbb{N}}$ denote respectively the eigenvalues and eigenstates of the one-particle hamiltonian $\widehat{H}$. It follows from (15)–(17) that

$$\chi(\omega) = \sum_{n,m\in\mathbb{N}} |\langle\varphi_n,\widehat{A}\varphi_m\rangle|^2\ \delta(\hbar\omega + E_n - E_m)\cdot[f(E_n) - f(E_m)]\,. \tag{18}$$

Now, using the analyticity of f:

$$f(E) - f(E+\hbar\omega) = -\sum_{k=1}^{\infty}\frac{(\hbar\omega)^k}{k!}\ f^{(k)}(E)$$

so that (18) is rewritten as

$$\chi(\omega) = -\sum_{n,m\in\mathbb{N}} |A_{nm}|^2\ \delta(\hbar\omega + E_n - E_m)\sum_{k=1}^{\infty}\frac{(\hbar\omega)^k}{k!}\ f^{(k)}(E_n) \tag{19}$$

where we use the notation

$$A_{nm} =: \langle\varphi_n,\widehat{A}\varphi_m\rangle\,. \tag{20}$$

After a careful justification of the commutation of various infinite summations, (19) yields:

$$\chi(\omega) = -\sum_{k=1}^{\infty}\frac{(\hbar\omega)^k}{k!}\int dE\ f^{(k)}(E)\sum_{n,m\in\mathbb{N}}|A_{nm}|^2\delta(E-E_n)\delta(\hbar\omega + E_n - E_m) \tag{21}$$

and we therefore have to study, semiclassically, the behaviour of distributions in E and ω of the form:

$$C(E,\omega) =: \sum_{n,m}|A_{nm}|^2\ \delta(E-E_n)\ \delta(\hbar\omega + E_n - E_m) \tag{22}$$

acting on suitable test functions, and in particular on derivatives of the Fermi-Dirac distribution f. However in all that follows, as in [7], we are only able to replace f_σ by $f_\sigma * \widetilde{\rho}_\tau$ for any fixed τ as large as we want, where

$$\sigma = \beta\hbar \tag{23}$$

is a fixed parameter which has the dimension of time and $\widetilde{\rho}_\tau$ is the Fourier transform of a $\mathcal{C}_0^\infty$ function ρ_τ:

$$\rho_\tau(t) = \rho(t/\tau)$$

where

$$\rho(t) \begin{cases} \equiv 1 & \text{if } |t| \leq 1 \\ \equiv 0 & \text{if } |t| \geq 2 \end{cases}$$

$\int \rho(t)dt = 1$.

Let g be a $\mathcal{C}^\infty$ function such that its Fourier transform $\widetilde{g}$ has compact support contained in $[-T, T]$, for some $T > 0$. Then denote:

$$\ell(E) =: \left(f_\sigma^{(k)} * \widetilde{\rho}_\tau\right)(E) \tag{24}$$

(if $\tau \to \infty$, $\ell\left(\frac{E-\mu}{\hbar}\right)$ would be simply the kth derivative of the Fermi-Dirac function $k \geq 1$)

$$\langle C(E,\omega), g(\omega)\ell\left(\frac{E-\mu}{\hbar}\right)\rangle = \sum_{n,m\in\mathbb{N}} |A_{nm}|^2\, g\left(\frac{E_n - E_m}{\hbar}\right)\, \ell\left(\frac{E_n-\mu}{\hbar}\right). \tag{25}$$

The RHS of (25) can be rewritten, using a Fourier transform, as

$$\begin{aligned} &\frac{1}{2\pi}\int_{-\infty}^{+\infty} dt\; \widetilde{g}(t) \sum_{n,m=0}^{\infty} |A_{nm}|^2\, e^{it(E_n-E_m)/\hbar}\ell\left(\frac{E_n-\mu}{\hbar}\right) \\ =&\frac{1}{2\pi}\int_{-\infty}^{+\infty} dt\; \widetilde{g}(t)\left\{\operatorname{Tr} \widehat{A}(t)\widehat{A}\,\ell\left(\frac{\widehat{H}-\mu}{\hbar}\right)\right\}. \end{aligned} \tag{26}$$

Now since $\widetilde{g}$ is of compact support, and so is $\widetilde{\ell}$ (Fourier transform of ℓ) due to its definition (24), we can adapt our treatment of semiclassical trace formulae using coherent state decomposition [6] to this situation. Here the quantum observable appearing inside the Trace is the product of observable $\widehat{A}$ with its Heisenberg time translated $\widehat{A}(t)$ given by (16). We thus have also to make use of Egorov's theorem [8] to recognize, as a dominant classical symbol of $\widehat{A}(t)\widehat{A}$ simply $(A \circ \phi_t)(z)A(z)$ where $z = (x,\xi) \in \mathbb{R}^n \times \mathbb{R}^n$ is the classical phase-space point.

We therefore define the various "classical objects" that will manifest themselves, in the limit as $\hbar \to 0$, in the semiclassical expansion of (26).

First of all, the classical flow induced by the classical Hamiltonian H is ϕ_t; since H does not depend explicitly on time, the classical flow lives on the energy surface

$$\sum\nolimits_{\mu} = \{z = (x,\xi) \in \mathbb{R}^{2n} : H(x,\xi) = \mu\}$$

and we denote by $d\sigma_\mu$ the Liouville measure living on $\sum_\mu$. We call γ a generic periodic orbit living on $\sum_\mu$, i.e.

$$\gamma = \{z : \exists\, T_\gamma > 0 : \phi_{T_\gamma}(z) = z\} .$$

T_γ is therefore the classical period of orbit γ, which can be a repetition of a primitive orbit γ^*. The classical action along the closed orbit γ is called S_γ, σ_γ is the corresponding Maslov index, and P_γ the corresponding Poincaré map. (See [6].)

$$C_\mu(t) =: \int_{\sum_\mu} d\sigma_\mu(z)\ A(z)\ A \circ \phi_t(z) \tag{27}$$

is the classical autocorrelation of A on the energy surface at the Fermi level μ;

$$c_\gamma(t) =: \int_0^{T_{\gamma^*}} ds\ A \circ \phi_s(z)\ A \circ \phi_{t+s}(z) \tag{28}$$

is the classical autocorrelation of A along the closed primitive orbit γ^*. It is independent of the point z of γ^* where we start the integration, and it is of course T_{γ^*} periodic, as a function of t. Therefore it can be expanded in Fourier series in the following form:

$$c_{\gamma^*}(t) = \sum_{p \in Z} c_{\gamma^*,p}\ e^{2i\pi p t/T_{\gamma^*}} . \tag{29}$$

Now we state the result

Theorem 1. *Assume (H1-2) together with*

(H.3) *μ is non-critical for H.*

(H.4) *On $\sum_\mu$, the set $(\Gamma_\mu)_\tau$ of classical periodic orbits γ with period smaller than τ is such that the corresponding Poincaré maps P_γ do not have eigenvalue 1.*

Then, as $\hbar \to 0$, (26) has a complete expansion in $\hbar$ of the following form:

$$\int dt\ \widetilde{g}(t) \Bigg\{ h^{-n}\ \widetilde{\ell}(0)\ C_\mu(t) + \sum_{j \geq 1} h^{-n+j}\ \lambda_j(\widetilde{\ell}, t) + \sum_{\gamma \in (\Gamma_\mu)_\tau} h^{-1} \frac{e^{iS_\gamma/\hbar + i\sigma_\gamma \pi/2}}{|\det(1 - P_\gamma)|^{1/2}}\ \widetilde{\ell}(T_\gamma)\ c_\gamma(t) + \sum_{j \geq 0} h^j\ d^j_\gamma(\widetilde{\ell}, t) \Bigg\}$$

where $\lambda_j(\cdot, t)$ and $d^j_\gamma(\cdot, t)$ are distributions supported respectively by $\{0\}$ and $\{T_\gamma\}$.

The proof of this theorem will be given elsewhere [5].

From this theorem, we can deduce an important result for $\chi(\omega)$, as a distribution, or rather to $\chi_\tau(\omega)$, a "regularized" version of it where the Fermi-Dirac function f is replaced by $f * \rho_{\tau/\hbar}$. This yields the following result:

Corollary 2. *Assume (H1-4) together with (H.5) and (H.6) below:*

(H.5) *the classical dynamics on $\sum_\mu$ is sufficiently mixing, in the sense that*

$$\int_{-\infty}^{+\infty} dt|C_\mu(t)| < \infty$$

(assuming A has been adjusted so that $\langle A\rangle_\mu = 0$).

(H.6) $\sum_{k=-\infty}^{+\infty} |c_{\gamma,k}| < \infty$ *(any γ so that $|T_\gamma| < \tau$).*

Then $\chi_\tau(\omega)$ admits, as a distribution, a complete asymptotic expansion of the form:

$$-\omega h^{1-n} \int dt\ C_\mu(t)\ e^{i\omega t} + \sum_{j\geq 2} h^{j-n}\ \mu_j(\omega)$$

$$+\omega \sum_{\gamma\in(\Gamma_\mu)_\tau} \frac{e^{iS_\gamma/\hbar + i\sigma_\gamma\pi/2}}{|\det(1-P_\gamma)|^{1/2}} \frac{\pi T_\gamma/\sigma}{sh\ \pi T_\gamma/\sigma} \left(\sum_{k=-\infty}^{+\infty} c_{\gamma,k}\ \delta\left(\omega - \frac{2\pi k}{T_{\gamma^*}}\right) + \sum_{j\geq 1} h^j\ \nu_j(\omega)\right)$$

where μ_j and ν_j are distributions.

References

[1] O. Agam, *The magnetic response of chaotic mesoscopic systems*, J. Phys. I (France), **4** (1994), 697–730.

[2] M. Aizenman and G. M. Graf, *Localization bounds for an electron gas*, J. Phys. A (Math. and Gen.), **31** (1998), 6783–6806.

[3] M. Berry, *Some quantum-to-classical asymptotics* in Chaos et Physique Quantique, Les Houches (1989) M. J. Giannoni, A. Voros and J. Zin-Justin, North Holland.

[4] J. D. Butler, *Semiclassical counting function with application to quantum current*, Orsay preprint (1999).

[5] M. Combescure and D. Robert, *Semiclassical results in linear response theory*, (in preparation).

[6] M. Combescure, J. Ralston and D. Robert, *A proof of the Gutzwiller semiclassical trace formula using coherent states decomposition*, Commun. Math. Phys., **202** (1999), 463–480.

[7] M. Combescure and D. Robert, *Rigorous semiclassical results for the magnetic response of an electron gas*, Orsay preprint (2000).

[8] Y. Egorov, *On canonical transformations of pseudodifferential operators*, Uspehi Mat. Nauk, **25** (1969), 235–236.

[9] S. Fournais, *Semiclassics of the quantum current*, Commun. in PDE, **23** (1998), 601–628.

[10] G. Gallavotti and E. G. Cohen, *Dynamical ensembles in stationary states*, J. Stat. Phys., **80** (1995), 931–970.

[11] G. Gallavotti and D. Ruelle, *SRB states and nonequilibrium statistical mechanics close to equilibrium*, Commun. Math. Phys., **190** (1997), 279–285.

[12] P. Gaspard and S. Jain, *Semiclassical theory for many-body fermionic systems*, Pramana Journal of Physics, **48** (1997), 503–516.

[13] R. Kubo, M. Toda and N. Hashitsume, *Statistical Physics II*, (Nonequilibrium Statistical Mechanics), Springer-Verlag, Berlin, Heidelberg, New York, Tokyo (1978).

[14] L. D. Landau and E. M. Lifshitz, *Statistical Physics*, Course of Theoretical Physics, vol. 5, Pergamon Press, Oxford, New York, Toronto, Sydney, Paris, Frankfurt (1980).

[15] B. Mehlig and K. Richter, *Semiclassical linear response : far-infrared absorption in ballistic quantum systems*, Phys. Rev. Lett., **80** (1998), 1936–1939.

[16] K. Richter, D. Ullmo and R. Jalabert, *Orbital magnetism in the ballistic regime: geometrical effects*, Phys. Rep., **276** (1996), 1–83.

[17] D. Ruelle, *General linear response formula in statistical mechanics and the fluctuation-dissipation theorem far from equilibrium*, Phys. Lett., **A245** (1998), 220–224.

[18] D. Ruelle, *Smooth dynamics and new theoretical ideas in nonequilibrium statistical mechanics*, J. Stat. Phys., **95** (1999), 393–468.

[19] D. Ruelle, *Natural nonequilibrium states in quantum statistical mechanics*, J. Stat. Phys., **98** (2000), 57–75.

[20] M. Wilkinson, *A semiclassical sum rule for matrix elements of classically chaotic systems*, J. Phys. A : (Math. and Gen.), **20** (1987), 2415–2423.

Monique Combescure
Laboratoire de Physique Théorique
Unité Mixte de Recherche - CNRS - UMR N° 8627
Université de Paris XI
Bâtiment 210
91405 Orsay Cedex, France
E-mail address: `monique.combescure@th.u-psud.fr`

Didier Robert
Département de Mathématiques
URA CNRS 758
Université de Nantes
2, rue de la Houssinière
F-44072 Nantes Cedex, France
E-mail address: `robert@math.univ-nantes.fr`

The Berry-Tabor Conjecture

Jens Marklof

Abstract. One of the central observations of *quantum chaology* is that statistical properties of quantum spectra exhibit surprisingly universal features, which seem to mirror the chaotic or regular dynamical properties of the underlying classical limit. I will report on recent studies of simple regular systems, where some of the observed phenomena can be established rigorously. The results discussed are intimately related to the distribution of values of quadratic forms, and in particular to a quantitative version of the Oppenheim conjecture.

Quantum Chaos

One of the main objectives of *quantum chaology* is to identify characteristic properties of quantum systems which, in the semiclassical limit, reflect the regular or chaotic features of the underlying classical dynamics.

Take for example the geodesic flow on the unit tangent bundle of a compact two-dimensional Riemannian surface M. The corresponding quantum system is described by the stationary Schrödinger equation

$$-\Delta\varphi_j = \lambda_j\varphi_j\,, \tag{1}$$

where Δ is the Laplacian of M, λ_j represent the quantum energy eigenvalues and φ_j the corresponding eigenfunctions. The spectrum of the negative Laplacian is a discrete ordered subset of the real line,

$$0 \le \lambda_1 < \lambda_2 \le \lambda_3 \le \cdots \to \infty\,. \tag{2}$$

According to Weyl's law, the number of eigenvalues below λ is asymptotically

$$\#\{j : \lambda_j \le \lambda\} \sim \frac{\operatorname{area}(M)}{4\pi}\lambda \tag{3}$$

as $\lambda \to \infty$. Hence the mean spacing between adjacent levels is asymptotically $4\pi/\operatorname{area}(M)$. For simplicity, we may assume in what follows that $\operatorname{area}(M) = 4\pi$.

One quantity measuring the "randomness" of such a sequence on the scale of the mean spacing is the *consecutive level spacing distribution*, which is defined as

$$P(s, N) = \frac{1}{N}\sum_{j=1}^{N}\delta(s - \lambda_{j+1} + \lambda_j)\,, \tag{4}$$

where $\delta(x)$ is the Dirac mass. If the deterministic sequence $\{\lambda_j\}$ is sufficiently "random" we may expect to find a limit distribution $P(s)$ of $P(s,N)$ for $N \to \infty$, that is,

$$\lim_{N\to\infty} \int_0^\infty P(s,N)\, h(s)\, ds = \int_0^\infty P(s)\, h(s)\, ds\,, \tag{5}$$

for any sufficiently nice test function h.

It was conjectured by Berry and Tabor in 1977 [1] that *if the corresponding classical dynamics is completely integrable, then $P(s)$ exists and is equal to the waiting time between consecutive events of a Poisson process,*

$$P(s) = \exp(-s)\,. \tag{6}$$

That is, $\{\lambda_j\}$ behaves like a sequence of independent random variables. There are a number of obvious (and less obvious) counter examples to this behaviour already known to Berry and Tabor [1], which is why one expects the above to hold only for "generic" systems. Let us compare this with the case of "generic" chaotic dynamical systems. Here, the statistical nature of the limit distribution is completely different: Bohigas, Giannoni, and Schmit [3] suggested in 1984 that *if the geodesic flow is hyperbolic, then $P(s)$ exists and is equal to the consecutive level spacing distribution of a suitable Gaussian ensemble of hermitian random matrices.* In the case of the Gaussian unitary ensemble, the distribution is for example (approximately)

$$P_{\mathrm{GUE}}(s) \approx \frac{32}{\pi^2}\, s^2 \mathrm{e}^{-\frac{4}{\pi}s^2}\,. \tag{7}$$

The characteristic level repulsion (i.e., $P(s) \to 0$ as $s \to 0$) was observed already by McDonald and Kaufman [16] and Berry [2] in the case of chaotic billiards.

There has been some recent progress on the above conjectures in the case of integrable systems, which will be reported here. The chaotic case is less well understood.

Lattice Point Problems

The eigenvalues λ_j of the Laplacian on a compact two-dimensional surface with integrable geodesic flow are asymptotically given by the values $F(\mathbf{m}) \sim \lambda_j$ at lattice points $\mathbf{m} \in \mathbb{Z}^2$, where the function $F = F_2 + F_0$ is piecewise analytic and F_2, F_0 homogeneous of degree 2 and 0, respectively [6]. The question of spectral statistics may therefore be reformulated as a lattice point problem of lattice points $\mathbf{m} \in \mathbb{Z}^2$ close to the curve $F(\mathbf{x}) = \lambda$.

Sinai [22] and Major [11] proved that the statistics of λ_j converge to the Poisson limit if F is generic in a certain function space. Unfortunately, a "generic" function in that space is not twice differentiable, and hence there is no corresponding smooth classical dynamical system. It is still open, if the Poisson limit exists for smooth F. Let us have a look at a few examples.

THE SPHERE AND OTHER ZOLL SURFACES We begin by discussing an obviously non-generic example with respect to the above conjectures. Consider the geodesic flow on the unit tangent bundle of the sphere, which clearly is completely integrable. The eigenvalues of $-\Delta$ are (up to some constant factor) $l(l+1)$ with integer $l = 0, 1, 2, \ldots$, each with multiplicity $2l+1$. Label these numbers in increasing order by $\lambda_1, \lambda_2, \ldots$; due to the high multiplicity one thus has $P(s) = \delta(s)$. The same result holds for all Zoll surfaces, since the eigenvalues are strongly clustered around the values $l(l+1)$, compare [8, 24, 7, 23]. The reason for this clustering is that all geodesics are closed and have the same period. Zoll surfaces are clearly quite non-generic in this respect, and it is therefore of no surprise that their eigenvalues do not follow the expected statistical behaviour.

FLAT TORI We now turn to a more generic class of surfaces, flat two-dimensional tori $\mathbb{T}^2 = \mathbb{R}^2/\mathfrak{L}$, where $\mathfrak{L}$ is some lattice in $\mathbb{R}^2$. The Laplacian on $\mathbb{T}^2$ is the standard Euclidean Laplacian acting on functions which are invariant under the lattice $\mathfrak{L}$. The eigenvalues λ_j are easily seen to be values at integers $(m,n) \in \mathbb{Z}^2$ of the positive definite quadratic form

$$Q(m,n) = \alpha m^2 + \beta mn + \gamma n^2 , \tag{8}$$

where the coefficients α, β, γ depend on the lattice $\mathfrak{L}$. In the special case when $\mathfrak{L} = \mathbb{Z}^2$ one has

$$Q(m,n) = m^2 + n^2 . \tag{9}$$

The spacings are in this case clearly integer, so $P(s, N)$ cannot converge to the conjectured exponential distribution. In fact, one can show that the limit distribution is again $P(s) = \delta(s)$, which is in disagreement to what is expected generically. This result follows directly from a classical result of Landau, which says that the number of ways of representing an integer as a sum of two squares grows on average logarithmetically. The same holds when Q is proportional to a form with rational coefficients [21].

Results which support the Berry-Tabor conjecture on a rigorous level have so far been obtained only for a statistic which is easier to handle than the consecutive spacing distribution: the pair correlation density.

Pair Correlation

The pair correlation density of the sequence $\{\lambda_j\}$ is defined as

$$R_2(s,N) = \frac{1}{N} \sum_{j,k=1}^{N} \delta(s - \lambda_j + \lambda_k) , \tag{10}$$

where one therefore considers the distribution of *all* spacings, instead of just nearest neighbours. According to the Berry-Tabor conjecture, we expect that

$$\lim_{N\to\infty} \int_{-\infty}^{\infty} R_2(s,N)\, h(s)\, ds = \int_{-\infty}^{\infty} R_2(s)\, h(s)\, ds \tag{11}$$

exists for generic integrable systems, where $R_2(s)$ is the pair correlation density of a Poisson process[1]

$$R_{2\ \mathrm{Poisson}}(s) = \delta(s) + 1\,. \tag{12}$$

Sarnak [21] was able to prove the Berry-Tabor conjecture for the pair correlation of almost all flat tori: almost all with respect to Lebesgue measure in the moduli space of two-dimensional flat tori. His proof uses averaging techniques to reduce the pair correlation problem to estimating the number of solutions of systems of diophantine equations. The almost-everywhere result then follows from a variant of the Borel-Cantelli argument.[2]

Eskin, Margulis and Mozes [10] recently strengthened Sarnak's result by giving explicit diophantine conditions on (α,β,γ) under which the Berry-Tabor conjecture holds. Admissible forms are for example

$$m^2 + \gamma n^2 \tag{13}$$

with γ diophantine. [γ is called diophantine if there exist constants $\kappa \geq 2, C > 0$ such that $|\alpha - \frac{p}{q}| > Cq^{-\kappa}$ for all rationals $\frac{p}{q}$.[3]] These diophantine conditions cannot be dropped, since one can construct an uncountable set of tori with area 4π (say), whose pair correlation density does not converge, as pointed out by Sarnak [21]. In some sense, such tori feel the degeneracies in the spectra of tori corresponding to rational forms. The above uncountable set is in fact a set of second Baire category, and hence "generic" in the topological sense. This illustrates the subtlety of the problem: "generic" tori —in the measure-theoretic sense— follow the Berry-Tabor conjecture, topologically "generic" tori do not.

The pair correlation problem for positive definite quadratic forms may be viewed as a special case of the quantitative version of the Oppenheim conjecture of quadratic forms of signature $(2,2)$, which is particularly difficult [9].

TORI WITH AHARONOV-BOHM FLUX We have seen that the eigenvalue statistics for the standard square torus with lattice $\mathfrak{L} = \mathbb{Z}^2$ are non-generic. In order to break

[1]Here, the delta mass $\delta(s)$ is a result of our definition which counts spacings between equal elements whose spacing is trivially zero. The interesting part is the "1", the mean density of the eigenvalue sequence.

[2]For further related examples of sequences whose pair correlation function converges to the uniform density almost everywhere in parameter space, see [17, 19, 25, 26, 28]. For recent results on higher correlations see [18, 20, 27].

[3]Almost all numbers are diophantine. An example of a diophantine number is $\gamma = \sqrt{2}$, where $\kappa = 2$.

the degeneracy, let us replace the periodicity conditions with quasi-periodicity conditions,

$$\varphi(x+k, y+l) = \mathrm{e}^{-2\pi\mathrm{i}(\alpha k+\beta l)}\varphi(x, y), \quad k, l \in \mathbb{Z}. \tag{14}$$

The eigenvalues of the negative Laplacian are then

$$(m-\alpha)^2 + (n-\beta)^2 . \tag{15}$$

Such an effect is observed for instance in the presence of Aharonov-Bohm flux lines threading the torus. This is a pure quantum-topological effect, the corresponding classical geodesic flow is not affected. Cheng and Lebowitz have studied such a system numerically, and found good agreement with the Berry-Tabor conjecture for generic α, β. Cheng, Lebowitz and Major [5] were able to prove the convergence of the pair correlation density to the expected Poisson distribution on average over $(\alpha, \beta) \in [0,1]^2$.

In [15] it is proved that the conjecture is in fact true for fixed α, β, provided α or β is diophantine (e.g. $\alpha = \sqrt{2}, \beta = \sqrt{3}$). This diophantine condition is necessary: there is a set of second Baire category of $(\alpha, \beta) \in [0,1]^2$ for which the pair correlation density does not converge. The idea of the proof is as follows. Consider the Fourier transform of the pair correlation density,

$$K_2(\tau, N) = \int_{-\infty}^{\infty} R_2(s, N)\, \mathrm{e}^{2\pi\mathrm{i}\tau s}\, ds \tag{16}$$

hence

$$K_2(\tau, N) = \left|N^{-1/2} \sum_{j=1}^{N} \mathrm{e}^{2\pi\mathrm{i}\lambda_j \tau}\right|^2 . \tag{17}$$

The function $K_2(\tau, N)$ is often called the *spectral form factor*. We are interested in the asymptotics of

$$\int_{-\infty}^{\infty} R_2(s, N)\, h(s)\, ds = \int_{-\infty}^{\infty} K_2(\tau, N)\, \hat{h}(\tau)\, d\tau$$

where $\hat{h}$ is the Fourier transform of the test function h. The trick is now that the exponential sum $\sum \mathrm{e}^{2\pi\mathrm{i}\lambda_j\tau}$ turns out to be a theta sum that can be identified with a function on a noncompact manifold Σ with finite measure, and that the integral over τ corresponds to an average over a unipotent orbit on Σ that becomes equidistributed on Σ, as $N \to \infty$. The equidistribution follows essentially from Ratner's classification of invariant measures of unipotent flows. The average over τ can therefore be replaced by the average over the whole space Σ, which luckily turns out to be exactly the expected limit. A crucial subtlety in the proof is that the theta sum is unbounded on Σ, which is where the diophantine conditions come in.

It is quite interesting to remark that, although the pair correlation statistics of the λ_j follow the Poisson prediction, the value distribution of the sum $\sum \mathrm{e}^{2\pi\mathrm{i}\lambda_j\tau}$ defining the form factor does not satisfy a central limit theorem in general. For example, in the case of the rectangular torus discussed above, where $\lambda_j = m^2+\gamma n^2$, a limiting distribution exists [14] but has a power-like tail, and is therefore not normal. The proof is based on the asymptotic distribution of values of theta sums [12, 13]. A similar deviation from the normal distribution is observed whenever the eigenvalues λ_j are values of quadratic forms at lattice points, e.g. also when $\lambda_j = (m-\alpha)^2+(n-\beta)^2$. If, however, the quadratic form is replaced by a more generic smooth function F homogeneous of degree 2, I expect the limiting distribution to be normal. This will be discussed in detail elsewhere.

References

[1] M. V. Berry and M. Tabor, Level clustering in the regular spectrum, *Proc. Roy. Soc.* A **356** (1977) 375–394.

[2] M. V. Berry, Quantizing a classically ergodic system: Sinai's billiard and the KKR method, *Ann. Phys.* **131** (1981) 163–216.

[3] O. Bohigas, M.-J. Giannoni and C. Schmit, Characterization of chaotic quantum spectra and universality of level fluctuation laws, *Phys. Rev. Lett.* **52** (1984) 1–4.

[4] Z. Cheng and J. L. Lebowitz, Statistics of energy levels in integrable quantum systems, *Phys. Rev.* A **44** (1991) 3399–3402.

[5] Z. Cheng, J. L. Lebowitz and P. Major, On the number of lattice points between two enlarged and randomly shifted copies of an oval, *Probab. Theory Related Fields* **100** (1994) 253–268.

[6] Y. Colin de Verdière, Quasi-modes sur les variétés Riemanniennes, *Invent. Math.* **43** (1977) 15–52.

[7] Y. Colin de Verdière, Sur le spectre des opérateurs elliptiques à bicaractéristiques toutes périodique, *Comment. Math. Helvetici* **54** (1979) 508–522.

[8] J. J. Duistermaat and V. W. Guillemin, The spectrum of positive elliptic operators and periodic bicharacteristics, *Invent. Math.* **29** (1975) 39–79.

[9] A. Eskin, G. Margulis and S. Mozes, Upper bounds and asymptotics in a quantitative version of the Oppenheim conjecture, *Ann. of Math.* **147** (1998) 93–141.

[10] A. Eskin, G. Margulis and S. Mozes, Quadratic forms of signature (2,2) and eigenvalue spacings on rectangular 2-tori, preprint.

[11] P. Major, Poisson law for the number of lattice points in a random strip with finite area, *Prob. Theo. Rel. Fields* **92** (1992) 423–464.

[12] J. Marklof, Limit theorems for theta sums, *Duke Math. J.* **97** (1999) 127–153.

[13] J. Marklof, Theta sums, Eisenstein series, and the semiclassical dynamics of a precessing spin, in: D. Hejhal et al. (eds.), *Emerging Applications of Number Theory*, IMA Vol. Math. Appl. **109** (Springer, New York, 1999) 405–450.

[14] J. Marklof, Spectral form factors of rectangle billiards, *Comm. Math. Phys.* **199** (1998) 169–202.

[15] J. Marklof, Pair correlation densities of inhomogeneous quadratic forms, preprint 2000.

[16] S. W. McDonald and A. N. Kaufman, Spectrum and eigenfunctions for a Hamiltonian with stochastic trajectories, *Phys. Rev. Lett.* **42** (1979) 1189–1191.

[17] Z. Rudnick and P. Sarnak, The pair correlation function for fractional parts of polynomials, *Comm. Math. Phys.* **194** (1998) 61–70.

[18] Z. Rudnick, P. Sarnak and A. Zaharescu, The distribution of spacings between the fractional parts of αn^2, preprint November 1999.

[19] Z. Rudnick and A. Zaharescu, A metric result on the pair correlation of fractional parts of sequences, *Acta Arith.* **LXXXIX** (1999) 283–293.

[20] Z. Rudnick and A. Zaharescu, The distribution of spacings between fractional parts of lacunary sequences, preprint December 1999.

[21] P. Sarnak, Values at integers of binary quadratic forms, *Harmonic Analysis and Number Theory* (Montreal, PQ, 1996), 181–203, CMS Conf. Proc. **21**, Amer. Math. Soc., Providence, RI, 1997.

[22] Ya. G. Sinai, Poisson distribution in a geometrical problem, *Adv. Sov. Math., AMS Publ.* **3** (1991) 199–215.

[23] A. Uribe and S. Zelditch, Spectral statistics on Zoll surfaces, *Comm. Math. Phys.* **154** (1993) 313–346.

[24] A. Weinstein, Asymptotics of eigenvalue clusters for the Laplacian plus a potential, *Duke Math. J.* **44** (1977) 883–892.

[25] J. M. VanderKam, Values at integers of homogeneous polynomials, *Duke Math. J.* **97** (1999) 379–412.

[26] J. M. VanderKam, Pair correlation of four-dimensional flat tori, *Duke Math. J.* **97** (1999) 413–438.

[27] J. M. VanderKam, Correlations of eigenvalues on multi-dimensional flat tori, *Comm. Math. Phys.* **210** (2000) 203–223.

[28] S. Zelditch, Level spacings for integrable quantum maps in genus zero, *Commun. Math. Phys.* **196** (1998) 289–329.

School of Mathematics
University of Bristol
Bristol BS8 1TW, U.K.
E-mail address: j.marklof@bristol.ac.uk

On Quantum Unique Ergodicity for Linear Maps of the Torus

Zeév Rudnick

Abstract. The problem of "quantum ergodicity" addresses the limiting distribution of eigenfunctions of classically chaotic systems. I survey recent progress on this question in the case of quantum maps of the torus. This example leads to analogues of traditional problems in number theory, such as the classical conjecture of Gauss and Artin that any (reasonable) integer is a primitive root for infinitely many primes, and to variants of the notion of Hecke operators.

1. Introduction

One of the few rigorous general results in the field of "Quantum Chaology" is *Quantum Ergodicity* [13, 2, 14]. To formulate this notion, recall that if the classical dynamics of a system are ergodic, then almost all trajectories of a particle cover the energy shell uniformly, that is to say that the time averages along the trajectory converge to the phase space average. The intuition afforded by the "Correspondence Principle" leads one to look for an analogous statement about the semiclassical limit of expectation values of observables in an energy eigenstate. As formulated by Schnirelman [13], the corresponding assertion is that when the classical dynamics is ergodic, for almost all eigenstates the expectation values of observables converge to the phase-space average.

A key question is: Under suitable assumptions on the system, can one say anything beyond "almost all"? For instance, when can one assert that *all* expectation values converge to the phase space average? Such behavior has sometimes been called quantum *unique* ergodicity [12].

Below is a survey of some recent attempts to understand question in the context of quantum maps of the torus. These are an important model for understanding the quantization of classically chaotic systems, first studied by Hannay and Berry [5]. We will devote special attention to linear hyperbolic automorphisms of the torus $\mathbf{T}^2$ – the so-called "cat maps".

Notation. *We will use the abbreviations* $e(z) := e^{2\pi i z}$, $e_N(z) = e(z/N)$.

2. Quantum Mechanics on the Torus

We review the basics of quantum mechanics on the torus $\mathbf{T}^2$, viewed as a phase space [5, 7, 3, 4].

2.1. Quantum states

We start with a description of the Hilbert space of states of such a system. In brief, Planck's constant is restricted to be an inverse integer: $h = 1/N$, and the Hilbert space of states $\mathcal{H}_N$ is N-dimensional, in keeping with the intuition that each state occupies a Planck cell of volume $h = 1/N$ and the constraint that the total phase-space $\mathbf{T}^2$ has volume one. The "state vectors" are distributions on the line which are periodic in both momentum and position representations: $\psi(q+1) = \psi(q)$, $[\mathcal{F}_h\psi](p+1) = [\mathcal{F}_h\psi](p)$, where $[\mathcal{F}_h\psi](p) = h^{-1/2}\int \psi(q)\, e(-pq/h)\, dq$. The space of such distributions is finite dimensional, of dimension precisely $N = 1/h$, and consists of periodic point-masses at the coordinates $q = Q/N$, $Q \in \mathbf{Z}$. We may then identify $\mathcal{H}_N$ with the N-dimensional vector space $L^2(\mathbf{Z}/N\mathbf{Z})$, with the inner product $\langle \cdot , \cdot \rangle$ defined by

$$\langle \phi, \psi \rangle = \frac{1}{N} \sum_{Q \bmod N} \phi(Q)\, \overline{\psi}(Q)\,.$$

2.2. Observables

Classical observables (i.e. functions $f \in C^\infty(\mathbf{T}^2)$) give rise to quantum observables, that is operators $\mathrm{Op}_N(f)$ on $\mathcal{H}_N$. To define these, one starts with the translation operators

$$[t_1\psi](Q) = \psi(Q+1)$$

and

$$[t_2\psi](Q) = e_N(Q)\, \psi(Q)\,,$$

which may be viewed as the analogues of differentiation and multiplication (respectively) operators. In fact in terms of the usual translation operators on the line $\hat{q}\psi(q) = q\psi(q)$ and $\hat{p}\psi(q) = \frac{h}{2\pi i}\frac{d}{dq}\psi(q)$, they are given by $t_1 = e(\hat{p})$, $t_2 = e(\hat{q})$. In this context, Heisenberg's commutation relations read

$$t_1^a t_2^b = t_2^b t_1^a e_N(ab) \quad \forall a, b \in \mathbf{Z}\,.$$

More generally, mixed translation operators are defined for $n = (n_1, n_2) \in \mathbf{Z}^2$ by

$$T_N(n) = e_N(\frac{n_1 n_2}{2}) t_2^{n_2} t_1^{n_1}\,.$$

These are unitary operators on $\mathcal{H}_N$, whose action on a wave-function $\psi \in \mathcal{H}_N$ is given by:

$$T_N(n)\psi(Q) = e^{\frac{i\pi n_1 n_2}{N}} e(\frac{n_2 Q}{N}) \psi(Q + n_1)\,.$$

For any smooth function $f \in \mathbf{C}^\infty(\mathbf{T}^2)$, define a *quantum observable* $\mathrm{Op}_N(f)$, called the *Weyl quantization of* f, by

$$\mathrm{Op}_N(f) = \sum_{n \in \mathbf{Z}^2} \widehat{f}(n) T_N(n)$$

where $\widehat{f}(n)$ are the Fourier coefficients of f.

The observables $\mathrm{Op}_N(f)$ satisfy: For any orthonormal basis of $\mathcal{H}_N$ we have

$$\frac{1}{N}\sum_{j=1}^{N} \langle \mathrm{Op}_N(f)\psi_j, \psi_j\rangle = \int_{\mathbf{T}^2} f + O_f\left(\frac{1}{N}\right). \tag{1}$$

That is, the *mean* of the expectation values is asymptotic to the classical average of the observable f.

2.3. Dynamics

To introduce dynamics, we consider a smooth, area-preserving (symplectic) map A of the torus. Iterating A we get a discrete dynamical system. For instance, if $A \in SL(2, \mathbf{Z})$ is a linear automorphism then the system is well known to be chaotic if A is hyperbolic, that is $|\operatorname{tr} A| > 2$ (such a map is called a "cat map" in the physics literature). Another example is the "Kronecker map",

$$\tau_\alpha : x \mapsto x + \alpha \mod 1, \quad \alpha = (\alpha_1, \alpha_2). \tag{2}$$

If $1, \alpha_1, \alpha_2$ are linearly independent over the rationals then this map is *uniquely ergodic*, i.e. the only τ_α-invariant probability measure on the torus is Lebesgue measure.

Definition 2.1. *A quantization of A is a sequence of unitary maps $U_N : \mathcal{H}_N \to \mathcal{H}_N$ such that*

$$U_N^* \mathrm{Op}_N(f) U_N - \mathrm{Op}_N(f \circ A) \to 0, \quad N \to \infty. \tag{3}$$

The operator U_N is called the *quantum propagator*, whose iterates give the evolution of the quantum system, and we require the quantum evolution to be asymptotic to the classical evolution as $N \to \infty$ (this is an analogue of "Egorov's theorem"). In this case we say that the map A is "quantizable". The eigenfunctions of U_N play the rôle of energy eigenstates.

In the example of the linear map A, if we further assume $A = \begin{pmatrix} a & b \\ c & d \end{pmatrix}$ with $ab \equiv cd \equiv 0 \mod 2$, then we can construct a unitary operator $U_N(A)$ which satisfies an *exact* version of Egorov's theorem:

$$U_N(A)^* \mathrm{Op}_N(f) U_N(A) = \mathrm{Op}_N(f \circ A). \tag{4}$$

A quantization of the Kronecker map τ_α (2) satisfying (3) was constructed in joint work with Jens Marklof (see [10] for the closely related case of a skew translation), by first doing so for rational α, in which case we have an exact Egorov theorem (4), and then for the general case by approximating α by rationals.

3. Quantum Ergodicity

For quantum maps, the form that quantum ergodicity assumes is the following:

Theorem 3.1. ([1, 15, 16]) *Let A be a quantizable area-preserving map of the torus. Assume A is* ergodic. *Then for any orthonormal basis ψ_j of $\mathcal{H}_N$ consisting of eigenfunctions of $U_N(A)$, there is a subset $J(N) \subset \{1, 2, \dots, N\}$, with $\frac{\#J(N)}{N} \to 1$, so that for $j \in J(N)$ we have:*

$$\langle \mathrm{Op}_N(f)\psi_j, \psi_j\rangle \to \int_{\mathbf{T}^2} f, \quad \text{as } N \to \infty$$

for all observables $f \in C^\infty(\mathbf{T}^2)$.

Theorem 3.1 is a consequence, using positivity and a standard diagonalization argument, of the following estimate for the variance due to Zelditch [15] (see the Appendix for the proof):

Theorem 3.2. ([15]) *Let A be a quantizable area-preserving, ergodic map of the torus. For any orthonormal basis ψ_j, $j = 1, \dots, N$ of of $\mathcal{H}_N$ consisting of eigenfunctions of $U_N(A)$, we have*

$$\frac{1}{N}\sum_{j=1}^{N} \left| \langle \mathrm{Op}_N(f)\psi_j, \psi_j\rangle - \int_{\mathbf{T}^2} f \right|^2 \to 0$$

for all observables $f \in C^\infty(\mathbf{T}^2)$.

A key problem is:

Problem 3.3. *Is it true that* all *eigenfunctions become equidistributed as $N \to \infty$?*

For the Kronecker map τ_α, Jens Marklof and I gave an affirmative answer (see [10] for skew translations):

Theorem 3.4. *If $1, \alpha_1, \alpha_2$ linearly independent over the rationals, then for all eigenfunctions ψ of $U_N(\tau_\alpha)$,*

$$\langle \mathrm{Op}_N(f)\psi, \psi\rangle \to \int_{\mathbf{T}^2} f, \quad N \to \infty .$$

This is a consequence of the quantization procedure for the map coupled with the fact that for such α, the map is classically uniquely ergodic.

In the most interesting case of *hyperbolic* maps (e.g. "cat maps"), a basic problem is the existence of several invariant measures.

4. Beyond Quantum Ergodicity for Cat Maps

I will now describe some recent attempts, joint with Pär Kurlberg, to improve on quantum ergodicity (Theorem 3.1) for cat maps.

4.1. Hecke operators [8]

It transpires that there is a commutative group of unitary operators on the state-space $\mathcal{H}_N$ which commute with the quantized map and therefore act on its eigenspaces. We call these "Hecke operators", in analogy with the setting of the modular surface.

To understand their origin, one needs to note that it is possible to define $U_N(A)$ so that it only depends on the remainder of $A \mod 2N$ and satisfies (4). One thus gets a *projective* representation $A \mapsto U_N(A)$ of the subgroup of "quantizable" elements in the finite modular group $SL(2, \mathbf{Z}/2N\mathbf{Z})$. It turns out that it can be made into an *ordinary* representation if we further restrict to the subgroup $\Gamma(4, 2N)$ given by $g = I \mod 4$ for N even, $g = I \mod 2$ for N odd. Thus for $A, B \in \Gamma(4, 2N)$ we have $U_N(AB) = U_N(A)U_N(B)$. Consequently, if $AB = BA \mod 2N$ then their propagators commute. This is the basic principle that we use to form the Hecke operators (see [6] for another application of this idea).

Remark 4.1. *The congruence $AB = BA \mod 2N$ is much less restrictive than the equation $AB = BA$. The latter has as its solutions in $SL(2, \mathbf{Z})$ essentially only $\pm$ powers of A (at least for A "primitive").*

4.2. Equidistribution of Hecke eigenfunctions

Since the Hecke operators commute with $U_N(A)$, they act on its eigenspaces, and since they commute with each other there is a basis of $\mathcal{H}_N$ consisting of joint eigenfunctions of $U_N(A)$ and the Hecke operators, whose elements we call Hecke eigenfunctions. We show

Theorem 4.2. ([8]) *Let $A \in SL(2, \mathbf{Z})$ be hyperbolic, $A = I \mod 4$, and $f \in C^\infty(\mathbf{T}^2)$ a smooth observable. Then for all normalized Hecke eigenfunctions $\phi \in \mathcal{H}_N$ of $U_N(A)$, the expectation values $\langle \mathrm{Op}_N(f)\phi, \phi\rangle$ converge to the phase-space average of f as $N \to \infty$. Moreover, for all $\epsilon > 0$ we have*

$$\langle \mathrm{Op}_N(f)\phi, \phi\rangle = \int_{\mathbf{T}^2} f(x)dx + O_{f,\epsilon}(N^{-1/4+\epsilon}), \quad \text{as } N \to \infty .$$

The exponent of $1/4$ in our theorem is certainly not optimal, and more likely the correct exponent is $1/2$. What we in fact show is that if ϕ_i, $i = 1, \dots, N$ is an orthonormal basis of $\mathcal{H}_N$ consisting of Hecke eigenfunctions then

$$\sum_{i=1}^{N} \left| \langle \mathrm{Op}_N(f)\phi_i, \phi_i\rangle - \int_{\mathbf{T}^2} f(x)dx \right|^4 \ll N^{-1+\epsilon} \tag{5}$$

(compare Theorem 3.2). We deduce Theorem 4.2 from (5) by taking an orthonormal basis with $\phi_1 = \phi$ and omitting all but one term on the LHS. If all terms on the LHS of (5) are of roughly the same size, then we would expect this to give the exponent $1/2$.

Remark 4.3. *The Hecke eigenspaces have small dimension (at most $O(\log\log N)$), while the eigenspaces of $U_N(A)$ may have large dimension. In fact, the* mean *degeneracy is $N/\operatorname{ord}(A, N)$ where $\operatorname{ord}(A, N)$ the* order *(or period) of A modulo N,* *that is the least integer $k \geq 1$ for which $A^k = I \mod N$. It can be shown that the mean degeneracy can be as large as $N/\log N$ for arbitrarily large N.*

4.3. Arbitrary eigenfunctions

Since not all eigenfunctions of $U_N(A)$ are Hecke eigenfunctions, we have not completely solved Problem 3.3 – whether *all* eigenfunctions become equidistributed, that is if we have quantum *unique* ergodicity. In [9], we show equidistribution of *all* eigenfunctions of $U_N(A)$ for *almost all* integers N:

Theorem 4.4. *Let $A \in SL(2, \mathbf{Z})$ be hyperbolic. There is a set of integers $\mathcal{N}^*$ of density one so that all eigenfunctions of $U_N(A)$ are equidistributed, as $N \to \infty$, $N \in \mathcal{N}^*$.*

Previously, the only result giving an infinite set of N for which all eigenfunctions of $U_N(A)$ become equidistributed is by Degli-Esposti, Graffi and Isola [4], which conditional on the Generalized Riemann Hypothesis gives an infinite set of *primes*.

A key step in the proof of Theorem 4.4 is an estimate for the fourth power moment of the expectation values, involving the order of A modulo N:

Theorem 4.5. ([9]) *There is a sequence of integers of density 1 so that for all observables $f \in C^\infty(\mathbf{T}^2)$ and any orthonormal basis $\{\psi_j\}_{j=1}^N$ of $\mathcal{H}_N$ consisting of eigenfunctions of $U_N(A)$ we have:*

$$\sum_{j=1}^{N} |\langle \mathrm{Op}_N(f)\psi_j, \psi_j\rangle - \int_{\mathbf{T}^2} f|^4 \ll \frac{N(\log N)^{14}}{\mathrm{ord}(A,N)^2} \,.$$

Thus for any subsequence of integers N such that

$$\frac{\mathrm{ord}(A,N)}{N^{1/2}(\log N)^7} \to \infty \tag{6}$$

(and satisfying an additional "genericity" assumption) we find that for all eigenfunctions of $U_N(A)$, $\langle \mathrm{Op}_N(f)\psi, \psi\rangle \to \int_{\mathbf{T}^2} f$ as $N \to \infty$.

4.4. Controlling the order of A modulo N

Theorem 4.5 reduces the problem of quantum ergodicity to that of finding sequences of integers satisfying (6), a problem closely related to the classical Gauss-Artin problem of showing that any integer, other than ± 1 or a perfect square, is a primitive root modulo infinitely many primes; see [11] for a survey. We show

Theorem 4.6. *Let $A \in SL(2, \mathbf{Z})$ be hyperbolic. Then there exist $\delta > 0$ and a density 1 subset S of the integers such that for all $N \in S$ we have*

$$\mathrm{ord}(A,N) \gg N^{1/2} \exp((\log N)^\delta) \,.$$

Combining Theorem 4.6 with Theorem 4.5 gives Theorem 4.4.

Remark 4.7. *We note that condition* (6) *fails infinitely often. In fact one can show that there are arbitrarily large integers so that* $\operatorname{ord}(A, N)$ *is smaller than* $\text{const.}\, N/\log N$.

Appendix A. Proof of Quantum Ergodicity

We review the proof of Quantum Ergodicity (Theorem 3.1) as given in [15], that is we show that given $f \in C^\infty(\mathbf{T}^2)$, for any orthonormal basis ψ_j, $j = 1, \dots, N$ of of $\mathcal{H}_N$ consisting of eigenfunctions of $U_N(A)$, we have

$$\frac{1}{N}\sum_{j=1}^{N}\left|\langle \mathrm{Op}_N(f)\psi_j, \psi_j\rangle - \int_{\mathbf{T}^2} f\right|^2 \to 0. \tag{7}$$

We do this to emphasize the difference between it and our results (Theorems 4.2, (5), 4.4 and 4.5), which while far stronger than what is given by (7), requires methods that are special to the arithmetic structure of the cat map. In contrast, the argument below uses nothing more than Egorov's theorem and the *ergodicity* of the map. Without loss of generality, we will in the sequel assume that $\int_{\mathbf{T}^2} f = 0$.

We first recall some basic properties of the quantized observables $\mathrm{Op}_N(f)$:

1. The adjoint is given by

$$\mathrm{Op}_N(f)^* = \mathrm{Op}_N(\bar{f}). \tag{8}$$

2. The composition of operators satisfies:

$$\mathrm{Op}_N(f)\,\mathrm{Op}_N(g) = \mathrm{Op}_N(fg) + O_{f,g}(\frac{1}{N}) \tag{9}$$

for $f, g \in C^\infty(\mathbf{T}^2)$.

We fix $T \geq 1$. By Egorov's theorem (3), as $N \to \infty$ we have

$$\frac{1}{T}\sum_{j=1}^{T}(U_N(A)^t)^*\,\mathrm{Op}_N(f)U_N(A)^t \sim \frac{1}{T}\sum_{t=1}^{T}\mathrm{Op}_N(f\circ A^t) = \mathrm{Op}_N(f^T)$$

where $f^T := \frac{1}{T}\sum_{t=1}^{T} f\circ A^t$ is the ergodic average of f. Moreover, if ψ_j is an eigenfunction: $U_N(A)\psi_j = e^{i\lambda_j}\psi_j$, then

$$\begin{aligned}\langle \mathrm{Op}_N(f)\psi_j, \psi_j\rangle &= \langle \mathrm{Op}_N(f)U_N(A)\psi_j, U_N(A)\psi_j\rangle \\ &= \langle U_N(A)^*\,\mathrm{Op}_N(f)U_N(A)\psi_j, \psi_j\rangle \\ &\sim \langle \mathrm{Op}_N(f\circ A)\psi_j, \psi_j\rangle.\end{aligned}$$

Consequently, if ψ_j is an eigenfunction then for all $T \geq 0$,

$$\langle \mathrm{Op}_N(f)\psi_j, \psi_j\rangle = \langle \mathrm{Op}_N(f^T)\psi_j, \psi_j\rangle.$$

Now we look at the sum (recall $\int f = 0$)

$$S_2(f,N) := \frac{1}{N}\sum_{j=1}^{N} |\langle \operatorname{Op}_N(f)\psi_j, \psi_j\rangle|^2 .$$

We will show that $\lim_{N\to\infty} S_2(f,N) = 0$.

By Egorov (3), we have $S_2(f,N) \sim S_2(f^T,N)$ as $N \to \infty$, for all $T \geq 1$. By Cauchy-Schwartz, we have

$$|\langle \operatorname{Op}_N(f^T)\psi_j, \psi_j\rangle|^2 \leq \|\operatorname{Op}_N(f^T)\psi_j\|^2 \|\psi_j\|^2 = \langle \operatorname{Op}_N(f^T)^* \operatorname{Op}_N(f^T)\psi_j, \psi_j\rangle .$$

Moreover, by (8), (9),

$$\operatorname{Op}_N(f^T)^* \operatorname{Op}_N(f^T) = \operatorname{Op}_N(|f^T|^2) + O_{f,T}(\frac{1}{N})$$

and so

$$S_2(f,N) \lesssim \frac{1}{N}\sum_{j=1}^{N} \langle \operatorname{Op}_N(|f^T|^2)\psi_j, \psi_j\rangle + O_{f,T}(\frac{1}{N}) .$$

By (1) we thus find that for fixed $T \geq 1$,

$$\limsup S_2(f,N) \leq \int_{\mathbf{T}^2} |f^T|^2 .$$

So far we have used nothing about the cat map except Egorov's theorem. Now we use the fact that it is *ergodic*, in particular the *mean ergodic theorem* holds: For $F \in L^2(\mathbf{T}^2)$, the ergodic averages F^T converge to $\int_{\mathbf{T}^2} F$ in L^2. Thus we have $\int_{\mathbf{T}^2} |f^T|^2 \to 0$ as $T \to \infty$. Therefore given $\epsilon > 0$, we can find $T = T(f,\epsilon)$ for which $\int_{\mathbf{T}^2} |f^T|^2 < \epsilon$ and consequently

$$\limsup S_2(f,N) < \epsilon$$

which shows that $S_2(f,N) \to 0$ as required. □

References

[1] A. Bouzouina and S. De Bièvre, *Equipartition of the eigenfunctions of quantized ergodic maps on the torus*, Comm. Math. Phys. **178** (1996), 83–105.

[2] Y. Colin de Verdière, *Ergodicité et fonctions propres du laplacien*, Comm. Math. Phys. **102** (1985), 497–502.

[3] M. Degli Esposti, *Quantization of the orientation preserving automorphisms of the torus*, Ann. Inst. Poincaré **58** (1993), 323–341.

[4] M. Degli Esposti, S. Graffi and S. Isola, *Classical limit of the quantized hyperbolic toral automorphisms*, Comm. Math. Phys. **167** (1995), 471–507.

[5] J. H. Hannay and M. V. Berry, *Quantization of linear maps on a torus – Fresnel diffraction by a periodic grating*, Physica D **1** (1980), 267–291.

[6] J. Keating and F. Mezzadri, *Pseudo-Symmetries of Anosov Maps and Spectral Statistics*, Nonlinearity **13** (2000), no. 3, 747–775.

[7] S. Knabe, *On the quantisation of Arnold's cat*, J. Phys. A: Math. Gen. **23** (1990), 2013–2025.

[8] P. Kurlberg and Z. Rudnick, *Hecke theory and equidistribution for the quantization of linear maps of the torus*, Duke Math. J. **103** (2000), no. 1, 47–77.

[9] P. Kurlberg and Z. Rudnick, *On quantum ergodicity for linear maps of the torus*, preprint math/9910145, to appear in Comm. Math. Phys.

[10] J. Marklof and Z. Rudnick, *Quantum unique ergodicity for parabolic maps*, Geom. and Funct. Analysis **10** (2000), no. 6, 1554–1578.

[11] M. Murty, *Artin's conjecture for primitive roots*, Math. Intelligencer **10** (1988), no. 4, 59–67.

[12] Z. Rudnick and P. Sarnak, *The behaviour of eigenstates of arithmetic hyperbolic manifolds*, Comm. Math. Phys. **161** (1994), 195–213.

[13] A. Schnirelman, *Ergodic properties of eigenfunctions*, Usp. Math. Nauk **29** (1974), 181–182.

[14] S. Zelditch, *Uniform distribution of eigenfunctions on compact hyperbolic surfaces*, Duke Math. J. **55** (1987), 919–941.

[15] S. Zelditch, *Quantum ergodicity of C^*-dynamical systems*, Comm. Math. Phys. **177** (1996), 507–528.

[16] S. Zelditch, *Index and dynamics of quantized contact transformations*, Ann. Inst. Fourier (Grenoble) **47** (1997), 305–363.

Raymond and Beverly Sackler School of Mathematical Sciences
Tel Aviv University
Tel Aviv 69978, Israel
E-mail address: `rudnick@math.tau.ac.il`

Bound Information: The Classical Analog to Bound Quantum Entanglement

Nicolas Gisin, Renato Renner and Stefan Wolf

Abstract. It was recently pointed out that there is a close connection between information-theoretic key agreement and quantum entanglement purification. This suggests that the concept of bound entanglement (entanglement which cannot be purified) has a classical counterpart: bound information, which cannot be used to generate a secret key by any protocol. We analyse a probability distribution which results when a specific bound entangled quantum state is measured. We show strong evidence for the fact that the corresponding mutual information is indeed bound. The probable existence of such information stands in contrast to previous beliefs in classical information theory.

1. Information-Theoretic Key Agreement from Classical and Quantum Information

Assume that two parties Alice and Bob, who are connected by an authentic but otherwise completely insecure channel, are willing to generate a secret key (allowing them to communicate securely). More precisely, Alice and Bob want to compute, after some rounds of communication (where the random variable C summarizes the communication carried out over the public channel), strings S_A and S_B, respectively, with the property that they are most likely both equal to a uniformly distributed string S about which the adversary Eve has virtually no information. More precisely,

$$\text{Prob}\,[S_A = S_B = S] \geq 1 - \varepsilon\ , \quad H(S) = \log_2 |\mathbf{S}|\ , \quad \text{and} \quad I(S;C) < \varepsilon \qquad (1)$$

(where $\mathbf{S}$ is the range of S and $|\mathbf{S}|$ is its cardinality) should hold for some small ε. Note that the security condition in (1) is information-theoretic (sometimes also called unconditional): Even an adversary with unlimited computer power must be unable to obtain useful information. The Diffie–Hellman protocol [1] for instance achieves the goal of key agreement by insecure communication only with respect to computationally bounded adversaries.

It is a straightforward generalization of Shannon's well-known impossibility result [13] that information-theoretic secrecy cannot be generated in this setting, i.e., from authenticity only: Public-key systems are never unconditionally secure.

Hence we have to assume some additional structure in the initial setting, for instance some pieces of information given to Alice and Bob (and also Eve), respectively.

1.1. Classical information

The general case where this information given to the three parties initially consists of the outcomes of some random experiment has been studied intensively [9, 10, 14]. Here, it is assumed that Alice, Bob, and Eve have access to realizations of random variables X, Y, and Z, respectively, jointly distributed according to P_{XYZ}. A special case is when all the parties receive noisy versions of a (binary) signal broadcast by some information source.

It was shown that if the setting is modified this way (where the secrecy condition in (1) must be replaced by $I(S;CZ) < \varepsilon$), then secret-key agreement is often possible. Shannon's pessimistic result now generalizes to the statement that the size of the resulting secret key S cannot (substantially) exceed the quantity

$$I(X;Y \downarrow Z) := \min_{XY \to Z \to \overline{Z}} I(X;Y|\overline{Z})$$

(where $XY \to Z \to \overline{Z}$ is a Markov chain) which was defined in [10] as the *intrinsic conditional information between* X *and* Y*, given* Z.

In the special case where the parties' initial information consists of the outcomes of many independent repetitions of the same random experiment given by P_{XYZ} (i.e., Alice knows $X^N := [X_1, X_2, \dots, X_N]$, and similarly for Bob and Eve), the *secret-key rate* $S(X;Y||Z)$ was defined as the maximal key-generation rate (measured with respect to the number of required realizations of P_{XYZ}) that is asymptotically achievable (for $N \to \infty$). The above-mentioned result then implies

$$S(X;Y||Z) \leq I(X;Y{\downarrow}Z) ,$$

and it was conjectured that intrinsic information can always be distilled into a secret key, i.e., that $I(X;Y{\downarrow}Z) > 0$ implies $S(X;Y||Z) > 0$ [10, 14]. This conjecture was supported by some evidence given in [10]; however, it is the objective of this paper to give much stronger evidence for the opposite, i.e., that there exist types of intrinsic information *not* allowing for secret-key agreement. The motivation for the corresponding considerations comes from quantum mechanics or, more precisely, from the concept of *bound entanglement* in quantum information theory.

1.2. Quantum information

When considering the model where certain pieces of information are given initially to the involved parties, it is a natural question where this information comes from. According to Landauer, information is always physical and hence ultimately quantum mechanical [7, 8]. Thus the random variables could come from measuring a certain quantum state $|\Psi\rangle$. In this case however it seems to be overly restrictive to force Alice and Bob to measure their quantum systems right at the beginning of the key-agreement process. It is possibly advantageous for them to carry out a protocol first (using classical communication and local quantum operations on their systems) after which they end up with a number of quantum bits in a maximally entangled state. Measuring them finally leads to a (classical) secret key. The first phase of this protocol is called *quantum (entanglement) purification.*

In order to understand what happens in a purification protocol and for which initial states such a protocol is at all possible, we recall some basic facts about quantum (information) theory[1]. In contrast to a classical bit (*Cbit* for short) which can take either of the values 0 or 1, a quantum bit (*Qbit*) can exist in a superposition of these two extremal states (with complex *probability amplitudes* a and b satisfying $|a|^2 + |b|^2 = 1$):

$$|\psi\rangle = a|0\rangle + b|1\rangle \ .$$

When measuring this state with respect to the basis $\{|0\rangle, |1\rangle\}$, we obtain $|0\rangle$ with probability $|a|^2$ and $|1\rangle$ otherwise. All (pure) states of one Qbit can be represented as unit vectors in the Hilbert space $\mathbf{C}^2$.

A possible state of a system of two Qbits can be

$$|\psi\rangle = |\psi_1\rangle \otimes |\psi_2\rangle =: |\psi_1\psi_2\rangle \ ,$$

which is simply the tensor product of the states $|\psi_1\rangle$ and $|\psi_2\rangle$ of the first and second Qbit, respectively. Such a state is called a *product state.* However, (normalized) linear combinations of quantum states lead to additional states; for instance,

$$|\psi^-\rangle := (|01\rangle - |10\rangle)/\sqrt{2}$$

is also a possible state of the two-Qbit system. This state is called *singlet state* and has the property that whenever the Qbits are measured with respect to the same basis, the outcomes are opposite bits. There is no classical explanation for this behavior which is called *(maximal) entanglement.* We conclude that two Qbits are not the same as "two times one Qbit."

As described above, the objective of Alice and Bob doing quantum purification is to generate two-Qbit systems in the state $|\psi^-\rangle$ (or in states very close to it) by classical communication and local quantum operations. The states they start with can for instance be their view of a pure state $|\Psi\rangle$ living in Alice's, Bob's, and a possible adversary Eve's (who is assumed to have total control over the entire environment) Hilbert spaces:

$$|\Psi\rangle \in H_{\text{Alice}} \otimes H_{\text{Bob}} \otimes H_{\text{Eve}} \ .$$

Then Alice and Bob's perspective

$$\rho_{AB} := \text{Tr}_{H_{\text{Eve}}}(|\Psi\rangle)$$

(the *trace* over Eve's space H_{Eve}) is generally a *mixed state.* In contrast to a pure state, which can be represented by a vector in a Hilbert space, a mixed state is described by a probability distribution over such a space[2]. A mixed state, such as ρ_{AB}, can be represented by a $(\dim H_{\text{Alice}}) \cdot (\dim H_{\text{Bob}}) \times (\dim H_{\text{Alice}}) \cdot (\dim H_{\text{Bob}})$

[1] For an introduction, see for example [11].

[2] Note that the notion of mixed state is actually rather information-theoretic than physical. Roughly speaking, a mixed state is a pure state that is only partially known.

matrix, namely the weighted sum (with respect to the probability distribution) of the projectors to the subspaces generated by the corresponding pure states. This matrix is called *density matrix.*

It is important to note that "purification," which transforms the mixed state ρ_{AB} into pure (singlet) states, actually means key agreement: Alice and Bob's final state is pure and hence not entangled with anything else, in particular not with anything under Eve's control. The adversary is out of the picture, whatever operations and measurements she performs.

Let us consider some properties of mixed states. A state ρ_{AB} which is *separable*, i.e., a mixture of product states, can be prepared remotely by purely classical communication. States that are not separable are called *entangled* and cannot be prepared this way. It is a natural question which states ρ_{AB} *can be purified* and which cannot. Separable states cannot be purified because of the property just described and because of the generalization of Shannon's Theorem mentioned at the beginning of this paper: No information-theoretic key agreement is possible from authentic but public (classical) communication. On the other hand, if Alice's and Bob's subsystems are two-dimensional (i.e., Qbits) and entangled, then purification is always possible [4]. However, the surprising fact was recently discovered that the same is not true for higher-dimensional systems: There exist entangled states which cannot be purified [5]. (This follows from the fact that the eigenvalues of the so-called partial transposition of certain entangled density matrices ρ_{AB} are non-negative [12].) This type of entanglement is called *bound* (in contrast to *free* entanglement, which *can* be purified). From the perspective of classical information theory, the interesting point is that bound entanglement seems to have a classical counterpart with unexpected properties.

2. Linking the Two Models

It was shown in [2] that there exists a close connection between the classical- and quantum-information-based secret-key-agreement scenarios, where the transition from quantum to classical information corresponds to certain measurements.

The following result was shown in [2]. Let $|\Psi\rangle$ be a pure quantum state of Alice, Bob, and Eve's systems, and let ρ_{AB} be the corresponding mixed state of Alice and Bob (obtained by tracing over Eve's space). Then the distribution P_{XYZ}, resulting from optimal measurement of $|\Psi\rangle$ by all the parties, has positive intrinsic conditional information, $I(X;Y\downarrow Z) > 0$, if and only if ρ_{AB} is an entangled state. By assuming "optimal measurement" we mean here the following statement. Whenever ρ_{AB} is entangled, but only then, there exist, for all possible measurements Eve can do (i.e., for all bases or, more generally, generating sets $\{|z\rangle\}$ she can choose to measure with respect to), measurement bases $\{|x\rangle\}$ and $\{|y\rangle\}$ for Alice and Bob, respectively, such that $I(X;Y\downarrow Z) > 0$ holds for the distribution $P_{XYZ}(x,y,z) := |\langle x,y,z|\Psi\rangle|^2$.

Given this result, it is natural to assume that the same is true also on the protocol level, i.e., that ρ_{AB} can be purified if and only if the corresponding distribution P_{XYZ} (again with respect to optimal measurements) satisfies $S(X;Y||Z) > 0$. However, this correspondence could not be generally proven so far. An interesting consequence of such a proof would be that bound entanglement has a classical counterpart, namely classical information which cannot be used for generating a secret key. We will call such information *bound.* The existence of bound information would contrast with previous beliefs in classical information theory.

Although the connection between classical key agreement and quantum purification has not been proven in generality, we give, in the next section, new independent evidence that bound information does exist. The arguments are based on the analysis of a distribution coming from the "translation" of a bound entangled state described in [6].

3. Bound Information and Binarizations

We analyse a distribution that results from measuring a bound entangled quantum state, described in [6], with respect to the standard bases. Without even having a closer look at the state or the optimality of the measurements performed, we give direct evidence for the fact that this distribution yields an example of bound intrinsic information.

The distribution we consider is the following. Let $0 \leq \alpha \leq 3$.

X Y (Z)	1	2	3
1	(0) $2/21$	(1) $(5-\alpha)/21$	(2) $\alpha/21$
2	(3) $\alpha/21$	(0) $2/21$	(4) $(5-\alpha)/21$
3	(5) $(5-\alpha)/21$	(6) $\alpha/21$	(0) $2/21$

(This table reads as follows: We have for instance $P_{XYZ}(1,1,0) = 2/21$ and $P_{XYZ}(1,1,z) = 0$ for $z \neq 0$.) The corresponding quantum state $|\Psi_\alpha\rangle$ is known to be free entangled for $\alpha \in [0,1)$, bound entangled for $\alpha \in [1,2)$, and separable for $\alpha \in [2,3]$. Not surprisingly, the above distribution satisfies $I(X;Y\downarrow Z) > 0$ for $\alpha \in [0,2)$, but a secret-key agreement protocol is known only for $\alpha \in [0,1)$. In the following, we give evidence for the fact that there does not exist such a protocol for $\alpha \in [1,2)$. More precisely, we show the following facts for this case.

First, we prove that whenever the random variable Y is "binarized," i.e., sent through a binary-output channel $P_{\overline{Y}|Y}$ (or a ternary-output channel but where only two symbols are actually considered in the computation of the mutual information), then the intrinsic information vanishes (Proposition 3.1).

Proposition 3.2 on the other hand suggests that intrinsic information which does not resist any binarization must be bound: Whenever secret-key agreement is possible with X and Y and with respect to Z, then there exist binarizations of a certain number of repetitions of X and Y such that the intrinsic information remains positive.

These two propositions together suggest that the given distribution is indeed bound entangled. (However, note that they do not prove this.)

Proposition 3.1. *Assume the above distribution with* $\alpha \in [1,2)$. *Let* $P_{\overline{Y}|Y}$ *be an arbitrary conditional distribution with* $\overline{\mathbf{Y}} = \{0,1,\Delta\}$, *and let* E *be the event that* $\overline{Y} \in \{0,1\}$. *Then* $I(X;\overline{Y} \downarrow Z \,|\, E) = 0$.

Proof. We only have to consider the case $\alpha = 1$. This implies the statement for all $\alpha \in [1,2)$. Let the following channel $P_{\overline{Y}|Y}$ be given (where $\overline{\mathbf{Y}} = \{0,1,\Delta\}$):

$$\begin{array}{rclrclrcl} P_{\overline{Y}|Y}(0,1) & = & x\,, & P_{\overline{Y}|Y}(0,2) & = & y\,, & P_{\overline{Y}|Y}(0,3) & = & z\,, \\ P_{\overline{Y}|Y}(1,1) & = & u\,, & P_{\overline{Y}|Y}(1,2) & = & v\,, & P_{\overline{Y}|Y}(1,3) & = & w\,. \end{array}$$

Here, we have $x,y,z,u,v,w,x+u,y+v,z+w \in [0,1]$. We get the following distribution $P_{X\overline{Y}Z|E}$ (to be normalized).

X / $\overline{Y}$ (Z)	1	2	3
0	(0) $2x$ (3) y (5) $4z$	(0) $2y$ (1) $4x$ (6) z	(0) $2z$ (2) x (4) $4y$
1	(0) $2u$ (3) v (5) $4w$	(0) $2v$ (1) $4u$ (6) w	(0) $2w$ (2) u (4) $4v$

The only symbol z of Z for which $I(X;\overline{Y}|Z=z,E) > 0$ holds is $z=0$. Let us now consider a channel $P_{\overline{Z}|Z}$ with $\overline{\mathbf{Z}} = \{\overline{0},\overline{1},\overline{2},\overline{3},\overline{4},\overline{5},\overline{6}\}$ and $P_{\overline{Z}|Z}(0,0) = 1$. Furthermore, $P_{\overline{Z}|Z}(\overline{0},1) = c$ and $P_{\overline{Z}|Z}(\overline{1},1) = 1-c$, and analogously for $Z = 2,3,4,5$, and 6 with transition probabilities e,a,f,b, and d, respectively. Then we get for the column vectors of the $P_{X\overline{Y}|\overline{Z}=\overline{0}}$ matrix:

$$\left[2\binom{x}{u} + a\binom{y}{v} + 4b\binom{z}{w}\,,\; 4c\binom{x}{u} + 2\binom{y}{v} + d\binom{z}{w}\,,\; e\binom{x}{u} + 4f\binom{y}{v} + 2\binom{z}{w}\right].$$

Clearly, the three vectors are linearly dependent. We can assume that

$$\binom{x}{u} = \lambda_1\binom{y}{v} + \lambda_2\binom{z}{w}$$

holds for some $\lambda_1, \lambda_2 \in [0,\infty)$. (The other cases are analogous.)

Let $\vec{s} := \binom{y}{v}$ and $\vec{t} := \binom{z}{w}$. We then get for the above matrix

$$\left[(a+2\lambda_1)\vec{s}+(4b+2\lambda_2)\vec{t},(2+4c\lambda_1)\vec{s}+(d+4c\lambda_2)\vec{t},(4f+e\lambda_1)\vec{s}+(2+e\lambda_2)\vec{t}\right].$$

The corresponding distribution satisfies $I(X;\overline{Y}|\overline{Z}=\overline{0},E)=0$ if

$$\frac{a+2\lambda_1}{4b+2\lambda_2}=\frac{2+4c\lambda_1}{d+4c\lambda_2}=\frac{4f+e\lambda_1}{2+e\lambda_2}$$

holds. This is equivalent to

$$\begin{aligned}\lambda_1(2d-16bc)+\lambda_2(4ac-4) &= 8b-ad\ ,\\ \lambda_1(4-4be)+\lambda_2(ae-8f) &= 16bf-2a\ .\end{aligned}$$

We show that this system is solvable, with $(a,b,c,d,e,f)\in[0,1]^6$, for all $\lambda_1,\lambda_2\in[0,\infty)$. For this, we prove that for all sufficiently large numbers $R>0$, the equations are solvable for all pairs (λ_1,λ_2) on the path $(0,0)$-$(R,0)$-(R,R)-$(0,R)$-$(0,0)$, and that the corresponding path in $[0,1]^6$ is homeomorphic to S^1. Then the claim follows by a simple topological argument.

We only sketch the remainder of the proof. For $(\lambda_1,\lambda_2)=(0,0)$, the equations are solvable by setting

$$d=f=1 \ \text{ and } \ 8b=d\ . \tag{2}$$

For $(\lambda_1,\lambda_2)=(R,0)$, where we assume R to be sufficiently large, a solution is given by

$$b\approx e\approx 1 \ \text{ and } \ d\approx 8c$$

(where additionally both equations of (2) should *not* be satisfied nor approximately satisfied). For $(\lambda_1,\lambda_2)=(R,R)$, the equalities are

$$\begin{aligned}R(2d-16bc+4ac-4) &= 8b-ad\\ R(4-4be+ae-8f) &= 16bf-2a\end{aligned}$$

with a possible approximate solution

$$b\approx e\approx 0\ ,\quad c\approx d\approx 1\ ,\quad a\approx f\approx 1/2\ .$$

Finally, the case $(\lambda_1,\lambda_2)=(0,R)$ can be solved by

$$a\approx c\approx 1 \ \text{ and } \ e\approx 8f\ .$$

When combining the solutions for the different cases, it is not difficult to see that there exists a path γ in $[0,1]^6$ that, mapped to the (λ_1,λ_2) plane, exactly corresponds to the square $(0,0)$-$(R,0)$-(R,R)-$(0,R)$-$(0,0)$. This is true for all sufficiently large R, and thus the argument is finished. □

Proposition 3.2. *Let X, Y, and Z satisfy $S(X;Y||Z)>0$. Then there exist a number N and ternary-output channels $P_{\overline{X}|X^N}$ and $P_{\overline{Y}|Y^N}$ (with ranges $\overline{\mathbf{X}}=\overline{\mathbf{Y}}=\{0,1,\Delta\}$ of $\overline{X}$ and $\overline{Y}$, respectively) such that the event E defined by $\overline{X}\neq\Delta\neq\overline{Y}$ has positive probability and $I(\overline{X};\overline{Y}\downarrow Z^N\,|\,E)>0$ holds.*

Proof sketch. Let $\varepsilon > 0$ to be determined later. Then, according to the definition of the secret-key rate $S(X;Y||Z)$, there exist N and a protocol that allows Alice and Bob to compute K-bit keys S_A and S_B for some $K \geq 1$ such that $S_A = S_B = S$ holds with probability at least $1-\varepsilon$, where S is a uniformly distributed K-bit string with $H(S|CZ^N) \geq K - \varepsilon$ if C is the communication exchanged over the public channel. We construct new random variables $X' := [X^N, R_A]$ and $Y' := [Y^N, R_B]$, where R_A and R_B are random strings independent from each other and from all the rest, in such a way that we can assume the protocol to take X' and Y' as inputs and to be deterministic. Then there must exist a communication string c which occurs with positive probability and is such that $\text{Prob}\,[S_A = S_B = S \,|\, C = c] = 1$ and $H(S \,|\, Z^N, C = c) \geq K - \varepsilon/(1-\varepsilon)$ hold. Let us consider the following mappings of the ranges $\mathbf{X}'$ and $\mathbf{Y}'$ of X' and Y', respectively, to $\{0, 1, \Delta\}$. If the communication c is possible from a particular value $x' \in \mathbf{X}'$, then x' is mapped to the first bit of the resulting secret key S_A; otherwise, x' is mapped to Δ. The map from $\mathbf{Y}'$ to $\{0, 1, \Delta\}$ is defined analogously. Then the event E corresponds to the event that the communication C actually equals c, and has hence positive probability. Since R_A and R_B are independent of each other and of all the rest, these mappings are (probabilistic) binarizations $\overline{X}$ and $\overline{Y}$ of X^N and Y^N, respectively, satisfying $I(\overline{X};\overline{Y} \downarrow Z^N \,|\, E) > 0$ if ε is small enough. □

4. Concluding Remarks

We have given evidence that a certain specific probability distribution has so-called bound intrinsic information, i.e., information that cannot be used for generating a secret key by any protocol. Although the motivation for considering this particular distribution is that it results from measuring a bound entangled quantum state, our arguments are purely classical.

The existence of bound information would stand in contrast to previous beliefs in the context of unconditionally secure key agreement and be additional support for the close connection between information-theoretic key agreement and quantum purification conjectured in [2]. It should be pointed out that bound information represents one of the very few examples where quantum information theory initiates a new concept in classical information theory.

References

[1] W. Diffie and M. E. Hellman, New directions in cryptography, *IEEE Transactions on Information Theory*, Vol. 22, No. 6, pp. 644–654, 1976.

[2] N. Gisin and S. Wolf, Linking classical and quantum key agreement: is there "bound information"?, *Advances in Cryptology – Proceedings of Crypto 2000*, Lecture Notes in Computer Science, Vol. 1880, pp. 482–500, 2000.

[3] N. Gisin and S. Wolf, Quantum cryptography on noisy channels: quantum versus classical key agreement protocols, *Phys. Rev. Lett.*, Vol. 83, pp. 4200–4203, 1999.

[4] M. Horodecki, P. Horodecki, and R. Horodecki, Inseparable 2 spin 1/2 density matrices can be distilled to a singlet form, *Phys. Rev. Lett.*, Vol. 78, p. 574, 1997.

[5] P. Horodecki, Separability criterion and inseparable mixed states with positive partial transposition, *Phys. Lett. A*, Vol. 232, p. 333, 1997.

[6] P. Horodecki, M. Horodecki, and R. Horodecki, Bound entanglement can be activated, *Phys. Rev. Lett.*, Vol. 82, pp. 1056–1059, 1999. quant-ph/9806058.

[7] R. Landauer, Information is inevitably physical, *Feynman and Computation 2*, Addison Wesley, Reading, 1998.

[8] R. Landauer, The physical nature of information, *Phys. Lett. A*, Vol. 217, p. 188, 1996.

[9] U. Maurer, Secret key agreement by public discussion from common information, *IEEE Transactions on Information Theory*, Vol. 39, No. 3, pp. 733–742, 1993.

[10] U. Maurer and S. Wolf, Unconditionally secure key agreement and the intrinsic conditional information, *IEEE Transactions on Information Theory*, Vol. 45, No. 2, pp. 499–514, 1999.

[11] A. Peres, *Quantum theory: concepts and methods*, Kluwer Academic Publishers, 1993.

[12] A. Peres, Separability criterion for density matrices, *Phys. Rev. Lett.*, Vol. 77, pp. 1413–1415, 1996.

[13] C. E. Shannon, Communication theory of secrecy systems, *Bell System Technical Journal*, Vol. 28, pp. 656–715, 1949.

[14] S. Wolf, *Information-theoretically and computationally secure key agreement in cryptography*, ETH dissertation No. 13138, ETH Zürich, 1999.

N. Gisin:
Group of Applied Physics
University of Geneva
CH-1211 Geneva, Switzerland
E-mail address: `Nicolas.Gisin@physics.unige.ch`

R. Renner:
Department of Computer Science
ETH Zürich
CH-8092 Zürich, Switzerland
E-mail address: `renner@inf.ethz.ch`

S. Wolf:
Centre for Applied Cryptographic Research
Department of Combinatorics and Optimization
University of Waterloo
Waterloo, ON N2L 3G1, Canada
E-mail address: `swolf@cacr.math.uwaterloo.ca`

D-Branes on Calabi–Yau Manifolds

Michael R. Douglas

Abstract. We give an overview of recent work on Dirichlet branes on Calabi–Yau threefolds which makes contact with Kontsevich's homological mirror symmetry proposal, proposes a new definition of stability which is appropriate in string theory, and provides concrete quiver categories equivalent to certain categories of branes on CY.

1. Introduction

Dirichlet branes, discovered by Dai, Leigh and Polchinski in 1989, have been a central part of the dramatic progress in superstring theory of recent years. They are the simplest solitons in string theory; the basic definition of Dirichlet brane (D-brane) is that it is an allowed boundary condition for a string [45].

Although they play many physical roles, from a mathematical point of view their most salient feature is that they provide the natural context in which Yang–Mills theory is embedded in string theory. This has allowed physicists to rederive and better understand some of the most beautiful mathematical constructions that have been discovered in this area, such as the ADHM construction of instantons and Nahm's construction of monopoles.

D-branes on Calabi–Yau (CY) manifolds have been the focus of a number of recent works (see [15, 18] for additional references). Early on it was realized that this theory had a close connection with mirror symmetry and especially Kontsevich's homological mirror symmetry proposal [36], which interprets mirror symmetry as an equivalence between the derived category of coherent sheaves (naturally associated to Yang–Mills on Kähler manifolds) and a derived category proposed by Fukaya, associated to isotopy classes of Lagrangian submanifolds. These two classes of objects (usually called "B"- and "A"-type branes respectively) correspond to the two classes of supersymmetry preserving (or BPS) branes on a CY, and thus the physical understanding of mirror symmetry indeed should imply such an equivalence.

In physical terms, Kontsevich's proposal addresses only the structure of topologically twisted open string theory (we explain this below), while one expects more structure to appear in discussing the full open string theory which governs the physics of D-branes.

One approach to this extra structure is to study the additional conditions on the D-brane world-volumes which follow from space-time supersymmetry. In the large volume limit, these two conditions are the hermitian Yang–Mills equations (for B branes) and the special Lagrangian condition (for A branes). In this context, Strominger, Yau and Zaslow formulated mirror symmetry as the statement that the mirror W to a CY M could be obtained as the moduli space of a particular A brane mirror to the B brane which is the point on M. [52]. This brane is expected to come in families which give a T^3 fibration structure to M, a claim which has been proven in many cases by Gross and by Ruan [59, 27, 49, 41].

Although this approach clearly captures some essential part of the truth, it should be realized that the specific additional conditions which were used, the hermitian Yang–Mills and special Lagrangian conditions, are not believed to be the correct physical conditions except in the large volume limit. This is related to the point that in string theory, CY manifolds are not really Ricci-flat manifolds except in the large volume limit; the Einstein equations are known to be corrected. Indeed, the stringy version of CY Kähler moduli space, which is best thought of as the complex structure moduli space of the mirror, looks quite different from the conventional geometric version.

The correct replacements for either Ricci-flatness or the supersymmetry conditions on D-branes are not known at this writing, but some things are known, which we summarize here. In particular, one has strong reasons to believe that the metric defined by the SYZ construction (which is the "D0-metric" of [14] on the mirror manifold) is *not* Ricci-flat.

Having cast doubt on the usual mathematical starting point for the discussion of D-branes, we are obliged to say something about what replaces it. Now the fundamental physical definition is quite clear – a D-brane is a boundary condition in the world-sheet conformal field theory (CFT); the A and B branes each correspond to boundary conditions preserving a particular half of world-sheet supersymmetry.

This definition has some conceptual advantages; notably, as in the discussion of Greene and Plesser [25], mirror symmetry is manifest. The problem with it is that it is rather hard to make precise contact with geometry except in the large volume limit. However, a number of general results have been obtained, which we outline. In keeping with the CFT framework, we freely use mirror symmetry in our considerations, while regarding the precise connection with geometry as following from subsequent considerations which are not yet completely understood.

An example of a geometrical statement which is on a sound footing in string theory is the statement that the holomorphic properties of B branes are independent of Kähler moduli, and thus are "the same" as in the large volume limit [4, 14]. The precise meaning of "same" turns out to be subtle however and this is the point where the derived category must enter the discussion. A recent development is a precise contact with the graded categories which appear in Kontsevich's proposal, which shows that (if we consider B branes), as the Kähler moduli are varied, the gradings undergo a "flow" determined by the variation of certain phases associated with the branes (in A brane language, the phase appearing in the special Lagrangian condition) [17, 19].

This puts us in a position to explain perhaps the simplest physical structure carried by D-branes which is not already implicit in [36]. This is the concept of "marginal stability" which governs the possible variation of the spectrum of BPS branes. This concept reduces to variation of μ-stability for B branes in the large volume limit and based on this analogy, the work of Joyce on such transitions for A branes [34], and many other considerations, a deformation of stability called Π-stability was proposed in [16] which governs the spectrum of BPS branes on arbitrary CY's in string theory. This proposal has undergone a number of tests; we also briefly describe work in progress which attempts to complete the proposal by associating natural t-structures to regions in Kähler moduli space.

Another front on which physical considerations have made progress is in the explicit construction of these BPS branes and their moduli spaces. A particularly simple class of constructions are the orbifold constructions. These can be exactly solved for orbifolds of flat space, and provide a physical context which contains the structures found both in orbifold resolution by quiver varieties [37] and in the study of the generalized McKay correspondence [3, 32, 48], including generalizations of Beilinson's construction of holomorphic bundles on $\mathbf{P}^n$ [1].

More recently, it has been found that results from the study of explicit Gepner model boundary states [47] can be rederived starting with a "Landau–Ginzburg orbifold theory." [18]. Geometrically, this uses an embedding of the CY in a weighted projective space, and describes bundles on CY by using the generalized McKay approach to describe bundles on the ambient space, and then restricting. One new point that emerges is that the physical construction naturally contains moduli which appear only on restriction. Another striking result from this approach is an explicit prediction for the connection formula which relates the natural bases of periods at the large volume and Gepner points.

2. Background on CFT on Calabi–Yau

The most general definition of a Dirichlet brane in weakly coupled type II string theory is that it is a conformal boundary condition in the world-sheet CFT. This definition is discussed at this level of generality in [45]. It is quite analogous to the discussion of branes in a geometrical framework as calibrated submanifolds carrying appropriate vector bundles in the sense that they are auxiliary objects; one requires some description of the ambient space (the CY for us) to get started on their definition.

In the most general case, where we take as the "ambient space" an abstract CFT defined in terms of a Hilbert space, Hamiltonian and operator product coefficients, our starting point is a priori non-geometric: it does not come with any local coordinates or other conventional geometric data. On general grounds, the

CFT's which are easy to describe in this abstract way are "highly stringy" – the conventional equations of physics such as the Einstein and Yang–Mills equations will not be valid (except in special cases with high symmetry) but will be drastically modified in string theory (we will say more about this below). In some cases, such as the Gepner models, we can argue that they are still connected to solutions of Einstein's equations by varying parameters (this is the "large volume limit"), and the interesting question is then what aspects of this conventional geometrical interpretation survive the stringy modifications and how do other aspects change. However, there are other cases such as asymmetric orbifolds [42] where we have no such large volume limit or other geometrical interpretation at present. In these cases, the abstract definition of D-brane makes sense, but the identification of D-branes with cycles carrying bundles on some space cannot even get started.

Although general, these remarks are intended to make the point that many of the statements about branes which are taken as definition or otherwise manifest in the mathematical treatments, actually require proof or at least justification in the physical discussion starting from string theory. While we do not want to belabor this point, it should be kept in mind.

We now specialize to the particular case of string compactification on a Calabi–Yau threefold, for which the CFT is a $(2,2)$ superconformal field theory with $\hat{c} = 3$. These have been discussed in [7, 9, 41, 55] and elsewhere. The notation $(2,2)$ means there are two independent $N = 2$ superconformal algebras describing left- and right-moving excitations on the string world-sheet. Each $N = 2$ algebra has two supercharges G^+ and G^- which anticommute to produce the Hamiltonian and a $U(1)$ charge J, which provides a grading on the Hilbert space.

A distinguishing feature of the $N = 2$ algebra is that it admits a topological twisting, a redefinition of the generators after which one of the supercharges can be used as a "BRST charge," a differential Q which squares to zero. The cohomology of the sum $Q_L + Q_R$ of these operators in the two $N = 2$ algebras is finite dimensional and is the Hilbert space of the topologically twisted theory. This Hilbert space also parameterizes a space of "operators" in the theory which correspond to linearized deformations of the theory; these always turn out to integrable in the $(2,2)$ case so these theories come in families with locally smooth moduli spaces.

Physicists are particularly interested in these models not because they admit topological twisting but because they lead to string theories with space-time supersymmetry. Seeing this requires discussing a bit more structure, namely the "bosonized $U(1)$." The main point here is that there are natural "spectral flow" operators denoted $e^{i\hat{c}\phi/2}$ (on left and right) which provide automorphisms between sectors of different $U(1)$ charge q and $q + \hat{c}/2$. Sectors with integral (resp. half-integral) $U(1)$ charge correspond to space-time bosons (fermions) and this one-to-one relation guarantees space-time supersymmetry. The topologically twisted world-sheet theories are important in this context mainly because their observables correspond to certain distinguished amplitudes in this space-time supersymmetric theory.

The case of threefolds with $\hat{c} = 3$ is further distinguished in that the correlator of three-operators (the "Yukawa coupling") is always well defined; one can further show that such correlators are the third derivatives of a function on moduli space F known in the physics literature as the "$N = 2$ prepotential" and in the math literature as the "Gromov–Witten potential." We will discuss the physical interpretation of this potential below.

One made a choice in this construction of whether to identify G^+ or G^- as the BRST operator Q; it turns out that only the relative choice between the two $N = 2$ algebras matters in the final result. This choice distinguishes "A" and "B" topologically twisted theories, which although they arise from the same CFT, need have little else in common. On the level of abstract CFT, one can also exchange G^+ and G^- in one of the $N = 2$ algebras to define a "new" CFT, all of whose physical predictions would be the same, but in which the A and B twisted theories would be exchanged. Thus there is no a priori difference between A and B, and this is the sense in which mirror symmetry is manifest in CFT.

This formalism arises from several more concrete constructions. Contact with geometry and the large volume limit is most direct in what is called the "nonlinear sigma model" definition. This starts from a Calabi–Yau threefold with Ricci-flat metric $g^{(0)}$, and maps of the string world-sheet into this "target space." The Ricci-flat metric is determined by a choice of complex structure (let the moduli space of these be $MC(M)$) and complexified Kähler class (moduli space $MK_0(M)$); physical arguments strongly suggest that this produces a $(2, 2)$ CFT determined by these moduli. The two gradings provided by left and right $U(1)$ charge essentially correspond to the bigrading of Dolbeault cohomology, and the B and A twisted theories as above correspond respectively to a theory describing variation of Hodge structure (the Yukawa couplings are the second derivatives of the periods of the holomorphic three-form) and to the "quantum cohomology theory" whose prepotential generates the Gromov–Witten invariants.

Mirror symmetry equates the A model on M, which sees only $MK(M)$, with the B model on its mirror W, which only sees $MC(W)$. In particular, the complexified Kähler cone $MK_0(M)$ is only a large volume approximation to the "true" stringy Kähler moduli space $MK(M)$, which is best defined simply as $MC(W)$. Much work has been done on the physical interpretation of this point. [24]. Since there is such a close connection between these topologically twisted theories and the complex geometry, the claim that these originate from physical CFTs for which mirror symmetry is manifest places this formulation of mirror symmetry essentially beyond doubt; various explicit arguments have confirmed this [41].

Beyond these results, which can be convincingly justified in the topologically twisted string theory, what one actually has in this definition is a way to compute general observables as a series expansion in powers of the curvature of the metric $g^{(0)}$ multiplied by a distinguished scale of length l_s, the "string length." One also has prescriptions for adding "instanton" corrections associated to non-trivial holomorphic maps from world-sheets of various genus into the CY. It should be realized that this expansion by itself does not provide a very convincing argument for existence of these models on finite size CY, because such series expansions often do not converge. However, given evidence from other directions that models do indeed exist, we can interpret them as valid descriptions at least of the asymptotics of these observables for large CYs. (Whenever we talk about "large" CYs in the following, we simply mean the case in which the leading terms are good approximations.)

Perhaps the most basic observation one can make from these results is that a CY is actually not Ricci flat in string theory [26]. One traditionally derives Ricci-flatness as the condition for "zero beta function" required for conformal invariance; however this is just the leading term and one sees the next correction at order l_s^6; the true condition takes the form

$$0 = \frac{\partial}{\partial\mu} g^{\beta}_{ij} = R_{ij} + l_s^6 (R^{(4)})_{ij} + \cdots , \tag{1}$$

with the term $R^{(4)}$ given in [26], and a higher order term shown to be nonzero in [33]. It has also been shown that the corrections are such that a solution can be found at each order in the series [43].

A conceptual argument that Ricci-flatness is not to be expected is simply that the structure of the Kähler moduli space predicted by mirror symmetry is quite different from the geometric Kähler cone or a naive complexification of this, and we would expect the "true" stringy metric to depend naturally on this true moduli space; the space of Ricci-flat metrics does not.

However, it is difficult to go further with this definition, simply because neither the equation (1) nor the metric g^{β} it describes are canonically defined; the theory of renormalization which led to it does not distinguish in any convincing way between g^{β} and any other metric $g^{\beta'}$ which would be produced by an arbitrary local functional redefinition $g \to g + \alpha_1 R + \alpha_2 R^2 + \cdots$.

The study of D-branes will provide a better definition of the metric, which we discuss in the next section. Let us turn now to alternatives to the non-linear sigma model. The most famous of these is the Gepner model, also described in the references above. This starts off from completely non-geometric data, but is believed to be equivalent to the "analytic continuation" of the CY nonlinear sigma model to a specific highly stringy point in the true Kähler moduli space. The fact that such models do exist and are described by the $N = 2$ formalism is our best reason to believe that the series expansions described earlier can indeed be summed in some way to define the models under discussion.

The best understanding we have at present of how the Gepner model is connected to the geometric picture comes through Witten's "linear sigma model" definition ([56]; see [23, 30] for work on D-branes in this framework). One can think of this as defining the CY through a specific embedding in $\mathbf{C}^6$. This embedding is described in two steps. First, one constructs a weighted projective space as a symplectic quotient of $\mathbf{C}^6$ by a $U(1)$ action, in coordinates $z^i \to e^{iw_i\theta} z^i$ for chosen weights w_i. One then defines the CY as a complex hypersurface in this ambient space. (One can similarly get more general toric varieties, etc.)

The symplectic quotient is the ordinary quotient by $U(1)$ of the stable points in the preimage $\pi^{-1}(0)$ of a moment map $\pi\colon \mathbf{C}^6 \to \mathbf{R}$ defined by

$$\{z^i\} \to -\mu + \sum w_i |z^i|^2.$$

To get weighted projective space with weights w_i, one takes $w_i \geq 0$ for $1 \leq i \leq 5$, w_6 such that $\sum_i w_i = 0$, and $\mu > 0$. Actually, the quotient is a line bundle $O_{WP}(w_6)$ over the weighted projective space; one then imposes the complex equations grad $W = 0$ with $W = z^6 f(z^i)$, which reduce to $z^6 = 0$ and $f(z^i) = 0$ within this zero section.

The important point is now that one can argue that varying the true Kähler parameter to the Gepner point corresponds to an analytic continuation to the "Landau–Ginzburg phase," the limit $\mu << 0$. Here the symplectic quotient is rather different: it is the orbifold $\mathbf{C}^5/\mathbf{Z}_{-w_5}$, and the equations grad $W = 0$ now constrain the coordinates to $z^i = 0$ for $1 \leq i \leq 5$. Although on the surface this looks quite different from the original nonlinear sigma model, since the orbifold is birational to the original line bundle, from the point of view of complex geometry the two phases are not so different, and further physical considerations bear this out, leading to a moduli space MK covering both phases with only isolated singularities, and no immediately evident distinction between the phases.

This discussion motivates the idea that a good starting point for defining CY is orbifold resolution, and indeed an even simpler class of CYs can be defined this way, as the resolution of orbifolds $\mathbf{C}^3/\Gamma$ with Γ a discrete subgroup of $SU(3)$. This produces non-compact CYs, which have also been much studied as local models of singularities in CYs, with the general conclusion being that mirror symmetry and almost all of the qualitative features of CY physics are visible in these simpler examples. The simplest case is $\mathbf{C}^3/\mathbf{Z}_3$, studied in [12, 13, 17].

3. Dirichlet Branes and Stability

The coarsest classification of boundary conditions in CFT is according to the portion of the superconformal algebra they preserve. One way to specify a boundary condition is in terms of a linear transformation implementing the "reflection" from left-moving to right-moving excitations; this transformation must contain an automorphism of the entire symmetry algebra and thus we need to know these.

In the $(2,2)$ context, there are two basic automorphisms one can use to relate left- and right-moving algebras. [44]. The trivial choice $G_L^\pm = G_R^\pm$ and $J_L = J_R$ is referred to as a "B-type" boundary condition as it is compatible with the B topological twisting.[1] Such boundary conditions can be obtained in the nonlinear sigma model by fixing the boundary to sit on a holomorphic submanifold of the CY carrying a holomorphic bundle. The other basic choice is $G_L^\pm = G_R^\mp$ and $J_L = -J_R$ or "A-type," which is obtained by fixing the boundary to sit on a Lagrangian submanifold, carrying a flat bundle.

[1]This twisting is defined in open string conventions.

This is all the data which is required to specify a boundary condition in the topologically twisted open string theory [57], and in this context mirror symmetry can be interpreted as a one-to-one equivalence between the two classes of boundary conditions and all topological correlators. This is the physical background behind Kontsevich's proposal; however there is much more to say about its physical underpinnings and meaning which has only been addressed recently.

From a physical point of view, the most basic question is what physical observables could one compute if one knew something about the derived category of coherent sheaves or its mirror. This question is not entirely trivial as the categorical framework is quite different from the usual physics language, which assumes that the objects involved come in moduli spaces with explicit coordinatizations, etc.

There are various physical realizations of the BPS branes we are talking about, depending on what string theory we use and how many flat dimensions the brane extends in. All are essentially identical in terms of the underlying string theory; the primary distinction is between those which can be treated purely classically (so, objects are solutions of equations of motion derived from some action), and those in which quantum effects are taken into account (so objects are zero energy wave functions or typically harmonic forms on the classical moduli space). In any case, they will have half of the supersymmetry available in the original type II theory on CY, because in that theory left- and right-movers led to independent supersymmetries, which are now related by the boundary conditions.

Two realizations seem to be particularly useful to keep in mind. In the first, the physical picture is of open strings free to move in the $4 = 10 - 6$ dimensions not taken up by the Calabi–Yau; this leads to a world-volume theory with $N = 1$ supersymmetry in four dimensions, which we treat classically. These theories are gauge theories and the basic specification of such a theory is a choice of gauge group (a semi-simple Lie group) G, a Kähler manifold X of "chiral matter" configurations admitting a G-action by isometries, a G-invariant holomorphic function $W\colon X \to \mathbf{C}$ called "superpotential," and finally moment maps for the G action.

Although all of this data is physical, for the primary question of finding the BPS branes and their moduli space (as a complex variety), the data only gets used in the following way: a configuration of the brane system is a solution of grad $W = 0$ in the symplectic quotient $X//G$. We refer to the set of these as the moduli space of brane configurations. Since quotient commutes with restriction, one can also think of this as the symplectic G-quotient of the variety grad $W = 0$.

The second realization takes the branes to be particles in the extra four dimensions. This description is related to the first as follows: given a moduli space of simple brane configurations (i.e., with endomorphism group $U(1)$); one identifies particles with zero energy states in a supersymmetric quantum mechanics on this moduli space. By standard arguments, these will be cohomology classes. (If the moduli space is singular, it turns out that the quotient/restriction definition gives it a natural embedding in a non-singular space, which can be used to make the definition.) These are the "BPS branes" of [29, 22] etc.

From now on we will consider only the first realization of brane, but use the second to define what we will call "BPS central charge" and explain the concept of marginal stability. It can be shown [9] that in an $N = 2$ supersymmetric theory (such as type II on CY), the mass of a particle (in the usual space-time sense) satisfies a bound

$$\begin{aligned} M &\geq |Z(Q)|; \\ Z(Q) &= q_i z^i + m^i \frac{\partial F}{\partial z_i} = \int_\Sigma \Omega \end{aligned} \tag{2}$$

with equality attained only for BPS particles. The quantity $Z(Q)$ is the "BPS central charge" and depends only on the "charge" or topological type of the brane and on the CY moduli. It does not depend on the particular point or cohomology class in the moduli space of brane configurations.

For A branes, $Z(Q)$ is the integral of the holomorphic three-form over the Lagrangian submanifold, and thus depends only on the homology class of the submanifold and the moduli MC. The BPS bound is the same as that coming from the calibrated geometry of special Lagrangian submanifolds [28].

For B branes, $Z(Q)$ is defined in terms of the prepotential F, a function on $MK(M)$, usually computed by invoking mirror symmetry, and depends on the K-theory class of the brane and the moduli MK. From now on, when we speak of "CY moduli space" we mean the part (MC or MK) which appears in these considerations, and "brane charge" means the homology or K-theory class.

The most basic physical role of this prepotential is its appearance in (2). Implicit in this setup is a flat (Gauss–Manin) connection on the CY moduli space, related to the choice of a fixed basis on homology. This allows transporting a BPS brane between points in moduli space and following the variation of its central charge.

Naively, this transport allows us to identify the complete spectrum (list) of BPS branes at each point in moduli space. However, this is not true, because of the existence of "lines of marginal stability" on which the spectrum changes. There is a strong constraint on these changes: a brane with charge Q can undergo a decay (and thus disappear from the spectrum) into constituents with charges Q_i (satisfying $Q = \sum Q_i$) only if the phases

$$\varphi(Q_i) \equiv \frac{1}{\pi} \text{Im } \log Z(Q_i) \tag{3}$$

are equal. (The normalization factor $1/\pi$ will be explained later.) This follows from energy conservation, which implies that $M \geq \sum M_i$, and (2), and no other assumptions. We will return to this point; for now we stress that the spectrum of BPS branes does change with moduli, in both realizations. In the first realization, it will turn out that this is the direct generalization of "wall crossing" phenomena related to the behavior of μ-stability under variations of Kähler class.

Once we know that moduli spaces of brane configurations exist, of course it would be quite interesting to know more about them and in particular to find their Kähler metrics. In particular, the problem of how to give a canonical definition of the metric of a CY in string theory has a natural answer in this context: it is just the moduli space of the "D0-brane," the B brane which embeds in a point [14]. This presupposes that such a brane exists and has a moduli space which is the CY, which although manifest at large volume cannot be taken for granted on a string-scale CY. Furthermore, there might not be a unique candidate; one can choose one by using the flat connection of Kähler moduli space to carry the large volume D0-brane along some path to the point of interest, a prescription which in general is path dependent.

Anyways, assuming that such a D0-brane exists, we stress that for each point in the true Kähler moduli space (the complex structure moduli space of the mirror), there exists a canonical Kähler metric on the CY, which away from the large volume limit is not expected to be Ricci-flat (we will mention some explicit results bearing this out below [13]). This metric can also be thought of as the metric on the moduli space of the D3-brane wrapped on the T^3 fibration of SYZ mirror symmetry; again from this point of view, except in special limits, it has no reason to be Ricci-flat. (In the explicit computations of [52], this would come about after instanton corrections.) It is not even clear that it has the same Kähler class as the original metric; indeed once one goes from $MK_0(M)$ to $MK(M)$ one loses any clear sense of what this would mean.

This metric or even the question of what equations determine it seems rather inaccessible at present and we return to the question of what Kontsevich's ideas should help physicists to compute. In some sense, the answer is the space X, group G and superpotential W; however it has taken some time for physicists to recognize any of this data in Kontsevich and Fukaya's rather abstract derived categories, and indeed this relation has not been completely spelled out in the literature. The most basic point is simply that brane configurations can naturally be thought of as objects in a category whose morphisms are the linearized variations of the gauge theory data described above (tangent vectors to a point in X modulo gauge directions) associated to the system which is the direct sum of two brane subsystems. The superpotential enters because such variations typically correspond to obstructed deformations; the non-trivial point seems to be that in cases associated to Calabi–Yau threefolds this obstruction theory can always be summarized in terms of gradients of a potential. This point has been recognized in numerous special cases – for example, the functional W for holomorphic bundles is the holomorphic Chern–Simons functional, and for holomorphic curves is the functional proposed in [10, 58] – but a mathematical argument as simple and general as the physical argument based on $N = 1$ supersymmetry does not seem to have appeared in the literature.

Having seen what Kontsevich's framework can describe, we also can see in this language what it does not describe: namely, the details of the symplectic quotient which are determined by the specific moment maps. In particular, not all holomorphic objects (solutions of grad $W = 0$) correspond to points in this quotient; only stable objects do (in the GIT sense, which depends on the specific moment maps). Thus, the question of characterizing stable BPS branes (in the physical sense) could be answered by proposing the mathematical stability condition which they satisfy. One can show by physical arguments that just as the derived category of coherent sheaves does not depend on the Kähler structure of the CY, the moment maps should only depend on the Kähler structure and not on the complex structure [19].

On large CYs, branes correspond to bundles satisfying the hermitian Yang–Mills equations, so the stability condition is just the one given by the Donaldson and Uhlenbeck–Yau theorems, which is slope or μ-stability: a bundle E (or coherent sheaf) is μ-stable if, for all subsheaves E', one has $\mu(E') < \mu(E)$ with the slope $\mu(E) = c_1(E)/\text{rank}(E)$. This condition depends on both Kähler and complex structure but in a clearly separated way: μ depends on Kähler structure, and the subobject relation on complex structure. This will be the prototype for our stringy stability condition.

A first step towards such a condition is to consider a deformed hermitian Yang–Mills equation derived from string theory by Marino et. al. [40] as the condition for unbroken supersymmetry from the "Born–Infeld action." Their equation is a deformation of the Yang–Mills action which involves higher powers of the Yang–Mills curvature F, and takes the form

$$0 = \mu_{d,\theta} \equiv \text{Im } e^{i\theta}\text{Tr } (\omega + il_s^2 F)^d \tag{4}$$

on a d-fold with Kähler form ω. Considerations of geometric invariant theory in [38, 53] show that such an equation will have a solution for "$\mu_\theta(E)$-stable" bundles; i.e., a stability condition with $\mu_\theta \equiv \int \mu_{d,\theta}$ playing the role of the slope.

Although very explicit, this equation only takes into account power-like corrections in l_s, while it is known that further instanton corrections are present. Related to this, it is not clear which Kähler form one should take for ω (the Ricci-flat or some stringy version.) This point could also be addressed by D-brane considerations, but it is not clear that this is the best approach to take as one would prefer to have a condition which depends not on ω but on the point in $MK(M)$, i.e., on the complex structure of the mirror.

One can identify the precise quantity which generalizes slope by appealing to mirror symmetry to relate this question to work of Joyce on invariants counting special Lagrangian manifolds. [34]. Among the many results of this work is an analysis of the local stability of a special Lagrangian manifold under variations of the complex structure; this is determined by an inequality involving (in physics language) the phase of the BPS central charge (3). This fits very well with the known physical considerations on marginal stability, and the sign of the inequality is new information from this point of view.

This brings us to the proposal of [16], that the BPS branes at a specific point in (true) Kähler moduli space are the "Π-stable" branes, i.e., each brane based on a holomorphic object E such that for all subobjects E' one has $\varphi(E') < \varphi(E)$. Compared to Joyce's criterion, the main difference is to use morphisms and subobjects in the definition. There is also a conceptual difference: instead of working with special Lagrangians, we take the stability condition back to the

B picture and apply it there. This has the great practical advantage that the definition of subobject is fairly clear in the B picture (though we will have to say more about this below) while computing the morphisms in the A picture is a harder and still not fully understood problem.

Besides the arguments we cited, in [16] the proposal is compared with more explicit results in the large volume and in orbifold limits (more below) and finds agreement. It was studied in the example of the $\mathbf{C}^3/\mathbf{Z}_3$ orbifold and its resolution in [17], about which more below, and seems to produce sensible results there. As we mentioned, on physical grounds it is impossible for the spectrum to change except on lines of marginal stability as defined earlier, and it is hard to come up with any competing proposal which satisfies this constraint. Having said this, the proposal has two ill-defined points. First, one must lift the phases from the interval $[0, 2)$ (which is what (3) gives us) to $\mathbf{R}$. The geometry underlying this (in the A picture) is explained in [50]; in our work we have only postulated this lifting ad hoc. Second, it requires objects to live in an abelian category (as do all GIT notions of stability), but the only universal category associated to a Calabi–Yau seems to be the derived category used by Kontsevich, which is not abelian. The two points are connected and we will come back to them below.

For completeness, we should say that (in our opinion) the assumption that A branes are in fact special Lagrangian submanifolds is on less firm a footing than most of the other elements of the picture. The basic problem is the one that we mentioned, that the appropriate definition of the metric and Kähler form of the CY in string theory has not in fact been settled, but is almost certainly not the Ricci flat metric. One possibility is that one still has the special Lagrangian condition but with respect to a preferred non-Ricci flat Kähler form, possibly the D0-metric described above. Some evidence for this idea can be found in the work of Leung, Yau and Zaslow [39] which shows that the special Lagrangian condition is related to the MMMS equation (4), in which the metric could be derived from D-brane considerations, by a Fourier–Mukai transform of the type which should describe T duality in string theory. At present however this is just a guess, and more physical work is needed to understand this point; for example it is clear that with some work a series expansion analogous to (1) could be found for the corrections to the special Lagrangian condition.

4. Flow of Gradings

We now turn to a point which emerges quite clearly from physics and CFT but does not seem to have appeared in the mathematical discussion, namely a "flow of gradings" which is induced under variation of Kähler moduli (in the B picture) or complex structure (in the A picture).

Consider the B branes for definiteness; in the large volume limit each brane corresponds to some stable coherent sheaf and thus for each pair of branes E

and F we have graded morphisms $\mathrm{Ext}^n(E,F)$. One often denotes this space as $\mathrm{Hom}(E,F[n])$, a notation which finds its justification in homological algebra.

Suppose we now follow a general path in Kähler moduli space between two points K and L. The new claim is that a morphism of degree n at the starting point K will undergo "flow of grading" determined by the phases (3): its degree at the point L will be

$$n' = n + \varphi_L(F) - \varphi_K(F) - \varphi_L(E) + \varphi_K(E)\,.$$

(This is the reason for the $1/\pi$ in (3), which first appeared in [46].) In particular, the gradings are $\mathbf{R}$-valued. This rule can be expressed more simply by saying that the grading of objects varies with φ, so that the starting $\mathrm{Hom}(E[\varphi_K(E)], F[n + \varphi_K(F)])$ turns into $\mathrm{Hom}(E[\varphi_L(E)], F[n + \varphi_L(F)])$.

In itself this is just a definition but it can be checked physically if we have more than one definition of the underlying CFT. As we discuss in the next section, this check has been made between large volume and orbifold points in various models. The phases are known from mirror symmetry results and flow by integers between these points, and the gradings of morphisms agree with these predictions.

The argument for the flow from CFT is quite simple. [19]. It relies on the identification of a morphism with a "winding string in the bosonized $U(1)$," and has the following intuitive picture. One can think of each brane as having location φ on a circle of circumference 2 (the bosonized $U(1)$) and a morphism as an open string stretched between the pair of branes; grading corresponds to length, so the flow is simply induced by motion of the branes on the circle.

Once we grant this point, we quickly see that no single abelian category (such as the category of coherent sheaves) could possibly describe the branes throughout Kähler moduli space. This is because a general path will cause gradings to flow below zero, and convert objects from even to odd grading, both of which lead to violations of the axioms of an abelian category. However, the derived category and its distinguished triangles still make sense with these flows. This is perhaps the fundamental reason why one is forced to the derived category in these considerations. Recent work of Seidel and Thomas and of Horja [51, 31] provides even more concrete motivation for this point; they show that the monodromies associated to general paths in true Kähler moduli space act naturally on the derived category, not the category of coherent sheaves.

One does not have a notion of subobject in the derived category and thus no GIT definition of stability can apply in this context. The most direct way out of this problem is to propose that each point in Kähler moduli space comes with a preferred abelian category. This could be defined by a choice of t-structure [2], which might be determined by the phases as well: one basically wants to keep all objects whose phases lie in the interval $(-1,0]$ and which are not involved in morphisms of negative degree. This point is presently under investigation.

5. Orbifolds

The most accessible physical definitions of Calabi–Yau manifolds are the orbifolds $\mathbf{C}^3/\Gamma$ and the Landau–Ginzburg orbifolds described in the introduction. As it turns out, the basic constructions involved in defining D-branes in these theories are already fairly standard in mathematics and thus we can be relatively brief.

For B branes, onc clearly wants to take some sort of Γ-equivariant bundles on $\mathbf{C}^n$, and this can be implemented in physical terms by starting with the gauge theory Lagrangian for branes in $\mathbf{C}^n$ and applying a quotient which acts simultaneously on $\mathbf{C}^n$ and in the gauge group. [11]. This leads directly to the quiver theories which appear in the constructions of Kronheimer and subsequent work. [37]. In particular, B branes are objects in a "McKay" quiver category, whose primitive objects (nodes) correspond to irreps of Γ, whose links correspond to terms in the tensor product with the representation defining the Γ action on $\mathbf{C}^n$, and with specified quadratic relations.

The physical application requires $\Gamma \subset SU(n)$ and such orbifold singularities can often be resolved (always for $n \leq 3$) to smooth manifolds. This resolution is visible as the moduli space of a D0-brane and one can even compute the metric (actually, a controlled approximation in the small blow-up limit); for $\mathbf{C}^3/\Gamma$ one obtains explicit non-Ricci-flat metrics of the type discussed earlier [13].

More generally, the question arises of how B branes as quiver objects or irreps of Γ are related to the large volume definition of B branes as coherent sheaves. In recent mathematical work [3, 32, 48], a generalization of the McKay correspondence has been developed, which in particular provides a natural basis for K theory with compact support (thus supported on the exceptional divisor) labelled by irreps of Γ. The natural conjecture is that this is the physical correspondence we are looking for [18].

This conjecture can be tested against results from mirror symmetry which determine the Gauss–Manin connection on Kähler moduli space and lead to a connection formula which relates the orbifold "charge" basis to the large volume basis. This gives an explicit expression for the Chern character of the bundle corresponding to each irrep, and this has been checked to agree with the conjecture for $\mathbf{C}^3/\mathbf{Z}_3$ [17]. A somewhat more intricate version of this applies to the Landau–Ginzburg orbifolds; one uses the McKay correspondence to get the natural basis for the K theory of the weighted projective space, and then restricts this to the Calabi–Yau. Again one can compare with results from mirror symmetry and find agreement [18].

These constructions are intimately related to and generalize Beilinson's construction of quiver categories from sheaves on $\mathbf{P}^n$ [1], and it is these categories of quiver representations which are compared to the category of coherent sheaves in the test of "flow of gradings" mentioned above. More generally, all this provides a fairly concrete relation between the problems of classifying coherent sheaves and of solving certain $N = 1$ supersymmetric gauge theories. Although neither problem is easy in general, the second problem is not only more familiar to physicists but is a more appropriate language for the subsequent considerations than an explicit classification of sheaves would have been. It would be even more valuable if such constructions could be found for all sheaves, not just those which arise on restriction from the ambient space.

The first step in this generalization already appears in the physical theories – it turns out that these include additional moduli which can be shown to correspond to linearized deformations of bundles which appear after restriction to the CY. [20]. Presumably one could go beyond this linearized level by computing exact superpotentials; this may be possible using physical methods.

It would be quite interesting to know more about the variation of the spectrum of BPS branes with Kähler moduli. Some simple results of this type were found for $\mathbf{C}^3/\mathbf{Z}_3$ in [17]; for example it was shown that there are lines of marginal stability arbitrarily close to the large volume limit, by constructing bundles which were ϵ away from violating slope stability.

It would clearly be important to find a microscopic derivation of the stability condition, either from geometry (see [54] for work in this direction) or perhaps CFT or string field theory.

Our general conclusion has to be that Kontsevich's proposal, coming as it did before the study of D-branes, has turned out to be remarkably prescient. Although it has taken physicists some time to catch up, we are finally extending this picture to provide both more concrete pictures of branes on Calabi–Yau manifolds and new concepts which should be of interest to mathematicians.

Acknowledgements

We would like to thank D.-E. Diaconescu, B. Fiol and C. Römelsberger for enjoyable collaborations, and F. Bogomolov, M. Kontsevich, M. Marino, D. Morrison, A. Polishchuk, P. Seidel, and R. P. Thomas for valuable discussions. We thank R. P. Thomas for comments on the manuscript.

This work was supported by the Clay Mathematics Institute and by DOE grant DE-FG02-96ER40959.

References

[1] A. A. Beilinson, *Coherent sheaves on P^n and problems of linear algebra,* Funct. Anal. Appl. **12** (1978) 214–216.

[2] A. A. Beilinson, Bernstein and P. Deligne, *Faiseaux Pervers*, Asterisque **100** (1982).

[3] T. Bridgeland, A. King, and M. Reid, *Mukai implies McKay,* math.AG/9908027.

[4] I. Brunner, M. R. Douglas, A. Lawrence, and C. Römelsberger, *D-branes on the quintic,* to appear in J. High Energy Physics, hep-th/9906200.

[5] P. Candelas, X. de la Ossa, A. Font, S. Katz, and D. R. Morrison, *Mirror symmetry for two parameter models – I,* Nucl. Phys. **B416** (1994) 481, hep-th/9308083.

[6] P. Candelas, X. C. de la Ossa, P. S. Green, and L. Parkes, *A pair of Calabi–Yau manifolds as an exactly soluble superconformal theory,* Nucl. Phys. **B359** (1991) 21.

[7] D. A. Cox and S. Katz, *Mirror Symmetry and Algebraic Geometry,* Mathematical Surveys **68**, 1999, AMS.

[8] J. Dai, R. G. Leigh and J. Polchinski, *New Connections Between String Theories,* Mod. Phys. Lett. **A4**, 2073 (1989).

[9] R. Dijkgraaf, *Fields, Strings and Duality,* in *Les Houches LXIV, Symetries Quantiques,* eds. A. Connes, K. Gawedzki and J. Zinn-Justin, Elsevier (1998).

[10] S. Donaldson and R. Thomas, *Gauge Theory in Higher Dimensions,* in *The Geometric Universe; Science, Geometry, and the Work of Roger Penrose*, Oxford University Press, 1998.

[11] M. R. Douglas and G. Moore, *D-branes, Quivers, and ALE Instantons,* hep-th/9603167.

[12] M. R. Douglas, B. R. Greene, and D. R. Morrison, *Orbifold resolution by D-branes,* Nucl. Phys. **B506** (1997) 84, hep-th/9704151.

[13] M. R. Douglas and B. R. Greene, *Metrics on D-brane orbifolds,* Adv. Theor. Math. Phys. **1**, 184 (1998) hep-th/9707214.

[14] M. R. Douglas, *Two Lectures on D-Geometry and Noncommutative Geometry,* in *Nonperturbative Aspects of Strings, Branes and Supersymmetry,* World Scientific (1999), hep-th/9901146.

[15] M. R. Douglas, *Topics in D-geometry,* Class. Quant. Grav. **17** (2000) 1057, hep-th/9910170.

[16] M. R. Douglas, B. Fiol, and C. Römelsberger, *Stability and BPS branes,* hep-th/0002037.

[17] M. R. Douglas, B. Fiol, and C. Römelsberger, *The spectrum of BPS branes on a noncompact Calabi–Yau,* hep-th/0003263.

[18] M. R. Douglas and D.-E. Diaconescu, *D-branes on Stringy Calabi–Yau Manifolds,* hep-th/0006224.

[19] M. R. Douglas, *D-branes, Categories and $N = 1$ Supersymmetry,* to appear.

[20] M. R. Douglas and D.-E. Diaconescu, to appear.

[21] B. Fiol and M. Marino, *BPS states and algebras from quivers,* JHEP 0007 (2000) 031, hep-th/0006189.

[22] R. Gopakumar and C. Vafa, *M-Theory and Topological Strings–I,* hep-th/9809187.

[23] S. Govindarajan, T. Jayaraman, and T. Sarkar, *World sheet approaches to D-branes on supersymmetric cycles,* hep-th/9907131.

[24] B. R. Greene, *String theory on Calabi–Yau manifolds,* hep-th/9702155.

[25] B. R. Greene and M. R. Plesser, *Duality in Calabi–Yau Moduli Space,* Nucl. Phys. **B338** (1990) 15.

[26] M. T. Grisaru, A. E. van de Ven and D. Zanon, *Two-Dimensional Supersymmetric Sigma Models On Ricci-Flat Kahler Manifolds Are Not Finite,* Nucl. Phys. **B277** (1986), 388.

[27] M. Gross, *Topological Mirror Symmetry,* math.AG/9909015.

[28] F. Reese Harvey, *Spinors and Calibrations,* Academic Press (1990).

[29] J.A. Harvey and G. Moore, *On the algebras of BPS states,* Comm. Math. Phys. **197** 489 (1998), hep-th/9609017.

[30] K. Hori, A. Iqbal, and C. Vafa, *D-branes and mirror symmetry,* hep-th/0005247.

[31] R. P. Horja, *Hypergeometric functions and mirror symmetry in toric varieties,* math.AG/9912109.

[32] Y. Ito and H. Nakajima, *McKay correspondence and Hilbert schemes in dimension three,* math.AG/9803120.

[33] I. Jack, D. R. Jones and J. Panvel, *Six loop divergences in the supersymmetric Kahler sigma model,* Int. J. Mod. Phys. **A8**, 2591 (1993); hep-th/9311117.

[34] D. Joyce, *On counting special Lagrangian homology 3-spheres,* hep-th/9907013.

[35] S. Kachru, S. Katz, A. Lawrence and J. McGreevy, *Mirror Symmetry for Open Strings,* hep-th/0006047.

[36] M. Kontsevich, *Homological Algebra of Mirror Symmetry,* Proceedings of the 1994 ICM, alg-geom/9411018.

[37] P. B. Kronheimer, *The construction of ALE spaces as hyper-Kähler quotients,* J. Diff. Geom. **29** (1989) 665.

[38] N. C. Leung, *Einstein Type Metrics and Stability on Vector Bundles,* J. Diff. Geom. **45** (1997) 514.

[39] N. C. Leung, S.-T. Yau and E. Zaslow, *From Special Lagrangian to Hermitian-Yang–Mills via Fourier–Mukai Transform,* math.DG/0005118.

[40] M. Marino, R. Minasian, G. Moore and A. Strominger, *Nonlinear Instantons from Supersymmetric p-Branes,* JHEP 0001 (2000) 005, hep-th/9911206.

[41] D. R. Morrison, *Geometric Aspects of Mirror Symmetry,* math.AG/0007090.

[42] K.S. Narain, M.H. Sarmadi and C. Vafa (Harvard U.), *Asymmetric orbifolds,* Nucl. Phys. B288 (1987), 551.

[43] D. Nemeschansky and A. Sen, *Conformal Invariance Of Supersymmetric Sigma Models On Calabi–Yau Manifolds,* Phys. Lett. **B178** (1986), 365.

[44] H. Ooguri, Y. Oz and Z. Yin, *D-branes on Calabi–Yau spaces and their mirrors,* Nucl. Phys. **B477**, 407 (1996), hep-th/9606112.

[45] J. Polchinski, *TASI Lectures on D-branes,* hep-th/9611050.

[46] A. Polishchuk and E. Zaslow, *Categorical Mirror Symmetry: The Elliptic Curve,* Adv. Theor. Math. Phys. **2** (1998) 443-470, math.AG/9801119.

[47] A. Recknagel and V. Schomerus, *D-branes in Gepner models,* Nucl. Phys. **B531** (1998) 185, hep-th/9712186.

[48] M. Reid, *La correspondance de McKay,* Séminaire Bourbaki (novembre 1999), no. 867, math.AG/9911165.

[49] W.-D. Ruan, *Lagrangian torus fibration and mirror symmetry of Calabi–Yau hypersurface in toric variety,* math.DG/0007028.

[50] P. Seidel, *Graded Lagrangian submanifolds,* math.SG/9903049.

[51] P. Seidel and R. P. Thomas, *Braid group actions on derived categories of coherent sheaves,* math.AG/0001043.

[52] A. Strominger, S.-T. Yau and E. Zaslow, *Mirror Symmetry is T-Duality,* Nucl.Phys. B479 (1996) 243-259, hep-th/9606040.

[53] R. P. Thomas, *D-Branes and Mirror Symmetry*, to appear in the proceedings of the 2000 Clay Mathematics Institute school on Mirror Symmetry.

[54] R. P. Thomas, to appear.

[55] C. Voisin, *Mirror Symmetry,* SMF/AMS Texts **1**, AMS 1999.

[56] E. Witten, *Phases of N=2 theories in two dimensions,* Nucl. Phys. **B403** (1993) 159, hep-th/9301042.

[57] E. Witten, *Chern–Simons gauge theory as a string theory,* hep-th/9207094.

[58] E. Witten, *Branes and the Dynamics of QCD,* Nucl. Phys. B507 (1997) 658; hep-th/9706109.

[59] I. Zharkov, *Torus Fibrations of Calabi–Yau Hypersurfaces in Toric Varieties and Mirror Symmetry,* math.AG/9806091.

Department of Physics and Astronomy
Rutgers University
Piscataway, NJ 08855–0849, USA
and
I.H.E.S.
Le Bois-Marie, Bures-sur-Yvette, 91440 France
E-mail address: `mrd@physics.rutgers.edu`

Knot Invariants and Chern-Simons Theory

José M. F. Labastida

Abstract. A brief review of the development of Chern-Simons gauge theory since its relation to knot theory was discovered in 1988 is presented. The presentation is done guided by a dictionary which relates knot theory concepts to quantum field theory ones. From the basic objects in both contexts the quantities leading to knot and link invariants are introduced and analysed. The quantum field theory approaches that have been developed to compute these quantities are reviewed. Perturbative approaches lead to Vassiliev or finite type invariants. Non-perturbative ones lead to polynomial or quantum group invariants. In addition, a brief discussion on open problems and future developments is included.

In 1988, Edward Witten established the connection between Chern-Simons gauge theory and the theory of knot and link invariants [34]. Since then the theory has been intensively studied, making important progress as a result of the application of standard field theory methods. The development of the theory of knot and link invariants has been also very impressive in the last fifteen years and at some stages has occurred parallel to Chern-Simons gauge theory. There is a natural correspondence between both developments. This has led to the construction of the dictionary introduced in [22], and reproduced in Table I, which will be used as a guide in this presentation.

Chern-Simons gauge theory was first analysed from a non-perturbative point of view. The original paper by Witten presented a series of non-perturbative methods which led him to establish the equivalence between vacuum expectation values (vevs) of Wilson loops and polynomial invariants like the Jones polynomial [16] and its generalizations. Perturbative studies started one year later and soon their connection to Vassiliev invariants [31, 8] was pointed out. It turned out that the coefficients of the perturbative series correspond to these invariants [6, 9, 24].

The perturbative series expansion has been studied for different gauge-fixings. The first analysis in the covariant Landau gauge [14, 7] was later extended in a general framework [2, 1], reobtaining the formulation by Bott and Taubes of their configuration space integral [10]. Their integral corresponds precisely to the perturbative series expansion of the vev of a Wilson loop in Chern-Simons gauge theory in the Landau gauge. Before the work by Bott and Taubes, Kontsevich presented a different integral [21] for Vassiliev invariants which turned out to correspond to the perturbative series expansion of the vev of a Wilson loop in the light-cone gauge [11, 23, 19].

Additional studies of the perturbative series expansion have been performed in the temporal gauge [11, 25]. This gauge has the important feature that the integrals which are present in the expressions for the coefficients of the perturbative series expansion can be carried out, leading to combinatorial expressions [25]. This has been shown to be the case up to order 4 and it seems likely that the approach can be generalized. In this analysis a crucial role is played by the factorization theorem for Chern-Simons gauge theory proved in [4]. The resulting expressions are better presented when written in terms of Gauss diagrams for knots [26]. Recent results demonstrate the existence of a combinatorial formula of this type [13]. Chern-Simons gauge theory has provided combinatorial expressions for all the Vassiliev invariants up to order 4 [25]. Further work is needed to obtain a general combinatorial expression.

The invariants obtained in the perturbative framework for each gauge-fixing are the same. This is guaranteed by the fact that the theory is gauge invariant and Wilson loops are gauge invariant operators. In fact, this property is responsible, from a field theory point of view, for the connection between Vassiliev invariants and polynomial invariants, as they appear in the non-perturbative approach, and for the existence of different representations of Vassiliev invariants. The correspondence between the Chern-Simons gauge theory description and the knot theory one is listed in the following table.

Table I

Knot Theory	**Chern-Simons Gauge Theory**
Knots and links	Wilson loops
Knot and link polynomial invariants	Vevs of products of Wilson loops
Singular knots	Operators for singular knots
Invariants for singular knots	Vevs of the new operators
Finite type or Vassiliev invariants	Coeffs. of the perturbative series
Chord diagram	First coeff. of the perturbative series
{1T,4T} and {1T,AS,IHX,STU}	Lie-algebra structure of group factors
Configuration space integral	Landau gauge
Kontsevich integral	Light-cone gauge
??	Temporal gauge

Before entering into the review of the developments of the last years guided by this table it is worth remarking two facts. First, the entry in the knot-theory column corresponding to the temporal gauge has not been filled in yet. Further work is needed to complete it. Second, recent developments based on the application of Maldacena's conjecture has led to introduction of a new context in which knot invariants are organized differently [29]. It is possible that some important new boxes are still missing in the table.

Chern-Simons gauge theory on a smooth three-manifold M is defined by the action,

$$S_{\rm CS}(A) = \frac{k}{4\pi}\int_M \mathrm{Tr}(A \wedge dA + \frac{2}{3} A \wedge A \wedge A)\,, \tag{1}$$

where k is an integer constant. The exponential of this action is gauge invariant. Wilson loops are some of the relevant operators of the theory. They are defined as:

$$W^R_\gamma(A) = \mathrm{Tr}_R\left(\mathrm{Hol}_\gamma(A)\right) = \mathrm{Tr}_R\, \mathrm{P}\exp\int_\gamma A\,, \tag{2}$$

where R denotes a representation of the gauge group G and γ a 1-cycle. Products of these operators are the natural candidates to obtain topological invariants after computing their vev respect to the action (1).

The Wilson loop (2) and the vevs of its products are the first two items of the right column of Table I. The first two in the left column are the basic ingredients of knot theory. Knot theory studies inequivalent embeddings $\gamma\colon S^1 \to M$. Each of these is a knot. Knots are classified by constructing knot invariants or quantities which can be computed taking a representative of a class and are invariant within the class. The problem of the classification of knots in $M = S^3$ can be reformulated in a two-dimensional framework using regular knot projections. The previous equivalence problem then translates into the equivalence of regular knot projections under Reidemeister moves.

The study of knot and link invariants resulted in important progress in the 1980s after the discovery of the Jones polynomial [16] and its generalizations like the HOMFLY [12] and the Kauffman [20] polynomial invariants. Witten showed in 1988 that the vevs of products of Wilson loops correspond to the Jones polynomial when one considers $SU(2)$ as the gauge group and all the Wilson loops entering in the vev are taken in the fundamental representation F. For example, if one considers a knot K, with Jones polynomial $V_K(t)$, he showed that, $V_K(t) = \langle W_K^F \rangle$, provided that one performs the identification $t = \exp(2\pi i/k + h)$ where $h = 2$ is the dual Coxeter number of the gauge group $SU(2)$. He also showed that if instead of $SU(2)$ one considers $SU(N)$ and the Wilson loop carries the fundamental representation, the resulting invariant is the HOMFLY polynomial. If, instead, one considers $SO(N)$ as gauge group and Wilson loops carrying the fundamental representation, one is led to the Kauffman polynomial. The case of $SU(2)$ as gauge group and Wilson loops carrying a representation of spin $j/2$ leads to the Akutsu-Wadati [5] polynomials. The framework generated by Chern-Simons gauge theory leads to an enormous variety of knot and link invariants. They can be also obtained from a quantum group approach [17], and from more general formalisms [18].

In our discussion of the next five items in Table I we will deal first with the left column. In 1990, V. A. Vassiliev [31] introduced new knot invariants based on singular knots which were reformulated later by Birman and Lin [9] from an axiomatic point of view. A singular knot with j double points consists of the image of a map from S^1 into S^3 with j simple self-intersections. The key ingredient in the construction by Birman and Lin is the observation that any knot invariant extends to generic singular knots by the Vassiliev resolution:

$$\nu(K^{j+1}) = \nu(K^j_+) - \nu(K^j_-), \tag{3}$$

where K^{j+1} is a singular knot with $j+1$ double points which differs from the knots K^j_+ and K^j_- only in the region where the double point is resolved by an overcrossing (+) and an undercrossing (−). Using this extension Birman and Lin [9] characterized the invariants of finite type or Vassiliev invariants introducing the following definition: a Vassiliev or finite type invariant of order m is a knot invariant which is zero on the unknot and that, after extending it to singular knots, it is zero on singular knots K^j with $j > m$ double points.

Besides introducing an axiomatic approach to Vassiliev invariants, Birman and Lin proved an important theorem in 1993 [9]. Any polynomial invariant $P_K(t)$ for a knot K can be expanded as:

$$Q_K(x) = P_K(\mathrm{e}^x) = \sum_{m=0}^{\infty} \nu_m(K) x^m \,. \tag{4}$$

Birman and Lin proved that if one extends the quantities $\nu_m(K)$ to Vassiliev invariants for singular knots using Vassiliev resolution (3), then $\nu_m(K)$ are Vassiliev invariants of order m. An immediate consequence of this theorem is that the coefficients of the perturbative expansion associated to the vev of a Wilson loop in Chern-Simons gauge theory are Vassiliev invariants. This property of the coefficients of the perturbative series expansion has been proved using standard quantum field theory methods [24].

From a singular knot with m double points one can construct a particular object which determines Vassiliev invariants of order m: its chord diagram [6]. Given a singular knot K^m, its chord diagram, $CD(K^m)$, is built in the following way. Take a base point and draw the preimages of the map associated to a given representative of K^m on a circle. Then join by straight lines the pairs of preimages which correspond to each singular point. If $\nu(K^m)$ is a Vassiliev invariant of order m then it is completely determined by $CD(K^m)$.

Chord diagrams play an important role in the theory of Vassiliev invariants [6]. Since Vassiliev invariants of order m for singular knots with m double points are codified by chord diagrams, one could ask if there are as many independent invariants of this kind as chord diagrams. The answer to this question is no. Chord diagrams are associated to knot diagrams and these diagrams must be considered modulo the equivalence relation dictated by the Reidemeister moves. These relations indeed impose some relations among chord diagrams, the so-called 1T and 4T relations [6]. The general expression for the dimensions of the spaces of chord diagrams is an open problem which has challenged many people. These dimensions correspond in fact to the dimensions of the spaces of primitive Vassiliev invariants.

The vector space of chord diagrams can be characterized in an equivalent way using trivalent diagrams and introducing a series of new relations. This characterization is very important because it corresponds to the one that naturally arises from the point of view of Chern-Simons gauge theory. It consists of the expansion of the set of chord diagrams to a new set in which trivalent vertices are allowed. This means that now the lines in the interior of the circle can join a point on the circle to a point on one of the internal lines. Bar-Natan showed [6] that the space of chord diagrams modulo 1T and 4T relations is equivalent to the new one after modding out by the so-called 1T, AS, IHX and STU relations.

The relations AS, STU and IHX are reminiscent of a Lie-algebra structure. If one assigns totally antisymmetric structure constants f_{abc} to the internal trivalent vertices, and group generators T_a to the vertices on the circle, the STU relation is just the defining Lie-algebra relation, while the IHX relation corresponds to the Jacobi identity. The group factors associated to the perturbative series expansion of the vev of a Wilson loop in Chern-Simons gauge theory correspond precisely to these spaces.

We will now turn our attention to the column on the right column in Table I. Singular knots play a central role in the theory of Vassiliev invariants. As shown in [24], they have an operator counterpart in Chern-Simons gauge theory. It has a rather simple form. Let us consider a singular knot K^n with n double points, and let us assign to each double point i a triple $\tau_i = \{s_i, t_i, T^{a_i}\}$ where s_i and t_i, $s_i < t_i$, are the values of the K^n-parameter at the double point, and T^{a_i} is a group generator. The gauge-invariant operator associated to the singular knot K^n is:

$$(\frac{4\pi i}{k})^n \,\mathrm{Tr}\Big[T^{\phi(w_1)}U(w_1,w_2)T^{\phi(w_2)}U(w_2,w_3)T^{\phi(w_3)}\dots$$
$$\dots U(w_{2n-1},w_{2n})T^{\phi(w_{2n})}U(w_{2n},w_1)\Big], \quad (5)$$

where $\{w_i; i=1,\dots,2n\}$, $w_i < w_{i+1}$, is the set that results from ordering the values s_i and t_i, for $i=1,\dots,n$, and ϕ is a map that assigns to each w_i the group generator in the triple to which it belongs.

Some immediate implications of the singular operators (5) are the following. First, they lead to a proof of the theorem by Birman and Lin discussed above; second, they allow direct contact with chord diagrams since these diagrams correspond to these operators at lowest order. This has been shown in [24]. The quantities which result after the assignment of Lie-algebra data to chord diagrams are called weight systems [6]. For each system one chooses a group and a representation. They correspond to the group theory factors in the context of Chern-Simons perturbation theory.

We will now describe the last three items of Table I starting with the right column. Perturbative studies of Chern-Simons gauge theory started with the works by Guadagnini, Martellini and Mintchev [14] and by Bar-Natan [7]. These studies were made in the covariant Landau gauge. Subsequent works [2, 1] in this gauge led to a framework linked to the theory of Vassiliev invariants, which constituted the configuration space integral approach [10]. The elements of the resulting Feynman rules in this gauge are:

$$\frac{i}{4\pi}\delta_{ab}\epsilon^{\mu\nu\rho}\frac{(x-y)_\rho}{|x-y|^3}, \quad -igf_{abc}\epsilon_{\mu\nu\rho}\int d^3x, \quad g(T^a)_i^j, \tag{6}$$

which correspond to internal line (gauge propagator), internal vertex, and vertex on the Wilson loop, respectively. In these equations $g^2 = 4\pi/k+h$. To these rules one must include the ones related to the ghost fields present in the Landau gauge. One of the consequences of their presence is that higher-loop corrections to two- and three-point functions can be ignored, at least in some of the standard regularization schemes.

In analysing the perturbative series one must deal with an important subtlety. If one computes the first order contribution to the perturbative series expansion of the vev of a Wilson loop, one finds that the resulting quantity is not a topological invariant. In the gauge fixing of the theory we have introduced a metric dependence that could lead to quantities which are not topological. This first order contribution is just a manifestation of it. Fortunately, only in this term, and in its propagation in higher order contributions, topological invariance is lost. The rest of the perturbative series expansion is truly topological. Thus, although vevs are not topological invariant quantities, they fail to be so in a controllable way. The non-topological terms factorize and multiply a term which is topological. The factor turns out to have a framing dependence equivalent to the one obtained in non-perturbative approaches.

The Feynman rules allow splitting the contributions to each order in two factors: a geometrical factor which includes all the space dependence, and a group factor which includes all the group theoretical dependence. The general form is:

$$\langle W_K^R\rangle = \dim R \sum_{i=0}^{\infty}\sum_{j=1}^{d_i}\alpha_{ij}(K)r_{ij}(R)x^i, \quad x=\frac{2\pi i}{k+h}=ig^2/2, \tag{7}$$

where R is the dimension of the representation R, $\alpha_{0,1} = r_{0,1} = 1$, and $d_0 = 1$. The factors $\alpha_{ij}(K)$ and $r_{ij}(R)$ appearing at each order i incorporate all the dependence dictated from the Feynman rules apart from the dependence on the coupling constant, which is contained in x. Of these two factors, in the $r_{ij}(R)$ all the group-theoretical dependence is collected. The rest is contained in the $\alpha_{ij}(K)$ or geometrical factors. They have the form of integrals over the Wilson loop corresponding to the knot K of products of propagators, as dictated by the Feynman rules. The first index in $\alpha_{ij}(K)$ denotes the order in the expansion and the second index labels the different geometrical factors which can contribute at the given order. Similarly, $r_{ij}(R)$ stands for the independent group structures which appear at order i, which are also dictated by the Feynman rules. The object d_i in (7) is the dimension of the space of invariants at a given order.

Among the basis of group factors which can be chosen there is a special class called canonical basis which turns out to be very useful. Basically, it consists of connected diagrams. If we denote by $r_{ij}^c(R)$ the group factors associated to this basis, and $\alpha_{ij}^c(K)$ the corresponding geometrical factors, the perturbative series expansion (7) can be written as [4]:

$$\langle W_K^R \rangle = \dim R \exp\left\{ \sum_{i=1}^{\infty} \sum_{j=1}^{\hat{d}_i} \alpha_{ij}^c(K)\, r_{ij}^c(R)\, x^i \right\}, \tag{8}$$

where $\hat{d}_i$ stands for the number of connected elements in the canonical basis at order i. The result (8) is known as the factorization theorem, and it holds for arbitrary gauges. The geometrical factors $\alpha_{ij}^c(K)$ are a selected set of Vassiliev invariants. They are called primitive Vassiliev invariants. They have been computed for general classes of knots as torus knots [3, 33] up to order 6.

The contribution at first order in (8) is precisely the framing factor. The rest of the terms in the exponent of (8) are knot invariants. The series contained in that exponent was analysed by Bott and Taubes [10] in their work on the configuration space for Vassiliev invariants (listed on the left column in Table I). They showed that the integral expression entering the geometrical factors $\alpha_{ij}^c(K)$ are convergent [10, 30]. Further work on the subject has led to a proof of their invariance [1, 35].

The explicit expression for the integrals entering in the second order contribution was first presented in [14]. It was later analysed in detail by Bar-Natan [7]. This invariant turns out to be the total twist of the knot and coincides mod 2 with the Arf invariant. The integral expression for the order 3 invariant, $\alpha_{31}^c(K)$ was first presented in [2]. Properties of the primitive Vassiliev invariants $\alpha_{21}^c(K)$ and $\alpha_{31}^c(K)$ have been studied in [15]. In these works the integral expressions for $\alpha_{21}^c(K)$ and $\alpha_{31}^c(K)$ were studied in the flat-knot limit and combinatorial expressions were obtained.

The perturbative analysis of Chern-Simons gauge theory in the light-cone gauge leads to the Kontsevich integral, which constitutes a particular representation of Vassiliev invariants. Non-covariant gauges are characterized by a unit constant vector n and have the form $n^\mu A_\mu = 0$. In the case of the light-cone gauge the unit vector n satisfies the condition $n^2 = 0$. In this gauge there is only one Feynman rule to be taken into account to compute the vevs of operators: the one associated to the propagator. The group factors that remain in this case correspond just to chord diagrams. The fact that in this gauge no group factors with trivalent vertices have to be taken into account is a quantum field theory ratification of the Bar-Natan theorem among the equivalence of the two representations of the space of diagrams. Non-covariant gauges share the problem of the presence of unphysical poles in their propagators [27]. Several prescriptions have been proposed to avoid these unphysical poles. Usually, a prescription is chosen so that some particular properties of the correlation functions are fulfilled. In the light-cone gauge there is a natural prescription which is motivated by the simple form that the elements of the perturbative expansion take after performing a Wick rotation. This prescription leads [23] to the Kontsevich integral.

The studies in the Landau and in the light-cone gauge provide integral expression for Vassiliev invariants. It is difficult to obtain information on these invariants from these expressions. Combinatorial formulas are much preferred. It is known that a general combinatorial formula for Vassiliev invariants exists [13, 28]. The search for an explicit construction of the combinatorial formula has led to the study of Chern-Simons gauge theory in the temporal gauge [25]. This turns out to be the more suitable gauge to carry out all the intermediate integrals and obtain combinatorial expressions. This approach has provided a combinatorial expression for the two primitive Vassiliev invariants at order 4. The temporal gauge has been also treated in [11, 32]. Previous studies of the configuration space integrals in the limit of flat knots [15] have also led to combinatorial expressions for Vassiliev invariants of order 2 and 3.

The starting point of the analysis in the temporal gauge is the same as in the light-cone gauge. The gauge-fixing condition is the same but now n is a unit vector of the form $n^\mu = (1,0,0)$. As before, the propagator presents unphysical poles, and a prescription to regulate it is needed. In this case a prescription-independent analysis is done splitting the propagators in two terms. It leads to the concept of kernel, as introduced in [25]. The kernels are quantities which depend on the knot projection chosen and therefore are not knot invariants. However, at a given order i a kernel differs from an invariant of type i by terms that vanish in signed sums of order i. The kernel contains the part of a Vassiliev invariant which is the last in becoming zero when performing signed sums, in other words, a kernel vanishes in signed sums of order $i+1$ but does not in signed sums of order i. Kernels plus the structure of the perturbative series expansion seem to contain enough information to reconstruct the full Vassiliev invariants [25]. The general expression for the kernels can be written in a universal form much in the spirit of the universal form which shares some resemblence with the Kontsevich integral.

Using the kernels and taking into account general properties of the perturbative series expansion one can reconstruct the complete perturbative coefficients obtaining combinatorial formulas. Vassiliev invariants up to order 4 were expressed in terms of these quantities and the crossing signatures in reference [25]. Here, we collect only the formula for the primitive Vassiliev invariant at second order. It has the following expression:

$$\alpha_{21}(K) = \alpha_{21}(U) + \langle \bigoplus, \bar{G}(\mathcal{K}) \rangle \,, \tag{9}$$

where $\alpha_{21}(U)$ stands for the value of α_{21} for the unknot and $\bar{G}(\mathcal{K}) = G(\mathcal{K}) - G(\alpha(\mathcal{K}))$, where $\alpha(\mathcal{K})$ denotes the ascending diagram of the knot projection $\mathcal{K}$. $G(\mathcal{K})$ is the Gauss diagram corresponding to $\mathcal{K}$. The inner product used in (9) consists of the sum over all the embeddings of the diagram $\bigoplus$ into $G(\mathcal{K})$, each one weighted by a factor, $\varepsilon_1\varepsilon_2$, where ε_1 and ε_2 are the signatures of the chords of $G(\mathcal{K})$ involved in the embedding.

The analysis presented in [25] up to order 4 should be generalized to arbitrary order, trying to obtain a general expression similar to the one existing in the light-cone corresponding to the Kontsevich integral. The resulting formula would allow us to fill the last box on the left column of Table I. Though this study seems promising, the problems inherent to the proper treatment of gauge theories in non-covariant gauges constitute an important barrier. Much work has to be done to understand the subtle issues related to the use of non-covariant gauges. The kernels plus the properties of the perturbative series expansion are probably enough to compute the explicit form of a given invariant but certainly it does not provide a systematic way of deriving the general universal formula.

The relation between knot theory and Chern-Simons gauge theory does not end here. Most likely additional boxes to Table I are waiting to be discovered. Quantum field theory is a very rich framework which is enlarged when regarded from the point of view of string theory. Recent work [29] indicates that new important connections can be established that could lead to entirely new approaches to the theory of knot invariants.

References

[1] D. Altschuler and L. Friedel, 'Vassiliev knot invariants and Chern-Simons perturbation theory to all orders," *Commun. Math. Phys.* **187** (1997) 261, and 'On universal Vassiliev invariants," **170** (1995) 41.

[2] M. Alvarez and J. M. F. Labastida, "Analysis of observables in Chern-Simons perturbation theory," *Nucl. Phys.* **B395** (1993) 198, hep-th/9110069, and 'Numerical knot invariants of finite type from Chern-Simons gauge theory," **B433** (1995) 555, hep-th/9407076; Erratum, ibid. **B441** (1995) 403.

[3] M. Alvarez and J. M. F. Labastida, "Vassiliev invariants for torus knots," *Journal of Knot Theory and its Ramifications* **5** (1996) 779; q-alg/9506009.

[4] M. Alvarez and J. M. F. Labastida, "Primitive Vassiliev invariants and factorization in Chern-Simons gauge theory," *Commun. Math. Phys.* **189** (1997) 641, q-alg/9604010.

[5] Y. Akutsu and M. Wadati, "Exactly solvable models and knot theory," *Phys. Rep.* **180** (1989) 247.

[6] D. Bar-Natan, "On the Vassiliev knot invariants," *Topology* **34** (1995) 423.

[7] D. Bar-Natan, "Perturbative aspects of Chern-Simons topological quantum field theory", Ph.D. Thesis, Princeton University, 1991.

[8] J. S. Birman, "New points of view in knot theory," *Bull. AMS* **28** (1993) 253.

[9] J. S. Birman and X. S. Lin, "Knot polynomials and Vassiliev's invariants," *Invent. Math.* **111** (1993) 225.

[10] R. Bott and C. Taubes, "On the self-linking of knots," *Jour. Math. Phys.* **35** (1994) 5247.

[11] A. S. Cattaneo, P. Cotta-Ramusino, J. Frohlich and M. Martellini, "Topological BF theories in three-dimensions and four-dimensions," *J. Math. Phys.* **36** (1995) 6137.

[12] P. Freyd, D. Yetter, J. Hoste, W. B. R. Lickorish, K. Millet and A. Ocneanu, "A new polynomial invariant of knots and links," *Bull. AMS* **12** (1985) 239.

[13] M. Goussarov, M. Polyak and O. Viro, "Finite Type Invariants of Classical and Virtual Knots", preprint, 1998, math.GT/9810073.

[14] E. Guadagnini, M. Martellini and M. Mintchev, 'Perturbative aspects of the Chern-Simons field theory," *Phys. Lett.* **B227** (1989) 111; "Chern-Simons model and new relations between the HOMFLY coefficients," **B228** (1989) 489, and "Wilson lines in Chern-Simons theory and link invariants," *Nucl. Phys.* **B330** (1990) 575.

[15] A. C. Hirshfeld and U. Sassenberg, "Derivation of the total twist from Chern-Simons theory," *Journal of Knot Theory and its Ramifications* **5** (1996) 489, and "Explicit formulation of a third order finite knot invariant derived from Chern-Simons theory," **5** (1996) 805.

[16] V. F. R. Jones, "Hecke algebras representations of braid groups and link polynomials," *Ann. of Math.* **126** (1987) 335.

[17] C. Kassel, M. Rosso and V. Turaev, "Quantum groups and knot invariants", Panoramas et syntheses 5, Societe Mathematique de France, 1997.

[18] C. Kassel and V. Turaev, "Chord diagram invariants of tangles and graphs," *Duke Math. J.* **92** (1998) 497–552.

[19] L. Kauffman, "Witten's Integral and Kontsevich Integral", Particles, Fields and Gravitation, Lodz, Poland 1998, Ed. Jakub Rembieliski; AIP Proceedings 453 (1998), 368.

[20] L. H. Kauffman, "An invariant of regular isotopy," *Trans. Am. Math. Soc.* **318** (1990) 417.

[21] M. Kontsevich, "Vassiliev's knot invariants," *Advances in Soviet Math.* **16**, Part 2 (1993) 137.

[22] J. M. F. Labastida, "Chern-Simons Gauge Theory: Ten Years After", *Trends in Theoretical Physics II*, H. Falomir, R. Gamboa, F. Schaposnik, eds., American Institute of Physics, New York, 1999, CP 484, 1–41, hep-th/9905057.

[23] J. M. F. Labastida and E. Pérez, "Kontsevich integral for Vassiliev invariants from Chern-Simons perturbation theory in the light-cone gauge," *J. Math. Phys.* **39** (1998) 5183; hep-th/9710176.

[24] J. M. F. Labastida and E. Pérez, "Gauge-invariant operators for singular knots in Chern-Simons gauge theory," *Nucl. Phys.* **B527** (1998) 499, hep-th/9712139.

[25] J. M. F. Labastida and E. Pérez, "Combinatorial Formulae for Vassiliev Invariants from Chern-Simons Perturbation Theory", *J. Math. Phys.* **41** (2000), 2658–2699.

[26] J. M. F. Labastida and E. Pérez, "Vassiliev Invariants in the Context of Chern-Simons Gauge Theory", Santiago de Compostela preprint, US-FT-18/98; hep-th/9812105.

[27] G. Leibbrandt, "Introduction to noncovariant gauges," *Rev. Mod. Phys.* **59** (1987) 1067.

[28] M. Polyak and O. Viro, "Gauss diagram formulas for Vassiliev invariants," *Int. Math. Res. Notices* **11** (1994) 445.

[29] H. Ooguri and C. Vafa, "Knot Invariants and Topological Strings", Harvard preprint, HUTP-99/A070, hep-th/9912123.

[30] D. Thurston, "Integral expressions for the Vassiliev knot Invariants", Harvard University senior thesis, April 1995; math/9901110.

[31] V. A. Vassiliev, "Cohomology of knot spaces", *Theory of singularities and its applications, Advances in Soviet Mathematics*, vol. 1, *Americam Math. Soc.*, Providence, RI, 1990, 23–69.

[32] J. F. W. H. van de Wetering, "Knot invariants and universal R-matrices from perturbative Chern-Simons theory in the almost axial gauge," *Nucl. Phys.* **B379** (1992) 172.

[33] S. Willerton, "On Universal Vassiliev Invariants, Cabling, and Torus Knots", University of Melbourne preprint (1998).

[34] E. Witten, "Quantum field theory and the Jones polynomial," *Commun. Math. Phys.* **121** (1989) 351.

[35] S.-W. Yang, "Feynman integral, knot invariant and degree theory of maps", National Taiwan University preprint, September 1997; q-alg/9709029.

Departamento de Física de Partículas
Universidade de Santiago de Compostela
E-15706 Santiago de Compostela, Spain
E-mail address: `labasti@fpaxp1.usc.es`

Topological Quantum Field Theory and Four-Manifolds

Marcos Mariño

Abstract. I review some recent results on four-manifold invariants which have been obtained in the context of topological quantum field theory. I focus on three different aspects: (a) the computation of correlation functions, which give explicit results for the Donaldson invariants of non-simply connected manifolds, and for generalizations of these invariants to the gauge group $SU(N)$; (b) compactifications to lower dimensions, and connections to three-manifold topology and to intersection theory on the moduli space of flat connections on Riemann surfaces; (c) four-dimensional theories with critical behaviour, which give some remarkable constraints on Seiberg-Witten invariants and new results on the geography of four-manifolds.

1. Introduction

One of the original motivations of Witten [19] to introduce topological quantum field theories (TQFT) was precisely to understand the Donaldson invariants of four-manifolds from a physical point of view. This approach proved its full power in 1994, when it was realized that all the information of Donaldson theory was contained in the Seiberg-Witten (SW) invariants. These new invariants led to a true revolution in four-dimensional topology, and they were introduced in [20] based on nonperturbative results in supersymmetric quantum field theory. The relation between Donaldson invariants and SW invariants was fully clarified in an important paper by G. Moore and E. Witten [15], where they introduced the so-called u-plane integral.

In this note, I review some recent results on four-manifold invariants which have been obtained through the use of u-plane integral techniques. I emphasize how these results are related to the physics of four-dimensional quantum field theories. First, I discuss Donaldson invariants as correlation functions in TQFT. I present some new results for non-simply connected manifolds (for product ruled surfaces, in particular), and for extensions of Donaldson theory to higher rank gauge groups. Second, I use compactifications of the field theory to make contact with results in three and two dimensions. In the two-dimensional case, I recover in fact Thaddeus' celebrated formula for the intersection pairings on the moduli space of flat connections on a Riemann surface. Finally, I consider qualitatively new physics (a field theory with critical behaviour) to obtain new relations between the SW invariants and classical invariants of four-manifolds. In the last section, I briefly consider some open problems.

2. Correlation Functions

2.1. General aspects

The Donaldson invariants of smooth, compact, oriented four-manifolds X [2] are defined by using intersection theory on the moduli space of anti-self-dual connections. The cohomology classes on this space are associated to homology classes of X through the slant product [2] or, in the context of topological field theory, by using the descent procedure [19]. Here we will restrict ourselves to the Donaldson invariants associated to zero, one- and two-homology classes[1]. Define

$$\mathbf{A}(X) = \mathrm{Sym}(H_0(X) \oplus H_2(X)) \otimes \wedge^* H_1(X) . \tag{1}$$

Then, the Donaldson invariants can be regarded as functionals

$$D_X^{w_2(E)} \colon \mathbf{A}(X) \to \mathbf{Q} , \tag{2}$$

where $w_2(E) \in H^2(X, \mathbf{Z})$ is the second Stiefel-Whitney class of the gauge bundle. It is convenient to organize these invariants as follows. Let $\{\delta_i\}_{i=1,\dots,b_1}$ be a basis of one-cycles, $\{\beta_i\}_{i=1,\dots,b_1}$ the corresponding dual basis of harmonic one-forms, and $\{S_i\}_{i=1,\dots,b_2}$ a basis of two-cycles. We introduce the formal sums

$$\delta = \sum_{i=1}^{b_1} \zeta_i \, \delta_i, \qquad S = \sum_{j=1}^{b_2} v_i \, S_i , \tag{3}$$

where v_i are complex numbers, and ζ_i are Grassmann variables. The generator of the 0-class will be denoted by $x \in H_0(X, \mathbf{Z})$. We then define the Donaldson-Witten generating function:

$$Z_{DW}(p, \zeta_i, v_i) = D_X^{w_2(E)}(\mathrm{e}^{px+\delta+S}) , \tag{4}$$

so that the Donaldson invariants can be read off from the expansion of the left-hand side in powers of p, ζ_i and v_i. The main result in [19] is that Z_{DW} can be understood as the generating functional of a twisted version of the $N = 2$ supersymmetric gauge theory – with gauge group $SU(2)$ – in four dimensions. In the twisted theory one can define observables $O(x)$, $I_1(\delta) = \int_\delta O_1$, $I_2(S) = \int_S O_2$ (where O_i are functionals of the fields of the theory) in one-to-one correspondence with the homology classes of X, and in such a way that the generating functional

$$\langle \mathrm{e}^{pO(x)+I_1(\delta)+I_2(S)} \rangle$$

is precisely $Z_{DW}(p, \zeta_i, v_i)$.

Based on the low-energy effective descriptions of $N = 2$ gauge theories obtained in [17], Witten obtained an explicit formula for (4) in terms of SW invari-

[1]The inclusion of three-classes has been considered in [11].

ants for manifolds of $b_2^+ > 1$ and simple type [20]. The general framework to give a complete evaluation of (4) was established in [15]. The main result of Moore and Witten is an explicit expression for the generating function Z_{DW}:

$$Z_{DW} = Z_u + Z_{SW} \tag{5}$$

which consists of two pieces. Z_{SW} is the contribution from the moduli space M_{SW} of solutions of the SW monopole equations. Z_u (the u-plane integral henceforth) is the integral of a certain modular form over the fundamental domain of the group $\Gamma^0(4)$, that is, over the quotient $\Gamma^0(4) \setminus H$, where H is the upper half-plane. The explicit form of Z_u was derived in [15] for simply connected four-manifolds, and extended to the non-simply connected case in [11]. Z_u is non-vanishing only for manifolds with $b_2^+ = 1$, and provides a simple physical explanation of the failure of topological invariance of the Donaldson invariants on those manifolds [15].

2.2. Donaldson invariants in the non-simply connected case

Most of the computations of Donaldson invariants have focused on simply connected manifolds. The study of the non-simply connected side was initiated in [15, 8], and finally a complete description of the invariants was given in [11]. Some additional results were obtained in [7]. The non-simply connected case presents some new features, mostly when $b_2^+ = 1$. Of particular interest are the Donaldson invariants of product ruled surfaces $\mathbf{S}^2 \times \Sigma_g$, which as far as I know have not been completely determined from a mathematical point of view. Recall that the invariants depend on the chamber chosen in the Kähler cone. The result gets simpler in the limiting chambers of very small or large volumes for $\mathbf{S}^2$. We will take a symplectic basis of one-cycles in Σ_g, δ_i, $i = 1, \cdots, 2g$, and consider the $\mathrm{Sp}(2g, \mathbf{Z})$-invariant element $\iota = -2\sum_{i=1}^{g} \delta_i \delta_{i+g}$. In the limit of small volume for $\mathbf{S}^2$, the generating functions $Z_g^{w_2(E)} = D^{w_2(E)}_{\mathbf{S}^2\times\Sigma_g}(\mathrm{e}^{px+r\iota+s\Sigma_g+t\mathbf{S}^2})$ are given by [11, 7]:

$$Z_g^{w_2(E)=0} = -\tfrac{i}{4}\Big[(h_\infty^2 f_{2\infty})^{-1}\mathrm{e}^{2pu_\infty+2stT_\infty}\Big(2f_{1\infty}h_\infty^2 s + 2r\Big)^g \coth\Big(\tfrac{is}{2h_\infty}\Big)\Big]_{q^0}, \tag{6}$$

$$Z_{g,\mathbf{S}^2}^{w_2(E)=[\mathbf{S}^2]} = -\tfrac{1}{4}\Big[(h_\infty^2 f_{2\infty})^{-1}\mathrm{e}^{2pu_\infty+2stT_\infty}\Big(2f_{1\infty}h_\infty^2 s + 2r\Big)^g \csc\Big(\tfrac{s}{2h_\infty}\Big)\Big]_{q^0}, \tag{7}$$

and they vanish for the other choices of Stiefel-Whitney class. For $g = 0$, one recovers the expressions for $\mathbf{S}^2 \times \mathbf{S}^2$ which were obtained in [15, 5]. The above equations involve the modular forms with q-expansions:

$$\begin{aligned}
u_\infty &= \frac{1}{2}\frac{\vartheta_2^4+\vartheta_3^4}{(\vartheta_2\vartheta_3)^2} = \frac{1}{8q^{1/4}}(1+20q^{1/2}-62q+\cdots),\\
T_\infty &= -\frac{1}{24}\left(\frac{E_2}{h_\infty^2}-8u_\infty\right) = q^{1/4}(1-2q^{1/2}+6q+\cdots),\\
h_\infty(\tau) &= \frac{1}{2}\vartheta_2\vartheta_3 = q^{1/8}(1+2q^{1/2}+q+\cdots),
\end{aligned}$$

$$
\begin{aligned}
f_{1\infty}(q) &= \frac{2E_2+\vartheta_2^4+\vartheta_3^4}{3\vartheta_4^8} = 1+24q^{1/2}+\cdots\,, \\
f_{2\infty}(q) &= \frac{\vartheta_2\vartheta_3}{2\vartheta_4^8} = q^{1/8}+18q^{5/8}+\cdots\,,
\end{aligned} \tag{8}
$$

and the subscript q^0 means that one has to extract the q^0 power in the q-expansion. The expressions for the other limiting chamber can be found in two ways: since the wall-crossing formula in the non-simply connected case was obtained in [11], one can sum to the above expression an infinite number of wall-crossing terms. Alternatively, one can perform a direct evaluation of the u-plane integral [7].

2.3. Extension to gauge group $SU(N)$

The Donaldson invariants are usually defined for the gauge group $SU(2)$. In principle, one can formally consider invariants of four-manifolds defined from anti-self dual $SU(N)$ gauge connections. Although this seems to be pretty difficult from a mathematical point of view, the evaluation of the would-be $SU(N)$ Donaldson invariants turns out to be tractable using quantum field theory [10]. The result is simpler for manifolds of simple type and with $b_2^+ > 1$. Not surprisingly, it can be expressed in terms of the cohomology ring of X and of SW invariants:

$$
\langle e^{pO(x)+I_2(S)}\rangle_{SU(N)} = \alpha_N^{\chi}\beta_N^{\sigma}\sum_{k=0}^{N-1}\omega^{k(N^2-1)\chi_h}\sum_{\lambda^I}\prod_{I=1}^{N-1}SW(\lambda^I)
$$

$$
\cdot\Big(\prod_{1\le I<J\le N-1} q_{IJ}^{-(\lambda^I,\lambda^J)}\Big)\exp\left[p\omega^{2k}N+2\omega^{2k}S^2+2\omega^k\sum_{I=1}^{N-1}(S,\lambda^I)\sin\frac{\pi I}{N}\right], \tag{9}
$$

where $\omega = \exp[i\pi/N]$, $\chi_h = (\chi+\sigma)/4$, and χ, σ are the Euler characteristic and the signature of X, respectively. The q_{IJ} are $\exp \pi i \tau_{IJ}$, where τ_{IJ}, $I \neq J$, are the leading terms of the off-diagonal effective couplings τ_{IJ}, which have been computed in [3]. The sum in (9) is over basic classes, and $(\,,\,)$ is the intersection form of X. Finally, α_N and β_N are universal constants. In the above expression we have considered only $SU(N)$ bundles with zero Stiefel-Whitney class. In addition, one can consider additional operators associated to higher Casimirs of the gauge group, that we have not included in (9). Notice that the above expression shows that the theory factorizes down to the "magnetic" Cartan torus $U(1)^{N-1}$, but there is an important mixing measured by the off-diagonal effective couplings.

3. Compactification

3.1. Down to three dimensions

To make contact with results in three-dimensional topology, one should consider four-manifolds of the form $X = \mathbf{S}^1 \times Y$. Donaldson theory on these manifolds has been explored in [12]. Using results from supersymmetric gauge theory, we would expect the partition function of Donaldson-Witten theory on $Y \times \mathbf{S}^1$ for gauge group G to agree with the Rozansky-Witten invariant $Z_{RW}(Y, X_G)$ [16], where X_G is a hyperKähler manifold. When $G = SU(2)$, $X_{SU(2)}$ is the Atiyah-Hitchin manifold and the Rozansky-Witten invariant is the Casson-Walker-Lescop invariant $\lambda_{\rm CWL}(Y)$. For $G = SU(N)$, $X_{SU(N)}$ is the reduced moduli space of N monopoles, which is a hyperKähler manifold of dimension $4(N-1)$.

This expectation can be partially checked. Using (9) and the Meng-Taubes theorem [14], one can prove that, for $b_1(Y) > 1$

$$Z_{DW}^{SU(N)}(Y \times \mathbf{S}^1) = N^2(\lambda_{\rm CWL}(Y))^{N-1}, \tag{10}$$

and the left-hand side is in fact (up to an overall constant) $Z_{RW}(Y, X_{SU(N)})$, which has been recently computed by Habegger and Thompson [6]. This gives an interesting non-trivial check of (9). For $b_1(Y) = 1$ there are important subtleties in the correspondence with Rozansky-Witten theory, which have been discussed in [12] when the gauge group is $SU(2)$.

3.2. Down to two dimensions

The connection to two-dimensional moduli problems appears when one considers product ruled surfaces $X = \mathbf{S}^2 \times \Sigma_g$. Anti-self dual connections on $X = \mathbf{S}^2 \times \Sigma_g$ with zero instanton number and $w_2(E) = [\mathbf{S}_2]$ are in one-to-one correspondence with flat connections on Σ_g with odd degree, which form a moduli space M_g. Donaldson invariants correspond to intersection pairings on M_g, which were determined by Thaddeus in [18]. The $\mathrm{Sp}(2g, \mathbf{Z})$-invariant cohomology ring of M_g is generated by cohomology classes α, β and γ, of degrees 2, 4 and 6, respectively. The relation between the intersection pairings and the Donaldson invariants of product ruled surfaces is given by:

$$\langle \alpha^m \beta^n \gamma^p \rangle_{M_g} = -D_{\mathbf{S}^2 \times \Sigma_g}^{w_2(E)=[\mathbf{S}^2]}((2\Sigma_g)^m(-4x)^n \iota^p), \tag{11}$$

where the overall minus sign is due to a different choice of orientation. On the other hand, we know the explicit expression for the Donaldson invariants, which is given in (7), and we can then rederive some important results about the intersection pairings [7]. The first thing that we can prove is the recursive relation for insertions of γ. One easily sees that

$$\frac{\partial}{\partial r} Z_g^{w_2(E)=[\mathbf{S}^2]} = 2g Z_{g-1}^{w_2(E)=[\mathbf{S}^2]}, \tag{12}$$

and this implies, using (11), that $\langle \alpha^m \beta^n \gamma^p \rangle_{M_g} = 2g \langle \alpha^m \beta^n \gamma^{p-1} \rangle_{M_{g-1}}$, which is precisely Thaddeus' recursive relation.

We now compute the intersection pairings $\langle \alpha^m \beta^n \rangle$. To do this, we use the expansion:

$$\csc z = \sum_{k=0}^{\infty} (-1)^{k+1}(2^{2k} - 2) B_{2k} \frac{z^{2k-1}}{(2k)!}, \tag{13}$$

where B_{2k} are the Bernoulli numbers. We have to extract now the powers $s^m p^n$ from the generating function (7). Fortunately, only the leading terms contribute in the q-expansion of the modular forms. Taking into account the comparison factors from (11), and the dimensional constraint $2m+4n=6g-6$, one finds

$$\langle \alpha^m \beta^n \rangle = (-1)^g \frac{m!}{(m-g+1)!} 2^{2g-2}(2^{m-g+1}-2)B_{m-g+1}\,, \tag{14}$$

which is exactly Thaddeus' formula for the intersection pairings.

The relation between topological Yang-Mills theory on $\mathbf{S}^2 \times \Sigma_g$ and two-dimensional moduli problems is in fact more interesting, since the Donaldson invariants in the limiting chamber of small volume for Σ_g correspond to the Gromov-Witten invariants of M_g. We refer the reader to [7] for results in this direction.

4. Critical Behaviour

4.1. Superconformal points

When one considers topological quantum field theories in four dimensions with qualitative new physics, one also finds a completely different kind of predictions from a mathematical point of view. In [13] we studied a quantum field theory with a critical behaviour on a four-manifold X of simple type and with $b_2^+>1$, namely twisted $N=2$ supersymmetric QCD with gauge group $SU(2)$ and one massive hypermultiplet with mass m. It is known [1] that the low-energy theory becomes superconformal for a certain critical value of the mass m_*, and that the quantities that characterize the theory (like the masses of the BPS particles) have a scaling behaviour near the critical point. The theory has the same BRST operators as topological Yang-Mills theory, although mathematically it describes equivariant intersection theory on the moduli space of $SU(2)$ monopoles (see [9] for a review). Using the results of [15] and some additional input, one can compute the analog of the generating function (4) for this theory, which now depends on the extra parameter m. To write the result, we need the family of Seiberg-Witten elliptic curves for the $N_f=1$ theory [17], parameterized by $(u,m)\in \mathbf{C}^2$ and given by:

$$y^2 = x^2(x-u)+2mx-1\,. \tag{15}$$

The curve is easily put into standard form $y^2=4x^3-g_2x-g_3$, with $g_2(u;m)=\frac{4}{3}(u^2-6m)$, $g_3(u;m)=\frac{1}{27}(8u^3-72mu+108)$, and discriminant $\Delta(u;m)=g_2^3-27g_3^2$. This discriminant is a cubic in u and has three roots $u_j(m)$, $j=1,2,3$. For generic, but fixed, values of m one of the periods of (15) goes to infinity as $u\to u_j$ while the other period, $\varpi_j \equiv \varpi(u_j(m);m)$ remains finite, and in fact is given by $(\varpi_j)^2 = g_2/(36g_3)$. The generating function of the critical theory is given by a sum over the singular fibers of the Weierstrass family (15) and over the basic classes of X:

$$Z(p,S;m) = k\sum_{j=1}^{3}\left(\frac{g_2^3(u_j(m);m)}{\Delta'(u_j(m);m)}\right)^{\chi_h}(\varpi_j(m))^{7\chi_h-c_1^2}$$
$$\cdot\sum_x SW(x)(-1)^{(v^2+v\cdot x)/2}\exp\left[2pu_j+S^2T_j-i\frac{(S,x)}{2\varpi_j}\right]. \quad (16)$$

Here $\Delta'=\frac{\partial}{\partial u}\Delta$, $T_j=-\frac{1}{24}\big((\varpi_j)^{-2}-8u_j\big)$, and k is a nonvanishing constant, independent of p,S,m. The topological data of the manifold X enter through v, which is an integral lifting of $w_2(X)$, the basic classes x and their SW invariants, and the numerical invariants χ_h and $c_1^2=2\chi+3\sigma$.

The critical behaviour of this theory is associated to the cusp singularity of (15) when $m_*=\frac{3}{2}$, $u_*=3$. Indeed, when $z=m-m_*\to 0$, two of the roots of $\Delta(u;m)=0$, call them $u_\pm(m)$, coincide, and the period $\varpi_\pm$ diverges as $z^{-1/4}$, while $g_2(u_\pm(m);m)\sim z$ and $\Delta'(u_\pm(m);m)\sim\delta u_\pm\sim z^{3/2}$. At the third singularity all the various factors in (16) are given by nonvanishing analytic series in z, but, evidentally, the contributions from $u_\pm(m)$ contain factors which are diverging or vanishing as $z\to 0$. What can we say about the behaviour of the complete function $Z(p,S,m)$ as $z\to 0$? For physical reasons, we do not expect any divergence in the correlation functions: there are no infrared divergences in spacetime, since X is compact, and since the moduli space of vacua is also compact for $b_2^+>1$, we do not expect any divergence from the target geometry. In more physical terms, since we are working at finite volume, correlation functions should still be finite near the critical point. This implies that $Z(p,S,m)$ *must be a regular analytic function of* z *near* $z=0$.

4.2. Mathematical implications

Let us now see what are the mathematical implications of this fact. We first define the "twisted" Seiberg-Witten series as follows.

$$SW_X^{w_2(X)}(z) := \sum_x (-1)^{\frac{v^2+v\cdot x}{2}} SW(x)e^{zx}\,. \quad (17)$$

This is a finite sum [20]. Notice that a change of lifting changes (17) only by a sign. We now make the key definition:

Definition 4.1. *Let X be a compact, oriented four-manifold of simple type with $b_2^+>1$. We say that "X is SST" if $SW_X^{w_2(X)}(z)$ has a zero at $z=0$ of order $\geq\chi_h-c_1^2-3$.*

One has the following result, whose proof can be found in [13]:

Theorem 4.2. *If X is SST, then $Z(p,S,m)$ is regular at $m=m_*$.*

It is interesting to notice that, for most of the SST manifolds, the contributions to $Z(p,S,m)$ from the colliding singularities $u_\pm$ go to infinity separately as $z \to 0$, but when we sum the two contributions (and we are forced to do that because the manifold is compact) the infinities cancel and we get a finite result.

All the simple type, four-manifolds we are aware of are in fact SST. Using the definition, one can check that SST manifolds satisfy the following remarkable property, which gives a relation between SW invariants and the problem of geography for four-manifolds:

Theorem 4.3. (Generalized Noether inequality) *Let X be SST. If X has B distinct basic classes and $B > 0$, then*

$$B \geq \left[\frac{\chi_h - c_1^2}{2} \right].$$

In particular, $c_1^2 \geq \chi_h - 2B - 1$.

Although being SST is only a sufficient condition for $Z(p, S, m)$ to be finite, the analysis of [13] leads naturally to the following conjecture:

Conjecture 4.4. *All compact four-manifolds of simple type and with $b_2^+ > 1$ are SST.*

In fact, Feehan, Kronheimer, Lenness and Mrowka have proved in [4] that the above conjecture is true under some mild assumptions, by using the $PU(2)$ monopole equations.

5. Conclusions and Open Problems

I think it is fair to say that we have a rather complete understanding of the relation between Donaldson theory and TQFT in four dimensions. There are however a few open problems that deserve investigation, both in physics and in mathematics:

1) There are many predictions from TQFT that should still be checked from the mathematical side, and I think that this is interesting by itself. For example, the results (6) and (7), as well as the wall-crossing formula of [11] for nonsimply connected manifolds, may be obtained by generalizing [5]. The extension to $SU(N)$ seems still out of reach mathematically, but it would be extremely interesting to check (9) in some detail. One can invert the logic and say that (9) gives a good reason *not* to study the $SU(N)$ Donaldson invariants, since it shows that these generalizations have the same topological information as the SW invariants!
2) In a different direction, it would be interesting to study the theory for four-manifolds with $b_2^+ = 0$. A motivation to do that would be to shed some four-dimensional light (via compactification on a circle) on the relation between the Casson invariant and the three-dimensional Seiberg-Witten invariant for homology three-spheres.
3) Finally, the twisted counterparts of superconformal field theories in four dimensions certainly deserve closer scrutiny.

References

[1] P. C. Argyres, M. R. Plesser, N. Seiberg and E. Witten, "New $N = 2$ superconformal field theories in four dimensions," hep-th/9511114, Nucl. Phys. **B 461** (1996) 71.

[2] S. K. Donaldson and P.B. Kronheimer, *The geometry of four-manifolds*, Oxford University Press, 1990.

[3] J. D. Edelstein and J. Mas, "Strong-coupling expansion and Seiberg-Witten-Whitham equations," hep-th/9901006, Phys. Lett. **B 452** (1999) 69.

[4] P. M. N. Feehan, P. B. Kronheimer, T. G. Lenness, and T. Mrowka, "$PU(2)$ monopoles and a conjecture of Mariño, Moore and Peradze," math.DG/9812125, Math. Res. Letters **6** (1999) 169.

[5] L. Göttsche, "Modular forms and Donaldson invariants for 4-manifolds with $b_+ = 1$," alg-geom/9506018; J. Am. Math. Soc. **9** (1996) 827. L. Göttsche and D. Zagier, "Jacobi forms and the structure of Donaldson invariants for 4-manifolds with $b_+ = 1$," alg-geom/9612020, Selecta. Math. (N.S.) **4** (1998) 69.

[6] N. Habegger and G. Thompson, "The universal perturbative quantum three-manifold invariant, Rozansky-Witten invariants and the generalized Casson invariant," math.DG/9911049.

[7] C. Lozano and M. Mariño, "Donaldson invariants of product ruled surfaces and two-dimensional gauge theories," hep-th/9907165, to appear in Commun. Math. Phys.

[8] A. Losev, N. Nekrasov, and S. Shatashvili, "Issues in topological gauge theory," hep-th/9711108, Nucl. Phys. **B 549** (1998); "Testing Seiberg-Witten solution," hep-th/9801061.

[9] M. Mariño, "The geometry of supersymmetric gauge theories in four dimensions," hep-th/9701128.

[10] M. Mariño and G. Moore, "The Donaldson-Witten function for gauge groups of rank larger than one," hep-th/9802185, Commun. Math. Phys. **199** (1998) 25.

[11] M. Mariño and G. Moore, "Donaldson invariants for non-simply connected manifolds," hep-th/9804114, Commun. Math. Phys. **203** (1999) 249.

[12] M. Mariño and G. Moore, "Three-manifold topology and the Donaldson-Witten partition function," Hep-th/9811214, Nucl. Phys. **B 547** (1999) 569.

[13] M. Mariño, G. Moore and G. Peradze, "Superconformal invariance and the geography of four-manifolds," hep-th/9812055, Commun. Math. Physics. **205** (1999) 691; "Four-manifold geography and superconformal symmetry," math.DG/9812042, Math. Res. Lett. **6** (1999) 429.

[14] G. Meng and C. H. Taubes, "$\underline{SW}$=Milnor torsion," Math. Res. Lett. **3** (1996) 661.

[15] G. Moore and E. Witten, "Integration over the u-plane in Donaldson theory," hep-th/9709193, Adv. Theor. Math. Phys. **1** (1997) 298.

[16] L. Rozansky and E. Witten, "HyperKähler geometry and invariants of three-manifolds," hep-th/9612216, Selecta. Math. (N.S.) **3** (1997) 401.

[17] N. Seiberg and E. Witten, "Electric-magnetic duality, monopole condensation, and confinement in $N = 2$ supersymmetric Yang-Mills theory," hep-th/9407087, Nucl. Phys. **B 426** (1994) 19. "Monopoles, duality and chiral symmetry breaking in $N = 2$ supersymmetric QCD," hep-th/9408099, Nucl. Phys. **B 431** (1994) 484.

[18] M. Thaddeus, "Conformal field theory and the cohomology of the moduli space of stable bundles," J. Diff. Geom. **35** (1992) 131.

[19] E. Witten, "Topological Quantum Field Theory," Commun. Math. Phys. **117** (1988) 353.

[20] E. Witten, "Monopoles and four-manifolds," hep-th/9411102; Math. Res. Letters **1** (1994) 769.

Department of Physics and Astronomy
Rutgers University
126 Frelinghuysen Road
Piscataway, NJ 08854, USA
E-mail address: `marcosm@physics.rutgers.edu`

D-Brane Conformal Field Theory and Bundles of Conformal Blocks

Christoph Schweigert and Jürgen Fuchs

Abstract. Conformal blocks form a system of vector bundles over the moduli space of complex curves with marked points. We discuss various aspects of these bundles. In particular, we present conjectures about the dimensions of sub-bundles. They imply a Verlinde formula for non-simple connected groups like $PGL(n, C)$.

We then explain how conformal blocks enter in the construction of conformal field theories on surfaces with boundaries. Such surfaces naturally appear in the conformal field theory description of string propagation in the background of a D-brane. In this context, the sub-bundle structure of the conformal blocks controls the structure of symmetry breaking boundary conditions.

1. Introduction

Two-dimensional conformal field theory plays a fundamental role in the theory of two-dimensional critical systems of classical statistical mechanics, in quasi one-dimensional condensed matter physics and in string theory. Moreover, this field of research has repeatedly contributed substantially to the interaction between mathematics and physics. The study of defects in systems of condensed matter physics, of percolation probabilities, and of (open) string perturbation theory in the background of certain string solitons, the so-called D-branes, has recently driven theoretical physicists to analyse conformal field theories on surfaces that may have boundaries and/or can be non-orientable.

In the present contribution, we would like to describe some mathematical aspects of this recent development and state some conjectures about the sub-bundle structure of the bundles of conformal blocks. In physics, this structure enters in the description of symmetry breaking boundary conditions. In the special case of WZW models our conjectures imply a Verlinde formula for non-simply connected groups like $PGL(n, C)$.

We are grateful to the organizers of the third European Congress of Mathematics for the invitation to present these results.

2. Chiral and Full Conformal Field Theory

The investigation of boundary conditions makes it clear that it is indispensable to distinguish carefully between *chiral* conformal field theory, CCFT, and *full* conformal field theory, CFT. Although a CFT is constructed from a CCFT, both types of theories are rather different in nature. Mathematicians typically have in mind *chiral* conformal field theory, CCFT, when they use the term conformal field theory; one goal of the present note is to draw their attention to this difference. In physics, both CCFT and CFT have found applications: CFT in string theory and the study of two-dimensional critical systems, CCFT in the description of quantum Hall fluids (for a recent review see [14]).

A CCFT is defined on a closed, compact complex curve $\hat{X}$; a CFT, in contrast, is defined on a real two-dimensional manifold X that is endowed with a conformal structure, i.e. a class of metrics modulo local rescalings. While $\hat{X}$ is oriented and has empty boundary, X is allowed to have a boundary and/or to be non-orientable.

To any such X, one finds a complex curve $\hat{X}$, the *double*, together with an anti-conformal involution σ of $\hat{X}$ such that X is isomorphic to the quotient of $\hat{X}$ by σ. (When X has empty boundary, $\hat{X}$ is just the total space of the orientation bundle of X.) $\hat{X}$ is not necessarily connected; also, σ does not necessarily act freely: fixed points of σ correspond to boundary points of $\hat{X}$.

A double can also be constructed when X has dimension higher than two, and it will be an oriented conformal manifold. However, only in two dimensions a conformal structure plus an orientation are equivalent to a holomorphic structure; in higher dimensions, additional integrability constraints have to be fulfilled. This simple fact implies that specifically in two dimensions the powerful tools of the theory of holomorphic functions are at one's disposal.

We will start with a discussion of chiral conformal field theory, exhibiting in particular some aspects of the bundles of conformal blocks. For instance, we will explain conjectures about the sub-bundle structure of these bundles. Full CFT will be discussed in Section 4. The general idea is to relate full CFT on X to chiral conformal field theory on the double $\hat{X}$.

3. Chiral Conformal Field Theory

3.1. Conformal vertex operator algebras

The definition of a chiral conformal field theory starts with a conformal vertex operator algebra (VOA) $(\mathcal{H}_\Omega, v_\Omega, Y, v_{\rm vir})$. Here $\mathcal{H}_\Omega$ is a complex vector space, v_Ω and $v_{\rm vir}$ are vectors in $\mathcal{H}_\Omega$, and

$$Y\colon \mathcal{H}_\Omega \to \mathrm{End}(\mathcal{H}_\Omega)[[z, z^{-1}]] \tag{1}$$

is a map into the space of Laurent series in a formal variable z with values in $\mathrm{End}(\mathcal{H}_\Omega)$, called a *field-state correspondence.* One imposes a weakened version of commutativity, called *locality*: for each pair of vectors $v, w \in \mathcal{H}_\Omega$, there is an integer $N = N(v, w)$ such that

$$(z-w)^N Y(v,z)Y(w,b) = (z-w)^N Y(w,b)Y(v,z) \tag{2}$$

in the sense of formal power series. For precise definitions and an introduction to VOAs, see e.g. [24].

One way to think of the field-state correspondence is as a family of products on the vector space $\mathcal{H}_\Omega$ that depend on the formal variable z:

$$v \star_z w = Y(v,z)w\,. \tag{3}$$

It is remarkable that "commutativity" (2) implies a kind of associativity: one has

$$a \star_z (b \star_w c) = (a \star_{z-w} b) \star_w c\,. \tag{4}$$

for all $a, b, c \in \mathcal{H}_\Omega$. The vector Ω is called the vacuum vector. Under field-state correspondence, it is mapped to the identity map, $Y(\Omega, z) = \mathrm{id}$. The Virasoro vector v_{vir} has the property that the modes L_n in the expansion of the corresponding field, the *stress energy tensor*,

$$Y(v_{\mathrm{vir}}, z) = \sum_{n\in\mathbb{Z}} L_n\, z^{-n-2} \tag{5}$$

span an infinite-dimensional Lie algebra that is isomorphic to the Virasoro algebra.

At first sight a VOA might seem a rather ad hoc kind of algebraic structure. We would like to stress, however, that it is a very natural concept indeed. Apart from their motivation from physics (for a review see [24]), VOAs can be naturally characterized in a category theoretic framework as singular commutative rings in certain categories [6]. Examples of VOAs can be constructed from various infinite-dimensional Lie algebras like the Heisenberg algebra, the Virasoro algebra or untwisted affine Lie algebras; the latter case gives rise to the so-called WZW models. Moreover, the chiral de Rham complex [25] allows us to associate to any complex variety a VOA as an invariant.

For every algebraic structure, it is natural to investigate its representation theory. Thus denote by I the set of equivalence classes of irreducible representations of a given VOA, and $\mathcal{H}_\mu$ a representative for $\mu \in I$. Elements of I will also be called labels. Any irreducible representation carries in particular a representation of the Virasoro algebra. Usually, one requires that L_0 acts by semi-simple operators.[1] The axioms of a VOA then imply that the spectrum of L_0 is integrally spaced and bounded from below. The lowest eigenvalue of L_0 in $\mathcal{H}_\mu$ is called the *conformal weight* Δ_μ.

A special representation is given by the vector space $\mathcal{H}_\Omega$ of the VOA itself; it is called the vacuum representation, and the corresponding label in I will be denoted by Ω. A VOA is called rational iff the set I is finite and every representation is completely reducible. The representation theory of VOAs is a model dependent

[1] This requirement is weakened in so-called logarithmic conformal field theories.

problem that can be quite intricate; see e.g. [17, 27] for the so-called coset conformal field theories.

3.2. Conformal blocks

VOAs formalize aspects of the 'observables' of chiral two-dimensional field theories. As a consequence, the algebraic structure of a VOA fits together very well with certain two-dimensional geometries – more precisely, since the theory is chiral, with the geometry of complex curves. To see how this works out, we fix a complex curve $\hat{X}=\hat{X}_g$ of genus g, and m distinct smooth points $\vec{z}=(z_1,\dots,z_m)$ on $\hat{X}$, the insertion points. A VOA should be thought of as being associated to a (formal) punctured disc with coordinate z; to the data $\hat{X},\vec{z}$ one can associate a global version of the VOA.[2] In the case of $\hat{X}=\mathbb{C}P^1$ with three marked points, this global version also leads to the coproduct for the VOA that was defined in [12].

When we fix additional data, labels $\vec{\lambda}=\lambda_1,\lambda_2,\dots,\lambda_m$ and local coordinates $\xi_1,\xi_2,\dots,\xi_m$ for each insertion point, this global algebra acts on the algebraic dual

$$(\mathcal{H}_{\lambda_1}\otimes_{\mathbb{C}}\cdots\otimes_{\mathbb{C}}\mathcal{H}_{\lambda_m})^*\,, \tag{6}$$

of the tensor product of the modules $\mathcal{H}_{\lambda_i}$. The vector space $V_{\vec{\lambda}}(\hat{X}_g)$ of conformal blocks is by definition the invariant subspace under this action. We call a complex curve with the additional data we just described an *extended curve.*

For the vertex operator algebra of the WZW model based on an untwisted affine Lie algebra $\mathfrak{g}=\bar{\mathfrak{g}}^{(1)}$ at level $k\in\mathbb{Z}_{\geq 0}$, the vector spaces $V_{\vec{\lambda}}(\hat{X}_g)$ also appear naturally in algebraic geometry. The space $V_\Omega(\hat{X}_g)$, e.g., can be identified with the space of holomorphic sections in the k-th power of a line bundle $\mathcal{L}$ over the moduli space of holomorphic G-connections over $\hat{X}$, where G is the simple, connected and simply connected complex Lie group with Lie algebra $\bar{\mathfrak{g}}$.

One can show that the vector spaces $V_{\vec{\lambda}}(\hat{X}_g)$ with fixed $\vec{\lambda}$ and g fit together into a vector bundle $\mathcal{V}_{\vec{\lambda},g}$ over the moduli space $\mathcal{M}_{g,m}$ of complex curves of genus g with m marked points. There is no reason why this vector bundle $\mathcal{V}_{\vec{\lambda},g}$ should be irreducible. Later, we will describe some sub-bundles which enter in the description of symmetry breaking boundary conditions.

The Virasoro element $v_{\rm vir}$ can also be used to endow this bundle with a projectively flat connection, the Knizhnik-Zamolodchikov connection. As a result, the dimension of the vector space $V_{\vec{\lambda}}(\hat{X}_g)$ depends only on the genus g and on $\vec{\lambda}$. The Knizhnik-Zamolodchikov connection implies a projective action of the fundamental group of $\mathcal{M}_{g,m}$, the mapping class group, on $V_{\vec{\lambda}}(\hat{X}_g)$. In the particular case of g=1 and with one insertion of the vacuum Ω, we obtain a projective representation of the modular group $PSL(2,\mathbb{Z})$ on a complex vector space of dimension $|I|$. In a natural basis, the generator $T\colon \tau\mapsto\tau{+}1$ of $PSL(2,\mathbb{Z})$ is represented by a unitary diagonal matrix $T_{\lambda,\mu}$, and $S\colon \tau\mapsto -1/\tau$ by a unitary symmetric matrix $S_{\lambda,\mu}$.

[2]This algebra has been named chiral algebra in [4]; unfortunately this does not agree with terminology in physics, where a chiral algebra is a VOA.

Up to this point, the structures we discussed are mathematically essentially under control (see e.g. [13]). There is, however, one further crucial aspect of conformal blocks, known as factorization. The curve $\hat{X}$ is allowed to have ordinary double points. Blowing up such a point p yields a new curve $\hat{X}'$ with a projection to $\hat{X}$ under which p has two pre-images $p'_\pm$. By factorization one means the existence of a canonical isomorphism

$$V_{\vec{\lambda}}(\hat{X}) \cong \bigoplus_{\mu\in I} V_{\vec{\lambda}\cup\{\mu,\mu^+\}}(\hat{X}') \tag{7}$$

between the blocks on $\hat{X}$ and $\hat{X}'$. This structure tightly links the system of bundles $\mathcal{V}_{\vec{\lambda},g}$ over the moduli spaces $\mathcal{M}_{g,m}$ for different values of g and m. One consequence is that one can express the rank of $\mathcal{V}_{\vec{\lambda},g}$ for all values of m and g in terms of the matrix S that we encountered in the description of the action of the modular group. This results in the famous *Verlinde formula*, which reads

$$\operatorname{rank} \mathcal{V}_{\vec{\lambda},g} = \sum_{\mu\in I} \prod_{i=1}^{m} \frac{S_{\lambda_i,\mu}}{S_{\Omega,\mu}} \, |S_{\Omega,\mu}|^{2-2g} \,. \tag{8}$$

The matrix S can be computed explicitly for concrete models; in the case of WZW models, S is given by the Kac-Peterson formula. The combination of the Kac-Peterson formula for S with the general Verlinde formula (8) then gives the Verlinde formula in the sense of algebraic geometry [1, 8].

3.3. Sub-bundles

We now discuss sub-bundles of the bundles $\mathcal{V}_{\vec{\lambda},g}$. To this end, we introduce the *fusion rules* $\mathcal{N}_{\lambda_1\lambda_2\lambda_3}$, which are dimensions of the three-point blocks at genus 0:

$$\mathcal{N}_{\lambda_1,\lambda_2,\lambda_3} = \operatorname{rank} \mathcal{V}_{\lambda_1,\lambda_2,\lambda_3} \,. \tag{9}$$

It turns out that $\mathcal{N}_{\lambda_1,\lambda_2,\Omega}$ describes a permutation of order 2 on I which we denote by $\mu\mapsto\mu^+$. Consider the free $\mathbb{Z}$-module $\mathcal{R} = \mathbb{Z}^{|I|}$ with a basis $\{\Phi_\mu\}$ labelled by $\mu\in I$. The definition

$$\Phi_{\mu_1}\star\Phi_{\mu_2} = \sum_{\mu_3\in I} \mathcal{N}_{\mu_1,\mu_2,\mu_3^+}\, \Phi_{\mu_3} \tag{10}$$

turns $\mathcal{R}$ into a ring, the *fusion ring*. Factorization and Knizhnik-Zamolodchikov connection imply that this ring is commutative, associative and semi-simple. Notice that by construction a fusion ring comes with a distinguished basis labelled by I.

Invertible elements of $\mathcal{R}$ that are elements of the distinguished basis are called simple currents. In the WZW case, simple currents correspond (with one exception, appearing for E_8 at level 2), to elements of the center of the Lie group G [15]. The group $\mathcal{G}$ of simple currents acts on $\mathcal{R}$, and also on I. This action is in general not free, and to each element $\mu\in I$ we associate its stabilizer $\mathcal{S}_\mu := \{\Phi_J\in\mathcal{G} \,|\, \Phi_J\star\Phi_\mu=\Phi_\mu\}$. Given an m-tuple of labels $\vec{\lambda}$, consider the subgroup $\mathcal{S}^1_{\vec{\lambda}}$ of $\mathcal{S}_{\lambda_1}\times\cdots\times\mathcal{S}_{\lambda_m}$ that consists of those elements $(J_1, J_2, \ldots, J_m)$ whose product equals the unit element of the fusion ring, $\prod_i J_i = \Omega$.

Henceforth we will for simplicity assume that $H^2(\mathcal{G},\mathbb{C}^*)=0$; for results in more general situations see [18, 20]. One expects that then the group $\mathcal{S}^1_{\vec\lambda}$ acts by bundle automorphisms on $\mathcal{V}_{\vec\lambda,g}$; for WZW models and $g=0$, see [20]. The invariant subspaces under this action form sub-bundles; once an action of $\mathcal{S}^1_{\vec\lambda}$ is fixed, they are characterized by their eigenvalues, i.e. by a character of $\mathcal{S}^1_{\vec\lambda}$.

A conjecture for the ranks of these sub-bundles was presented in [18, 20]. It is actually more convenient to give the Fourier transforms of these ranks which can be interpreted as traces of the action of $\mathcal{S}^1_{\vec\lambda}$ on the conformal blocks. For the trace of $(J_1, J_2, \dots, J_m)\in\mathcal{S}^1_{\vec\lambda}$ we obtain

$$\mathrm{Tr}_{V_{\vec\lambda,g}}\Theta_{J_1,\dots,J_m} = \sum_{\mu\in I}\prod_{i=1}^{m}\frac{S^{J_i}_{\lambda_i,\mu}}{S_{\Omega,\mu}}\,|S_{\Omega,\mu}|^{2-2g}\,. \tag{11}$$

Here S^J is the matrix that describes the transformation under the modular group of the one-point blocks on an elliptic curve with insertion J. In particular, for $J=\Omega$, S^J equals the ordinary matrix S and we recover the Verlinde formula (8) for $\Theta=\mathrm{id}$.

In the case of WZW models, there is a conjecture [18] for the explicit form of the matrix S^J: it is given by the S-matrix of Kac-Peterson form for another Lie algebra, the so-called orbit Lie algebra [17, 16] associated to the data $(\mathfrak{g}, J)$.[3] With this conjecture, formula (11) is as concrete as the Verlinde formula and can be implemented in a computer program, see `http://www.nikhef.nl/~t58/kac.html`. In the case of WZW models, these sub-bundles also enter in a Verlinde formula, in the sense of algebraic geometry, for non-simple connected groups like $PGL(n,\mathbb{C})$, see [2].

4. Full Conformal Field Theory

Having discussed some structures of CCFT, our next goal is to construct full CFT on a real two-dimensional manifold X with conformal structure, using CCFT on the double $\hat X$ of X. To this end, we have to prescribe some data for each insertion point, which can lie either in the interior of X or on the boundary. For each point on the boundary of X, we choose a label and a local coordinate ξ such that the lift to the double $\hat X$ gives a local holomorphic coordinate. We also choose an orientation for each component of the boundary.

For each point in the interior of X, a so-called bulk point, we choose a local coordinate such that on the double we obtain local holomorphic coordinates around the two pre-images. Notice that possible choices of such coordinates come in two

[3]Orbit Lie algebras also enter in the study of moduli spaces of flat connections with non-simple connected structure group on elliptic curves [26], and in the related problem of determining almost commuting elements in Lie groups [7].

disjoint sets which are related by a complex conjugation of the coordinate. We call them a local orientation of the coordinate.

To specify the labels of the bulk fields, we need to fix one more datum: an automorphism $\dot{\omega}$ of the fusion ring that preserves conformal weight modulo integers, $T_{\dot{\omega}\lambda}=T_{\lambda}$. The labels for bulk fields are now defined as equivalence classes of triples (λ, or', or), where λ is a label of the chiral CCFT, or' is a local orientation of the coordinate, and or is a local orientation of X around the insertion point. The triples (λ, or', or) and $(\dot{\omega}\lambda, -or', -or)$ with opposite orientations are identified. We call a surface with this structure a *labelled surface* (cf. [11], where the notion was introduced in the category of topological manifolds for the case when $\dot{\omega}$ is the conjugation).

Some thought shows that these data precisely allow us to give the double, with the pre-images of the insertion points, the structure of an extended curve, so that we can define conformal blocks. While the moduli space $\mathcal{M}_X$ of a labelled surface cannot be embedded into the moduli space $\mathcal{M}_{\hat{X}}$ of its double as an extended curve, an embedding of the corresponding Teichmüller spaces $\mathcal{T}$ does exist. This is sufficient to give us conformal blocks as vector bundles $\mathcal{V}_{\vec{\lambda},g}$ over $\mathcal{M}_X$. The involution σ on $\hat{X}$ induces an involution σ_* on $\mathcal{T}_{\hat{X}}$. The subgroup of the mapping class group of $\hat{X}$ that commutes with σ_* is called the relative modular group [5]. (In the case of a closed and oriented surface, $\hat{X}$ has two connected components of opposite orientation, each of which is isomorphic to X. The relative modular group is then isomorphic to the mapping class group of X.)

The central problem in the construction of a CFT from a CCFT is to specify, for all choices of $\vec{\lambda}$ and all genera g, *correlation functions* of the CFT as sections in $\mathcal{V}_{\vec{\lambda},g}$. For fixed labels $\vec{\lambda}$ and fixed g, these sections depend on the positions of the insertion points and on the moduli of the conformal structure. We impose the locality requirement: these sections have to be genuine functions (rather than multivalued sections) of these parameters. This includes in particular invariance under the relative modular group. Moreover, we require compatibility with factorization (for a formulation of the latter requirement in the category of topological manifolds, see [11]).

Due to factorization, the calculation of any correlation function can be reduced to four basic cases: the sphere with three points, the disc with three boundary points, the disc with one boundary point and one interior point, and the real projective plane with one point. Except for the last case, they involve three marked points on each connected component of the double.

Constraints on these particular correlators follow by considering situations where four marked points appear on each connected component of $\hat{X}$. In particular, from the consideration of four points on the boundary of the disc one can deduce that the correlator of three boundary fields is proportional to the so-called fusing matrix [3, 9]. We stress that this is a model independent statement. Indeed, the construction of a CFT from a CCFT can be formulated in a completely model independent way; see [10, 11], where a general prescription for the correlators has been given for the case when $\dot\omega$ is charge conjugation.

Similarly, one can consider two bulk fields on a disc to derive constraints on the correlator of one bulk field and one boundary field on a disc with boundary condition a. To be more explicit, introduce a basis $\{e^{(\alpha)}_{\lambda,\mu,\nu}\}$ for the space $V_{\lambda,\mu,\nu}$ of three-point blocks on the sphere; we are looking for a linear combination

$$\sum_{\alpha=1}^{\mathcal{N}_{\lambda,\dot\omega\lambda,\mu}} {}^{\alpha}R^{a}_{\lambda,\dot\omega\lambda,\mu}\, e^{(\alpha)}_{\lambda,\dot\omega\lambda,\mu} \tag{12}$$

that is compatible with factorization. It turns out that, with suitable normalizations, the factorization constraints imply that

$$R^{a}_{\lambda_1,\dot\omega\lambda_1,\Omega} R^{a}_{\lambda_2,\dot\omega\lambda_2,\Omega} = \sum_{\lambda_3} \tilde{\mathcal{N}}_{\lambda_1,\lambda_2}^{\lambda_3} R^{a}_{\lambda_3,\dot\omega\lambda_3,\Omega}\,. \tag{13}$$

This equation shows that the *reflection coefficients* $R^{a}_{\lambda,\dot\omega\lambda,\Omega}$ form one-dimensional representations of a certain algebra, called the *classifying algebra* [19]. The problem of classifying boundary conditions is therefore reduced to the representation theory of the classifying algebra.

It turns out that, in the case when $\dot\omega$ is charge conjugation, the structure constants $\tilde{\mathcal{N}}_{\lambda_1,\lambda_2}^{\lambda_3}$ are just the fusion rules, i.e. the dimensions of the spaces of three-point blocks. This result has been generalized both to more general choices of the fusion rule automorphism $\dot\omega$ [19] and to the case when the boundary conditions break some of the symmetries of the theory [21, 22, 23]. The structure constants of the classifying algebra can be expressed in terms of the modular transformation matrices S^J that appear in (11) and of characters of simple current groups [22, 23]. The resulting expressions can be shown to coincide with traces of the form (11) for appropriate twisted intertwiner maps on the spaces of three-point blocks. Via this relationship, the sub-bundle structure of the bundles of conformal blocks controls boundary conditions in these more general cases, in particular boundary conditions that break some of the bulk symmetries. Solitonic solutions in field and string theory typically do not respect all symmetries, and D-brane solutions are no exception to this. The sub-bundle structure of conformal blocks is therefore an essential ingredient for the CFT approach to D-brane physics.

References

[1] A. Beauville, in: *Hirzebruch 65 Conference on Algebraic Geometry* [Israel Math. Conf. Proc. 9], M. Teicher, ed. (Bar-Ilan University, Ramat Gan 1996), p. 75–96.

[2] A. Beauville, in: *The Mathematical Beauty of Physics*, J.M. Drouffe and J.-B. Zuber, eds. (World Scientific, Singapore 1997), p. 141.

[3] R. E. Behrend, P. A. Pearce, V. B. Petkova, and J.-B. Zuber, *Boundary conditions in rational conformal field theories*, Nucl. Phys. B 570 (2000) 525–589.

[4] A. A. Beilinson and V. G. Drinfeld, *Chiral algebras I*, preprint.

[5] M. Bianchi and A. Sagnotti, *Open strings and the relative modular group*, Phys. Lett. B 231 (1989) 389–396.

[6] R. E. Borcherds, *Quantum vertex algebras*, preprint math.QA/9903038.

[7] A. Borel, R. Friedman, and J. W. Morgan, *Almost commuting elements in compact Lie groups*, preprint math.GR/9907007.

[8] G. Faltings, *A proof for the Verlinde formula*, J. Algebraic Geom. 3 (1994) 347–374.

[9] G. Felder, J. Fröhlich, J. Fuchs, and C. Schweigert, *The geometry of WZW branes*, preprint hep-th/9909030, J. Geom. Phys. 34 (2000) 162–190.

[10] G. Felder, J. Fröhlich, J. Fuchs, and C. Schweigert, *Conformal boundary conditions and three-dimensional topological field theory*, Phys. Rev. Lett. 84 (2000) 1659–1662.

[11] G. Felder, J. Fröhlich, J. Fuchs, and C. Schweigert, *Correlation functions and boundary conditions in RCFT and three-dimensional topology*, preprint hep-th/9912239., to appear in Comp. Math.

[12] G. Felder, J. Fröhlich, and G. Keller, *On the structure of unitary conformal field theory II: Representation theoretic approach*, Comm. Math. Phys. 130 (1990) 1–49.

[13] E. Frenkel, D. Ben-Zvi, *Vertex algebras and algebraic curves*, AMS, to appear.

[14] J. Fröhlich, B. Pedrini, C. Schweigert, and J. Walcher, Universality in quantum Hall systems: coset construction of incompressible states, J. Stat. Phys. 103 (2001) 527–567.

[15] J. Fuchs, *Simple WZW currents*, Comm. Math. Phys. 136 (1991) 345–356.

[16] J. Fuchs, U. Ray, and C. Schweigert, *Some automorphisms of Generalized Kac-Moody algebras*, J. Algebra 191 (1997) 518–540.

[17] J. Fuchs, A. N. Schellekens, and C. Schweigert, *The resolution of field identification fixed points in diagonal coset theories*, Nucl. Phys. B 461 (1996) 371–404.

[18] J. Fuchs, A. N. Schellekens, and C. Schweigert, *A matrix S for all simple current extensions*, Nucl. Phys. B 473 (1996) 323–366.

[19] J. Fuchs and C. Schweigert, *A classifying algebra for boundary conditions*, Phys. Lett. B 414 (1997) 251–259.

[20] J. Fuchs and C. Schweigert, *The action of outer automorphisms on bundles of chiral blocks*, Comm. Math. Phys. 206 (1999) 691–736.

[21] J. Fuchs and C. Schweigert, *Orbifold analysis of broken bulk symmetries*, Phys. Lett. B 447 (1999) 266–276.

[22] J. Fuchs and C. Schweigert, *Symmetry breaking boundaries I. General theory*, Nucl. Phys. B 558 (1999) 419–483.

[23] J. Fuchs and C. Schweigert, *Symmetry breaking boundaries II. More structures; examples*, Nucl. Phys. B 568 (2000) 543–593.

[24] V. G. Kac, *Vertex Algebras for Beginners* (American Mathematical Society, Providence 1996.)

[25] F. Malikov, V. Schechtman, and A. Vaintrob, *Chiral de Rham complex*, Comm. Math. Phys. 204 (1999) 439–473.

[26] C. Schweigert, *On moduli spaces of flat connections with non-simply connected structure group*, Nucl. Phys. B 492 (1997) 743–755.

[27] F. Xu, *Algebraic coset conformal field theories I,II*, preprint math.OA/9810035, math.OA/9903096.

C. Schweigert
Lpthe
Université Paris 6
4, Place Jussieu
F–75 252 Paris Cedex 05, France
E-mail address: `Schweige@Lpthe.Jussieu.Fr`

J. Fuchs
Institutionen För Fysik
Karlstads Universitet
Universitetsgatan 1
S–651 88 Karlstad, Sweden
E-mail address: `Jfuchs@Fuchs.Tekn.Kau.Se`

From Quiver Diagrams to Particle Physics

Angel M. Uranga

Abstract. Recent scenarios of phenomenologically realistic string compactifications involve the existence of gauge sectors localized on D-branes at singular points of Calabi–Yau threefolds. The spectrum and interactions in these gauge sectors are determined by the local geometry of the singularity, and can be encoded in quiver diagrams. We discuss the physical models arising for the simplest case of orbifold singularities, and generalize to non-orbifold singularities and orientifold singularities (a generalization naturally arising in string theory). Finally we show that relatively simple singularities lead to gauge sectors surprisingly close to the standard model of elementary particles.

1. Introduction

The construction of string theory models which at low energies reproduce the basic features of the standard model of elementary particles (or some extension thereof) is a non-trivial task from the physical point of view. Interestingly, it also involves beautiful mathematics, and leads to a rich interplay between the physical and mathematical points of view. To quote a traditional example, the construction of four-dimensional $\mathcal{N} = 1$ supersymmetric heterotic string vacua involves the study of stable G-bundles on Calabi–Yau threefolds, with G a subgroup of $E_8 \times E_8$ or $SO(32)$.

Recent develoments in string theory have led to new kinds of string theory vacua with potential phenomenological interest. A particular class corresponds to compactification on a Calabi–Yau threefold X with gauge bundles defined on subvarieties of the internal space (times non-compact spacetime). In physical terms, the compactification includes a set of D-branes [18], which are dynamical extended objects partially wrapped on the internal manifold, and filling non-compact spacetime. Their dynamics is controlled by a gauge theory defined on their world-volume. The rank of the gauge bundle they carry is usually referred to as the number of D-branes. In the following we center on the simplest case of D3-branes in type IIB string theory, where each D-brane spans four-dimensional spacetime times a point $P \in X$. The properties of the corresponding gauge theory sector in the low-energy theory are then determined in terms of the local geometry of X around P.

For a set of D3-branes sitting at a smooth point in X, the low-energy gauge field theory on the world-volume has $\mathcal{N} = 4$ supersymmetry, and is therefore non-chiral and phenomenologically uninteresting. Chiral gauge sectors arise when D3-branes sit at singular points in X. Singularities preserving at least $\mathcal{N} = 1$ supersymmetry, and at a finite distance in Calabi–Yau moduli space must be Gorenstein canonical singularities.

In Section 2 we discuss several systems of branes at singular points in non-compact Calabi–Yau spaces. In Section 2.1 we center on the simple case of threefold quotient singularities, which leads to an implementation of the McKay correspondence in string theory. In Sections 2.2. and 2.3 we point out several generalizations suggested by string theory. In Section 3 we discuss how an extremely simple $\mathbf{Z}_3$ quotient singularity leads to gauge groups and particle contents remarkably similar to those of the (minimal supersymmetric) standard model.

I am grateful to G. Aldazabal, A. Hanany, L. E. Ibáñez, J. Park, F. Quevedo and R. Rabadán for collaboration and useful discussion on these issues. I also thank M. González for encouragement and support.

2. Branes at Singularities

2.1. Branes at orbifold singularities

Let Γ be a discrete group of $SU(3)$, and $\{\mathbf{r}_i\}$ the set of its unitary irreducible representations. We want to consider a set of D3-branes at the origin of $\mathbf{C}^3/\Gamma$, where Γ acts on $\mathbf{C}^3$ through a three-dimensional representation $\mathcal{R}_\mathbf{3}$.

Consider first a set of N D3-branes at the origin in $\mathbf{C}^3$, labeled by an index $a = 1, \dots, N$ referred to as Chan–Paton index. Quantization of open strings with endpoints of the D3-branes lead to a set of dynamical fields propagating on the D3-brane world-volume. In terms of $\mathcal{N} = 1$ supersymmetry multiplets, the corresponding gauge field theory contains a set of vector multiplets (each containing one gauge field and one complex fermion) with gauge group $U(N)$, and three chiral multiplets Φ^a, $a = 1, 2, 3$ (each containing one complex scalar and one complex fermion). The latter transform in the adjoint representation of $U(N)$ and form a triplet under the $SU(3)$ action on $\mathbf{C}^3$. The interactions are encoded in the superpotential function $W = \epsilon_{abc}\mathrm{tr}\,(\Phi^a\Phi^b\Phi^c)$, where ϵ_{abc} correspond to the components of the $SU(3)$ invariant tensor.

Following [5] (see [9, 4] for generalizations) the field theory on D3-branes at the origin in $\mathbf{C}^3/\Gamma$ is obtained from the above field theory associated to N D3-branes in flat space, by keeping the states which are invariant under the combined action of Γ on $\mathbf{C}^3$ (as determined by $\mathcal{R}_\mathbf{3}$) and on the space of Chan–Paton indices (through an N-dimensional representation $\mathcal{R}$, with decomposition $\mathcal{R} = \bigoplus_i N_i\mathbf{r}_i$). Following [12], we regard fields in the adjoint of $U(N)$ as $\mathrm{Hom}(\mathbf{C}^N, \mathbf{C}^N)$. The projection on the $\mathcal{N} = 1$ vector multiplets leaves the fields

$$\mathrm{Hom}(\mathbf{C}^N, \mathbf{C}^N)^\Gamma = \bigoplus_i \mathrm{Hom}(\mathbf{C}^{N_i}, \mathbf{C}^{N_i}) \tag{1}$$

corresponding to a gauge group $\prod_i U(N_i)$. The projection on the $SU(3)$ triplet of $\mathcal{N}=1$ chiral supermultiplets leaves the fields

$$(\mathcal{R}_\mathbf{3} \otimes \mathrm{Hom}(\mathbf{C}^N, \mathbf{C}^N))^\Gamma = \bigoplus_{i,j} a_{ij}^\mathbf{3} \mathrm{Hom}(\mathbf{C}^{N_i}, \mathbf{C}^{N_j}) \tag{2}$$

where $a_{ij}^\mathbf{3}$ are defined by $\mathcal{R}_\mathbf{3} \otimes \mathbf{r}_i = \bigoplus_j a_{ij}^\mathbf{3} \mathbf{r}_j$. Hence we obtain $a_{ij}^\mathbf{3}$ $\mathcal{N}=1$ chiral multiplets transforming in the representation $(N_i, \overline{N}_j)$ of the gauge group. The superpotential is obtained by resticting the above one to the surviving fields.

The field content on the D3-brane world-volume can be encoded in a quiver diagram, where the ith node represents the $U(N_i)$ factor in the gauge group, and $a_{ij}^\mathbf{3}$ oriented arrows from the ith to the jth node correspond to the $\mathcal{N}=1$ chiral multiplets in the $(N_i, \overline{N}_j)$ representation. Finally, closed triangles of oriented arrows are associated to superpotential couplings of the corresponding chiral multiplets.

Several interesting mathematical connections arise at this point. For instance, the quiver diagrams encoding the field theory content and interactions coincide with the McKay quivers associated to the singularity, which are related to the homology of the resolved space by the McKay correspondence [15]. The correspondence arises in the string theory context since branes giving rise to a specific gauge factor have the geometrical interpretation of higher-dimensional branes wrapped on homology cycles of the space. Hence, gauge theory data (the quiver diagram) are related to the homology of the ambient space.

Also, if $\mathcal{R}$ is chosen to be the regular representation of Γ, the moduli space of vacua of the gauge theory corresponds to the space of possible locations of the D3-brane, which is isomorphic to the transverse space $\mathbf{C}^3/\Gamma$. The construction of the moduli space amounts to performing a symplectic quotient in the subspace of fields subject to relations $\frac{\partial W}{\partial \Phi_i} = 0$ (F-term constraints) determined by the superpotential [4]. It provides the string theory counterpart of the construction of $\mathbf{C}^3/\Gamma$ as the moduli of representations of a quiver diagram with relations [20]. In the particular case $\gamma \subset SU(2)$, studied in [5, 9], one recovers the hyperkähler quotient construction of ALE spaces [11]. String theory also provides a description of the resolved spaces, by a suitable modification of the symplectic quotient due to a non-zero moment map (D-term).

We conclude this section with some physical considerations. In string theory D-branes are sources of certain p-form gauge fields from the closed string sector, whose equations of motion may impose certain consistency conditions on the D-brane configuration. In the particular case of D3-branes at $\mathbf{C}^3/\Gamma$ singularities, the equations of motion for fields located at the singularity impose the so-called twisted tadpole cancellation conditions, which for quotient singularities amount to the vanishing of the character of the representation $\mathcal{R}$. This constraint is equivalent to the cancellation of non-abelian anomalies in the gauge field theory on the D3-branes world-volume [13]. They also imply that the remaining mixed $U(1)$-non-abelian anomalies have a factorized form and are cancelled by a version of the

Green–Schwarz mechanism [8]. The anomalous $U(1)$ factors become massive and disappear from the low-energy dynamics.

2.2. Generalizations

2.2.1. NON-ORBIFOLD SINGULARITIES There is no simple recipe to obtain the field theory on the world-volume of stacks of D3-branes at a general singularity. However, the requirement that its moduli space should correspond to the space of possible locations of D-branes in the transverse space is enough to determine the field theory in the simple example of the conifold singularity X_{con} [10], defined by the hypersurface $x^2 + y^2 + z^2 + w^2 = 0$ in $\mathbf{C}^4$. A set of N D3-branes at a conifold singularity yields an $\mathcal{N} = 1$ supersymmetric field theory with gauge group $U(N) \times U(N)$ and chiral multiplets A_i, B_i, $i = 1, 2$ in the representations $(N, \overline{N})$ and $(\overline{N}, N)$ under the gauge group, and in the representations $(2,1)_{1/2}$, $(1,2)_{1/2}$ under the $SU(2)^2 \times U(1)$ symmetry group of X_{con}. The interactions are determined by a quartic superpotential $W = \epsilon^{ij}\epsilon^{kl}\mathrm{tr}\, A_i B_j A_k B_l$.

This example provides a whole new family of models [22] (see [3] for a related discussion) corresponding to D3-branes at quotients of the conifold, X_{con}/Γ, with Γ a subgroup of $SU(2) \times SU(2)$ in order to preserve $\mathcal{N} = 1$ supersymmetry. The strategy is, as in Section 2.1, to start with D3-branes at X_{con}, embed the action of Γ on the Chan–Paton indices, and keep only fields which are invariant under the combined geometrical and Chan–Paton action of Γ. The resulting field theories can be encoded in a quiver diagram, with nodes and arrows corresponding to gauge factors and chiral multiplets, and superpotential terms correspond to closed polygons formed by four arrows. When Γ acts on the Chan–Paton indices in the regular representation, the moduli space of the gauge theory is a symplectic quotient subject to F-term contraints from the superpotential, as obtained in explicit examples [21], generalizing the orbifold result. The string theory construction also suggests that these quiver diagrams also encode the homology of the resolution of X_{con}/Γ.

A general recipe to obtain the field theory on D3-branes at a general toric singularity X was proposed in [16] (see [6] for a more general discussion). It is based on realizing X as a partial resolution of a threefold quotient singularity $\mathbf{C}^3/\Gamma$ for suitable $\Gamma \subset SU(3)$. The construction requires a precise identification of the effect of the resolution on the field theory, which is a specific Higgs mechanism in which several gauge factors break to their diagonal subgroup, and some chiral multiplets become massive and disappear from the light spectrum. The detailed map has been worked out in some explicit examples, e.g., in [16, 1], and provides an explanation for the existence of quiver diagrams for non-orbifold singularities. They are obtained by joining nodes and deleting arrows in the quiver of the initial quotient singularity, as dictated by the Higgs mechanism in the field theory. It would be interesting to obtain a more precise characterization of these operations on quiver diagrams.

Another interesting question is related to the uniqueness of the field theory corresponding to a given singularity [6]. In some cases, different field theories may have isomorphic moduli spaces. Mathematically, the representation moduli of the corresponding quivers with relations are isomorphic. From the physical point of view, the equality of moduli spaces may in some cases be a reflection of Seiberg duality, a non-trivial infrared equivalence of seemingly different field theories.

2.2.2. ORIENTIFOLD PROJECTIONS Type IIB string theory is invariant under an operation Ω which reverses the orientation on the world-sheet. Hence it is possible to consider modding out type IIB configurations by Ω, possibly accompanied by a geometric involution g, also leaving the theory invariant [19]. The new configurations thus obtained are called orientifolds, and are characterized by the inclusion of non-orientable world-sheets in the string theory perturbative expansion.

We are interested in studying D3-branes at orientifold singularities, i.e., Ωg quotients of the system of D3-branes at singularities. The best studied case is again that of quotient singularities $\mathbf{C}^3/\Gamma$, and for the sake of clarity here we center on $\Gamma = \mathbf{Z_N}$ with the action of the generator θ of $\mathbf{Z}_N$ on $\mathbf{C}^3$ represented by $\gamma = \mathrm{diag}(e^{2\pi i t_1/N}, e^{2\pi i t_2/N}, e^{2\pi i t_3/N})$, with t_i integers defined modulo N, and $\sum_{a=1}^{3} t_a = 0 \bmod N$. Let us embed the action of θ on the Chan–Paton indices by a diagonal matrix $\gamma_{\theta,3}$ with N_k entries $e^{2\pi i\, k/N}$. The field theory one obtains has vector multiplets with gauge group $\prod_{i=1}^{N} U(N_i)$, and chiral multiplets $\Phi^a_{i,i+t_a}$, $a = 1,\dots,3$, $i = 1,\dots,N$ in the representation $(N_i, \overline{N}_{i+t_a})$.

The orientifold action may also be embedded in the space of Chan–Paton indices, through a matrix $\gamma_{\Omega g,3}$. The representations of Γ and Ωg are usually constrained from mutual consistency requirements [7]. Several solutions to these conditions are known (quite exhaustively in the two-fold case [2]), but a complete classification is lacking. In the following we center on a concrete example [8] where g inverts all coordinates in $\mathbf{C}^3$, and exchanges closed string fields in oppositely twisted sectors, i.e., has a non-trivial action on the homology cycles shrunk at the singularity. We also choose $\gamma_{\Omega g,3}$ symmetric and such that they exchange the eigenspaces of conjugate eigenvalues in $\gamma_{\theta,3}$. The action of $\overline{Z}_N$ on Chan–Paton indices is therefore constrained to form a real representation, $N_k = N_{-k}$.

The effect of the orientifold projection on the spectrum is as follows. Gauge factors associated to conjugate irreducible representations of $\overline{Z}_N$ are identified, and unitary gauge factors associated to real representations are reduced to their maximal orthogonal subgroups. Correspondingly, the chiral multiplets $\Phi^a_{i,i+t_a}$ and $\Phi^a_{-i-t_a,-i}$ are identified. When $i + t_a = -i$ the bifundamental field is projected down to a two-index antisymmetric representation of the unitary gauge factor after the projection.

The effect on the quiver diagram of $\mathbf{C}^3/\Gamma$ is an identification of nodes and arrows related by a $\mathbf{Z}_2$ action, with a specific prescription for nodes and arrows which are mapped to themselves (it would be interesting to characterize these mappings more precisely in order to allow the classification of resulting models). One may wonder about the meaning of the resulting quiver. String theory suggests it encodes the information about the homology of the resolved orientifold singularity. In fact, the string theory counting of homology cycles in the resolved space (by counting of twisted sector blow-up moduli) gives roughly speaking half the number encountered before the orientifold projection, due to the non-trivial action of g by exchanging oppositely twisted sectors. This agrees with the counting of nodes in the orientifolded quiver, which also gives half the number encountered before the orientifold projection.

We conclude by pointing out that orientifolds of non-orbifold singularities can be constructed as partial resolutions of orientifolds of quotient singularities. Preliminary results on some simple examples [17] indicate that the effect of the orientifold projection on the quiver of the non-orbifold singularity is also a $\mathbf{Z}_2$ involution, and that the orientifold singularity can be constructed as a symplectic quotient on the space of fields constrained by the F-terms conditions.

3. Particle Physics

In this section we discuss particular examples of singularities leading to interesting low-energy field theories, in that they resemble the structure of the (minimal supersymmetric) standard model (or some extension thereof) which we now review. Such models would provide phenomenologically interesting string theory vacua when embedded in a compact Calabi–Yau context.

3.1. Review of the (minimal supersymmetric) standard model

All known non-gravitational interactions between elementary particles are described by the quantum field theory known as the standard model. The simplest supersymmetric extension of this model contains vector multiplets with gauge group $SU(3) \times SU(2) \times U(1)$. It also contains a set of chiral multiplets transforming in three copies of the representation $(3,2)_{1/6}+(\overline{3},1)_{-2/3}+(\overline{3},1)_{1/3}+(1,2)_{1/2}+(1,1)_1$, where subscripts denote $U(1)$ charges. Successful breaking of the electroweak interactions requires also at least one chiral multiplet in the representation $(1,2)_{1/2} + (1,2)_{-1/2}$.

Interactions are encoded in a complicated superpotential, which in principle is the most general gauge invariant function of these chiral multiplets. We will skip these details since realistic phenomenology usually involves additional model-dependent assumptions, like additional global symmetries.

3.2. Realistic models from $\mathbf{Z}_3$ singularities

The replication of families is an intriguing feature of the above field theory. To reproduce it from branes at singularities, the case of $\mathbf{C}^3/\mathbf{Z}_3$ with the $\mathbf{Z}_3$ action defined by $(z_1,z_2,z_3) \to (e^{2\pi i/3}z_1, e^{2\pi i/3 z_2}, e^{2\pi i/3}z_3)$, is singled out, in that it leads to natural triplication. The two examples we are to consider are based on this singularity.

The first example we consider forms a subsector of the model considered in [14]. It is constructed by placing eleven D3-branes on the orientifold of the $\mathbf{C}^3/\mathbf{Z}_3$ singularity, introduced in Section 2.2.2. We choose $\gamma_{\theta,3}$=diag$(1,e^{2\pi i/3}\mathbf{1}_5,e^{-2\pi i/3}\mathbf{1}_5)$. Following our rules above, we obtain vector multiplets with gauge group $SU(5)$

(the $U(1)$ factor disappears from the light spectrum as explained in Section 2.1), and there are chiral multiplets in three copies of the representation $5 + \overline{10}$. This spectrum resembles the structure of $SU(5)$ grand unified theories, which reproduce the spectrum of the minimal supersymmetric standard model when an additonal field in the adjoint representation is present to break $SU(5)$ to the standard model group through the Higgs mechanism (an additional $5 + \overline{5}$ pair is further required for electroweak symmetry breaking). Unfortunately, all Higgs fields are absent in our field theory, which therefore is suggestive but not truly realistic.

The second example is based on the $\mathbf{C}^3/\mathbf{Z}_3$ orbifold (rather than orientifold) singularity, with Chan–Paton embedding given by $\gamma_{\theta,3}$=diag($\mathbf{1_3}$,$e^{2\pi i/3}\mathbf{1}_2$,$e^{-2\pi i/3}$). This embedding does not satisfy the tadpole cancellation condition stated in Section 2.1 (the character of the representation of $\mathbf{Z}_3$ is non-zero). The problem can be solved by introducing an additional set of D-branes, for instance D7$_a$-branes, with $a = 1, 2, 3$, wrapped on the complex surfaces $z_a = 0$. For non-compact spaces these additional branes are non-dynamical, but they contribute additional fields in the four-dimensional world-volume of the D3-branes, arising from open strings stretched between D3- and D7-branes. The $\mathbf{Z}_3$ acts on the space of D7-brane Chan–Paton indices, through matrices $\gamma_{\theta,7_a}$. The modified tadpole cancellation conditions are satisfied for the very symmetric choice $\mathrm{tr}\,\gamma_{\theta,7_a} = -\mathrm{tr}\,\gamma_{\theta,3}$, $a = 1, 2, 3$. Choosing e.g., $\gamma_{\theta,7_a} = \mathrm{diag}(e^{2\pi i/3}, e^{-2\pi i/3}\mathbf{1}_2)$ we obtain the following spectrum:

$$\begin{array}{rrl} 3-3 \text{ strings} & \text{Gauge group} & SU(3)\times SU(2)\times U(1) \\ & \text{Chiral multiplets} & 3\,[\,(3,2)_{1/6} + (1,2)_{1/2} + (\overline{3},1)_{-2/3}\,] \\ 3-7_a \text{ strings} & \text{Chiral multiplets} & (3,1)_{-1/3} + 2(1,2)_{-1/2} + \\ a=1,2,3 & & +2(\overline{3},1)_{1/3} + (1,1)_1\,. \end{array} \tag{3}$$

We have included the charges under the only non-anomalous linear combinations of the $U(1)$ factors in the original $U(3) \times U(2) \times U(1)$ gauge group. This $U(1)$ does not become massive and plays the role of hypercharge in the standard model we have just constructed.

Notice how close to the spectrum in Section 3.1 one can get using very simple singularities. Since such constructions would provide a rationale for the existence of three generations (one per complex transverse dimensions) and hypercharge assignments, we believe these examples illustrate the phenomenological interest of string theory compactifications with branes at singularities.

References

[1] C. Beasley, B. R. Greene, C. I. Lazaroiu, M. R. Plesser, *D3-branes on partial resolutions of abelian quotient singularities of Calabi–Yau threefolds*, Nucl.Phys. B566 (2000) 599-6.

[2] J. D. Blum, K. Intriligator, *New phases of string theory and 6-D RG fixed points via branes at orbifold singularities*, Nucl. Phys. B506 (1997) 199-22; J. Park,

A. M. Uranga, *A Note on superconformal N=2 theories and orientifolds*, Nucl. Phys. B542 (1999) 139-15.
[3] K. Dasgupta, S. Mukhi, *Brane constructions, conifolds and M theory*, Nucl. Phys. B551 (1999) 204-22.
[4] M. R. Douglas, B. R. Greene, D. R. Morrison, *Orbifold resolution by D-branes*, Nucl. Phys. B506 (1997) 84-10.
[5] M. R. Douglas, G. Moore, *D-branes, quivers, and ALE instantons*, hep-th/9603167.
[6] B. Feng, A. Hanany, Y.-H. He, *D-brane gauge theories from toric singularities and toric duality*, hep-th/0003085.
[7] E. G. Gimon, J. Polchinski, *Consistency conditions for orientifolds and d manifolds*, Phys. Rev. D54 (1996) 1667; E. G. Gimon, C. V. Johnson, *K3 orientifolds*, Nucl.Phys. B477 (1996) 715; A. Dabholkar, J. Park, *Strings on orientifolds*, Nucl. Phys. B477 (1996) 701.
[8] L. E. Ibáñez, R. Rabadán, A. M. Uranga, *Anomalous U(1)'s in type I and type IIB $D = 4$, N=1 string vacua*, Nucl. Phys. B542 (1999) 112-13.
[9] C. V. Johnson, R. C. Myers, *Aspects of type IIB theory on ALE spaces*, Phys. Rev. D55 (1997) 6382.
[10] I. R. Klebanov, E. Witten, *Superconformal field theory on three-branes at a Calabi–Yau singularity*, Nucl. Phys. B536 (1998) 199-21.
[11] P. B. Kronheimer, *The construction of ALE spaces as hyper-Kähler quotients*, J. Diff. Geom. 29 (1989) 665.
[12] A. Lawrence, N. Nekrasov, C. Vafa, *On conformal field theories in four-dimensions*, Nucl. Phys. B533 (1998) 199-20.
[13] R. G. Leigh, M. Rozali, *Brane boxes, anomalies, bending and tadpoles*, Phys. Rev. D59 (1999) 026004.
[14] J. Lykken, E. Poppitz, S. P. Trivedi, *Branes with GUTs and supersymmetry breaking*, Nucl. Phys. B543 (1999) 105.
[15] J. McKay, *Graphs, singularities and finite groups*, Proc. Symp. Pure Math. 37 (1980) 183; M. Reid, *McKay correspondence*, alg-geom/9702016.
[16] D. R. Morrison, M. R. Plesser, *Nonspherical horizons*, Adv. Theor. Math. Phys. 3 (1999) 1-8.
[17] J. Park, R. Rabadán, A. M. Uranga, *Orientifolding the conifold*, hep-th/9907086.
[18] J. Polchinski, *TASI lectures on D-branes*, hep-th/9611050.
[19] A. Sagnotti in Cargese'87, *Non-perturbative quantum field theory*, ed. G. Mack *et al* (Pergamon Press 1988), pag. 521 *Some properties of open string theories.*
[20] A. V. Sardo Infirri, *Resolutions of orbifold singularities and the transportation problem on the McKay quiver*, alg-geom/9610005.
[21] R. von Unge, *Branes at generalized conifolds and toric geometry*, J. High E. Phys. 9902 (1999) 023; K. Oh, R. Tatar, *Branes at orbifolded conifold singularities and supersymmetric gauge field theories*, J. High E. Phys. 9910 (1999) 031.
[22] A. M. Uranga, *Brane configurations for branes at conifolds*, J. High Energy Phys. 9901(1999)022.

Theory Division,
C.E.R.N.,
CH-1211 Geneve 23, Switzerland
E-mail address: `Angel.Uranga@cern.ch`

From Symplectic Packing to Algebraic Geometry and Back

Paul Biran

Abstract. In this paper we survey various aspects of the symplectic packing problem and its relations to algebraic geometry, going through results of Gromov, McDuff, Polterovich and the author.

1. Introduction

1.1. Symplectic packing

Let (M^{2n}, Ω) be a $2n$-dimensional symplectic manifold with finite volume. Fix an integer $N \geq 1$ and consider *symplectic packing of* (M, Ω) *by* N *equal balls* of radius λ, that is symplectic embeddings

$$\varphi\colon \underbrace{B(\lambda) \coprod \cdots \coprod B(\lambda)}_{N \text{ times}} \to (M, \Omega) \tag{1}$$

of N disjoint copies of $B(\lambda)$ – the closed $2n$-dimensional ($2n = \dim M$) Euclidean ball of radius λ, endowed with the standard symplectic structure of $\mathbb{R}^{2n}$, $\omega_{\text{std}} = dx_1 \wedge dy_1 + \cdots + dx_n \wedge dy_n$.

While Darboux's theorem assures that such packings always exist for small enough λ's, when trying to increase the radii one runs into an obvious *volume obstruction*: φ being a symplectic embedding must also be volume preserving. We thus have:

$$N \operatorname{Vol} B(\lambda) = \operatorname{Vol}(\operatorname{Image}(\varphi)) \leq \operatorname{Vol}(M, \Omega), \quad \text{or equivalently} \quad \lambda^{2n} \leq \frac{\int_M \Omega^n}{\pi^n N}. \tag{2}$$

The symplectic packing problem is the following question: *Does the symplectic structure impose other obstructions on symplectic packings, beyond inequality (2)?*

1.2. Singularities of plane algebraic curves

Let $p_1, \ldots, p_N \in \mathbb{C}P^2$ be N general points in the complex projective plane. Consider *irreducible* algebraic curves $C \subset \mathbb{C}P^2$ which pass through $p_1, \ldots, p_N$ with given multiplicities $m_1, \ldots, m_N$. The following is a classical question in singularity theory: *What is the minimal possible degree of a curve* C *with the above prescribed singularities?* An old (and still open) conjecture of Nagata asserts that when $N \geq 9$ the degree of C must satisfy

$$\deg(C) \geq \frac{m_1 + \cdots + m_N}{\sqrt{N}}.$$

Surprisingly, this purely algebro-geometric problem is intimately related to the symplectic packing problem.

In this paper we shall survey the symplectic packing problem and its mutual relations with algebraic geometry and the conjecture of Nagata.

2. Packing Obstructions and Algebraic Geometry

First results in this direction were established by Gromov in 1985. Using his theory of pseudo-holomorphic curves he proved the following:

Theorem 2.1. (Gromov [14]) *If $B(\lambda_1) \coprod B(\lambda_2)$ embeds symplectically into $B^{2n}(R)$ then $\lambda_1^2 + \lambda_2^2 < R^2$.*

Notice that for $\lambda_1 = \lambda_2 = \lambda$ this inequality becomes $\lambda^2 < \frac{1}{2}R^2$, or when written differently:

$$2 \operatorname{Vol} B(\lambda) < \frac{1}{2^{n-1}} \operatorname{Vol} B^{2n}(R).$$

In other words, the maximal portion of the volume of $B^{2n}(R)$ that can be filled via two disjoint symplectic balls of equal radius is smaller than $\frac{1}{2^{n-1}}$. This is a purely symplectic phenomenon! Indeed it is not hard to prove that if one replaces the condition "symplectic" on the embedding φ by "volume preserving" then the problem becomes trivial in the sense that no obstructions beyond inequality (2) exist, namely an arbitrarily large portion of the volume of (M, Ω) can be filled by N disjoint copies of equal balls embedded via a volume preserving map. This holds for every manifold M and any $N \geq 1$.

Trying to formalize the study of the preceding phenomenon one introduces the following quantity:

$$v_N(M, \Omega) = \sup_{\lambda} \frac{N \operatorname{Vol} B(\lambda)}{\operatorname{Vol}(M, \Omega)},$$

where λ passes over all possible radii for which there exists a symplectic embedding φ as in (1) above. The number $v_N(M, \Omega)$ measures the "maximal" portion of the volume of M that can be symplectically packed by N equal balls. When $v_N(M, \Omega) = 1$ we say that (M, Ω) admits a *full symplectic packing* by N equal balls, while the case $v_N(M, \Omega)$ is referred to as a *packing obstruction.*

Generalizing Gromov's work, McDuff and Polterovich discovered the following interesting result:

Theorem 2.2. (McDuff-Polterovich [26]) *For each of the manifolds $B^4(1)$ and $\mathbb{C}P^2$ (both endowed with their standard symplectic structures) we have:*

N	1	2	3	4	5	6	7	8	9
v_N	1	$\frac{1}{2}$	$\frac{3}{4}$	1	$\frac{20}{25}$	$\frac{24}{25}$	$\frac{63}{64}$	$\frac{288}{289}$	1

Moreover, $v_N = 1$ *when* $N = k^2$.

2.1. Relations to algebraic geometry

McDuff and Polterovich discovered that the symplectic packing problem is intimately related to algebraic geometry. The first ingredient in building up this connection is the symplectic blowing-up construction. We shall only give a general overview of it here referring the reader to [26, 27] for more details.

The symplectic blowing-up construction is a symplectic version of the classical blowing-up procedure in algebraic geometry. It was discovered by Gromov [15] and further developed by Guillemin and Sternberg [16]. Topologically symplectic blowing-up is the same as in the algebraic framework, but it also defines a symplectic form on the blown-up manifold. The deep connection between this operation and symplectic packing was first established by McDuff in [23].

Topologically, blowing-up amounts to removing a point p from M and replacing it by the (complex) projectivization of the tangent space of M at p. The same construction also allows us to blow-up M at N distinct points $p_1, \ldots, p_N \in M$.

It turns out that in order to define a symplectic form on the blow-up of M at $p_1, \ldots, p_N$ one has to specify a symplectic embedding

$$\varphi\colon B(\lambda_1) \coprod \cdots \coprod B(\lambda_N) \to (M, \Omega)$$

which sends the centre of the ith ball to p_i for every i. Roughly speaking, once such an embedding is given, one cuts out from M the images of the embedded balls and collapses their boundaries to copies of $\mathbb{C}P^{n-1}$ (called exceptional divisors) using the Hopf map. If we denote by $\Theta\colon \widetilde{M} \to M$ the blow-up of M at $p_1, \ldots, p_N$, then the symplectic blowing-up construction defines a symplectic form $\widetilde{\Omega}$ on $\widetilde{M}$ such that $\widetilde{\Omega}|_{T(\Sigma_i)} = \lambda_i^2 \sigma_{\text{std}}$ and $\widetilde{\Omega} = \Theta^*\Omega$ outside the images of the embedded balls. Here we have denoted by Σ_i the exceptional divisor corresponding to p_i, $\Sigma_i = \Theta^{-1}(p_i) \approx \mathbb{C}P^{n-1}$, and by σ_{std} the standard symplectic structures of $\mathbb{C}P^{n-1}$ normalized so that the area of a projective line $\mathbb{C}P^1 \subset \mathbb{C}P^{n-1}$ is π. Writing $E_1, \ldots, E_N$ for the homology classes of the exceptional divisors in $H_{2n-2}(\widetilde{M})$ and $e_1, \ldots, e_N \in H^2(\widetilde{M})$ for their Poincaré duals we thus have:

$$[\widetilde{\Omega}] = \Theta^*[\Omega] - \pi\lambda_1^2 e_1 \ldots - \pi\lambda_1^2 e_N\,. \tag{3}$$

Conversely, given a symplectic form $\widetilde{\Omega}$ on the blow-up $\widetilde{M}$ of a manifold one can, by *blowing down*, obtain a symplectic form Ω on M, and a symplectic packing of (M, Ω) by balls. The radii of these balls are determined by $\widetilde{\Omega}|_{\Sigma_i}$ as can be easily seen from (3) above.

In the language of symplectic blow-ups, the problem of existence of symplectic packings with balls of given radii, is equivalent to determining how much it is possible to blow-up symplectically the manifold. In view of the blowing down construction this amounts to determination of those cohomology classes on the blow-up of the manifold which can be represented by symplectic forms. For example, in order to show that $\mathbb{C}P^2$ admits a symplectic packing by N equal balls of radius λ one has to show that the cohomology class

$$a = \pi l - \pi\lambda^2(e_1 + \cdots + e_N) \tag{4}$$

on the blow-up of $\mathbb{C}P^2$ has a symplectic representative. Here $l \in H^2(\mathbb{C}P^2, \mathbb{Z})$ stands for the (positive) generator.

Although there are no general tools to attack this type of problem on a general symplectic manifold, some information can be still obtained using tools from classical algebraic geometry. The idea is to produce a *Kähler* form (rather than just symplectic) in the given cohomology class. The main tool for this purpose is the following criterion which is due independently to Nakai and to Moishezon (see [17]). Henceforth we shall denote by $H^{(1,1)}(S, \mathbb{R}) = H^{(1,1)}(S) \cap H^2(S, \mathbb{R})$ the space of cohomology classes which can be represented by *real* $(1,1)$-forms on a given smooth algebraic variety S.

Theorem 2.3. (Nakai-Moishezon criterion) *Let S be a smooth algebraic surface and $a \in H^{(1,1)}(S, \mathbb{R})$. The class a can be represented by a Kähler form if and only if the following two conditions are satisfied:*

1. $\int_S a \cup a > 0$.
2. $\int_C a > 0$ *for every algebraic curve $C \subset S$.*

Note that in order to apply this criterion one has to have some information on *all* the possible homology classes $[C] \in H_2(S)$ which can be represented by algebraic curves $C \subset S$. In general this might be hard, however when S has a positive anti-canonical class a great deal of information is available on the relevant classes $[C] \in H_2(S)$. Such surfaces S are called *Fano* (or *del-Pezzo*) and it is not hard to check that blow-ups of $\mathbb{C}P^2$ at no more than eight points belong to this category. It turns out that for Fano surfaces one can replace the second condition in the Nakai-Moishezon criterion by the following much weaker one: "$\int_C a > 0$ *for every exceptional rational curve $C \subset S$*". By an *exceptional rational curve* we mean a rational algebraic curve with self intersection -1. Furthermore, it is a classical fact that Fano surfaces have only finitely many such curves and a complete list of their homology classes is available (see [8]). Therefore in this case the Nakai-Moishezon criterion boils down to finitely many explicit inequalities. All this leads to a precise computation of the maximal λ for which the class a in (4) is Kähler. In particular this gives a lower bound on the supremum of all λ for which the class in (4) is symplectic, hence a lower bound on $v_N(\mathbb{C}P^2)$ when $N \leq 8$.

To obtain an upper bound on λ (that is, a packing obstruction) it is not possible to use algebraic geometry any longer since a priori it is not clear that the symplectic forms obtained by the symplectic blowing-up operation are always Kähler. Nevertheless, using the theory of pseudo-holomorphic curves McDuff and Polterovich managed to obtain limitations on the possible radii λ. Roughly speaking, the idea is that the symplectic forms obtained from blowing-up can be deformed to Kähler forms via symplectic deformations. The point is that certain types of algebraic curves (such as exceptional rational curves) persist under deformations of the symplectic structure and so one can still find symplectic surfaces in the same homology classes as the rational exceptional curves for every symplectic blow-up of $\mathbb{C}P^2$. The blown-up form $\widetilde{\Omega}$ must have positive area on such surfaces and so one gets from each such symplectic surface an inequality which bounds λ from above. In other words, the same set of inequalities on λ as in the algebraic case applies here also to the (more general) symplectic case. As a result, the upper and lower bounds mentioned above coincide. This leads to the *precise* values of v_N which appear in Theorem 2.2.

The case $N > 9$ could not be attacked by algebraic geometry since the blow-up of $\mathbb{C}P^2$ at $N > 9$ is not Fano anymore and much less is known on the possible homology classes of irreducible algebraic curves on such surfaces.

Notice that in order to prove that $\mathbb{C}P^2$ admits a full symplectic packing by N equal balls one has to show that the cohomology class

$$a_\epsilon = l - (\frac{1}{\sqrt{N}} - \epsilon)(e_1 + \cdots + e_N) \tag{5}$$

has a symplectic representative for every $\epsilon > 0$. McDuff and Polterovich discovered that surprisingly this would follow from the old conjecture of Nagata made in the early 1950s which (in an equivalent formulation) asserts that for every $N \geq 9$ the cohomology class a_ϵ has a Kähler representative for every $\epsilon > 0$.

We shall discuss Nagata's conjecture in some more detail in Section 5 below. For the moment let us only mention that Nagata's conjecture holds true in the case $N = k^2$ due to a very simple argument, hence $v_N(\mathbb{C}P^2) = 1$ when $N = k^2$.

3. Existence of Full Packings: First Iteration

Although Nagata's conjecture is still not settled the following was proved in [2]:

Theorem 3.1. *For every $N \geq 9$ we have $v_N(B^4(1)) = v_N(\mathbb{C}P^2) = 1$. That is, both $B^4(1)$ and $\mathbb{C}P^2$ admit a full symplectic packing by N equal balls for every $N \geq 9$.*

Let us outline the main ideas leading to this theorem. As already mentioned, the existence of symplectic packing is equivalent to existence of symplectic forms representing certain cohomology classes on the blow-up of the manifold. We therefore need a version of the Nakai-Moishezon criterion which will be valid in the symplectic category. The main tool for establishing such a criterion (in dimension 4) is the theory of *Seiberg-Witten invariants* and their interpretation, due to Taubes, as *Gromov* invariants in the language of pseudo-holomorphic curves. The reader is referred to [31] and [24] for excellent presentations of this theory and its applications to symplectic geometry.

In order to state our criterion we need to introduce a class $\mathcal{C}$ of symplectic 4-manifolds for which it is valid. The precise definition of the class $\mathcal{C}$ involves knowledge of the the theory of Seiberg-Witten invariants which we shall not attempt to give here, but for the purpose of symplectic packing it is enough to mention that the class $\mathcal{C}$ contains the following types of symplectic 4-manifolds:

- Manifolds with $b_2^+ = 1$ and $b_1 = 0$.
- Ruled surfaces.
- If $(M, \Omega) \in \mathcal{C}$ then any (symplectic) blow-up of (M, Ω) also belongs to $\mathcal{C}$.

Next we need the notion of *exceptional classes.* In analogy to algebraic geometry we call a homology class $A \in H_2(M, \mathbb{Z})$ exceptional if it can be represented by a symplectic 2-sphere $\Sigma \subset (M, \Omega)$ with $\Sigma \cdot \Sigma = -1$. It is a standard fact (that can be proved using Gromov's compactness theorem) that if Ω_0 and Ω_1 are symplectic forms that can be joined by a path of symplectic forms $\{\Omega_t\}$ then the set of exceptional classes for Ω_0 and and for Ω_1 coincide. Thus the set of Ω-exceptional classes $\mathcal{E}_\Omega \subset H_2(M, \mathbb{Z})$ depends only on the deformation class of Ω.

Our symplectic version of the Nakai-Moishezon criterion is (see [1]):

Theorem 3.2. *Let M be a closed 4-manifold and $\alpha \in H^2(M, \mathbb{R})$. Suppose that M admits a symplectic form Ω such that the following conditions are satisfied:*

1. *(M, Ω) is in the class $\mathcal{C}$.*
2. *$\int_M [\Omega] \cup \alpha > 0$ and $\int_M \alpha \cup \alpha > 0$.*
3. *$\alpha(E) \geq 0$ for every exceptional class $E \in \mathcal{E}_\Omega$.*

Then arbitrarily close to α in $H^2(M, \mathbb{R})$ there exist cohomology classes which can be represented by symplectic forms. Moreover, these symplectic forms may be assumed to be in the deformation class of Ω.

The proof of this theorem consists of two main ingredients. The first one is the *inflation* procedure which was introduced into symplectic geometry by Lalonde and McDuff (see [20, 21]) in the context of the problem of symplectic isotopies (see Section 4.1 below). This procedure can be formulated as follows: *If $C \subset (M^4, \Omega)$ is a 2-dimensional symplectic submanifold with $C \cdot C \geq 0$ then there exists a closed 2-form ρ, supported arbitrarily close to C, whose cohomology class is Poincaré dual to $[C]$ and such that for every $t > 0$, $s \geq 0$ the form $\Omega_{t,s} = t\Omega + s\rho$ is symplectic.*

In our context this is an important tool for obtaining symplectic cohomology classes (i.e. classes which can be represented by symplectic forms). Indeed by taking $t > 0$ smaller and smaller the cohomology class $[\Omega_{t,s}]$ remains symplectic and becomes closer and closer to the class Poincaré dual to $s[C]$. Therefore, in order to prove that a given cohomology class, say a, can be approximated by symplectic cohomology classes it is enough to produce a symplectic submanifold $C \subset (M, \Omega)$ with non-negative self intersection whose homology class is Poincaré dual to a positive multiple of a. We would like to point out that a similar principle applies in the algebraic category as well (see Section 5.1 below).

At present the problem of existence of symplectic hypersurfaces in given homology classes is in general out of reach, however in dimension 4 for manifolds which belong to the class $\mathcal{C}$ one can apply the machinery of Taubes-Seiberg-Witten theory of Gromov invariants to obtain the wanted submanifold C as a pseudo-holomorphic curve. Condition 2 of the theorem is exactly the technical assumption needed in order to force these invariants to be non-trivial on manifolds in the class $\mathcal{C}$. In general, non-triviality of the Gromov invariants gives a reducible pseudo-holomorphic curve C which may have some components of negative self intersection. If such components appear the inflation cannot be performed. It is this point at which condition 3 of Theorem 3.2 comes into play. This assumption, it turns out, assures that C has only one irreducible component. We refer the reader to [24, 25] for more details on the structure of the components of curves arising from non-zero Seiberg-Witten Gromov invariants (see also [2]).

Returning to symplectic packings, Theorem 3.2 implies that the "main" obstruction (beyond having positive volume) for a cohomology class α to carry a symplectic representative comes from the exceptional classes $E \in \mathcal{E}_\Omega$ (on which α must be positive). Applying this to the blow-up of $\mathbb{C}P^2$ at $N \geq 9$ points, an easy computation shows that when α is the class a_ϵ from (5) the restrictions coming from exceptional classes (i.e. $a_\epsilon(E) > 0$ for every $E \in \mathcal{E}_\Omega$) are *weaker* than the volume inequality $\int_M a_\epsilon \cup a_\epsilon > 0$. Consequently $v_N(\mathbb{C}P^2) = 1$ for every $N \geq 9$.

4. Other Directions

Apart from the problem of existence of packings there are other important related aspects on which extensive research has been carried out. We shall give here a brief overview of some of them without any attempt to be complete.

4.1. Isotopies of balls

Two symplectic packings

$$\varphi_0, \varphi_1 \colon B(\lambda_1) \coprod \cdots \coprod B(\lambda_N) \to (M, \Omega) \tag{6}$$

are called symplectically isotopic if there exists a smooth family $\{\varphi_t\}$ of symplectic packings

$$\varphi_t \colon B(\lambda_1) \coprod \cdots \coprod B(\lambda_N) \to (M, \Omega), \quad 0 \leq t \leq 1$$

which interpolates between φ_0 and φ_1.

The main question in this context is: *Given N and radii $\lambda_1, \dots, \lambda_N$, are every two symplectic packings of (M, Ω) by balls of these radii symplectically isotopic?* Or put in a different way, *Is the space of symplectic packings of (M, Ω) by balls of these radii connected?*

This problem can be easily translated to an equivalent question on uniqueness of the symplectic blowing-up construction: *Does symplectic blowing-up really depend on the symplectic packing φ used to define it, or in fact solely on the "weights" $\lambda_1, \dots, \lambda_N$?* In other words, *If $\widetilde{\Omega}_0$ and $\widetilde{\Omega}_1$ are two symplectic forms on the blow-up $\Theta \colon \widetilde{M} \to M$ of M, constructed using symplectic packings φ_0 and φ_1 as in (6), are $\widetilde{\Omega}_0$ and $\widetilde{\Omega}_1$ symplectomorphic? Are they isotopic?*

First results in this direction were obtained by McDuff [23] for the case of $\mathbb{C}P^2$. These results have been later generalized to other cases by Lalonde [20] and later on by the author [5]. At present, the most general result concerning the uniqueness problem is the following theorem due to McDuff [25]:

Theorem 4.1. (McDuff) *Let (M^4, Ω) be a closed symplectic 4-manifold in the class $\mathcal{C}$. Then any two cohomologous symplectic forms $\widetilde{\Omega}_0$ and $\widetilde{\Omega}_1$ obtained from symplectic blowing up are isotopic. In particular, any two symplectic packings of such a manifold by balls of given radii are symplectically isotopic.*

The arguments used to prove this theorem are somewhat parallel to those used for the proof of Theorem 3.2, the main techniques being the inflation procedure and Taubes-Seiberg-Witten theory.

At present the problem of isotopies is still open for symplectic manifolds which *do not* belong to the class $\mathcal{C}$ (e.g. $\mathbb{T}^4$). In higher dimension than 4 the problem is completely open for *all* manifolds (see Section 7.1 below).

4.2. Explicit packing constructions

Another interesting direction deals with explicit packing constructions that realize the maximal values of v_N. The methods proving existence of symplectic packings are very indirect, thus explicit constructions are important in order to gain intuition about how these symplectic embeddings really look.

First such constructions were obtained by Karshon [18] for $\mathbb{C}P^n$ with $1 \leq N \leq n+1$ balls. Traynor [32] found explicit constructions realizing full packings of $\mathbb{C}P^n$ by k^n equal balls, the maximal packings of $\mathbb{C}P^2$ by $1 \leq N \leq 6$ equal balls, as well as some constructions for other manifolds (see also [19] for some generalizations and [28] for other types of packings). All these constructions are based on the fact that $\mathbb{C}P^n$ is a toric manifold and make use of its moment map.

To the best of the author's knowledge no explicit constructions for the *maximal* packings of $\mathbb{C}P^2$ by 7, 8 or any $N > 9$ ($N \neq k^2$) equal balls are known.

5. Nagata's Conjecture

In the early 1950's Nagata made the following conjecture regarding singularities of plane algebraic curves (see [29]):

Conjecture 5.1. (Nagata) *Let $p_1, \dots, p_N \in \mathbb{C}P^2$ be $N \geq 9$ very general[1] points. Then for every algebraic curve $C \subset \mathbb{C}P^2$ the following inequality holds:*

$$\deg(C) \geq \frac{\operatorname{mult}_{p_1} C + \cdots + \operatorname{mult}_{p_N} C}{\sqrt{N}}.$$

[1] By very general we mean that $(p_1, \dots, p_N)$ is allowed to vary in a subset of the configuration space $\mathcal{C}_N = \{(x_1, \dots, x_N) \mid x_i \neq x_j \ \forall i \neq j\}$ whose complement contains at most a countable union of proper subvarieties.

This conjecture can be easily translated via the Nakai-Moishezon criterion to a statement about the *Kähler cone* of blow-ups of $\mathbb{C}P^2$. By the Kähler cone of an algebraic manifold S we mean the subset $\mathcal{K}_S \subset H^{(1,1)}(S,\mathbb{R})$ of all cohomology classes a which admit a Kähler representative. It is also relevant to consider the closure $\overline{\mathcal{K}}_S$ of $\mathcal{K}_S$, and we shall call classes in $\overline{\mathcal{K}}_S$ by the name *semi-Kähler.*

Returning to blow-ups $\Theta\colon V_N \to \mathbb{C}P^2$ of $\mathbb{C}P^2$ at N points we have:

$$H^{(1,1)}(V_N,\mathbb{R}) = H^2(V_N,\mathbb{R}) = \mathbb{R}l \oplus \mathbb{R}e_1 \oplus \cdots \oplus \mathbb{R}e_N \quad \text{and}$$
$$H_2(V_N,\mathbb{R}) = \mathbb{R}L \oplus \mathbb{R}E_1 \oplus \cdots \oplus \mathbb{R}E_N\,.$$

Here, the E_i's denote the homology classes of the exceptional divisors, L the homology class of the proper transform of a projective line which does not pass through the blown-up points, and $l, e_1, \dots, e_N$ the Poincaré duals to $L, E_1, \dots, E_N$. Note that if $C \subset \mathbb{C}P^2$ is a an algebraic curve, then its proper transform $\overline{C} \subset V_N$ lies in the homology class

$$[\overline{C}] = \deg(C)L - \mathrm{mult}_{p_1}(C)E_1 - \cdots - \mathrm{mult}_{p_N}(C)E_N\,.$$

With these notations, Nagata's conjecture is equivalent to saying that the cohomology class

$$a = l - \frac{1}{\sqrt{N}}(e_1 + \cdots + e_N)$$

is non-negative when evaluated on each homology class $[\overline{C}] \in H_2(V_N)$ that can be represented by an algebraic curve. Adding this to the fact that $a \cdot a = 0$, it follows from the Nakai-Moishezon criterion that Nagata's conjecture is equivalent to the class a being semi-Kähler.

5.1. The special case $N = k^2$

Interestingly, Nagata's conjecture can be easily confirmed in case the number of points N is a square. This is based on the following simple observation: *Let V be a smooth surface and $D \subset V$ an irreducible curve with non-negative self intersection. Then the cohomology class Poincaré dual to $[D]$ lies in $\overline{\mathcal{K}}_V$.* This follows easily from Nakai-Moishezon criterion and the irreducibility of C.

Now let $C \subset \mathbb{C}P^2$ be a smooth curve of degree $k > 0$ and choose $N = k^2$ points $p_1, \dots, p_N$ on C. Let $\Theta\colon V_N \to \mathbb{C}P^2$ be the blow-up of $\mathbb{C}P^2$ at $p_1, \dots, p_N$ and denote by D the proper transform of C in V_N. Since D is irreducible and $D \cdot D = 0$, the class Poincaré dual to D (namely $kl - (e_1 + \cdots + e_N)$) is semi-Kähler, which is the statement of Nagata's conjecture for $N = k^2$. The fact that we have chosen the points $p_1, \dots, p_N$ to lie in a specific position rather than a very general one is not restrictive. Indeed, it is not hard to show that if Nagata's conjecture holds for a specific choice of points $p_1, \dots, p_N$, then it continues to hold also for a very general choice of points.

5.2. Numerical invariants of the Kähler cone

Let X be an algebraic manifold and $a \in H^{(1,1)}(X, \mathbb{R})$. Given real numbers $\lambda_1, \ldots, \lambda_N$ we say that the vector $(X, a; \lambda_1, \ldots, \lambda_N)$ is Kähler (resp. semi-Kähler) if there exist N points $p_1, \ldots, p_N \in X$ such that the cohomology class $\Theta^* a - \lambda_1 e_1 - \cdots - \lambda_N e_N$ lies in $\mathcal{K}_{\widetilde{X}}$ (resp. $\overline{\mathcal{K}}_{\widetilde{X}}$), where $\Theta\colon \widetilde{X} \to X$ is the blow-up of X at $p_1, \ldots, p_N$. Now, given a class $a \in \mathcal{K}_X$ one can introduce the following quantity which we call the *Nth remainder* of a:

$$\mathcal{R}_N(a) = 1 - \frac{\sup_\lambda \{N\lambda^n \mid a_\lambda = (X, a; \lambda, \ldots, \lambda) \text{ is Kähler}\}}{\int_X a^n},$$

where $n = \dim_{\mathbb{C}} X$. The numbers $\mathcal{R}_N(a)$ are analogous to the quantities introduced in Section 2 and play the role of $1 - v_N$ in the algebraic category. Similar quantities have been previously introduced in algebraic geometry and are commonly called *Seshadri constants* (see [7, 11] and [12]). Denoting by $l \in H^2(\mathbb{C}P^2, \mathbb{Z})$ the positive generator, Nagata's conjecture can now be reformulated as:

$$\mathcal{R}_N(l) = 0 \quad \textit{for every} \quad N \geq 9.$$

As already mentioned, Nagata's conjecture is still far from being settled (except for the case $N = k^2$). This makes it interesting to try to bound $\mathcal{R}_N(l)$ from above, or to find asymptotics on $R_N(l)$. For example:

Theorem 5.2. **(Xu** [33]**)** $\mathcal{R}_N(l) \leq \frac{1}{N}$ *for every* N.

At present, this seems to be the best general asymptotic on $\mathcal{R}_N(l)$, however for some special families of Ns it can be improved quadratically. The following was proved in [3]:

Theorem 5.3. *Let* $a, r \in \mathbb{N}$.

1. *For* $N = a^2r^2 + 2r$, $\mathcal{R}_N(l) \leq \frac{1}{(a^2r+1)^2}$.
2. *For* $N = a^2l^2 - 2l$, $\mathcal{R}_N(l) \leq \frac{1}{(a^2r-1)^2}$.
3. *For* $N = a^2r^2 + r$ *with* $ar \geq 3$, $\mathcal{R}_N(l) \leq \frac{1}{(2a^2r+1)^2}$.

This result was obtained using a recursive algorithmic procedure for constructing Kähler classes on algebraic manifolds. Although this algorithm has purely algebro-geometric foundations it originated from very simple ideas coming from symplectic geometry which we shall now explain.

5.3. An algorithm for constructing Kähler classes on algebraic manifolds

The first ingredient in our algorithm is the following general theorem from [3]:

Theorem 5.4. *Let* X^n *be a smooth algebraic variety of (complex) dimension* n *and let* $a \in H^{(1,1)}(X, \mathbb{R})$. *Suppose that:*

1. $(X, a; m)$ *is Kähler (resp. semi-Kähler) for some* $m \in \mathbb{R}$.
2. $(\mathbb{C}P^n, ml; \alpha_1, \ldots, \alpha_r)$ *is semi-Kähler.*

Then $(X, a; \alpha_1, \ldots, \alpha_r)$ *is Kähler (resp. semi-Kähler).*

As a corollary we get:

Corollary 5.5. *Suppose that* $(\mathbb{C}P^n, dl; m_1, \ldots, m_k, m)$ *is Kähler (resp. semi-Kähler) and* $(\mathbb{C}P^n, ml; \alpha_1, \ldots, \alpha_r)$ *is semi-Kähler. Then* $(\mathbb{C}P^n, dl; m_1, \ldots, m_k, \alpha_1, \ldots, \alpha_r)$ *is Kähler (resp. semi-Kähler).*

This means that one can reduce questions about the Kähler cone of blow-ups of $\mathbb{C}P^n$ to the same questions on blow-ups of $\mathbb{C}P^n$ at *fewer* points.

Before we describe the algorithm itself let us explain the symplectic rationale behind Theorem 5.4. We remark in advance that the following is not a rigorous mathematical argument but just a sequence of intuitive ideas which are supposed to explain how one can guess that such a statement should be true. The mathematical proof can be carried out using purely algebro-geometric techniques (see [3]).

To start with, note that the first condition of Theorem 5.4 means that if Ω is a symplectic form with $[\Omega] = a$, then (X, Ω) admits a symplectic embedding of a ball of radius $\sqrt{\frac{m}{\pi}}$,

$$\varphi \colon B\left(\sqrt{\frac{m}{\pi}}\right) \to (X, \Omega).$$

The second assumption means that $\mathbb{C}P^n$ endowed with the symplectic form $m\sigma_{\text{std}}$ admits a symplectic packing by r balls of radii $\sqrt{\alpha_1}, \ldots, \sqrt{\alpha_r}$. Here, σ_{std} stands for the standard symplectic form of $\mathbb{C}P^n$, normalized so that $[\sigma_{\text{std}}] = \pi l \in H^2(\mathbb{C}P^n, \mathbb{R})$. By a standard procedure in symplectic geometry (see [26]) we may assume that all these r balls lie in the complement of some complex hyperplane $H \approx \mathbb{C}P^{n-1}$. Now, it is well known that $(\mathbb{C}P^n \setminus \mathbb{C}P^{n-1}, m\sigma_{\text{std}})$ is symplectomorphic to $\operatorname{Int} B(\sqrt{m})$. Therefore we get a symplectic packing of $B(\sqrt{m})$ by r balls of radii $\sqrt{\alpha_1}, \ldots, \sqrt{\alpha_r}$. Rescaling everything we obtain a symplectic packing

$$\psi \colon B\left(\sqrt{\frac{\alpha_1}{\pi}}\right) \coprod \cdots \coprod B\left(\sqrt{\frac{\alpha_r}{\pi}}\right) \to B\left(\sqrt{\frac{m}{\pi}}\right).$$

The composition $\varphi \circ \psi$ will give us a symplectic packing of (X, Ω) by r balls of radii $\sqrt{\frac{\alpha_1}{\pi}}, \ldots, \sqrt{\frac{\alpha_1}{\pi}}$. This in turn implies that the cohomology class

$$\Theta^* a - \alpha_1 e_1 - \cdots - \alpha_r e_r \tag{7}$$

on the blow-up $\Theta \colon \widetilde{X} \to X$ at r points has a symplectic representative. Of course, this still does not mean that this class admits a Kähler representative, however since everything in this "argument" consisted of Kähler classes it seems reasonable to expect the class in (7) to be also Kähler. It turns out that this intuitive argument can be translated to a rigorous algebro-geometric proof using a so-called *degeneration* argument (see [3]. See also [30] and [6] for more about degenerations and their applications).

In this context it is worth mentioning here Lazarsfeld's work [22] on Seshadri constants of Abelian varieties which is also based on a simple "symplectic packing" argument.

The second ingredient in our algorithm is the *Cremona* group action on the Kähler cone of blow-ups of $\mathbb{C}P^2$. Let $\Theta \colon V_N \to \mathbb{C}P^2$ be a blow-up of $\mathbb{C}P^2$

at $N \geq 3$ points. The Cremona group is a group of reflections which acts on $H^{(1,1)}(V_N, \mathbb{R})$. A very important feature of this action is that it preserves the Kähler cone $\mathcal{K}_{V_N} \subset H^{(1,1)}(V_N, \mathbb{R})$. The reader is referred to [10] and to [3] for the definition and basic properties of the action of this group.

In order to detect Kähler (or semi-Kähler) classes on V_N we use Theorem 5.4 and the Cremona action successively until we arrive at a vector on which we can easily verify Kählerness. Here is a simple example which illustrates how the algorithm works. For brevity let us write

$$(\mathbb{C}P^2, dl; \alpha_1^{\times r_1}, \dots, \alpha_k^{\times r_k}) = (\mathbb{C}P^2, dl; \underbrace{\alpha_1, \dots, \alpha_1}_{r_1 \text{ times}}, \dots, \underbrace{\alpha_k, \dots, \alpha_k}_{r_k \text{ times}}).$$

Consider the case $N = 15$. We shall show now that the vector $v = (\mathbb{C}P^2, 15l; 4^{\times 14})$ is semi-Kähler. Note that this would imply that

$$\mathcal{R}_{15}(l) \leq 1 - \frac{14 \cdot 4^2}{15^2} = \frac{1}{225}.$$

To start with, recall that for every $k > 0$ the vector $(\mathbb{C}P^2, kl; 1^{\times k^2})$ is semi-Kähler, hence for every $m > 0$, $(\mathbb{C}P^2, mkl; m^{\times k^2})$ is also semi-Kähler. In particular, $(\mathbb{C}P^2, 8l; 4^{\times 4})$ is semi-Kähler. Therefore according to Corollary 5.5, in order to show that our vector v is semi-Kähler it is enough to show that the vector $v_1' = (\mathbb{C}P^2, 15l; 4^{\times 10}, 8)$ is semi-Kähler. Applying suitable Cremona transformations to this vector we obtain the vector $v_1 = (\mathbb{C}P^2, 10l; 3^{\times 11})$. Since the Cremona group preserves semi-Kähler classes, v_1 is semi-Kähler if and only if v_1' is. Note that the vector v_1 is "simpler" then v_1' in the sense that its entries are smaller than those of v_1'. Now, in order to show that v_1 is semi-Kähler we decompose it into $v_2' = (\mathbb{C}P^2, 10l; 3^{\times 7}, 6)$ and $(\mathbb{C}P^2, 6l; 3^{\times 4})$. The latter being semi-Kähler, it is enough by Corollary 5.5 to show that v_2' is semi-Kähler. Applying suitable Cremona transformations to v_2' we get the vector $v_2 = (\mathbb{C}P^2, l; 0^{\times 8})$. This vector is obviously semi-Kähler since it stands for the cohomology class that is Poincaré dual to a projective line on V_8 not passing through the exceptional divisors. All the above imply now that our original vector v is semi-Kähler.

Similar applications of the algorithm lead to the asymptotics of Theorem 5.3. In fact this algorithm gives rise to a mysterious pattern which links upper bounds on $\mathcal{R}_N(l)$ with *continued fractions* expansions of the number $\sqrt{N}$ (see [3]).

Another result of a successive application of Corollary 5.5 is the following, somewhat amusing, corollary:

Corollary 5.6. *If Nagata's conjecture holds for* N_1 *and* N_2 *then it holds also for* $N_1 N_2$.

It is of course a pity that the product of two squares is also a square.

5.4. Nagata's conjecture via ellipsoids

Recall that our algorithm for producing Kähler classes on blow-ups of $\mathbb{C}P^2$ (or more generally on any algebraic manifold) originated from the very simple symplectic picture that if we can embed symplectically a ball B into (X, Ω), then any packing of this ball would give rise also to a packing of M by balls of the same sizes. Note that the role played by the ball B is merely auxiliary and in the end we are interested only in the balls which pack (X, Ω).

In view of this there is no a priori reason not to replace B by a different manifold or shape. In fact it is reasonable to expect that doing so may lead to sharper estimates on $\mathcal{R}_N(l)$. Of course balls have the important advantage in that they have a nice characteristic foliation on the boundary thus giving rise to the symplectic blowing-up construction. Therefore, good candidates to replace the ball B seem to be ellipsoids. By a *symplectic ellipsoid with multi-radius* $\underline{r} = (r_1, \dots, r_n)$ we mean the following subset of $\mathbb{C}^n$

$$E(\underline{r}) = \left\{ z \in \mathbb{C}^n \;\middle|\; \sum_{i=1}^{n} \frac{|z_i|^2}{r_i^2} \leq 1 \right\},$$

endowed with the standard symplectic form of $\mathbb{C}^n$. Now let $\underline{w} = (w_1, \dots, w_n)$ be a vector of integral weights and $m \in \mathbb{R}$. As before, if we can embed symplectically the ellipsoid $E(m\underline{w})$ into (X, Ω), then any symplectic packing

$$\varphi \colon B(\alpha_1) \coprod \cdots \coprod B(\alpha_r) \to E(m\underline{w}) \tag{8}$$

would give rise to a symplectic packing of (X, Ω) by the same balls.

It seems reasonable that this scheme would translate well into pure algebro-geometric terms. The idea is that precisely as $\mathbb{C}P^n$ is the algebro-geometric analogue of a symplectic ball, *weighted projective spaces* should correspond to symplectic ellipsoids. Weighted projective spaces are defined as follows: given a vector of integral weights $\underline{w} = (w_1, \dots, w_n)$, define a $\mathbb{C}^*$-action on $\mathbb{C}^{n+1} \setminus 0$ by $\lambda \cdot (z_0, \dots, z_n) = (z_0, \lambda^{w_1} z_1, \dots, \lambda^{w_n} z_n)$ for every $\lambda \in \mathbb{C}^*$. The quotient space is a singular (unless $w_i = 1$ for all i) toric algebraic variety which we denote by $\mathbb{C}P^n(\underline{w})$. Existence of a packing φ as in (8) should translate here to *ampleness* of some divisor class[2] on blow-ups of $\mathbb{C}P^n(\underline{w})$ at r points.

The next step is a bit more involved. We have to translate the fact that (X, Ω) admits a symplectic embedding of an ellipsoid $E(m\underline{w})$. For this end we would like to define a weighted blowing-up of X at a point $p \in X$ with weights $\underline{w}$. The problem is that this procedure is not well defined since there does not exist any canonical weighted-action of $\mathbb{C}^*$ on the tangent space $T_p(X)$. In other words, such an action depends on a choice of coordinates (unless $\underline{w} = (1, \dots, 1)$ of course). Nevertheless, when X is a toric variety this difficulty should be tractable and one expects to obtain a weighted blown-up variety $\mathrm{Bl}^{(\underline{w})}(X)$.

If all the above is indeed feasible then one would expect a theorem of the following type to hold:

[2] We cannot work here with Kähler classes for these varieties are not smooth manifolds. Nevertheless the notion of ample divisor exits and essentially the same theory (with some modifications) applies to them as in the smooth case.

If $(\mathrm{Bl}^{(\underline{w})}(X), a; m)$ *is ample (resp. semi-ample) and* $(\mathbb{C}P^n(\underline{w}), ml; \alpha_1, \dots, \alpha_r)$ *is ample (resp. semi-ample) then* $(X, a; \alpha_1, \dots, \alpha_r)$ *is also ample (resp. semi-ample).* Here a is a given divisor class on X and l is a suitable hyperplane in $\mathbb{C}P^n(\underline{w})$.

A successive application of this splitting procedure in combination with the Cremona group action (on blow-ups of weighted projective planes) would hopefully improve the existing asymptotics on the remainders $\mathcal{R}_N(l)$ of $\mathbb{C}P^2$ and perhaps shed some new light on Nagata's conjecture.

6. Back to Symplectic Packing: a Stability Phenomenon

Let S be a smooth algebraic surface and $C \subset S$ a smooth irreducible curve. Consider the ruled surface

$$P_{C/S} = \mathbb{P}(N_{C/S} \oplus \mathcal{O}_C) \xrightarrow{p_C} C,$$

where $N_{C/S} \to C$ is the normal bundle of C in S. Inside the surface $P_{C/S}$ we have the curve $Z_{C/S} = \mathbb{P}(N_{C/S} \oplus 0)$. Denote by $z_C, f \in H^{(1,1)}(P_{C/S}, \mathbb{R})$ the Poincaré duals to $[Z_{C/S}]$ and to the fibre of $p_C\colon P_{C/S} \to C$ respectively.

The following theorem allows us to reduce questions on the Kähler cone of blow-ups of an arbitrary algebraic surface to the same questions on blow-ups of ruled surfaces. It can be proved using a degeneration argument similarly to Theorem 5.4 (see [3]).

Theorem 6.1. *Let S and C be as above and denote by $c \in H^{(1,1)}(S, \mathbb{R})$ the class Poincaré dual to $[C]$. Let $a \in H^{(1,1)}(S, \mathbb{R})$, and suppose that the following two conditions are satisfied:*

1. *The class $a - mc$ is Kähler (resp. semi-Kähler) for some $m \in \mathbb{R}$.*
2. *The vector $\left(P_{C/S}, (\int_C a) f + m z_C; \alpha_1, \dots, \alpha_k\right)$ is Kähler (resp. semi-Kähler).*

Then the vector $(S, a; \alpha_1, \dots, \alpha_k)$ is Kähler (resp. semi-Kähler).

Remark 6.2. *When the class a of the theorem is Poincaré dual itself to a smooth curve in S, we may take C to be this curve and $m = 1$. Then the class $a - mc$ in condition 1 of the theorem is just* 0 *hence semi-Kähler. Thus in this case only condition 2 has to be checked.*

With this remark in mind, let us return to the problem of packing of a symplectic 4-manifold (M, Ω). It turns out that the degeneration argument (see [3]) used to prove Theorem 6.1 can be neatly translated to symplectic geometry using a construction called *Gompf surgery.* The ability to find a smooth curve C which represents the Poincaré dual to the class $a = [\Omega]$ as in Remark 6.2 follows from Donaldson's symplectic hypersurface theorem (see [9]). The upshot of all this is that in order to obtain symplectic packings of an arbitrary symplectic 4-manifold it is enough to pack a symplectic ruled surface constructed in an analogous way to $P_{C/S}$ above. The main advantage is that symplectic ruled surfaces belong to the class of manifolds $\mathcal{C}$ described in Section 3 and so our symplectic version of Nakai-Moishezon criterion 3.2 applies to them. An analysis of the possible exceptional classes on blow-ups of ruled surfaces shows that similarly to $\mathbb{C}P^2$ ruled surfaces admit full symplectic packing by N equal balls for *every* large enough N. The conclusion is that the same thing holds also for (M, Ω). More precisely we have the following (see [4]):

Theorem 6.3. *Let (M, Ω) be a closed symplectic 4-manifold with $[\Omega] \in H^2(M, \mathbb{Q})$. Then, there exists N_0 such that for every $N \geq N_0$, (M, Ω) admits a full symplectic packing by N equal balls. In fact, if for some $k_0 \in \mathbb{Q}$ the Poincaré dual to $k_0[\Omega]$ can be represented by a symplectic submanifold of genus at least 1, then one can assume that $N_0 = 2k_0^2 \operatorname{Vol}(M, \Omega)$, where $\operatorname{Vol}(M, \Omega) = \frac{1}{2}\int_M \Omega \wedge \Omega$.*

This means that packing obstructions occur for at most a finite number of values of N. Thus on the phenomenological level, essentially on every symplectic manifold we have stability of the process of symplectic packing: starting from some N all obstructions disappear and the quantities $v_N(M, \Omega)$ stablize on the value 1.

Theorem 6.3 also allows us to bound from above the number N starting from which the process becomes stable. Here are some concrete examples:

Corollary 6.4. *In each of the following cases N_0 is a number (not necessarily minimal) for which the relevant symplectic manifold admits full symplectic packing by N equal balls for every $N \geq N_0$:*

1. *Let $S \subset \mathbb{C}P^n$ be an irrational smooth projective surface of degree d, and let Ω be the restriction of the standard Kähler form of $\mathbb{C}P^n$ to S. Then for (S, Ω) we have $N_0 = d$.*
2. *For $(\mathbb{T}^2 \times \mathbb{T}^2, \sigma \oplus \sigma)$ the 4-dimensional symplectic split-torus, where σ is an area form on $\mathbb{T}^2$, we have $N_0 = 2$.*
3. *Let (C_1, σ_1), (C_2, σ_2) be (real) symplectic surfaces with $\int_{C_1} \sigma_1 = \int_{C_2} \sigma_2$, and let $a, b \in \mathbb{N}$. Then, for $(C_1 \times C_2, a\sigma_1 \oplus b\sigma_2)$ we have $N_0 = 8ab$.*

7. Open Problems, New Directions and Some Speculations

7.1. Symplectic packing in higher dimensions

Most of the results mentioned up to now are special to dimension 4. This is not a coincidence. While (part) of the algebro-geometric results we mentioned can be generalized to any dimension, the results on the stability of symplectic packing rely strongly on Seiberg-Witten theory which is very special to dimension 4, and it is not clear how to generalize it (if possible at all) to higher dimensions.

The only known results about symplectic packings in higher dimensions are Gromov's Theorem 2.1 and a result due to McDuff and Polterovich [26] which states that $v_N(B^{2n}(1)) = v_N(\mathbb{C}P^n) = 1$ whenever $N = k^n$ (the latter can be easily generalized to any algebraic manifold).

The following two problems seem to be of particular interest:

1. Does the process of symplectic packing stabilize starting from some number of balls N_0, as in dimension 4?
2. Given a symplectic manifold (M, Ω), does there exist a radius λ_0 such that for every $\lambda \leq \lambda_0$ any two symplectic embeddings of a ball of radius λ are symplectically isotopic?

7.2. What is the symplectic nature of the phenomenon of packing obstructions? Although powerful tools such as pseudo-holomorphic curves allow us to obtain quite a lot of information on symplectic packings there is at least one aspect of the problem which remains mysterious: the *geometric nature* of this phenomenon.

It seems reasonable to categorize packing obstructions (for $N \geq 2$) as a *symplectic intersection phenomenon* (e.g. if one tries to embed symplectically two large enough balls into $\mathbb{C}P^n$ then their images must intersect). Intersections play a fundamental role in symplectic geometry and appear in many forms and disguises. Probably the most fundamental known such phenomenon is that of *Lagrangian intersections.* This phenomenon has to do with non-removable intersections between Lagrangian submanifolds that cannot be detected by means of classical topology or differential geometry. Lagrangian intersections have been studied intensively over the last decade by numerous people and nowadays there is both a systematic machinery to study them (see [13] for example) as well as a fair geometric intuition about their symplectic nature. In fact, this feel or intuition preceded the invention of the first mathematical tools used to study Lagrangian intersections by more than a decade. Let us also mention that several symplectic rigidity phenomena can be well understood and studied in the framework of Lagrangian intersections, most notably Arnold conjecture on fixed points of Hamiltonian diffeomorphisms.

It seems to be a question of conceptual relevance to figure out whether packing obstructions are in fact a "hidden" Lagrangian intersection phenomenon or something else, of a genuinely different nature.

A perhaps naive, but still worth exploring, idea would be to try to approach this question using Floer homology. For example, given a symplectic embedding of a ball $\phi\colon B(\lambda) \to \mathbb{C}P^n$ with $2\lambda^2 > 1$ one may consider a Lagrangian submanifold L which lies on the boundary of $\text{Image}(\phi)$ (a Lagrangian torus for example). If it can be shown that the Floer homology $HF_*(L, L)$ does not vanish, then it would follow that any two such symplectic balls $\varphi_1, \varphi_2\colon B(\lambda) \to \mathbb{C}P^n$ which are Hamiltonian isotopic *must have non-empty intersection.* Although this would not recover Gromov's packing Theorem 2.1 in its full generality[3], it would strongly indicate that packing obstructions are in fact a Lagrangian intersection phenomenon.

At present the implementation of this plan using Floer homology unfortunately runs into (technical) difficulties due to analytic problems which cause Floer homology not to be well defined in our setting. Hopefully the techniques will be refined in the future so as to enable us to settle this "dilemma".

[3]Since, except in dimension 4 (see [23]), we do not know whether or not any two symplectic balls in $\mathbb{C}P^n$ are Hamiltonian isotopic.

References

[1] P. Biran, *Geometry of symplectic packing,* PhD Thesis, Tel-Aviv University (1997).

[2] P. Biran, *Symplectic packing in dimension* 4, Geom. Funct. Anal. **7** (1997), 420–437.

[3] P. Biran, *Constructing new ample divisors out of old ones,* Duke Math. J. **98** (1999), 113–135.

[4] P. Biran, *A stability property of symplectic packing,* Invent. Math. **136** (1999), 123–155.

[5] P. Biran, *Connectedness of spaces of symplectic embeddings,* Internat. Math. Res. Notices **10** (1996), 487–491.

[6] C. Ciliberto and R. Miranda, *Linear systems of plane curves with base points of equal multiplicity,* Trans. Amer. Math. Soc. **352** (2000), no. 9, 4037–4050.

[7] J.-P. Demailly, *L^2-vanishing theorems for positive line bundles and adjunction theory,* in Transcendental methods in Algebraic Geometry. F. Catanese and C. Ciliberto eds., Lect. Notes in Math. **1646** (1996), 1–97, Springer-Verlag, Berlin.

[8] M. Demazure, *Surfaces de del Pezzo II-V,* in Séminaire sur les singularités de surfaces (1976–1977), Lecture Notes in Mathematics **777** (1980), Springer-Verlag, Berlin.

[9] S. K. Donaldson, *Symplectic submanifolds and almost-complex geometry,* J. Differential Geom. **44** (1996), 666–705.

[10] I. Dolgachev and I. Ortland, *Point sets in projective spaces and theta functions,* Astérisque **165** (1988).

[11] L. Ein and R. Lazarsfeld, *Seshadri constants on smooth surfaces,* Astérisque **218** (1993), 177–186.

[12] L. Ein, O. Küchle and R. Lazarsfeld, *Local positivity of ample line bundles,* J. Differential Geom. **42** (1995), 193–219.

[13] A. Floer, *Morse theory for Lagrangian intersections,* J. Differential Geom. **28** (1988), 513–547.

[14] M. Gromov, *Pseudoholomorphic curves in symplectic manifolds,* Invent. Math. **82** (1985), 307–347.

[15] M. Gromov, *Partial differential relations,* Ergebnisse der Mathematik und ihrer Grenzgebiete (3) **9** (1986), Springer-Verlag, Berlin-New York.

[16] V. Guillemin and S. Sternberg, *Birational equivalence in the symplectic category,* Invent. Math. **97** (1989), 485–522.

[17] R. Hartshorne, *Algebraic geometry,* Graduate Texts in Mathematics **52** (1977), Springer-Verlag, New York-Heidelberg.

[18] Y. Karshon, Appendix to *Symplectic packings and algebraic geometry,* Invent. Math. **115** (1994), 405–434.

[19] B. Kruglikov, *A remark on symplectic packings,* (Russian) Dokl. Akad. Nauk **350** (1996), 730–734.

[20] F. Lalonde, *Isotopy of symplectic balls, Gromov's radius and the structure of ruled symplectic* 4*-manifolds,* Math. Ann. **300** (1994), 273–296.

[21] F. Lalonde and D. McDuff, *The classification of ruled symplectic* 4*-manifolds,* Math. Res. Lett. **3** (1996), 769–778.

[22] R. Lazarsfeld, *Lengths of periods and Seshadri constants of abelian varieties,* Math. Res. Lett. **3** (1996), 439–447.

[23] D. McDuff, *Blow ups and symplectic embeddings in dimension* 4, Topology **30** (1991), 409–421.

[24] D. McDuff, *Lectures on Gromov invariants for symplectic* 4*-manifolds,* in Gauge theory and symplectic geometry (Montreal, PQ, 1995), 175–210, Kluwer Acad. Publ., Dordrecht, 1997.

[25] D. McDuff, *From symplectic deformation to isotopy,* in Topics in symplectic 4-manifolds (Irvine, CA, 1996), 85–99, Internat. Press, Cambridge, MA, 1998.

[26] D. McDuff and L. Polterovich, *Symplectic packings and algebraic geometry,* Invent. Math. **115** (1994), 405–434.

[27] D. McDuff and D. Salamon, *Introduction to symplectic topology,* Oxford Mathematical Monographs, (1998), Oxford University Press.

[28] F. M. Maley, J. Mastrangeli and L. Traynor, *Symplectic packings in cotangent bundles of tori,* Experimental Mathematics **9** (2000), no. 3, 435–455.

[29] M. Nagata, *Lecture notes on the 14'th problem of Hilbert,* Tata Institute of Fundamental Research, (1965), Bombay.

[30] Z. Ran, *Enumerative geometry of singular plane curves,* Invent. Math. **97** (1989), 447–465.

[31] C. H. Taubes, *The Seiberg-Witten and Gromov invariants,* Math. Res. Lett. **2** (1995), 221–238.

[32] L. Traynor, *Symplectic packing constructions,* J. Differential Geom. **42** (1995), 411–429.

[33] G. Xu, *Curves in P^2 and symplectic packings,* Math. Ann. **299** (1994), 609–613.

School of Mathematical Sciences
Tel-Aviv University
Ramat-Aviv, Tel-Aviv 69978, Israel
E-mail address: biran@math.tau.ac.il

New Invariants of Legendrian Knots

Yuri Chekanov

Abstract. We discuss two different ways to construct new invariants of Legendrian knots in the standard contact $\mathbf{R}^3$. These invariants are defined combinatorially, in terms of certain planar projections, and (sometimes) allow us to distinguish Legendrian knots which are not Legendrian isotopic but have the same classical invariants.

1. Introduction

A smooth knot L in the standard contact space $(\mathbf{R}^3, \alpha) = (\{(p, q, u)\}, du - pdq)$ is called Legendrian if the restriction of α to L vanishes (or, in other words, if L is everywhere tangent to the 2-plane distribution given by $\alpha = 0$). The general, open, problem of the theory of Legendrian knots is to give a classification of Legendrian knots up to Legendrian isotopy. The latter is defined as follows: two Legendrian knots are said to be Legendrian isotopic if they can be connected by a path in the space of Legendrian embeddings (or, equivalently, if one can be sent to another by a diffeomorphism g of $\mathbf{R}^3$ such that $g^*\alpha = \varphi\alpha$, where φ is a positive function). It is easy to show that every smooth knot is isotopic to a Legendrian one. However, two different Legendrian knots belonging to the same smooth isotopy class may be not Legendrian isotopic.

The so-called classical invariants of an oriented Legendrian knot L are defined as follows. First of them is just the smooth isotopy type of L. The Thurston-Bennequin number $\beta(L)$ of L is the linking number (with respect to the orientation defined by $\alpha \wedge d\alpha$) between L and $s(L)$, where s is a small shift along the u direction. The Maslov number $m(L)$ (which actually is an invariant of Legendrian immersion) is twice the rotation number of the projection of L to the (p, q) plane (or, equivalently, the value of the Maslov 1-cohomology class on the fundamental class of L). That m and β are indeed invariant under Legendrian isotopy follows since vectors tangent to a Legendrian curve are never parallel to the u axis. The change of orientation on L changes the sign of $m(L)$ and preserves $\beta(L)$.

One can ask whether there exists a pair of Legendrian knots which have the same classical invariants but are not Legendrian isotopic. Eliashberg and Fraser showed that this cannot happen when the knots are trivial as smooth knots [4, 6]. However, it turns out that there exist Legendrian knots with the same classical invariants but not Legendrian isotopic to each other. In the present talk, we discuss two entirely different new constructions of invariants of Legendrian knots [1, 2] that sometimes allow us to distinguish Legendrian knots with the same classical invariants. These new invariants are combinatorially defined in terms of projections of L to certain planes. They do not change when the orientation of the knot changes, so essentially they are invariants of non-oriented Legendrian knots. There is no visible relation between the two constructions. Neither of them is known to provide a priori stronger invariants; there are examples where the first construction works better. Both constructions extend, with minor modifications, to the case of Legendrian links, which we do not discuss here.

In order to visualize a knot in $\mathbf{R}^3$, it is convenient to use its projection to a plane. In the Legendrian case, the character of the resulting picture will depend on the choice of the projection. We shall use two of them: $\pi\colon \mathbf{R}^3 \to \mathbf{R}^2$, $(p,q,u) \mapsto (p,q)$ and $\sigma\colon \mathbf{R}^3 \to \mathbf{R}^2$, $(p,q,u) \mapsto (q,u)$.

We say that a Legendrian knot $L \subset \mathbf{R}^3$ is generic with respect to π, or π-generic, if all self-intersections of the immersed curve $\pi(L)$ are transverse double points. We can represent a π-generic Legendrian knot L by its (π-)diagram: the curve $\pi(L) \subset \mathbf{R}^2$, at every crossing of which the overpassing branch (the one with the greater value of u) is marked. Of course, not every abstract knot diagram in $\mathbf{R}^2$ is a diagram of a Legendrian knot, or is oriented diffeomorphic to such. Thus it requires a bit of extra work (which we are going to skip here) to check that the diagrams we draw indeed correspond to Legendrian knots. The Thurston-Bennequin number of a π-generic Legendrian knot can be computed by counting the crossings of its diagram with signs:

$\beta(L) = \#($  $) - \#($  $)$,

(where the p axis is horizontal and the q axis vertical). The Maslov number $m(L)$ is twice the rotation number of $\pi(L)$. Properties of the projection σ are discussed in Section 3.

Theorem 1.1. ([1, 2]) *Legendrian knots L, L' whose π-diagrams are given in Figure 1 have the same classical invariants (smooth knot type 5_2, $m = 0$, $\beta = 1$) but are not Legendrian isotopic.*

It is easy to check that the classical invariants are the same. Two different constructions of invariants distinguishing L and L' are given in Section 2 and Section 3.

2. First Construction: Differential Algebra

2.1. Introduction

In order to construct new invariants of Legendrian knots, we associate with every π-generic Legendrian knot L a differential graded algebra (A, ∂) over $\mathbf{Z}_2$. Our constructions can be viewed as an algebraically refined combinatorial version of a certain particular case of a more general Morse-Witten-type theory, an outline of

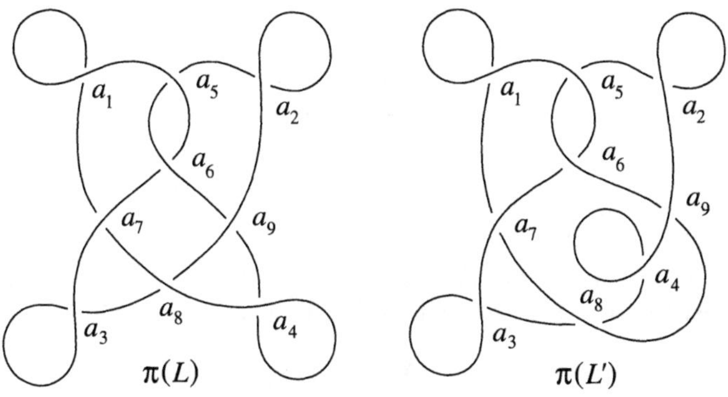

FIGURE 1

which was given by Eliashberg, Givental, and Hofer [5]. Similar results were also announced by Eliashberg.

2.2. Definitions and results

Let $\{a_1, \ldots, a_n\}$ be the set of crossings of $Y = \pi(L)$. Define A to be the tensor algebra (free associative unital algebra) $T(a_1, \ldots, a_n)$ generated by $a_1, \ldots, a_n$. The grading on A, which takes values in the group $\mathbf{Z}/m(L)\mathbf{Z}$, is defined as follows. Given a crossing a, consider the points $z_+, z_- \in L$ such that $\pi(z_+) = \pi(z_-) = a$ and $u(z_+) > u(z_-)$ (where $u(z)$ denotes the u coordinate of a point $z \in \mathbf{R}^3$). These points divide L into two pieces, γ_1 and γ_2, which we orient from z_- to z_+. We can assume, without loss of generality, that the intersecting branches are orthogonal at a. Then, for $\varepsilon \in \{1, 2\}$, the rotation number of the curve $\pi(\gamma_\varepsilon)$ has the form $N_\varepsilon/2 + 1/4$, where $N_\varepsilon \in \mathbf{Z}$. Clearly, $N_1 - N_2$ is equal to $\pm m(L)$. Hence N_1 and N_2 represent the same element of the group $\Gamma = \mathbf{Z}/m(L)\mathbf{Z}$, which we define to be the degree of a.

We are going to define the differential ∂. For every natural k, fix a (curved) convex k-gon $\Pi_k \subset \mathbf{R}^2$ whose vertices $x_0^k, \ldots, x_{k-1}^k$ are numbered anticlockwise. The form $dp \wedge dq$ defines an orientation on $\mathbf{R}^2$. Denote by $W_k(Y)$ the collection of smooth orientation-preserving immersions $f\colon \Pi_k \to \mathbf{R}^2$ such that $f(\partial \Pi_k) \subset Y$. Note that $f \in W_k(Y)$ implies $f(x_i^k) \in \{a_1, \ldots, a_n\}$. Consider the set of non-parametrized immersions $\widetilde{W}_k(Y)$, which is the quotient of $W_k(Y)$ by the action of the group $\{g \in \mathrm{Diff}_+(\Pi_k) \mid g(x_i^k) = x_i^k\}$.

The diagram Y divides a neighbourhood of each of its crossings into four sectors. We call positive two of them which are swept out by the underpassing curve rotating anticlockwise, and negative the other two (see Figure 2). For each vertex x_i^k of the polygon Π_k, a smooth immersion $f \in \widetilde{W}_k(Y)$ maps its neighbourhood in Π_k to either a positive or a negative sector; we shall say that x_i^k is, respectively, a positive or negative vertex for f. Define the set $W_k^+(Y)$ to consist of immersions $f \in \widetilde{W}_k(Y)$ such that the vertex x_0^k is positive for f, and all other vertices are negative. Let $W_k^+(Y, a_j) = \{f \in W_k^+(Y) \mid f(x_0^k) = a_j\}$.

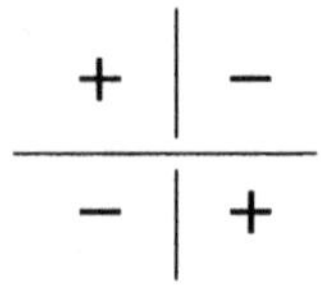

FIGURE 2

Denote $A_1 = \{a_1, \dots, a_n\} \otimes \mathbf{Z}_2 \subset A$, $A_k = (A_1)^{\otimes k}$. Then $A = \oplus_{l=0}^{\infty} A_l$. There is a decomposition $\partial = \sum_{k \geq 0} \partial_k$, where $\partial_k(A_i) \in A_{i+k-1}$. Define

$$\partial_k(a_j) = \sum_{f \in W_{k+1}^+(Y, a_j)} f(x_1) \cdots f(x_k),$$

which, in particular, means $\partial_0(a_j) = \#(W_1^+(Y, a_j))$, and extend ∂ to A by linearity and the Leibniz rule.

Theorem 2.1. *The differential ∂ is well defined. We have* $\deg(\partial) = -1$ *and* $\partial^2 = 0$.

The above theorem allows us to consider the homology ring $\ker(\partial)/\operatorname{im}(\partial)$, which turns out to be a Legendrian isotopy invariant:

Theorem 2.2. *Let (A, ∂), (A', ∂') be the differential graded algebras associated with Legendrian isotopic (π-generic) Legendrian knots L, L'. Then the homology rings of (A, ∂) and (A', ∂') are isomorphic as graded rings.*

The hard part in the proof of Theorem 2.1 is to show that $\partial^2 = 0$. The proof of this fact mimics, in a combinatorial way, the classical gluing-compactness argument of the Floer theory (cf. [7]). The proof of Theorem 2.2 involves a careful study of the behaviour of the differential graded algebra associated with a Legendrian knot when the diagram goes through elementary bifurcations (Legendrian Reidemeister moves).

2.3. Examples

2.3.1. Consider the Legendrian knots L_1, L_2, L_3 whose diagrams are given in Figure 3. The diagram $Y_1 = \pi(L_1)$ is the simplest possible diagram of a Legendrian knot. The classical invariants of L_1 are as follows: $m(L_1) = 0$, $\beta(L_1) = -1$, the knot is an unknot in the smooth category. Since $m(L_1) = 0$, the grading on the algebra $A = T(a)$ takes values in $\mathbf{Z}$. We have $\deg(a) = 1$. The set of immersions $\widetilde{W}_k(Y_1)$ is empty for $k > 1$ and consists of two elements $f_1, f_2 \in W_1^+(Y_1, a)$, whose images are the closures of the two bounded components of $\mathbf{R}^2 \setminus Y_1$. Hence $\partial(a) = 1 + 1 = 0$.

The Legendrian knot L_2 is a right-handed (with respect to the orientation defined by $\alpha \wedge d\alpha$) trefoil as a smooth knot. We have $m(L_2) = 0$, $\beta(L_2) = 1$, $A = T(a_1, \dots, a_5)$, where $\deg(a_1) = \deg(a_2) = 1$, $\deg(a_3) = \deg(a_4) = \deg(a_5) = 0$, $\partial(a_1) = 1 + a_3 + a_5 + a_3 a_4 a_5$, $\partial(a_2) = 1 + a_3 + a_5 + a_5 a_4 a_3$, $\partial(a_3) = \partial(a_4) = \partial(a_5) = 0$.

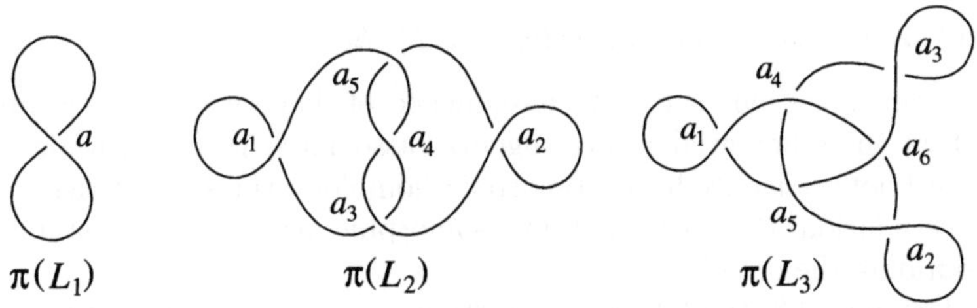

FIGURE 3

The Legendrian knot L_3 is a left-handed trefoil as a smooth knot. We have $m(L_3) = \pm 2$, $\beta(L_3) = -6$. $A = T(a_1, \dots, a_6)$, where all generators have degree $1 \in \mathbf{Z}_2$, $\partial(a_1) = 1 + a_4a_6$, $\partial(a_2) = 1 + a_5a_4$, $\partial(a_3) = 1 + a_6a_5$, $\partial(a_4) = \partial(a_5) = \partial(a_6) = 0$.

2.3.2. Let $(A, \partial) = (T(a_1, \dots, a_9), \partial)$ be the differential graded algebra associated with the Legendrian knot L given in Figure 1. We have $m(L) = 0$, $\beta(L) = 1$, $\deg(a_i) = 1$ for $i \leq 4$, $\deg(a_5) = 2$, $\deg(a_6) = -2$, $\deg(a_i) = 0$ for $i \geq 7$, $\partial(a_1) = 1 + a_7 + a_7a_6a_5$, $\partial(a_2) = 1 + a_9 + a_5a_6a_9$, $\partial(a_3) = 1 + a_8a_7$, $\partial(a_4) = 1 + a_8a_9$, $\partial(a_i) = 0$ for $i \geq 5$.

Let $(A', \partial) = (T(a_1, \dots, a_9), \partial)$ be the differential graded algebra associated with the Legendrian knot L' given in Figure 1. We have $m(L') = 0$, $\beta(L') = 1$, $\deg(a_i) = 1$ for $i \leq 4$, $\deg(a_i) = 0$ for $i \geq 5$, $\partial(a_1) = 1 + a_7 + a_5 + a_7a_6a_5 + a_9a_8a_5$, $\partial(a_2) = 1 + a_9 + a_5a_6a_9$, $\partial(a_3) = 1 + a_8a_7$, $\partial(a_4) = 1 + a_8a_9$, $\partial(a_i) = 0$ for $i \geq 5$.

2.4. Poincaré polynomials

Homology rings of differential graded algebras are rather hard to handle. That is why we introduce another, a bit more subtle, invariant of Legendrian knots. This invariant is a finite subset of the group monoid $\mathbf{N}_0[\Gamma]$, where $\mathbf{N}_0 = \{0, 1, \dots\}$, $\Gamma = \mathbf{Z}/m(L)\mathbf{Z}$. Assume that $\partial_0 = 0$. Then $\partial_1^2 = 0$. Since $\partial(A_1) \subset A_1$, we can consider the homology $H(A_1, \partial_1) = \ker(\partial_1|_{A_1}) / \operatorname{im}(\partial_1|_{A_1})$, which is a vector space graded by the cyclic group Γ. Define the Poincaré polynomial $P_{(A,\partial)} \in \mathbf{N}_0[\Gamma]$ by

$$P_{(A,\partial)}(t) = \sum_{\lambda \in \Gamma} \dim\bigl(H_\lambda(A_1, \partial_1)\bigr) t^\lambda ,$$

where $H_\lambda(A_1, \partial_1)$ is the degree λ homogeneous component of $H(A_1, \partial_1)$.

Define the group $\operatorname{Aut}_0(A)$ to consist of graded automorphisms of A such that for each $i \in \{1, \dots, n\}$ we have $g(a_i) = a_i + c_i$, where $c_i \in A_0 = \mathbf{Z}_2$. (of course $c_i = 0$ when $\deg(a_i) \neq 0$). Consider the set $U_0(A, \partial)$ consisting of automorphisms $g \in \operatorname{Aut}_0(A)$ such that $(\partial^g)_0 = 0$ (where $\partial^g = g^{-1} \circ \partial \circ g$). Define

$$I(A, \partial) = \{P_{(A,\partial^g)} \mid g \in U_0(A, \partial)\} .$$

Since $\operatorname{Aut}_0(A)$ has at most 2^n elements, this invariant is not hard to compute. We can associate with every (π-generic) Legendrian knot L the set $I(L) = I(A_L, \partial_L)$. The set $I(L)$ can be empty (see Section 4) but no examples are known where $I(L)$ contains more than one element.

Theorem 2.3. *If L is Legendrian isotopic to L' then $I(L) = I(L')$.*

Theorem 2.3 is not a direct consequence of Theorem 2.2. It is proved by showing that the differential graded algebras associated with Legendrian isotopic Legendrian knots not only have isomorphic homology rings but satisfy a stricter equivalence relation: they become tame isomorphic after (iterated) stabilizations. We are skipping the details here.

By Theorem 2.3, the following assertion, which is straightforward to verify, implies Theorem 1.1:

Proposition 2.4. *For the knots L, L' given in Figure 1, we have*

$$I(L) = \{t^{-2} + t^1 + t^2\}, \quad I(L') = \{2t^0 + t^1\}.$$

As an obvious corollary of the definitions we get

Proposition 2.5. *If $P \in I(L)$ then $P(-1) = \beta(L)$.*

Thus the relation between the Thurston-Bennequin number and the coefficients of the polynomial P is the same as between the Casson invariant of a homology 3-sphere and its Floer homology. This analogy was the starting point for developing the theory described above.

3. Second Construction: Decompositions of Fronts

In this section, we present the invariants of Legendrian knots constructed in [2]. These invariants are defined in terms of the σ-projection.

3.1. Fronts of Legendrian knots

Given a Legendrian knot $L \subset \mathbf{R}^3$, its σ-projection, or front, $\sigma(L) \subset \mathbf{R}^2$ is a singular curve with nowhere vertical tangent vectors (the q axis is horizontal and the u axis vertical). Its singularities, generically, are semi-cubic cusps. We say that L is σ-generic if, moreover, all self-intersections of $\sigma(L)$ are transverse double points with different q coordinates. Every closed planar curve with these types of singularities and nowhere vertical tangent vectors is a front of a Legendrian knot. Note that there is no need to explicitly indicate the type of a crossing of $\sigma(L)$ (which was necessary for $\pi(L)$): the overpassing branch (the one with the greater value of p) is always the one with the greater slope.

The Maslov number a σ-generic oriented Legendrian knot L can be computed by counting the right cusps of the front $\sigma(L)$ with signs depending on the orientations:

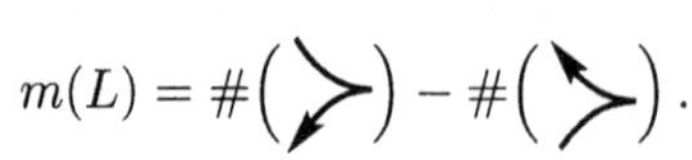

The Thurston-Bennequin number of L equals the sum of crossings of $\sigma(L)$ with signs minus half the total number of cusps:

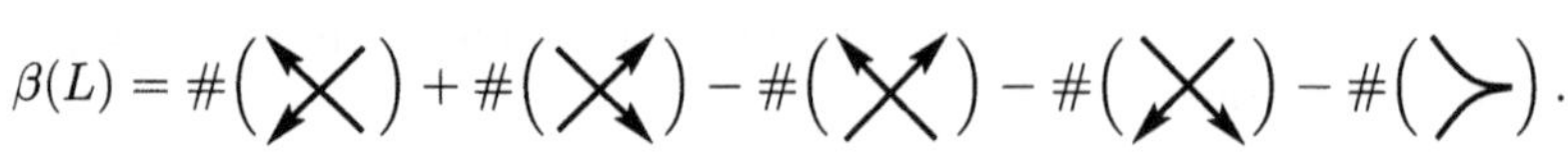

Given a σ-generic oriented Legendrian knot L, denote by $C(L)$ the set of points that correspond to cusps of $\sigma(L)$. Define the Maslov index $r\colon L \setminus C(L) \to \Gamma = \mathbf{Z}/m(L)\mathbf{Z}$ to be a locally constant function, uniquely defined up to adding a constant, whose value changes near points of $C(L)$ as shown in Figure 4. We say that a crossing of $\Sigma = \sigma(L)$ is Maslov if r takes the same value on both its branches.

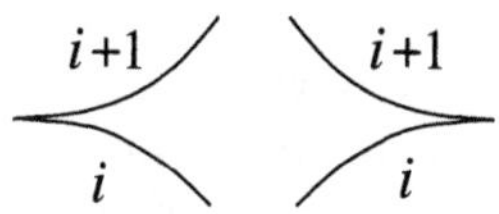

FIGURE 4

3.2. Admissible decompositions

Assume that $\Sigma = \sigma(L)$ is a union of closed curves $X_1, \dots, X_n$ that have finitely many self-intersections and meet each other at finitely many points. Then we call the set $\{X_1, \dots, X_n\}$ a decomposition of Σ.

We are going to formulate the main definition. A decomposition $\{X_1, \dots, X_n\}$ is called admissible if it satisfies the four conditions stated below. The first two of them are as follows:

(1) Each curve X_i bounds a topological disk: $X_i = \partial B_i$.
(2) For each $i \in \{1, \dots, n\}$, $q \in \mathbf{R}$, the set $B_i(q) = \{u \in \mathbf{R} \mid (q,u) \in B_i\}$ is either a segment, or consists of a single point u such that (q,u) is a cusp of Σ, or is empty.

Conditions (1) and (2) imply that each curve X_i has exactly two cusps (and hence the total number of cusps is $2n$). These cusps divide X_i into two pieces, on which the coordinate q is a monotone function. Near a crossing point $x \in X_i \cap X_j$, the curves X_i and X_j may look in one of three ways represented in Figure 5, (one drawn by solid lines and the other by dashed ones). Conditions (1) and (2), in particular, forbid the case shown in Figure 5a. We call the crossing point x switching if X_i and X_j are not smooth near x (Figure 5b), and non-switching otherwise (Figure 5c).

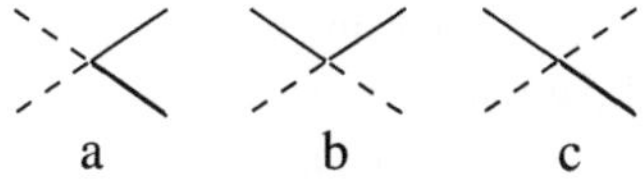

FIGURE 5

We can now formulate the remaining two conditions:

(3) If $(q_0, u) \in X_i \cap X_j$ $(i \neq j)$ is switching then for each $q \neq q_0$ sufficiently close to q_0 the set $B_i(q) \cap B_j(q)$ either coincides with $B_i(q)$ or $B_j(q)$, or is empty.
(4) Every switching crossing is Maslov.

Denote by $\mathrm{Adm}(\Sigma)$ the set of admissible decompositions of Σ. Given $D \in \mathrm{Adm}(\Sigma)$, denote by $\mathrm{Sw}(D)$ the set of its switching points. Define $\theta(D) = \#(D) - \#(\mathrm{Sw}(D))$.

Theorem 3.1. *If σ-generic Legendrian knots $L, L' \subset \mathbf{R}^3$ are Legendrian isotopic then there exists a one-to-one mapping $g\colon \operatorname{Adm}(\sigma(L)) \to \operatorname{Adm}(\sigma(L'))$ such that $\theta(g(D)) = \theta(D)$ for all $D \in \operatorname{Adm}(\sigma(L))$. In particular, the number $\#(\operatorname{Adm}(\sigma(L)))$ is an invariant of Legendrian isotopy.*

3.3. Remarks

3.3.1. Decompositions of fronts were first considered by Eliashberg in [3]. He stated a theorem, a particular case of which asserts that if a σ-generic Legendrian knot L is Legendrian isotopic to the one whose front is shown in Figure 6a then $\sigma(L)$ admits a decomposition satisfying conditions (1) and (2) (the proof has not yet appeared). The results of [3] in their general formulation cannot be proved by the methods of [2].

3.3.2. In the definition of admissible decomposition one can skip condition (4), or weaken it by replacing the group Γ with its quotient $\mathbf{Z}/\widetilde{m}\mathbf{Z}$ (the condition becomes void when $\widetilde{m} = 1$). A complete analogue of Theorem 3.1 will hold for thus defined larger sets $\operatorname{Adm}_{\widetilde{m}}(\sigma(L))$. However, it seems that the corresponding invariants are weaker.

3.3.3. The proof of Theorem 3.1 goes as follows: we connect L with L' by a generic path in the space of Legendrian knots and define a canonical way to extend admissible decompositions through the points where the front is not σ-generic. The mapping g depends on the choice of the path: going along a loop in the space of Legendrian knots we can produce a non-trivial automorphism of $\operatorname{Adm}(\sigma(L))$ even when the loop is contractible. It is not clear whether this phenomenon has some geometrical meaning, or is due to imperfections of the construction.

3.4. Examples

3.4.1. Before considering examples illustrating the definition of an admissible decomposition, notice that every admissible decomposition D of a front Σ is uniquely defined by its set of switching points. Indeed, denote by $X(\Sigma)$ the set of crossings of Σ, then each subset $E \subset X(\Sigma)$ defines a decomposition $D(E)$ of Σ which near $x \in X(\Sigma)$ has the form shown in Figure 5b if $x \in E$, and the form shown in Figure 5c otherwise. Clearly, if $E = \operatorname{Sw}(D)$ then $D = D(E)$.

FIGURE 6

3.4.2. Each of the fronts shown in Figure 6abc (the corresponding Legendrian knots are Legendrian isotopic to each other) admits a single admissible decomposition, all crossings being switching. Consider the front Σ_0 given in Figure 6d. All of its three crossings a, b, c are Maslov. It is easy to see that if $E \subset X(\Sigma_0)$ is empty or consists of two elements, then the decomposition $D(E)$ consists of a single curve, and condition (1) is not satisfied. It is straightforward to check that each of the subsets $\{a\}$, $\{c\}$, $\{a,b,c\}$ defines an admissible decomposition. The decomposition $D(\{b\})$ is not admissible since condition (3) is not satisfied at the point b.

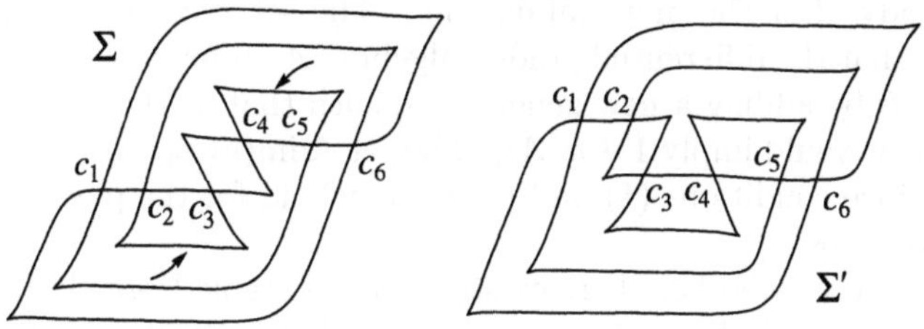

FIGURE 7

3.4.3. The Legendrian knots corresponding to the fronts Σ, Σ' represented in Figure 7 are respectively Legendrian isotopic to the Legendrian knots L, L' defined in Figure 1. We are going to show that $\#(\mathrm{Adm}(\Sigma)) \neq \#(\mathrm{Adm}(\Sigma'))$ and hence Theorem 1.1 is a consequence of Theorem 3.1.

Assume that $D \in \mathrm{Adm}(\Sigma)$. Consider the curve $X_1 \in D$ containing the piece of Σ indicated by the lower arrow. Being applied to X_1, conditions (1) and (2) imply that $c_2, c_3 \in \mathrm{Sw}(D)$. Similarly, looking at the curve $X_2 \in D$ containing the piece of Σ indicated by the upper arrow, one concludes that $c_4, c_5 \in \mathrm{Sw}(D)$. Since c_1 and c_6 are not Maslov, we have $\mathrm{Sw}(D) = \{c_2, c_3, c_4, c_5\}$. It is not hard to check that this decomposition is indeed admissible, and hence $\#(\mathrm{Adm}(\Sigma)) = 1$. Arguing similarly, one can find that $\#(\mathrm{Adm}(\Sigma')) = 2$, the admissible decompositions D_1, D_2 being defined by $\mathrm{Sw}(D_1) = \{c_2, c_3, c_4, c_5\}$, $\mathrm{Sw}(D_2) = \{c_1, c_2, c_3, c_4, c_5, c_6\}$.

4. Instability of Invariants

There are two stabilizing operations, S_- and S_+, on Legendrian isotopy classes of oriented Legendrian knots, defined as follows. Given an oriented Legendrian knot L, we perform one of the operations described in Figure 8 in a small neighbourhood of a point on L. One can check that, up to Legendrian isotopy, the resulting Legendrian knot $S_\pm(L)$ does not depend on the choices involved, and the operations S_-, S_+ commute. An important observation is that two Legendrian knots L, L' have the same classical invariants if and only if they are stable Legendrian isotopic in the sense that there exist $n_-, n_+ \in \mathbf{N}_0$ such that $S_-^{n_-}(S_+^{n_+}(L))$ is Legendrian isotopic to $S_-^{n_-}(S_+^{n_+}(L'))$ (see e. g. [8]).

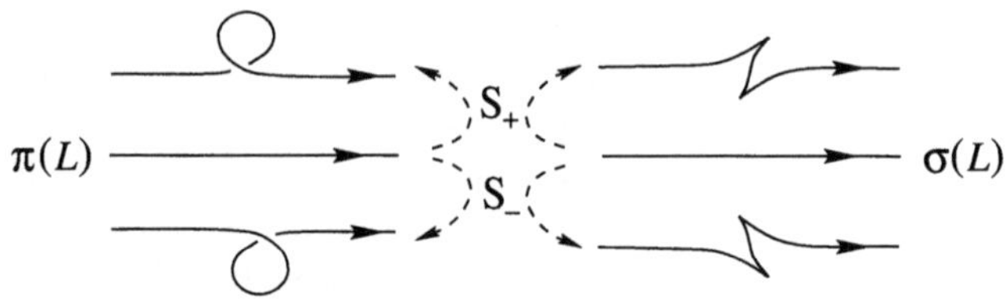

FIGURE 8

Thus the invariants constructed in Sections 1 and 2 cannot be stable. In fact, they fail already after the first stabilization. The set $I(S_\pm(L))$ is always empty. The reason is that the differential graded algebra (A, ∂) for $S_\pm(L)$ can be obtained from that for L by adding a new generator a such that $\partial(a) = 1$ (while $I(S_\pm(L))$ being nonempty would imply $1 \notin \partial(A)$). The set $\mathrm{Adm}(\sigma(S_\pm(L)))$ is empty as well. This follows since conditions (1) and (2) cannot hold for the piece containing the newly created cusps.

It is not known whether there exists a pair of stable Legendrian isotopic but not Legendrian isotopic Legendrian knots such that at least one of them has the form $S_\pm(L)$.

References

[1] Yu. V. Chekanov, Differential algebra of Legendrian links, to appear.

[2] Yu. V. Chekanov and P. E. Pushkar, Arnold's four cusp conjecture and invariants of Legendrian knots, in preparation.

[3] Ya. Eliashberg, A theorem on the structure of wave fronts and its application in symplectic topology, Funct. Anal. Appl. **21** 1987, 227–232.

[4] Ya. Eliashberg, Legendrian and transversal knots in tight contact 3-manifolds, In: Topological methods in modern mathematics (Stony Brook, NY, 1991), Publish or Perish, 1993, 171–193.

[5] Ya. Eliashberg, Invariants in contact topology, In: Proceedings ICM 1998 II, Doc. Math., Extra volume, 327–338.

[6] Ya. Eliashberg and M. Fraser, Classification of topologically trivial Legendrian knots, In: Geometry, topology, and dynamics (Montreal, PQ, 1995), CRM Proc. Lecture Notes, 15, AMS, Providence, 1998, 17–51.

[7] A. Floer, Symplectic fixed points and holomorphic spheres, Comm. Math. Phys. **120** 1989, 575–611.

[8] D. Fuchs and S. Tabachnikov, Invariants of Legendrian and transverse knots in the standard contact space, Topology **36** (1997), 1025–1053.

Moscow Center for Continuous Mathematics Education
M. Znamensky per. 7/10–5
Moscow 121019 Russia
E-mail address: `chekanov@mccme.ru`

Contact Topology in Dimension Greater than Three

Hansjörg Geiges

Abstract. The aim of this talk is to give a survey of the known methods for constructing contact structures on manifolds of dimension greater than 3. We give an extensive list of contact manifolds that can be constructed via these methods, including some recent examples of contact structures on 5-dimensional manifolds found by a combination of contact surgery and cobordism theoretic techniques.

1. Introduction

Let M be a differentiable manifold of odd dimension $2n-1$. A codimension 1 distribution ξ on M (i.e. a smooth tangent hyperplane field) is called a **contact structure** if any differential 1-form α that locally defines ξ as $\xi = \ker\alpha$ satisfies the condition that $\alpha \wedge (d\alpha)^{n-1}$ is nowhere zero. This condition is independent of the choice of α: Replacing α by $f\alpha$ with f some nowhere zero function, we have $f\alpha \wedge (d(f\alpha))^{n-1} = f^n\alpha \wedge (d\alpha)^{n-1}$.

If ξ is coorientable, such an α exists globally (by a partition of unity argument) and is then called a **contact form**. For convenience we restrict attention to orientable manifolds and coorientable contact structures.

In the present survey we concentrate on the question as to what can be said about the existence of contact structures, in particular on manifolds of dimension greater than 3. For a discussion of contact topology in a wider context of its recent and not so recent history see [8, 14].

2. Some Classical Constructions

In this section we briefly recall the constructions of contact manifolds known before the early 1990s. According to the usage of C. T. C. Wall, as related to me by C. B. Thomas, the attribute 'classical' may be employed with reference to any result prior to one's thesis.

2.1. Hypersurfaces in symplectic manifolds

Let (V,ω) be a symplectic manifold of dimension $2n$. This means that ω is a closed 2-form ($d\omega = 0$) of maximal rank ($\omega^n \neq 0$). A vector field X on V is called a **Liouville vector field** if $L_X\omega = \omega$. With the help of Cartan's formula for the Lie derivative $L_X = d \circ i_X + i_X \circ d$ this may be rewritten as $d(i_X\omega) = \omega$. Then the 1-form $\alpha = i_X\omega$ defines a contact form on any hypersurface M in V transverse to X. Indeed,

$$\alpha \wedge (d\alpha)^{n-1} = (i_X\omega) \wedge (d(i_X\omega))^{n-1} = i_X\omega \wedge \omega^{n-1} = \frac{1}{n} i_X(\omega^n),$$

which restricts to a volume form on $M \subset V$ provided M is transverse to X. This is how contact structures arise on suitable energy hypersurfaces in Hamiltonian systems.

Example 2.1. *$M = S^{2n-1} \subset \mathbb{R}^{2n}$ with $\omega = \sum_{i=1}^n dx_i \wedge dy_i$ and Liouville vector field $X = \frac{1}{2}\sum_{i=1}^n (x_i\partial_{x_i} + y_i\partial_{y_i})$, whence $\alpha = \frac{1}{2}\sum_i (x_i dy_i - y_i dx_i)$. Notice that if we regard S^{2n-1} as the unit sphere in $\mathbb{C}^n$ with complex structure J, then $\xi = \ker\alpha$ defines at each point $x \in S^{2n-1}$ the $(n-1)$-dimensional complex subspace of T_xS^{2n-1}, since $\alpha = -\frac{1}{2} r\, dr \circ J$ in terms of the radial coordinate r. The hermitian form $d\alpha(\,\cdot\,, J\,\cdot\,)$ on ξ is called the Levi form of the hypersurface $S^{2n-1} \subset \mathbb{C}^n$, and the contact condition for α corresponds to the positive definiteness of that Levi form, or the strict pseudoconvexity of the hypersurface.*

Example 2.2. *The cotangent bundle T^*N of an arbitrary manifold N of dimension n carries a canonical symplectic form ω which in local coordinates q_i on N and dual coordinates p_i can be written as $\sum_{i=1}^n dp_i \wedge dq_i$. The radial vector field in the fibre direction, $X = \sum_{i=1}^n p_i\partial_{p_i}$, is a Liouville vector field on (T^*N, ω). So the unit cotangent bundle of N (with respect to any Riemannian metric on N) inherits a contact form, given in local coordinates by $\alpha = \sum_{i=1}^n p_i dq_i$.*

2.2. S^1-bundles

Let (B, ω) be a symplectic manifold such that ω defines an integral de Rham cohomology class, i.e. $[\omega]$ lies in the image of the natural homomorphism $H^2(B;\mathbb{Z}) \hookrightarrow H^2(B;\mathbb{R}) \cong H^2_{dR}(B)$. Let M be the total space of the S^1-bundle over B with Euler class $e = [\omega]$. Then, as observed by Boothby and Wang [4], one can find a connection 1-form α for this bundle such that the curvature equation reads $d\alpha = \pi^*\omega$, where $\pi\colon M \to B$ is the bundle projection. This easily implies that α is a contact form on M.

Example 2.3. *We can recover Example 2.1 by considering the generalised Hopf fibration $\pi\colon S^{2n-1} \to \mathbb{C}P^{n-1}$, where $\mathbb{C}P^{n-1}$ is equipped with its standard Fubini-Study symplectic form. By computing the Betti numbers of S^1-bundles over certain Kähler manifolds with integral Kähler class, Boothby and Wang produced the first systematic examples of contact manifolds that are not spheres or cotangent bundles.*

This construction was extended by C. B. Thomas [31] to certain Seifert fibrations. As an application, he could show that most simply connected, indecomposable 5-dimensional spin manifolds carry a contact structure, since they can be realised as Seifert S^1-bundles over $\mathbb{C}P^2$.

2.3. Brieskorn manifolds

In the mid 1970s several teams of authors (see [26, 30, 1]) observed that Brieskorn manifolds

$$\Sigma(a_0, \ldots, a_n) = \{(z_0, \ldots, z_n) \in \mathbb{C}^{n+1} |\, z_0^{a_0} + \cdots + z_n^{a_n} = 0\} \cap S^{2n+1}$$

(here the a_j are natural numbers ≥ 2) admit a contact structure. One of the several equivalent ways of writing this contact structure is as the kernel of the contact form

$$\alpha = \frac{i}{4} \sum_{j=0}^{n} (z_j d\overline{z}_j - \overline{z}_j dz_j) .$$

Example 2.4. *Every odd dimensional homotopy sphere that bounds a parallelisable manifold can be realised as a Brieskorn manifold, see* [22], *and hence admits a contact structure. This includes all 28 seven-dimensional and all 992 eleven-dimensional homotopy spheres.*

Example 2.5. *Using the Brieskorn examples, and the connected sum construction for contact manifolds due to Meckert* [28] *(this will be discussed below in the context of contact surgery), C. B. Thomas* [32] *proved the existence of contact structures on a range of* $(n-1)$*-connected* $(2n+1)$*-manifolds.*

2.4. 3-dimensional constructions

Our main emphasis is on higher dimensions, but for completeness we mention some 3-dimensional constructions, so that below we can see to what extent they have higher-dimensional analogues. For more detailed surveys see [8, 17, 33].

The existence of contact structures on every closed, orientable 3-manifold was first shown by Martinet [27], using the surgery presentation for 3-manifolds due to Lickorish and Wallace. In dimension 3, surgery preserving a contact structure can be performed on any S^1 embedded *transversely* to a given contact structure (and with any framing), and the standard contact structure on S^3 provides the starting point.

Thurston and Winkelnkemper [34] gave a short alternative proof using the Alexander open book decomposition for 3-manifolds, and Gonzalo [20] obtained the same result from a branched covering description of 3-manifolds due to Hilden, Montesinos, and Thickstun.

A further proof of the existence of contact structures on every closed, orientable 3-manifold M was given by Altschuler [2]. His proof is based on a parabolic deformation of 1-forms. Starting from a nowhere zero 1-form α which satisfies $\alpha \wedge d\alpha \geq 0$ but not $\alpha \wedge d\alpha \equiv 0$ (plus some additional restrictions), a heat equation is used to diffuse the 'positivity' of the form over the whole 3-manifold. Eliashberg and Thurston [10] considerably extended these ideas, using more geometric arguments.

3. Contact Surgery and Cobordism Theory

In 1990 Eliashberg [6] proved a remarkable theorem about the topology of Stein manifolds (affine complex analytic manifolds). As was known from the work of Lefshetz, Serre, Frankel-Andreotti and Milnor, a Stein manifold of real dimension $2n$ admits a proper Morse function with all critical points of index $\leq n$, so in particular it has the homotopy type of an n-dimensional CW complex. Eliashberg's theorem says that the converse is true provided $n \geq 3$.

Theorem 3.1. (Eliashberg) *Let (W, J) be a $2n$-dimensional compact almost complex manifold with boundary ∂W. If $n \geq 3$ and W admits a Morse function constant on the boundary and with all critical points of index $\leq n$, then J is homotopic to an integrable almost complex structure J' such that ∂W is strictly J'-convex and the interior of W is Stein.*

The situation for $n = 2$ is more complicated, see [19].

3.1. Contact surgery

As mentioned in Example 2.5, in 1982 Meckert had given a proof that the connected sum of two contact manifolds also admits a contact structure, but her construction did not seem to allow any simple generalisation to surgeries along higher-dimensional spheres (observe that the connected sum of two manifolds may be thought of as a surgery along a zero-dimensional sphere S^0, one each of the two points constituting S^0 being embedded in the two manifolds).

Recall from Example 2.1 that the strict J-convexity of a hypersurface M in an almost complex manifold (V, J) is equivalent to saying that the J-invariant hyperplanes in the tangent spaces of M define a contact structure on M. So Theorem 3.1 may be interpreted as a statement about contact surgery. Weinstein [36] gave an alternative description of this aspect of Eliashberg's theorem, using hypersurfaces transverse to a Liouville vector field. The basic idea in Weinstein's construction is strikingly simple.

We illustrate this idea for connected sums of 3-manifolds, see Figure 1. Consider $\mathbb{R}^4$ with coordinates x, y, z, t and standard symplectic form $\omega = dx \wedge dy + dz \wedge dt$. The vector field $X = \frac{1}{2}x\partial_x + \frac{1}{2}y\partial_y + 2z\partial_z - t\partial_t$ is a Liouville vector field for ω. We can find a smooth handle, with boundary transverse to X, joining the hyperplanes $\{t = 1\}$ and $\{t = -1\}$. Thus $\alpha = \omega(X, \cdot)$ induces a contact form on these hyperplanes and the (boundary of) the smooth handle joining them.

On $\{t = \pm 1\}$, this 1-form α restricts to $\frac{1}{2}x\,dy - \frac{1}{2}y\,dx \pm dz$. By the Darboux theorem for contact forms, any contact form can in suitable local coordinates be written exactly like that. Thus, by choosing the handle in Figure 1 thin enough, we can glue its ends into two given contact 3-manifolds to perform a connected sum (or a 0-surgery on one component).

By replacing the coordinate axes in this picture by higher-dimensional euclidean spaces, one obtains the corresponding handle for surgeries along higher-dimensional **isotropic** spheres (i.e. spheres tangent to a given contact structure). Notice that an isotropic submanifold in (M^{2n-1}, ξ) has dimension at most $n - 1$.

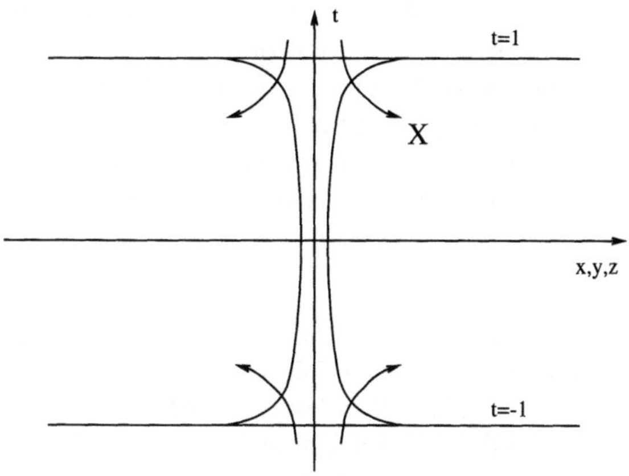

FIGURE 1. Handle used in contact surgery.

Thanks to an h-principle for isotropic embeddings (cf. [21, 6, 11]), and a neighbourhood theorem for isotropic submanifolds in place of the Darboux theorem, the question which kinds of surgeries can be performed as contact surgeries now essentially becomes a problem in algebraic topology.

It is important to bear in mind that there are two questions that need to be addressed when trying to perform a contact surgery:

(1) Can a given embedding $S^k \hookrightarrow (M, \xi)$ be approximated by an isotropic embedding (i.e. can the h-principle be applied)?
(2) Which framings (i.e. trivialisations of the normal bundle of $S^k \hookrightarrow M$) allow contact surgery?

These issues are discussed at length in [12, 15] for various applications. The introductory sections in those papers might provide a useful guide.

Example 3.2. *Using contact surgery, an essentially complete solution to the existence problem for contact structures on $(n-1)$-connected $(2n+1)$-manifolds was given in* [11, 12]. *The essential point is that these manifolds can be obtained from a sphere by doing surgery below the middle dimension. The precise existence statement is complicated by the presence of exotic spheres and certain exceptional invariants in the classification of these highly connected manifolds. Here are some examples that are easy to state:*

(a) *Any simply-connected* 5*-manifold M whose second Stiefel-Whitney class $w_2(M) \in H^2(M;\mathbb{Z}_2)$ admits an integral lift $c \in H^2(M;\mathbb{Z})$ carries a contact structure ξ such that the conformally symplectic (and hence complex) bundle $(\xi, d\alpha)$ has first Chern class $c_1 = c$.*
(b) *Every* 2*-connected* 7*-manifold and every* 4*-connected* 11*-manifold admits a contact structure (cf. Example 2.4).*

A coorientable codimension 1 distribution η on M with a complex bundle structure or, equivalently, a reduction of the structure group of TM^{2n+1} to $\mathrm{U}(n) \times 1$, is called an **almost contact structure**. Clearly, the existence of an almost contact structure is a necessary condition for the existence of a coorientable contact structure. In dimension 5 (without any assumption on the fundamental group of M), almost contact structures are classified by the integral lifts of $w_2(M)$.

3.2. Plumbing

By arguments related to those in the proof of Theorem 3.1, Eliashberg [7] showed the following: Given a symplectic manifold (W, ω) and $L \looparrowright W$ an immersed Lagrangian submanifold (i.e. $\omega|TL \equiv 0$ and $\dim L = (\dim W)/2$) with only transverse double points, one can find a small neighbourhood of L whose boundary inherits a contact structure. In particular – recall the remarks in Example 2.2 – this allows the plumbing of cotangent disc bundles (for a discussion of plumbing see for instance [22]).

Example 3.3. *In [7] this was used by Eliashberg to construct exotic contact structures on S^{2n-1}, i.e. contact structures that are not diffeomorphic to the standard structure of Example 2.1. The exoticity of the contact structure is detected by a symplectic filling not containing any symplectic spheres, but different from a disc D^{2n}.*

Two contact structures on a manifold are called **homotopically equivalent** if they induce homotopic almost contact structures. A contact structure ξ on $S^{2n-1} = \partial D^{2n}$ is called **homotopically standard** if it is homotopically equivalent to the standard contact structure, or equivalently, if the stable almost complex structure on the stable tangent bundle $TS^{2n-1} \oplus \varepsilon^1 \cong TD^{2n}|S^{2n-1}$ defined by ξ extends as an almost complex structure over D^{2n}.

Example 3.4. *For n odd, in which case the obstruction group $\pi_{2n-1}(\mathrm{SO}_{2n}/\mathrm{U}_n)$ detecting homotopical standardness is finite, Eliashberg obtained exotic but homotopically standard contact structures on S^{2n-1} by taking the connected sum of spheres as in Example 3.3. By refining the homotopical considerations and making direct use of contact surgery, this argument was extended in [12] to prove the existence of exotic but homotopically standard contact structures also on S^7 and S^{8k+3}. The case S^{8k+7} with $k \geq 1$ still seems to be open (in spite of the result attributed to me in [9]).*

A more effective way of detecting homotopically equivalent but nondiffeomorphic contact structures is provided by the newly developed contact homology, see [9]. For instance, Ustilovsky [35] has used this to show that on S^{4n+1} there exist in fact infinitely many nondiffeomorphic, but homotopically standard contact structures (constructed from different descriptions of S^{4n+1} as Brieskorn manifold).

3.3. Spin bordism

One strategy to produce examples of contact manifolds is to find classes $\mathcal{C}$ of manifolds with the following properties:

(i) $\mathcal{C}$ contains a subclass $\mathcal{C}_0$ of manifolds which carry a contact structure by some explicit construction (e.g. Brieskorn manifolds).

(ii) Any manifold in $\mathcal{C}$ is obtained from one in $\mathcal{C}_0$ by surgery below the middle dimension, and all the necesary surgeries can be performed as contact surgeries.

For instance, in the results from [11, 12] illustrated in Example 3.2, this strategy is applied to the class $\mathcal{C}$ of all highly connected manifolds carrying an almost contact structure, starting from the class $\mathcal{C}_0$ of homotopy spheres bounding parallelisable manifolds (cf. Example 2.4). The main difficulty lies in distilling from Wall's classification of highly connected manifolds the information that is necessary to answer questions (1) and (2) from Section 3.1.

As explained in [15, Lemma 3], these questions are easy to answer in dimension 5. Here we only have to deal with questions (1) and (2) in the case of 1- and 2-spheres. There are no obstructions to approximating an embedding $S^1 \hookrightarrow (M,\xi)$ by an isotropic one, and for an embedding $i\colon S^2 \hookrightarrow (M,\xi)$ one only has to require $i^*c_1(\xi) = 0$ to be able to find an isotropic approximation. In the case of 1-spheres all framings can be realised by contact surgeries (this is false for 3-manifolds). In the case of 2-spheres there is no problem with framings because they are controlled by the group $\pi_2(\mathrm{SO}_3) = 0$.

In [15, 16] this has been used to prove the existence of contact structures on classes $\mathcal{C}_\pi$ of 5-dimensional spin manifolds M with fundamental group π_1 isomorphic to some given group π. The key steps in the argument are as follows, for details see [16].

1. Bordism theorem: Let $f\colon M \to B\pi$ be the classifying map of the universal covering $\widetilde{M} \to M$ and σ a spin structure on M. If the bordism class $[f\colon M \to B\pi, \sigma] \in \Omega_5^{\mathrm{spin}}(B\pi)$ contains a contact manifold (M_0, ξ_0) with $c_1(\xi_0) = 0$, then M admits a contact structure.

   This is analogous to results about metrics of positive scalar curvature, cf. [29]. What the assumptions of the theorem really imply (in any dimension) is that M can be obtained from M_0 by surgery in codimension ≥ 3. This information is sufficient for constructing a positive scalar curvature metric on M, given one on M_0, but for contact structures this does not allow going beyond dimension 5.

2. Reduction theorem: To solve the problem for the class $\mathcal{C}_\pi$, it suffices to solve it for all the classes $\mathcal{C}_{\pi_p}$, where π_p denotes a p-Sylow subgroup of π. This is formally analogous to the result of Kwasik and Schultz [23] for positive scalar curvature metrics.

3. It remains to find a class $\mathcal{C}_{\pi_p,0}$ of contact manifolds (with $c_1 = 0$) generating $\Omega_5^{\mathrm{spin}}(B\pi_p)$. In [16], step 3 is completed for certain groups π having periodic cohomology (periodic groups, for short). The result of [15] for $\pi = \mathbb{Z}_2$ can also be interpreted in this framework, but by *ad hoc* arguments a somewhat stronger result was achieved there (both concerning the existence of contact structures and the topological structure of 5-manifolds with fundamental group of order 2). We summarise the main existence results of these two papers in the following example.

Example 3.5. *Let M be a closed oriented 5-manifold with universal cover $\widetilde{M}$. The manifold M admits a contact structure in either of the following cases:*

(a) *$\pi = \pi_1(M)$ is periodic, $|\pi|$ is odd, $9 \not| \, |\pi|$, and M is spin, i.e. $w_2(M) = 0$.*
(b) *$\pi_1(M) = \mathbb{Z}_2$ and $\widetilde{M}$ is spin.*

By the reduction theorem (and the structure of periodic groups of odd order), case (a) reduces to the study of cyclic groups. This causes the problems at the prime 3, since the spin bordism group $\Omega_5^{\mathrm{spin}}(B\pi_{p^k})$ has a different structure for $p = 3$ or $p \geq 5$. The class $\mathcal{C}_{\mathbb{Z}_{p^k},0}$ may be taken to consist of lens spaces, i.e. quotients of S^5 under some fixed point free representation $\rho\colon \mathbb{Z}_{p^k} \rightarrow \mathrm{U}(3)$. The standard contact structure on S^5 descends to these quotients.

In (b), for the class $\mathcal{C}_{\mathbb{Z}_2,0}$ one needs ten manifolds, constructed explicitly as $\mathbb{Z}_2$-quotients of Brieskorn manifolds.

An interesting illustrative example is the following:

Example 3.6. *Let $D_{s,3}$ denote the metacyclic group*

$$\{x, y \,|\, x^s = y^3 = 1, yxy^{-1} = x^r\},$$

where s is an odd integer, r is a primitive third root of $1 \bmod s$, *and furthermore* $\gcd(3(r-1), s) = 1$. *This group acts freely and smoothly, but not freely and linearly, on S^5. The quotient $S^5/D_{s,3}$ admits a contact structure by the results above, see also* [12] *for a more direct argument. The contact structure given by this construction lifts to an exotic contact structure on S^5. Indeed, there is evidence that $D_{s,3}$ cannot act compatibly with the standard contact structure.*

4. Other Constructions and Examples

We conclude this survey with a summary of other methods for constructing contact manifolds. For a more detailed discussion of the first three of the following constructions see [13].

4.1. Branched covers

As mentioned above, Gonzalo had introduced a branched cover construction for 3-dimensional contact manifolds. Gromov [21, p. 343] observed that contact structures lift to branched covers in arbitrary dimension, provided the branching locus is a codimension two contact submanifold. For a proof see [13].

Example 4.1. *Let M^3 be a closed, orientable 3-manifold and Σ_g a closed, orientable surface of genus g. Then $M^3 \times \Sigma_g$ admits a contact structure. Indeed, M^3 is parallelisable, so by Example 2.2 we have a contact form on $M^3 \times S^2$, thought of as the unit cotangent bundle of M^3. A contact structure on $M^3 \times \Sigma_g$ can now be obtained by writing this manifold as a branched cover of $M^3 \times S^2$, branched along $2 + 2g$ suitable sections of the unit cotangent bundle.*

Contact structures on more general *torus* bundles over 3-manifolds had been found earlier by Lutz [25].

4.2. Fibre connected sum

Again this is a construction that was first observed by Gromov [21, p. 343]. In the symplectic case this construction was worked out by Gompf [18] and used to remarkable effect. For the contact case a proof can again be found in [13].

Let $j_i \colon (N, \alpha_N) \hookrightarrow (M, \alpha)$, $i = 1, 2$, be two disjoint contact embeddings ($j_i^* \alpha = \alpha_N$). Let $\psi \colon V_1 - j_1(N) \to V_2 - j_2(N)$ be an orientation preserving diffeomorphism of punctured tubular neighbourhoods (reversing the radial coordinate), induced by an orientation reversing isomorphism of the two normal bundles. Then the fibre connected sum of M along the $j_i(N)$, obtained from $M - (j_1(N) \cup j_2(N))$ by gluing with the help of ψ, admits a contact structure.

Example 4.2. *The manifold $M^3 \times \Sigma_g$ can also be obtained by starting with $M^3 \times S^2$ and performing g fibre connected sums along suitable sections of the unit cotangent bundle of M^3.*

4.3. Contact reduction

This construction is the contact analogue of symplectic reduction, which has been studied much more intensively. A concise description is given in [13], but the possibility of contact reduction has been observed independently by several authors. This construction has recently attracted a great deal of attention, see for instance [5, 24]. It has already led to some interesting results, but its potential for the construction of higher-dimensional contact manifolds remains to be explored.

4.4. Heat flow

Recently, Altschuler, in collaboration with L. F. Wu [3], has extended his construction (alluded to in Section 2.4) from dimension 3 to higher dimensions. Independently, R. Gulliver, M. Schwarz and the author have obtained results in a similar direction. The set-up, however, becomes more restrictive, and so far the applications of this method remain scarce. One of the concrete examples of Altschuler and Wu is yet another construction of contact structures on trivial surface bundles over 3-manifolds.

Acknowledgements

Part of the work on this survey was done during an enjoyable stay at the Max-Planck-Institute for Mathematics in the Sciences, Leipzig. I thank Matthias

Schwarz for the kind invitation and his hospitality, and the MPI for its support. I also thank the London Mathematical Society for supporting my attendance of the ECM with a travel grant.

References

[1] K. Abe, *On a generalization of the Hopf fibration I. Contact structures on the generalized Brieskorn manifolds*, Tôhoku Math. J. (2), **29** (1977), 335–374.

[2] S. J. Altschuler, *A geometric heat flow for one-forms on three-dimensional manifolds*, Illinois J. Math., **39** (1995), 98–118.

[3] S. J. Altschuler and L. F. Wu, *On deforming confoliations*, preprint.

[4] W. M. Boothby and H. C. Wang, *On contact manifolds*, Annals of Math. (2), **68** (1958), 721–734.

[5] C. P. Boyer and K. Galicki, *A note on toric contact geometry*, J. Geom. Phys., **35** (2000), 288–298.

[6] Y. Eliashberg, *Topological characterization of Stein manifolds of dimension* > 2, Internat. J. Math., **1** (1990), 29–46.

[7] Y. Eliashberg, *On symplectic manifolds with some contact properties*, J. Differential Geom., **33** (1991), 233–238.

[8] Y. Eliashberg, *Symplectic topology in the nineties*, Differential Geom. Appl., **9** (1998), 59–88.

[9] Y. Eliashberg, *Invariants in contact topology*, in: Proc. International Congress of Mathematicians (Berlin, 1998), Doc. Math., Extra Vol. II (1998), 327–338.

[10] Y. Eliashberg and W. P. Thurston, *Confoliations*, Univ. Lecture Ser., **13**, Amer. Math. Soc., Providence, 1998.

[11] H. Geiges, *Contact structures on* 1*-connected* 5*-manifolds*, Mathematika, **38** (1991), 303–311.

[12] H. Geiges, *Applications of contact surgery*, Topology, **36** (1997), 1193–1220.

[13] H. Geiges, *Constructions of contact manifolds*, Math. Proc. Cambridge Philos. Soc., **121** (1997), 455–464.

[14] H. Geiges, *A brief history of contact geometry and topology,* Expo. Mat., **19** (2001), 25–53.

[15] H. Geiges and C. B. Thomas, *Contact topology and the structure of* 5*-manifolds with* $\pi_1 = \mathbb{Z}_2$, Ann. Inst. Fourier (Grenoble), **48** (1998), 1167–1188.

[16] H. Geiges and C. B. Thomas, *Contact structures, equivariant spin bordism and periodic fundamental groups*, Math. Ann., to appear.

[17] E. Giroux, *Topologie de contact en dimension* 3 *(autour des travaux de Yakov Eliashberg)*, Séminaire Bourbaki, vol. 1992/93, Astérisque, **216** (1993), 7–33.

[18] R. E. Gompf, *A new construction of symplectic manifolds*, Ann. of Math. (2), **142** (1995), 527–595.

[19] R. E. Gompf, *Handlebody construction of Stein surfaces*, Ann. of Math. (2), **148** (1998), 619–693.

[20] J. Gonzalo, *Branched covers and contact structures*, Proc. Amer. Math. Soc., **101** (1987), 347–352.

[21] M. Gromov, *Partial Differential Relations*, Ergeb. Math. Grenzgeb. (3), **9**, Springer-Verlag, Berlin, 1986.

[22] F. Hirzebruch and K. H. Mayer, $O(n)$*-Mannigfaltigkeiten, exotische Sphären und Singularitäten*, Lecture Notes in Math., **57**, Springer-Verlag, Berlin, 1968.

[23] S. Kwasik and R. Schultz, *Positive scalar curvature and periodic fundamental groups*, Comment. Math. Helv., **65** (1990), 271–286.

[24] E. Lerman, *Contact cuts*, preprint (2000).

[25] R. Lutz, *Sur la géométrie des structures de contact invariantes*, Ann. Inst. Fourier (Grenoble), **29** (1979), 283–306.

[26] R. Lutz and C. Meckert, *Structures de contact sur certaines sphères exotiques*, C. R. Acad. Sci. Paris Sér. A-B, **282** (1976), A591–A593.

[27] J. Martinet, *Formes de contact sur les variétés de dimension* 3, in: Proc. Liverpool Singularities Sympos. II, Lecture Notes in Math., **209** (Springer-Verlag, Berlin) (1971), 142–163.

[28] C. Meckert, *Forme de contact sur la somme connexe de deux variétés de contact de dimension impaire*, Ann. Inst. Fourier (Grenoble), **32** (1982), 251–260.

[29] J. Rosenberg and S. Stolz, *Manifolds of positive scalar curvature*, in: Algebraic Topology and its Applications (G.E. Carlsson et al., eds.), Math. Sci. Res. Inst. Publ., **27** (Springer-Verlag, New York) (1994), 241-267.

[30] S. Sasaki and C. J. Hsu, *On a property of Brieskorn manifolds*, Tôhoku Math. J. (2), **28** (1976), 67–78.

[31] C. B. Thomas, *Almost regular contact manifolds*, J. Differential Geom., **11** (1976), 521–533.

[32] C. B. Thomas, *Contact structures on* $(n-1)$*-connected* $(2n+1)$*-manifolds*, Banach Center Publ., **18** (1986), 255–270.

[33] C. B. Thomas (based on lectures of Y. Eliashberg and E. Giroux), 3*-dimensional contact geometry*, in: Contact and Symplectic Geometry (C.B. Thomas, ed.), Publ. Newton Inst., **8** (Cambridge University Press) (1996), 48–65.

[34] W. P. Thurston and H.E. Winkelnkemper, *On the existence of contact forms*, Proc. Amer. Math. Soc., **52** (1975), 345–347.

[35] I. Ustilovsky, *Infinitely many contact structures on* S^{4m+1}, Internat. Math. Res. Notices (1999), 781–791.

[36] A. Weinstein, *Contact surgery and symplectic handlebodies*, Hokkaido Math. J., **20** (1991), 241–251.

Mathematisch Instituut
Universiteit Leiden
Postbus 9512
2300 RA Leiden, The Netherlands
E-mail address: `geiges@math.leidenuniv.nl`

The Hamiltonian Seifert Conjecture: Examples and Open Problems

Viktor L. Ginzburg

Abstract. Hamiltonian dynamical systems tend to have infinitely many periodic orbits. For example, for a broad class of symplectic manifolds almost all levels of a proper smooth Hamiltonian carry periodic orbits. The Hamiltonian Seifert conjecture is the existence problem for regular compact energy levels without periodic orbits.

Very little is known about how large the set of regular energy values without periodic orbits can be. For instance, in all known examples of Hamiltonian flows on $\mathbb{R}^{2n}$ such energy values form a discrete set, whereas "almost existence theorems" only require this set to have zero measure. We describe constructions of Hamiltonian flows without periodic orbits on one energy level and formulate conjectures and open problems.

1. Introduction

Hamiltonian flows of proper smooth functions on $\mathbb{R}^{2n}$ or on many other symplectic manifolds are known to have periodic orbits on almost all energy levels; see [2, 15, 17]. On the other hand, there exist examples of proper smooth Hamiltonians on $\mathbb{R}^{2n}$ with one regular level carrying no periodic orbits, [4, 7, 14].

In this notice we address the question of how large the set $\mathcal{AP}_H$ of regular values of H without periodic orbits can be. We refer to this problem as the Hamiltonian Seifert conjecture (see [23, 24]). Below we recall some results and notions concerning the existence of periodic orbits and constructions of examples without periodic orbits. We also formulate conjectures on the size of the aperiodic set $\mathcal{AP}_H$. Very little is known on this problem and the conjectures made below are based exclusively on supporting evidence rather than on the lack of examples.

Another subject we briefly touch upon is Hamiltonian systems describing the motion of a charge in a magnetic field. (See [5] for an introduction and a survey of results.) These systems play an important role in the constructions of "counterexamples" to the Hamiltonian Seifert conjecture. A detailed review of these counterexamples in the context of dynamical systems can be found in [8].

The work is partially supported by the NSF and by the faculty research funds of the University of California, Santa Cruz.

Two relevant questions are entirely omitted from our discussion: the existence of periodic orbits on contact type energy levels (see, e.g., [2, 15, 17, 26, 34]) and the dynamics of the transition from low to high energy levels for twisted geodesic flows (see, e.g., [29] and [5]).

2. General Existence Results for Periodic Orbits of Hamiltonian Systems

Let W be a symplectic manifold and $H: W \to \mathbb{R}$ a proper smooth Hamiltonian. One cannot guarantee that a given regular level of H carries a periodic orbit even when $W = \mathbb{R}^{2n}$, as we will see in Section 3. However, for a broad class of symplectic manifolds it is true that periodic orbits exist on almost all levels of H. (Here "almost all" is understood in the sense of measure theory.)

Theorem 2.1. (Almost Existence of Periodic Orbits) *Let P be a compact symplectic manifold with $\pi_2(P) = 0$. Then for $W = P \times \mathbb{R}^{2l}$ with the split symplectic structure, almost all energy levels of a proper smooth Hamiltonian $H: W \to \mathbb{R}$ carry periodic orbits.*

This theorem is proved for $W = \mathbb{R}^{2n}$ by Hofer, Zehnder, and Struwe (see [16, 17, 33]) and in the general case by Floer, Hofer, and Viterbo, [2].

Analogues of Theorem 2.1 hold for many other manifolds W, for example, for $W = \mathbb{CP}^n$, [15]. (See also [26].) The theorem also holds for $W = P \times D^2(r)$, where $\pi_2(P)$ is not necessarily trivial, provided that H is proper as a function to its image and that the radius $r > 0$ is less than a certain constant $m(P, \omega)$ depending on the symplectic topology of P, [15, 28].

The Hofer-Zehnder capacity c_{HZ} provides a convenient setting for working with the almost existence problem. Let (W, ω) be a symplectic manifold. Consider the class $\mathcal{H}$ of smooth non-negative functions on W such that

1. a function $H \in \mathcal{H}$ vanishes on some open set, depending on H;
2. a function $H \in \mathcal{H}$ assumes its maximal value $m(H)$ on a complement to a compact set, which again may depend on H; and
3. the Hamiltonian flow of $H \in \mathcal{H}$ does not have non-constant periodic orbits of period $T \leq 1$.

By definition, $c_{HZ}(W, \omega) = \sup m(H)$ over all $H \in \mathcal{H}$. Clearly, c_{HZ} takes values in $(0, \infty]$. The capacity c_{HZ} is monotone with respect to inclusions and homogeneous of degree 1 with respect to conformal symplectic diffeomorphisms (in contrast with the symplectic volume, which is homogeneous of degree n, where $2n$ is the dimension). In addition, it satisfies the following normalization condition: $c_{HZ}(D^{2n}(r)) = c_{HZ}(D^2(r) \times \mathbb{R}^{2n-2}) = \pi r^2$, where $D^{2k}(r)$ is the ball of radius r in $\mathbb{R}^{2k}$ equipped with the standard symplectic structure. (The proof of the normalization condition is non-trivial. See [17] for more details and further references.)

In what follows, we say that W has bounded capacity if for every open set $U \subset W$ with compact closure, $c_{HZ}(U) < \infty$. Note that a non-compact manifold W with

$c_{HZ}(W) = \infty$ can have bounded capacity (e.g., $W = \mathbb{R}^{2n}$). However, $c_{HZ}(W) < \infty$ does imply, by monotonicity, that W has bounded capacity.

Proposition 2.2. (Hofer-Zehnder, [17]) *Assume that W has bounded capacity. Then the almost existence theorem holds for every smooth function H on W such that the sets $\{H \leq a\}$ are compact.*

This proposition is a relatively straightforward consequence of the definition of the Hofer-Zehnder capacity, see [17, Section 4.2]. However, proving that W has bounded capacity essentially amounts to showing that periodic orbits exist on a dense set of levels, which is quite non-trivial already for $\mathbb{R}^{2n}$. The manifolds $W = P \times \mathbb{R}^{2n}$ (with $\pi_2(P) = 0$), $\mathbb{CP}^n$, and also $P \times D^2(r)$ (with $r < m(P, \omega)$) have bounded capacity, [2, 16, 17, 15]), as do all manifolds for which the almost existence theorem has been established. (It seems to be unknown whether or not $\mathbb{T}^{2n}$ with the standard symplectic structure has finite capacity for $n > 1$. However, almost existence has been proven for $H\colon \mathbb{T}^{2n} \to \mathbb{R}$ satisfying some additional conditions; see, e.g., [18].)

Almost existence of periodic orbits (Theorem 2.1) does not extend to all compact symplectic manifolds:

Example 2.3. (Zehnder's torus, [36]) Let $2n \geq 4$. Consider the torus $W = \mathbb{T}^{2n}$ with an irrational translation-invariant symplectic structure ω_{irr}. Choose a Hamiltonian H on W so that every level $\{H = c\}$ with $c \in (0.5, 1.5)$ is the union of two standard embedded tori $\mathbb{T}^{2n-1} \subset \mathbb{T}^{2n}$. Since ω_{irr} is irrational, the Hamiltonian flow of H on $\mathbb{T}^{2n-1}$ is quasiperiodic. Thus, none of the levels $\{H = c\}$ with $c \in (0.5, 1.5)$ carries a periodic orbit.

According to M. Herman, [12, 13], this phenomenon is stable: the flow of a sufficiently $C^{2n+\delta}$-small perturbation of H still has no periodic orbits in the energy range $(0.6, 1.4)$, provided that ω_{irr} satisfies a certain Diophantine condition.

As a consequence of Proposition 2.2, $c_{HZ}(\mathbb{T}^{2n}, \omega_{\text{irr}}) = \infty$, even though $\mathbb{T}^{2n}$ is compact. Note also that in Zehnder's example the symplectic form is not exact on the energy levels without periodic orbits. However, many flows for which almost existence has been established (e.g., on $P \times \mathbb{R}^2$) also have this property. Thus, although it feels that non-existence of periodic orbits in Zehnder's example is forced by the topology of ω_{irr}, it is not clear how to make rigorous sense of this statement.

Now we are in a position to state the main question considered in this paper. We call a value a of H *aperiodic* if the level $\{H = a\}$ carries no periodic orbits.

> Let W be a symplectic manifold of bounded capacity and H a smooth proper function on W. How large can the set $\mathcal{AP}_H$ of regular aperiodic values of H be?

As we show in the next section, the set $\mathcal{AP}_H$ can be non-empty for an appropriately chosen function H on $\mathbb{R}^{2n}$, and hence, by Darboux's theorem, on any symplectic manifold.

3. The Hamiltonian Seifert Conjecture

3.1. Results

The *Seifert conjecture* is the question, posed by Seifert in [32], whether or not every non-vanishing vector field on S^3 has a periodic orbit. Of course, in this question the sphere S^3 can be replaced by another manifold, and additional restrictions can be imposed on the vector field. For example, the Seifert conjecture can be formulated for C^1- or C^∞-smooth or real analytic vector fields, divergence-free vector fields, etc. Note that the Seifert conjecture, when interpreted as above, is a question rather than a conjecture. Yet, we will refer to negative answers to this question as counterexamples to the Seifert conjecture.

The Seifert conjecture has a rich history extending for more than forty years. Here we mention only some of the relevant results. The first breakthrough was due to Wilson, [35], who constructed a C^∞-smooth vector field without periodic orbits on S^{2n+1}, $2n+1 \geq 5$. A C^1-smooth non-vanishing vector field without periodic orbits on S^3 was found by Schweitzer, [31], and a C^2-vector field by Harrison, [10]. Finally, a real analytic non-vanishing vector field on S^3 without periodic orbits was constructed by K. Kuperberg, [22]. The reader interested in a detailed discussion and further references should consult [8, 20, 21, 23, 24].

For Hamiltonian flows, the Seifert conjecture can be formulated in a number of ways; see [8]. For example, one may ask if there is a proper smooth function on a given symplectic manifold (e.g., $\mathbb{R}^{2n}$), having a regular aperiodic value, or, more generally, as in the previous section, if there exists a smooth proper function H with a given set $\mathcal{AP}_H$. Here we state and discuss known results and make some conjectures.

Recall that a *characteristic* of a two-form η of rank $(2n-2)$ on a $(2n-1)$-dimensional manifold is an integral curve of the line field formed by the null-spaces $\ker\eta$. The Hamiltonian Seifert conjecture (in a form slightly weaker than considered above) can be stated as whether or not in a given symplectic manifold there exists a regular compact hypersurface without closed characteristics.

Let $i\colon M \hookrightarrow W$ be an embedded smooth compact hypersurface without boundary in a $2n$-dimensional symplectic manifold (W, ω).

Theorem 3.1. ([4, 7, 14]) *Assume that $2n \geq 6$ and that $i^*\omega$ has a finite number of closed characteristics. Then there exists a C^∞-embedding $i'\colon M \hookrightarrow W$, which is C^0-close and isotopic to i, such that $i'^*\omega$ has no closed characteristics.*

An irrational ellipsoid M in the standard symplectic vector space $\mathbb{C}^n$ is the unit energy level of the quadratic Hamiltonian $H_0 = \sum \lambda_j |z_j|^2$, where the eigenvalues λ_j are positive and rationally independent. The level M carries exactly n periodic orbits. Applying Theorem 3.1, we obtain the following

Corollary 3.2. *For $2n \geq 6$, there exists a C^∞-function $H\colon \mathbb{R}^{2n} \to \mathbb{R}$, C^0-close and isotopic (with a compact support) to H_0, such that the Hamiltonian flow of H has no closed trajectories on the level set $\{H = 1\}$.*

Remark 3.3. The method used in [4, 7] to construct such a function H relies on a Hamiltonian version of Wilson's plug. The horocycle flow (see Example 4.1) is employed to interrupt periodic orbits of the flow of H_0 on the level $\{H_0 = 1\}$. The function H differs from H_0 only in n small balls with centers on these periodic orbits. A simple outline of the proof of Theorem 3.1, some generalizations, and the list of all known Hamiltonian flows with $\mathcal{AP}_H \neq \emptyset$ can be found in [8]. Here we just mention that irrational ellipsoids are not the only hypersurfaces in $\mathbb{R}^{2n}$ with a finite number of closed characteristics. Examples of such non-simply connected hypersurfaces are found by Laudenbach, [25].

The construction given in [14] uses plugs arising as symplectizations of Wilson's plug for $2n \geq 8$ and Harrison's plug [10] for $2n = 6$. (Hence, the $C^{3-\epsilon}$-smoothness of the flow obtained by this method in dimension 6.)

3.2. Conjectures and open problems

Returning to the question of the possible size of the set $\mathcal{AP}_H$, we primarily focus in this section on proper Hamiltonians H on $\mathbb{R}^{2n}$. (See the next section and [8] for results and conjectures for some other symplectic manifolds.) Recall that an upper bound on this set is given by Theorem 2.1: the set $\mathcal{AP}_H$ must have zero measure. By Theorem 3.1 and Corollary 3.2, $\mathcal{AP}_H$ can be non-empty.

There is almost no doubt that the method used in [4, 7] to modify the function H_0 can be applied, literally without any change, to a sequence of energy levels a_k converging to some value a_0. The resulting function H can then be smoothened along the level $\{H = a_0\}$ at the expense of turning a_0 into a critical value of infinite order degeneration. Thus, $\mathcal{AP}_H$ can contain a sequence converging to a singular, infinitely degenerate, value of H.

The next step is more subtle. It appears plausible that the function H_0 can be modified as in the proof of Theorem 3.1 simultaneously for all energy values from a compact zero measure set $K \subset (0, \infty)$. The resulting function H will then be smoothly isotopic to H_0 and have K as the set of aperiodic values. This leads to the following

Conjecture 3.4. *For $2n \geq 6$, there exists a C^∞-function $H\colon \mathbb{R}^{2n} \to \mathbb{R}$, C^0-close and isotopic (with a compact support) to H_0, such that the Hamiltonian flow of H has no closed trajectories for energy values in the set K.*

Remark 3.5. This method is unlikely to be applicable to the construction of a function H with a dense set $\mathcal{AP}_H$. The difficulty can be best seen by considering the function H from Corollary 3.2. The characteristic foliation of ω on the level $\{H = a\}$, where $a \neq 1$ and $a > 0$, is diffeomorphic to that on $\{H_0 = a\}$. (In fact, the forms $\omega|_{\{H=a\}}$ and $\omega|_{\{H_0=a\}}$ are conformally equivalent.) As a consequence, the flow of H on $\omega|_{\{H=a\}}$ has exactly n periodic orbits and these orbits are non-degenerate and hence stable under small perturbations. The "bifurcation" that happens at $a = 1$ is simply that the periods of these orbits go to infinity as $a \to 1$. Other systems obtained by this method from any function H_0 with non-degenerate orbits will exhibit a similar behavior: periodic orbits on $\{H = a\}$, where $a \notin \mathcal{AP}_H$, will be non-degenerate. On the other hand, for a system with a dense set $\mathcal{AP}_H$ every energy level must have only degenerate periodic orbits.

Remark 3.6. Neither the method from [4, 7] nor from [14] apply to four-dimensional manifolds. There are also serious difficulties in adapting Kuperberg's plug, [22], to the symplectic setting. A C^1-smooth divergence-free vector field on S^3 was found by G. Kuperberg, [20]. It is not known whether or not this vector field can be obtained by a C^2- (or even C^1-) embedding of S^3 into $\mathbb{R}^4$. Such an embedding would give a C^2-smooth "counterexample" to the Hamiltonian Seifert conjecture in dimension 4.

4. A Charge in a Magnetic Field

The first example of a Hamiltonian flow with a compact energy level carrying no periodic orbits was found by Hedlund in 1936 when he proved that the horocycle flow is minimal, i.e., every integral curve is dense, [11]. To put Hedlund's result in the context of the Hamiltonian Seifert conjecture, consider a compact Riemannian manifold M equipped with a closed two-form σ. Denote by ω_0 the standard symplectic form on T^*M and by $\pi\colon T^*M \to M$ the natural projection. The form $\omega = \omega_0 + \pi^*\sigma$ on T^*M is symplectic (a twisted symplectic structure). The flow of the standard metric Hamiltonian H (the kinetic energy) on (T^*M, ω) describes the motion of a charge on M in the magnetic field σ. We will refer to this flow as a *twisted geodesic flow.* The behavior of twisted geodesic flows seems to depend on whether σ is allowed to degenerate or not. In what follows we focus entirely on the case where σ is non-degenerate, i.e., symplectic.

The study of twisted geodesic flows by methods of symplectic topology originates from [1], where V. Arnold showed that for $M = \mathbb{T}^2$ with a flat metric and a non-vanishing form σ every level of H carries periodic orbits. This theorem is extended to low energy levels on other compact surfaces in [3, 6]. (See [5] for a survey of related results.) The existence of periodic orbits on low energy levels has also been established for many manifolds M, when $2m = \dim M > 2$. For example, when σ is symplectic and compatible with the metric on M, every low energy level carries at least $\mathrm{CL}(M) + m$ periodic orbits, [19], and at least $\mathrm{SB}(M)$ periodic orbits when the orbits are non-degenerate, [9]. (Here $\mathrm{CL}(M)$ is the cup-length of M and $\mathrm{SB}(M)$ is the sum of Betti numbers of M.) Note also that this problem of existence of periodic orbits can be extended to a broader class of Hamiltonian flows to include the Weinstein-Moser theorem, [9, 19].

The following example shows that the above results do not generalize to all energy levels.

Example 4.1. (The horocycle flow) Let M be a compact surface of genus $g \geq 2$ equipped with a metric of constant negative curvature $K = -1$. Let σ be the area form on M. Then the Hamiltonian flow on the energy level $\{H = 1\}$ has no periodic orbits. In fact, the flow on this energy level is smoothly equivalent to the horocycle flow (see, e.g., [6]), which is minimal by a theorem of Hedlund, [11]. This example has (Hamiltonian) analogues in all even dimensions; see [8, Example 4.2].

It is also known that a neighborhood of M in T^*M with a twisted symplectic structure has finite Hofer-Zehnder capacity under some additional assumptions on M, [27]. Moreover, for some manifolds M, the twisted cotangent bundle has bounded capacity, [9, 18, 27]. This implies almost existence of periodic orbits for a certain class of twisted geodesic flows (e.g., on $\mathbb{T}^{2m}$ with any form σ).

Conjecture 4.2. *For every M and any symplectic form σ, a neighborhood of M in T^*M has finite Hofer-Zehnder capacity. Moreover, almost all levels of H carry periodic orbits. When σ is symplectic, every low energy level of H has a contractible periodic orbit.*

Remark 4.3. This conjecture is supported by some additional evidence; see, e.g., [30]. Interestingly, there seems to be no sufficient evidence that T^*M has bounded Hofer-Zehnder capacity in general. For example, it is not known whether in the setting of Example 4.1 the sets $\{H < a\}$ with $a > 1$ have finite capacity.

Example 4.1 is the only known "naturally arising" example of a Hamiltonian flow with a regular compact aperiodic energy level. Furthermore, this is the only known example of a twisted geodesic flow with an aperiodic energy level. It is not known if there exists a twisted geodesic flow with more than one aperiodic energy level.

Acknowledgments

The author is grateful to Yael Karshon and Ely Kerman for their advice and numerous useful suggestions.

References

[1] V. I. Arnold, *On some problems in symplectic topology*, in *Topology and Geometry – Rochlin Seminar*, Ed.: O. Viro, Lecture Notes in Math. **1346**, Springer-Verlag, Berlin-New York, 1988, 1–5.

[2] A. Floer, H. Hofer and C. Viterbo, *The Weinstein conjecture in* $P \times \mathbb{C}^l$, Math. Z., **203** (1990), 469–482.

[3] V. L. Ginzburg, *New generalizations of Poincaré's geometric theorem*, Funct. Anal. Appl., **21** (2) (1987), 100–106.

[4] V. L. Ginzburg, *An embedding* $S^{2n-1} \to \mathbb{R}^{2n}$, $2n - 1 \geq 7$, *whose Hamiltonian flow has no periodic trajectories*, IMRN (1995), no. 2, 83–98.

[5] V. L. Ginzburg, *On closed trajectories of a charge in a magnetic field. An application of symplectic geometry*, in *Contact and Symplectic Geometry*, C. B. Thomas (Editor), Publications of the Newton Institute, Cambridge University Press, Cambridge, 1996, p. 131–148.

[6] V. L. Ginzburg, *On the existence and non-existence of closed trajectories for some Hamiltonian flows*, Math. Z., **223** (1996), 397–409.

[7] V. L. Ginzburg, *A smooth counterexample to the Hamiltonian Seifert conjecture in* $\mathbb{R}^6$, IMRN (1997), no. 13, 641–650.

[8] V. L. Ginzburg, *Hamiltonian dynamical systems without periodic orbits*, in *Northern California Symplectic Geometry Seminar*, Eds.: Ya. Eliashberg et al., Amer. Math. Soc. Transl. (2), **196** (1999), 35–48.

[9] V. L. Ginzburg and E. Kerman, *Periodic orbits in magnetic fields in dimensions greater than two*, in *Geometry and Topology in Dynamics*, Eds.: M. Barge and K. Kuperberg, Contemporary Mathematics, **246** (1999), 113–121.

[10] J. Harrison, *A* C^2 *counterexample to the Seifert conjecture*, Topology, **27** (1988), 249–278.

[11] G. A. Hedlund, *Fuchsian groups and transitive horocycles*, Duke Math. J., **2** (1936), 530–542.

[12] M.-R. Herman, *Examples de flots hamiltoniens dont aucune perturbations en topologie* C^∞ *n'a d'orbites périodiques sur ouvert de surfaces d'énergies*, C. R. Acad. Sci. Paris Sér. I Math., **312** (1991), 989–994.

[13] M.-R. Herman, *Différentiabilité optimale et contre-exemples à la fermeture en topologie* C^∞ *des orbites récurrentes de flots hamiltoniens*, C. R. Acad. Sci. Paris Sér. I Math., **313** (1991), 49–51.

[14] M.-R. Herman, *Communication at the conference in honor of A. Douady* Géométrie Complexe et Systèmes Dynamiques, Orsay, France, 1995.

[15] H. Hofer and C. Viterbo, *The Weinstein conjecture in the presence of holomorphic spheres*, Comm. Pure Appl. Math., **45** (1992), 583–622.

[16] H. Hofer and E. Zehnder, *Periodic solution on hypersurfaces and a result by C. Viterbo*, Invent. Math., **90** (1987), 1–9.

[17] H. Hofer and E. Zehnder, *Symplectic Invariants and Hamiltonian Dynamics*, Birkhäuser, Advanced Texts; Basel-Boston-Berlin, 1994.

[18] M.-Y. Jiang, *Periodic solutions of Hamiltonian systems on hypersurfaces in a torus*, Manuscripta Math., **85** (1994), 307–321.

[19] E. Kerman, *Periodic orbits of Hamiltonian flows near symplectic critical submanifolds*, IMRN (1999) no. 17, 953–969.

[20] G. Kuperberg, *A volume-preserving counterexample to the Seifert conjecture*, Comment. Math. Helv., **71** (1996), 70–97.

[21] G. Kuperberg and K. Kuperberg, *Generalized counterexamples to the Seifert conjecture*, Ann. Math., **144** (1996), 239–268.

[22] K. Kuperberg, *A smooth counterexample to the Seifert conjecture in dimension three*, Ann. Math., **140** (1994), 723–732.

[23] K. Kuperberg, *Counterexamples to the Seifert conjecture*, Proceedings of the International Congress of Mathematicians, Vol. II (Berlin, 1998). Doc. Math. (1998) Extra Vol. II, 831–840.

[24] K. Kuperberg, *Aperiodic dynamical systems*, Notices Amer. Math. Soc., **46** (1999), 1035–1040.

[25] F. Laudenbach, *Trois constructions en topologie symplectique*, Ann. Fac. Sci. Toulouse Math., **6** (1997), 697–709.

[26] G. Liu and G. Tian, *Weinstein conjecture and GW invariants*, Commun. Cont. Math., **2** (2000), 405–459.

[27] G. Lu, *Periodic motion of a charge on a manifold in the magnetic field*, Preprint 1999, math.DG/9905146.

[28] R. Ma, *Symplectic capacity and Weinstein conjecture in certain cotangent bundles and Stein manifolds*, NoDEA Nonlinear Differential Equations Appl., **2** (1995), 341–356.

[29] G. P. Paternain and M. Paternain, *First derivative of topological entropy for Anosov geodesic flow in the presence of magnetic field*, Nonlinearity, **10** (1997), 121–131.

[30] L. Polterovich, *Geometry on the group of Hamiltonian diffeomorphisms*, Proceedings of the International Congress of Mathematicians, Vol. II (Berlin, 1998). Doc. Math. 1998, Extra Vol. II, 401–410.

[31] P. A. Schweitzer, *Counterexamples to the Seifert conjecture and opening closed leaves of foliations*, Ann. Math., **100** (1970), 229–234.

[32] H. Seifert, *Closed integral curves in the 3-space and 2-dimensional deformations*, Proc. Amer. Math. Soc., **1** (1950), 287–302.

[33] M. Struwe, *Existence of periodic solutions of Hamiltonian systems on almost every energy surfaces*, Bol. Soc. Bras. Mat., **20** (1990), 49–58.

[34] C. Viterbo, *A proof of Weinstein's conjecture in* $\mathbb{R}^{2n}$, Ann. Inst. H. Poincaré, Anal. Non Linéaire, **4** (1987), 337–356.

[35] F. Wilson, *On the minimal sets of non-singular vector fields*, Ann. Math., **84** (1966), 529–536.

[36] E. Zehnder, *Remarks on periodic solutions on hypersurfaces*, in *Periodic Solutions of Hamiltonian Systems and Related Topics* Eds.: P. Rabinowitz et al., Reidel Publishing Co. (1987), 267–279.

Department of Mathematics
UC Santa Cruz
Santa Cruz, CA 95064, USA
E-mail address: `ginzburg@math.ucsc.edu`

The Fine Geometry of the Cantor Families of Invariant Tori in Hamiltonian Systems

Àngel Jorba and Jordi Villanueva

Abstract. This work focuses on the dynamics around a partially elliptic, lower dimensional torus of a real analytic Hamiltonian system. More concretely, we investigate the abundance of invariant tori in the directions of the phase space corresponding to elliptic modes of the torus. Under suitable (but generic) non-degeneracy and non-resonance conditions, we show that there exist plenty of invariant tori in these elliptic directions, and that these tori are organized in manifolds that can be parametrized on suitable Cantor sets. These manifolds can be seen as "Cantor centre manifolds", obtained as the nonlinear continuation of any combination of elliptic linear modes of the torus. Moreover, for each family, the density of the complementary of the set filled up by these tori is exponentially small with respect to the distance to the initial torus. These results are valid in the limit cases when the initial torus is an equilibrium point or a maximal dimensional torus. It is remarkable that, in the case in which the initial torus is totally elliptic, we can derive Nekhoroshev-like estimates for the diffusion time around the torus. Due to the use of weaker non-resonance conditions, these results are an improvement on previous results [7].

1. Introduction

The study of the dynamics in a neighbourhood of a given invariant manifold is a classical topic in Dynamical Systems. Here we will focus on the description of the phase space around an invariant torus of a real analytic Hamiltonian system.

Let H be a real analytic Hamiltonian system with ℓ degrees of freedom, having an r-dimensional invariant torus, $0 \leq r \leq \ell$. We assume that this torus is isotropic (the canonical symplectic form of $\mathbb{R}^{2\ell}$ vanishes on the tangent bundle of the torus). Hence, there exists a set of canonical coordinates $(\theta, x, I, y) \in \mathbb{T}^r \times \mathbb{R}^m \times \mathbb{R}^r \times \mathbb{R}^m$, $m = \ell - r$, in such a way that the torus corresponds to the manifold $\{x = y = 0,\, I = 0\}$, and it is parameterized by the angular variables θ. Moreover, we will also assume that the dynamics on the torus is a linear quasi-periodic flow with respect to θ, given by the vector of basic frequencies $\omega^{(0)} \in \mathbb{R}^r$. So, in these (canonical) coordinates the Hamiltonian can be written as

$$H(\theta, x, I, y) = \langle \omega^{(0)}, I \rangle + \frac{1}{2}\langle z, B(\theta) z \rangle + H_1(\theta, x, I, y)\,, \tag{1}$$

where $z = (x, y)$ and the Taylor expansion of H_1, with respect to (z, I), starts with monomials of degree 3 (as is usual in this context, the degree of a monomial $z^l I^s$, $l \in \mathbb{N}^{2m}$, $s \in \mathbb{N}^r$, is defined as $|l|_1 + 2|s|_1$, being $|l|_1 = \sum_j |l_j|$).

As usual in KAM theory for lower dimensional tori, we will restrict to the context of the so-called reducible tori ([3]). So, we will assume that the normal variational equations of the torus, which are given by $\dot{z} = J_m B(\omega^{(0)} t) z$ (J_m denotes the matrix of the canonical 2-form of $\mathbb{R}^{2m}$), can be reduced to constant coefficients by means of a linear change of variables in z, depending on time in a quasi-periodic way (with $\omega^{(0)}$ as vector of basic frequencies). Then, we can perform a canonical transformation on the initial Hamiltonian system (depending on θ in a 2π-periodic way), such that (1) takes the form

$$H(\theta, x, I, y) = \langle \omega^{(0)}, I \rangle + \frac{1}{2} \langle z, Bz \rangle + H_1(\theta, x, I, y), \tag{2}$$

where B is a symmetric matrix with constant coefficients. To simplify the notations, here we have also kept the names H and H_1 for the Hamilton function and for the higher order terms. Moreover, in order to have a Hamiltonian given by a real function, we restrict our set-up to the case in which the linear transformation is given by a real change of variables, and hence, we assume that we have a real reduced matrix B. The linear normal behavior around the torus is now given by the system $\dot{z} = J_m B z$. For definiteness, we denote the eigenvalues of $J_m B$ as $(\lambda, -\lambda) = (\lambda_1, \dots, \lambda_m, -\lambda_1, \dots, -\lambda_m)$. As usual, we will call the imaginary part of these eigenvalues the normal frequencies of the torus.

We are concerned with the local behaviour around this lower dimensional torus. Let us start by looking at the linear approximation to this dynamics. As the dynamics in the hyperbolic directions (eigenvalues with nonzero real part) is locally very simple – trajectories either escape or tend asymptotically to the torus – we will focus on the elliptic directions. For simplicity, let us start by assuming that $\pm\lambda_1$ are a conjugate pair of purely imaginary eigenvalues, $\lambda_1 = i\omega^{(0)}_{r+1}$. It is clear that the dynamics in the corresponding two-dimensional real invariant space is that of a linear oscillator, with frequency $\omega^{(0)}_{r+1}$. More concretely, this linear subspace is filled by a 1-parametric family of periodic orbits whose frequency is fixed, i.e., it does not depend on the amplitude. Hence, taking into account that these linear oscillations take place around an r-dimensional torus carrying a linear dynamics with frequency $\omega^{(0)}$, we have that the (linear approximation to the) dynamics in the (real) eigenspace corresponding to $\pm\lambda_1$ is described by a 1-parametric family of $(r+1)$-dimensional tori, having $(\omega^{(0)}, \omega^{(0)}_{r+1})$ as a vector of basic frequencies. This family is degenerate, in the sense that the frequencies are constant. A natural question is the effect that the nonlinear terms have on this picture. We will show that, under generic conditions, this family "survives" the effect of the nonlinear terms, but it becomes Cantorian (and now, the frequencies depend on the amplitudes), see [4, 17, 20, 7, 21]. The dense set of holes that conform the Cantor structure is due to the resonances between the basic and normal frequencies. We will also show that the size of those holes is exponentially small with respect to the distance to the initial torus.

The above description has been done using a single normal frequency and a single r-dimensional torus. A similar discussion can be made using several normal frequencies: if we select a set of m_1 normal frequencies then, in the corresponding directions, the dynamics of the linear system is described by an m_1-parametric family of m_1-dimensional tori, all of them having the same frequencies. Moreover, if we skip the term H_1 in (2), it is clear that the initial torus is embedded in an r-parametric family of r-dimensional tori (parameterized by I), that also becomes Cantorian when we consider the nonlinear part (see [4, 17, 3, 7, 21]). Combining this family with the families in the normal directions we obtain an $(r+m_1)$-parametric family of $(r+m_1)$-dimensional tori. As we will see, most of these tori survive the effect of H_1, and are organized as Cantor manifolds. As before, the measure of the holes of the Cantor structure is exponentially small with respect to the distance to the initial torus.

In the next sections we will discuss these issues in more detail, and we will give some hints about the scheme used in the proofs (full details can be found in [7] and [9]). Moreover, in the case in which the torus is totally elliptic, we will derive lower bounds on the escaping time from a neighbourhood of the torus.

2. Hypotheses and Set-Up

We will assume that the Hamiltonian (2) is a real analytic function, defined on

$$\mathcal{D}_{r,m}(\rho_0, R_0) = \left\{(\theta, x, I, y) \in \mathbb{C}^{2\ell} \colon |\mathrm{Im}(\theta)| \leq \rho_0,\ |z| \leq R_0,\ |I| \leq R_0^2\right\}, \qquad (3)$$

for certain $0 < \rho_0 < 1$ and $0 < R_0 < 1$, where $|\cdot|$ refers to the sup norm of complex vectors. We will denote by $\|\cdot\|_{\rho,R}$ the sup norm of a function on the set $\mathcal{D}_{r,m}(\rho, R)$ and we will also assume that $\|H\|_{\rho_0,R_0} < +\infty$.

As we are interested in the dynamics in the elliptic directions of the torus, let us define m_e as the number of elliptic directions of the initial torus, and let us write $\lambda = (\lambda^e, \lambda^h)$, where $\pm\lambda^e$ denotes the $2m_e$ elliptic eigenvalues, and $\pm\lambda^h$ the $2m_h$ hyperbolic ones ($m = m_e + m_h$). Let m_1 be any number of elliptic directions ($0 \leq m_1 \leq m_e$) that, without loss of generality, we will assume correspond to the first m_1 elliptic eigenvalues. So, we will write $\lambda^e = (\widetilde{\lambda}^e, \widehat{\lambda}^e)$, where $\widetilde{\lambda}^e$ are these first m_1 components and $\widehat{\lambda}^e$ the remaining ones. Moreover, we will denote $\widetilde{\lambda}^e = i\widetilde{\omega}^{(0)}$, and $\Omega^{(0)} = (\omega^{(0)}, \widetilde{\omega}^{(0)}) \in \mathbb{R}^{r+m_1}$. We want to construct the family of invariant tori for the elliptic directions corresponding to the frequency $\widetilde{\omega}^{(0)}$ (so, the basic frequencies of this family will be close to $\Omega^{(0)}$). These tori will have $m_e - m_1$ elliptic directions and m_h hyperbolic ones.

2.1. Non-resonance conditions

An important point for the quantitative analysis is the control of the small divisors. So, we will assume that some linear combinations of basic frequencies and normal eigenvalues satisfy a standard Diophantine condition. More concretely, we suppose that, for certain $c > 0$ and $\gamma > r + m_1 - 1$, we have

$$\left| i\langle k, \Omega^{(0)} \rangle + \langle \widehat{\lambda}^e, \widehat{l}^e \rangle \right| \geq \frac{c}{(|k|_1 + |\widehat{l}^e|_1)^\gamma}, \tag{4}$$

for any $k \in \mathbb{Z}^{r+m_1}$ and any $\widehat{l}^e \in \mathbb{Z}^{m_e - m_1}$, with $|\widehat{l}^e|_1 \leq 2$, $|k|_1 + |\widehat{l}^e|_1 \neq 0$. It is well known that if we take a fixed $\gamma > r + m_1 - 1$, the set of $(\Omega^{(0)}, \widehat{\lambda}^e)$ for which these conditions are not fulfilled, for any $c > 0$, has zero measure.

3. Normal Form

This is the first step of the methodology used. The purpose is to apply a (canonical) change of variables to bring Hamiltonian (2) into a more convenient form. So, we will perform a finite number of normal form transformations, in order to kill some specific monomials of the Taylor-Fourier expansion of the Hamiltonian. In order to enumerate these monomials, we introduce the following notation. Let us split z as (z^e, z^h), where z^e denotes the variables corresponding to the elliptic normal directions, and z^h to the hyperbolic ones. Moreover, we write $z^e = (\widetilde{z}^e, \widehat{z}^e)$, where $\widetilde{z}^e$ refers to the variables related to the first m_1 elliptic directions (the corresponding frequencies are $\widetilde{\omega}^{(0)}$), and $\widehat{z}^e$ refers to the remaining elliptic directions. Finally, z^n is defined as $(\widehat{z}^e, z^h)$ (z^n will be the normal directions to the new family of invariant tori). Hence, the monomials we would like to remove are:

1. all the monomials of order one in z^n; then, $z^n = 0$ is an invariant manifold,
2. all the monomials that do not contain z^n (except the unavoidable resonances); then, the manifold $z^n = 0$ is foliated by invariant tori,
3. all the monomials of order two in z^n that involve either one elliptic and one hyperbolic direction or two elliptic directions (except, of course, the unavoidable resonances); then, the elliptic part of the normal variational equation of these tori is reduced to constant coefficients.

As is well known, killing these monomials at all orders is a (generically) divergent process, so we will remove them up to a suitable finite degree. This degree will be selected depending on the distance to the initial torus, in order to have an exponentially small remainder with respect to that distance (see Theorem 3.1). To write this normal form, let $\widetilde{I}_j$ be $\frac{1}{2}((x_j^e)^2 + (y_j^e)^2)$ $(j = 1, \dots, m_1)$, and $\bar{I} = (I, \widetilde{I})$. Then, skipping the higher order terms that do not belong to the normal form, we have that

$$H(\theta, x, I, y) = F(\bar{I}) + \frac{1}{2}\langle z^n, Q^n(\theta, I, \widetilde{z}^e) z^n \rangle + O_3(z^n), \tag{5}$$

where the matrix Q^n takes the following form:

$$\langle z^n, Q^n(\theta, I, \widetilde{z}^e) z^n \rangle = \langle \hat{z}^e, Q^e(I) \hat{z}^e \rangle + \langle z^h, Q^h(\theta, I, \widetilde{z}^e) z^h \rangle,$$

and $\partial_{\bar{I}} F(0) = \Omega^{(0)}$. The action $\widetilde{I}$ can be taken as a parameter in the normal direction. It is easy to check that we have the following quasi-periodic solutions for the canonical equations associated to (5):

$$\theta(t) = \partial_I F(\bar{I}^{(0)})t + \theta^{(0)}, \quad \widetilde{x}_j^e(t) = \sqrt{2\widetilde{I}_j^{(0)}} \sin\left(\partial_{\widetilde{I}_j} F(\bar{I}^{(0)})t + \widetilde{\theta}_j^{(0)}\right), \quad z^n(t) = 0,$$

$$I(t) = I^{(0)}, \qquad \widetilde{y}_j^e(t) = \sqrt{2\widetilde{I}_j^{(0)}} \cos\left(\partial_{\widetilde{I}_j} F(\bar{I}^{(0)})t + \widetilde{\theta}_j^{(0)}\right), \tag{6}$$

where $\bar{I}^{(0)} = (I^{(0)}, \widetilde{I}^{(0)})$ and $\widetilde{\theta}_j^{(0)}$ are arbitrary initial conditions. The following result contains the estimates for the normal form.

Theorem 3.1. *For any $R > 0$ small enough, there exists a real analytical canonical transformation $\Psi^R\colon \mathcal{D}_{r,m}(\rho_0/2, R\,\mathrm{e}^{-\rho_0/2}) \mapsto \mathcal{D}_{r,m}(\rho_0, R)$, with $\Psi^R - \mathrm{Id}$ 2π-periodic in θ and small. More concretely, if we write $\Psi^R - \mathrm{Id} = (\Theta^R, \mathcal{X}^R, \mathcal{I}^R, \mathcal{Y}^R)$, and $\mathcal{Z}^R = (\mathcal{X}^R, \mathcal{Y}^R)$, we have*

$$\|\Theta^R\| \le \rho_0/2, \quad \|\mathcal{I}^R\| \le R^2(1 - \mathrm{e}^{-4\rho_0}), \quad \|\mathcal{Z}^R\| \le R(1 - \mathrm{e}^{-2\rho_0}).$$

The transformed Hamiltonian $H^R = H \circ \Psi^R$ takes the form

$$H^R(\theta, x, I, y) = F^R(\bar{I}) + \frac{1}{2}\langle z^n, Q^{R,n}(\theta, I, \widetilde{z}^e) z^n\rangle + S^R(\theta, x, I, y) + T^R(\theta, x, I, y),$$

where, skipping the term S^R, we have the normal form (5), with $T^R \equiv O_3(z^n)$. Moreover, we have the following estimate:

$$\|S^R\| \le c_1\,\mathrm{e}^{-(c_1/R)^{\frac{2}{\gamma+1}}},$$

where $c_1 > 0$ is a constant. All the norms are taken on the set $\mathcal{D}_{r,m}(\rho_0/2, R\,\mathrm{e}^{-\rho_0/2})$.

We point out that the kind of dependence of Ψ^R on R does not matter for our purposes. What follows from the proof is that it is piecewise analytic.

Remark 3.2. *If the torus is totally elliptic ($m_h = 0$), the bound on the remainder S^R immediately implies a Nekhoroshev estimate on the diffusion time around the torus (see [14, 16, 12, 7, 15]). More concretely, it can be shown that the time needed to go from a distance of R to a distance of $2R$ has to be bigger than $c_2 \exp((c_2/R)^{2/(\gamma+1)})$, for some constant $c_2 > 0$. This includes the limit cases of elliptic equilibrium points as well as of maximal dimensional tori.*

4. Cantor Manifolds of Lower Dimensional Tori

The usual schemes to prove the existence of invariant tori are based on an iterative process with quadratic convergence (Newton method), to overcome the effect of the small divisors ([10, 1]). These methods are usually constructed by means of a sequence of canonical transformations, that puts the initial Hamiltonian into a form where the existence of the tori follows immediately.

The equations that define the canonical transformations used at each step of the Newton method are given by a linear system of partial differential equations. In order to simplify the resolution of these equation we only deal with tori that are reducible in the elliptic directions: in this way, we directly solve this part of the equations by reducing them to constant coefficients while the hyperbolic part can be easily solved by means of a standard fixed point scheme ([5]). The reducibility has been imposed in the construction of the normal form in Section 3 and in the Diophantine condition (4) by asking for $|\hat{l}^e|_1 = 2$. Following the ideas in [2], it should be possible to construct invariant tori without asking for reducibility, that is, only asking for $|\hat{l}^e|_1 \leq 1$ in (4).

As usual in these situations, we need to control the frequencies along the iterative process, to avoid divisors that are too small. To this end we will use suitable non-degeneracy conditions to control the dependence of the frequencies with respect to a suitable set of parameters. The internal frequencies of the tori in the family can be easily controlled by the corresponding actions $\bar{I}$ (see hypothesis H1 in Theorem 4.1). The control of the frequencies in the elliptic directions is more involved, since we do not have free parameters available (this constitutes the so-called lack-of-parameters problem, [13, 6, 22, 3, 8]). Here we also use the actions $\bar{I}$, by asking for a suitable transversality condition with respect to the set of resonant frequencies (see hypothesis H2 in Theorem 4.1).

Note that the lower bound (4) for the small divisors appears in the denominators of the coefficients of the series involved in the Newton process. As this iterative process will be applied to the Hamiltonian obtained from Theorem 3.1, the remainder to be killed is exponentially small. Then, we can select the value of c in (4) to be of the same order as this remainder, such that the measure of the frequencies that do not satisfy (4) is also exponentially small. Then, the above-mentioned non-degeneracy conditions allows us to complete the proof, except for a set of frequencies of exponentially small measure.

Theorem 4.1. *Assume the same hypotheses as in Theorem 3.1, and compute the normal form (5) up to degree 2. Let us define $\mu(\bar{I})$ as the normal frequencies of the elliptic directions of the family of lower dimensional tori in this normal form. Then, we assume the following non-degeneracy conditions:*

H1: $\det \partial^2_{\bar{I},\bar{I}} F(0) \neq 0$,

H2: $\mathrm{Re}(l^\top \partial_{\bar{I}}\mu(0)(\partial^2_{\bar{I},\bar{I}}F(0))^{-1}) \notin \mathbb{Z}^{r+m_1}$, *for any* $l \in \mathbb{Z}^{m_e-m_1}$, $0 < |l|_1 \leq 2$.

Then, there exists $c_0 > 0$ such that for any $R > 0$ small enough, there is a Cantor set $\mathcal{W}_R \subset \mathcal{V}_R = \{\Omega \in \mathbb{R}^{r+m_1} : |\Omega - \Omega^{(0)}| \leq c_0 R^2\}$ such that, for any $\Omega \in \mathcal{W}_R$, the Hamiltonian system H has an invariant torus of dimension $r + m_1$, having Ω as vector of basic frequencies, that is close to the one predicted by the normal form. The elliptic directions of the normal variational equations of these tori are reducible. Moreover, $\mathcal{W}_R$ can be characterized in terms of the Lebesgue measure:

$$\mathrm{meas}(\mathcal{V}_R \setminus \mathcal{W}_R) \leq c_0\, \mathrm{e}^{-(c_0/R)^{\frac{2}{\gamma+1}}} .$$

Conditions H2 are usually called first order Melnikov conditions (see [11, 4]).

Remark 4.2. *Condition H1 is a standard (and generic) non-degeneracy condition on the dependence of the basic frequencies on the actions ([10, 1]). A similar result can be proved replacing H1 by a higher order non-degeneracy condition ([18, 19]).*

Remark 4.3. *It can also be proved that the invariant Cantor manifold can be extended to a (non-invariant) C^∞ manifold that contains all the invariant tori in the family (see [23]). The measure of the complementary of the set of tori on this manifold is also exponentially small.*

Remark 4.4. *Some of the invariant tori provided by Theorem 4.1 are "complex tori", in the sense that the corresponding quasi-periodic trajectories (for real time) live in $\mathbb{C}^{2\ell}$. This follows immediately from (6) by selecting some $\widetilde{I}_j$ negative.*

Acknowledgements

The authors are indebted to M. Sevryuk for his remarks. This research has been supported by the Spanish grant DGICYT PB94–0215, the Catalan grant CIRIT 1998S0GR–00042 and the INTAS project 97–10771.

References

[1] V. I. Arnold. Proof of A. N. Kolmogorov's theorem on the preservation of quasi-periodic motions under small perturbations of the Hamiltonian. *Russian Math. Surveys*, 18(5):9–36, 1963.

[2] J. Bourgain. On Melnikov's persistency problem. *Math. Res. Lett.*, 4(4):445–458, 1997.

[3] H. W. Broer, G. B. Huitema, and M. B. Sevryuk. *Quasi-Periodic Motions in Families of Dynamical Systems: Order amidst Chaos*, volume 1645 of *Lecture Notes in Math.* Springer-Verlag, New York, 1996.

[4] L. H. Eliasson. Perturbations of stable invariant tori for Hamiltonian systems. *Ann. Sc. Norm. Super. Pisa, Cl. Sci.*, 15(1):115–147, 1988.

[5] S. Graff. On the conservation of hyperbolic invariant tori for Hamiltonian systems. *J. Differential Equations*, 15(1):1–69, 1974.

[6] À. Jorba and C. Simó. On the reducibility of linear differential equations with quasiperiodic coefficients. *J. Differential Equations*, 98:111–124, 1992.

[7] À. Jorba and J. Villanueva. On the normal behaviour of partially elliptic lower dimensional tori of Hamiltonian systems. *Nonlinearity*, 10:783–822, 1997.

[8] À. Jorba and J. Villanueva. On the persistence of lower dimensional invariant tori under quasi-periodic perturbations. *J. Nonlinear Sci.*, 7:427–473, 1997.

[9] À. Jorba and J. Villanueva. On the density of quasi-periodic motions around a partially elliptic lower dimensional torus. In preparation.

[10] A. N. Kolmogorov. On the persistence of conditionally periodic motions under a small change of the Hamilton function. *Dokl. Acad. Nauk. SSSR*, 98(4):527–530, 1954.

[11] V. K. Melnikov. On some cases of the conservation of conditionally periodic motions under a small change of the Hamiltonian function. *Soviet Math. Dokl.*, 6:1592–1596, 1965.

[12] A. Morbidelli and A. Giorgilli. Superexponential stability of KAM tori. *J. Statist. Phys.*, 78(5–6):1607–1617, 1995.

[13] J. Moser. Convergent series expansions for quasi-periodic motions. *Math. Ann.*, 169:136–176, 1967.

[14] N. N. Nekhoroshev. An exponential estimate of the time of stability of nearly-integrable Hamiltonian systems. *Russian Math. Surveys*, 32:1–65, 1977.

[15] L. Niederman. Nonlinear stability around an elliptic equilibrium point in a Hamiltonian system. *Nonlinearity*, 11(5):1465–1479, 1998.

[16] A. D. Perry and S. Wiggins. KAM tori are very sticky: rigorous lower bounds on the time to move away from an invariant Lagrangian torus with linear flow. *Phys. D*, 71:102–121, 1994.

[17] J. Pöschel. On elliptic lower dimensional tori in Hamiltonian systems. *Math. Z.*, 202(4):559–608, 1989.

[18] H. Rüssmann. Non-degeneracy in the perturbation theory of integrable dynamical systems. In *Number theory and dynamical systems*, volume 134 of *Lond. Math. Soc. Lect. Note Ser.*, pages 5–18. 1989.

[19] M. B. Sevryuk. Invariant tori of hamiltonian systems that are nondegenerate in Rüssman's sense. *Dokl. Akad. Nauk, Ross. Akad. Nauk*, 346(5):590–593, 1996.

[20] M. B. Sevryuk. Excitation of elliptic normal modes of invariant tori in Hamiltonian systems. In A. G. Khovanskii, A. N. Varchenko, and V. A. Vassiliev, editors, *Topics in Singularity Theory. V. I. Arnold's 60th Anniversary Collection*, volume 180 of *American Mathematical Society Translations-Series 2, Advances in the Mathematical Sciences*, pages 209–218. American Mathematical Society, Providence, Rhode Island, 1997.

[21] M. B. Sevryuk. Invariant tori of intermediate dimensions in Hamiltonian systems. *Regul. Chaotic Dyn.*, 3(1):39–48, 1998.

[22] M. B. Sevryuk. The lack-of-parameters problem in the KAM theory revisited. In C. Simó, editor, *Hamiltonian Systems with Three or More Degrees of Freedom*, NATO Adv. Sci. Inst. Ser. C Math. Phys. Sci., pages 568–572. Held in S'Agaró, Spain, 19–30 June 1995. Kluwer Acad. Publ., Dordrecht, Holland, 1999.

[23] H. Whitney. Analytic extensions of differentiable functions defined in closed sets. *Trans. Amer. Math. Soc.*, 36(1):63–89, 1934.

À. Jorba
Departament de Matemàtica Aplicada i Anàlisi
Universitat de Barcelona
Gran Via 585, 08007 Barcelona, Spain
E-mail address: `angel@maia.ub.es`

J. Villanueva
Departament de Matemàtica Aplicada I
Universitat Politècnica de Catalunya
Diagonal 647, 08028 Barcelona, Spain
E-mail address: `jordi@vilma.upc.es`

Symplectic and Contact Geometry and Hamiltonian Dynamics

Mikhail B. Sevryuk

Abstract. This is an introduction to the contributions by the lecturers at the mini-symposium on symplectic and contact geometry. We present a very general and brief account of the prehistory of the field and give references to some seminal papers and important survey works.

Symplectic geometry is the geometry of a closed nondegenerate two-form on an even-dimensional manifold. Contact geometry is the geometry of a maximally nondegenerate field of tangent hyperplanes on an odd-dimensional manifold. The symplectic structure is fundamental for Hamiltonian dynamics, and in this sense symplectic geometry (and its odd-dimensional counterpart, contact geometry) is as old as classical mechanics. However, the science the present mini-symposium is devoted to is usually believed to date from H. Poincaré's "last geometric theorem" [70] concerning fixed points of area-preserving mappings of an annulus:

Theorem 1. (Poincaré-Birkhoff) *An area-preserving diffeomorphism of an annulus $\mathbb{S}^1\times[0;1]$ possesses at least two fixed points provided that it rotates two boundary circles in opposite directions.*

This theorem proven by G. D. Birkhoff [16] was probably the first statement describing the properties of symplectic manifolds and symplectomorphisms "in large", thereby giving birth to *symplectic topology.*

In the mid 1960s [2, 3] and later in the 1970s ([4, 5, 6] and [7, Appendix 9]), V. I. Arnol'd formulated his famous conjecture generalizing Poincaré's theorem to higher dimensions. This conjecture reads as follows:

Conjecture 2. (Arnol'd) *A flow map A of a (possibly nonautonomous) Hamiltonian system of ordinary differential equations on a closed symplectic manifold M possesses at least as many fixed points as a smooth function on M must have critical points, both "algebraically" and "geometrically".*

The weakened "algebraic" version of this conjecture asserts that the number of fixed points of A counting multiplicities is no less than the sum of the Betti numbers (over $\mathbb{Z}$) of manifold M. The weakened "geometric" version states that the number of geometrically distinct fixed points of A is no less than the Lyusternik-Schnirel'man category of manifold M. For instance, a flow map of a Hamiltonian system on the torus $\mathbb{T}^{2n}$ possesses at least $2n+1$ geometrically distinct fixed points, and at least 4^n fixed points counting multiplicities. It is worthwhile to emphasize here that the "correct" higher-dimensional generalization of area-preserving two-dimensional mappings in this theory is symplectomorphisms rather than volume-preserving diffeomorphisms. More precisely, Conjecture 2 considers symplectomorphisms that are flow maps of (possibly nonautonomous) Hamiltonian systems (such symplectomorphisms are said to be *homological to the identity*).

The Arnol'd conjecture has affected greatly the development of the theory of symplectic manifolds in the subsequent years. The first noticeable step here was Ya. M. Èliashberg's proof [23] of this conjecture for all the two-dimensional surfaces. Of other important achievements in symplectic geometry and topology in the 1970s and early 1980s, one should mention A. Weinstein's results on Lagrangian submanifolds [80] and Èliashberg's theorem [24] on the so-called *rigidity*, or *hardness*, of symplectomorphisms (discussed previously by him and M. L. Gromov since the late 1960s):

Theorem 3. **(Èliashberg-Gromov)** *The group of symplectomorphisms of a closed symplectic manifold is C^0-closed in the group of all diffeomorphisms.*

Theorem 3 is often referred to as "the existence theorem of symplectic topology" [8]. It shows that symplectic geometry is an intrinsically topological science.

In their milestone paper [19], C. C. Conley and E. Zehnder proved Conjecture 2 for tori $\mathbb{T}^{2n}$ of all the even dimensions with the standard symplectic structure. They introduced a new technique of constructing a certain action functional on the space of contractible loops on the manifold. This technique can be regarded as a hyperbolic analogue of the Morse theory for positive functionals. Work [19], together with Gromov's celebrated paper [41] on the so-called pseudo-holomorphic curves (two-dimensional submanifolds that are symplectic analogues of geodesics) in a symplectic manifold, marked the beginning of the modern period of symplectic and contact topology, cf. [8]. In particular, Gromov [41] gave a new proof of the rigidity of symplectomorphisms and proved the following fundamental *nonsqueezing theorem.* Let $B^{2n}(R)$ denote the closed ball with center 0 and radius R in $\mathbb{R}^{2n}$ equipped with the standard symplectic structure.

Theorem 4. **(Gromov)** *There is no symplectic embedding $B^{2n}(R) \hookrightarrow B^2(r)\times\mathbb{R}^{2n-2}$ for $R > r$.*

This theorem shows that the symplectic invariants (called *symplectic capacities*) are essentially two-dimensional.

By now, symplectic/contact geometry/topology and the related aspects of Hamiltonian dynamics have turned into a vast and flourishing branch of mathematics which can definitely not be surveyed during 4.5 hours of the mini-symposium. The contributions collected here should therefore be thought of as just a certain "snapshot" of several active studies and interesting results in the field. Instead of trying to trace the development of the theory of symplectic and contact manifolds since 1985 or reviewing the state of the art, we will present here a brief account of each of the topics selected for the mini-symposium to give them a unity.

Two lectures are devoted to the Arnol'd conjecture discussed above. In the late 1980s, A. Floer published a series of very important papers (of which we cite here only three, [30, 31, 32]) where he, apart from other achievements, combined the variational approach by Conley and Zehnder [19] with Gromov's elliptic methods [41] and defined what has become known as the Floer (co)homology theory. This enabled him to prove Conjecture 2 for the so-called *positive*, or *monotone*, symplectic manifolds [32]. Afterwards, Floer's landmark result was generalized by H. Hofer and D. A. Salamon [45] and by K. Ono [69] to *semi-positive*, or *weakly monotone*, manifolds (in particular, to all the symplectic manifolds of dimensions ≤ 6), and by G. C. Lu [58, 59], to products of weakly monotone manifolds (and Calabi-Yau manifolds). Finally, a further extension of Floer's ideas and the theory of the so-called Gromov-Witten invariants have led K. Fukaya-K. Ono, H. Hofer-D. A. Salamon, J. Li-G. Liu-G. Tian, Y. B. Ruan, and B. Siebert to a proof of the weakened Arnol'd conjecture for every closed symplectic manifold (for the case where all the fixed points are nondegenerate), we would confine ourselves to four references [34, 35, 56, 57]. The lecture by Salamon surveys this stream of studies in symplectic topology.

A quite different approach to the Arnol'd conjecture was proposed by B. Fortune [33] who proved it for projective spaces $\mathbb{C}\mathrm{P}^n$ with the standard symplectic structure [7, Appendix 3]. This proof was based on the fact that $\mathbb{C}\mathrm{P}^n$ is the reduced symplectic manifold of $\mathbb{C}^{n+1}$ under the Hopf $\mathbb{S}^1$-action and any Hamiltonian system on $\mathbb{C}\mathrm{P}^n$ is the Marsden-Weinstein reduction of an appropriate Hamiltonian system on $\mathbb{C}^{n+1}$. L. A. Ibort and C. Martínez Ontalba [49] showed that Fortune's method is in fact universal: the fixed point problem for a symplectomorphism (homological to the identity) of every closed symplectic manifold can be translated into a critical point problem with symmetry on loops in the space $\mathbb{R}^{2N}$ (for suitable N) endowed with the standard symplectic structure. All these questions are treated in Ibort's talk.

The lecture by P. Biran considers the interesting problem of *symplectic packing*: given a closed symplectic manifold M of dimension $2n$, what is the supremum $\nu_k(M)$ of volumes that can be filled by symplectic embeddings of k equal disjoint balls $B^{2n}(R)$ into M? This question was first addressed by Gromov [41] as an extension of the nonsqueezing phenomenon: whereas volume-preserving packing is obvious, there do exist obstructions to symplectic packing, and the latter turns out to be highly nontrivial already for the case $n = 2$ [14, 15, 60]. However, for every closed symplectic 4-manifold M with the symplectic structure representing a rational cohomology class, there exists an integer N such that for $k \geq N$, this manifold has a full packing: $\nu_k(M) = \mathrm{Volume}(M)$ [15].

In contrast to these three talks, the lecture by V. M. Zakalyukin is devoted to the *local* problem of generalized caustics. Let M be a symplectic manifold of dimension $2n$, and let functions $f_i \colon M \to \mathbb{R}$, $1 \leq i \leq m$, be independent and pairwise in involution ($m \leq n$). Their common level sets $f = c \in \mathbb{R}^m$ are coisotropic $(2n - m)$-dimensional submanifolds of M. Given a Lagrangian submanifold $L \hookrightarrow M$, the set of values c for which L is not transversal to the fiber $f = c$ is called a *coisotropic caustic.* The singularities of "conventional" caustics ($m = n$) are well-studied [9], and the talk treats the case $m < n$. The singularities of caustics for $m < n$ were first examined in [81].

The next two lectures pertain to contact topology. H. Geiges' talk considers various constructions of contact manifolds, cf. [36]. A progress in constructing symplectic manifolds is exemplified by R. E. Gompf's method [40]. The lecture by Yu. V. Chekanov deals with Legendrian knots and their invariants. Here the problem is to determine when two topologically isotopic Legendrian knots in a contact 3-space are isotopic through contactomorphisms. For instance, two topologically trivial Legendrian knots can be transformed to each other by contact isotopies if and only if their Thurston-Bennequin invariants and Maslov numbers coincide respectively [27]. For an analogous problem for Lagrangian (two-dimensional) knots in symplectic 4-manifolds see, e.g., [25].

Finally, two lectures are devoted to the "core" Hamiltonian dynamics, to be more precise, to periodic and quasi-periodic motions in autonomous Hamiltonian systems. In the talk by V. L. Ginzburg, the speaker describes his constructions of smooth Hamilton functions $H\colon \mathbb{R}^{2n} \to \mathbb{R}$ such that the Hamiltonian flow afforded by H on the compact energy hypersurface $H = 1$ has no periodic trajectories (the symplectic structure on $\mathbb{R}^{2n}$ is assumed to be standard), see [37, 38, 39]. Such Hamiltonian systems provide counterexamples to the so-called *Hamiltonian Seifert conjecture.* Finally, À. Jorba's and J. Villanueva's lecture studies the complicated "exponential" structure of the set of invariant tori (carrying quasi-periodic motions) near a given one in an analytic autonomous Hamiltonian flow, the relevant reference being [50]. This topic is within the framework of the KAM (Kolmogorov-Arnol'd-Moser) theory concerning quasi-periodic motions in generic dynamical systems.

For the basic ideas of the KAM theory, see [7, Appendix 8]; volume [17] presents a modern survey. Here we would like only to remark that whereas the KAM theory is always local with respect to the *action* variables, its global character with respect to the *angle* variables is best pronounced while considering coisotropic invariant tori of dimensions greater than the number n of degrees of freedom (see a bibliography and discussion in [17]). Indeed, an invariant torus of a Hamiltonian flow or symplectic diffeomorphism is automatically *isotropic* provided that this torus carries a quasi-periodic motion and the symplectic structure on the phase space is exact [17, 42]. Thus, coisotropic invariant KAM tori of dimensions $> n$ can occur for non-exact symplectic structures only (in particular, they are impossible in the local theory, e.g., near equilibrium/fixed points of Hamiltonian systems).

An interplay between a) Gromov's theory of pseudoholomorphic curves and Floer's homology theory and b) examining periodic orbits of Hamiltonian vector fields within energy surfaces is exemplified by paper [46], see also Hofer's plenary lecture [47] at the 23rd International Congress of Mathematicians.

As was already emphasized, the present mini-symposium covers unavoidably only a small fraction of modern symplectic and contact geometry and topology. Of the significant missed achievements, we would mention here only C. H. Taubes' results on a precise relation between the Seiberg-Witten invariants and the Gromov invariants for closed symplectic 4-manifolds [73, 74, 75, 76, 77, 78] (see also [20]) and S. K. Donaldson's works on symplectic Lefschetz pencils [21, 22] (see also [11]).

Of survey monographs on the field, one should first of all mention influential books [43, 62] as well as advanced textbooks [1, 12]. Monographs [26, 61] are devoted to special topics. Important contributions can be found in collections [10, 13, 18, 28, 29, 44, 48, 51, 55, 71, 72, 79]. Finally, we would like specifically to draw the reader's attention to the stimulating reviews of the field by F. Lalonde [52, 53, 54] and D. McDuff [63, 64, 65, 66, 67, 68].

References

[1] B. Aebischer, M. Borer, M. Kälin, Ch. Leuenberger and H. M. Reimann, *Symplectic Geometry,* Progr. Math., **124** (Birkhäuser, Basel, 1994).

[2] V. I. Arnol'd, *Sur une propriété topologique des applications globalement canoniques de la mécanique classique,* C. R. Acad. Sci. Paris, **261** (1965), 3719–3722.

[3] V. I. Arnol'd, *The stability problem and ergodic properties of classical dynamical systems,* in: Proceedings of the Intern. Congress of Mathematicians (Moscow, 1966) (Mir, Moscow, 1968), 387–392 (in Russian).

[4] V. I. Arnol'd, *A comment to H. Poincaré's paper "Sur un théorème de géométrie",* in: H. Poincaré, Selected Works in Three Volumes, Vol. II (Nauka, Moscow, 1972), 987–989 (in Russian).

[5] V. I. Arnol'd, *Fixed points of symplectic diffeomorphisms,* in: F. E. Browder, Ed., Mathematical Developments Arising from Hilbert Problems, Proc. Symp. Pure Math., **28** (A.M.S., Providence, RI, 1976), 66.

[6] V. I. Arnol'd, *Some problems in the theory of differential equations,* in: Unsolved Problems in Mechanics and Applied Mathematics (Moscow State Univ. Press, Moscow, 1977), 3–9 (in Russian).

[7] V. I. Arnol'd, *Mathematical Methods of Classical Mechanics,* Graduate Texts in Math., **60** (Springer-Verlag, New York, 1978) [the Russian original is of 1974, the 3rd Russian edition is of 1989].

[8] V. I. Arnol'd, *The first steps of symplectic topology,* Russian Math. Surveys, **41** (1986), no. 6, 1–21.

[9] V. I. Arnol'd, *Singularities of Caustics and Wave Fronts,* Math. Appl. (Soviet Ser.), **62** (Kluwer, Dordrecht, 1990).

[10] M. Audin and J. Lafontaine, Eds., *Holomorphic Curves in Symplectic Geometry,* Progr. Math., **117** (Birkhäuser, Basel, 1994).

[11] D. Auroux, *Asymptotically holomorphic families of symplectic submanifolds,* Geom. Funct. Anal., **7** (1997), 971–995.

[12] R. Berndt, *Einführung in die Symplektische Geometrie* (Friedr. Vieweg & Sohn, Braunschweig, 1998).

[13] E. Bierstone, B. A. Khesin, A. G. Khovanskiĭ and J. E. Marsden, Eds., *The Arnol'dfest,* Proceedings of a Conference in Honour of V. I. Arnol'd for his Sixtieth Birthday, Fields Inst. Comm., **24** (A.M.S., Providence, RI, 2000).

[14] P. Biran, *Symplectic packing in dimension* 4, Geom. Funct. Anal., **7** (1997), 420–437.

[15] P. Biran, *A stability property of symplectic packing,* Inv. Math., **136** (1999), 123–155 [see also Featured Review 2000b:57039 by M. Schwarz of this paper in Math. Reviews].

[16] G. D. Birkhoff, *Proof of Poincaré's geometric theorem,* Trans. Amer. Math. Soc., **14** (1913), 14–22.

[17] H. W. Broer, G. B. Huitema and M. B. Sevryuk, *Quasi-Periodic Motions in Families of Dynamical Systems: Order amidst Chaos,* Lecture Notes in Math., **1645** (Springer-Verlag, Berlin, 1996).

[18] R. Budzyński, S. Janeczko, W. Kondracki and A. F. Künzle, Eds., *Symplectic Singularities and Geometry of Gauge Fields,* Banach Center Publ., **39** (Polish Acad. Sci., Inst. Math., Warsaw, 1997).

[19] C. C. Conley and E. Zehnder, *The Birkhoff-Lewis fixed point theorem and a conjecture of V. I. Arnol'd,* Inv. Math., **73** (1983), 33–49.

[20] S. K. Donaldson, *The Seiberg-Witten equations and 4-manifold topology,* Bull. Amer. Math. Soc. (N.S.), **33** (1996), 45–70 [see also Featured Review 96k:57033 by D. S. Freed of this paper in Math. Reviews].

[21] S. K. Donaldson, *Symplectic submanifolds and almost-complex geometry,* J. Differential Geom., **44** (1996), 666–705 [see also Featured Review 98h:53045 by D. Pollack of this paper in Math. Reviews].

[22] S. K. Donaldson, *Lefschetz fibrations in symplectic geometry,* in: G. Fischer and U. Rehmann, Eds., Proceedings of the Intern. Congress of Mathematicians, Vol. II (Berlin, 1998), Doc. Math., **1998**, Extra Vol. II, 309–314 (electronic).

[23] Ya. M. Èliashberg, *An estimate of the number of fixed points of area-preserving transformations,* Preprint (Syktyvkar, 1978) (in Russian).

[24] Ya. M. Èliashberg, *Rigidity of symplectic and contact structures,* Preprint (1981) (in Russian); see also in: Abstracts of reports to the 7th Intern. Topology Conference in Leningrad (1982).

[25] Ya. M. Èliashberg and L. V. Polterovich, *The problem of Lagrangian knots in four-manifolds,* in [51], 313–327.

[26] Ya. M. Èliashberg and W. P. Thurston, *Confoliations,* Univ. Lecture Series, **13** (A.M.S., Providence, RI, 1998).

[27] Ya. M. Èliashberg and M. Fraser, *Classification of topologically trivial Legendrian knots,* in [55], 17–51.

[28] Ya. M. Èliashberg and L. Traynor, Eds., *Symplectic Geometry and Topology,* IAS/Park City Math. Series, **7** (A.M.S., Providence, RI, 1999).

[29] Ya. M. Èliashberg, D. B. Fuchs, T. Ratiu and A. Weinstein, Eds., *Northern California Symplectic Geometry Seminar,* Amer. Math. Soc. Transl. Ser. 2, **196** (A.M.S., Providence, RI, 1999).

[30] A. Floer, *An instanton-invariant for 3-manifolds,* Comm. Math. Phys., **118** (1988), 215–240.

[31] A. Floer, *Morse theory for Lagrangian intersections,* J. Differential Geom., **28** (1988), 513–547.

[32] A. Floer, *Symplectic fixed points and holomorphic spheres,* Comm. Math. Phys., **120** (1989), 575–611.

[33] B. Fortune, *A symplectic fixed point theorem for* $\mathbb{C}P^n$, Inv. Math., **81** (1985), 29–46.

[34] K. Fukaya and K. Ono, *Arnol'd conjecture and Gromov-Witten invariant,* Topology, **38** (1999), 933–1048 [see also Featured Review 2000j: 53116 by D. E. Hurtubise of this paper in Math. Reviews].

[35] K. Fukaya and K. Ono, *Arnol'd conjecture and Gromov-Witten invariant for general symplectic manifolds,* in [13], 173–190.

[36] H. Geiges, *Constructions of contact manifolds,* Math. Proc. Cambridge Philos. Soc., **121** (1997), 455–464.

[37] V. L. Ginzburg, *An embedding* $S^{2n-1} \to \mathbb{R}^{2n}$, $2n-1 \geq 7$, *whose Hamiltonian flow has no periodic trajectories,* Internat. Math. Res. Notices, **1995**, no. 2, 83–97 (electronic).

[38] V. L. Ginzburg, *A smooth counterexample to the Hamiltonian Seifert conjecture in* $\mathbb{R}^6$, Internat. Math. Res. Notices, **1997**, no. 13, 641–650.

[39] V. L. Ginzburg, *Hamiltonian dynamical systems without periodic orbits,* in [29], 35–48.

[40] R. E. Gompf, *A new construction of symplectic manifolds,* Ann. of Math. (2), **142** (1995), 527–595 [see also Featured Review 96j:57025 by M. Ue of this paper in Math. Reviews].

[41] M. L. Gromov, *Pseudo holomorphic curves in symplectic manifolds,* Inv. Math., **82** (1985), 307–347.

[42] M. R. Herman, *Inégalités "a priori" pour des tores lagrangiens invariants par des difféomorphismes symplectiques,* Inst. Hautes Études Sci. Publ. Math., **70** (1989), 47–101.

[43] H. Hofer and E. Zehnder, *Symplectic Invariants and Hamiltonian Dynamics* (Birkhäuser, Basel, 1994) [see also Featured Review 96g:58001 by D. M. Burns, Jr. of this book in Math. Reviews].

[44] H. Hofer, C. H. Taubes, A. Weinstein and E. Zehnder, Eds., *The Floer Memorial Volume,* Progr. Math., **133** (Birkhäuser, Basel, 1995).

[45] H. Hofer and D. A. Salamon, *Floer homology and Novikov rings,* in [44], 483–524.

[46] H. Hofer, K. Wysocki and E. Zehnder, *The dynamics on three-dimensional strictly convex energy surfaces,* Ann. of Math. (2), **148** (1998), 197–289 [see also Featured Review 99m:58089 by M. Schwarz of this paper in Math. Reviews].

[47] H. Hofer, *Dynamics, topology, and holomorphic curves,* in: G. Fischer and U. Rehmann, Eds., Proceedings of the Intern. Congress of Mathematicians, Vol. I (Berlin, 1998), Doc. Math., **1998**, Extra Vol. I, 255–280 (electronic).

[48] J. Hurtubise, F. Lalonde and G. Sabidussi, Eds., *Gauge Theory and Symplectic Geometry,* NATO Adv. Sci. Inst. Ser. C Math. Phys. Sci., **488** (Kluwer, Dordrecht, 1997).

[49] L. A. Ibort and C. Martínez Ontalba, *Arnol'd's conjecture and symplectic reduction,* J. Geom. Phys., **18** (1996), 25–37.

[50] À. Jorba and J. Villanueva, *On the normal behaviour of partially elliptic lower-dimensional tori of Hamiltonian systems,* Nonlinearity, **10** (1997), 783–822.

[51] W. H. Kazez, Ed., *Geometric Topology,* AMS/IP Stud. Adv. Math., **2.1** (A.M.S., Providence, RI; Intern. Press, Cambridge, MA, 1997).

[52] F. Lalonde, *Energy and capacities in symplectic topology,* in [51], 328–374.

[53] F. Lalonde, *J-holomorphic curves and symplectic invariants,* in [48], 147–174.

[54] F. Lalonde, *New trends in symplectic geometry,* C. R. Math. Rep. Acad. Sci. Canada, **19** (1997), 33–50.

[55] F. Lalonde, Ed., *Geometry, Topology, and Dynamics,* CRM Proceedings & Lecture Notes, **15** (A.M.S., Providence, RI, 1998).

[56] J. Li and G. Tian, *Virtual moduli cycles and Gromov-Witten invariants of general symplectic manifolds,* in [72], 47–83.

[57] G. Liu and G. Tian, *Floer homology and Arnol'd conjecture,* J. Differential Geom., **49** (1998), 1–74 [see also Featured Review 99m:58047 by J.-C. Sikorav of this paper in Math. Reviews].

[58] G. C. Lu, *The Arnol'd conjecture for a product of weakly monotone manifolds,* Chinese J. Math., **24** (1996), 145–157.

[59] G. C. Lu, *The Arnol'd conjecture for a product of monotone manifolds and Calabi-Yau manifolds,* Acta Math. Sinica (N.S.), **13** (1997), 381–388.

[60] D. McDuff and L. V. Polterovich, *Symplectic packings and algebraic geometry,* Inv. Math., **115** (1994), 405–434.

[61] D. McDuff and D. A. Salamon, *J-Holomorphic Curves and Quantum Cohomology,* Univ. Lecture Series, **6** (A.M.S., Providence, RI, 1994) [see also Featured Review 95g:58026 by B. Hunt of this book in Math. Reviews].

[62] D. McDuff and D. A. Salamon, *Introduction to Symplectic Topology* (The Clarendon Press, Oxford Univ. Press, New York, 1995 [1st ed.], 1998 [2nd ed.]).

[63] D. McDuff, *Lectures on Gromov invariants for symplectic 4-manifolds,* in [48], 175–210.

[64] D. McDuff, *Recent developments in symplectic topology,* in: A. Balog, G. O. H. Katona, A. Recski and D. Szász, Eds., Proceedings of the Second European Congress of Mathematics, Vol. II (Budapest, 1996), Progr. Math., **169** (Birkhäuser, Basel, 1998), 28–42.

[65] D. McDuff, *Fibrations in symplectic topology,* in: G. Fischer and U. Rehmann, Eds., Proceedings of the Intern. Congress of Mathematicians, Vol. I (Berlin, 1998), Doc. Math., **1998**, Extra Vol. I, 339–357 (electronic).

[66] D. McDuff, *Symplectic structures-a new approach to geometry,* Notices Amer. Math. Soc., **45** (1998), 952–960.

[67] D. McDuff, *Introduction to symplectic topology,* in [28], 5–33.

[68] D. McDuff, *A glimpse into symplectic geometry,* in: V. I. Arnol'd, M. Atiyah, P. Lax and B. Mazur, Eds., Mathematics: Frontiers and Perspectives (A.M.S., Providence, RI, 2000), 175–187.

[69] K. Ono, *On the Arnol'd conjecture for weakly monotone symplectic manifolds,* Inv. Math., **119** (1995), 519–537.

[70] H. Poincaré, *Sur un théorème de géométrie,* Rend. Circ. Mat. Palermo, **33** (1912), 375–407.

[71] D. A. Salamon, Ed., *Symplectic Geometry,* London Math. Soc. Lecture Note Series, **192** (Cambridge Univ. Press, Cambridge, 1993).

[72] R. J. Stern, Ed., *Topics in Symplectic 4-Manifolds,* First Intern. Press Lecture Series, **I** (Intern. Press, Cambridge, MA, 1998).

[73] C. H. Taubes, *The Seiberg-Witten invariants and symplectic forms,* Math. Res. Lett., **1** (1994), 809–822.

[74] C. H. Taubes, *The Seiberg-Witten and Gromov invariants,* Math. Res. Lett., **2** (1995), 221–238.

[75] C. H. Taubes, SW $\Rightarrow$ Gr: *from the Seiberg-Witten equations to pseudo-holomorphic curves,* J. Amer. Math. Soc., **9** (1996), 845–918 [see also Featured Review 97a:57033 by D. A. Salamon of this paper in Math. Reviews].

[76] C. H. Taubes, *Counting pseudo-holomorphic submanifolds in dimension* 4, J. Differential Geom., **44** (1996), 818–893 [see also Featured Review 97k:58029 by T. H. Parker of this paper in Math. Reviews].

[77] C. H. Taubes, *The structure of pseudo-holomorphic subvarieties for a degenerate almost complex structure and symplectic form on* $S^1 \times B^3$, Geom. Topol., **2** (1998), 221–332 (electronic) [see also Featured Review 99m:57029 by F. Lalonde of this paper in Math. Reviews].

[78] C. H. Taubes, *The geometry of the Seiberg-Witten invariants,* in: G. Fischer and U. Rehmann, Eds., Proceedings of the Intern. Congress of Mathematicians, Vol. II (Berlin, 1998), Doc. Math., **1998**, Extra Vol. II, 493–504 (electronic).

[79] C. B. Thomas, Ed., *Contact and Symplectic Geometry,* Publ. Newton Inst., **8** (Cambridge Univ. Press, Cambridge, 1996).

[80] A. Weinstein, *Lectures on Symplectic Manifolds,* CBMS Regional Conference Series in Math., **29** (A.M.S., Providence, RI, 1977).

[81] V. M. Zakalyukin and O. M. Myasnichenko, *Lagrangian singularities in symplectic reduction,* Functional Anal. Appl., **32** (1998), 1–9.

Institute of Energy Problems of Chemical Physics
The Russia Academy of Sciences
Lenin prospect 38, Bldg. 2
Moscow 117829, Russia
E-mail address: `sevryuk@mccme.ru, rusin@chph.ras.ru`

Simple Coisotropic Projections and Caustics

Vladimir Zakalyukin

Abstract. A generalization of Arnold's simple Lagrangian singularities is presented. Consider a fibration of a symplectic space by coisotropic fibers. A generic Lagrangian submanifold can meet certain fibers non-transversally. We call the set of corresponding points from the base of the fibration – the coisotropic caustic. It turn out that the classification of simple local singularities of coisotropic caustics is related to all Coxeter crystallographic groups generated by reflections A, B, C, D, E, F (except G_2).

Similar answers arise in the parallel theory of Legendre singularities in contact spaces.

1. Introduction

Various applications of singularity theory in geometry and physics involve symplectic geometry and in particular Arnold's construction [1] of Lagrangian projections.

A bundle with a symplectic total space and Lagrangian fibers is called a Lagrangian bundle. The main examples are: the cotangent bundle and the foliation by common level sets of integrals of a completely integrable Hamiltonian system. The restriction to an immersed Lagrangian submanifold L of the bundle projection is called a Lagrangian mapping. Its critical value locus (in other words, the base point set of nontransversal intersections of L with the corresponding fibers) is called caustic.

The simple stable classes of caustic germs are related to A, D, E simple Lie groups.

Here we describe a rather natural generalization of this construction corresponding to the case of noncompletely integrable Hamiltonian systems.

Consider a collection of independent functions $h_1, \ldots, h_k$ $\;k \leq n$ defined on a symplectic space (M^{2n}, ω), which are pairwise in involution. Their common level sets C_c^{2n-k}, $c \in \mathbf{R}^k$ are coisotropic (at each point the tangent space to C_c contains its symplectic orthogonal) and foliate M. Each fiber is itself foliated by characteristic (isotropic submanifolds of dimension k being the integral submanifolds of the distributions spanned by h_i Hamiltonian vector fields). The space of these characteristics is a symplectic (reduced) space of dimension $2(n-k)$.

The Lagrangian submanifold $L \subset M^{2n}$ can meet certain fiber C_c nontransversally. The isotropic variety $L \cap C_c$ in this case projects to a Lagrangian subvariety (which in general is singular) in the corresponding space of characteristics. The set of such values of $c \in \mathbf{R}^k$ is called a *coisotropic caustic* of L.

Our result is the classification of simple (having no continuous invariants) stable coisotropic caustics. The underlying equivalence relation is provided by the group of symplectomorphisms of the ambient space, which preserve the coisotropic fibration.

The answer was unexpected. Simple stable coisotropic caustics are diffeomorphic (with one exception) to irregular orbit hypersurfaces of Weil groups of A, B, C, D, E, F types.

The same list of normal forms remains for simple stable (with respect to the group of contactomorphisms) projections of Legendre submanifolds along the coisotropic fibration of a contact space.

The proofs (see [6]) are based on the classification of singularities of the contact of a Lagrangian submanifold with only one coisotropic subspace [5].

We determine also the range of nice dimensions n and k, for which simple stable coisotropic singularities are dense. In particular, if $k = 1$ then generic Morse nontransversal intersections of a Lagrangian submanifold with the level hypersurfaces of regular Hamiltonian function are stable.

All constructions are local, and the initial objects are supposed to be C^∞-smooth.

The work was supported by RFFI 99-01-00147 and INTAS 1644 grants. It was finished during the author's visit to Purdue University.

2. Definitions

Let $h\colon \mathbf{R}^{2n} \to \mathbf{R}^k$, $k \le n$ be a germ (at the origin) of a fibration with coisotropic fibers C_c, $c \in \mathbf{R}^k$ of the standard symplectic space $(\mathbf{R}^{2n}, \omega)$. The Darboux theorem implies that any two such fibrations (of equal dimensions) are symplectomorphic.

Consider a pair (L^n, h), where L is a germ of a Lagrangian submanifold of $(\mathbf{R}^{2n}, \omega)$. Two pairs will be called *equivalent*, if one of them can be transformed into the other by some local symplectomorphism of $\mathbf{R}^{2n}$.

A pair is called *stable* if its orbit of the (pseudo) group of equivalencies is an open subset (in the space of pairs equipped with the appropriate topology).

A pair is called *simple*, if germs of L and h have such representatives that the germs of their small deformations at any nearby point define a pair, which is equivalent to one of the finite list of normal forms.

Denote by h_0 a distinguished coordinate coisotropic fibration defined in Darboux coordinates $\mathbf{R}^{2n} = \{(x, y, u, v)\}$, $\omega = dy \wedge dx + dv \wedge du$, $x = (x_1, \ldots, x_k) \in \mathbf{R}^k$, $y = (y_1, \ldots, y_k) \in \mathbf{R}^k$, $u = (u_1, \ldots, u_{n-k}) \in \mathbf{R}^{n-k}$, $v = (v_1, \ldots, v_{n-k}) \in \mathbf{R}^{n-k}$ by the projection $h_0\colon (x, y, u, v) \mapsto y$.

The associated isotropic characteristics are the subspaces parallel to x coordinate subspace and the reduction mapping is the projection $\rho_c\colon C_c \to \mathbf{R}^{2(n-k)}$, $\rho_c\colon (x, c, u, v) \mapsto (u, v)$.

Denote by Γ_k the (pseudo)group of symplectomorphism germs at the origin of the standard symplectic space $\mathbf{R}^{2n}$, which commute with the projection h_0 (map fibers to fibers). It contains the germs of the products of symplectomorphisms of $(x, y), dy \wedge dx$ space, preserving Lagrangian fibration $(y, x) \mapsto y$, and symplectomorphisms of $(u, v), dv \wedge du$ space.

If k is odd the group is not connected. In this case we denote by Γ_k^+ its connected component of the identity, whose elements preserve the orientation of the fiber $C_0 = \{y = 0\}$ The group Γ_k (for odd k) is isomorphic to the semidirect product of Γ_k^+ and of $\mathbf{Z}_2$, generated by reflection

$$I\colon (x_1, \dots, x_k, y_1, \dots, y_k, u, v) \mapsto (-x_1, \dots, x_k, -y_1, \dots, y_k, u, v),$$

which changes the orientation of C_0.

Let r be the rank of the projection h_0 restricted to the tangent space T_0L at the origin of the Lagrangian submanifold L.

There exists a Lagrange coordinate subspace L_* transversal to T_0L having r-dimensional intersection with x isotropic coordinate subspace. A permutation of subset $(1, \dots, k)$ of indices and a symplectic permutation of (u, v) coordinate space induce a symplectomorphism from Γ_k^+. Hence, without loss of generality, suppose that this is the subspace $L_* = \{x_{r+1} = \dots = x_k = y_1 = \dots = y_r = 0, u = 0\}$. Then the Lagrangian germ L is defined by a (generating) function S in the variables $x_{r+1}, \dots, x_k, y_1 \dots, y_r,\ u$ by standard formulas

$$L = \left\{ (x, y, u, v) \,\middle|\, y_j = \frac{\partial S}{\partial x_j}, \, j = r+1, \dots, k; \, x_i = -\frac{\partial S}{\partial y_i} \quad j = 1, \dots, r; \, v = \frac{\partial S}{\partial u} \right\}.$$

Such a Lagrangian germ, its generating function and its variables will be called *(r, k)-adapted.*

If $r = k$ (that is L^n is transversal to C_0 at the origin), then the intersection $L \cap C_0$ is transversal to characteristic fibers of C and ρ_c is nonsingular.

All such transversal pairs are equivalent to each other. Really, in this case function S depends only on y and u and the symplectomorphism

$$(x, y, u, v) \mapsto \left(x + \frac{\partial S}{\partial y}, \, y, \, v - \frac{\partial S}{\partial u}, \, u \right)$$

belongs to Γ_k^+ and maps L to the coordinate Lagrangian submanifold L_* defined by the zero generating function.

Denote by G_k the subgroup of symplectomorphisms, which preserve only one distinguished coisotropic fiber C_0.

3. Classification Theorems

If $k = n$ we are in the classical Lagrangian projection setting without any v, u coordinates. The simple stable pairs are classified by generating functions S in x and y only, which are versal (with respect to the R^+ group of right diffeomorphisms and additions with constants) deformations with y parameters of functions in x having A, D, E simple singularities.

Suppose now $k < n$.

Theorem 3.1. **i.** *Any simple stable nontransversal pair has corank $k-r$ equal to 1.* **ii.** *Any simple stable nontransversal germ (L^n, h) is equivalent to a pair germ (L_S, h_0), where Lagrangian submanifold L_S is defined in $(k-1,k)$-adapted Darboux coordinates by the generating function (with $k-r=1$) $S(y_1,\dots,y_{k-1},t,u)$ (here t stands for x_k) of one of the following forms:*

(In the following $Q(u_1,\dots,u_k)$ denotes a nondegenerate quadratic form $Q = \pm u_1^2 \pm \dots \pm u_k^2$.)

(1) *$S = t^3 + tf(y,u)$, where $f(y,u)$ is a restricted versal deformation with parameters y of one of the simple singularities of functions in u:*

$$A_m: \quad k \geq m \geq 1 \quad f(u) = \pm u_1^{m+1} + Q(u_2,\dots,u_{n-k}) + y_1u_1 + \dots + y_{m-1}u_1^{m-1}$$

(for even m the singularities with $\pm$ signs are equivalent);

$$D_m: \quad k \geq m \geq 4 \quad f(y,u) = u_1^2u_2 \pm u_2^{m-1} + Q(u_3,\dots,u_{n-k}) + y_1u_2^1 + \dots + y_{m-2}u_2^{m-2} + y_{m-1}u_1$$

(for odd m the singularities with $\pm$ signs are equivalent);

$$E_6: \quad k \geq 5 \quad f(y,q) = u_1^3 \pm u_2^4 + Q(u_3,\dots,u_{n-k}) + y_1u_1 + y_2u_2 + y_3u_1u_2 + y_4u_2^2 + y_5u_1u_2^2;$$

$$E_7: \quad k \geq 6 \quad f(q) = u_1^3 + u_1u_2^3 + Q(u_3,\dots,u_{n-k}) + y_1u_1 + y_2u_2 + y_3u_1u_2 + y_4u_2^2 + y_5u_1^2 + y_6u_1^2u_2;$$

$$E_8: \quad k \geq 7 \quad f(q) = u_1^3 + u_2^5 + Q(u_3,\dots,u_{n-k}) + y_1u_1 + y_2u_2 + y_3u_1u_2 + y_4u_2^2 + y_5u_1u_2^2 + y_6u_2^3 + y_7u_2^3u_1.$$

(2) *Classes, corresponding to boundary singularities*

$$C_m: \quad k \geq m \geq 2 \quad S = t^{2m+1} + y_{m-1}t^{2m-1} + \dots + y_1t^3 + t^2u_1 + tQ(u_2,\dots,u_{n-k});$$

$$B_m: \quad k \geq m \geq 2 \quad S = t^3u_2 + t^2u_1 + t(\pm u_2^m + Q(u_3,\dots,u_{n-k}) + y_1u_2^1 + \dots + y_{m-1}u_2^{m-1});$$

$$F_4: \quad k \geq 4 \quad S = t^5 + y_3t^3 + t^2u_1 + t(u_2^3 + Q(u_3,\dots,u_{n-k}) + y_2u_2^2 + y_1u_2).$$

(3) *Exceptional class*

$$U_{n-k+2}: \quad k \geq n-k+2 \quad S = \pm t^4 + t^2(y_1u_1 + \dots + y_{n-k}u_{n-k} + y_{n-k+1}) + tQ(u_1,\dots,u_{n-1}).$$

Remark 3.2. *Classes C_k, F_4 have an alternative equivalent form (when $n-k>2$), $S=t^3u_2+t^2u_1+tf$, where f is a versal deformation (with parameters y) of the corresponding simple boundary singularity of functions in $u_2,\dots,u_{n-k}$ with the hypersurface $u_2=0$ as the boundary. In particular, B_2 is equivalent to C_2.*

Remark 3.3. *The adjacency table of these simple classes differs from that of boundary singularities by the term $U_{n-k+2}\to B_2$ only.*

Generically the coisotropic caustic of a pair consists of two components: one (Σ_s) is formed by those values of $y=(y_1,\dots,y_k)$, for which the corresponding reduced Lagrangian variety $\rho(L\cap C_y)$ has singular points, and the other (Σ_i) – is formed by those values of y, for which the corresponding reduced Lagrangian variety $\rho(L\cap C_y)$ has nontransversal intersections of smooth branches. The straightforward calculations imply the following

Proposition 3.4. *The hypersurface (Σ_s) in y space for A, D, E classes from Theorem 3.1 is defined by the equations $f(u,y)=y_k, \frac{\partial f}{\partial u}=0$, for the corresponding family $f=f(u,y_1,\dots,y_{k-1})$. Thus it is the irregular orbit hypersurface in the space of orbits of one of the A, D, E Coxeter finite groups, generated by reflections.*

The hypersurface (Σ_i) for these classes is a cylinder (with the line generator parallel to y_k axis) over the ordinary A, D, E caustic in $y_1,\dots,y_{k-1}$ space, defined by equations

$$\frac{\partial f}{\partial u}=0,\quad \det\left(\frac{\partial^2 f}{\partial u_i\partial u_j}\right)=0\,.$$

Remark 3.5. *Using methods of [4, 3] one can verify that the Lagrangian projections defined by the fibration $(u,v)\mapsto u$ of reduced singular Lagrangian varieties $\rho(L\cap C_y)$ for A, D, E classes of Theorem 3.1 are stable in the sense of [3] (with respect to perturbations of symplectic structure and Lagrangian projections) for any y.*

Proposition 3.6. *The coisotropic caustics of normal forms B, C, F are diffeomorphic to the bifurcation sets of zeros (alternatively called wave fronts) of the corresponding boundary singularities, that is to the irregular orbit hypersurface Σ of the corresponding reflection group.*

Amazingly, one irreducible component of Σ coincides with Σ_s and the other – with Σ_i.

For example, the curve Σ_s of the normal form $C_2: S=t^5+y_1t^3+t^2u_1$ (here we put $k=2, n=4$) is the set of parameters y_1, y_2 where zero is a multiple root of the polynomial $P(t)=5t^4+3y_1t^2+2tu_1-u_2$, while the curve Σ_i is determined by the condition that $P(t)$ has an arbitrary multiple root provided that $u_1=0$.

Thus the coisotropic caustic in the (y_1,y_2)-plane is the union of a line and a half of a parabola tangent to this line (that is the C_2 bifurcation diagram).

Similar answers arise in the parallel problem of classifying the simple stable pairs of Legendre submanifold germs and coisotropic contact fibrations with respect to the group of contactomorphisms of contact space.

Let (K^{2n+1}, α) be a contact space. A submanifold $M \subset K$ is called *coisotropic* if at each point $m \in M$ the subspace $T_m M$ is transversal to the contact distribution $A_m = \{v \in T_m K : \alpha(v) = 0\}$ and subspace $ST_m M \subset A_m$ of vectors skew-orthogonal to $A_m \cap T_m M$ with respect to non-degenerate two-form $\beta = d\alpha\,|_A$ (β is defined up to a multiplication by a non-zero factor) belong to $A_m \cap T_m M$ (therefore $ST_m M$ determines an integrable distribution on M).

A fibration $\rho \colon K \to E$ with isotropic fibers is called *isotropic*. Locally all isotropic fibrations are contactomorphic to standard one $\rho \colon (z, x, y, u, v,) \mapsto (z, y)$, $z \in \mathbf{R}$, $\alpha = dz - ydx - udv$.

A pair consisting of a germ of Legendre submanifold $\mathcal{L} \in K$ and a coisotropic fibration is contactomorphic to a pair of the standard fibration and a Legendre germ $\mathcal{L} = \{z, (x, y, u, v) | z = S, (x, y, u, v) \in L\}$ determined in appropriate adapted coordinates by a generating function S of the associated Lagrangian submanifold $L \subset \mathbf{R}^{2n}$.

The contact counterparts of definitions of simple stable pairs are straightforward.

Theorem 3.7. *Non-transversal contact pair $\mathcal{L}, \rho$ is simple and stable if it is contactomorphic to a pair determined by one of the generating functions from Theorem 3.1.*

The following theorem characterizes the range of nice dimensions (n, k), for which any generic pair of Lagrangian submanifold and collection of functions in involution has stable and simple singularities.

However, involutive collections of functions $h_1, \dots, h_k$ on $\mathbf{R}^{2n}$ form a subset with very complicated singularities in the space of collections of arbitrary functions, if at some point the rank of the mapping h is less than $k-1$. In this case it is not clear what "generic" means. To avoid this difficulty consider only collections (called *1-generic*) without such points.

Theorem 3.8. *For an open and dense subset in the space (equipped with the fine Whitney topology) of pairs formed by a proper Lagrangian submanifold in $\mathbf{R}^2 n$ and a 1-generic coisotropic mapping h, the germs of pairs (L, h) are stable at each point if and only if $n = k$ and $k < 6$, or $n > k$ and $k < 4$ except for the pair $(k, n) = (3, 4)$.*

Corollary 3.9. *In the range of nice dimensions $k < n$ only the following classes generically appear: for $k = 1 : A_1$ (at isolated points); for $k = 2 : A_1$ (on curves), $A_2, B_2 \approx C_2$ (at isolated points) and for $k = 3, n > 4 : A_1$ (on surfaces), $A_2, B_2 \approx C_2$ (on curves), B_3, C_3 (at isolated points).*

4. Infinitesimal Stability

Time-depending families of local symplectomorphisms from Γ_k are phase flows of time-dependent Hamiltonian vector fields, whose $-\frac{\partial H}{\partial x}\frac{\partial}{\partial y}$ components may depend only on y. Similarly for G_k-families these components vanish on C_0. This observation proves the following

Lemma 4.1. i. *Infinitesimal transformations from Γ_k^+ are defined by Hamiltonian vector fields with Hamiltonian functions H of the form*

$$H = \sum_{i=1}^{k} x_i H_i(y) + H_0(y,u,v)$$

with certain smooth functions H_i and H_0.

In particular, this lemma implies that Γ_k-symplectomorphisms preserve a well-defined affine structure on isotropic characteristic fibers.

The image of an (r,k)-adapted Lagrangian germ under a symplectomorphism φ from a neighbourhood of the identity in Γ_k^+ or in G_k^+ remains transversal to L_*. Thus φ acts on the adapted generating families.

Let for (r,k)-adapted coordinates $p = (p_1,\ldots,p_{k-r}) = (y_{r+1},\ldots,y_k)$, $q = (q_1,\ldots,q_{k-r}) = (x_{r+1},\ldots,x_k)$, $w = (w_1,\ldots,w_r) = (y_1,\ldots,y_r)$ and $z = (z_1,\ldots z_r)$ $= (x_1,\ldots,x_r)$.

Lemma 4.2. i. *The tangent space $T_S\Gamma$ to the Γ_k^+-orbit of generating function S consists of all functions of the form*

$$\tilde{S} = \sum_{i=1}^{k-r} q_i H_i\,(\frac{\partial S}{\partial q}, w) + \sum_{j=1}^{r} \frac{\partial S}{\partial w} H_j\,(\frac{\partial S}{\partial q}, w) + H_0\,(\frac{\partial S}{\partial q}, w, u, \frac{\partial S}{\partial u}),$$

with smooth functions H_i, H_j and H_0.

ii. *The tangent space $T_S G$ to the G_k^+-orbit of generating function S consists of all functions of the form*

$$\tilde{S} = \sum_{i=1}^{k-r} \frac{\partial S}{\partial q_i} H_i(q,w,u) + \sum_{j=1}^{r} w_j H_j(q,w,u) + H_0(\frac{\partial S}{\partial u}, u),$$

with smooth functions H_i, H_j and H_0.

Remark 4.3. *In the contact case the tangent space to the orbit of the group of contactomorphisms is determined by the set of contact Hamiltonians. The counterparts of the formulas from this lemma are similar: only each summand can contain the functions S itself as an extra argument.*

The (r,k)-adapted Lagrangian germ $L(S)$ with generating function germ $S(q,w,u)$ is called *infinitesimally stable* if its tangent space $T_S\Gamma$ coincides with the total space $C^\infty(q,w,u)$ of germs of smooth functions in q, w and u. The infinitesimally stable germ is stable.

Lemma 4.4. *Germ S is infinitesimally stable if and only if any function $\tilde{S}$ in x, y, u variables has a decomposition*

$$\tilde{S} = \sum_{i=1}^{k-r} \frac{\partial S}{\partial q_i} H_i(q, y, u) + \sum_{j=1}^{k} \frac{\partial S}{\partial y_j} H_j(y, u) + H_0(y, \frac{\partial S}{\partial u}, u).$$

Similarly to the classical result of J.Mather on right-left equivalence of smooth mappings, the classes, which are stable with respect to the group Γ_k, are deformations transversal to the orbits of the bigger group G_k.

Lemma 4.5. *If classes of* $\left.\frac{\partial(S(q,w,u)-q_1p_1-\cdots-q_{k-r}p_{k-r})}{\partial y}\right|_{y=o}$ *span $\mathcal{O}$, then the germ S is infinitesimally stable (and therefore stable).*

Remark 4.6. *The transversality to the G_k orbit persists under the transformations from Γ_k. Thus any stable adapted germ S of (r,k)-finite type is infinitesimally stable. Otherwise it would be not equivalent to nearby germs which are transversal to the corresponding G_k-orbit.*

Since a Γ_k-simple germ ought to be G_k simple, it remains to classify simple G_k orbits (see[5, 6]).

References

[1] V. I. Arnold, *Normal forms of functions near degenerate critical points, A_k, D_k, E_k Weil groups and Lagrangian singularities,* Functional analysis and applic.,**6** (1972) n.4, 3–25.

[2] V. I. Arnold, A. N. Varchenko, S. M. Gusein-Zade, *Singularities of differentiable mapping 1.* Moscow,"Nauka", 1982.

[3] A. B. Givental, *Singular Lagrangian varieties and their Lagrangian mappings,* Sovremennye probl. matem. Noveishie dostigenia, ViNITI. **.33** (1988), 55–112.

[4] R. M. Roberts, V. M. Zakalyukin, *On singular Lagrangian varieties,* Functional analysis and applic., **26** (1992) n.3, 28–34.

[5] O. M. Myasnichenko, V. M. Zakalyukin, *Lagrange Singularities under Symplectic Reduction,* Functional analysis and applic., **32**, (1998) n.1, 1–9.

[6] V. M. Zakalyukin, *Simple coisotropic caustics*, Preprint 189, Laboratoire de Topologie, Université de Bourgogne, (1999), 1–16.

Moscow Aviation Institute
125871, Moscow, Russia
E-mail address: `vladimir@zakal.mccme.ru`

Reassigned Scalograms and Singularities

Eric Chassande-Mottin and Patrick Flandrin

Abstract. Reassignment is a general nonlinear technique aimed at increasing the localization of time-frequency and time-scale distributions. Its principle consists in supplementing an energy distribution with a suitable vector field, thanks to which energy contributions are moved on the plane so as to sharpen the initial distribution. Reassignment can be performed in an efficient way and, in the case of scalograms (i.e., wavelet-based energy densities), it can be equipped with fast algorithms too. When applied to isolated Hölder singularities, scalogram reassignment acts as a squeezing operator upon the influence cone of the underlying wavelet transform, thus permitting a sharper localization and a higher contrast as compared to conventional scalograms. Closed form expressions can be obtained for the specific family of Klauder wavelets, with the Morlet wavelet as a limiting case. When considered as a function of scale at the time instant of the singularity, reassigned scalograms are shown to undergo a power-law evolution from which the Hölder exponent can be estimated.

1. Introduction

Time-frequency analysis of nonstationary signals can be performed in many different ways, with techniques ranging from short-time Fourier or wavelet transforms to Wigner-type methods [7, 14]. Whereas the former approaches exhibit poor localization properties, the latter ones are impaired by interference phenomena which limit the effectiveness of their sharper localization in multicomponent situations [10]. A nonlinear technique, referred to as *reassignment*, has been proposed to overcome both limitations. In a nutshell, reassignment is a two-step process which consists in smoothing out oscillating interference terms while squeezing the localized terms which have been spread over the plane by the smoothing operation. Originally proposed for the only spectrograms (short-time Fourier energy densities) [12, 13], the method has since been generalized far beyond [1], with a possible application to time-scale techniques such as scalograms (wavelet energy densities). The purpose of this paper is to address some specific issues related to scalogram reassignment, and especially to investigate how reassigned scalograms may be used for characterizing isolated Hölder singularities.

2. Reassigning Scalograms

Given an admissible wavelet $\psi(t) \in L^2(\mathbb{R})$ and a scale factor $a > 0$, the continuous wavelet transform (CWT) of a signal $x(t) \in L^2(\mathbb{R})$ is classically defined by [14]

$$T_x^{(\psi)}(t,a) := \frac{1}{\sqrt{a}} \int_{-\infty}^{+\infty} x(s)\, \overline{\psi\left(\frac{s-t}{a}\right)}\, ds\,. \tag{1}$$

As written in (1), the CWT is a function of time and scale but one can remark that, under mild conditions on the analyzing wavelet $\psi(t)$ (namely that its spectrum[1] energy density $|\Psi(\omega)|^2$ is unimodal and characterized by some reference (angular) frequency $\omega_0 > 0$), it can also be expressed as a function of time and frequency $\omega > 0$, with the identification $\omega := \omega_0/a$.

As it is well known, a by-product of the admissibility condition imposed on $\psi(t)$ (essentially that it is zero-mean, together with a proper normalization) is that an energetic interpretation can be attached to a CWT, thanks to the relation

$$\int_{-\infty}^{+\infty} \int_{0}^{+\infty} |T_x^{(\psi)}(t,a)|^2 \, \frac{dt\, da}{a^2} = \|x\|_2^2\,. \tag{2}$$

The energy distribution $S_x^{(\psi)}(t,a) := |T_x^{(\psi)}(t,a)|^2$ associated to a CWT is referred to as a *scalogram* and, whereas it is usually defined as the squared modulus of a CWT, it turns out [16] that it can be expressed as well as

$$S_x^{(\psi)}(t,a) = \int_{-\infty}^{+\infty} \int_{0}^{+\infty} W_x(s,\xi)\, W_\psi\left(\frac{s-t}{a}, a\xi\right) \frac{ds\, d\xi}{2\pi}, \tag{3}$$

by introducing the *Wigner-Ville distribution* (WVD) of $x(t)$ [7] :

$$W_x(t,\omega) := \int_{-\infty}^{+\infty} x(t+s/2)\, \overline{x(t-s/2)}\, e^{-i\omega s}\, ds\,. \tag{4}$$

The above relation (3) makes explicit the fact that a scalogram results from an affine smoothing of the WVD, with the consequence that the scalogram value at a given point (t,a) of the plane cannot be considered as pointwise. In fact, thanks to the smoothing operation (3), it rather results from the summation of all WVD contributions within some time-frequency domain defined as the essential time-frequency support of the wavelet, properly shifted by t and scaled by a. A whole distribution of values is therefore summarized by a single number, and this number is assigned to the geometrical center of the domain over which the distribution is considered. Reasoning with a mechanical analogy, the situation is as if the total mass of an object was assigned to its geometrical center, an arbitrary point which – except in the very specific case of a homogeneous distribution over the domain – has no reason to suit the actual distribution. A much more meaningful choice is to assign the total mass to the *center of gravity* of the distribution within

[1] Throughout this paper, we will use the convention of using lower case symbols for representing signals in the time domain, while labelling by the corresponding upper case symbols their spectra in the frequency domain.

the domain, and this is precisely what reassignment does: at each point where a scalogram value is computed, we also compute the local centroïd $(\hat{t}_x(t,a), \hat{\omega}_x(t,a))$ of the WVD W_x, as seen through the time-frequency "window" W_ψ centered in $(t, \omega = \omega_0/a)$, and the scalogram value is moved from the point where it has been computed to a new time-scale point deduced from this centroïd. Operating directly on the time-scale plane, this leads to defining the reassigned scalogram as [1]:

$$\check{S}_x^{(\psi)}(t,a) := \int_{-\infty}^{+\infty}\int_0^{+\infty} S_x^{(\psi)}(s,b)\,\delta\left(t - \hat{t}_x(s,b), a - \hat{a}_x(s,b)\right) \frac{b^2}{\hat{a}_x^2(s,b)}\, ds\, db\,, \quad (5)$$

with the centroïd in time given by

$$\hat{t}_x(t,a) = \frac{1}{S_x^{(\psi)}(t,a)} \int_{-\infty}^{+\infty}\int_0^{+\infty} s\, W_x(s,\xi)\, W_\psi\left(\frac{s-t}{a}, a\xi\right) \frac{ds\, d\xi}{2\pi}\,, \quad (6)$$

whereas the associated centroïd in scale $\hat{a}_x(t,a)$ needs some intermediate step in frequency. Precisely, it is necessary to first compute the quantity:

$$\hat{\omega}_x(t,a) = \frac{1}{S_x^{(\psi)}(t,a)} \int_{-\infty}^{+\infty}\int_0^{+\infty} \xi\, W_x(s,\xi)\, W_\psi\left(\frac{s-b}{a}, a\xi\right) \frac{ds\, d\xi}{2\pi}\,, \quad (7)$$

from which a frequency-to-scale conversion is then achieved according to:

$$\hat{a}_x(t,a) = \frac{\omega_0}{\hat{\omega}_x(t,a)}\,, \quad (8)$$

where ω_0 is the mean frequency of the wavelet spectrum energy density $|\Psi(\omega)|^2$, assumed to be of unit integral over $\mathbb{R}_+$. Whereas the proposed displacement rule is natural from the viewpoint of the mechanical analogy mentioned above, it is worth noting that other prescription rules can be derived from group theory arguments [5].

From a practical point of view, the centroïds (6) and (8) can be evaluated in a much more efficient way. Given a wavelet $\psi(t)$, this requires the introduction of the two additional wavelets $(\mathbf{T}\psi)(t) := t\,\psi(t)$ and $(\mathbf{D}\psi)(t) := (d\psi/dt)(t)$, thanks to which we have [1]:

$$\hat{t}_x(t,a) = t + a\,\mathrm{Re}\left\{T_x^{(>\mathbf{T}\psi)}(t,a)/T_x^{(\psi)}(t,a)\right\}\,; \quad (9)$$

$$\hat{a}_x(t,a) = -\frac{a\,\omega_0}{\mathrm{Im}\left\{T_x^{(>\mathbf{D}\psi)}(t,a)/T_x^{(\psi)}(t,a)\right\}}\,. \quad (10)$$

As compared to conventional scalograms based on one single CWT, reassigned scalograms involve three different CWT's (and even only two when using a Morlet wavelet $g(t)$ for which $(\mathbf{T}g)(t)$ and $(\mathbf{D}g)(t)$ are proportional). The computational burden remains therefore of the same order, with moreover the possibility of an efficient implementation based on fast algorithms [6].

3. Reassignment as Squeezing

Whereas there is no time-scale curve onto which conventional scalograms can perfectly localize, reassigned scalograms inherit automatically the localization properties of the WVD on straight lines of the time-frequency plane. More precisely, it is well known that, for unimodular signals of the form $x(t) = \exp\{i(\omega_0 t + \alpha t^2/2)\}$, we have $W_x(t,\omega) = \delta\left(\omega - (\omega_0 + \alpha t)\right)$ for any α (linear "chirp"), and not only for $\alpha = 0$ (pure tone) as is the case for the Fourier transform [7]. Letting α go to $\pm\infty$ leads formally to idealized impulses for which we also have:

$$x(t) = \delta(t - t_0) \Rightarrow W_x(t,\omega) = \delta\left(t - t_0\right) . \tag{11}$$

In this case, it immediately follows from (9) that $\hat{t}_x(t,a) = t_0$ for any a and any $\psi(t)$, thus guaranteeing that $\check{S}_x^{(\psi)}(t,a) \propto \delta(t - t_0)$, as does the WVD in (11). This situation contrasts with that of the ordinary scalogram for which

$$x(t) = \delta(t - t_0) \Rightarrow S_x^{(\psi)}(t,a) = \frac{1}{a}\left|\psi\left(\frac{t_0 - t}{a}\right)\right|^2 . \tag{12}$$

In this latter case, the essential support of the scalogram corresponds to a time-scale domain (referred to as its *influence cone*) that is limited by the two lines of equations $t = t_0 \pm a\,\Delta t_\psi/2$, where Δt_ψ stands for a measure of effective width of the wavelet $\psi(t)$. Reassignment acts therefore as a squeezing operator that permits us to end up with a perfectly localized distribution in the case of a perfectly localized impulse[2]. For the picture to be complete, one has to also consider the reassignment operator acting on scale and it turns out, from (8), that

$$\hat{a}_x(t,a) = \frac{\omega_0}{\omega_{\psi_a}(t)}, \tag{13}$$

where $\omega_{\psi_a}(t)$ refers to the instantaneous frequency of $\psi_a(t) := \psi(t/a)/\sqrt{a}$, i.e., of the wavelet at scale a. Reassignment in scale depends therefore on the chosen wavelet and, except in Morlet-type cases for which the instantaneous frequency $\omega_g(t)$ is constant, reassignment trajectories will in general be curves that are not parallel to the time axis.

4. Singularity Characterization from Reassigned Scalograms

Time-scale techniques – including scalograms – are known to provide a tool simultaneously adapted for the detection and the characterization of singularities [14, 15]. Inspired by the impulse example sketched above, it is therefore natural to consider reassigned scalograms of singularities, the underlying motivation being that reassignment methods may improve the contrast in the representation of singularities and therefore their detection.

[2]As such, reassignment shares much with related techniques such as those discussed in [4, 8, 9].

We will, here, limit ourselves to isolated Hölder singularities. Any isolated Hölder singularity can be written as (details about the construction of this family can be found in [3]):

$$X(\omega) = A_\nu |\omega|^{-\nu-1}, \tag{14}$$

where the amplitude function A_ν is parameterized by the Hölder exponent ν according to:

$$A_\nu = \begin{cases} 2\Gamma(\nu+1)\,(-\sin(\nu\pi/2))\,, & \text{if } \nu \in \mathbb{R} - \mathbb{Z}\,, \\ 2(\nu\,!)\,(-1)^{(\nu+1)/2}\,, & \text{if } \nu \in \mathbb{N}\,, \\ \pi(-1)^{\nu/2}/|\nu+1|\,, & \text{if } \nu \in \mathbb{Z}_-^*\,. \end{cases} \tag{15}$$

These functions are obviously a particular case of singularities. Nevertheless, they are good local approximations for a wide variety of observable singular behaviours.

Defining the fractional derivative of order α of a signal $x(t)$ as

$$x^{(\alpha)}(t) := \int_0^{+\infty} (i\omega)^\alpha\, X(\omega)\, e^{it\omega}\, \frac{d\omega}{2\pi}, \tag{16}$$

it turns out [9] that isolated Hölder singularities of the type (14) have the property that their CWT is equal to a rescaled version of their fractional derivative of order $\alpha = -\nu - 1$:

$$T_x^{(\psi)}(t,a) = A_{-\alpha-1}\, a^{-(\alpha+1/2)}\, i^\alpha\, \overline{\psi^{(\alpha)}(-t/a)}\,. \tag{17}$$

One can see in (17) two important characteristics of the scalogram structure of Hölder singularities. First, the energy is almost entirely concentrated in an influence cone defined by the support of $|\psi^{(\alpha)}(-t/a)|$. Second, from the restriction of (17) at time $t = 0$, namely

$$\log|T^{(\psi)}(0,a)|^2 = \log|A_\nu \psi^{(-\nu-1)}(0)|^2 + (2\nu+1)\log a\,, \tag{18}$$

one can get a simple estimate of ν by measuring the scalogram slope along the scale axis in a log-log diagram [15].

In order to go further, the wavelet $\psi(t)$ has to be specified, and it proves convenient to make use of the so-called *Klauder* [11] (or Cauchy) wavelet, defined in the time domain as

$$\kappa_{\beta,\gamma}(t) = \frac{C_{\beta,\gamma}}{(\gamma - it)^{\beta+1}}, \tag{19}$$

where the constant $C_{\beta,\gamma} = (2\gamma)^{\beta+1/2}\Gamma(\beta+1)/\sqrt{2\pi\Gamma(2\beta+1)}$ ensures a unit energy normalization. From the Fourier transform of $\kappa_{\beta,\gamma}(t)$,

$$K_{\beta,\gamma}(\omega) = C_{\beta,\gamma}\, \frac{2\pi}{\Gamma(\beta+1)}\, \omega^\beta\, e^{-\gamma\omega}\, U(\omega)\,, \tag{20}$$

with the convergence conditions $\beta > -1/2$ and $\gamma > 0$, and where $U(\cdot)$ is the Heaviside step function, one can see that the Klauder wavelet family is covariant to fractional differentiation:

$$\kappa_{\beta,\gamma}^{(\alpha)}(t) = \left(\frac{i}{2\gamma}\right)^{\alpha} \sqrt{\frac{\Gamma(2(\alpha+\beta)+1)}{\Gamma(2\beta+1)}}\, \kappa_{\alpha+\beta,\gamma}(t)\,. \tag{21}$$

This last equation gives us the possibility of obtaining the following closed form expression for the (Klauder) CWT of a Hölder singularity:

$$T_x^{(\kappa_{\beta,\gamma})}(t,a) = A_{-\alpha-1}\,(2\gamma a)^{\beta+1/2}\,\frac{\Gamma(\alpha+\beta+1)}{\sqrt{\Gamma(2\beta+1)}}\,(\gamma a - it)^{-(\alpha+\beta+1)}\,, \tag{22}$$

whence that of the corresponding scalogram, by taking the squared modulus.

Concerning the computation of the centroïds (9) and (10), the two wavelet transforms $T_x^{(\mathbf{T}\psi)}$ and $T_x^{(\mathbf{D}\psi)}$ need to be expressed. For this, we can use the property that the Klauder wavelet is stable by multiplication by t and by differentiation (the second property resulting from (21) with $\alpha = 1$):

$$\begin{aligned} (d\kappa_{\beta,\gamma}/dt)(t) &= (i/2\gamma)\sqrt{(2\beta+3)(2\beta+2)}\,\kappa_{\beta+1,\gamma}(t), \\ t\,\kappa_{\beta,\gamma}(t) &= i\gamma\sqrt{(2\beta)/(2\beta-1)}\,\kappa_{\beta-1,\gamma}(t) - i\gamma\kappa_{\beta,\gamma}(t)\,. \end{aligned} \tag{23}$$

Combining (17), (21) and (23), we can get the algebraic form of the three wavelet transforms involved in (9) and (10), leading finally to (see [3] for details):

$$\hat{t}(t,a) = \frac{\alpha}{\alpha+\beta}\,t\,; \tag{24}$$

$$\hat{a}(t,a) = \frac{\omega_0}{\alpha+\beta+1}\left(\gamma a + \frac{t^2}{\gamma a}\right). \tag{25}$$

with the reference frequency of the Klauder wavelet equal to $\omega_0 = (\beta+1/2)/\gamma$.

A remarkable feature of the above equation is that, although the Klauder wavelet is of infinite support in time (with an effective width $\Delta t_{\kappa_{\beta,\gamma}} \propto \gamma/\sqrt{2\beta-1}$), it leads to a reassigned scalogram whose time-scale support is *strictly* limited to a cone centered at the time of occurrence $t = 0$ of the singularity. Border lines of this cone are given by

$$\hat{t} = \pm\,\frac{\alpha\gamma\,(\alpha+\beta+1)}{(\alpha+\beta)(2\beta+1)}\,\hat{a}\,, \tag{26}$$

thus showing that the sharpness of the cone is controlled by both the singularity strength (through α) and the chosen wavelet (through β and γ). More precisely, it is easy to check that, for any fixed β and γ (i.e., for any fixed Klauder wavelet), the angle θ of the cone goes to zero as $\theta \sim [(\gamma/\beta)(\beta+1)/(\beta+1/2)] \times \alpha$ when $\alpha \to 0$, as expected from the result we obtained in the impulse case. Conversely, we also get that θ goes to zero as $\theta \sim \gamma\alpha/\beta$ when α is fixed and $\beta \to \infty$. In such a limiting case, the Klauder wavelet tends to become a Morlet wavelet, thus evidencing that a Morlet scalogram perfectly localizes Hölder singularities, even in cases where $\alpha \neq 0$.

The explicit evaluation of the reassigned scalogram can finally be achieved by inverting the reassignment operators (24) and (25), so as to identify which quantities sum up at a given reassignment point $(\hat{t}, \hat{a})$, leading to

$$\check{S}_x^{(\kappa)}(\hat{t}, \hat{a}) = \frac{\hat{C}\,\hat{a}^{-(\alpha+\beta-1)}}{(2C)^{\alpha+\beta+1}} \sum_{\epsilon=\pm 1} \left(C\hat{a} + \epsilon\sqrt{(C\hat{a})^2 - (1+\beta/\alpha)^2\hat{t}^2} \right)^{-\alpha+\beta-2} \tag{27}$$

where the constants are given by $\hat{C} = A^2_{-\alpha-1} 2^{2\beta+1} \gamma^{\alpha+\beta+3} \Gamma^2(\alpha+\beta+1)/\Gamma(2\beta+1)$ and $C = \gamma(\alpha+\beta+1)/(2\beta+1)$.

At the time of occurrence $t = 0$ of the singularity, (27) simplifies to

$$\check{S}_x^{(\kappa)}(0, \hat{a}) = \frac{\hat{C}}{(2C)^{2\alpha-3}}\,\hat{a}^{-(2\alpha+1)}, \tag{28}$$

an equation from which we conclude that (i), as for the scalogram, the reassigned scalogram undergoes a power-law behaviour with respect to scales, and (ii) the exponent $-(2\alpha+1) = 2\nu+1$ of this power-law is the same as in the scalogram case (see (18)). This means that the measurement of the Hölder exponent ν can be done possibly with a reassigned scalogram.

5. Conclusion

Reassigned scalograms have been shown to be candidates for the detection and characterization of isolated Hölder singularities, and by extension for the measurement of the local regularity index of an arbitrary signal. Some technical details (such as existence conditions for the Klauder wavelet, or the consideration of special cases where the reassignment operators cannot be inverted) and questions concerning practical implementation (influence of the time-frequency grid resolution) have been omitted here but can be found in [3].

Software — Matlab codes for computing reassigned time-frequency and time-scale distributions are available as part of a Toolbox [2], freely distributed on the Internet.

References

[1] F. Auger, P. Flandrin. Improving the readability of time-frequency and time-scale representations by the reassignment method. *IEEE Trans. Signal Proc.*, SP-43(5): 1068–1089, 1995.

[2] F. Auger, P. Flandrin, P. Gonçalvès, O. Lemoine. Time-frequency toolbox for MATLAB, user's guide and reference guide. Available on the Internet at the URL `http://iut-saint-nazaire.univ-nantes.fr/~auger/tftb.html`.

[3] E. Chassande-Mottin. *Méthodes de réallocation dans le plan temps-fréquence pour l'analyse et le traitement de signaux non stationnaires.* Thèse de Doctorat, Université de Cergy-Pontoise (France), 1998.

[4] I. Daubechies, S. Maes. A nonlinear squeezing of the continuous wavelet transform based on auditory nerve models. In *Wavelets in Medicine and Biology* (A. Aldroubi and M. Unser, *eds.*), 527–546. CRC Press, Boca Raton (FL), 1996.
[5] L. Daudet, M. Morvidone, B. Torresani. Time-frequency and time-scale vector fields for deforming time-frequency and time-scale representations. Preprint, 1999.
[6] P. Flandrin, E. Chassande-Mottin, P. Abry. Reassigned scalograms and their fast algorithms. *Proc. SPIE-95*, 2569: 152–158, San Diego (USA), 1995.
[7] P. Flandrin. *Time-Frequency/Time-Scale Analysis.* Academic Press, San Diego (CA), 1999.
[8] P. Guillemain, R. Kronland-Martinet. Horizontal and vertical ridges associated to continuous wavelet transforms. *Proc. of the IEEE Int. Symp. on Time-Frequency and Time-Scale Analysis*, 63–66, Victoria (Canada), 1992.
[9] P. Guillemain, R. Kronland-Martinet. Characterization of acoustic signals through continuous linear time-frequency representations. *Proc. IEEE*, 84(4): 561–587, 1996.
[10] F. Hlawatsch, P. Flandrin. The interference structure of Wigner distributions and related time-frequency signal representations. In *The Wigner Distribution: Theory and Applications in Signal Processing* (W. Mecklenbräuker and F. Hlawatsch, *eds.*), 59–133, Elsevier, Amsterdam (NL), 1998.
[11] J. R. Klauder. Path integrals for affine variables. In *Functional Integration: Theory and Applications* (J.-P. Antoine and E. Tirapegui, *eds.*), 101–119. Plenum Press, New York (NY), 1980.
[12] K. Kodera, C. De Villedary, R. Gendrin. A new method for the numerical analysis of nonstationary signals. *Phys. Earth and Plan. Int.*, 12: 142–150, 1976.
[13] K. Kodera, R. Gendrin, C. De Villedary. Analysis of time-varying signals with small BT values. *IEEE Trans. on Acoust., Speech, and Signal Proc.*, ASSP-26(1): 64–76, 1978.
[14] S. Mallat. *A Wavelet Tour of Signal Processing.* Academic Press, New York (NY), 1998.
[15] S. Mallat, W. L. Hwang. Singularity detection and processing with wavelets. *IEEE Trans. on Info. Theory*, IT-38(2): 617–643, 1992.
[16] O. Rioul, P. Flandrin. Time-scale energy distributions : a general class extending wavelet transforms. *IEEE Trans. on Signal Proc.*, SP-40(7): 1746–1757, 1992.

E. Chassande-Mottin
Max Planck Institut für Gravitationsphysik
Albert Einstein Institut
Am Mühlenberg, 1
D-14424 Golm, Germany
E-mail address: `eric@aei-potsdam.mpg.de`

P. Flandrin
Laboratoire de Physique (UMR 5672 CNRS)
École Normale Supérieure de Lyon
46, allée d'Italie
F-69364 Lyon Cedex 07, France
E-mail address: `flandrin@ens-lyon.fr`

The Impact of Wavelet Coefficient Correlations on Fractionally Differenced Process Estimation

Peter F. Craigmile, Donald B. Percival and Peter Guttorp

Abstract. The discrete wavelet transform (DWT) approximately decorrelates a fractionally differenced (FD) process, allowing for simple maximum likelihood estimation of the FD process parameters using the wavelet coefficients. In previous work we have established limit theorems for the parameters based on a model where scales are uncorrelated and two simple models for within-scale correlation, namely, white noise and a first order autoregressive (AR) process. Here we assess the adequacy of these simple models for handling between- and within-scale correlations. We compare the performance of these simple models for estimating the FD process parameters against procedures that use longer wavelet filters (to reduce between-scale correlations) and use AR models of higher order (to more precisely model within-scale correlations).

1. Introduction

Time series collected in areas such as atmospheric sciences, geosciences and hydrology often exhibit a long range dependence, i.e., slowly decaying auto-correlations or, equivalently, a spectral density function (SDF) that is proportional to $|f|^\alpha$ at low frequencies for some $\alpha < 0$. A convenient model for such series is a fractionally differenced (FD) process [5, 6]. Specifically, for $d \in [-\frac{1}{2}, \frac{1}{2})$ and $\sigma_\epsilon^2 > 0$, $\{X_t\}_{t \in \mathcal{Z}}$ is an FD(d, σ_ϵ^2) process if its SDF is given by

$$S_X(f) \quad = \quad \sigma_\epsilon^2 |2\sin(\pi f)|^{-2d} \quad f \in [-\tfrac{1}{2}, \tfrac{1}{2}]. \tag{1}$$

d is known as the *difference parameter* and σ_ϵ^2 is the *innovation variance.* When $d = 0$, $\{X_t\}$ is a white noise (i.e. uncorrelated) process. Extending this model by letting $d \geq \frac{1}{2}$ in equation (1), we obtain a class of non-stationary FD processes that are stationary if we difference $\lfloor d + \frac{1}{2} \rfloor$ times.

Given a time series that is a realisation of a portion $\{X_t\}_{t=0}^{N-1}$ of a stationary FD process, McCoy and Walden [9] extended earlier work by Wornell [13] to obtain effective approximate maximum likelihood (ML) estimators of the FD parameters d and σ_ϵ^2. The basis of their scheme was to formulate the likelihood function in terms of the discrete wavelet transform (DWT) of $\{X_t\}$ by making use of the assumption that the DWT of a Gaussian FD process yields approximately independent deviates. In previous work [10, 3], we extended the McCoy and Walden estimator to handle both stationary and non-stationary FD processes observed in the presence of a trend; i.e., the observed time series is taken to be a realisation of

$$Y_t = T_t + X_t \qquad t = 0, \dots, N-1. \tag{2}$$

Here $\{T_t\}$ is a deterministic polynomial trend of order K, and $\{X_t\}$ is a realisation of a Gaussian FD(d, σ_ϵ^2) process. As with the McCoy and Walden scheme, the key assumption behind this extension is the independence of certain wavelet coefficients across and between scales. In this paper we re-examine this assumption. After a review of background material in Section 2, we argue in Section 3 that the correlation between scales can be made arbitrarily small by increasing the length of the wavelet filter. This increase in filter length, however, does not help reduce the correlation within scales, so we consider in Section 4 modelling this correlation using autoregressive (AR) models whose coefficients are scale-dependent but are solely determined by d. We conclude that a first order AR model is adequate for modelling the correlation structure within scales.

2. Definitions and Background on Wavelet Coefficients

For an even integer L, let $\{h_l\}_{l=0}^{L-1}$ denote a Daubechies [4] wavelet filter. By definition this filter has squared gain function

$$\mathcal{H}_{1,L}(f) \equiv 2\sin^L(\pi f) \sum_{l=0}^{L/2-1} \binom{L/2-1+l}{l} \cos^{2l}(\pi f). \tag{3}$$

Associated with the wavelet filter we define the scaling filter by $g_l \equiv (-1)^{l+1} h_{L-1-l}$ (with a squared gain function of $\mathcal{G}_{1,L}(f) = \mathcal{H}_{1,L}(\frac{1}{2} - f)$). Assume for convenience that $N = 2^J$ for some integer J, and let $N_j \equiv N2^{-j}$. The level j wavelet coefficients can be computed using the level j wavelet filter $\{h_{j,l}\}_{l=0}^{L_j-1}$:

$$W_{j,k} = \sum_{l=0}^{L_j-1} h_{j,l} X_{2^j(k+1)-l-l \bmod N_{j-1}} \qquad j = 1, \dots, J, \quad k = 0, \dots, N_j - 1$$

where $L_j \equiv (2^j - 1)(L-1) + 1$ and $\{h_{j,l}\}$ has squared gain function

$$\mathcal{H}_{j,L}(f) \equiv \mathcal{H}_{1,L}(2^{j-1}f) \prod_{k=0}^{j-2} \mathcal{G}_{1,L}(2^k f). \tag{4}$$

These coefficients are associated with changes in averages on scale $\tau_j \equiv 2^{j-1}$ and with times spaced $\lambda_j \equiv 2^j$ units apart. In practice we use the pyramid algorithm (Mallat [8]) to calculate these DWT coefficients efficiently (see Percival and Walden [11]). Since the DWT handles filtering operations periodically, the first $B_j \equiv \lceil (L-2)(1-2^{-j}) \rceil$ wavelet coefficients are explicitly affected by the circularity assumption. We call these coefficients the *boundary dependent (BD) coefficients.* We call the remaining $M_j \equiv N_j - B_j$ which are unaffected by boundaries the *boundary independent (BI) coefficients.* Let $M \equiv \sum_{j=1}^{J} M_j$.

In Craigmile *et al.* [3] we noted that the BI wavelet coefficients are unaffected by the polynomial trend if $K \leq \frac{L}{2}$, and thus we can estimate the parameters of $\{X_t\}$ via Gaussian likelihood using these coefficients (if $\frac{L}{2} \geq \lfloor d - \frac{1}{2} \rfloor$). We further assumed that the BI wavelet coefficients are uncorrelated between scales, and either a white noise or AR(1) model was a good fit for these coefficients on each level. We now investigate this further.

3. Between-Scale Decorrelation

Let $(W_w)_{j,k}$ denote the BI wavelet coefficients ($j = 1, \ldots, J$, $k = 0, \ldots, M_j$). From chapter 9 of Percival and Walden [11] we have that

$$\begin{aligned} \mathrm{Cov}((W_w)_{j,k}, (W_w)_{j',k'}) &= 2^{1-2d}\sigma_\epsilon^2 \int_0^{\frac{1}{2}} \cos(2\pi f(2^j(k+1) - 2^{j'}(k'+1))) \\ &\qquad \times H_{j,L}(f)\, H^*_{j',L}(f)\, \sin^{-2d}(\pi f)\, df\,, \end{aligned}$$

where $H_{j,L}(f)$ is the Fourier transform of $\{h_{j,l}\}$ and denotes * the complex conjugation operator. An extension of Theorem 3.2 in Craigmile *et al.* [3] shows that this integral is finite for $d < \frac{L+1}{2}$. $\mathcal{H}_{j,L}(f)$ corresponds to an approximate band-pass filter with pass-band $[2^{-(j+1)}, 2^{-j}]$ (see, e.g., Daubechies [4]). This approximate filter has a squared gain function given by $\mathcal{H}_{1,bp}(f) \equiv 2^j 1_{[2^{-(j+1)},2^{-j}]}(f)$. Lai [7] defines the following squared gain function

$$\mathcal{H}_{1,I}(f) \equiv \begin{cases} 0, & f \in [0, \frac{1}{4})\,; \\ 1, & f = \frac{1}{4}\,; \\ 2, & f \in (\frac{1}{4}, \frac{1}{2}]\,, \end{cases}$$

and shows that $\mathcal{H}_{1,L}(f) \to \mathcal{H}_{1,I}(f)$ as $L \to \infty$ for all $f \in [0, \frac{1}{2}]$. Thus if we define $\mathcal{H}_{j,I}(f) \equiv \mathcal{H}_{1,I}(2^{j-1}f) \prod_{k=0}^{j-2} \mathcal{H}_{1,I}(\frac{1}{2} - 2^k f)$, we have $\mathcal{H}_{j,L}(f) \to \mathcal{H}_{j,I}(f)$ as $L \to \infty$ for all $f \in [0, \frac{1}{2}]$ and $j \geq 1$. $\mathcal{H}_{j,I}(f)$ differs from $\mathcal{H}_{1,bp}(f)$ on a countable set of points and thus an integral involving either of these two squared gain functions will be the same. By the spectral representation theorem we can therefore see that the BI wavelet coefficients at different scales are asymptotically uncorrelated for large L, since the pass-bands of these squared gain functions do not intersect (see Craigmile [2] for additional details).

Lai [7] also proves that convergence of $\mathcal{H}_{1,L}(f)$ is monotone in the following sense. For all even L, $\mathcal{H}_{1,L}(\frac{1}{4}) = 1$,

$$\begin{aligned} \mathcal{H}_{1,L}(f) \geq \mathcal{H}_{1,L+2}(f) \geq \mathcal{H}_{1,I}(f)\,, &\quad f \in [0, \tfrac{1}{4})\,; \\ \mathcal{H}_{1,L}(f) \leq \mathcal{H}_{1,L+2}(f) \leq \mathcal{H}_{1,I}(f)\,, &\quad f \in (\tfrac{1}{4}, \tfrac{1}{2}]\,. \end{aligned}$$

For $j > 1$ this translates into

$$\mathcal{H}_{j,L}(f) \geq \mathcal{H}_{j,L+2}(f) \geq \mathcal{H}_{j,I}(f)\,, \quad f \in [2^{-j}, \tfrac{1}{2}]$$

meaning that the side lobe behaviour of $\mathcal{H}_{j,L}(f)$ reduces with increasing L. Also the decorrelation between higher and lower scales is rapid with increasing L because

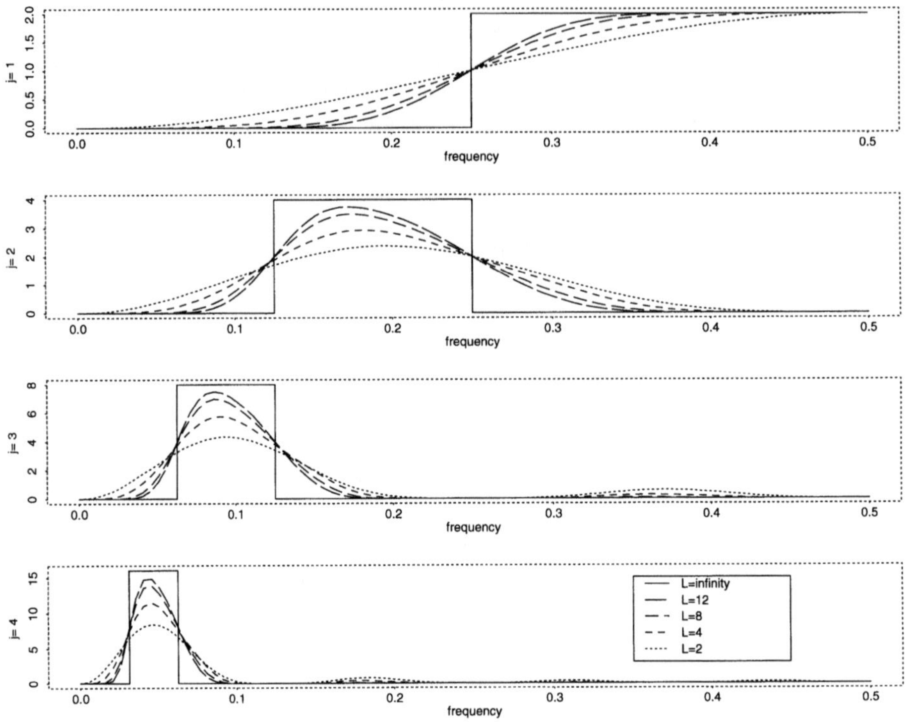

FIGURE 1. Squared gain functions for various filter lengths, L, and $j = 1, \ldots, 4$ ($L = \infty$ denotes the ideal wavelet filter). For example, with $j = 3$ the side-lobes for $f > \frac{1}{4}$ decrease with increasing L.

there is less intersection of the squared gain functions. Figure 1 illustrates this decay for a number of wavelet filter lengths and $j = 1, \ldots, 4$.

See Tewfik and Kim [12] for a related discussion on the correlation structure of a DWT of fractional Brownian motion.

4. An AR(p) Wavelet Model

We now consider the within-scale dependence. On level j we can write the lag τ auto-covariance as

$$\sigma_\epsilon^2 \sigma_{j,\tau}(d) \equiv \int_{-\frac{1}{2}}^{\frac{1}{2}} e^{i2\pi f\tau} S_j(f)\, df\,, \tag{5}$$

where we define the SDF of the level j BI wavelet coefficients by

$$S_j(f) \equiv \sigma_\epsilon^2\, 2^{-j-2d} \sum_{k=0}^{2^j-1} \mathcal{H}_{j,L}(2^{-j}(f+k))\, \sin^{-2d}(2^{-j}(f+k))\,. \tag{6}$$

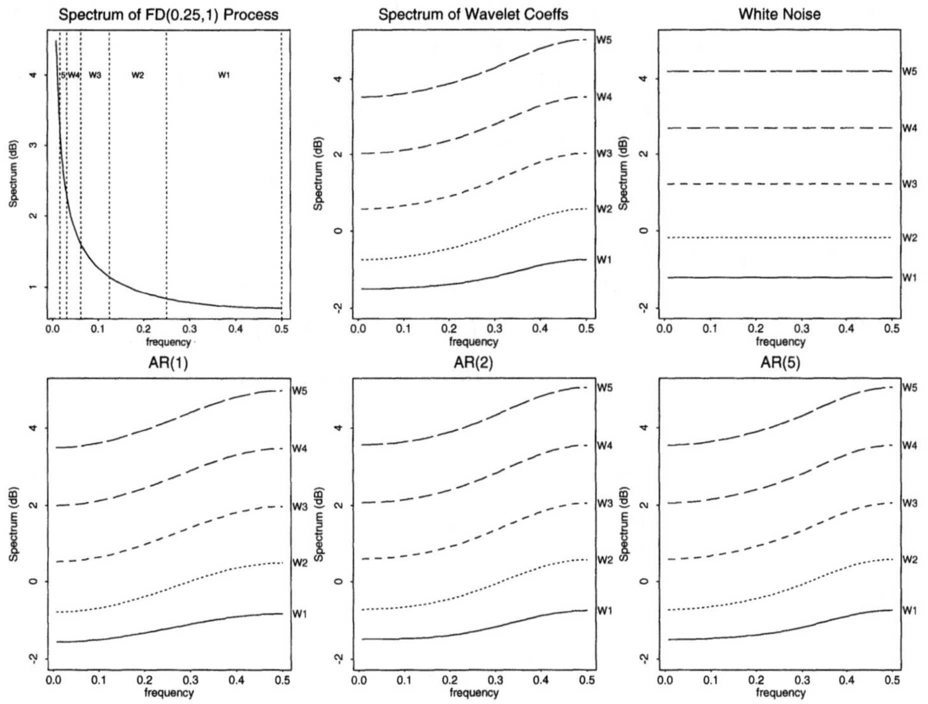

FIGURE 2. Going from left to right, top to bottom, plots show the SDF of a FD(0.25,1) process (dotted vertical lines indicate the approximate bandpasses for the first five wavelet levels), the SDF of the BI wavelet coefficients with $L = 8$, and the SDF assumed in the white noise and AR(p) models for $p = 1, 2, 5$.

In Craigmile *et al.* [3] we showed that, if we assume that the BI wavelet coefficients are either a portion of a white noise or AR(1) process, then the estimate of d is asymptotically normal, and the estimate of σ_ϵ^2 follows a scaled chi-squared distribution. Both estimates are consistent. Since we can approximate any continuous SDF by an AR(p) SDF for large enough p (Anderson [1]), a better approximation is given by supposing that the BI wavelet coefficients $\{(W_w)_{j,k} : k = 0, \ldots, M_j - 1\}$ are a portion of an AR(p) process, i.e.,

$$(W_w)_{j,k} = \sum_{r=1}^{p} \phi_{r,p}(j,d)(W_w)_{j,k-r} + Z_{j,k} \qquad (k = p, \ldots, M_j - 1) \qquad (7)$$

where $\{Z_{j,k} \sim \text{i.i.d. } N(0, \eta(j,d)\sigma_\epsilon^2) : j = 1 \ldots J,\ k = 0, \ldots, M_j - 1\}$. Figure 2 illustrates this for an FD(0.25,1) process analysed using an wavelet filter with $L = 8$. The top left panel shows the SDF of the process along with the approximate band-passes that correspond to the first five wavelet levels. The top middle panel shows the actual SDF of the BI wavelet coefficients (equation 6) and the right-hand panel is the SDF if we assume that that the BI coefficients are uncorrelated per each wavelet level. If we assume the AR(p) model as above for $p = 1, 2, 5$ we have an SDF given in the lower panels of Figure 2. The SDF looks better for higher p, but not by too much (the AR(1) approximation is very good as it stands). We now employ the Levinson-Durbin (LD) recursions. Let $\phi_{1,1}(j,d) \equiv \sigma_1(j,d)/\sigma_0(j,d)$, $P_0(j,d) \equiv \sigma_0(j,d)$ and $P_1(j,d) \equiv \sigma_0(j,d)(1-\phi_{1,1}^2(j,d))$. Then for $s > 1$,

$$\begin{aligned}
\phi_{s,s}(j,d) &= P_{s-1}^{-1}(j,d)\left(\sigma_s(j,d) - \textstyle\sum_{r=1}^{s-1}\phi_{r,s-1}(j,d)\,\sigma_{s-r}(j,d)\right);\\
\phi_{r,s}(j,d) &= \phi_{r,s-1}(j,d) - \phi_{s,s}(j,d)\,\phi_{s-r,s-1}(j,d) \qquad (r = 1,\dots,s-1);\\
P_s(j,d) &= \sigma_0(j,d) - \textstyle\sum_{r=1}^{s}\phi_{r,s}(j,d)\,\sigma_r(j,d) = P_{s-1}(j,d)(1-\phi_{s,s}^2(j,d)).
\end{aligned}$$

The Yule-Walker equations show that $P_p(j,d) = \eta(j,d)$. Letting $M_{jp} \equiv M_j - p$, the likelihood for one wavelet level ($j = 1 \dots J$) is

$$\begin{aligned}
l(j,d,\sigma_\epsilon^2) \equiv\; & -\tfrac{M_j}{2}\left[\log(2\pi\sigma_\epsilon^2)\right] - \tfrac{1}{2}\left[M_{jp}\log(P_p(j,d)) + \textstyle\sum_{k=0}^{p-1}\log(P_k(j,d))\right]\\
& -\tfrac{1}{2\sigma_\epsilon^2}\left[\textstyle\sum_{k=0}^{p-1}\frac{(\vec{Z}_{j,k}^{(k)}(d))^2}{P_k(j,d)} + \sum_{k=p}^{M_j-1}\frac{Z_{j,k}^2}{P_p(j,d)}\right]
\end{aligned}$$

where

$$\vec{Z}_{j,k}^{(k)}(d) \equiv \begin{cases} (W_w)_{j,0}, & k = 0;\\ (W_w)_{j,k} - \sum_{r=1}^{k}\phi_{r,k}(j,d)\,(W_w)_{j,k-r}, & k = 1\dots p-1, \end{cases}$$

and hence assuming that wavelet levels are uncorrelated

$$l_N(d,\sigma_\epsilon^2) \equiv \textstyle\sum_{j=1}^{J} l(j,d,\sigma_\epsilon^2). \tag{8}$$

Maximizing with respect to σ_ϵ^2 yields the ML estimate

$$\widehat{\sigma}_{\epsilon,N,p}^2(d) = \tfrac{1}{M}\textstyle\sum_{j=1}^{J}\left[\sum_{k=0}^{p-1}\frac{(\vec{Z}_{j,k}^{(k)}(d))^2}{P_k(j,d)} + \sum_{k=p}^{M_j-1}\frac{Z_{j,k}^2}{P_p(j,d)}\right]. \tag{9}$$

The profile likelihood with respect to d is

$$\begin{aligned}
l_N(d,\widehat{\sigma}_{\epsilon,N,p}^2(d)) \equiv\; & -\tfrac{M}{2}\left[\log(2\pi\widehat{\sigma}_{\epsilon,N,p}^2(d)) + 1\right]\\
& -\tfrac{1}{2}\textstyle\sum_{j=1}^{J}\left[M_{jp}\log(P_p(j,d)) + \sum_{k=0}^{p-1}\log(P_k(j,d))\right].
\end{aligned} \tag{10}$$

We maximize this expression to obtain $\hat{d}_{N,p}$. Now let $\hat{\theta}^T \equiv (\hat{d}_{N,p}, \widehat{\sigma}_{\epsilon,N,p}^2(d))$ denote the vector of estimates. We can extend the results of Craigmile *et al.* [3] as follows (see [2] for a proof).

Theorem 4.1. *For a differentiable function $g(\cdot)$, let $\Delta_1(g(x)) \equiv [\frac{\partial}{\partial y}g(y)|_{y=x}]/g(x)$. Suppose that equation (7) holds. For $d < \frac{L+1}{2}$, as $N \to \infty$,*

(a) $(\hat{\theta} - \theta) \to_p 0$;
(b) $\sqrt{N}(\hat{\theta} - \theta) \to_d N(0, \Gamma^{-1}(\theta))$;
(c) $\sqrt{N}(\hat{d}_{N,p} - d) \to_d N(0, \sigma_{d,p}^2)$,

		d				
L	p	0	0.25	0.50	0.75	1.00
2	0	1.260	1.036	0.896	0.781	0.664
2	1	1.260	1.052	0.906	0.767	0.664
2	2	1.260	1.055	0.908	0.762	0.664
2	5	1.260	1.056	0.912	0.754	0.664
4	0	1.060	0.982	0.921	0.867	0.816
4	1	1.060	1.005	0.960	0.909	0.839
4	2	1.060	1.004	0.963	0.916	0.836
4	5	1.060	1.004	0.963	0.917	0.829
8	0	0.991	0.956	0.923	0.893	0.864
8	1	0.991	0.977	0.965	0.952	0.934
8	2	0.991	0.975	0.963	0.953	0.937
8	5	0.991	0.975	0.964	0.953	0.937
∞	0	0.944	0.931	0.919	0.906	0.894
∞	1	0.944	0.944	0.945	0.944	0.944
∞	2	0.944	0.944	0.944	0.944	0.944
∞	5	0.944	0.944	0.945	0.945	0.945

TABLE 1. Calculation of $\sigma^2_{d,p}$ for various filter lengths, L, ($L = \infty$ refers to using the ideal wavelet filter $\mathcal{H}_{j,I}(f)$), AR(p) wavelet model ($p = 0$ refers to the white noise model of Craigmile *et al.* [3]) and difference parameter d.

where

$$2\Gamma(\theta) \quad \equiv \quad \begin{bmatrix} \sum_j \Delta_1^2(P_p(j,d))\, 2^{-j} & \sigma_\epsilon^{-2} \sum_j \Delta_1(P_p(j,d))\, 2^{-j} \\ \sigma_\epsilon^{-2} \sum_j \Delta_1(P_p(j,d))\, 2^{-j} & \sigma_\epsilon^{-4} \end{bmatrix},$$

and $\sigma^2_{d,p} \equiv 2[\sum_j \Delta_1^2(P_p(j,d))\, 2^{-j} - (\sum_j \Delta_1(P_p(j,d))\, 2^{-j})^2]^{-1}$.

For the same range of d and any N, $\widehat{\sigma}^2_{\epsilon,N,p}(d) =_d M^{-1}\sigma^2_\epsilon \chi^2_M$.

Table 1 shows $\sigma^2_{d,p}$ for various values of filter length L ($L = \infty$ refers to using the ideal wavelet filter $\mathcal{H}_{j,I}(f)$) and difference parameter d under different AR(p) wavelet models ($p = 0$ refers to the white noise wavelet model of Craigmile *et al.* [3]). We analyse to J=6. In general, keeping L fixed, the asymptotic variance decreases with increasing d (especially for shorter values of L). It also decreases with increasing L for stationary $d < \frac{1}{2}$, but increases with L for non-stationary $d \geq \frac{1}{2}$. The limit variance changes only slightly for $L = 2$ as we increase p. For $L > 2$ there is little change in the asymptotic variance with $p \geq 1$. In fact Monte Carlo studies to estimate d for various samples sizes, filter lengths and values of the difference parameter showed that an AR(1) model in this case was sufficient. An AR(p) ($p > 1$) gave no improvement to estimation.

5. Conclusions

In this paper we have further examined the estimation of the parameters of a trend contaminated FD process using the DWT. We have demonstrated that as we increase the filter length we can decorrelate between wavelet scales and decrease side-lobe behaviour. By extending the white noise and AR(1) wavelet models to the AR(p) ($p > 1$) case, we do not improve the estimation of d from that of the AR(1) model. Clearly these results give an attractive framework in which to model other short and long dependent processes.

Acknowledgments

The authors gratefully acknowledge support for this research from an STTR grant from AFOSR (MathSoft, Inc., and the University of Washington), an NSF grant (MathSoft, Inc., and the University of Washington) and an EPA grant (National Research Center for Statistics and the Environment).

References

[1] T. W. Anderson, *The Statistical Analysis of Time Series (Classics Edition)*, Wiley-Interscience, (1994).

[2] P. F. Craigmile, *Wavelet-Based Parameter Estimation for Trend Contaminated Long Memory Processes*, Ph.D. Thesis, University of Washington, (2000).

[3] P. F. Craigmile, D. B. Percival and P. Guttorp, *Wavelet-Based Parameter Estimation for Trend Contaminated Fractionally Differenced Processes*, Technical Report, National Research Center for Statistics and the Environment, University of Washington, (2000).

[4] I. Daubechies, *Ten Lectures on Wavelets*, CBMS-NSF Series in Applied Mathematics, 61, SIAM, Philadelphia, (1992).

[5] C. W. J. Granger, and R. Joyeux, *An Introduction to Long-memory Time Series Models and Fractional Differencing*, Journal of Time Series Analysis, **1** (1980) 15–29.

[6] J. R. M. Hosking, *Fractional Differencing*, Biometrika, **68(1)** (1981) 165–176.

[7] M.-J. Lai, *On the digital filter associated with Daubechies' wavelets*, IEEE Trans. on Signal Processing, **43** (1995) 2203–2205.

[8] S. Mallat, *A theory for multiresolution signal decomposition: The wavelet representation*, IEEE Trans. on Pattern Analysis and Machine Intelligence, **11** (1989) 674–693.

[9] E. McCoy and A. Walden, *Wavelet Analysis and Synthesis of Stationary Long-Memory Processes*, J. Computational and Graphical Statistics, **5(1)** (1996) 26–56.

[10] D. B. Percival, and A. G. Bruce, *Wavelet-Based Approximate maximum likelihood estimates for Trend-Contaminated Fractional Difference Processes*, MathSoft Research Report, **67** (1998).

[11] D. B. Percival and A. Walden, *Wavelet Methods for Time Series Analysis*, Cambridge University Press, (2000).

[12] A. H. Tewfik and M. Kim *Correlation structure of the DWT of fractional Brownian motion*, IEEE Trans. Information Theory, **38(2)** (1992) 904–909.

[13] G. W. Wornell, *Signal Processing with Fractals: A Wavelet-Based Approach*, Prentice Hall, Upper Saddle River, New Jersey, (1995).

P. F. Craigmile, P. Guttorp
Department of Statistics
Box 354322
University of Washington
Seattle. WA 98195–4322, USA
E-mail address: `pfc@stat.washington.edu`
E-mail address: `peter@stat.washington.edu`

D. B. Percival
Applied Physics Laboratory
Box 355640
University of Washington
Seattle, WA 98195–5640, USA
E-mail address: `dbp@apl.washington.edu`

D. B. Percival
MathSoft Inc.
1700 Westlake Avenue North
Suite 500
Seattle, WA 98109–3044, USA
E-mail address: `dbp@statsci.com`

Wavelet-Based Modelling of Persistent Periodicities

Emma J. McCoy

Abstract. The k-factor Gegenbauer ARMA (Auto Regressive Moving Average) models of Woodward, Zhang and Gray (*J. Time Ser. Anal.* 19 (1998), 485–504) can model persistent periodic behaviour associated with each of k frequencies in $[0, 0.5]$. Current methods of parameter estimation of such processes is not automatic, and often involve many approximations, in addition to problems of model order identification. We propose a wavelet packet based analysis of such processes which utilizes a reversible jump Markov Chain Monte Carlo (MCMC) algorithm. This algorithm automatically accounts for model order determination of the number of AR and MA parameters as well as the number of persistent periodicities. From a Bayesian perspective, there is no fundamental difference between observables and parameters, and so missing values and forecasts could easily be incorporated into the algorithm. We demonstrate the method by applying the model to the Mauna Loa atmospheric CO_2 data.

1. Introduction

Long-memory processes are found in a wide variety of sources and have been analysed by a number of authors. Wavelet transforms have been recognized as a useful tool for studying such processes e.g., [2, 10]. The simplest models exhibiting long-memory are fractionally differenced (FD) processes, the discrete wavelet transform (DWT) which maps slowly decaying autocorrelations to approximate uncorrelation. This decorrelation property can be exploited to obtain simple likelihood based inference of the parameters of such models, e.g., [6, 1].

Time series exhibiting both short and long-memory are harder to deal with in that the DWT no longer maintains the decorrelation property. However, choice of an appropriate transform obtained from a wavelet packet library can reinstate this property, thus allowing extensions of the previous likelihood methods to more complicated models allowing for both long and short-memory.

2. GARMA Models

The models we shall study in this paper are extensions of the ARMA(p, q) model given by

$$\phi(B)x_t = \theta(B)\epsilon_t ,$$

where

$$\phi(B) = 1 - \phi_1 B - \cdots - \phi_p B^p ,$$
$$\theta(B) = 1 - \theta_1 B - \cdots - \theta_q B^q ,$$

ϵ_t is zero-mean white noise with variance σ_ϵ^2, and B^k is the backward shift operator defined by $B^k x_t = x_{t-k}$.

Stationary versions of such processes exhibit short-memory behaviour.

Hosking [4, 5] defined an extension of this model: the fractional ARMA (FARMA) model, which also allows for incorporation of long-memory, namely

$$\phi(B)(1-B)^d x_t = \theta(B)\epsilon_t ,$$

where in this case d can take on fractional values. A natural, further extension has been proposed by Woodward et al [9]: the Gegenbauer ARMA (GARMA) model, which allows for long-term dependency in stationary models associated with any k frequencies $f_j = \cos^{-1} u_j/(2\pi) \in [0, 0.5]$, $j = 1, \ldots, k$,

$$\phi(B) \prod_{j=1}^{k} (1 - 2u_j B + B^2)^{\lambda_j} x_t = \theta(B)\epsilon_t . \tag{1}$$

This GARMA process is stationary if the u_i are distinct and $\lambda_i < \frac{1}{2}$ whenever $|u_i| \neq 1$, $\lambda_i < \frac{1}{4}$ when $|u_i| = 1$, and, in addition, the roots of the characteristic equation $\phi(z) = 0$ lie outside the unit circle. This model will exhibit long memory when $\lambda_i > 0$, $i = 1, \ldots, k$.

2.1. Spectral considerations

The spectrum associated with (1) is

$$S_x(f) = \sigma_\epsilon^2 \frac{|\theta(e^{i2\pi f})|^2}{|\phi(e^{i2\pi f})|^2} \prod_{j=1}^{k} [4\{\cos(2\pi f) - u_j\}^2]^{-\lambda_j} . \tag{2}$$

The frequencies $f_j = (\cos^{-1} u_j)/(2\pi)$ at which the spectrum becomes unbounded are called the Gegenbauer frequencies. When $k = 1$ and $u_1 = 0$, this model reduces to a FARMA model.

3. Discrete Wavelet Packet Transforms

In an attempt to obtain the decorrelating properties of the DWT for more general processes, we consider selecting a transform from a wavelet packet (WP) table. Figure 1 shows an example of a three level wavelet packet table, and the route through the table corresponding to the conventional DWT.

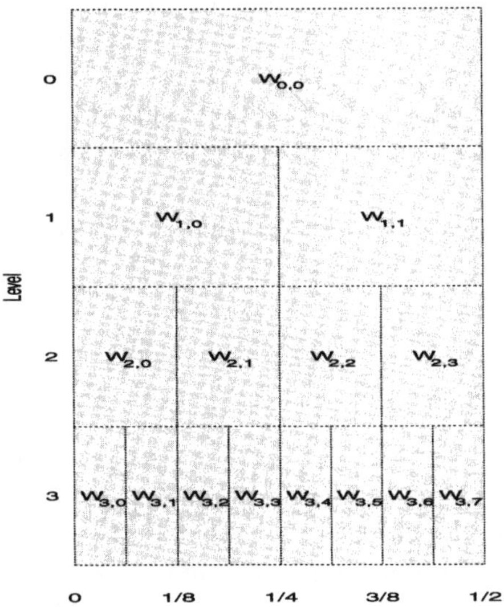

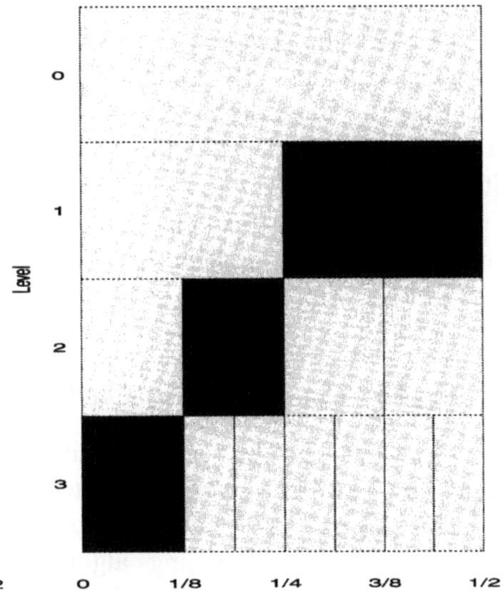

FIGURE 1. 3-level WP table (left), DWT in the table (right)

The DWPT (discrete wavelet packet transform) is an orthonormal transform, and as for the DWT, can be formulated using filtering operations involving the low and high pass filters associated with the conventional DWT, see e.g., [7]. From the table, a suitable transform can be chosen which partitions the interval $[0, \frac{1}{2}]$ in such a way that $S_x(f)$ is fairly flat over these intervals, and with this criterion, the DWPT should be approximately uncorrelated.

4. Likelihood Considerations

With the assumption of decorrelation from the DWPT, we can model our wavelet coefficients as forming an independent sample with,

$$\mathbf{W}_{j,k} \sim \mathrm{N}(0, \sigma_{j,k}^2),$$

where $\sigma_{j,n}^2$ is the appropriate band-pass variance,

$$\sigma_{j,k}^2 = \int_{\frac{k}{2^{j+1}}}^{\frac{k+1}{2^{j+1}}} 2^j S_x(f)\,\mathrm{d}f, \tag{3}$$

with associated likelihood

$$L(\gamma \mid \mathbf{W}) = \prod_{j,k} (2\pi\sigma_{j,k})^{-\frac{1}{2}} \exp\left\{-\frac{\mathbf{W}_{j,k}^2}{2\sigma_{j,k}^2}\right\}.$$

Where γ is the full set of parameters involved in the GARMA model.

The aim of maximum likelihood estimation is to determine the values of the parameters which maximize this function. Despite the simplification gained by the decorrelating properties of the DWPT, we are nonetheless still faced with a complicated likelihood to maximize, which involves an intermediate integral. We note that the band-pass integral given in equation (3) is approximate, the exact bandpass variance can be obtained using the square gain function for the wavelet filters (see [1]).

Our approach to solving this problem is to move away from analytical evaluation towards a simulation based approach.

5. Markov Chain Monte Carlo (MCMC)

The idea behind MCMC and reversible jump MCMC e.g., [3], is to set up a Markov Chain (MC) which converges to the distribution of interest, in our case the posterior distribution (likelihood times prior) of the parameters. One way of doing this is via the Metropolis–Hastings algorithm, which involves sampling proposal values of the parameters and then either accepting or rejecting these values. Then, we can run the MC, and after the chain has converged, we will be sampling parameter values from the posterior and can trivially obtain values of the parameter which maximize this posterior.

5.1. The Metropolis algorithm

Given initial values for the parameters, we propose a new parameter configuration, by either changing the current value of one of the parameters, introducing a new parameter (a birth – equivalent to an increase in model order), or deleting one of the parameters, or pair of parameters (a death – equivalent to a reduction in model order). If the posterior is increased by the proposal we accept the parameter set; otherwise we accept with probability

$$\frac{L(\gamma_{\text{new}} \mid \mathbf{W})\, p(\gamma_{\text{new}})\, q(\text{new} \to \text{old})}{L(\gamma_{\text{old}} \mid \mathbf{W})\, p(\gamma_{\text{old}})\, q(\text{new} \to \text{old})},$$

where $q(\cdot)$ accounts for the dimensionality change between proposals, and γ is the parameter currently under study (see [3]).

5.2. Latent root parameterization

In order to sample values of the parameters from the stationarity region of our model, we consider a latent (or characteristic) root parameterization. We know that for stationarity and invertibility, the roots of the characteristic polynomials must lie outside the unit circle. So, instead of parameterizing the ARMA part of the model in terms of ϕ and θ, for simulation purposes we parameterize in terms of the reciprocal roots, α_1 and α_2. These roots are either real or occur as complex conjugate pairs.

$$\phi(u) = 1 - \sum_{j=1}^{p} \phi_j u^j = \prod_{j=1}^{p} (1 - \alpha_{1,j} u),$$

$$\theta(u) = 1 - \sum_{j=1}^{q} \theta_j u^j = \prod_{j=1}^{q} (1 - \alpha_{2,j} u),$$

where

$$\alpha_{1,j} = r_{1,j} e^{\pm i\beta_{1,j}}, \qquad \alpha_{2,j} = r_{2,j} e^{\pm i\beta_{2,j}},$$

represent root pairs, or real roots when $\beta = 0$.

5.3. Prior specification

In order to simulate values for the parameters, we adopt the following priors, which are all trivial to simulate,

$$p(\beta) \sim U(-\pi, \pi); \qquad p(r) \sim U(0, 1),$$
$$p(\lambda) \sim U\left(0, \frac{1}{2}\right), \; (|u| \neq 1); \quad p(u) \sim U(-1, 1).$$

Once we have simulated proposal values for parameters we can either accept or reject based on the Metropolis algorithm. The integral (3) in the likelihood is carried out using Monte Carlo integration as part of the simulation.

6. Application to CO_2 Data

The monthly atmospheric CO_2 measurements collected at the summit of Mauna Loa in Hawaii between 1968 and 1990 are shown in Figure 2, along with the periodogram of the second differences (to remove the trend). The periodogram is suggestive of two or perhaps three-component GARMA behaviour.

6.1. DWPT

In order to determine the "route" through the wavelet packet table, we carry out significance tests at each level based on the cumulative periodogram of the wavelet coefficients at that level, to determine whether the spectrum is varying or not. If the spectrum is flat, the periodogram ordinates should follow a chi-squared distribution with equal mean and variance determined solely by the variance of the wavelet coefficients at that level. If the test is failed at a certain level we progress to the next level; see [8] for a full discussion of this and alternative methods of determining the best decorrelating transform.

The DWPT chosen using this method for the CO_2 data is shown in Figure 3.

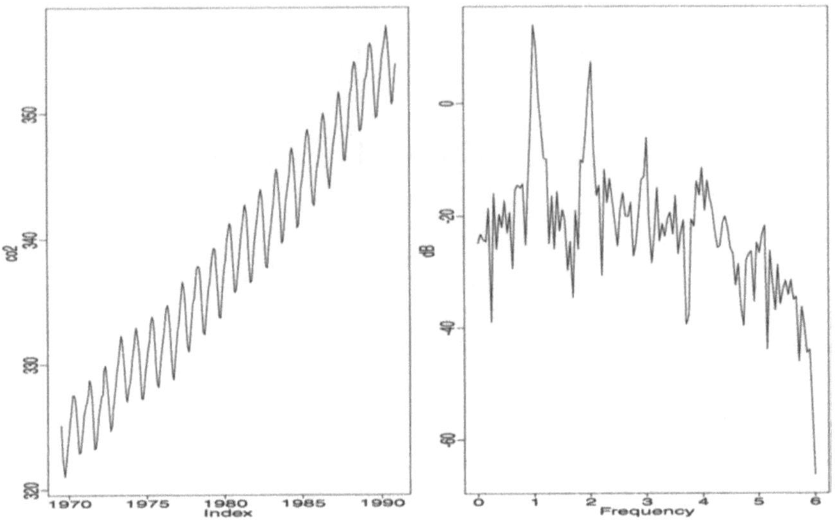

FIGURE 2. Monthly CO_2 measurements (left) and the periodogram of the second differences (right)

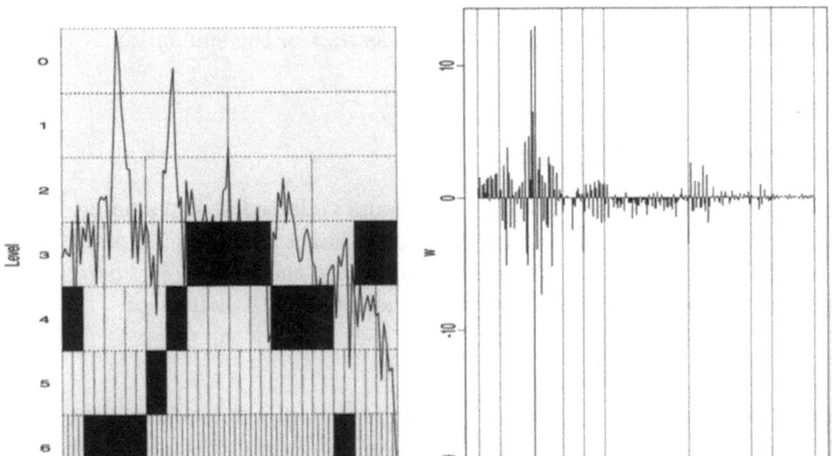

FIGURE 3. Route through the wavelet packet table and periodogram (left) and the associated wavelet coefficients (right)

6.2. Results

Histograms displaying the posterior distributions for the model order parameters are shown in Figure 4.

Fixing the model order as a 2-factor GARMA(1, 1), we can then ascertain the optimal values of the parameters.

A sample from the posterior distribution of the theoretical spectrum with this fixed model order (as given in equation (2)) is shown in Figure 5, and matches the periodogram well.

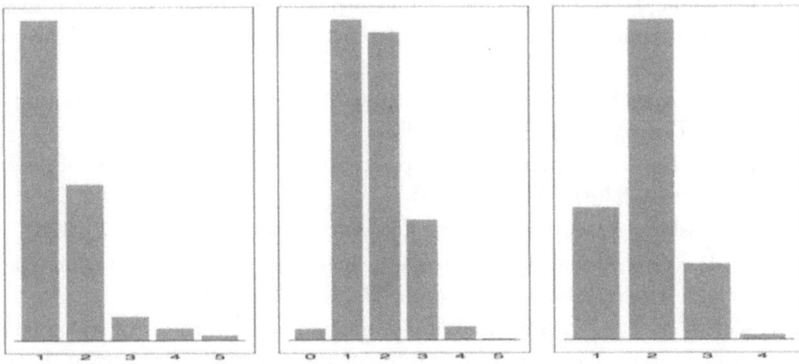

FIGURE 4. Histograms of the posterior dists. for the GARMA model order: AR (left), MA (middle), Gegenbauer (right).

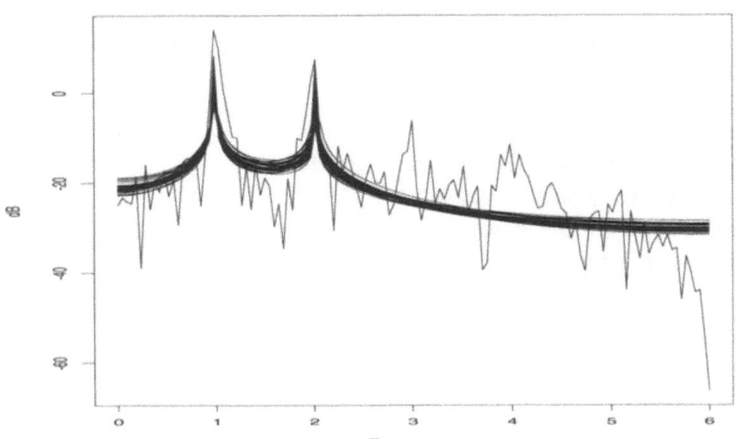

FIGURE 5. Theoretical 2-factor GARMA(1, 1) spectra and periodogram for CO_2 data

7. Conclusions

The use of MCMC combined with the DWPT enables efficient analysis of processes involving both long and short-memory. We can obtain parameter estimates for GARMA models which enables us to determine the location of persistent seasonal periodicities. In addition, these parameter estimates could be used for forecasting.

References

[1] P. F. Craigmile, D. B. Percival and P. Guttorp. Wavelet-based parameter estimation for trend contaminated fractionally differenced processes. Preprint, University of Washington, 1999.

[2] P. Flandrin. Wavelet analysis and synthesis of fractional Brownian-motion. *IEEE Trans. Inf. Theory*, 38(2): 910–917, 1992.

[3] P. J. Green. Reversible jump MCMC computation and Bayesian model determination. *Biometrika*, 82(4): 711–732, 1995.

[4] J. R. M. Hosking. Fractional differencing. *Biometrika*, 61(1): 165–176, 1981.

[5] J. R. M. Hosking. Modeling persistence in hydrological time series using fractional differencing. *Water Resources Research*, 20(12): 1898–1908, 1984.

[6] E. J. McCoy and A. T. Walden. Wavelet analysis and synthesis of stationary long-memory processes. *Journal of Computational and Graphical Statistics*, 5, 1–31, 1996.

[7] D. B. Percival and A. T. Walden. *Wavelet methods for time series analysis.* Cambridge University Press, 2000.

[8] D. B. Percival, S. Sardy and A. C. Davison. Wavestrapping time series: adaptive wavelet-based bootstrapping. Preprint, University of Washington, 2000.

[9] W. A. Woodward, Q. C. Cheng and H. L. Grey. A k-factor GARMA long-memory model. *Journal of Time Series Analysis*, 19(4): 485–504, 1998.

[10] G. W Wornell. Wavelet-based representations for the $1/f$ family of fractal processes. *Proceedings of the IEEE*, 81(10): 1428–1450, 1993.

Department of Mathematics
Imperial College of Science, Technology and Medicine
Huxley Building, 180 Queen's Gate
London, SW7 2BZ, United Kingdom
E-mail address: `e.mccoy@ic.ac.uk`

Network Traffic Modeling Using a Multifractal Wavelet Model

Rudolf H. Riedi, Vinay J. Ribeiro, Matthew S. Crouse, and Richard G. Baraniuk

Abstract. In this paper, we develop a simple and powerful multiscale model for synthesizing nonGaussian, long-range dependent (LRD) network traffic. Although wavelets effectively decorrelate LRD data, wavelet-based models have generally been restricted by a Gaussianity assumption that can be unrealistic for traffic. Using a multiplicative superstructure on top of the Haar wavelet transform, we exploit the decorrelating properties of wavelets while simultaneously capturing the positivity and "spikiness" of nonGaussian traffic. This leads to a swift $O(N)$ algorithm for fitting and synthesizing N-point data sets. The resulting model belongs to the class of multifractal cascades, a set of processes with rich statistical properties. We elucidate our model's ability to capture the covariance structure of real data and then fit it to real traffic traces. Queueing experiments demonstrate the accuracy of the model for matching real data.

1. Introduction

Fractal models arise frequently in a variety of scientific disciplines, such as physics, chemistry, astronomy, and biology. More recently, fractal models have had a major impact on the analysis of data communication networks such as the Internet. In their landmark paper [1], Leland et al. demonstrated that network traffic loads exhibit fractal properties such as self-similarity, "burstiness," and long-range dependence (LRD) that are inadequately described by classical traffic models. Characterization of these fractal properties, particularly LRD, has provided exciting new insights into network behavior and performance.

As the pre-eminent random fractal model, fractional Brownian motion (fBm) has played a central rôle in many fields [1, 2]. FBm is the unique Gaussian process with stationary increments and the following scaling property for all $a > 0$

$$B(at) \overset{fd}{=} a^H B(t), \tag{1}$$

This work was supported by the National Science Foundation, grant no. CCR-9973188, by ONR, grant no. N00014-99-10813, by DARPA/AFOSR, grant no. F49620-97-1-0513, and by Texas Instruments. Email: {riedi, vinay, mcrouse, richb}@rice.edu. URL: www.dsp.rice.edu.

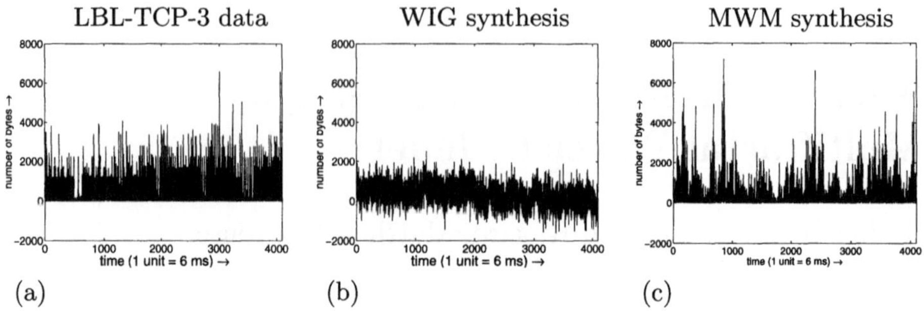

FIGURE 1. Bytes-per-time arrival process for (a) wide-area TCP traffic at the Lawrence Berkeley Laboratory (trace LBL-TCP-3) [4], (b) one realization of the state-of-the-art wavelet-domain independent Gaussian (WIG) model [3], and (c) one realization of the multifractal wavelet model (MWM) synthesis. The MWM traces closely resemble the real data, while the WIG traces (with their large number of negative values) do not.

with the equality in (finite-dimensional) distribution. The parameter H, $0 < H < 1$, is known as the *Hurst parameter*. It rules the LRD of fBm, as we will see later, but it also governs its local "spikiness." In particular, for all t

$$B(t+s) - B(t) \simeq s^H , \tag{2}$$

meaning that, for $0 < H < 1$, fBm has "infinite slope" everywhere. Gaussian processes with a more flexible scaling relation than (1) can be synthesized by the wavelet-domain independent Gaussian (WIG) model [3].

Real network traffic traces, however, do not exhibit the strict self-similarity of (1) and are positive and nonGaussian in nature thus limiting the use of fBm as a traffic model. The transmission control protocol (TCP) traffic we study in this paper exhibits local scaling similar to (2), but with an exponent H_t that depends on t. This has been termed *multifractal behavior* and was reported for the first time in [5] and subsequently in [6, 7, 8, 9]. Amazingly, the statistical properties of H_t as a random variable in t can be described compactly through a function $T(q)$ that controls the scaling behavior of the sample moments of order q. This powerful relation, called the *multifractal formalism*, ties burstiness, higher-order dependence structure, and moments of marginals together in one unified theory.

In this paper, we propose a new non-linear model for network traffic data. The *multifractal wavelet model* (MWM) is based on a multifractal cascade in the wavelet domain that by design guarantees a positive output. Each sample of the MWM process is obtained as a product of several positive independent random variables resulting in an approximately lognormal marginal density.

Fitting the MWM to real traffic traces results in an excellent match, far better than the Gaussian WIG model, visually (see Figure 1) and, as we will see, in the multifractal partition function $T(q)$, the burstiness as measured by the multifractal spectrum, the marginals, and the queueing behavior.

In this paper, we describe LRD and its relationship with fBm in Section 2. After introducing the wavelet transform and describing the WIG model in Section 3, we derive the MWM in Section 4. Section 6 reports on the results of simulation experiments with real data traces. We give an intuitive introduction to multifractal cascades in Section 5 and close with conclusions in Section 7.

2. fBm and LRD

Although we analyse fBm from a continuous-time point of view, for practical computations and simulations, we often work with sampled continuous-time fBm. The increments process of sampled fBm

$$X[n] := B(n) - B(n-1) \tag{3}$$

defines a stationary Gaussian sequence known as discrete *fractional Gaussian noise* (fGn) with covariance behavior [10]

$$r_X[k] \simeq |k|^{2H-2}, \text{ for } |k| \text{ large}. \tag{4}$$

For $1/2 < H < 1$, the covariance of fGn is strictly positive and decays so slowly that it is non-summable (i.e., $\sum_k r_X[k] = \infty$). This property is called LRD.

The LRD of fGn can be equivalently characterized in terms of how the aggregated processes

$$X^{(m)}(n) := \frac{1}{m} \sum_{i=(k-1)m+1}^{km} X(i) \tag{5}$$

behave. It follows from (1) that $X(n) \stackrel{fd}{=} m^{1-H} X^{(m)}(n)$.

Hence, a log-log plot of the variance of $X^{(m)}(n)$ as a function of m – known as a *variance-time plot* – will have a slope of $2H-2$. The variance-time plot can characterize LRD in non-Gaussian, non-zero-mean data as well [1].

3. Wavelets and LRD Processes

3.1. Wavelet transform

The discrete wavelet transform is a multi-scale signal representation of the form [11]

$$c(t) = \sum_k u_k \, 2^{-J_0/2} \, \phi(2^{-J_0}t - k) + \sum_{j=-\infty}^{J_0} \sum_k w_{j,k} \, 2^{-j/2} \psi\left(2^{-j}t - k\right), \ j,k \in \mathbf{Z}$$

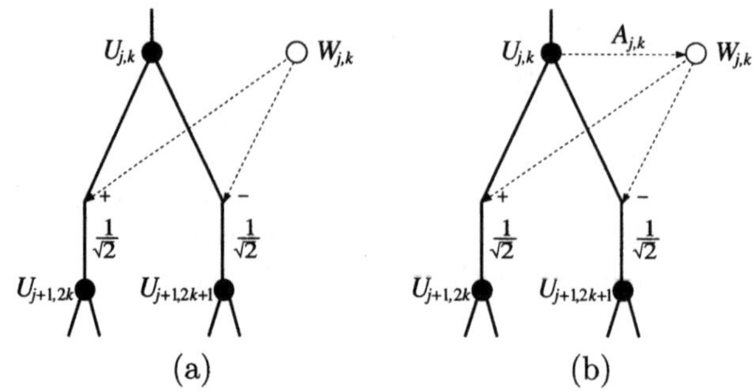

FIGURE 2. (a) WIG construction: Generate the $W_{j,k}$'s as mutually independent and identically distributed within scale according to $W_{j,k} \sim N(0, \sigma_j^2)$. Then compute the Haar scaling coefficients $U_{j+1,2k}$ and $U_{j+1,2k+1}$ at scale $j+1$ as sums and differences of the scaling and wavelet coefficients $U_{j,k}$ and $W_{j,k}$ at scale j (normalized by $1/\sqrt{2}$). (b) MWM construction: At scale j, generate the multiplier $A_{j,k} \sim \beta(p_j, p_j)$, form the wavelet coefficient as the product $W_{j,k} = A_{j,k} U_{j,k}$ and them form scaling coefficients at scale $j+1$ as in (a).

with J_0 the coarsest scale and u_k and $w_{j,k}$ the scaling and wavelet coefficients, respectively. The scaling coefficients may be viewed as providing a coarse approximation of the signal, with the wavelet coefficients providing higher-frequency "detail" information. Using filter bank techniques, the wavelet transform and inverse wavelet transform can be computed in $O(N)$ operations for a length-N

In the *Haar* wavelet transform (see Figure 2), the prototype scaling and wavelet functions are given by

$$\phi(t) = \begin{cases} 1, & 0 \le t < 1 \\ 0, & \text{else} \end{cases} \quad \text{and} \quad \psi(t) = \begin{cases} 1, & 0 \le t < 1/2 \\ -1, & 1/2 \le t < 1 \\ 0, & \text{else}. \end{cases}$$

The Haar scaling and wavelet coefficients can be recursively computed via [11]

$$u_{j-1,k} = 2^{-1/2}(u_{j,2k} + u_{j,2k+1}), \ w_{j-1,k} = 2^{-1/2}(u_{j,2k} - u_{j,2k+1}). \qquad (6)$$

3.2. Modeling LRD data

Wavelets serve as an approximate Karhunen-Loève or decorrelating transform for fBm [2], fGn, and more general LRD signals [12]. Hence, modeling and processing of these signals in the wavelet domain is often more efficient and powerful than in the time domain.

Gaussian LRD processes can be approximately synthesized by generating wavelet coefficients as independent zero-mean Gaussian random variables, identically distributed within scale according to $W_{j,k} \sim N(0, \sigma_j^2)$,[1] with σ_j^2 the wavelet-coefficient variance at scale j [3]. We call the resulting model the wavelet-domain independent Gaussian (WIG) model [3] (see Figure 2(a)). A power-law decay for the σ_j^2's leads to approximate wavelet synthesis of fBm or fGn [2].

The WIG model assumes Gaussianity even though network traffic signals (such as loads and interarrival times) are positive and can be highly nonGaussian. We seek a more accurate marginal characterization for these spiky, non-negative LRD processes.

3.3. Modeling non-negative data with the Haar wavelet

It is easily shown that the Haar scaling coefficients $u_{j,k}$ are non-negative if and only if the signal itself is non-negative; that is, $c(t) \geq 0 \Leftrightarrow u_{j,k} \geq 0,\ \forall\, j,k$. Solving (6) for $u_{j,2k}$ and $u_{j,2k+1}$, we find

$$u_{j,2k} = 2^{-1/2}(u_{j-1,k} + w_{j-1,k}),\ u_{j,2k+1} = 2^{-1/2}(u_{j-1,k} - w_{j-1,k}). \qquad (7)$$

Combining (7) with the constraint $u_{j,k} \geq 0$, we obtain the condition

$$c(t) \geq 0 \Leftrightarrow |w_{j,k}| \leq u_{j,k},\ \forall\, j,k. \qquad (8)$$

4. Multifractal Wavelet Model

The positivity constraints (8) on the Haar wavelet coefficients suggest a very simple multiscale, multiplicative signal model for positive processes. In the *multifractal wavelet model* (MWM) – a name which we will explain in a moment – we compute the wavelet coefficients recursively by

$$W_{j,k} = A_{j,k}\, U_{j,k}, \qquad (9)$$

with $A_{j,k}$ a random variable supported on the interval $[-1,1]$. Together with (7), we obtain (see Figure 2(b))

$$U_{j,2k} = 2^{-1/2}(1 + A_{j+1,k})\, U_{j-1,k},\ U_{j,2k+1} = 2^{-1/2}(1 - A_{j+1,k})\, U_{j-1,k}. \qquad (10)$$

Thus, to generate a MWM realization we first synthesize $U_{0,0}$ and then compute scaling coefficients at finer scales recursively using (10) till we reach a desired finest scale. This algorithm synthesizes an N-point using $O(N)$ computations.

The synthesized signal $C[k]$ is a discrete time approximation of the continuous time signal $c(t)$. Decomposing each shift k into a binary expansion $k = \sum_{i=0}^{n-1} k_i' 2^{n-1-i}$, we can write

$$C[k] = 2^{-n/2} U_{n,k} = 2^{-n}\, U_{0,0} \prod_{i=0}^{n-1} \frac{(1 + (-1)^{k_i'} A_{i,k_i})}{2}, \qquad (11)$$

[1] We use capital letters when we consider the underlying variables to be random.

with

$$k_0 \equiv 0, \text{ and } k_i = \sum_{j=0}^{i-1} k_i' 2^{i-1-j}, \ i = 1, \ldots, n-1. \tag{12}$$

In our experiments we choose the *symmetric beta distribution, $\beta(p,p)$* for the $A_{j,k}$'s

$$A_{j,k} \sim \beta(p_j, p_j), \tag{13}$$

with p_j the beta parameter at scale j. We set the p_j's to get the desired decay of the variances of $W_{j,k}$'s.

5. MWM is a Cascade

The MWM is a special case of the rich class of *multiplicative cascades*. Cascades provide a natural framework for producing positive "bursty" processes and offer greater flexibility and richer scaling properties than fractal models such as fGn and fBm. The subtle structure of cascades is best understood in terms of the powerful theory of *multifractals*, a statistical tool for measuring "burstiness" superior to LRD which merely measures "high variability".

Identifying the MWM algorithm with a multiplicative cascade allows us to benefit from the accumulated theoretical and practical knowledge of the field of *multifractals*, including a precise understanding of the convergence of the MWM algorithm, properties of the marginal distributions, advantages over monofractal fGn models, and a range of possible refinements and extensions [5, 9]. For these reasons, we find it useful to examine the MWM within the context of cascades and multifractals.

5.1. Cascades

The backbone of a cascade is a construction where one starts at a coarse scale and develops details of the process on finer scales iteratively in a multiplicative fashion. The MWM, e.g., is a multiplicative cascade: as (10) and (11) reveal we may write

$$C_{\mathrm{MWM}}[k] = 2^{-n} M_0^0 \prod_{i=1}^{n} M_{i,k_i}, \text{ with } M_{k_i}^i = \frac{(1 + (-1)^{k'_{i-1}} A_{i-1,k_{i-1}})}{2}. \tag{14}$$

This construction procedure naturally results in a process that "sits" just above the zero line and emits occasional positive jumps or spikes. In contrast, additive self-similar models such as fGn and the WIG "hover" around the mean with occasional outbursts in both positive and negative directions.

5.2. Multifractal analysis

Intuitively, multifractal analysis measures the frequency with which bursts of different strengths occur in a signal. Consider a positive process $Y(t)$. The strength of the burst of Y at time t, also called the degree of *Hölder continuity,* can be characterized by

$$\alpha(t) = \lim_{k_n 2^{-n} \to t} \alpha_{k_n}^n \text{ where } \alpha_{k_n}^n := -\frac{1}{n} \log_2 \left| Y((k_n+1)2^{-n}) - Y(k_n 2^{-n}) \right| \quad (15)$$

where $k_n 2^{-n} \to t$ means that $t \in [k_n 2^{-n}, (k_n+1)2^{-n})$ and $n \to \infty$. The smaller the $\alpha(t)$, the larger the increments of Y around time t, and the "burstier" it is at time t. The frequency of occurrence of a given strength α, can be measured by the *multifractal spectrum:*

$$f(\alpha) := \lim_{\varepsilon \to 0} \lim_{n \to \infty} \tfrac{1}{n} \log_2 \#\{k_n = 0, \dots, 2^n - 1 : \alpha_{k_n}^n \in (\alpha - \varepsilon, \alpha + \varepsilon)\}. \quad (16)$$

By definition, f takes values between 0 and 1 and is often shaped like a $\cap$ and concave. The smaller the $f(\alpha)$, the "fewer" points t will exhibit $\alpha(t) \approx \alpha$. If α_0 denotes the value $\alpha(t)$ assumed by "most" points t, then $f(\alpha_0) = 1$. See Figure 3(b) for the multifractal spectrum of the LBL-TCP-3 data set and of synthetic MWM data. We observe that the MWM captures the spectrum of the real data except for large values of α. This means that the MWM does not generate as many small values as the signal possesses.

5.3. Multifractal spectrum and higher-order moments

Though (16) gives us a simple measure of burstiness in data, in practice it is impossible to compute the right side of (16). However, $f(\alpha)$ can be obtained through the use of high and low-order moments of the signal $Y(t)$.

Define the *partition function* that captures the scaling of different moments of Y as

$$T(q) := \lim_{n \to \infty} \frac{1}{-n} \log_2 \mathbb{E}\left[S_n(q)\right], \quad (17)$$

with

$$S_n(q) := \sum_{k_n=0}^{2^n-1} \left| Y((k_n+1)2^{-n}) - Y(k_n 2^{-n}) \right|^q = \sum_{k_n=0}^{2^n-1} 2^{-qn\alpha_{k_n}^n}. \quad (18)$$

The multifractal spectrum $f(\alpha)$ and $T(q)$ are closely related, as the following hand-waving argument shows. Grouping in the sum $S_n(q)$ of (18) the terms behaving as $\alpha_{k_n}^n \approx \alpha$, and using (16) we get

$$S_n(q) = \sum_{\alpha} \sum_{\alpha_n \sim \alpha} \left(2^{-n\alpha}\right)^q \approx \sum_{\alpha} 2^{nf(\alpha)} 2^{-nq\alpha} \approx 2^{-n \inf_\alpha (q\alpha - f(\alpha))}. \quad (19)$$

We conclude that we must "expect" $T(q)$ to equal $\inf_\alpha(q\alpha - f(\alpha))$, the so-called *Legendre transform* of $f(\alpha)$. For the special case of an MWM process, i.e., $Y(t) =$

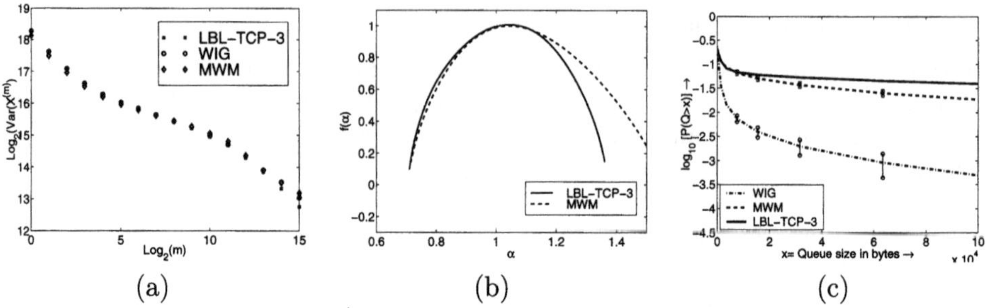

FIGURE 3. (a) Variance-time plot of the LBL-TCP-3 data "×", the WIGdata "◇", and one realization of the MWM synthesis "∘". (b) Multifractal spectra of the LBL-TCP-3 data and one realization of the MWM synthesis. (c) Comparison of the queuing performance of real data traces with those of synthetic WIG and MWM traces. Observe that the MWM synthesis matches the queuing behavior of the LBL-TCP-3 data closely, while the WIG synthesis does not.

$\int_0^t c(t)\ dt$, it can be shown (see [13]) that the inverse relation holds, called the *multifractal formalism*

$$f(\alpha) = T^*(\alpha) := \inf_q (q\alpha - T(q)) \ . \tag{20}$$

This relation makes the multifractal spectrum more accessible for practical purposes.

6. Experimental Results

In order to compare the MWM and WIG models we train them on a well-known real data trace, the LBL-TCP-3 [4]. We compare the correlation matching abilities through the variance-time plots of the real data, the MWM traces, and the WIG traces in Figure 3(a). Observe that, as expected, both the MWM and WIG models perform well in matching the correlation structure of the real data.

We plot the multifractal spectra (see Section 5) of the LBL-TCP-3 data and the synthetic MWM trace in Figure 3(b) (calculations for the negative moments of the WIG data become numerically unstable and hence the spectra for the WIG is not included). We observe that the spectra match extremely well except for large values of α. This corresponds to a close match of the scaling of higher-order moments, but a somewhat less accurate match of the scaling of the negative moments.

In Figure 3(c) we compare the average queuing behavior of the MWM and WIG traces to that of the real trace LBL-TCP-3. We observe that the MWM traces match the queuing behavior of the real data trace much better than the WIG traces.

7. Conclusions

The multiplicative wavelet model (MWM) combines the power of multifractals with the efficiency of the wavelet transform to form a flexible framework natural for characterizing and synthesizing positive-valued data with LRD. As our numerical experiments have shown, the MWM is particularly suited to the analysis and synthesis of network traffic data. In addition, the model could find application in areas as diverse as financial time-series characterization, geophysics (using 2-d and 3-d wavelets), and texture modeling. The parameters of the MWM are simple enough to be easily inferred from observed data or chosen a priori. Computations involving the MWM are extremely efficient – synthesis of a trace of N sample points requires only $O(N)$ computations. Finally, several extensions to the MWM are straightforward. The choice of β-distributed wavelet multipliers $A_{j,k}$ is not essential. Alternatively, we can employ mixtures of β's or even purely discrete distributions to fit higher-order multifractal moments.

References

[1] W. Leland, M. Taqqu, W. Willinger, and D. Wilson, "On the self-similar nature of Ethernet traffic (extended version)," *IEEE/ACM Trans. Networking*, pp. 1–15, 1994.

[2] P. Flandrin, "Wavelet analysis and synthesis of fractional Brownian motion," *IEEE Trans. Inform. Theory*, vol. 38, pp. 910–916, Mar. 1992.

[3] S. Ma and C. Ji, "Modeling video traffic in the wavelet domain," in *Proc. of 17th Annual IEEE Conf. on Comp. Comm., INFOCOM*, pp. 201–208, Mar. 1998.

[4] V. Paxson and S. Floyd, "Wide-area traffic: The failure of Poisson modeling," *IEEE/ACM Transactions on Networking*, vol. 3, pp. 226–244, 1995.

[5] R. Riedi and J. L. Véhel, "Multifractal properties of TCP traffic: A numerical study," *Technical Report No 3129, INRIA Rocquencourt, France*, Feb, 1997. Available at `www.dsp.rice.edu`.

[6] P. Mannersalo and I. Norros, "Multifractal analysis of real ATM traffic: A first look," *COST257TD*, 1997.

[7] A. Feldmann, A. C. Gilbert, and W. Willinger, "Data networks as cascades: Investigating the multifractal nature of Internet WAN traffic," *Proc. ACM/Sigcomm 98*, vol. 28, pp. 42–55, 1998.

[8] A. C. Gilbert, W. Willinger, and A. Feldmann, "Scaling analysis of random cascades, with applications to network traffic," *IEEE Trans. on Info. Theory, (Special issue on multiscale statistical signal analysis and its applications)*, vol. 45, April 1999.

[9] R. H. Riedi, M. S. Crouse, V. Ribeiro, and R. G. Baraniuk, "A multifractal wavelet model with application to network traffic," *IEEE Trans. Info. Theory*, vol. 45, pp. 992–1018, April 1999. Available at `www.dsp.rice.edu`.

[10] T. Lundahl, W. Ohley, S. Kay, and R. Siffert, "Fractional Brownian motion: A maximum likelihood estimator and its application to image texture," *IEEE Trans. on Medical Imaging*, vol. 5, pp. 152–161, Sep. 1986.

[11] I. Daubechies, *Ten Lectures on Wavelets.* New York: SIAM, 1992.

[12] L. Kaplan and C.-C. Kuo, "Extending self-similarity for fractional Brownian motion," *IEEE Trans. Signal Proc.*, vol. 42, pp. 3526–3530, Dec. 1994.

[13] R. H. Riedi, "Multifractal processes," *Technical Report, ECE Dept. Rice Univ., TR 99-06.* To appear in "Long range dependence: Theory and applications," eds. Doukhan, Oppenheim and Taqqu (2000). Available at `www.dsp.rice.edu/publications`.

Department of Electrical and Computer Engineering
Rice University
6100 South Main Street
Houston, TX 77005, USA
E-mail address: `mcrouse, riedi, vinay, richb@rice.edu`
URL: `www.dsp.rice.edu`.

Denoising Via Block Wiener Filtering in Wavelet Domain

Vasily Strela

Abstract. In this paper we describe a new method for image denoising. We analyse statistical properties of the wavelet coefficients of natural images. It turns out that there is a strong local covariance structure introduced by the edges. We suggest a model for this covariance which allows us to estimate it from the noisy image. Then Wiener filter is employed in order to remove the noise.

We compare our approach to other noise removal techniques. Wiener-wavelet denoising produces superior results both visually and in terms of mean square error.

1. Introduction

A good model of the signal statistics is essential in many applications. This paper describes a simple and effective model for the covariance structure of natural images. We use this model for noise removal.

There are two powerful techniques to reduce the noise level in a signal: Wiener filtering [3] and wavelet thresholding [2]. Wiener filtering is a linear procedure. Wavelet thresholding is nonlinear. Classical versions of both methods tend to blur edges in images. We try to blend these two approaches in order to improve the performance. Our idea is to apply Wiener filter to blocks of wavelet coefficients. This requires an estimate of the covariance matrix of each block. We obtain it adaptively which makes the whole procedure nonlinear (similar to thresholding).

Many authors argue that the distribution of wavelet coefficients of images is strongly non-Gaussian. In particular the histogram has a sharp peak at zero. A common approach is to model the distribution as a generalized Gaussian $e^{-|y/\lambda|^p}$ [1, 4, 6, 7, 8]. Instead of modeling statistics of all wavelet coefficients together we suggest that each sub-band has slightly different statistics. Moreover, it is important to concentrate on blocks of wavelet coefficients.

Why do we choose to work with blocks? It is known that the wavelet transform acts as an edge detector. In images edges represent features. Each feature corresponds to a block of wavelet coefficients. Working with these blocks is natural in order to preserve the edges (features) better.

It can be proved that if both the signal and the noise are Gaussian then Wiener filtering is the best possible mean square estimator. Our approach is to assume that in each sub-band, blocks of wavelet coefficients are Gaussian vectors with slowly changing covariance matrix. We concentrate on creating a model for the covariance of the blocks and hope that Wiener filtering will be close to optimal. Experimental results presented in Section 5 confirm our point of view.

2. Wiener Filter

Suppose a vector $\mathbf{S}$ is corrupted by Gaussian white noise with variance σ^2 and mean 0, $\mathbf{X} = \mathbf{S} + \sigma\mathbf{Z}$. Wiener filtering is the following linear procedure:

$$\widehat{\mathbf{X}} = \sum_m \frac{\beta_m^2}{\beta_m^2 + \sigma^2} \langle \mathbf{X}, \mathbf{g}_m \rangle \mathbf{g}_m \,. \tag{1}$$

Here β_m and $\mathbf{g}_m$ are eigenvalues and eigenvectors of the covariance matrix (Karhuen-Loeve transform) of $\mathbf{S}$. If $\mathbf{S}$ is Gaussian then $\widehat{\mathbf{X}}$ is the best mean square estimate of $\mathbf{S}$ (see for example [3]). In order to apply a Wiener filter one needs to estimate the covariance matrix (Karhuen-Loeve transform) of the signal. We develop a model for the block covariance structure in the next section.

3. Model of Local Covariance

It is known that wavelet transform has good decorrelating properties and is a reasonable approximation to the Karhuen-Loeve basis of images. Nevertheless, wavelet coefficients are highly structured. Figure 1 shows a 512 by 512 fingerprint image and its 256 by 256 horizontal detail wavelet sub-band.

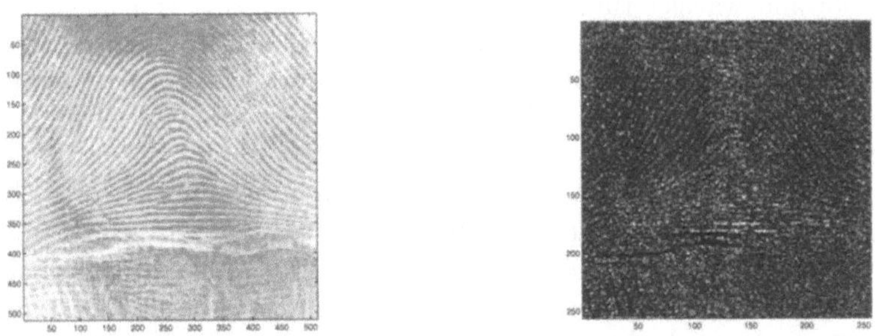

FIGURE 1. Fingerprint image and its horizontal detail wavelet sub-band.

One can see that large wavelet coefficients (lighter in color) correspond to edges of the original image. Also, because it is the horizontal detail sub-band, large coefficients tend to align horizontally. It is reasonable to assume that there

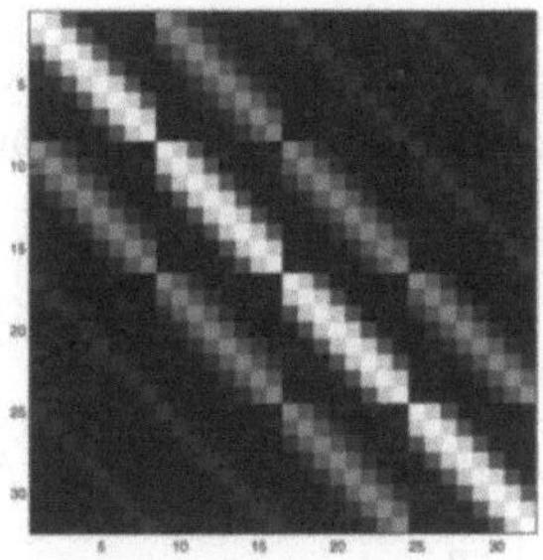

FIGURE 2. Covariance matrix of 4 by 8 block of wavelet coefficients estimated from the horizontal detail sub-band of the fingerprint image.

is correlation within the rows. On the other hand, edges are mostly local and only a few adjacent coefficients should be correlated.

Figure 2 presents covariance matrix of 4 by 8 block of wavelet coefficients estimated from the horizontal detail sub-band of the fingerprint image. Lighter color corresponds to the elements with larger absolute value. One can see that there is a strong correlation between adjacent coefficients within rows. This confirms our hypothesis.

Figure 2 gives an idea how the block covariance structure of the horizontal detail sub-band should look. Nevertheless, it would be an oversimplification to use the same covariance matrix for all blocks. Coefficients are correlated along and decorrelated across the edges. A block without edges should be weakly correlated while a block with many edges should exhibit strong correlation structure of the form similar to the one shown in Figure 2. We would like to distinguish between these two types of blocks.

Suppose a vector $\mathbf{X}_j$ corresponds to a block of noisy wavelet coefficients belonging to a given sub-band. We suggest the following model for its covariance matrix $\mathbf{C}_j$:

$$\mathbf{C}_j = \sigma^2 \mathbf{I} + \gamma_j \mathbf{B} . \tag{2}$$

Here σ^2 is the variance of the noise, γ_j is a parameter which has to be estimated from the data, $\mathbf{I}$ is the identity matrix. The covariance matrix $\mathbf{C}_j$ has two parts. $\sigma^2\mathbf{I}$ represents the contribution of the noise (the transform is orthogonal and the wavelet coefficients of the noise are decorrelated). $\gamma_j\mathbf{B}$ is the contribution of the edges. Matrix $\mathbf{B}$ characterizes the local covariance structure of noise-free wavelet coefficients of the particular sub-band. Figure 2 gives an example of such a matrix. We assume that each block $\mathbf{X}_j$ has the same "amount" of noise but a different "amount" of edges proportional to γ_j. This parameter changes from block to block.

Parameter γ_j can be estimated using the maximum likelihood method. Assuming that $\mathbf{X}_j$ is Gaussian with zero mean and covariance $\mathbf{C}_j$,

$$\mathbf{X}_j \sim \frac{1}{(2\pi)^{n/2}|\mathbf{C}_j|^{1/2}} e^{-\mathbf{X}^T\mathbf{C}_j^{-1}\mathbf{X}/2} = \frac{1}{(2\pi)^{n/2}|\mathbf{C}_j|^{1/2}} e^{-\mathbf{X}^T(\sigma^2\mathbf{I}+\gamma_j\mathbf{B})^{-1}\mathbf{X}/2},$$

we obtain the likelihood functional

$$\begin{aligned} L(\mathbf{Y}_j) &= -\frac{1}{2}\ln|\mathbf{C}_j| - \frac{1}{2}\mathbf{Y}_j(\sigma^2\mathbf{I} + \gamma_j\Lambda)^{-1}\mathbf{Y}_j \\ &= -\frac{1}{2}\sum_{k=1}^{n}\ln(\sigma^2+\gamma_j\lambda_k) - \frac{1}{2}\sum_{k=1}^{n}\frac{y_k^2}{\sigma^2+\gamma_j\lambda_k}. \end{aligned} \tag{3}$$

Here $|\mathbf{C}| = \det \mathbf{C}$, y_k, $k = 1,\dots,n$ are components of the vector $\mathbf{Y}_j = \mathbf{Q}\mathbf{X}_j$, λ_k are eigenvalues of $\mathbf{B}$, and $\mathbf{Q}$ is the matrix of eigenvectors of $\mathbf{B}$. Differentiating (3) by γ_j and setting the derivative equal to zero we obtain an equation for γ_j:

$$\frac{1}{\sigma^2}\sum_{k=1}^{n}\frac{\lambda_k y_k^2}{(1+\gamma_j\lambda_k)^2} - \sum_{k=1}^{n}\frac{\lambda_k}{1+\gamma_j\lambda_k} = 0. \tag{4}$$

This equation can be solved numerically.

3.1. Estimate of matrix B

In order to solve (4), eigenvalues and eigenvectors of the matrix $\mathbf{B}$ are needed. By our model, $\mathbf{B}$ characterizes average covariance structure of the given sub-band. We estimate it from the mean of $\mathbf{X}_j\mathbf{X}_j^T$ using model (2).

$$\frac{1}{n}\sum_j \mathbf{X}_j\mathbf{X}_j^T = \sigma^2\mathbf{I} + \beta\mathbf{B}.$$

Multiplication of $\mathbf{B}$ by a constant changes only normalization and is not important. We set $\mathbf{B}$ to be the difference between $\frac{1}{n}\sum_j \mathbf{X}_j\mathbf{X}_j^T$ and $\sigma^2\mathbf{I}$:

$$\mathbf{B} = \frac{1}{n}\sum_j \mathbf{X}_j\mathbf{X}_j^T - \sigma^2\mathbf{I}. \tag{5}$$

4. Denoising Algorithm

Using results from previous sections we suggest the following procedure for image denoising.

1. *Perform orthogonal wavelet decomposition of an image corrupted by Gaussian white noise.*
2. *For each detail (high pass) sub-band*
    a) *Estimate general block covariance matrix* $\mathbf{B}$ *using (5).*
    b) *Split the sub-band into non-intersecting blocks* $\mathbf{X}_j$. *Estimate covariance matrix* $\mathbf{C}_j$ *of each block in the form (2). Compute coefficients* γ_j *by solving equations (4) numerically.*
    c) *Apply Wiener filter (1) to each block* $\mathbf{X}_j$ *using covariance matrix* $\mathbf{C}_j$.

3. *Keep scaling (low pass) coefficients unchanged.*
4. *Reconstruct the image from denoised wavelet coefficients.*

We call this algorithm Wiener-wavelet denoising.

5. Numerical Results

We implemented Wiener-wavelet denoising and compared it to two other methods, MATLAB's `wiener2.m` routine and *wavelet soft thresholding*.

`wiener2.m` uses a pixel-wise adaptive Wiener method based on statistics estimated from a local neighborhood of each pixel. In all experiments we chose a neighborhood such that the root mean square error (rmse) was minimal.

The value of the threshold in wavelet thresholding was optimized in order to produce the smallest possible rmse.

We used Daubechies least asymmetric orthogonal wavelets with eight coefficients both in Wiener-wavelet and thresholding methods.

Table 1 compares peak signal to noise ratios (PSNR) obtained by these three methods for different images (PSNR= $20 \log_{10} \frac{255}{\text{rmse}}$). Wiener-wavelet denoising always gave the best results.

	Lenna	Barbara	Boats	Yogi
noise	20.16db	20.16db	14.14db	14.14db
thresholding	27.93db	24.00db	23.90db	21.38db
`wiener2.m`	28.13db	24.28db	23.94db	22.18db
Wiener-wavelet	29.53db	25.19db	24.80db	22.81db

TABLE 1. PSNR obtained by different denoising methods.

From Figure 3 one can see that, in the presence of strong noise, the Wiener-wavelet method preserves edges better. Careful examination shows that some visual information is lost in the noisy image and can be recovered only from Wiener-wavelet denoised image.

6. Conclusions

This paper presents a new technique for image denoising. We assume that blocks of wavelet coefficients are Gaussian with slowly changing covariance matrix. We develop a model for the covariance and use it for Wiener filtering.

In almost all experiments our method produces rmse lower than other methods. Visually, the Wiener-wavelet method preserves the edges better.

In case of white Gaussian noise with unknown variance, Wiener-wavelet denoising allows an easy and accurate estimate of the noise variance. If the covariance structure of the noise is known it can be easily taken into account.

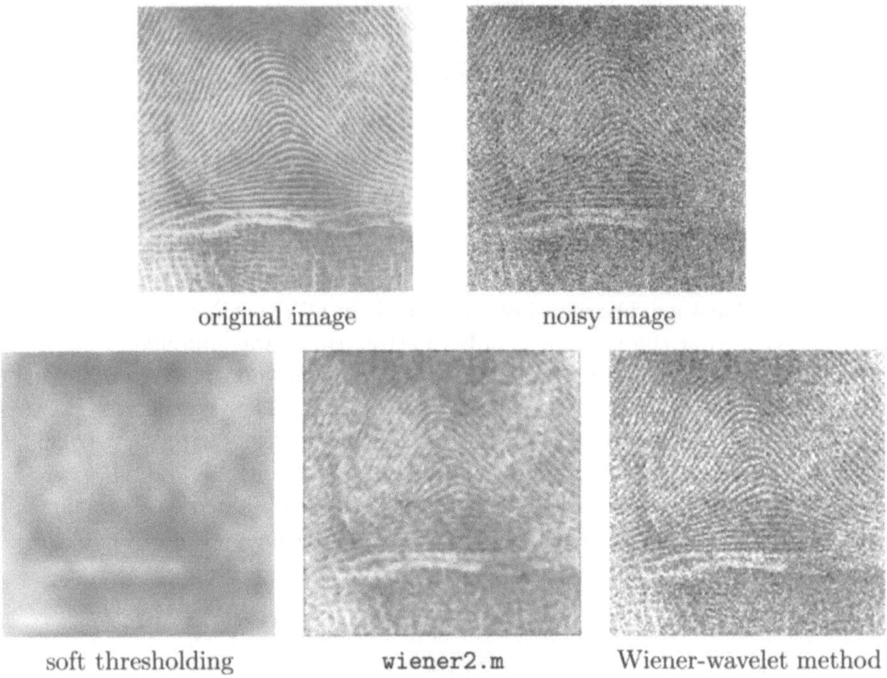

FIGURE 3. Visual comparison of different denoising methods.

There is another possible modification of our covariance model. Instead of estimating γ_j as a deterministic parameter one can model it as a random variable. A similar approach is adopted in [8]. We also would like to mention an independent work [5] in which image wavelet coefficients are modeled as a scalar Gaussian variable with spatially changing variance.

Acknowledgements

The author would like to thank Geoffrey Davis and Eero Simoncelli for useful discussions and comments.

References

[1] G. Chang, B Yu and M. Vetterli, *Spatially adaptive wavelet thresholding with context modeling for image denoising*, preprint (1998).

[2] D. Donoho, *De-noising by soft-thresholding*, IEEE Trans. on Info. Theory, **43** (1995), 613–627.

[3] S. Mallat, *Wavelet Tour of Signal Processing*, Academic Press, San Diego (1998).

[4] S. Mallat, *A theory for multiresolution signal decomposition: the wavelet representation*, IEEE Trans. Patttern Anal. Machine Intell., **11** (1989), 674–693.

[5] K. Mihcak, I. Kozintsev and K. Ramchadndran, *Spatially adaptive statistical modeling of wavelet image coefficients and its application to denoising*, preprint, (1999).

[6] P. Moulin and J. Liu, *Analysis of multiresolution image denoising schemes using a generalized Gaussian and complexity priors*, IEEE Trans. Info. Theory, **45** (1999), 909–919.

[7] E. Simoncelli and E. Adelson, *Noise removal via Bayesian wavelet coring*, Proc. IEEE ICIP, **I** (1996), 379–382.

[8] M. Wainwright and E. Simoncelli, *Scale mixtures of Gaussians and the statistics of natural images* in: S. A. Solla, T. K. Leen, and K.-R. Müller, Eds., Advances in Neural Information Processing Systems 12, (MIT Press, Cambridge MA) (2000).

Department of Mathematics & Computer Science
Drexel University
Philadelphia, PA 19104, USA
E-mail address: `vstrela@mcs.drexel.edu`

Wavelet Analysis of Discrete Time Series

Andrew T. Walden

Abstract. We give a brief review of some of the wavelet-based techniques currently available for the analysis of arbitrary-length discrete time series. We discuss the maximal overlap discrete wavelet packet transform (MODWPT), a non-decimated version of the usual discrete wavelet packet transform, and a special case, the maximal overlap discrete wavelet transform (MODWT). Using least-asymmetric or coiflet filters of the Daubechies class, the coefficients resulting from the MODWPT can be readily shifted to be aligned with events in the time series. We look at several aspects of denoising, and compare MODWT and cycle-spinning denoising. While the ordinary DWT basis provides a perfect decomposition of the autocovariance of a time series on a scale-by-scale basis, and is well suited to decorrelating a stationary time series with 'long-memory' covariance structure, a time series with very different covariance structure can be decorrelated using a wavelet packet 'best-basis' determined by a series of white noise tests.

1. Introduction

One scientific area where wavelet methods have been finding many applications is that of the analysis of discrete time series. A time series is defined to be a sequence of observations associated with an ordered independent variable t. Here we consider only the case of discrete values of t, but note that t could represent time, depth or distance along a line.

In this paper we seek to give a brief review of a few discrete wavelet techniques which have proven useful for analysing time series with a view to answering scientific questions. The fundamental ideas are introduced, but many more details and examples can be found in the references and in the comprehensive book [15] which includes many other useful approaches and aspects not touched on here.

In §2 we particularly concentrate attention on the 'maximal overlap' versions of the discrete wavelet transform and discrete wavelet packet transform which are applicable for arbitrary sample sizes. The ability to 'time align' the transform coefficients for certain choices of filters is also stressed. In §3 we look at the progress of wavelet denoising from its 'universal threshold' roots, while in §4 we discuss the scale-by-scale decomposition of the autocovariance sequence of a stationary process via the discrete wavelet transform. As discussed in §5 the correlations of the discrete wavelet transform coefficients of a time series from a 'long-memory' process are mostly negligible, because the transform is well matched to the covariance structure of the process; recent research suggests that a time series with very different covariance structure can be successfully decorrelated using a wavelet packet 'best-basis' determined by a series of white noise tests.

2. The Pyramid Algorithms

2.1. The discrete wavelet transform

For discrete compactly supported filters of the Daubechies class, ([4, Chapter 6]), we denote the even-length scaling (low-pass) filter by $\{g_l : l = 0, \dots, L-1\}$ and the wavelet (high-pass) filter $\{h_l : l = 0, \dots, L-1\}$. The low-pass filter satisfies

$$\sum_{l=0}^{L-1} g_l^2 = 1, \quad \sum_{l=0}^{L-1} g_l g_{l+2n} = \sum_{l=-\infty}^{\infty} g_l g_{l+2n} = 0\,, \tag{1}$$

for all nonzero integers n, so that the filter has unit energy and is orthogonal to its even shifts. The high-pass filter is also required to satisfy equation (1) but additionally the high- and low-pass filters are chosen to be quadrature mirror filters (QMFs) satisfying:

$$h_l = (-1)^l g_{L-l-1} \quad \text{or} \quad g_l = (-1)^{l+1} h_{L-l-1} \quad \text{for} \quad l = 0, \dots, L-1\,.$$

Denote the series to be transformed by $\{X_t : t = 0, \dots, N-1\}$. With $V_{0,t}^{(D)} \equiv X_t$, the jth stage input to the pyramid algorithm is $\{V_{j-1,t}^{(D)} : t = 0, \dots, N_{j-1} - 1\}$, where $N_j \equiv N/2^j$. For the discrete wavelet transform (DWT) pyramid algorithm the jth stage outputs are the jth level wavelet and scaling coefficients given by, respectively,

$$W_{j,t}^{(D)} = \sum_{t=0}^{L-1} h_l V_{j-1,(2t+1-l) \bmod N_{j-1}}^{(D)}, \quad V_{j,t}^{(D)} = \sum_{t=0}^{L-1} g_l V_{j-1,(2t+1-l) \bmod N_{j-1}}^{(D)},$$

$t = 0, \dots, N_j - 1$. If we write $\{W_{j,t}^{(D)} : t = 0, \dots, N_j - 1\}$ as $\mathbf{W}_j^{(D)}$ and $\{V_{j,t}^{(D)} : t = 0, \dots, N_j - 1\}$ as $\mathbf{V}_j^{(D)}$, then if $N = 2^J$ the pyramid algorithm is complete after J repetitions yielding $\mathbf{W}_1^{(D)}, \dots, \mathbf{W}_J^{(D)}, \mathbf{V}_J^{(D)}$, where the latter two vectors contain only one coefficient each. This defines the *full DWT*. If however N is an integer multiple of 2^{J_0} say, then we can carry out a *partial DWT* to level J_0. The DWT is an orthonormal transform of $\{X_t : t = 0, \dots, N-1\}$.

The jth level wavelet and scaling coefficients may be linked directly to the series $\{X_t\}$. We define the jth level wavelet filter $\{h_{j,l}\}$ formed by convolving together

$$\begin{array}{ll} \text{filter } 1 & g_0, g_1, \dots, g_{L-2}, g_{L-1}; \\ \vdots & \vdots \\ \text{filter } j-1 & g_0, \underbrace{0,\dots,0}_{2^{j-2}-1 \text{ zeros}}, g_1, \underbrace{0,\dots,0}_{2^{j-2}-1 \text{ zeros}}, \dots, g_{L-2}, \underbrace{0,\dots,0}_{2^{j-2}-1 \text{ zeros}}, g_{L-1}; \text{ and} \\ \text{filter } j & h_0, \underbrace{0,\dots,0}_{2^{j-1}-1 \text{ zeros}}, h_1, \underbrace{0,\dots,0}_{2^{j-1}-1 \text{ zeros}}, \dots, h_{L-2}, \underbrace{0,\dots,0}_{2^{j-1}-1 \text{ zeros}}, h_{L-1}. \end{array} \quad (2)$$

Filter $\{h_{j,l}\}$ has $L_j = (2^j - 1)(L-1) + 1$ terms. To obtain the jth level scaling filter $\{g_{j,l}\}$ the only difference is that filter j in the list above is replaced by

$$\text{filter } j \quad g_0, \underbrace{0,\dots,0}_{2^{j-1}-1 \text{ zeros}}, g_1, \underbrace{0,\dots,0}_{2^{j-1}-1 \text{ zeros}}, \dots, g_{L-2}, \underbrace{0,\dots,0}_{2^{j-1}-1 \text{ zeros}}, g_{L-1}. \quad (3)$$

Then

$$W_{j,t}^{(D)} = \sum_{l=0}^{L_j-1} h_{j,l} X_{(2^j(t+1)-1-l) \bmod N} \quad \text{and} \quad V_{j,t}^{(D)} = \sum_{l=0}^{L_j-1} g_{j,l} X_{(2^j(t+1)-1-l) \bmod N}.$$

The jth level filters have the following properties:

$$\sum_{l=0}^{L_j-1} g_{j,l} = 2^{j/2}; \ \sum_{l=0}^{L_j-1} g_{j,l}^2 = 1; \ \sum_{l=0}^{L_j-1} g_{j,l} h_{j,l} = 0; \ \sum_{l=0}^{L_j-1} h_{j,l} = 0; \ \sum_{l=0}^{L_j-1} h_{j,l}^2 = 1. \quad (4)$$

At level j the nominal frequency band to which the corresponding wavelet coefficients $\{W_{j,t}^{(D)}\}$ are associated is given by $|f| \in (1/2^{j+1}, 1/2^j]$; an 'octave' band. For example $\{W_{1,t}^{(D)}\}$, $\{W_{2,t}^{(D)}\}$ and $\{W_{3,t}^{(D)}\}$ have nominal pass-bands of $(1/4, 1/2], (1/8, 1/4]$ and $(1/16, 1/8]$ respectively.

2.2. The maximal overlap discrete wavelet transform

The DWT has several limitations:

- It requires the sample size to be an integer multiple of 2^{J_0} for a partial DWT, or to be exactly a power of 2 for the full transform.
- It is sensitive to where we 'break into' a time series. The wavelet and scaling coefficients are not circularly shift equivariant, i.e., circularly shifting the time series by some amount will not circularly shift the DWT wavelet and scaling coefficients by the same amount.
- The number of wavelet and scaling coefficients, N_j, decreases by a factor of 2 for each increasing level of the transform, limiting the ability to carry out statistical analyses on the coefficients.

These deficiencies can be overcome if the downsampling in the DWT can be avoided. This can be achieved by using the maximal overlap discrete wavelet transform (MODWT) ([12, 13]); see also the undecimated discrete wavelet transform ([19]), and stationary discrete wavelet transform ([10]). A rescaling of the defining filters is required to conserve energy: $\tilde{g}_l = g_l/\sqrt{2}$ and $\tilde{h}_l = h_l/\sqrt{2}$ so that now for example, $\sum_{l=0}^{L-1} \tilde{g}_l^2 = 1/2$, while the filters are still QMFs. Now the DWT pyramid algorithm consists of filter-subsample steps repeated for each level. If we define $V_{0,t}^{(M)} = X_t$, then the MODWT pyramid algorithm generates the MODWT wavelet coefficients $\{W_{j,t}^{(M)}\}$ and the MODWT scaling coefficients $\{V_{j,t}^{(M)}\}$ from $\{V_{j-1,t}^{(M)}\}$ using the 'new filters' (2) and (3) with non-zero coefficients divided by $\sqrt{2}$; the circular filterings (convolutions) can be written as

$$W_{j,t}^{(M)} = \sum_{l=0}^{L-1} \tilde{h}_l V_{j-1,(t-2^{j-1}l) \bmod N}^{(M)}, \quad V_{j,t}^{(M)} = \sum_{l=0}^{L-1} \tilde{g}_l V_{j-1,(t-2^{j-1}l) \bmod N}^{(M)},$$

$t = 0, \ldots, N-1$.

These coefficients can also be formulated in terms of a filtering of $\{X_t\}$, using the filters $\{\tilde{h}_{j,l} = h_{j,l}/2^{j/2}\}$ and $\{\tilde{g}_{j,l} = g_{j,l}/2^{j/2}\}$:

$$W_{j,t}^{(M)} = \sum_{l=0}^{L_j-1} \tilde{h}_{j,l} X_{(t-l) \bmod N} \quad \text{and} \quad V_{j,t}^{(M)} = \sum_{l=0}^{L_j-1} \tilde{g}_{j,l} X_{(t-l) \bmod N}.$$

Notice that the DWT of $\{X_t\}$ can be extracted from the MODWT via a rescaling and subsampling:

$$W_{j,t}^{(D)} = 2^{j/2} W_{j,2^j(t+1)-1}^{(M)} \quad \text{and} \quad V_{j,t}^{(D)} = 2^{j/2} V_{j,2^j(t+1)-1}^{(M)}. \tag{5}$$

The MODWT coefficients at level j are associated to the same nominal frequency band $|f| \in (1/2^{j+1}, 1/2^j]$ as for the DWT, but there are always N of them at each level. The overdetermined MODWT is not an orthonormal transform of $\{X_t : t = 0, \ldots, N-1\}$.

2.3. The discrete wavelet packet transform

The discrete wavelet packet transform (DWPT) is a generalization of the DWT which at level j of the transform partitions the frequency axis into 2^j equal width frequency bands, often labelled $n = 0, \ldots, 2^j - 1$. Increasing the transform level increases frequency resolution, but, starting with a series of length N, at level j there are only $N/2^j$ DWPT coefficients for each frequency band n.

We denote the time series to be transformed, $\{X_t : t = 0, \ldots, N-1\}$ by $\mathbf{X}$, thought of as a column vector. Initially we set $\mathbf{W}_{0,0}^{(D)} \equiv \mathbf{X}$. At the first step of the algorithm $\mathbf{X}$ is circularly filtered by the low-pass filter $\{g_l\}$, with corresponding transfer function $G(f) = \sum_{l=0}^{L-1} g_l e^{-i2\pi fl}$, and then downsampled by 2, to give first level coefficients $\mathbf{W}_{1,0}^{(D)} = \{W_{1,0,t}^{(D)}, t = 0, \ldots, (N/2)-1\}$ and $\mathbf{X}$ is also circularly filtered by the high-pass filter $\{h_l\}$, with corresponding transfer function $H(f)$, and then downsampled by 2, to give $\mathbf{W}_{1,1}^{(D)} = \{W_{1,1,t}^{(D)}, t = 0, \ldots, (N/2)-1\}$. Both $\mathbf{W}_{1,0}^{(D)}$ and $\mathbf{W}_{1,1}^{(D)}$ are of length $N_1 = N/2$ due to the downsampling.

Subsequent levels j of the transform repeat these steps: entries for the DWPT table up to level $j = 3$ are processed as shown in Figure 1; the figure emphasizes that circular filtering is used by showing that the jth level coefficients are obtained by filtering the $(j-1)$th level coefficients with the circular filter having discrete Fourier transform $\{H(\frac{k}{N_{j-1}})\}$ or $\{G(\frac{k}{N_{j-1}})\}$, followed by downsampling by 2.

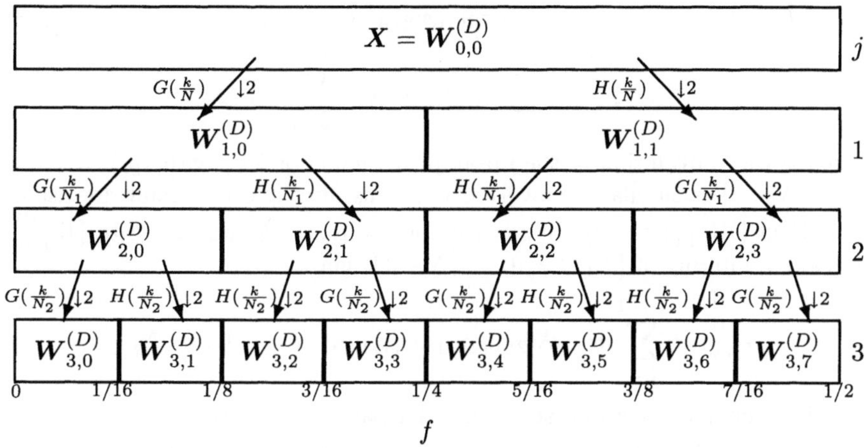

FIGURE 1. Wavelet packet table showing the DWPT of **X**.

Figure 1 can be summarized as follows. To compute the DWPT coefficients for levels $j = 1, \dots, J_0$, we circularly filter the wavelet packet coefficients at the previous stage and downsample by 2. Given the series $\{W^{(D)}_{j-1,\lfloor n/2\rfloor,t}\}$ of length N_{j-1} we calculate $\{W^{(D)}_{j,n,t}\}$ using

$$W^{(D)}_{j,n,t} \equiv \sum_{l=0}^{L-1} r_{n,l} W^{(D)}_{j-1,\lfloor n/2\rfloor,(2t+1-l) \bmod N_{j-1}}, \quad t = 0, \dots, N_j - 1,$$

where

$$r_{n,l} = \begin{cases} g_l, & \text{if } n \bmod 4 = 0 \text{ or } 3; \\ h_l, & \text{if } n \bmod 4 = 1 \text{ or } 2. \end{cases} \tag{6}$$

Here $\lfloor\cdot\rfloor$ denotes 'the integer part' operator. The ordering of the filters means that at level j band n is nominally associated with frequencies in the interval $(\frac{n}{2^{j+1}}, \frac{n+1}{2^{j+1}}]$. Such a DWPT would be said to be *sequency ordered* [22].

The transform that takes $\mathbf{X}$ to $\mathbf{W}^{(D)}_{j,n}$, $n = 0, \dots, 2^j - 1$, for any j between 0 and J_0 is called a DWPT, and is orthonormal.

We can also write $\{W^{(D)}_{j,n,t}\}$ in terms of filtering $\{X_t : t = 0, \dots, N-1\}$. Suppose we let

$$\{u_{1,0,l}\} = \{g_l : l = 0, \dots, L-1\} \quad \text{and} \quad \{u_{1,1,l}\} = \{h_l : l = 0, \dots, L-1\},$$

and for general (j,n) define

$$u_{j,n,l} = \sum_{k=0}^{L-1} r_{n,k}\, u_{j-1,\lfloor n/2\rfloor, l-2^{j-1}k}, \quad l = 0,\ldots,L_j - 1, \tag{7}$$

where $L_j = (2^j - 1)(L-1) + 1$ is the length of $\{u_{j,n,l}\}$. For example, for $j = 2$,

$$u_{2,1,l} = \sum_{k=0}^{L-1} r_{1,k}\, u_{1,0,l-2k} = \sum_{k=0}^{L-1} h_k g_{l-2k}$$

which is the convolution of the upsampled version of $\{h_l\}$ with $\{g_l\}$; if $\{X_t\colon t = 0,\ldots,N-1\}$ is circularly convolved with this filter and the result downsampled by 4, then $\{W^{(D)}_{2,1,t}\}$ results. In general, for $j = 1,\ldots,J_0$, we can write $\{W^{(D)}_{j,n,t}\}$ in terms of a filtering of $\{X_t\colon t = 0,\ldots,N-1\}$, via

$$W^{(D)}_{j,n,t} = \sum_{l=0}^{L_j-1} u_{j,n,l} X_{(2^j[t+1]-1-l) \bmod N}, \quad t = 0,1,\ldots,N_j - 1 .$$

2.4. The maximal overlap discrete wavelet packet transform

The downsampling step in the DWPT can also be removed. The maximal overlap (undecimated, stationary) discrete wavelet packet transform (MODWPT) as developed in [20] can be briefly summarized as follows.

Let $\widetilde{G}(f) = \sum_{l=0}^{L-1} \tilde{g}_l e^{-i2\pi f l}$, be the transfer function of $\{\tilde{g}_l\}$, and let the transfer function corresponding to $\{\tilde{h}_l\}$ be defined similarly. Initially we set $\mathbf{W}^{(M)}_{0,0} \equiv \mathbf{X}$. At the first step of the algorithm $\mathbf{X}$ is circularly filtered by the low-pass filter $\{\tilde{g}_l\}$, with corresponding transfer function $\widetilde{G}(f)$, to give first level coefficients $\mathbf{W}^{(M)}_{1,0} = \{W^{(M)}_{1,0,t},\ t = 0,\ldots,N-1\}$ and $\mathbf{X}$ is also circularly filtered by the high-pass filter $\{\tilde{h}_l\}$, with corresponding transfer function $\widetilde{H}(f)$, to give $\mathbf{W}^{(M)}_{1,1} = \{W^{(M)}_{1,1,t},\ t = 0,\ldots,N-1\}$. Both $\mathbf{W}^{(M)}_{1,0}$ and $\mathbf{W}^{(M)}_{1,1}$ are of length N since there is no downsampling.

For subsequent levels j of the transform we again insert $2^{j-1} - 1$ zeros, $j \geq 1$, between the elements of $\{\tilde{g}_l\}$ as in equation (3); the resulting filter has a transfer function given by $\widetilde{G}(2^{j-1}f)$. We can do likewise for $\{\tilde{h}_l\}$ to obtain $\widetilde{H}(2^{j-1}f)$. The set of entries for the MODWPT table up to level $j = 3$ are processed as shown in Figure 2. The jth level coefficients are obtained by filtering the level $j-1$ coefficients with the circular filter having discrete Fourier transform $\{\widetilde{H}(2^{j-1}\frac{k}{N})\}$ or $\{\widetilde{G}(2^{j-1}\frac{k}{N})\}$, as appropriate. Figure 2 can be summarized as follows. To compute the MODWPT coefficients for levels $j = 1,\ldots,J_0$, we circularly filter the wavelet packet coefficients at the previous stage. Given the series $\{W^{(M)}_{j-1,\lfloor n/2\rfloor,t}\}$ of length N we calculate $\{W^{(M)}_{j,n,t}\}$ using

$$W^{(M)}_{j,n,t} \equiv \sum_{l=0}^{L-1} r_{n,l} W^{(M)}_{j-1,\lfloor n/2\rfloor,(t-2^{j-1}l) \bmod N}, \quad t = 0,\ldots,N-1,$$

where now

$$r_{n,l} = \begin{cases} \tilde{g}_l, & \text{if } n \bmod 4 = 0 \text{ or } 3\,; \\ \tilde{h}_l, & \text{if } n \bmod 4 = 1 \text{ or } 2\,. \end{cases} \tag{8}$$

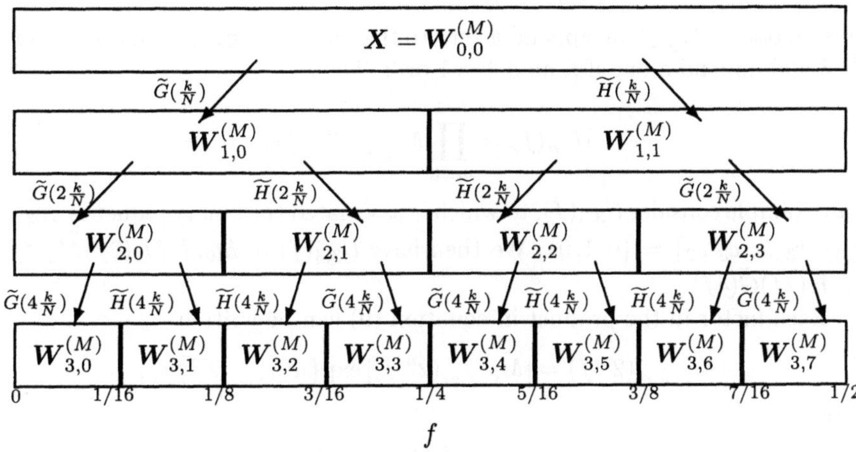

FIGURE 2. Undecimated wavelet packet table showing the MODWPT of **X**.

We can also write $\{W^{(M)}_{j,n,t}\}$ in terms of filtering $\{X_t\colon t = 0, \ldots, N-1\}$. Suppose we now let

$$\{u_{1,0,l}\} = \{\tilde{g}_l\colon l = 0, \ldots, L-1\} \quad \text{and} \quad \{u_{1,1,l}\} = \{\tilde{h}_l\colon l = 0, \ldots, L-1\}\,,$$

and for general (j, n) define $\{u_{j,n,l}\}$ as in (7) so for example, for $j = 2$,

$$u_{2,1,l} = \sum_{k=0}^{L-1} r_{1,k}\, u_{1,0,l-2k} = \sum_{k=0}^{L-1} \tilde{h}_k \tilde{g}_{l-2k}$$

which is the convolution of the upsampled version of $\{\tilde{h}_l\}$ with $\{\tilde{g}_l\}$; if $\{X_t\colon t = 0, \ldots, N-1\}$ is circularly convolved with this filter then $\{W^{(M)}_{2,1,t}\}$ results. In general, for $j = 1, \ldots, J_0$, we can write $\{W^{(M)}_{j,n,t}\}$ in terms of a filtering of $\{X_t\colon t = 0, \ldots, N-1\}$, via

$$W^{(M)}_{j,n,t} = \sum_{l=0}^{L_j-1} u_{j,n,l} X_{(t-l) \bmod N}, \quad t = 0, \ldots, N-1\,.$$

The transfer function $U_{j,n}(f) = \sum_{l=0}^{L_j-1} u_{j,n,l} e^{-i2\pi fl}$, of $\{u_{j,n,l}\colon l = 0, \ldots, L_j - 1\}$ is defined entirely by the two filters $\{\tilde{g}_l\}$ and $\{\tilde{h}_l\}$. Suppose we let $\widetilde{M}_0 = \widetilde{G}(f)$, say, and $\widetilde{M}_1 = \widetilde{H}(f)$, say, as in [3]. Further, suppose we attach to each pair (j, n) in the wavelet packet tree a sequence of ones and zeros according to the following rule. Let $\{\mathbf{c}_{1,0}\} = \{0\}$ and $\{\mathbf{c}_{1,1}\} = \{1\}$ and let $\{\mathbf{c}_{j,n}\} =$

$\{c_{j,n,0},\dots,c_{j,n,j-1}\}$ where $c_{j,n,k}\in\{0,1\}$ for $k=0,\dots,j-1$. Define, for $n=0,\dots,2^j-1$,

$$\mathbf{c}_{j,n}=\begin{cases}\{\mathbf{c}_{j-1,\lfloor n/2\rfloor},0\}, & \text{if } n \bmod 4 = 0 \text{ or } 3\,;\\ \{\mathbf{c}_{j-1,\lfloor n/2\rfloor},1\}, & \text{if } n \bmod 4 = 1 \text{ or } 2\,.\end{cases}$$

Hence to obtain $\{\mathbf{c}_{j,n}\}$ we append a zero to the sequence $\{\mathbf{c}_{j-1,\lfloor n/2\rfloor}\}$ if $n \bmod 4 = 0$ or 3 and append a one if $n \bmod 4 = 1$ or 2. Then,

$$U_{j,n}(f)=\prod_{m=0}^{j-1}\widetilde{M}_{c_{j,n,m}}(2^m f)\,. \tag{9}$$

As an example consider $U_{3,3}(f)$, to which is associated the binary sequence $\{\mathbf{c}_{3,3}\}=\{c_{3,3,0},c_{3,3,1},c_{3,3,2}\}=\{0,1,0\}$. We then have $U_{3,3}(f)=\widetilde{M}_0(f)\widetilde{M}_1(2f)\widetilde{M}_0(4f)=\widetilde{G}(f)\widetilde{H}(2f)\widetilde{G}(4f)$.

Each factor in the product in equation (9) can be written

$$\widetilde{M}_{c_{j,n,m}}(2^m f)=|\widetilde{M}_{c_{j,n,m}}(2^m f)|\exp\{i\theta_{c_{j,n,m}}(2^m f)\}\,,$$

where

$$\theta_{c_{j,n,m}}(2^m f)=\begin{cases}\theta^{(\tilde G)}(2^m f), & \text{if } c_{j,n,m}=0\,;\\ \theta^{(\tilde H)}(2^m f), & \text{if } c_{j,n,m}=1\,,\end{cases}$$

where $\theta^{(\tilde G)}(f)$ and $\theta^{(\tilde H)}(f)$ are the phase functions of the filters $\{\tilde g_l\}$ and $\{\tilde h_l\}$, respectively.

For the Daubechies least asymmetric filters of lengths $L=8(2)20$, denoted here LA(L), and the Daubechies coiflet filters of lengths $L=6(6)30$, denoted here C(L), it may be shown that $\theta^{(\tilde G)}(f)\approx 2\pi f\nu$ and $\theta^{(\tilde H)}(f)\approx -2\pi f(L-1+\nu)$, where, for the LA($L$) filters

$$\nu=\begin{cases}-(L/2)+1, & \text{if } L=8,12,16 \text{ or } 20\,;\\ -(L/2), & \text{if } L=10 \text{ or } 18\,;\\ -(L/2)+2, & \text{if } L=14\,;\end{cases}$$

and for the C(L) filters, $\nu=-(2L/3)+1$, $L=6,12,18,24$ and 30. The overall phase function of $U_{j,n}(f)$ is thus

$$\begin{aligned}\sum_{m=0}^{j-1}\theta_{c_{j,n,m}}(2^m f) &\approx 2\pi f\left(\nu\sum_{\substack{m=0\\ \{m:c_{j,n,m}=0\}}}^{j-1}2^m-(L-1+\nu)\sum_{\substack{m=0\\ \{m:c_{j,n,m}=1\}}}^{j-1}2^m\right)\\ &=2\pi f\nu_{j,n},\quad\text{say}\,.\end{aligned}$$

Since $\nu_{j,n}<0$ for the filters discussed above, we obtain the closest approximation to zero phase filtering if we associate the coefficient $W^{(M)}_{j,n,(t+|\nu_{j,n}|) \bmod N}$ with the input X_t. (For further details, including the necessity of reversing some of the filters listed in [4] for phase correction purposes, see [20] and [15].)

3. Denoising

3.1. The DWT and universal thresholding

Suppose that our observed time series consists of a signal plus noise, i.e.,

$$\mathbf{X} = \mathbf{D} + \epsilon\,, \tag{10}$$

where $\mathbf{D}$ is a deterministic signal and ϵ represents an N dimensional vector of independent and identically distributed (IID) Gaussian noise, each random variable having variance σ_ϵ^2.

For the purpose of denoising via thresholding Donoho and Johnstone ([5]) recommend firstly computing a level J_0 partial DWT giving coefficient vectors $\mathbf{W}_1^{(D)}, \ldots, \mathbf{W}_{J_0}^{(D)}$ and $\mathbf{V}_{J_0}^{(D)}$. Component-wise, we have, say,

$$W_{j,t}^{(D)} = d_{j,t} + e_{j,t} \quad j = 1, \ldots, J_0; \;\; t = 0, \ldots, N_j - 1\,.$$

(J_0 must be specified by the user.) Then, only the coefficients in the $\mathbf{W}_k^{(D)}$ vectors are subjected to thresholding; i.e., the elements of $\mathbf{V}_{J_0}^{(D)}$ are untouched, so that portion of $\mathbf{X}$ is automatically assigned to the signal $\mathbf{D}$.

Next a threshold level must be chosen. A key property about an orthonormal transform, (such as the partial DWT), of IID Gaussian noise is that the transformed noise has the same statistical properties as the untransformed noise, so that the $\{e_{j,t}\}$ are also IID Gaussian with mean zero and variance σ_ϵ^2. The *universal threshold* was defined in [5] as

$$\delta_U \equiv \sqrt{[2\sigma_\epsilon^2 \log(N)]}\,.$$

To understand the rationale for this threshold, suppose that the signal $\mathbf{D}$ is in fact a vector of zeros, so that the transform coefficients $\{W_{j,t}^{(D)}\}$ are a portion of an IID Gaussian sequence $\{e_{j,t}\}$ with zero mean and variance σ_ϵ^2. Then, as $N \to \infty$, we have

$$\mathbf{P}\big[\max\{|W_{j,t}^{(D)}|\} \le \delta_U\big] \equiv \mathbf{P}\big[\max\{|e_{j,t}|\} \le \delta_U\big] \to 1\,,$$

so that asymptotically we will correctly estimate the signal vector. Universal thresholding typically removes all the noise, but, in doing so, it can mistakenly set some small signal transform coefficients to zero. Universal thresholding thus ensures, with high probability, that the reconstruction is at least as smooth as the true deterministic signal. If σ_ϵ^2 is unknown as is frequently the case in applications, a practical procedure is to estimate it based upon the *median absolute deviation* (MAD) standard deviation estimate using just the $N/2$ level $j = 1$ coefficients in $\mathbf{W}_1^{(D)}$. By definition, this standard deviation estimator is

$$\hat{\sigma}_{\mathrm{MAD}} \equiv \frac{\mathrm{median}\{|W_{1,0}^{(D)}|, |W_{1,1}^{(D)}|, \ldots, |W_{1,\frac{N}{2}-1}^{(D)}|\}}{0.6745}\,.$$

The factor 0.6745 rescales so that $\hat{\sigma}_{\mathrm{MAD}}$ is also a suitable estimator for the standard deviation for Gaussian white noise. $\hat{\sigma}_{\mathrm{MAD}}$ is calculated from the elements of $\mathbf{W}_1^{(D)}$ because the smallest scale wavelet coefficients should be noise dominated, with the possible exception of the largest values. The MAD standard deviation estimate is designed to be robust against large deviations and hence should reflect the noise variance rather than the signal variance.

Finally, for $W_{j,t}^{(D)}, j = 1, \ldots, J_0$ and $t = 0, \ldots, N_j - 1$, we apply a chosen thresholding rule, such as hard thresholding defined by

$$\widehat{W}_{j,t}^{(D)} = \begin{cases} 0, & \text{if } |W_{j,t}^{(D)}| \leq \delta_U\,; \\ W_{j,t}^{(D)}, & \text{otherwise}, \end{cases}$$

to obtain the thresholded coefficients $\{\widehat{W}_{j,t}^{(D)}\}$, which are then used to form $\widehat{\mathbf{W}}_j^{(D)}$, $j = 1, \ldots, J_0$. $\mathbf{D}$ is estimated as $\widehat{\mathbf{D}}$ obtained by inverse transforming $\widehat{\mathbf{W}}_1^{(D)}, \ldots, \widehat{\mathbf{W}}_{J_0}^{(D)}$ and $\mathbf{V}_{J_0}^{(D)}$.

3.2. Level-dependent thresholding

In the event that the noise wavelet coefficients $\{e_{j,t}\}$ are uncorrelated, but not identically distributed, as will be the case if the additive noise is still IID but is non-Gaussian, a level-dependent (universal thresholding) can be used where now

$$\delta_{U,j} \equiv \sqrt{[2\sigma_j^2 \log(N)]}\,,$$

and σ_j^2 is the variance of the jth level coefficients. As pointed out in [7] in the general case of unknown variances, we could use the MAD estimate at each level j, namely,

$$\hat{\sigma}_{\mathrm{MAD},j} \equiv \frac{\mathrm{median}\{|W_{j,0}^{(D)}|, |W_{j,1}^{(D)}|, \ldots, |W_{j,N_j-1}^{(D)}|\}}{0.6745}\,,$$

and then use the universal threshold

$$\hat{\delta}_{U,j} \equiv \sqrt{[2\hat{\sigma}_{\mathrm{MAD},j}^2 \log(N)]}$$

at each level. The estimate $\hat{\sigma}_{\mathrm{MAD},j}$ is appropriate only for small j, where there is a considerable number of coefficients in a given level, and the signal is sparse.

Scale-dependent thresholding was investigated in the context of spectrum estimation in [21]; in this application the variances σ_j^2 at each level were known, and did not need to be estimated.

3.3. Cycle-spinning denoising and the MODWT

It was noted in Section 2.2 that one potential problem with the DWT is its sensitivity to where we 'break into' a time series. As a consequence the result of denoising using the DWT will depend somewhat on the starting point of the series. To try to alleviate this problem the idea of *cycle spinning* was introduced in [2]. Consider a partial DWT of level J_0 calculated from a time series of length N, an integer multiple of 2^{J_0}. The idea of denoising via cycle spinning is to apply denoising not only to $\mathbf{X}$, but also to all possible unique circularly shifted versions of $\mathbf{X}$, and to average the results. Consider the model in (10). Let $\mathcal{T}\mathbf{X} = [X_{N-1}, X_0, X_1, \ldots, X_{N-2}]$, i.e., $\mathcal{T}$ (circularly) delays $\mathbf{X}$ by one time unit, and let $\mathcal{T}^2\mathbf{X} = \mathcal{T}\mathcal{T}\mathbf{X}$ etc. Also we take $\mathcal{T}^{-1}$ to be a circular advance operator. Suppose that $\widehat{\mathbf{D}}_n$ is the estimate of $\mathbf{D}$

resulting from applying a denoising procedure to $\mathcal{T}^n\mathbf{X}$. Then the cycle spinning denoising estimate of $\mathbf{D}$ is given by

$$\overline{\mathbf{D}} = \frac{1}{2^{J_0}} \sum_{n=0}^{2^{J_0}-1} \mathcal{T}^{-n}\widehat{\mathbf{D}}_n .$$

(Shifts greater than or equal to 2^{J_0} are redundant.)

However, we know that the DWT of each $\mathcal{T}^n\mathbf{X}$ can be extracted readily from the MODWT of $\mathbf{X}$: see equation (5) where we note that the variance of the DWT coefficients is 2^j times that of the MODWT coefficients. Hence cycle-spinning can be implemented efficiently in terms of the MODWT as follows ([15]):

- Compute a level J_0 partial MODWT giving coefficient vectors $\mathbf{W}_1^{(M)},\ldots,$ $\mathbf{W}_{J_0}^{(M)}$ and $\mathbf{V}_{J_0}^{(M)}$.
- For each $j = 1,\ldots,J_0$ apply a chosen thresholding rule to each element of $\mathbf{W}_j^{(M)}$ using the level-dependent universal threshold $\delta_{U,j} \equiv \sqrt{[2\sigma_j^2 \log(N)]}$, with $\sigma_j^2 = \sigma_\epsilon^2/2^j$.
- The thresholded coefficients $\{\widehat{W}_{j,t}^{(M)}\}$ are then used to form $\widehat{\mathbf{W}}_j^{(M)}$, $j = 1,\ldots,J_0$. $\mathbf{D}$ is estimated as $\widehat{\mathbf{D}}$ obtained by applying the inverse MODWT to $\widehat{\mathbf{W}}_1^{(M)},\ldots,\widehat{\mathbf{W}}_{J_0}^{(M)}$ and $\mathbf{V}_{J_0}^{(M)}$.
-

The advantages of this approach to cycle spinning are two-fold. Cycle spinning originated assuming that the sample size N is an integer multiple of 2^{J_0}, whereas the MODWT-based approach above is valid for general N. Secondly, if σ_ϵ^2 is unknown, the MAD scale estimate can be adapted by taking

$$\tilde{\sigma}_{\mathrm{MAD}} \equiv \frac{2^{1/2}\,\mathrm{median}\{|W_{1,0}^{(M)}|, |W_{1,1}^{(M)}|,\ldots,|W_{1,N-1}^{(M)}|\}}{0.6745}.$$

The scaling factor of '$2^{1/2}$' in the numerator above, reflects the fact that $W_{1,t}^{(D)} = 2^{1/2}W_{1,2t+1}^{(M)}$.

4. Scale-Based Decomposition

Let $\{X_t, t \in \mathbb{Z}\}$ denote a real-valued second-order stationary process for which $\{X_t\colon t = 0,\ldots,N-1\}$ would represent a realization. The autocovariance sequence $\{s_{X,m}\}$ is defined as

$$s_{X,m} \equiv \mathrm{cov}\{X_t, X_{t+m}\} = E\{[X_t - \mu_X][X_{t+m} - \mu_X]\} = s_{X,-m},$$

where μ_X is the mean of $\{X_t, t \in \mathbb{Z}\}$.

The stationary stochastic processes resulting from applying the filters $\{\tilde{h}_{j,l}\}$, $j = 1,\ldots,J_0$, and $\{\tilde{g}_{J_0,l}\}$, to $\{X_t\}$ are given by the jth level wavelet coefficients

$\{\mathcal{W}_{j,t}^{(M)}\}$, $j = 1, \ldots, J_0$, and J_0th scaling coefficients $\{\mathcal{V}_{J_0,t}^{(W)}\}$, calculated by non-circular filtering, where

$$\mathcal{W}_{j,t}^{(M)} = \sum_{l=0}^{L_j-1} \tilde{h}_{j,l} X_{t-l},\ j = 1, \ldots, J_0, \quad \text{and} \quad \mathcal{V}_{J_0,t}^{(M)} = \sum_{l=0}^{L_{J_0}-1} \tilde{g}_{J_0,l} X_{t-l},\ t \in \mathbb{Z}.$$

The autocovariance of the wavelet coefficient process at transform level j, $\{\mathcal{W}_{j,t}^{(M)}\}$, is given by

$$s_{\mathcal{W}_j,m} = E\{\mathcal{W}_{j,t}^{(M)} \mathcal{W}_{j,t+m}^{(M)}\}.$$

Note that $\{\mathcal{W}_{j,t}^{(M)}\}$ has a mean of zero since by design $\sum_{l=0}^{L_j-1} \tilde{h}_{j,l} = 0$.

The autocovariance of the scaling coefficient process at transform level J_0, $\{\mathcal{V}_{J_0,t}^{(M)}\}$, is given by

$$s_{\mathcal{V}_{J_0},m} = E\{\mathcal{V}_{J_0,t}^{(M)} \mathcal{V}_{J_0,t+m}^{(M)}\} - E^2\{\mathcal{V}_{J_0,t}^{(M)}\} = E\{\mathcal{V}_{J_0,t}^{(M)} \mathcal{V}_{J_0,t+m}^{(M)}\} - \mu_X^2,$$

since the mean of $\{\mathcal{V}_{j,t}^{(M)}\}$ is μ_X, because by design $\sum_{l=0}^{L_j-1} \tilde{g}_{j,l} = 1$. Then ([16]),

$$s_{X,m} = \sum_{j=1}^{J_0} s_{\mathcal{W}_j,m} + s_{\mathcal{V}_{J_0},m} \tag{11}$$

i.e., the autocovariance sequence of the process $\{X_t, t \in \mathbb{Z}\}$ can be decomposed in terms of the autocovariance sequences of the wavelet coefficient processes, and the autocovariance sequence of the single scaling coefficient process. (This type of decomposition extends to cross-covariances (see [16]); it was noted for sequence variance in [11].) Most usefully, we can interpret this result as a scale-by-scale decomposition.

A standard measure of effective width of the jth level scaling filter is the 'autocorrelation width,' ([1]), defined as $\text{width}_a\{\tilde{g}_{j,l}\} = (\sum_l \tilde{g}_{j,l})^2 / \sum_l \tilde{g}_{j,l}^2$. But, from (4),

$$\left(\sum_l \tilde{g}_{j,l}\right)^2 = 1 \quad \text{and} \quad \sum_l \tilde{g}_{j,l}^2 = 2^{-j},$$

so that $\text{width}_a\{\tilde{g}_{j,l}\} = 2^j$. The averaging effect of $\{\tilde{g}_{j,l}\}$ thus extends over a scale of 2^j. Moreover, because the wavelet filter $\{\tilde{h}_{j,l}\}$ has the same length L_j as $\{\tilde{g}_{j,l}\}$, is orthogonal to $\{\tilde{g}_{j,l}\}$ and sums to zero, it represents the difference between two generalized averages, each occupying half the width of 2^j. Thus the wavelet coefficients at transform level j are associated with a scale of 2^{j-1}. For processes with sample interval Δt the jth level scaling coefficients are associated with a physical scale of $2^j\,\Delta t$ and the jth level wavelet coefficients with a physical scale of $2^{j-1}\,\Delta t$. We are thus able to think of the decomposition in (11) as a scale-by-scale decomposition of the autocovariance sequence of $\{X_t, t \in \mathbb{Z}\}$. For estimation considerations and examples of practical applications of such scale-based decompositions see [17] and [18].

5. Decorrelation

It has become well recognized in recent years ([6, 8, 23]) that wavelet transforms are particularly well suited to the analysis of long-memory or '$1/f$-type' processes; these have power spectra which plot as straight lines on log-frequency/log-power scales over many octaves of frequency and tend to infinity as the frequency tends to zero. The application of the DWT to a particular discrete-time long-memory model class, namely the stationary fractionally differenced (FD) processes, was investigated in [9]. A long-memory FD(d) process can be written as an infinite order moving average process:

$$X_t = \sum_{k=0}^{\infty} \frac{\Gamma(k+d)}{\Gamma(k+1)\Gamma(d)} \varepsilon_{t-k} ,$$

where $0 < d < 1/2$, $\Gamma(\cdot)$ is the gamma function, and $\{\varepsilon_t\}$ is a sequence of uncorrelated random variables (white noise) with mean zero and variance σ_ε^2. The spectrum of such a process is of the form $S(f) = \sigma_\varepsilon^2 (4\sin^2 \pi f)^{-d}$, $|f| \leq 1/2$. Using a full DWT for $N = 2^J = 2^5 = 32$ the exact 32×32 correlation matrix, (including boundary effects), of the wavelet coefficients $\{W_{j,t}^{(D)}\}$, $j = 1, \ldots, 5$ and the single scaling coefficient $V_{5,0}^{(D)}$ was computed, (for each of three different Daubechies scaling and wavelet filters). The only large off-diagonal correlations occurred between, rather than within, levels of the transform, the largest being due to boundary effects. The lack of correlation within a level can be explained by computing the covariance of the wavelet coefficients, ignoring boundary effects; from [15],

$$\operatorname{cov}\{W_{j,t}^{(D)}, W_{j,t+\tau}^{(D)}\} = \int_{-1/2}^{1/2} e^{i2^{j+1}\pi f\tau} |H_j(f)|^2 S(f) df ,$$

where $H_j(f) = \sum_{l=0}^{L_j-1} h_{j,l} e^{-i2\pi fl}$ is the transfer function of $\{h_{j,l}\}$. The jth level wavelet filter has an (ideal) pass band $|f| \in (1/2^{j+1}, 1/2^j]$; hence if in addition to $|H_j(f)|^2$ being approximately constant over this band, $S(f)$ is also, then approximately

$$\operatorname{cov}\{W_{j,t}^{(D)}, W_{j,t+\tau}^{(D)}\} \propto \int_{-1/2^j}^{-1/2^{j+1}} e^{i2^{j+1}\pi f\tau} df + \int_{1/2^{j+1}}^{1/2^j} e^{i2^{j+1}\pi f\tau} df = 0 \quad \text{for } \tau \neq 0 .$$

But the basis underlying the DWT ensures that $S(f)$ for an FD process is indeed approximately constant over $|f| \in (1/2^{j+1}, 1/2^j]$; although $S(f)$ rises rapidly towards zero frequency, the interval $(1/2^{j+1}, 1/2^j]$ becomes increasingly narrow towards zero frequency as j increases, 'tracking' the change in the spectrum. Hence the covariance, and correlation, is negligible within a level. Covariances between levels can be considered using similar reasoning – see [15].

The ability to decorrelate a process is a powerful weapon in statistical analysis as it enables the use of statistical tools which will fail in the presence of strong correlations (see for example [14]). For a stochastic process with a spectrum very different from that of a long-memory process, the DWT can be very sub-optimal as a decorrelator. A more flexible basis is of course provided by the DWPT; we can define a disjoint dyadic decomposition of the table in Figure 1 by noting that at each potential parent node (j, n) we can either carry out the splitting using both $G(\cdot)$ and $H(\cdot)$ or not split at all. Such a resulting decomposition is also orthonormal, being associated with a nonoverlapping partition of $(0, 1/2]$. A way of determining a 'best basis,' suitable for decorrelating, from all the possible orthonormal partitions, is given in [14]. The algorithm looks at each node (j, n) and carries out a statistical white noise test on $W^{(D)}_{j,n,t}$, $t = 0, \ldots, N_j - 1$. If we fail to reject the hypothesis that the DWPT coefficients are from a white noise process, then we retain $W^{(D)}_{j,n,t}$, $t = 0, \ldots, N_j - 1$, i.e., this sequence is processed no further. If we reject the hypothesis, then $W^{(D)}_{j,n,t}$, $t = 0, \ldots, N_j - 1$, is further filtered and downsampled to produce $W^{(D)}_{j+1,2n,t}$, $t = 0, \ldots, N_{j+1} - 1$ and $W^{(D)}_{j+1,2n+1,t}$, $t = 0, \ldots, N_{j+1} - 1$. Further details and application examples can be found in [14].

6. Conclusions

We have shown that a number of very useful wavelet-based tools are currently available for the analysis of discrete time series from both algorithmic and statistical perspectives. We expect many more to be developed in the coming years.

References

[1] R. N. Bracewell (1978). *The Fourier Transform and Its Applications (Second Edition)*. New York: McGraw-Hill.

[2] R. R. Coifman and D. L. Donoho (1995). Translation-invariant denoising. In *Wavelets and Statistics (Lecture Notes in Statistics, Volume 103)*, (A. Antoniadis and G. Oppenheim, *eds.*), 125–50, New York: Springer-Verlag.

[3] R. R. Coifman, Y. Meyer and M. V. Wickerhauser (1992). Size properties of wavelet packets. In *Wavelets and Their Applications*, (M. B. Ruskai et al, *eds.*), 453–79, Boston: Jones and Bartlett.

[4] I. Daubechies (1992). *Ten Lectures on Wavelets*. Philadelphia: SIAM.

[5] D. L. Donoho and I. M. Johnstone (1994). Ideal spatial adaptation by wavelet shrinkage. *Biometrika*, 81, 425–55.

[6] P. Flandrin (1992). Wavelet analysis and synthesis of fractional Brownian motion. *IEEE Transactions on Information Theory*, 38, 910–7.

[7] I. M. Johnstone and B. W. Silverman (1997). Wavelet threshold estimators for data with correlated noise. *Journal of the Royal Statistical Society, Series B*, 59, 319–51.

[8] E. Masry (1993). The wavelet transform of stochastic processes with stationary increments and its application to fractional Brownian motion. *IEEE Transactions on Information Theory*, 39, 260–4.

[9] E. J.McCoy and A. T. Walden (1996). Wavelet analysis and synthesis of stationary long-memory processes. *Journal of Computational and Graphical Statistics*, 5, 26–56.

[10] G. P. Nason and B. W. Silverman (1994). The discrete wavelet transform in *S*. *Journal of Computational and Graphical Statistics*, 3, 163–91.

[11] D. B. Percival (1995). On estimation of the wavelet variance. *Biometrika*, 82, 619–31.

[12] D. B. Percival and P. Guttorp (1994). Long-memory processes, the Allan variance and wavelets. In *Wavelets in Geophysics*, (E. Foufoula-Georgiou and P. Kumar, *eds.*), 325–44, San Diego: Academic Press.

[13] D. B. Percival and H. Mofjeld (1997). Analysis of subtidal coastal sea level fluctuations using wavelets. *Journal of the American Statistical Association*, 92, 868–80.

[14] D. B. Percival, S. Sardy and A. C. Davison (2001). Wavestrapping time series: adaptive wavelet-based bootstrapping. In *Nonlinear and Nonstationary Signal Processing*, (W. Fitzgerald, R. L. Smith, A. T. Walden and P. Young, *eds.*), Cambridge University Press, 2001.

[15] D. B. Percival and A. T. Walden (2000). *Wavelet Methods for Time Series Analysis*. Cambridge UK: Cambridge University Press.

[16] A. Serroukh and A. T. Walden (2000). Wavelet scale analysis of bivariate time series I: motivation and estimation. *Journal of Nonparametric Statistics*, 13, 1–36.

[17] A. Serroukh and A. T. Walden (2000). Wavelet scale analysis of bivariate time series II: statistical properties for linear processes. *Journal of Nonparametric Statistics*, 13, 37–56.

[18] A. Serroukh, A. T. Walden and D. B. Percival (2000). Statistical properties and uses of the wavelet variance estimator for the scale analysis of time series. *Journal of the American Statistical Association*, 95, 184–96.

[19] M. J. Shensa (1992). The discrete wavelet transform: wedding the à trous and Mallat algorithms. *IEEE Transactions on Signal Processing*, 40, 2464–82.

[20] A. T. Walden and A. Contreras Cristan (1998). The phase-corrected undecimated discrete wavelet packet transform and its application to interpreting the timing of events. *Proceedings of the Royal Society of London, Series A.*, 454, 2243–66.

[21] A. T. Walden, D. B. Percival and E. J. McCoy (1998). Spectrum estimation by wavelet thresholding of multitaper estimators. *IEEE Transactions on Signal Processing*, 46, 3153–65.

[22] M. V. Wickerhauser (1994). *Adapted Wavelet Analysis from Theory to Software Algorithms.* A K Peters, Wellesley, Mass.

[23] G. W. Wornell (1993). Wavelet-based representations for the $1/f$ family of fractal processes. *Proceedings of the IEEE*, 81, 1428–50.

Department of Mathematics
Imperial College of Science, Technology and Medicine
Huxley Building, 180 Queen's Gate
London SW7 2BZ, UK
E-mail address: a.walden@ic.ac.uk

MIX
Papier aus verantwortungsvollen Quellen
Paper from responsible sources
FSC® C105338

If you have any concerns about our products,
you can contact us on
ProductSafety@springernature.com

In case Publisher is established outside the EU,
the EU authorized representative is:
Springer Nature Customer Service Center GmbH
Europaplatz 3, 69115 Heidelberg, Germany

Printed by Libri Plureos GmbH
in Hamburg, Germany